ST. MARY'S CITY, MARYLAND 20686

Fluorine Chemistry: A Comprehensive Treatment

Fluorine Chemistry: A Comprehensive Treatment

ENCYCLOPEDIA REPRINT SERIES

Mary Howe-Grant, Editor

A WILEY-INTERSCIENCE PUBLICATION

JOHN WILEY & SONS

NEW YORK · CHICHESTER · BRISBANE · TORONTO · SINGAPORE

This text is printed on acid-free paper.

Copyright © 1995 by John Wiley & Sons, Inc.

All rights reserved. Published simultaneously in Canada.

Reproduction or translation of any part of this work
beyond that permitted by Section 107 or 108 of the
1976 United States Copyright Act without the permission
of the copyright owner is unlawful. Requests for
permission or further information should be addressed to
the Permissions Department, John Wiley & Sons, Inc.,
605 Third Avenue, New York, NY 10158-0012.

Library of Congress Catalog Number: 94-39173
ISBN: 0-471-12031-6

Printed in the United States of America

10 9 8 7 6 5 4 3 2 1

CONTENTS

Fluorine, 1
Fluorine compounds, inorganic
   Introduction, 27
   Aluminum, 33
   Ammonium, 47
   Antimony, 50
   Arsenic, 55
   Barium, 58
   Boron, 60
      Boron trifluoride, 60
      Fluoroboric acid and fluoroborates, 69
   Calcium, 83
   Cobalt, 96
   Copper, 98
   Germanium, 100
   Halogens, 102
   Hydrogen, 115
   Iron, 136
   Lead, 139
   Lithium, 141
   Magnesium, 143
   Mercury, 146
   Molybdenum, 148
   Nickel, 150
   Nitrogen, 152
   Oxygen, 158
   Phosphorus, 165
   Potassium, 176
   Rhenium, 180
   Silver, 182
   Sodium, 186
   Sulfur, 188
      Sulfur fluorides, 188
      Fluorosulfuric acid, 202
   Tantalum, 210
   Tin, 211
   Titanium, 214
   Tungsten, 217
   Zinc, 221
   Zirconium, 223
Fluorine compounds, organic
   Introduction, 227
   Direct fluorination, 242
   Fluorinated aliphatic compounds, 259
   Fluoroethanols, 281
   Fluoroethers and fluoroamines, 285
   Perfluoroepoxides, 294
   Fluorinated acetic acids, 304
   Fluorinated higher carboxylic acids, 311
   Perfluoroalkanesulfonic acids, 318
   Fluorinated aromatic compounds, 325
   Polytetrafluoroethylene, 381
   Perfluorinated ethylene-propylene copolymers, 404
   Tetrafluoroethylene-ethylene copolymers, 417
   Tetrafluoroethylene-perfluorovinyl ether copolymers, 431
   Poly(vinyl fluoride), 443
   Poly(vinylidene fluoride), 454
   Polychlorotrifluoroethylene, 472
   Bromotrifluoroethylene, 479
   Poly(fluorosilicones), 482
   Fluorocarbon elastomers, 490

EDITORIAL STAFF

Executive Editor: **Jacqueline I. Kroschwitz**
Editor: **Mary Howe-Grant**
Associate Managing Editor: **Lindy Humphreys**
Copy Editors: **Christine Punzo**
                Lawrence Altieri

CONTRIBUTORS

Webb I. Bailey, *Air Products and Chemicals, Inc., Allentown, Pennsylvania,* Halogens (under Fluorine compounds, inorganic)

William X. Bajzer, *Dow Corning Corporation, Midland, Michigan,* Introduction; Poly(fluorosilicones) (both under Fluorine compounds, organic)

viii CONTRIBUTORS

Max M. Boudakian, *Chemical Consultant, Pittsford, New York,* Fluorinated aromatic compounds (under Fluorine compounds, organic)

Julius E. Dohany, *Consultant, Berwyn, Pennsylvania,* Poly(vinylidene fluoride) (under Fluorine compounds, organic)

S. Ebnesajjad, *E. I. du Pont de Nemours & Co., Inc., Wilmington, Delaware,* Poly(vinyl fluoride) (under Fluorine compounds, organic)

Arthur J. Elliot, *Halocarbon Products Corporation, North Augusta, South Carolina,* Bromotrifluoroethylene; Fluorinated acetic acids; Fluoroethanols (all under Fluorine compounds, organic)

Francis E. Evans, *Hamburg, New York,* Boron, boron trifluoride; Sulfur (both under Fluorine compounds, inorganic)

Richard E. Fernandez, *E. I. du Pont de Nemours & Co., Inc., Wilmington, Delaware,* Fluorinated aliphatic compounds (under Fluorine compounds, organic)

Richard M. Flynn, *3M Company, St. Paul, Minnesota,* Fluoroethers and fluoroamines (under Fluorine compounds, organic)

Subhash V. Gangal, *E. I. du Pont de Nemours & Co., Inc., Wilmington, Delaware,* Perfluorinated ethylene–propylene copolymers; Polytetrafluoroethylene; Tetrafluoroethylene–ethylene copolymers; Tetrafluoroethylene–perfluorovinyl ether copolymers (all under Fluorine compounds, organic)

Werner M. Grootaert, *3M Company, St. Paul, Minnesota,* Fluorocarbon elastomers (under Fluorine compounds, organic)

Philip B. Henderson, *Air Products and Chemicals, Inc., Allentown, Pennsylvania,* Nitrogen; Tungsten (both under Fluorine compounds, inorganic)

Yung K. Kim, *Dow Corning Corporation, Midland, Michigan,* Introduction; Poly(fluorosilicones) (both under Fluorine compounds, organic)

Jack L. Kosmala, *3M Company, St. Paul, Minnesota,* Polycholorotrifluoroethylene (under Fluorine compounds, organic)

Richard J. Lagow, *University of Texas at Austin,* Direct fluorination (under Fluorine compounds, organic)

Charles B. Lindahl, *Ozark-Mahoning Co., Tulsa, Oklahoma,* Introduction; Antimony; Arsenic; Barium; Calcium; Germanium; Phosphorus; Tantalum; Tin; Zinc (all under Fluorine compounds, inorganic)

Tariq Mahmood, *Ozark-Mahoning Co., Tulsa, Oklahoma,* Introduction; Antimony; Arsenic; Calcium; Germanium; Phosphorus; Tantalum; Tin; Zinc (all under Fluorine compounds, inorganic)

Ganpat Mani, *AlliedSignal Inc., Morristown, New Jersey,* Boron, boron trifluoride; Sulfur (both under Fluorine compounds, inorganic)

Dayal T. Meshri, *Advance Research Chemicals, Inc., Catoosa, Oklahoma,* Aluminum; Cobalt; Copper; Iron; Lead; Mercury; Molybdenum; Nickel; Rhenium; Silver; Titanium; Zirconium (all under Fluorine compounds, inorganic)

George H. Millet, *3M Company, St. Paul, Minnesota,* Fluorocarbon elastomers; Polychlorotrifluoroethylene (both under Fluorine compounds, organic)

Werner H. Mueller, *Hoechst-Celanese Corporation, Charlotte, North Carolina,* Sodium (under Fluorine compounds, inorganic)

John R. Papcun, *Atotech, Cleveland, Ohio,* Ammonium; Boron, fluoroboric acid and fluoroborates; Lithium; Magnesium; Potassium (all under Fluorine compounds, inorganic)

Paul R. Resnick, *E. I. du Pont de Nemours & Co., Inc., Wilmington, Delaware,* Perfluoroepoxides (under Fluorine compounds, organic)

Patricia Savu, *3M Company, St. Paul, Minnesota,* Fluorinated higher carboxylic acids; Perfluoroalkanesulfonic acids (both under Fluorine compounds, organic)

George Shia, *AlliedSignal, Buffalo, New York,* Fluorine

Jean'ne M. Shreeve, *University of Idaho, Moscow,* Oxygen (under Fluorine compounds, inorganic)

Bruce E. Smart, *E. I. du Pont de Nemours & Co., Inc., Wilmington, Delaware,* Fluorinated aliphatic compounds (under Fluorine compounds, organic)

Robert A. Smith, *AlliedSignal, Morristown, New Jersey,* Hydrogen (under Fluorine compounds, inorganic)

L. G. Snow, *E. I. du Pont de Nemours & Co., Inc., Wilmington, Delaware,* Poly(vinyl fluoride) (under Fluorine compounds, organic)

I. J. Solomon, *IIT Research Institute, Chicago, Illinois,* Oxygen (under Fluorine compounds, inorganic)

Kenneth Wachter, *Olin Corporation, Stamford, Connecticut,* Sodium (under Fluorine compounds, inorganic)

Allan T. Worm, *3M Company, St. Paul, Minnesota,* Fluorocarbon elastomers (under Fluorine compounds, organic)

Andrew J. Woytek, *Air Products and Chemicals, Inc., Allentown, Pennsylvania,* Halogens; Nitrogen; Tungsten (all under Fluorine compounds, inorganic)

PREFACE

This volume is one of a series of carefully selected reprints from the world-renowned *Kirk-Othmer Encyclopedia of Chemical Technology* designed to provide specific audiences with articles grouped by a central theme. Although the 27-volume Kirk-Othmer Encyclopedia is widely available, many readers and users of this key reference tool have expressed interest in having selected articles in their specialty collected for handy desk reference or teaching purposes. In response to this need, we have chosen original, complete articles related to fluorine and fluorine compounds to make up this new volume. The full texts, tables, figures, and reference materials from the original work have been reproduced here unchanged. All articles are by industrial or academic experts in the various fields of fluorine chemistry and its application to chemical technology. The final work represents the result of careful review by competent specialists and the thorough editorial processing of the professional Wiley staff. Introductory information from the Encyclopedia concerning nomenclature, SI units and conversion factors, and related information has been provided as a further guide to the contents.

This volume includes comprehensive coverage of fluorine, synthetic fluorine chemistry, and fluorine-containing materials, eg, fluorine superacids, inorganic fluorine glasses, monomeric and polymeric fluorocarbons, perfluorocompounds, and poly(fluorosilicones). Articles on direct fluorination and such high performance materials as fluorocarbon elastomers may also be found herein. Chemical Abstracts Service (CAS) Registry Numbers and chemical structures of the various fluorine compounds, alphabetical organization, extensive cross-referencing, and a complete index further enhance the utility of this comprehensive treatment. The 55 articles in this work, prepared by 31 leading authorities from chemical corporations, universities, and research institutes, vary in length from one to 57 pages depending on commercial importance and current scientific interest in the topic. The contents should be of interest to all those engaged in the design, synthesis, analysis, and manufacture of fluorine-containing intermediates and compounds as well as to those specifically engaged in the application of these materials. The book should be an important research reference tool, desk-top information resource, and supplementary reading asset for teaching professionals and their students.

M. Howe-Grant

NOTE ON CHEMICAL ABSTRACTS SERVICE REGISTRY NUMBERS AND NOMENCLATURE

Chemical Abstracts Service (CAS) Registry Numbers are unique numerical identifiers assigned to substances recorded in the CAS Registry System. They appear in brackets in the *Chemical Abstracts* (CA) substance and formula indexes following the names of compounds. A single compound may have synonyms in the chemical literature. A simple compound like phenethylamine can be named β-phenylethylamine or, as in *Chemical Abstracts*, benzeneethanamine. The usefulness of the *Encyclopedia* depends on accessibility through the most common correct name of a substance. Because of this diversity in nomenclature careful attention has been given to the problem in order to assist the reader as much as possible, especially in locating the systematic CA index name by means of the Registry Number. For this purpose, the reader may refer to the CAS Registry Handbook—Number Section which lists in numerical order the Registry Number with the *Chemical Abstracts* index name and the molecular formula; eg, **458-88-8**, Piperidine, 2-propyl-, (*S*)-, $C_8H_{17}N$; in the *Encyclopedia* this compound would be found under its common name, coniine [*458-88-8*]. Alternatively, this information can be retrieved electronically from CAS Online. In many cases molecular formulas have also been provided in the *Encyclopedia* text to facilitate electronic searching. The Registry Number is a valuable link for the reader in retrieving additional published information on substances and also as a point of access for on-line data bases.

In all cases, the CAS Registry Numbers have been given for title compounds in articles and for all compounds in the index. All specific substances indexed in *Chemical Abstracts* since 1965 are included in the CAS Registry System as are a large number of substances derived from a variety of reference works. The CAS Registry System identifies a substance on the basis of an unambiguous computer-language description of its molecular structure including stereochemical detail. The Registry Number is a machine-checkable number (like a Social Security number) assigned in sequential order to each substance as it enters the registry system. The value of the number lies in the fact that it is a concise and unique means of substance identification, which is independent of, and therefore bridges, many systems of chemical nomenclature. For polymers, one Registry Number may be used for the entire family; eg, polyoxyethylene (20) sorbitan monolaurate has the same number as all of its polyoxyethylene homologues.

Cross-references are inserted in the index for many common names and for some systematic names. Trademark names appear in the index. Names that are incorrect, misleading, or ambiguous are avoided. Formulas are given very frequently in the text to help in identifying compounds. The spelling and form used, even for industrial names, follow American chemical usage, but not always the usage of *Chemical Abstracts* (eg, *coniine* is used instead of *(S)-2-propylpiperidine*, *aniline* instead of *benzenamine*, and *acrylic acid* instead of *2-propenoic acid*).

There are variations in representation of rings in different disciplines. The dye industry does not designate aromaticity or double bonds in rings. All double bonds and aromaticity are shown in the *Encyclopedia* as a matter of course. For example, tetralin has an aromatic ring and a saturated ring and its structure

appears in the *Encyclopedia* with its common name, Registry Number enclosed in brackets, and parenthetical CA index name, ie, tetralin [*119-64-2*] (1,2,3,4-tetrahydronaphthalene). With names and structural formulas, and especially with CAS Registry Numbers, the aim is to help the reader have a concise means of substance identification.

CONVERSION FACTORS, ABBREVIATIONS, AND UNIT SYMBOLS

SI Units (Adopted 1960)

The International System of Units (abbreviated SI), is being implemented throughout the world. This measurement system is a modernized version of the MKSA (meter, kilogram, second, ampere) system, and its details are published and controlled by an international treaty organization (The International Bureau of Weights and Measures) (1).

SI units are divided into three classes:

BASE UNITS

length	meter[†] (m)
mass	kilogram (kg)
time	second (s)
electric current	ampere (A)
thermodynamic temperature[‡]	kelvin (K)
amount of substance	mole (mol)
luminous intensity	candela (cd)

SUPPLEMENTARY UNITS

plane angle	radian (rad)
solid angle	steradian (sr)

[†]The spellings "metre" and "litre" are preferred by ASTM; however, "-er" is used in the *Encyclopedia*.

[‡]Wide use is made of Celsius temperature (t) defined by

$$t = T - T_0$$

where T is the thermodynamic temperature, expressed in kelvin, and $T_0 = 273.15$ K by definition. A temperature interval may be expressed in degrees Celsius as well as in kelvin.

FACTORS, ABBREVIATIONS, AND SYMBOLS

DERIVED UNITS AND OTHER ACCEPTABLE UNITS

These units are formed by combining base units, supplementary units, and other derived units (2–4). Those derived units having special names and symbols are marked with an asterisk in the list below.

Quantity	Unit	Symbol	Acceptable equivalent
*absorbed dose	gray	Gy	J/kg
acceleration	meter per second squared	m/s^2	
*activity (of a radionuclide)	becquerel	Bq	1/s
area	square kilometer	km^2	
	square hectometer	hm^2	ha (hectare)
	square meter	m^2	
concentration (of amount of substance)	mole per cubic meter	mol/m^3	
current density	ampere per square meter	A//m^2	
density, mass density	kilogram per cubic meter	kg/m^3	g/L; mg/cm^3
dipole moment (quantity)	coulomb meter	C·m	
*dose equivalent	sievert	Sv	J/kg
*electric capacitance	farad	F	C/V
*electric charge, quantity of electricity	coulomb	C	A·s
electric charge density	coulomb per cubic meter	C/m^3	
*electric conductance	siemens	S	A/V
electric field strength	volt per meter	V/m	
electric flux density	coulomb per square meter	C/m^2	
*electric potential, potential difference, electromotive force	volt	V	W/A
*electric resistance	ohm	Ω	V/A
*energy, work, quantity of heat	megajoule	MJ	
	kilojoule	kJ	
	joule	J	N·m
	electronvolt[†]	eV[†]	
	kilowatt-hour[†]	kW·h[†]	
energy density	joule per cubic meter	J/m^3	
*force	kilonewton	kN	
	newton	N	kg·m/s^2

[†]This non-SI unit is recognized by the CIPM as having to be retained because of practical importance or use in specialized fields (1).

FACTORS, ABBREVIATIONS, AND SYMBOLS

Quantity	Unit	Symbol	Acceptable equivalent
*frequency	megahertz	MHz	
	hertz	Hz	1/s
heat capacity, entropy	joule per kelvin	J/K	
heat capacity (specific), specific entropy	joule per kilogram kelvin	J/(kg·K)	
heat transfer coefficient	watt per square meter kelvin	W/(m²·K)	
*illuminance	lux	lx	lm/m²
*inductance	henry	H	Wb/A
linear density	kilogram per meter	kg/m	
luminance	candela per square meter	cd/m²	
*luminous flux	lumen	lm	cd·sr
magnetic field strength	ampere per meter	A/m	
*magnetic flux	weber	Wb	V·s
*magnetic flux density	tesla	T	Wb/m²
molar energy	joule per mole	J/mol	
molar entropy, molar heat capacity	joule per mole kelvin	J/(mol·K)	
moment of force, torque	newton meter	N·m	
momentum	kilogram meter per second	kg·m/s	
permeability	henry per meter	H/m	
permittivity	farad per meter	F/m	
*power, heat flow rate, radiant flux	kilowatt	kW	
	watt	W	J/s
power density, heat flux density, irradiance	watt per square meter	W/m²	
*pressure, stress	megapascal	MPa	
	kilopascal	kPa	
	pascal	Pa	N/m²
sound level	decibel	dB	
specific energy	joule per kilogram	J/kg	
specific volume	cubic meter per kilogram	m³/kg	
surface tension	newton per meter	N/m	
thermal conductivity	watt per meter kelvin	W/(m·K)	
velocity	meter per second	m/s	
	kilometer per hour	km/h	
viscosity, dynamic	pascal second	Pa·s	
	millipascal second	mPa·s	
viscosity, kinematic	square meter per second	m²/s	
	square millimeter per second	mm²/s	

xviii FACTORS, ABBREVIATIONS, AND SYMBOLS

Quantity	Unit	Symbol	Acceptable equivalent
volume	cubic meter	m^3	
	cubic decimeter	dm^3	L (liter) (5)
	cubic centimeter	cm^3	mL
wave number	1 per meter	m^{-1}	
	1 per centimeter	cm^{-1}	

In addition, there are 16 prefixes used to indicate order of magnitude, as follows:

Multiplication factor	Prefix	Symbol	Note
10^{18}	exa	E	
10^{15}	peta	P	
10^{12}	tera	T	
10^{9}	giga	G	
10^{6}	mega	M	
10^{3}	kilo	k	
10^{2}	hecto	h^a	^{a}Although hecto, deka, deci, and centi are SI prefixes, their use should be avoided except for SI unit-multiples for area and volume and nontechnical use of centimeter, as for body and clothing measurement.
10	deka	da^a	
10^{-1}	deci	d^a	
10^{-2}	centi	c^a	
10^{-3}	milli	m	
10^{-6}	micro	μ	
10^{-9}	nano	n	
10^{-12}	pico	p	
10^{-15}	femto	f	
10^{-18}	atto	a	

For a complete description of SI and its use the reader is referred to ASTM E 380 (4) and the article UNITS AND CONVERSION FACTORS which appears in Vol. 24.

A representative list of conversion factors from non-SI to SI units is presented herewith. Factors are given to four significant figures. Exact relationships are followed by a dagger. A more complete list is given in the latest editions of ASTM E 380 (4) and ANSI Z210.1 (6).

Conversion Factors to SI Units

To convert from	To	Multiply by
acre	square meter (m^2)	4.047×10^3
angstrom	meter (m)	$1.0 \times 10^{-10\dagger}$
are	square meter (m^2)	$1.0 \times 10^{2\dagger}$

†Exact.

FACTORS, ABBREVIATIONS, AND SYMBOLS xix

To convert from	To	Multiply by
astronomical unit	meter (m)	1.496×10^{11}
atmosphere, standard	pascal (Pa)	1.013×10^{5}
bar	pascal (Pa)	$1.0 \times 10^{5\dagger}$
barn	square meter (m^2)	$1.0 \times 10^{-28\dagger}$
barrel (42 U.S. liquid gallons)	cubic meter (m^3)	0.1590
Bohr magneton (μ_B)	J/T	9.274×10^{-24}
Btu (International Table)	joule (J)	1.055×10^{3}
Btu (mean)	joule (J)	1.056×10^{3}
Btu (thermochemical)	joule (J)	1.054×10^{3}
bushel	cubic meter (m^3)	3.524×10^{-2}
calorie (International Table)	joule (J)	4.187
calorie (mean)	joule (J)	4.190
calorie (thermochemical)	joule (J)	$4.184^\dagger$
centipoise	pascal second (Pa·s)	$1.0 \times 10^{-3\dagger}$
centistokes	square millimeter per second (mm^2/s)	$1.0^\dagger$
cfm (cubic foot per minute)	cubic meter per second (m^3/s)	4.72×10^{-4}
cubic inch	cubic meter (m^3)	1.639×10^{-5}
cubic foot	cubic meter (m^3)	2.832×10^{-2}
cubic yard	cubic meter (m^3)	0.7646
curie	becquerel (Bq)	$3.70 \times 10^{10\dagger}$
debye	coulomb meter (C·m)	3.336×10^{-30}
degree (angle)	radian (rad)	1.745×10^{-2}
denier (international)	kilogram per meter (kg/m)	1.111×10^{-7}
	tex‡	0.1111
dram (apothecaries')	kilogram (kg)	3.888×10^{-3}
dram (avoirdupois)	kilogram (kg)	1.772×10^{-3}
dram (U.S. fluid)	cubic meter (m^3)	3.697×10^{-6}
dyne	newton (N)	$1.0 \times 10^{-5\dagger}$
dyne/cm	newton per meter (N/m)	$1.0 \times 10^{-3\dagger}$
electronvolt	joule (J)	1.602×10^{-19}
erg	joule (J)	$1.0 \times 10^{-7\dagger}$
fathom	meter (m)	1.829
fluid ounce (U.S.)	cubic meter (m^3)	2.957×10^{-5}
foot	meter (m)	$0.3048^\dagger$
footcandle	lux (lx)	10.76
furlong	meter (m)	2.012×10^{-2}
gal	meter per second squared (m/s^2)	$1.0 \times 10^{-2\dagger}$
gallon (U.S. dry)	cubic meter (m^3)	4.405×10^{-3}
gallon (U.S. liquid)	cubic meter (m^3)	3.785×10^{-3}
gallon per minute (gpm)	cubic meter per second (m^3/s)	6.309×10^{-5}
	cubic meter per hour (m^3/h)	0.2271

†Exact.
‡See footnote on p. xiii.

To convert from	To	Multiply by
gauss	tesla (T)	1.0×10^{-4}
gilbert	ampere (A)	0.7958
gill (U.S.)	cubic meter (m^3)	1.183×10^{-4}
grade	radian	1.571×10^{-2}
grain	kilogram (kg)	6.480×10^{-5}
gram force per denier	newton per tex (N/tex)	8.826×10^{-2}
hectare	square meter (m^2)	$1.0 \times 10^{4\dagger}$
horsepower (550 ft·lbf/s)	watt (W)	7.457×10^2
horespower (boiler)	watt (W)	9.810×10^3
horsepower (electric)	watt (W)	$7.46 \times 10^{2\dagger}$
hundredweight (long)	kilogram (kg)	50.80
hundredweight (short)	kilogram (kg)	45.36
inch	meter (m)	$2.54 \times 10^{-2\dagger}$
inch of mercury (32°F)	pascal (Pa)	3.386×10^3
inch of water (39.2°F)	pascal (Pa)	2.491×10^2
kilogram-force	newton (N)	9.807
kilowatt hour	megajoule (MJ)	$3.6^\dagger$
kip	newton(N)	4.448×10^3
knot (international)	meter per second (m/S)	0.5144
lambert	candela per square meter (cd/m^3)	3.183×10^3
league (British nautical)	meter (m)	5.559×10^3
league (statute)	meter (m)	4.828×10^3
light year	meter (m)	9.461×10^{15}
liter (for fluids only)	cubic meter (m^3)	$1.0 \times 10^{-3\dagger}$
maxwell	weber (Wb)	$1.0 \times 10^{-8\dagger}$
micron	meter (m)	$1.0 \times 10^{-6\dagger}$
mil	meter (m)	$2.54 \times 10^{-5\dagger}$
mile (statute)	meter (m)	1.609×10^3
mile (U.S. nautical)	meter (m)	$1.852 \times 10^{3\dagger}$
mile per hour	meter per second (m/s)	0.4470
millibar	pascal (Pa)	1.0×10^2
millimeter of mercury (0°C)	pascal (Pa)	$1.333 \times 10^{2\dagger}$
minute (angular)	radian	2.909×10^{-4}
myriagram	kilogram (kg)	10
myriameter	kilometer (km)	10
oersted	ampere per meter (A/m)	79.58
ounce (avoirdupois)	kilogram (kg)	2.835×10^{-2}
ounce (troy)	kilogram (kg)	3.110×10^{-2}
ounce (U.S. fluid)	cubic meter (m^3)	2.957×10^{-5}
ounce-force	newton (N)	0.2780
peck (U.S.)	cubic meter (m^3)	8.810×10^{-3}
pennyweight	kilogram (kg)	1.555×10^{-3}
pint (U.S. dry)	cubic meter (m^3)	5.506×10^{-4}
pint (U.S. liquid)	cubic meter (m^3)	4.732×10^{-4}

†Exact.

To convert from	To	Multiply by
poise (absolute viscosity)	pascal second (Pa·s)	$0.10^\dagger$
pound (avoirdupois)	kilogram (kg)	0.4536
pound (troy)	kilogram (kg)	0.3732
poundal	newton (N)	0.1383
pound-force	newton (N)	4.448
pound force per square inch (psi)	pascal (Pa)	6.895×10^3
quart (U.S. dry)	cubic meter (m³)	1.101×10^{-3}
quart (U.S. liquid)	cubic meter (m³)	9.464×10^{-4}
quintal	kilogram (kg)	$1.0 \times 10^{2\dagger}$
rad	gray (Gy)	$1.0 \times 10^{-2\dagger}$
rod	meter (m)	5.029
roentgen	coulomb per kilogram (C/kg)	2.58×10^{-4}
second (angle)	radian (rad)	$4.848 \times 10^{-6\dagger}$
section	square meter (m²)	2.590×10^6
slug	kilogram (kg)	14.59
spherical candle power	lumen (lm)	12.57
square inch	square meter (m²)	6.452×10^{-4}
square foot	square meter (m²)	9.290×10^{-2}
square mile	square meter (m²)	2.590×10^6
square yard	square meter (m²)	0.8361
stere	cubic meter (m³)	$1.0^\dagger$
stokes (kinematic viscosity)	square meter per second (m²/s)	$1.0 \times 10^{-4\dagger}$
tex	kilogram per meter (kg/m)	$1.0 \times 10^{-6\dagger}$
ton (long, 2240 pounds)	kilogram (kg)	1.016×10^3
ton (metric) (tonne)	kilogram (kg)	$1.0 \times 10^{3\dagger}$
ton (short, 2000 pounds)	kilogram (kg)	9.072×10^2
torr	pascal (Pa)	1.333×10^2
unit pole	weber (Wb)	1.257×10^{-7}
yard	meter (m)	$0.9144^\dagger$

†Exact.

Abbreviations and Unit Symbols

Following is a list of common abbreviations and unit symbols used in the *Encyclopedia*. In general they agree with those listed in *American National Standard Abbreviations for Use on Drawings and in Text (ANSI Y1.1)* (6) and *American National Standard Letter Symbols for Units in Science and Technology (ANSI Y10)* (6). Also included is a list of acronyms for a number of private and government organizations as well as common industrial solvents, polymers, and other chemicals.

Rules for Writing Unit Symbols (4):

1. Unit symbols are printed in upright letters (roman) regardless of the type style used in the surrounding text.
2. Unit symbols are unaltered in the plural.
3. Unit symbols are not followed by a period except when used at the end of a sentence.
4. Letter unit symbols are generally printed lower-case (for example, cd for candela) unless the unit name has been derived from a proper name, in which case the first letter of the symbol is capitalized (W, Pa). Prefixes and unit symbols retain their prescribed form regardless of the surrounding typography.
5. In the complete expression for a quantity, a space should be left between the numerical value and the unit symbol. For example, write 2.37 lm, *not* 2.37lm, and 35 mm, *not* 35mm. When the quantity is used in an adjectival sense, a hyphen is often used, for example, 35-mm film. *Exception:* No space is left between the numerical value and the symbols for degree, minute, and second of plane angle, degree Celsius, and the percent sign.
6. No space is used between the prefix and unit symbol (for example, kg).
7. Symbols, not abbreviations, should be used for units. For example, use "A," not "amp," for ampere.
8. When multiplying unit symbols, use a raised dot:

$$N \cdot m \quad \text{for} \quad \text{newton meter}$$

In the case of W·h, the dot may be omitted, thus:

$$Wh$$

An exception to this practice is made for computer printouts, automatic typewriter work, etc, where the raised dot is not possible, and a dot on the line may be used.

9. When dividing unit symbols, use one of the following forms:

$$m/s \quad or \quad m \cdot s^{-1} \quad or \quad \frac{m}{s}$$

In no case should more than one slash be used in the same expression unless parentheses are inserted to avoid ambiguity. For example, write:

$$J/(mol \cdot K) \quad or \quad J \cdot mol^{-1} \cdot K^{-1} \quad or \quad (J/mol)/K$$

but *not*

$$J/mol/K$$

10. Do not mix symbols and unit names in the same expression. Write:

$$\text{joules per kilogram} \quad or \quad \text{J/kg} \quad or \quad \text{J·kg}^{-1}$$

but *not*

$$\text{joules/kilogram} \quad nor \quad \text{joules/kg} \quad nor \quad \text{joules·kg}^{-1}$$

ABBREVIATIONS AND UNITS

A	ampere	AOAC	Association of Official Analytical Chemists
A	anion (eg, HA)		
A	mass number	AOCS	Americal Oil Chemists' Society
a	atto (prefix for 10^{-18})		
AATCC	American Association of Textile Chemists and Colorists	APHA	American Public Health Association
		API	American Petroleum Institute
ABS	acrylonitrile–butadiene–styrene		
		aq	aqueous
abs	absolute	Ar	aryl
ac	alternating current, *n*.	*ar*-	aromatic
a-c	alternating current, *adj*.	*as*-	asymmetric(al)
ac-	alicyclic	ASHRAE	American Society of Heating, Refrigerating, and Air Conditioning Engineers
acac	acetylacetonate		
ACGIH	American Conference of Governmental Industrial Hygienists		
		ASM	American Society for Metals
ACS	American Chemical Society		
		ASME	American Society of Mechanical Engineers
AGA	American Gas Association		
Ah	ampere hour	ASTM	American Society for Testing and Materials
AIChE	American Institute of Chemical Engineers		
		at no.	atomic number
AIME	American Institute of Mining, Metallurgical, and Petroleum Engineers	at wt	atomic weight
		av(g)	average
		AWS	American Welding Society
		b	bonding orbital
AIP	American Institute of Physics	bbl	barrel
		bcc	body-centered cubic
AISI	American Iron and Steel Institute	BCT	body-centered tetragonal
		Bé	Baumé
alc	alcohol(ic)	BET	Brunauer-Emmett-Teller (adsorption equation)
Alk	alkyl		
alk	alkaline (not alkali)	bid	twice daily
amt	amount	Boc	*t*-butyloxycarbonyl
amu	atomic mass unit	BOD	biochemical (biological) oxygen demand
ANSI	American National Standards Institute		
		bp	boiling point
AO	atomic orbital	Bq	becquerel

C	coulomb	DIN	Deutsche Industrie Normen
°C	degree Celsius		
C-	denoting attachment to carbon	*dl*-; DL-	racemic
		DMA	dimethylacetamide
c	centi (prefix for 10^{-2})	DMF	dimethylformamide
c	critical	DMG	dimethyl glyoxime
ca	circa (approximately)	DMSO	dimethyl sulfoxide
cd	candela; current density; circular dichroism	DOD	Department of Defense
		DOE	Department of Energy
CFR	Code of Federal Regulations	DOT	Department of Transportation
cgs	centimeter-gram-second	DP	degree of polymerization
CI	Color Index	dp	dew point
cis-	isomer in which substituted groups are on same side of double bond between C atoms	DPH	diamond pyramid hardness
		dstl(d)	distill(ed)
		dta	differential thermal analysis
cl	carload	(*E*)-	entgegen; opposed
cm	centimeter	ϵ	dielectric constant (unitless number)
cmil	circular mil		
cmpd	compound		
CNS	central nervous system	e	electron
CoA	coenzyme A	ECU	electrochemical unit
COD	chemical oxygen demand	ed.	edited, edition, editor
coml	commercial(ly)	ED	effective dose
cp	chemically pure	EDTA	ethylenediaminetetra-acetic acid
cph	close-packed hexagonal		
CPSC	Consumer Product Safety Commission	emf	electromotive force
		emu	electromagnetic unit
cryst	crystalline	en	ethylene diamine
cub	cubic	eng	engineering
D	debye	EPA	Environmental Protection Agency
D-	denoting configurational relationship		
		epr	electron paramagnetic resonance
d	differential operator		
d	day; deci (prefix for 10^{-1})	eq.	equation
d-	*dextro*-, dextrorotatory	esca	electron spectroscopy for chemical analysis
da	deka (prefix for 10^1)		
dB	decibel	esp	especially
dc	direct current, *n*.	esr	electron-spin resonance
d-c	direct current, *adj*.	est(d)	estimate(d)
dec	decompose	estn	estimation
detd	determined	esu	electrostatic unit
detn	determination	exp	experiment, experimental
Di	didymium, a mixture of all lanthanons	ext(d)	extract(ed)
		F	farad (capacitance)
dia	diameter	*F*	faraday (96,487 C)
dil	dilute	f	femto (prefix for 10^{-15})

FAO	Food and Agriculture Organization (United Nations)	hyd	hydrated, hydrous
		hyg	hygroscopic
		Hz	hertz
fcc	face-centered cubic	^{i} (eg, Pr^i)	iso (eg, isopropyl)
FDA	Food and Drug Administration	i-	inactive (eg, i-methionine)
		IACS	International Annealed Copper Standard
FEA	Federal Energy Administration	ibp	initial boiling point
FHSA	Federal Hazardous Substances Act	IC	integrated circuit
		ICC	Interstate Commerce Commission
fob	free on board		
fp	freezing point	ICT	International Critical Table
FPC	Federal Power Commission	ID	inside diameter; infective dose
FRB	Federal Reserve Board		
frz	freezing	ip	intraperitoneal
G	giga (prefix for 10^9)	IPS	iron pipe size
G	gravitational constant = 6.67 × 10^{11} N·m^2/kg^2	ir	infrared
		IRLG	Interagency Regulatory Liaison Group
g	gram		
(g)	gas, only as in H$_2$O(g)	ISO	International Organization Standardization
g	gravitational acceleration		
gc	gas chromatography	ITS-90	International Temperature Scale (NIST)
gem-	geminal		
glc	gas–liquid chromatography	IU	International Unit
		IUPAC	International Union of Pure and Applied Chemistry
g-mol wt; gmw	gram-molecular weight		
GNP	gross national product	IV	iodine value
gpc	gel-permeation chromatography	iv	intravenous
		J	joule
GRAS	Generally Recognized as Safe	K	kelvin
		k	kilo (prefix for 10^3)
grd	ground	kg	kilogram
Gy	gray	L	denoting configurational relationship
H	henry		
h	hour; hecto (prefix for 10^2)	L	liter (for fluids only) (5)
ha	hectare	l-	$levo$-, levorotatory
HB	Brinell hardness number	(l)	liquid, only as in NH$_3$(l)
Hb	hemoglobin	LC$_{50}$	conc lethal to 50% of the animals tests
hcp	hexagonal close-packed		
hex	hexagonal	LCAO	linear combination of atomic orbitals
HK	Knoop hardness number		
hplc	high performance liquid chromatography	lc	liquid chromatography
		LCD	liquid crystal display
HRC	Rockwell hardness (C scale)	lcl	less than carload lots
		LD$_{50}$	dose lethal to 50% of the animals tested
HV	Vickers hardness number		

FACTORS, ABBREVIATIONS, AND SYMBOLS

LED	light-emitting diode	N-	denoting attachment to nitrogen
liq	liquid		
lm	lumen	n (as n_D^{20})	index of refraction (for 20°C and sodium light)
ln	logarithm (natural)		
LNG	liquefied natural gas	^{n} (as Bu^n),	
log	logarithm (common)	n-	normal (straight-chain structure)
LPG	liquefied petroleum gas		
ltl	less than truckload lots	n	neutron
lx	lux	n	nano (prefix for 10^9)
M	mega (prefix for 10^6); metal (as in MA)	na	not available
		NAS	National Academy of Sciences
$\overline{M}$	molar; actual mass		
$\overline{M}_w$	weight-average mol wt	NASA	National Aeronautics and Space Administration
$\overline{M}_n$	number-average mol wt		
m	meter; milli (prefix for 10^{-3})	nat	natural
		ndt	nondestructive testing
m	molal	neg	negative
m-	meta	NF	*National Formulary*
max	maximum	NIH	National Institutes of Health
MCA	Chemical Manufacturers' Association (was Manufacturing Chemists Association)	NIOSH	National Institute of Occupational Safety and Health
MEK	methyl ethyl ketone	NIST	National Institute of Standards and Technology (formerly National Bureau of Standards)
meq	milliequivalent		
mfd	manufactured		
mfg	manufacturing		
mfr	manufacturer		
MIBC	methyl isobutyl carbinol	nmr	nuclear magnetic resonance
MIBK	methyl isobutyl ketone		
MIC	minimum inhibiting concentration	NND	New and Nonofficial Drugs (AMA)
min	minute; minimum	no.	number
mL	milliliter	NOI-(BN)	not otherwise indexed (by name)
MLD	minimum lethal dose		
MO	molecular orbital	NOS	not otherwise specified
mo	month	nqr	nuclear quadruple resonance
mol	mole		
mol wt	molecular weight	NRC	Nuclear Regulatory Commission; National Research Council
mp	melting point		
MR	molar refraction		
ms	mass spectrometry	NRI	New Ring Index
MSDS	material safety data sheet	NSF	National Science Foundation
mxt	mixture		
μ	micro (prefix for 10^{-6})	NTA	nitrilotriacetic acid
N	newton (force)	NTP	normal temperature and pressure (25°C and 101.3 kPa or 1 atm)
N	normal (concentration); neutron number		

NTSB	National Transportation Safety Board	qv	quod vide (which see)
O-	denoting attachment to oxygen	R	univalent hydrocarbon radical
		(R)-	rectus (clockwise configuration)
o-	ortho		
OD	outside diameter	r	precision of data
OPEC	Organization of Petroleum Exporting Countries	rad	radian; radius
		RCRA	Resource Conservation and Recovery Act
o-phen	o-phenanthridine		
OSHA	Occupational Safety and Health Administration	rds	rate-determining step
		ref.	reference
owf	on weight of fiber	rf	radio frequency, n.
Ω	ohm	r-f	radio frequency, adj.
P	peta (prefix for 10^{15})	rh	relative humidity
p	pico (prefix for 10^{-12})	RI	Ring Index
p-	para	rms	root-mean square
p	proton	rpm	rotations per minute
p.	page	rps	revolutions per second
Pa	pascal (pressure)	RT	room temperature
PEL	personal exposure limit based on an 8-h exposure	RTECS	Registry of Toxic Effects of Chemical Substances
		^{s} (eg, Bu^s); sec-	secondary (eg, secondary butyl)
pd	potential difference		
pH	negative logarithm of the effective hydrogen ion concentration	S	siemens
		(S)-	sinister (counterclockwise configuration)
phr	parts per hundred of resin (rubber)	S-	denoting attachment to sulfur
p-i-n	positive-intrinsic-negative		
pmr	proton magnetic resonance	s-	symmetric(al)
p-n	positive-negative	s	second
po	per os (oral)	(s)	solid, only as in $H_2O(s)$
POP	polyoxypropylene	SAE	Society of Automotive Engineers
pos	positive		
pp.	pages	SAN	styrene-acrylonitrile
ppb	parts per billion (10^9)	sat(d)	saturate(d)
ppm	parts per million (10^6)	satn	saturation
ppmv	parts per million by volume	SBS	styrene–butadiene–styrene
ppmwt	parts per million by weight	sc	subcutaneous
PPO	poly(phenyl oxide)	SCF	self-consistent field; standard cubic feet
ppt(d)	precipitate(d)		
pptn	precipitation	Sch	Schultz number
Pr (no.)	foreign prototype (number)	sem	scanning electron microscope(y)
pt	point; part		
PVC	poly(vinyl chloride)	SFs	Saybolt Furol seconds
pwd	powder	sl sol	slightly soluble
py	pyridine	sol	soluble

soln	solution	trans-	isomer in which substituted groups are on opposite sides of double bond between C atoms
soly	solubility		
sp	specific; species		
sp gr	specific gravity		
sr	steradian		
std	standard	TSCA	Toxic Substances Control Act
STP	standard temperature and pressure (0°C and 101.3 kPa)		
		TWA	time-weighted average
		Twad	Twaddell
sub	sublime(s)	UL	Underwriters' Laboratory
SUs	Saybolt Universal seconds	USDA	United States Department of Agriculture
syn	synthetic		
^{t} (eg, Bu^t), t-, tert-	tertiary (eg, tertiary butyl)	USP	*United States Pharmacopeia*
		uv	ultraviolet
T	tera (prefix for 10^{12}); tesla (magnetic flux density)	V	volt (emf)
		var	variable
		vic-	vicinal
t	metric ton (tonne)	vol	volume (not volatile)
t	temperature	vs	versus
TAPPI	Technical Association of the Pulp and Paper Industry	v sol	very soluble
		W	watt
		Wb	weber
TCC	Tagliabue closed cup	Wh	watt hour
tex	tex (linear density)	WHO	World Health Organization (United Nations)
T_g	glass-transition temperature		
tga	thermogravimetric analysis	wk	week
		yr	year
THF	tetrahydrofuran	(Z)-	zusammen; together; atomic number
tlc	thin layer chromatography		
TLV	threshold limit value		

Non-SI (Unacceptable and Obsolete) Units		Use
Å	angstrom	nm
at	atmosphere, technical	Pa
atm	atmosphere, standard	Pa
b	barn	cm^2
bar[†]	bar	Pa
bbl	barrel	m^3
bhp	brake horsepower	W
Btu	British thermal unit	J
bu	bushel	m^3; L
cal	calorie	J
cfm	cubic foot per minute	m^3/s
Ci	curie	Bq
cSt	centistokes	mm^2/s
c/s	cycle per second	Hz

[†]Do not use bar (10^5 Pa) or millibar (10^2 Pa) because they are not SI units, and are accepted internationally only for a limited time in special fields because of existing usage.

FACTORS, ABBREVIATIONS, AND SYMBOLS

Non-SI (Unacceptable and Obsolete) Units		Use
cu	cubic	exponential form
D	debye	C·m
den	denier	tex
dr	dram	kg
dyn	dyne	N
dyn/cm	dyne per centimeter	mN/m
erg	erg	J
eu	entropy unit	J/K
°F	degree Fahrenheit	°C; K
fc	footcandle	lx
fl	footlambert	lx
fl oz	fluid ounce	m^3; L
ft	foot	m
ft·lbf	foot pound-force	J
gf den	gram-force per denier	N/tex
G	gauss	T
Gal	gal	m/s^2
gal	gallon	m^3; L
Gb	gilbert	A
gpm	gallon per minute	(m^3/s); (m^3/h)
gr	grain	kg
hp	horsepower	W
ihp	indicated horsepower	W
in.	inch	m
in. Hg	inch of mercury	Pa
in. H_2O	inch of water	Pa
in.-lbf	inch pound-force	J
kcal	kilo-calorie	J
kgf	kilogram-force	N
kilo	for kilogram	kg
L	lambert	lx
lb	pound	kg
lbf	pound-force	N
mho	mho	S
mi	mile	m
MM	million	M
mm Hg	millimeter of mercury	Pa
mμ	millimicron	nm
mph	miles per hour	km/h
μ	micron	μm
Oe	oersted	A/m
oz	ounce	kg
ozf	ounce-force	N
η	poise	Pa·s
P	poise	Pa·s
ph	phot	lx
psi	pounds-force per square inch	Pa
psia	pounds-force per square inch absolute	Pa
psig	pounds-force per square inch gage	Pa
qt	quart	m^3; L
°R	degree Rankine	K
rd	rad	Gy
sb	stilb	lx
SCF	standard cubic foot	m^3
sq	square	exponential form
thm	therm	J
yd	yard	m

BIBLIOGRAPHY

1. The International Bureau of Weights and Measures, BIPM (Parc de Saint-Cloud, France) is described in Appendix X2 of Ref. 4. This bureau operates under the exclusive supervision of the International Committee for Weights and Measures (CIPM).
2. *Metric Editorial Guide (ANMC-78-1)*, latest ed., American National Metric Council, 5410 Grosvenor Lane, Bethesda, Md. 20814, 1981.
3. *SI Units and Recommendations for the Use of Their Multiples and of Certain Other Units (ISO 1000-1981)*, American National Standards Institute, 1430 Broadway, New York, N.Y. 10018, 1981.
4. Based on *ASTM E 380-89a (Standard Practice for Use of the International System of Units (SI))*, American Society for Testing and Materials, 1916 Race Street, Philadelphia, Pa. 19103, 1989.
5. *Fed. Regist.*, Dec. 10, 1976 (41 FR 36414).
6. For ANSI address, see Ref. 3.

R. P. LUKENS
ASTM Committee E-43 on SI Practice

FLUORINE

Fluorine [*7782-42-4*], F_2, is a diatomic molecule existing as a pale yellow gas at ordinary temperatures. Its name is derived from the Latin word *fleure*, meaning to flow, alluding to the well-known fluxing power of the mineral fluorite [*7789-75-5*], CaF_2, which is the most abundant naturally occurring compound of the element. Although radioactive isotopes between atomic weight 17 and 22 have been artificially prepared and have half-lives between 4 s for ^{22}F and 110 min for ^{18}F, fluorine has a single naturally occurring isotope, ^{19}F, and has an atomic weight of 18.9984 (1). Fluorine, the most electronegative element and the most reactive nonmetal, is located in the upper right corner of the Periodic Table. Its electron configuration is $1s^2 2s^2 2p^5$.

The only commercially feasible method of preparing elemental fluorine is by the electrolysis of molten fluoride-containing salts. Fluorine was first isolated in 1886 by the French chemist Moissan (2) who applied a method originally suggested and unsuccessfully tried by Davey and Ampere in 1810–1812. Moissan used potassium fluoride in anhydrous hydrogen fluoride resulting in an electrically conductive electrolyte. The only chemical route, which does not rely on compounds derived from F_2, has more recently been discovered (3). Both starting materials are easily prepared from HF, and react at 150°C evolving fluorine gas.

$$K_2MnF_6 + 2\,SbF_5 \rightarrow 2\,KSbF_6 + MnF_3 + 1/2\,F_2$$

Fluorine was first produced commercially ca 50 years after its discovery. In the intervening period, fluorine chemistry was restricted to the development of various types of electrolytic cells on a laboratory scale. In World War II, the demand for uranium hexafluoride [7783-81-5], UF_6, in the United States and United Kingdom, and chlorine trifluoride [7790-91-2], ClF_3, in Germany, led to the development of commercial fluorine-generating cells. The main use of fluorine in the 1990s is in the production of UF_6 for the nuclear power industry (see NUCLEAR REACTORS). However, its use in the preparation of some specialty products and in the surface treatment of polymers is growing.

Fluorine, which does not occur freely in nature except for trace amounts in radioactive materials, is widely found in combination with other elements, accounting for ca 0.065 wt % of the earth's crust (4). The most important natural source of fluorine for industrial purposes is the mineral fluorspar [14542-23-5], CaF_2, which contains about 49% fluorine. Detailed annual reports regarding the worldwide production and reserves of this mineral are available (5). A more complete discussion of the various sources of fluorine-containing minerals is given elsewhere (see FLUORINE COMPOUNDS, INORGANIC).

Physical Properties

Fluorine is a pale yellow gas that condenses to a yellowish orange liquid at $-188°C$, solidifies to a yellow solid at $-220°C$, and turns white in a phase transition at $-228°C$. Fluorine has a strong odor that is easily detectable at concentrations as low as 20 ppb. The odor resembles that of the other halogens and is comparable to strong ozone (qv).

Because of the extreme difficulty in handling fluorine, reported physical properties (Table 1) show greater than normal variations among investigators. A detailed summary and correlation of the physical, thermodynamic, transport, and electromagnetic properties of fluorine is given in Reference 20.

Chemical Properties

Fluorine is the most reactive element, combining readily with most organic and inorganic materials at or below room temperature. Many organic and hydrogen-containing compounds, in particular, can burn or explode when exposed to pure fluorine. With all elements except helium, neon, and argon, fluorine forms compounds in which it shows a valence of -1. Fluorine reacts directly with the heavier helium-group gases xenon, radon, and krypton to form fluorides (see HELIUM-GROUP GASES, COMPOUNDS).

Fluorine is the most electronegative element and thus can oxidize many other elements to their highest oxidation state. The small size of the fluorine atom facilitates the arrangement of a large number of fluorines around an atom of another element. These properties of high oxidation potential and small size allow

Table 1. Physical Properties of Fluorine

Property	Value	References
melting point, °C	-219.61^a	6
	-217.9	7
	-223	8
boiling point, °C	-188.13^a	6
	-187.7	7
	-187.0	8
	-188.22	9
	-188.03	10
solid transition temperature, °C	-227.60	6
critical temperature, °C	-129.2^a	11
	-129.00	9
critical pressure, kPa^b	5571	9,11
heat of vaporization, ΔH_{vap}, at -188.44°C and 98.4 kPa, J/mol^c	6544	6
heat of fusion, ΔH_{fus}, J/mol^c	510	6
heat of transition, J/mol^c	727.6	6
heat capacities, J/(mol·K)^c		
solid at -223°C	49.338	6
at -238°C	31.074	6
	23.267	7
at -253°C	12.987	6
	9.372	7
liquid	57.312^a	6
	45.35	7
gas, C_p	31.46^a	12
	31.456	13
	31.380	14
	31.325	15
density of liquid at bp, kg/m^3	1516^a	16
	1514	17
density of solid, kg/m^3	1900^d	18
refractive index		
liquid at bp	1.2	19
gas at 0°C and 101.3 kPa^b	1.000214	20
surface tension, liquid, mN/m(=dyn/cm)		
at -193.26°C	14.81	21
at -192.16°C	14.60	16
viscosity, mPa·s(=cP)		
liquid at -187.96°C	0.257	16
at -203.96°C	0.414	16
gas at 0°C and 101.3 kPa^b	0.0218	22
	0.0209	7
thermal conductivity, gas at 0°C and 101.3 kPa, W/(m·K)	0.02477	23
dielectric constant, ϵ		
at -189.95°C	1.517	7
at -215.76°C	1.567	7
vapor pressure, kPa^b		
at 53.56 K	0.22	6
at 63.49 K	2.79	6
at 72.56 K	18.62	6
at 83.06 K	80.52	6
at 89.40 K	162.11	6

^{a}Generally accepted value. ^{b}To convert kPa to mm Hg, multiply by 7.5. ^{c}To convert J to cal, divide by 4.184. ^{d}Mean estimate value.

the formation of many simple and complex fluorides in which the other elements are at their highest oxidation states.

The reactivity of fluorine compounds varies from extremely stable, eg, compounds such as sulfur hexafluoride [2551-62-4], nitrogen trifluoride [7783-54-2], and the perfluorocarbons (see FLUORINE COMPOUNDS, ORGANIC); to extremely reactive, eg, the halogen fluorides. Another unique property of nonionic metal fluorides is great volatility. Volatile compounds such as tungsten hexafluoride [7783-82-6], WF_6, and molybdenum hexafluoride [7783-77-9], MoF_6, are produced by the reaction of the particular metal with elemental fluorine.

Fluorine is the first member of the halogen family. However, many of its properties are not typical of the other halogens. Fluorine has only one valence state, -1, whereas the other halogens also form compounds in which their valences are $+1$, $+3$, $+5$, or $+7$. Fluorine also has the lowest enthalpy of dissociation relative to the other halogens, which is in part responsible for its greater reactivity. Furthermore, the strength of the bond fluorine forms with other atoms is greater than those formed by the other halogens.

Table 2 shows bond energies for the four diatomic molecular halogens, as well as for the halides of hydrogen, carbon, boron, and aluminum. Examination of these data indicates that the enthalpies of fluorination are much greater than those of other halogenations. Less energy is required to form fluorine molecules than that needed for chlorine or bromine molecules, and much more energy is evolved in the formation of the fluorides. Therefore, fluorination reactions occur more readily, generating intense heat, and these frequently occur in situations where other halogenations do not.

Reactions. *Metals.* At ordinary temperatures, fluorine reacts vigorously with most metals to form fluorides. A number of metals, including aluminum, copper, iron, and nickel, form an adherent and protective surface film of the metal fluoride salt thus allowing the metal's use in the storage and handling of the gas. A metal's susceptibility to reaction with fluorine depends, to a great extent, on its physical state. For example, powdered iron of 0.84-mm size (20 mesh) is not attacked by liquid fluorine, whereas in the 0.14-mm size (100 mesh) it ignites and burns violently. There is no apparent reaction between liquid fluorine and powdered nickel as fine as 0.14 mm. Massive copper burns at 692°C, whereas copper

Table 2. Average Bond Energies, kJ[a,b]

Halogen	XX[c]	HX[d]	BX_3[d]	AlX_3[e]	CX_4[d]
F	157.8	569	645	582	456
Cl	243.6	431	444	427	327
Br	193.0	368	368	360	272
I	151.1	297	272	285	239

[a]To convert J to cal, divide by 4.184.
[b]X = halogen.
[c]Ref. 24.
[d]Ref. 25.
[e]Ref. 26.

wool ignites at a much lower temperature. Nickel burns in fluorine at 1147°C, and aluminum burns above its melting point (27).

Tin reacts completely with fluorine above 190°C to form tin tetrafluoride [7783-62-2], SnF_4. Titanium reacts appreciably above 150°C at a rate dependent on the size of the particles; the conversion to titanium tetrafluoride [7783-63-3], TiF_4, is complete above 200°C. Fluorine reacts with zirconium metal above 190°C. However, the formation of a coating of zirconium tetrafluoride [7783-64-4], ZrF_4, prevents complete conversion, the reaction reaching only 90% completion even at 420°C (28). Tungsten powder reacts with fluorine at a temperature above 250°C to produce the volatile tungsten hexafluoride, WF_6.

Nonmetals. Sulfur reacts with fluorine to yield the remarkably stable sulfur hexafluoride, SF_6. Operating conditions must be controlled because a mixture of the lower fluorides such as disulfur difluoride [13709-35-8], S_2F_2, disulfur decafluoride [5714-22-7], S_2F_{10}, and sulfur tetrafluoride [7783-60-0], SF_4, may also be formed. When this reaction is carried out between 310 and 340°C, SF_4 is primarily obtained and essentially no SF_6 and only trace amounts of lower fluorides. Below 300°C, and preferably at ca 275°C, SF_6 is the primary product. At 450–500°C, a mixture comprising ca 50% SF_4 and the lower sulfur fluorides is formed (see FLUORINE COMPOUNDS, INORGANIC–SULFUR).

Silicon and boron burn in fluorine forming silicon tetrafluoride [7783-61-1], SiF_4, and boron trifluoride [7637-07-2], BF_3, respectively. Selenium and tellurium form hexafluorides, whereas phosphorus forms tri- or pentafluorides. Fluorine reacts with the other halogens to form eight interhalogen compounds (see FLUORINE COMPOUNDS, INORGANIC–HALOGENS).

Water. Fluorine reacts with water to form hydrofluoric acid [7664-39-3], HF, and oxygen difluoride [7783-41-7], OF_2. In dilute (<5%) caustic solutions, the reaction proceeds as follows:

$$2\ F_2 + 2\ NaOH \rightarrow OF_2 + 2\ NaF + H_2O$$

In the presence of excess caustic, the oxygen difluoride is gradually reduced to oxygen and fluoride:

$$OF_2 + 2\ NaOH \rightarrow 2\ NaF + O_2 + H_2O$$

The overall reaction under controlled conditions provides a method for the disposal of fluorine by conversion to a salt:

$$2\ F_2 + 4\ NaOH \rightarrow 4\ NaF + O_2 + 2\ H_2O$$

Oxygen. Oxygen does not react directly with fluorine under ordinary conditions, although in addition to oxygen difluoride, three other oxygen fluorides are known (29). Dioxygen difluoride [7783-44-0], O_2F_2, trioxygen difluoride [16829-28-0], O_3F_2, and tetraoxygen difluoride [12020-93-8], O_4F_2, are produced in an electric discharge at cryogenic temperatures by controlling the ratio of fluorine to oxygen.

Nitrogen. Nitrogen usually does not react with fluorine under ordinary conditions and is often used as a diluent to moderate fluorinations. However, nitrogen can be made to produce nitrogen trifluoride, NF_3, by radiochemistry (30), glow discharge (31), or plasma (32) synthesis (see PLASMA TECHNOLOGY).

Noble Gases. Fluorine has the unique ability to react with the heavier noble gases to form binary fluorides. Xenon reacts at room temperature under uv radiation forming xenon difluoride [13709-36-9], XeF_2 (16–18,21,33,34). Xenon tetrafluoride [13709-61-0], XeF_4, is obtained by mixing an excess of fluorine with xenon and heating the mixture to 400°C (33,35). Fluorine and xenon at 300°C under 6–6.8 MPa (60–67 atm) yield the xenon hexafluoride [13693-09-9], XeF_6 (36,37). Fluorine reacts with radon at 400°C to yield a compound of low volatility, probably the difluoride (38). Krypton reacts with fluorine in an electric discharge at liquid air temperatures to yield krypton difluoride [13773-81-4], KrF_2 (39).

Hydrogen. The reaction between fluorine and hydrogen is self-igniting and extremely energetic. It occurs spontaneously at ambient temperatures as evidenced by minor explosions which sometimes occur in fluorine-generating cells from the mixing of the H_2 and F_2 streams. The controlled high temperature reaction of fluorine atoms, whether generated thermally or photolytically from fluorine gas, with hydrogen or deuterium is an energy source for high power chemical lasers (qv) (40). However, NF_3 has become the preferred fluorine source because it is easier to handle (41).

Ammonia. Ammonia (qv) reacts with excess fluorine in the vapor phase to produce N_2, NF_3, N_2F_2, HF, and NH_4F. This reaction is difficult to control in the vapor phase because of the intense heat of reaction, and in some cases only N_2 and HF are produced. Nitrogen trifluoride was obtained in 6% yields in a gas-phase reaction over copper (42). Yields of ca 60% are achieved by the reaction of fluorine and ammonia in a molten ammonium acid fluoride solution (43,44).

Organic Compounds. The reaction of pure or undiluted fluorine and organic compounds is usually accompanied by either ignition or a violent explosion of the mixture because of the very high heat of reaction. However, useful commercial-scale syntheses using fluorine are undertaken. Volatile compounds may be fluorinated in the gas phase by moderating the reaction using an inert gas such as nitrogen, by reducing reaction temperatures ($\leq -78°C$), and/or by the presence of finely divided packing materials. Solutions or dispersions of higher boiling materials may be fluorinated in inert solvents such as 1,1,2-trichloro-1,2,2-trifluoroethane [76-13-1] or some perfluorocarbon fluids, eg, Fluorinert FC-27 or FC-75 (3M) (45,46). Efficient removal of the very high reaction heat, which leads to molecular fragmentation and runaway reactions, is the underlying principle in any of the aforementioned approaches.

Saturated hydrocarbons (qv) under controlled conditions react with elemental fluorine to produce perfluorocarbons; the reaction is usually accompanied by some fragmentation and polymerization (47,48). The fluorination of aromatic compounds gives degradation products, polymers, unstable unsaturated compounds, or highly fluorinated cyclohexane derivatives, but no aromatic compounds. Methanol (qv) and acetone (qv) produce a variety of fluorinated carbonyl compounds (49–51). Various polyethers have been successfully fluorinated to give perfluoropolyethers in high yields (45,46) (see FLUORINE COMPOUNDS, ORGANIC–DIRECT FLUORINATION). Fluorine is also used in the preparation of cobalt trifluoride [10026-18-3], CoF_3, and other higher oxidation state metal fluorides, which can be used in high temperature fluorinations of aliphatic compounds (52).

Selective fluorination, where only one or two fluorines are introduced into a molecule, is becoming more prominent. One well-known example is the direct fluorination of uracil (2,4-pyrimidinedione), $C_4N_2H_4O_2$, in aqueous solution to produce 5-fluorouracil [51-21-8] (5-fluoro-2,4-pyrimidinedione), $C_4N_2H_3FO_2$ (53). Certain nitrogen- and oxygen-containing organics react with fluorine to yield a group of compounds known as electrophilic fluorinating agents (54–57). These reagents are used in regiospecific fluorinations of other organic substrates, particularly those used in pharmaceuticals (qv). For more information regarding selective fluorination, see References 58 and 59.

Polymers. The dilution of fluorine using an inert gas significantly reduces the reactivity, thus allowing controlled reactions to take place with hydrocarbon polymers, even at elevated temperatures. High density polyethylene containers can be blow-molded using 1–10% fluorine in nitrogen mixtures to produce barrier layers on the inside of the containers (see BARRIER POLYMERS; OLEFIN POLYMERS) (60). The permeation rate of nonpolar solvents such as n-pentane can be reduced by a factor of almost 500 when containers are blow-molded with a 1% fluorine-in-nitrogen mixture (61,62). Mixtures of 1–10% fluorine in nitrogen have been used to improve the surface properties, ie, moisture transport, soil release, and soil redeposition, of polyester, polyamide, polyolefin, and polyacrylonitrile fibers (63,64) (see FIBERS, POLYESTER; POLYAMIDES, FIBERS).

Fluorine may also be used in conjunction with other reactive gases, eg, oxygen and water vapor, to activate polymer surfaces in order to improve chemical bonding and adhesion (65). For example, ethylene–propylene–diene monomer (EPDM) rubber moldings exposed to a very dilute fluorine mixture, eg, 1 to 5% F_2 in N_2, and subsequently to atmospheric oxygen and moisture have higher surface energy and can be bonded with adhesives (qv) (66). Adhesives do not wet and thus do not adhere to the untreated rubber. Fluorine activates the polymer surface by initiating reactions which form reactive C—OH, C—OOH, or C—OF surface groups. Also, contaminates such as plasticizers (qv) and polymer processing aids are oxidized and removed from the surface. Similarly, other materials including polyolefins (67), polyethylene–vinyl acetate foams (68), and rubber tire scrap (69), can be treated with fluorine.

Carbon and Graphite. Fluorine reacts with amorphous forms of carbon, such as wood charcoal, to form carbon tetrafluoride [75-73-0], CF_4, and small amounts of other perfluorocarbons. The reaction initiates at ambient conditions, but proceeds to elevated temperatures as the charcoal burns in fluorine.

Fluorine reacts with high purity carbon or graphite at elevated temperatures under controlled conditions to produce fluorinated carbon, $(CF_x)_n$. Compounds having colors ranging from black to white have been prepared with fluorine contents ranging from $x = 0.1$ to $x = 1.3$ (70–74). The material was first obtained (70) in 1934, when graphite was heated to 420–460°C in a stream of fluorine to produce a gray product of composition $CF_{0.92}$. Subsequently, a white material of composition $CF_{1.12}$ was obtained (71) by accurate control of the reaction temperature to 627 ± 3°C. Applications utilizing the unique properties of these materials began to be developed in the late 1960s. Fluorinated carbon is a specialty product used in lithium batteries (qv) (75) and lubricants (76). For a complete review of these materials see References 77 and 78.

Manufacture

Fluorine is produced by the electrolysis of anhydrous potassium bifluoride [7789-29-9], KHF_2 or $KF \cdot HF$, which contains various concentrations of free HF. The fluoride ion is oxidized at the anode to liberate fluorine gas, and the hydrogen ion is reduced at the cathode to liberate hydrogen. Anhydrous HF cannot be used alone because of its low electrical conductivity (see ELECTROCHEMICAL PROCESSING, INORGANIC).

Fluorine-generating cells are classified into three distinct types, based on operating temperatures: low (−80 to 20°C) temperature cells, medium (60–110°C) temperature cells, and high (220–300°C) temperature cells. Figure 1 is a melting point diagram for the KF–HF system showing the three distinct areas of temperature operation and the corresponding HF concentrations in the electrolyte. Reference 79 presents a complete description of various laboratory and industrial fluorine cells worldwide.

Cells operating at low (2,80,81) and high (79,82) temperatures were developed first, but discontinued because of corrosion and other problems. The first medium temperature cell had an electrolyte composition corresponding to $KF \cdot 3HF$, and operated at 65–75°C using a copper cathode and nickel anodes. A later cell operated at 75°C and used $KF \cdot 2.2HF$ or $KF \cdot 2HF$ as electrolyte (83,84), and nickel and graphite as anode materials.

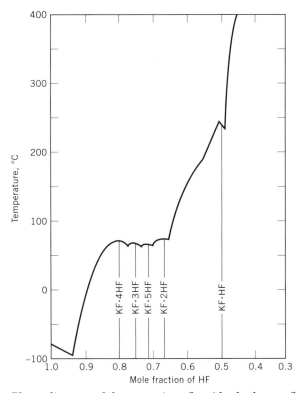

Fig. 1. Phase diagram of the potassium fluoride–hydrogen fluoride system.

Commercial Cells. All commercial fluorine installations employ medium temperature cells having operating currents of ≥ 5000 A. The medium temperature cell offers the following advantages over low and high temperature cells: (*1*) the vapor pressure of HF over the electrolyte is less; (*2*) the composition of the electrolyte can vary over a relatively wide range for only a small variation in the operation of the cell; (*3*) less corrosion or deterioration of the anode occurs; (*4*) tempered water can be used as cell coolant; and (*5*) the formation of a highly resistant film on the anode surface is considerably reduced compared to the high temperature cell.

The C and E type of the Atomic Energy Commission (AEC) (now the Department of Energy) cell designs (85–88) predominate in the United States and Canada. These were developed by Union Carbide Nuclear Co., under the auspices of the AEC, as part of the overall manufacturing process for uranium hexafluoride, UF_6. Large fluorine-generating plants using these designs were installed at the Paducah, Oak Ridge, and Portsmouth gaseous diffusion plants (which are no longer in operation) (see DIFFUSION SEPARATION METHODS). The AEC designs have been made available to industry and are used by several commercial producers. The other cell type used in the United States is a proprietary design developed by Allied Chemical, Corp. (now AlliedSignal, Inc.). This latter cell has a capacity of 5000 A and is used by AlliedSignal Inc. at its Metropolis, Illinois, plant. Table 3 gives the operating characteristics of a typical commercial size cell (AEC E-type).

AEC Cell. A diagram of the AEC cell is shown in Figure 2. The main components are the cell tank, cell head, anode assembly, cathode assembly, screen diaphragm, and packing gland (86,88,89). The E-type and C-type are similar in design and are both rated at 6000 A maximum capacity. The E-type incorporated design changes to improve heat removal efficiency and to prolong life of the anode-contact connection. The cell tank is constructed of 95-mm Monel plate surrounded by a thin-gauge Monel jacket to overcome the corrosion problems encountered by using a steel jacket. A recessed bolt carbon-plug fastens the anodes to a copper support bar. This design gives improved cell life over the C-type copper pressure-plate design by using a steel anode support bar.

The cell head is fabricated from a 2.54-cm steel plate and has separate compartments for fluorine and hydrogen. The outlet-gas manifolds, hydrogen fluoride feed and purge lines, and electrical connections are on top of the head. The gas

Table 3. AEC E-Type Cell Operating Characteristics

Characteristic	Value
current, A	6000
operating voltage, V	9–12
cell operating temperature, °C	90–105
hydrogen fluoride in electrolyte, %	40–42
effective anode area, m^2	3.9
anode current density, A/m^2	1500
anodes	32
anode life, A·h	$40-80 \times 10^6$

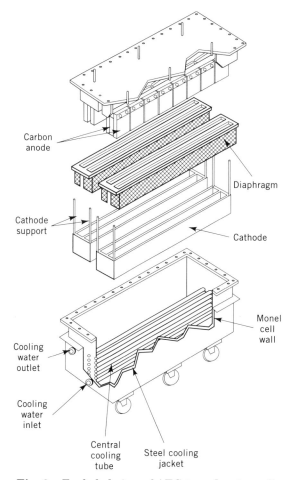

Fig. 2. Exploded view of AEC-type fluorine cell.

separation skirt is made of Monel. An insulating gasket maintains the seal between the tank and the head. The anode assembly consists of 32 carbon blades bolted onto a copper bar, each of which contains three copper conductor posts. The cathode assembly consists of three vertical, 0.6-cm parallel steel plates. The plates surround the anode assembly and are supported by three steel posts which also serve as conductors.

AlliedSignal Cell. A cross section of AlliedSignal's cell is shown in Figure 3. The cell body is a rectangular steel box on wheels with an outside water-cooling jacket. A central partition divides the box lengthwise. The negative side of the d-c bus is connected directly to the cell container. The interior of the box and the central partition act as the cathodic area of the cell.

Other Cell Designs. Although not used in the United States, another important cell is based on designs developed by ICI (90). Cells of this type are used by British Nuclear Fuels plc and differ from the cells shown in Figures 2 and 3

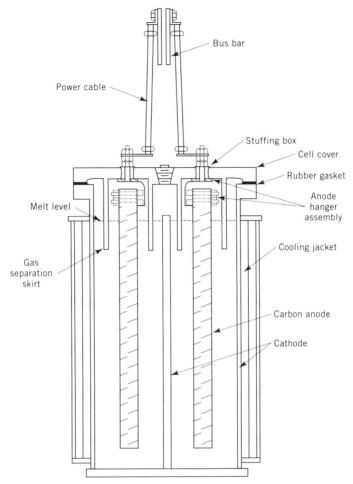

Fig. 3. Fluorine generator. Courtesy of AlliedSignal, Inc.

in two ways: (*1*) the anodes used are made of the same hard, nongraphitized carbon, but are more porous; and (*2*) the cathodes are formed from coiled tubes and provide additional cooling (91).

Anodes. Fluorine cell anodes are the most important cell component, and their design and materials of construction are key factors in determining productivity and cell life. Nickel and graphite, used in early cells, were abandoned when it was discovered that hard, nongraphitized carbon gave longer life. In the 1990s, anodes are made from petroleum coke and a pitch binder which is calcined at temperatures below that needed to convert the material to graphite. The anode carbon has low electrical resistance, high physical strength, and is resistant to reaction with fluorine. Historically, Union Carbide's YBD-grade carbon made the best anodes. More recently other carbon manufacturers have begun to offer improved anodes for fluorine service, eg, Carbone of America Ind. Corp.; Sociètè Des

Electrodes et Refractaires Savoié, part of the Pechiney group; and Toyo Tanso USA, Inc. The nature and quality of the starting materials (petroleum coke and pitch) and careful control of the calcining process are generally responsible for the improvements (92).

About 30% of the cell's operating voltage (9–12 V) is consumed by the anode overvoltage. Although some electrode overvoltage is common in all electrochemical processes, 3 to 4 V is exceptional. There are three generally accepted reasons for the high anode overvoltage: ohmic overvoltage, bubble overvoltage, and inhibition of charge transfer. Ohmic losses are generally small and result from the resistivity of the anode carbon and contact resistance between the anode and the metal current carriers. However, severe losses can arise if the contact between the anode and the current carrier becomes corroded. All commercial fluorine producers have proprietary designs for these connections, which mitigate this problem.

Bubble overvoltage and inhibition of charge transfer are related problems that are caused by the formation of a layer of fluorinated carbon, $(CF_x)_n$, on the anode surface (93,94). Because this material has very low electrical conductivity and is strongly nonwetting, electronic charge transfer is impeded and contact with the electrolyte is lost. As a result of the nonwetting character of the $(CF_x)_n$, bubbles of fluorine gas cling to the anode and grow in size, thereby reducing its effective surface area. Localized hot spots within the cell usually form under these conditions as the remaining working portions of the anode are subjected to higher than normal current densities. The locally excessive temperature also leads to the deterioration or burning of the anode. Signs of this problem are higher than normal cell voltage and higher levels of CF_4 in the fluorine product stream. In extreme cases the working surfaces of the anodes may become so restricted that fluorine production ceases.

Studies of anode electrochemistry have shown that water (>500 ppm) in the electrolyte can increase the formation of $(CF_x)_n$ on the anode surface (95). Under normal operating voltages, carbon can be electrochemically oxidized in the presence of water to form a graphite oxide, C_xO, film on the electrode surface, which then readily reacts with fluorine to form $(CF_x)_n$. Highly graphitic forms of carbon are much more susceptible to this problem. By starting out new cells at lower than normal operating voltages (6 V or less) trace amounts of water may be electrolyzed to H_2 and O_2 and removed from the cell, while avoiding the formation of C_xO.

Further improvements in anode performance have been achieved through the inclusion of certain metal salts in the electrolyte, and more recently by direct incorporation into the anode (92,96,97). Good anode performance has been shown to depend on the formation of carbon–fluorine intercalation compounds at the electrode surface (98). These intercalation compounds resist further oxidation by fluorine to form $(CF_x)_n$, have good electrical conductivity, and are wet by the electrolyte. The presence of certain metals enhance the formation of the intercalation compounds. Lithium, aluminum, or nickel fluoride appear to be the best salts for this purpose (92,98).

Other Cell Components. American fluorine manufacturers use Monel or steel cathodes. The early German investigators used magnesium cathodes without excessive corrosion in the high temperature cells. Welded steel or Monel construction is used for the cell body. Skirts are used to separate the hydrogen and fluorine above the electrolyte. The solid metal skirt is welded to the cell cover plate and extends vertically downward 10–15 cm into the electrolyte. In the AEC cells, the skirt extends further into the electrolyte as a wire mesh. This extension is called the diaphragm and acts to direct the flow of gases as they are liberated. The AEC cells use Monel skirts and diaphragms. The AlliedSignal cell uses a magnesium alloy skirt and does not employ a diaphragm.

Polytetrafluoroethylene (PTFE) provides the most satisfactory electrical insulation. Concentric rings of PTFE and PTFE impregnated with calcium fluoride are used for the packing glands which support the anode and cathode posts. Rubber is used as the gasket material to form a seal between the cover and the cell body.

Cells must be fitted with mild steel jackets and/or coils to remove heat during cell operation and to provide heat to maintain the electrolyte molten during shutdown. All commercial cells are totally jacketed. However, the accumulation of corrosive products can cause flow restrictions, resulting in decreased heat-transfer capacity. This problem was overcome in the AEC E-type cell by using a water jacket constructed of thin-gauge Monel with vertical corrugations to provide strength. The AEC cells also use internal Monel tubes, manifolded to the external jacket, for additional heat-transfer area. Because one of the operating limits is heat removal, improvements in jacket and tube design were a key factor in increasing the current-handling capacity of commercial cells.

Heat Transfer. A large portion of cell operating voltage is consumed in ohmic processes which generate heat and are a result of the large separation between anode and cathode and the resistivity of the electrolyte. Approximately 34.8 MJ (33,000 Btu) must be removed per kilogram of fluorine produced from any fluorine cell. This is accomplished by jacketing the cell and/or by using cooling tubes. The temperature of the cooling water should not drop below 58°C in order to avoid crystallization of bifluoride on the cell wall. The overall heat-transfer coefficient, U (thermal conductance), depends on operating current and inlet cooling water temperature (86). At 3000 A and an inlet water temperature of 40°C, an average U value is 109.5 W/(m^2·K) (19.3 Btu/(h·ft^2·°F)); for an operating current of 4000 A and a water temperature of 57°C, an average U value is 177.6 W/(m^2·K) (31.3 Btu/(h·ft^2·°F)).

Raw Material. The principal raw material for fluorine production is high purity anhydrous hydrofluoric acid. Each kilogram of fluorine generated requires ca 1.1 kg HF. Only a small portion of the hydrofluoric acid produced in the United States is consumed in fluorine production. The commercial grade is acceptable for use as received, provided water content is less than 0.02%. Typical specifications for hydrofluoric acid are

Assay	Wt %
HF, min	99.95
SO_2, max	0.005
H_2SiF_6, max	0.001
H_2O, max	0.02
nonvolatile acid (as H_2SO_4), max	0.01

Potassium bifluoride, KF·HF, is used as a raw material to charge the cells initially and for makeup when cells are rebuilt. A newly charged cell requires about 1400 kg KF·HF. Overall consumption of KF·HF per kilogram of fluorine generated is small. Commercial-grade flake potassium bifluoride is acceptable. Its specifications are

Assay	Wt %
KF·HF, min	99.3
Cl, max	0.01
K_2SiF_6	0.50
SO_4, max	0.01
Fe, max	0.02
Pb, max	0.005
H_2O, max	0.10

Process. The generation of fluorine on an industrial scale is a complex operation (89,99). The basic raw material, anhydrous hydrogen fluoride, is stored in bulk and charged to a holding tank from which it is continuously fed to the cells. Electrolyte for the cells is prepared by mixing KF·HF with HF to form KF·2HF. The newly charged cells are started up at a low current, which is gradually increased at a conditioning station separate from the cell operating position until full current is obtained at normal voltages. After conditioning, cells are connected in series using ca 12 V provided for each cell by a low voltage, 6000 A d-c rectifier. Hydrogen fluoride content is maintained between 40 and 42% by continuous additions. The electrolyte level must be set and controlled at a certain level below the cell head in order to maintain a seal between the fluorine and hydrogen compartments. The cells are operated at 95–105°C and cooled with water at 75°C.

Approximately 142.3 MJ/h (135,000 Btu/h) must be removed at an operating current of 6000 A. The hydrogen and fluorine gas leaving the cell contains ca 10 vol % HF. The individual gas streams from each cell are joined into separate hydrogen and fluorine headers for further processing in the plant. Demisters and filters are provided in the product gas streams to remove entrained electrolyte. The gas streams are then cooled to −110°C in refrigerant-cooled condensers to reduce the HF concentration to approximately 3 mol %. The condensed HF is recycled, and hydrogen stream is scrubbed with a caustic solution and vented or burned. Several possibilities are available for utilizing the fluorine stream. If lower than 3% HF levels are required, sodium fluoride towers or further cooling are employed to freeze out the HF and reduce the concentration to less than 0.2 mol %. Compressors or exhausters are normally required in both fluorine and hydrogen streams at some point in the system beyond the HF condensers.

Figure 4 presents the equipment flow sheet for a 9-t/d fluorine plant.

Equipment

Fluorine can be handled using a variety of materials (100–103). Table 4 shows the corrosion rates of some of these as a function of temperature. System clean-

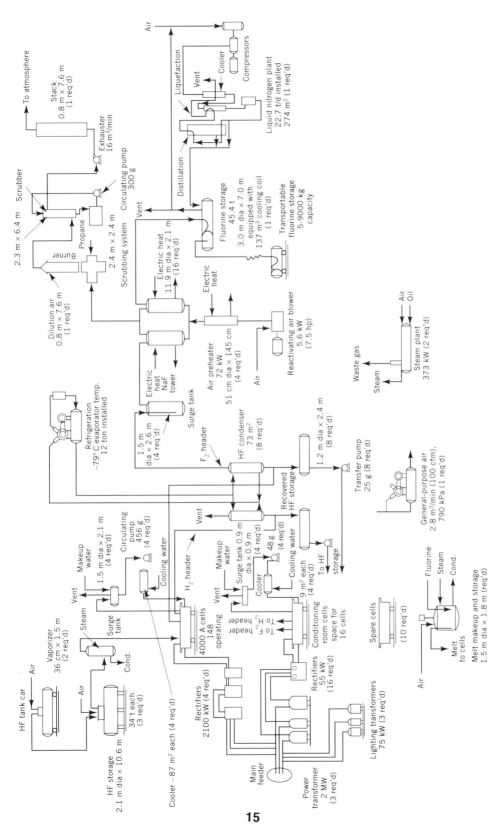

Fig. 4. Equipment flow sheet of elemental fluorine production and liquefaction plant, 9 t/d capacity. Step 1: purging residual F_2 at rates indicated; all but a trace of residual F_2 is removed in 15 min; N_2 purge is maintained for 1 h to remove last traces. Step 2: HF removal at rates indicated; all but a trace of HF is removed in 10 h; air purge is maintained for 10 h. To convert kPa to psi, multiply by 0.145.

FLUORINE

Table 4. Corrosion of Metals at Various Temperatures, mm/mo[a]

Material	Temperature, °C					
	200	300	400	500	600	700
nickel			0.018	0.129	0.74	0.86
Monel			0.013	0.051	1.5	3.8
Inconel			0.96	1.6	4.3	13
copper			4.1	3.0	25	74
aluminum			0	0.33	0.46	
magnesium	0	0				
iron	0	0.23	0.61	295		
steel						
0.27% carbon	0.051	0.23	0.38	503		
stainless, 310	0	0.79	14.2			

[a] Table abstracted from Reference 104.

liness and passivation are critical to success. Materials such as nickel, Monel, aluminum, magnesium, copper, brass, stainless steel, and carbon steel are commonly used. More information is available in the literature (20,104).

Copper, brass, and steel are generally used for gaseous service at temperatures below 200°C; Monel and nickel are used at elevated temperatures. For critical applications and where there is any danger of temperature buildup, only nickel or Monel should be used because of the stability of the nickel fluoride film. Only highly fluorinated polymers, such as polytetrafluoroethylene, are resistant to the gas under nominally static conditions. The PTFE must be free of any impurities, dirt, or foreign materials that could initiate ignition with fluorine. However, fluoropolymers are more susceptible to ignition than metals, and thus should be used as little as possible in fluorine service, especially in contact with flowing streams at high velocities or pressures.

All equipment, lines, and fittings intended for fluorine service must be leaktight, dry, and thoroughly cleansed of all foreign matter before use. The system should be checked for leaks, at least to its working pressure. It should be flushed with a nonaqueous degreasing solvent, such as methylene chloride, thoroughly purged with a stream of dry nitrogen, and evacuated to make certain no volatiles are present. Any foreign matter, particularly organics, not removed could burn with fluorine and initiate the burning of the metal equipment. After cleaning, the system should be filled with dry nitrogen.

The corrosion resistance of all materials used with fluorine depends on the passivation of the system. This is a pickling operation intended to remove the last traces of foreign matter, and to form a passive fluoride film on the metal surface. The dry nitrogen in the system is slowly replaced with gaseous fluorine in small increments until the concentration and pressure approach working conditions. Carbon steel is most commonly used for pipe and fittings (2.5 cm or greater) in gaseous fluorine service at ambient temperature conditions and pressures up to 2.86 MPa (415 psi).

Carbon steel or bronze-body gate valves are commonly used in gaseous fluorine service at low pressure. Plug valves, having Monel bodies and plugs, are recommended for moderate pressure service below 500 kPa (<5 atm). For valve-stem packing PTFE polymer is recommended and it must be maintained leak-tight. Valves lubricated or packed with grease or other organics should never be used. Bellows-type valves having Monel or stainless steel bellows are recommended for high pressure service, but not ball valves.

Compressors and blowers for gaseous fluorine service vary in design from multistage centrifugal compressors to diaphragm and piston types. Standard commercial instrumentation and control devices are used in fluorine systems. Pressure is measured using Bourdon-type gauges or pressure transducers. Stainless steel or Monel construction is recommended for parts in contact with fluorine. Standard thermocouples are used for all fluorine temperature-measuring equipment, such as the stainless-steel shielded type, inserted through a threaded compression fitting welded into the line. For high temperature service, nickel-shielded thermocouples should be used.

Dilute mixtures (eg, 10 or 20% F_2 in N_2) are generally less hazardous than pure fluorine, but the same precautions and procedures should be employed.

Economic Aspects

Availability and Shipping. Fluorine gas is packaged and shipped in steel cylinders conforming to Department of Transportation (DOT) specifications 3A1000 and 3AA1000 under a pressure of 2.86 MPa (415 psi). Table 5 lists world fluorine producers. Cylinders containing 2.2 and 0.7 kg are available from Air Products and Chemicals, Inc. All cylinders are equipped with special fluorine valves, the outlets of which have a left-hand thread conforming to the Compressed Gas Association (CGA) Specification 679 or 670. DOT regulations stipulate that cylinders must be shipped without safety relief devices and be equipped with valve protection caps. The total quantity of pure, gaseous fluorine in any size container is limited to 2.7 kg and the pressure must not exceed 2.86 MPa (415 psi) at 21°C.

Mixtures of 10 and 20% fluorine in nitrogen or other inert gases are commercially available in cylinders and tube trailers from Air Products and Chemicals, Inc. Blends can be safely packaged and stored at high, eg, 13.8 MPa (2000 psi), pressure. Filled with a 20% fluorine blend, tube trailers can contain up to 500 kg of fluorine. Such high pressure mixtures permit larger quantities of fluorine to be safely shipped.

Price. The 1993 U.S. price for fluorine in cylinders was $109/kg for 2.2 kg and $260/kg for 0.7 kg cylinders. The price in large volumes is determined by (1) the price of hydrofluoric acid; (2) power costs, ca 4.5 kW·h electricity is required for each kilogram of fluorine produced; (3) labor costs; (4) costs to maintain and rebuild cells; and (5) amortization of fixed capital. Fluorine production is highly capital intense. In addition, purification, compression, packaging, and distribution in cylinders increase the cost significantly.

Manufacturers. Besides manufacturers in the United States, commercial fluorine plants are operating in Canada, France, Germany, Italy, Japan, and the United Kingdom (see Table 5). Fluorine is also produced in the Commonwealth of Independent States (former Soviet Union); however, details regarding its manufacture, production volumes, etc, are regarded as secret information. The total

Table 5. Fluorine Producers

Name	Location	Products[a]	Capacity[b]
Air Products & Chemicals	Allentown, Pa.	SF_6, NF_3, WF_6, SF_4, ClF_3, BrF_3, IF_5, perfluorinated hydrocarbons, F_2 gas	
AlliedSignal, Inc.	Morristown, N.J.	UF_6, SF_6, IF_5, SbF_5, $(CF_x)_n$	large
Asahi Glass Co., Ltd.	Tokyo	SF_6	
Ausimont SpA	Milan, Italy	SF_6, fluoropolymer fluids, fluoromonomers, F_2 gas	large
British Nuclear Fuels plc	Preston, U.K.	UF_6, F_2 gas	large
Cameco Corp.	Port Hope, Ontario, Canada	UF_6	large
Central Glass Co., Ltd.	Tokyo	WF_6, NF_3, ClF_3, F_2 gas	moderate
Comurhex (Pechiney group)	Paris	UF_6, ClF_3, N_2F_2, WF_6, F_2 gas	large
Daikin Industries	Osaka, Japan	$(CF_x)_n$, IF_5, perfluorinated hydrocarbons	moderate
Kanto Denka Kogyo Co., Ltd.	Tokyo	SF_6, CF_4, CHF_3, WF_6, C_2F_6, NF_3, F_2 gas	moderate
Sequoyah Fuels Corp.	Gore, Okla.	UF_6	large
Solvay Fluor und Derivate, GmbH	Hannover, Germany	SF_6, IF_5, perfluorinated hydrocarbons, CF_4, WF_6, F_2 gas	large

[a]Not all companies that produce fluorine sell F_2 gas.
[b]Large, >1000 t/yr; moderate, >100 but < 1000 t/yr.

commercial production capacity of fluorine in the United States and Canada is estimated at over 5000 t/yr, of which 70–80% is devoted to uranium hexafluoride production. Most of the gas is used in captive uranium-processing operations.

Analysis

Direct analysis of gaseous fluorine is not possible by conventional methods because of its reactivity, therefore fluorine is converted (105) quantitatively to chlorine and the effluent stream analyzed. First, the fluorine sample is passed through a bed of sodium fluoride which retains the hydrogen fluoride. This step can be eliminated if the fluorine is free of HF as received. The sample is then passed over granular sodium chloride which reacts quantitatively with fluorine to release chlorine. This latter is reduced to chloride, which is determined by the Volhard method. Impurities such as oxygen, nitrogen, carbon tetrafluoride, sulfur hexafluoride, and carbon dioxide are determined by conventional gas chromatography techniques. Hydrogen fluoride is determined by infrared analysis of a separate sample. For processes using fluorine, the concentration may be monitored using uv analyzers such as those manufactured by Du Pont Instruments.

Fluorine in the atmosphere can be detected by chemical methods involving the displacement of halogens from halides. Dilute fluorine leaks are easily detected by passing a damp piece of starch iodide paper around the suspected area. The paper should be held with metal tongs or forceps to avoid contact with the gas stream and immediately darkens when fluorine is present.

Continuous monitoring for the presence of fluorine gas in the workplace may be accomplished using detectors available from Mine Safety Appliances (Pittsburgh, Pa.) or EIT (Exton, Pa.).

Specifications. Fluorine, having a dewpoint of 40°C and containing a maximum of 1.0 ppm water, is sold in cylinders according to the following specifications:

Assay	Mol %
fluorine, min	98
oxygen, max	0.5
nitrogen, max	1.0
carbon tetrafluoride, max	0.1
sulfur hexafluoride, max	0.1
hydrogen fluoride, max	0.2
carbon dioxide, max	0.1

Health and Safety

Fluorine, the most reactive element known, is a dangerous material but may be handled safely using proper precautions. In any situation where an operator may come into contact with low pressure fluorine, safety glasses, a neoprene coat, boots, and clean neoprene gloves should be worn to afford overall body protection. This protection is effective against both fluorine and the hydrofluoric acid which may form from reaction of moisture in the air.

In addition, face shields made of conventional materials or, preferably, transparent, highly fluorinated polymers, should be worn whenever operators approach equipment containing fluorine under pressure. A mask having a self-contained air supply or an air helmet with fresh air supply should always be available. Leaks in high pressure systems usually result in a flame from the reaction of fluorine with the metal. Shields should be provided for valves, pressure-reducing stations, and gauges. Valves are a particularly susceptible area for fluorine fires that can be initiated by foreign material accumulated at the valve seat. High pressure cylinders and valves should be remotely located with proper personnel protection and the latter should be operated using handle extensions. An excellent guide for the safe handling of fluorine in the laboratory is available (106).

Toxicity. Fluorine is extremely corrosive and irritating to the skin. Inhalation at even low concentrations irritates the respiratory tract; at high concentrations fluorine inhalation may result in severe lung congestion.

The American Conference of Governmental Industrial Hygienists (ACGIH) has established the 8-hour time-weighted average TLV as 1 ppm or 1.6 mg/m^3, and the short-term exposure limit TLV as 2 ppm or 3.1 mg/m^3. Fluorine has a sharp, penetrating odor detectable at levels well below the TLV. Manifestations of overexposure to fluorine include irritation or burns of the eyes, skin, and respiratory tract. The following emergency exposure limits (EEL) for humans have been suggested (107): 15.0 ppm for 10 min; 10 ppm for 30 min; and 7.5 ppm for 60 min.

Toxicity studies (108–110) established tolerance levels and degrees of irritations, indicating that the eye is the area most sensitive to fluorine. Comprehensive animal studies (111–113) determined a rat LC_{50} value of 3500 ppm·min for a single 5-min exposure and of 5850 ppm·min for a 15-min exposure. A no-effect concentration corresponded to a concentration-time value of ca 15% of the LC_{50} levels.

Because of the corrosive effects and discomfort associated with inhalation of fluorine, chronic toxicity does not occur. Although the metabolic fate of fluorine is not clear, it does not seem that much is converted to fluoride ion in the body (107). Therefore comparisons to effects of fluoride ion poisoning, known as fluorosis, are probably incorrect.

Burns. Skin burns resulting from contact with pure fluorine gas are comparable to thermal burns and differ considerably from those produced by hydrogen fluoride (114). Fluorine burns heal much more rapidly than hydrofluoric acid burns.

Disposal. Fluorine can be disposed of by conversion to gaseous perfluorocarbons or fluoride salts. Because of the long atmospheric lifetimes of gaseous perfluorocarbons (see ATMOSPHERIC MODELS), disposal by conversion to fluoride salts is preferred. The following methods are recommended: scrubbing with caustic solutions (115,116); reaction with solid disposal agents such as alumina, limestone, lime, and soda lime (117,118); and reaction with superheated steam (119). Scrubbing with caustic solution and, for dilute streams, reaction with limestone, are practiced on an industrial scale.

In a caustic scrubbing system, caustic potash, KOH, is preferred to caustic soda, NaOH, because of the higher solubility of the resulting potassium fluoride. Adequate solution contact and residence time must be provided in the scrub tower to ensure complete neutralization of the intermediate oxygen difluoride, OF_2. Gas residence times of at least one minute and caustic concentrations in excess of 5% are recommended to prevent OF_2 emission from the scrub tower.

Uses

Elemental fluorine is used captively by most manufacturers for the production of various inorganic fluorides (Table 5). The market for gaseous fluorine is small, but growing. The main use of fluorine is in the manufacture of uranium hexafluoride, UF_6, by

$$UF_4 + F_2 \rightarrow UF_6$$

Uranium hexafluoride is used in the gaseous diffusion process for the separation and enrichment of uranium-235, which exists in low concentration in natural uranium. The enriched UF_6 is converted back into an oxide and used as fuel for the nuclear power industry.

Another large use for elemental fluorine is in production of sulfur hexafluoride, SF_6, a gaseous dielectric for electrical and electronic equipment (see ELECTRONIC MATERIALS; FLUORINE COMPOUNDS, INORGANIC–SULFUR). Its high dielectric strength, inertness, thermal stability, and ease of handling have led to increased use as an electrically insulating medium, permitting reductions in size, weight, and cost of high voltage electrical switch gear, breakers, and substations. Elemental fluorine is also used to produce sulfur tetrafluoride, SF_4, by the reaction of sulfur and fluorine under controlled conditions. Sulfur tetrafluoride is a selective fluorinating agent used to produce fluorochemical intermediates in the pharmaceutical and herbicide industry (see HERBICIDES).

Fluorine reacts with the halogens and antimony to produce several compounds of commercial importance: antimony pentafluoride [*7783-70-2*], bromine trifluoride [*7787-71-5*], chlorine trifluoride [*7790-91-2*], and iodine pentafluoride [*7783-66-6*]. Chlorine trifluoride is used in the processing of UF_6 (see URANIUM AND URANIUM COMPOUNDS). Bromine trifluoride is used in chemical cutting by the oil well industry (see PETROLEUM). Antimony and iodine pentafluorides are used as selective fluorinating agents to produce fluorochemical intermediates (see FLUORINE COMPOUNDS, INORGANIC).

Fluorination of tungsten and rhenium produces tungsten hexafluoride, WF_6, and rhenium hexafluoride [*10049-17-9*], ReF_6, respectively. These volatile metal fluorides are used in the chemical vapor deposition industry to produce metal coatings and intricately shaped components (see THIN FILMS, FILM FORMATION TECHNIQUES).

Fluorine reacts with ammonia in the presence of ammonium acid fluoride to give nitrogen trifluoride, NF_3. This compound can be used as a fluorine source in the high power hydrogen fluoride–deuterium fluoride (HF/DF) chemical lasers and in the production of microelectronic silicon-based components.

Fluorine is used by a number of manufacturers to produce polyolefin containers that are resistant to permeation by organic liquids. In one application, the air which is normally used to blow-mold containers is replaced by a low concentration of fluorine in a mixture with nitrogen. In another approach, the containers are placed in a large enclosure subsequently flooded with very dilute fluorine–nitrogen or fluorine–air mixtures. Containers may vary in size from small bottles to automotive fuel tanks and show an outstanding resistance to nonpolar solvents and fuels. However, fuels containing polar additives, eg, alcohols, have been more difficult to contain and tank manufacturers are modifying the polyolefin and the fluorination process in an effort to meet Environmental Protection Agency (EPA) mandated fuel loss guidelines.

An important newer use of fluorine is in the preparation of a polymer surface for adhesives (qv) or coatings (qv). In this application the surfaces of a variety of polymers, eg, EPDM rubber, polyethylene–vinyl acetate foams, and rubber tire scrap, that are difficult or impossible to prepare by other methods are easily and quickly treated. Fluorine surface preparation, unlike wet-chemical surface treatment, does not generate large amounts of hazardous wastes and has been demonstrated to be much more effective than plasma or corona surface treatments. Figure 5 details the commercially available equipment for surface treating plastic components. Equipment to continuously treat fabrics, films, sheet foams, and other web materials is also available.

Fig. 5. Equipment for surface treating plastic components. Parts are loaded into one of the two lower chambers which is then evacuated to remove most of the air. This chamber is then flooded with a dilute mixture of fluorine and nitrogen which is made and stored in the upper chamber. After the treatment is completed, the fluorine mixture is pumped back up to the upper chamber for storage and the lower chamber repeatedly flooded with air and evacuated to remove any traces of fluorine gas. Two treatment chambers are cycled between the loading/unloading operation and the treatment step to increase equipment output. The fluorine–nitrogen blend may be used several times before by-products from the treatment process begin to interfere. All waste gases are purged through the scrubber shown to the right. Courtesy of FluorTec GmbH, Marksuhl, Germany.

Fluorine is used in the production of fluorinated organics both for the manufacture of perfluorinated materials and for the selective and regiospecific introduction of fluorine. Perfluorinated aliphatics, cycloaliphatics, and polyethers are made by fluorination of the hydrogen-containing analogue using F_2 or CoF_3. The superior chemical and thermal stability of perfluorocarbons has led to uses in high temperature lubrication, thermal testing of electronic components, and as specialty fluids for vacuum pumps, liquid seals, and hydraulic applications. Because of the high solubility of oxygen in these materials, perfluorinated aliphatics and cycloaliphatics have also been used as synthetic blood substitutes (see BLOOD, ARTIFICIAL) (120).

Although the selective introduction of fluorine into biologically important molecules has been demonstrated to provide dramatic improvements in efficacy and toxicity as compared to unfluorinated analogues, commercial methods using fluorine gas are uncommon. A notable exception is the production of 5-fluorouracil, made by the direct reaction of fluorine and uracil, which is used in cancer chemotherapy (see CHEMOTHERAPEUTICS, ANTICANCER). To overcome the difficulties encountered in direct fluorinations, electrophilic fluorinating agents have been commercialized by AlliedSignal, Inc.; Air Products & Chemicals, Inc.; and the Onoda Cement Co. These reagents are made from nitrogen-containing compounds, which when reacted with F_2 yield compounds that act as positive fluorine, F^+, sources. The selective introduction of fluorine into many types of organic molecules, including steroids, nucleosides, heterocycles, and aromatic compounds, can be effected using these reagents (54–57).

BIBLIOGRAPHY

"Fluorine" in *ECT* 1st ed., Vol. 6, pp. 656–667, by H. C. Miller and F. D. Lommis, Pennsylvania Salt Manufacturing Co.; in *ECT* 2nd ed., Vol. 9, pp. 506–525, by H. R. Neumark and J. M. Siegmund, Allied Chemical Corp.; in *ECT* 3rd ed., Vol. 10, pp. 630–654, by A. J. Woytek, Air Products & Chemicals, Inc.

1. D. N. Lapedes, ed., *Encyclopedia of Science and Technology*, Vol. 5, McGraw-Hill Book Co., Inc., New York, 1977, pp. 389–393.
2. H. Moissan, *Comp. Rend.* **102**, 1534 (1886); **103**, 202, 256 (1886); *Gmelins, Hanbuch der Anorganischen Chemie, System 5*, 8th ed., Deutsche Chemische Gesellschaft, Verlag Chemie, Berlin, 1926, pp. 4–16.
3. K. O. Christe, *Inorg. Chem.* **25**, 3721 (1986).
4. G. C. Finers, in M. Stacey, J. C. Tatlow, and A. G. Sharpe, eds., *Advances in Fluorine Chemistry*, Vol. 2, Butterworths, London, 1962.
5. M. M. Miller, *Fluorspar, Annual Report*, U.S. Dept. of Interior, Bureau of Mines, Washington, D.C., Sept. 1991.
6. J. H. Hu, D. White, and H. Johnson, *J. Am. Chem. Soc.* **75**, 5642 (1953).
7. E. Kanda, *Bull. Chem. Soc. Jpn.* **12**, 473 (1937).
8. J. D. Collins, L. S. Stone, and P. A. Juvner, *Background Chemistry for Development of Liquid Rocket Oxidizers, AD-18-283*, Callery Chemical Co., Callery, Pa., 1953.
9. G. H. Cady and J. H. Hildenbrand, *J. Am. Chem. Soc.* **52**, 3829 (1930).
10. W. H. Claussen, *J. Am. Chem. Soc.* **56**, 614 (1934).
11. D. Horovitz, *A Review of the Physical and Chemical Properties of Fluorine and Certain of Its Compounds, Report No. RMI-293-85*, Reaction Motors, Inc., Rockaway, N.J., 1950.
12. G. M. Murphy and J. E. Vance, *J. Chem. Phys.* **7**, 806 (1939).
13. V. N. Huff and S. Gordon, *Tables of Thermodynamics Functions for Analysis of Aircraft-Propulsion Systems, Tech. No. 2161*, National Advisory Committee for Aeronautics, Washington, D.C., Aug. 1950.
14. K. V. Butkov and R. B. Bozenbaum, *J. Phys. Chem. USSR* **24**, 706 (1950).
15. L. G. Cole, M. Farber, and G. W. Eluerum, Jr., *J. Chem. Phys.* **20**, 586 (1952).
16. G. W. Elverum, Jr., and R. N. Doeschev, *J. Chem. Phys.* **20**, 1834 (1952).
17. R. L. Jarry and H. C. Miller, *J. Am. Chem. Soc.* **78**, 1553 (1956).
18. W. T. Ziegler and J. C. Mullins, *Calculation of the Vapor Pressure and Heats of Vaporization and Sublimation of Liquids and Solids, Especially Below One Atmosphere, IV, Nitrogen and Fluorine. Technical Report no. 1*, Engineering Experiment Station, Georgia Institute of Technology, to Cryogenic Engineering Laboratory, National Bureau of Standards, Boulder, Colo., under NBS contract CST-7404, Apr. 1963.
19. E. U. Franck, *Naturwissenschaften* **41**, 37 (1954).
20. J. F. Tompkins and co-workers, *The Properties and Handling of Fluorine, Technical Report no. ASD-TDR-62-273*, Air Products & Chemicals, Inc., Allentown, Pa., 1963.
21. J. L. Weeks, C. L. Chernick, and M. S. Matheson, *J. Am. Chem. Soc.* **84**, 4612 (1962).
22. E. U. Franck and W. Stober, *Z. Naturforsch.* **7a**, 822 (1952).
23. E. U. Franck and E. Wicke, *Z. Elektrochem.* **55**, 643 (1951).
24. I. G. Stamper and R. F. Barrow, *Trans. Faraday Soc.* **54**, 1592 (1958).
25. D. A. Johnson, *Some Thermodynamic Aspects of Inorganic Chemistry*, Cambridge University Press, New York, 1968, p. 158.
26. E. L. Meutterties and C. W. Tullock, in W. L. Jolly, ed., *Preparative Inorganic Reactions*, Vol. 2, Interscience Publishers, New York, 1965, p. 243.
27. T. W. Godwin and C. F. Lorenzo, "Ignition of Seven Metals in Fluorine," paper no. 740, *American Rocket Society, 13th Annual Meeting*, New York, Nov. 17–21, 1958.
28. H. M. Haendler and co-workers, *J. Am. Chem. Soc.* **76**, 2177 (1954).

29. A. G. Streng, *Chem. Rev.* **63**, 607 (1963).
30. V. A. Dmitrievskii, V. N. Cherednikow, and E. K. Illin, *Khim Vys. Energ.* **7**(3), 206 (1973).
31. I. V. Nikitin and V. Ya. Rosolovski, *Izv. Adad, Nauk USSR Ser. Khim* **7**, 1464 (1970).
32. U.S. Pat. 3,304,248 (Feb. 14, 1967), H. T. Fullan and H. V. Scklemain (to Stauffer Chemical Co.).
33. C. L. Chernick and co-workers, *Science* **138**, 136 (1962).
34. J. H. Holloway, *Chem. Commun.*, 22 (1966).
35. H. H. Claassen, H. Selig, and J. G. Malm, *J. Am. Chem. Soc.* **84**, 3593 (1962).
36. J. G. Malm, I. Sheft, and C. L. Chernick, *J. Am. Chem. Soc.* **85**, 110 (1963).
37. E. E. Weaver, B. Weinstock, and C. P. Knop, *J. Am. Chem. Soc.* **85**, 111 (1963).
38. P. R. Fields, L. Stein, and M. H. Zirin, *J. Am. Chem. Soc.* **84**, 4164 (1962).
39. A. V. Grosse and co-workers, *Chem. Eng. News* **41**, 47 (Jan. 7, 1963).
40. G. C. Pimentel and J. H. Packer, *J. Chem. Phys.* **51**, 91 (1961).
41. M. C. Lin, M. E. Umstead, and N. Djeu, *Ann. Rev. Phys. Chem.* **34**, 557 (1983).
42. S. I. Morrow and co-workers, *J. Am. Chem. Soc.* **82**, 5301 (1960).
43. U.S. Pat. 4,091,081 (May 23, 1978), A. J. Woytek and J. T. Likeck (to Air Products and Chemicals, Inc.).
44. *Chem. Eng.* **84**(26), 116 (1977).
45. WO Pat. 90/06296 (June 14, 1990), M. G. Costello and G. I. Moore (to Minnesota Mining and Manufacturing Co.).
46. Eur. Pat. 0,332,601 (Sept. 13, 1989), F. R. Feher, and co-workers (to Monsanto Co.).
47. E. H. Hadley and L. A. Bigelow, *J. Am. Chem. Soc.* **62**, 3302 (1940).
48. E. A. Tyczkowski and L. A. Bigelow, *J. Am. Chem. Soc.* **77**, 3007 (1955).
49. K. B. Kellogg and G. H. Cady, *J. Am. Chem. Soc.* **70**, 3968 (1948).
50. N. Fukuhara and L. A. Bigelow, *J. Am. Chem. Soc.* **63**, 778 (1941).
51. W. D. Clark and R. J. Lagow, *J. Fluor. Chem.* **52**, 37 (1991).
52. M. Stacey and J. C. Tatlow, in M. Stacey, J. C. Tatlow, and A. G. Sharpe, eds., *Advances in Fluorine Chemistry*, Butterworths Publications, London, 1960, pp. 166–198.
53. Can. Pat. 3,954,749 (Mar. 16, 1976), P. D. Schuman and co-workers (to PCR, Inc.).
54. L. German and S. Zemskov, eds., *New Fluorinating Agents in Organic Synthesis*, Springer-Verlag, New York, 1989.
55. A. J. Poss and co-workers, *J. Org. Chem.* **56**, 5962 (1991).
56. E. Differding and co-workers, *Synlett*, 187 (1991); *Synlett*, 395 (1991).
57. T. Umemoto and co-workers, *J. Am. Chem. Soc.* **112**, 8563 (1990).
58. S. T. Purrington and B. S. Kagen, *Chem. Rev.* **86**, 997 (1986).
59. V. Grakauskas, *Intra-Science Chem. Rep.* **5**, 85 (1971).
60. U.S. Pat, 3, 862,284 (Jan. 21, 1975), D. D. Dixson, D. G. Manly, and G. W. Recktenwald (to Air Products and Chemicals, Inc.).
61. A. J. Woytek and J. F. Gentilecore, "A New Blow Molding Process to Reduce Solvent Permeation of Polyolefin Containers," paper no. 13 presented at *Advances in Blow Molding Conference*, Rubber and Plastics Institute, London, Dec. 6, 1977.
62. J. F. Gentilecore, M. A. Triolo, and A. J. Woytek, *Plast. Eng.* **34**(9), 40 (1978).
63. U.S. Pat. 4,020,223 (Apr. 26, 1977), D. D. Dixson and L. J. Hayes (to Air Products & Chemicals, Inc.).
64. U.S. Pat. 3,988,491 (Oct. 26 1976), D. D. Dixson and L. J. Hayes (to Air Products & Chemicals, Inc.).
65. R. Milker and A. Koch, in D. Satas, ed., *Coatings Technology Handbook*, Marcel Dekker, Inc., New York, 1990.
66. R. Milker and A. Koch, *Coating* **1**, 8 (1988).

67. B. D. Bauman, "Novel Polyurethane Composites with Surface-Modified Polymer Particles," paper presented at *SPI 32nd Annual Technical/Marketing Conference*, 1989.
68. R. Milker and A. Koch, *Kunststoffberater* **7/8**, 56 (1989).
69. B. D. Bauman, "Scrap Tire Reuse Through Surface-Modification Technology," paper presented at *International Symposium on Research and Development for Improving Solid Waste Management*, Cincinnati, Ohio, Feb. 7, 1991.
70. O. Ruff, D. Bretschneider, and F. Elert, *Z. Anorg. Chem.* **217**, 1 (1934).
71. U.S. Pat. 3,674,432 (July 4, 1972), R. J. Lagow and co-workers (to R. I. Patents, Inc.).
72. W. Rudorff, *Adv. Inorg. Chem. Radiochem.* **1**, 230 (1959).
73. G. R. Hennig, *Prog. Inorg. Chem.* **1**, 125 (1959).
74. N. Watanabe and K. Kumon, *Denki Kagaku* **35**, 19 (1967).
75. R. L. Fusaro and H. E. Sliney, *NASA Tech. Note D-5097*, National Aeronatics & Space Administration, Washington, D.C., 1969; *ALSE Trans.* **13**, 56 (1970).
76. M. Fukuda and T. Iijima, in J. P. Gabano, ed., *Lithium Batteries*, Academic Press, Inc., New York, 1983.
77. N. Watanabe, T. Nakajima, and H. Touhara, *Studies in Inorganic Chemistry 8, Graphite Fluorides*, Elsevier Science Publishers, New York, 1988.
78. G. A. Shia and G. Mani, in R. E. Banks, B. Smart, and J. C. Tatlow, eds., *Organofluorine Chemistry: Principles and Commercial Applications*, Plenum Publishing Corp., New York, in press.
79. R. J. Ring and D. Royston, *A Review of Fluorine Cells and Fluorine Production Facilities*, Australian Atomic Energy Commission, AAEC/E 281/, Sept. 1973.
80. A. J. Rudge, in A. Kuhn, ed., *Industrial Electrochemical Processes*, Elsevier Publishing Co., Amsterdam, the Netherlands, 1971, Chapt. 1.
81. A. J. Rudge, *Chem. Ind.* **22**, 504 (1956).
82. R. D. Fowler and co-workers, *Ind. Eng. Chem.* **39**, 3, 266 (1947).
83. G. H. Cady and H. S. Booth, eds., *Inorganic Synthesis*, Vol. 1, McGraw-Hill Book Co., Inc., New York, 1939, pp. 136–137.
84. C. H. Cady, D. A. Rogers, and C. A. Carlson, *Ind. Eng. Chem.* **34**, 4, 443 (1942).
85. S. H. Smiley and D. C. Brater, *USAEC Report TID-5295*, U.S. Atomic Energy Commission (USAEC), Washington, D.C., 1956.
86. J. Dykstra and co-workers, *Ind. Eng. Chem.* **47**, 5, 883 (1955).
87. B. W. Clark, *USAEC Report KY-326*, USAEC, Washington, D.C., 1960.
88. R. C. Kelley and W. E. Clark, eds., *USAEC Report TID4100, Suppl. 42, CAPE-55*, 1967, and *Suppl. 45, CAPE-486*, 1968, USAEC, Washington, D.C.
89. J. Dykstrra, A. P. Huber, and B. H. Thompson, "Multi-Ton Production of Fluorine for Manufacture of Uranium Hexafluoride," paper presented at *Second United Nations International Conference on the Peaceful Use of Atomic Energy*, A/CONF.15/P/524, June 1958.
90. Brit. Pat. 861,978 (1961), A. J. Rudge and A. Davies (to ICI).
91. J. F. Ellis and G. F. May, *J. Fluor. Chem.* **33** 133 (1986).
92. U.S. Pat. 4,312,718 (Jan. 26, 1982), N. Watanabe, M. Aramaki, and Y. Kita (to N. Watanabe, Central Glass., Ltd., and Toyo Tanso Co.).
93. L. Bai and B. E. Conway, *J. Appl. Electrochem.* **18**, 839 (1988).
94. L. Bai and B. E. Conway, *J. Appl. Electrochem.* **20**, 916 (1990).
95. T. Nakajima, T. Ogawa, and N. Watanabe, *J. Electrochem. Soc.* **134**, 8 (1987).
96. N. Watanabe, *Proc. Int. Symp. Molten Salt Chem. Technol., Molten Salt Comm. of the Electrochem. Soc. of Japan*, 21 (1983).
97. U.S. Pat. 4,915,809 (Apr. 10, 1990), O. Brown and M. Wilmott (to British Nuclear Fuels plc).
98. T. Nakajima and M. Touma, *J. Fluor. Chem.* **57**, 83 (1992).
99. J. Jacobson and co-workers, *Ind. Eng. Chem.* **47**, 5, 878 (1955).

100. J. R. McGuffey, R. Paluzelle, and W. E. Muldrew, *Ind. Eng. Chem.* **54**, 5, 46 (1962).
101. J. M. Siegmund, *Chem. Eng. Prog.* **63**, 6, 88 (1967).
102. W. C. Robinson, ed., *Fluorine Systerms Handbook*, Douglas Aircraft Co., Inc., Long Beach, Calif., 1967.
103. D. L. Endicott and L. H. Donahue, *Development and Demonstration of Criteria for Liquid Fluorine Feed System Components*, Report AFRPL-TR-65-133, McDonnell Douglas Astronautics Co., St. Louis, Mo., 1965.
104. W. R. Meyers and W. B. DeLong, *Chem. Eng. Progress* **44**, 359 (1948).
105. N. S. Nikolaev and co-workers, *Analytical Chemistry of Fluorine*, trans. by J. Schmorak, Halsted Press, New York, 1972, Chapt. 8.
106. E. A. Ranken and C. V. Borzileri, "The Safe Handling of Fluorine," *Health and Safety Manual, Supplement 21.12*, University of California, Lawrence Livermore National Laboratory, Berkeley, Apr. 1987.
107. Board on Toxicology and Environmental Health Hazards, National Research Council, *Emergency and Continuous Exposure Limits for Selected Airborne Contaminants*, Vol. 1, National Academy Press, Washington, D.C., Apr. 1984.
108. M. L. Keplinger and L. W. Suissa, *Am. Ind. Hyg. Assoc. J.* **29**, 10 (1968).
109. P. M. Ricca, *Am. Ind. Hyg. Assoc. J.* **31**, 22 (1970).
110. J. S. Lyon, *J. Occup. Med.* **4**, 199 (1962).
111. H. E. Stockinger, in C. Voegtlin and H. C. Hodge, eds., *The Pharmacology and Toxicology of Uranium*, McGraw-Hill Book Co., New York, 1949, Chapt. 17.
112. N. Ericksen and co-workers, *A Study of the Toxicological Effects of the Inhalation of Gaseous Fluorine at Concentrations of Approximately 25, 8, 3 and 0.7 mg/m³*, United States Atomic Energy Report 397, 407, 427, and 429, University of Rochester, New York, 1945.
113. N. Ericksen and co-workers, *A Study of the Lethal Effect of the Inhalation of Gaseous Fluorine at Concentrations from 100 ppm to 10,000 ppm*, United States Atomic Energy Report 435, University of Rochester, New York, 1945.
114. R. Y. Eagers, *Toxic Properties of Inorganic Fluorine Compounds*, Elsevier Publishing Co., Ltd., Amsterdam, the Netherlands, 1969, p. 43.
115. R. Landau and R. Rosen, *Ind. Eng. Chem.* **40**, 1239 (1948).
116. J. B. Ruch, *USAEC Report CF-60-4-38*, USAEC, Washington, D.C., 1960.
117. R. C. Liimatainer and W. L. Merchan, *Report ANL-5429*, Argonne National Laboratory, Northbrook, Ill., 1955.
118. J. D. Davratel, *USAEC Report RFP-1200*, USAEC, Washington, D.C., 1968.
119. S. H. Smiley and C. R. Schmitt, *Ind. Eng. Chem.* **46**, 244 (1954).
120. K. C. Lowe, *Adv. Mater.* **3**, 87 (1991).

General References

F. A. Cotton, ed., *Progress in Inorganic Chemistry*, Vol. 2, Interscience Publishers, New York, 1960.
R. Y. Eagers, *Toxic Properties of Inorganic Fluorine Compounds*, Elsevier Publishing Co. Ltd., London, 1969.
H. J. Emeleus, *J. Chem. Soc.*, 441 (1942).
Gmelins Handbuch der Anorganishcen Chemie, 8th ed., Suppl., Verlag Chemie, Weinheim, Germany, 1959, pp. 66–79.
M. Hudlicky, *Chemistry of Organic Fluorine Compounds*, Ellis Harwood Limited, Sussex, UK, 1976.
A. T. Kuhn, ed., *Industrial Electrochemical Processes*, Elsevier Publishing Co. Ltd., Amsterdam, the Netherlands, 1971.
Mellor's Comprehensive Treatise on Inorganic and Theoretical Chemistry, Vol. 2, Suppl., I. Longmans Green, New York, 1956, pp. 15–45.

N. S. Nikolaev and co-workers, *Analytical Chemistry of Fluorine*, trans. by J. Schmorak, Halsted Press, New York, 1972.

A. J. Rudge, *The Manufacture and Use of Fluorine and Its Compounds*, Oxford University Press, Inc., New York, 1962.

O. Ruff, *Chem. Ber.* **69A**, 181 (1936).

J. H. Simons, ed., *Fluorine Chemistry*, Vol. 1, Academic Press, Inc., New York, 1950.

C. Slesser and S. R. Schram, *Preparation Properties and Technology of Fluorine and Organic Fluoro-Compounds*, National Nuclear Energy Series, Div. VII, Vol. 1, McGraw-Hill Book Co., Inc., New York, 1951.

M. Stacey, J. C. Tatlow, and A. G. Sharpe, eds., *Advances in Fluorine Chemistry*, Vol. 2, Butterworths Inc., Washington, D.C., 1961.

Handbook of Compressed Gases, 3rd ed., Van Nostrand Reinhold, New York, 1990.

GEORGE SHIA
AlliedSignal, Inc.

FLUORINE COMPOUNDS, INORGANIC

Introduction, **27**
Aluminum, **33**
Ammonium, **47**
Antimony, **50**
Arsenic, **55**
Barium, **58**
Boron, **60**
Calcium, **83**
Cobalt, **96**
Copper, **98**
Germanium, **100**
Halogens, **102**
Hydrogen, **115**
Iron, **136**
Lead, **139**
Lithium, **141**
Magnesium, **143**
Mercury, **146**
Molybdenum, **148**
Nickel, **150**
Nitrogen, **152**
Oxygen, **158**
Phosphorus, **165**
Potassium, **176**
Rhenium, **180**
Silver, **182**
Sodium, **186**
Sulfur, **188**
Tantalum, **210**
Tin, **211**
Titanium, **214**
Tungsten, **217**
Zinc, **221**
Zirconium, **223**

INTRODUCTION

Fluorine (qv), the most electronegative element, is much more reactive than the other elements. On the Pauling scale of electronegativities, fluorine (value 4.0) lies well above oxygen (3.5), chlorine (3.0), and nitrogen (3.0) (1). Indeed, fluorine reacts with virtually every other element, including the helium group elements. These last were commonly called inert gases until 1962 when xenon, radon, and krypton were shown to react with fluorine (see HELIUM GROUP, GASES). Because of unique properties, fluorine has been called a superhalogen (1) and several of its compounds called superacids. The term superacid (2) is used for systems having higher acidities than anhydrous sulfuric or fluorosulfuric acid. A number of fluorine species exhibit superacid properties in HSO_3F or HSO_3F-SO_3 solutions. The SbF_5-HSO_3F system is an example.

The basic fluorine-containing minerals are fluorite [14542-23-5], commonly called fluorspar, CaF_2; and fluorapatite [1306-05-4], commonly called phosphate rock. The reaction of calcium fluoride and sulfuric acid produces hydrogen fluoride. Fluorosilicic acid is produced from fluorapatite as a by-product in the production of phosphoric acid. The boiling point of hydrogen fluoride, 19.54°C, is much higher than that of HCl, -84.9°C, owing to extensive molecular association via hydrogen bonding in the former. Hydrogen fluoride is the most common reagent for production of fluorine compounds. The first pure sample of anhydrous hydrogen fluoride was produced from thermal decomposition of KF·HF (3,4). Elemental fluorine, a pale greenish yellow gas, is produced by electrolysis of anhydrous potassium fluoride–hydrogen fluoride melts (see also ELECTROCHEMICAL PROCESSING, INORGANIC). IUPAC has recommended the prefix *fluoro* rather than the frequently used *fluo* for inorganic fluorine compounds. Terms such as fluoborate, fluosilicate, and silicofluoride are frequently used, however, rather than the preferred nomenclature, fluoroborate and fluorosilicate.

The fluoride ion is the least polarizable anion. It is small, having a diameter of 0.136 nm, 0.045 nm smaller than the chloride ion. The isoelectronic F^- and O^{2-} ions are the only anions of comparable size to many cations. These anions are about the same size as K^+ and Ba^{2+} and smaller than Rb^+ and Cs^+. The small size of F^- allows for high coordination numbers and leads to different crystal forms and solubilities, and higher bond energies than are evidenced by the other halides. Bonds between fluorine and other elements are strong whereas the fluorine–fluorine bond is much weaker, 158.8 kJ/mol (37.95 kcal/mol), than the chlorine–chlorine bond which is 242.58 kJ/mol (57.98 kcal/mol). This bond weakness relative to the second-row elements is also seen in O–O and N–N single bonds and results from electronic repulsion.

A number of elements exhibit the highest oxidation state only because fluorides and oxidation states of $+6$ and $+7$ are not uncommon. Examples of volatile fluorides of high oxidation state include VF_5, CrF_5, TaF_5, WF_6, MoF_6, PtF_6, ReF_6, ReF_7, and IF_7. Many complex fluorides demonstrate coordination not found for the other larger halides, eg, NiF_6^{2-}, SiF_6^{2-}, TaF_7^{2-}, TaF_8^{3-}, PF_6^-, AsF_6^-, ZrF_7^{3-}, and PbF_7^{3-}.

Fluorine forms very reactive halogen fluorides. Reaction of Cl_2 and F_2 at elevated temperatures can produce ClF, ClF_3, or ClF_5; BrF_3 and BrF_5 can be obtained from the reaction of Br_2 and F_2. These halogen fluorides react with all nonmetals, except for the noble gases, N_2, and O_2 (5). Fluorine also forms a class of compounds known as hypofluorites, eg, CF_3OF (6). Fluorine peroxide [7783-44-0], O_2F_2, has also been reported (6).

Fluorine's special properties lead to many applications. Its complexing properties account for its use as a flux in steelmaking and as an intermediate in aluminum manufacture. The reaction of fluorides with hydroxyapatite, $Ca_5(PO_4)_3OH$, which is found in tooth enamel, to form less soluble and/or more acid-resistant compounds, led to the incorporation of fluorides in drinking water and dentifrices (qv) to reduce dental caries. Many fluorides are volatile and in many cases are the most volatile compounds of an element. This property led to the use of UF_6 for uranium isotope enrichment, critical to the nuclear industry (see NUCLEAR REACTORS), and the use of metal fluorides in chemical vapor deposition (WF_6, MoF_6, ReF_6), in ion implantation (qv) for semiconductors (qv) (BF_3, PF_3, AsF_5, etc), and as unreactive dielectrics (SF_6). Because fluorine forms stable bonds, its compounds can be both extremely reactive (F_2, HF, interhalogens, hypofluorides, fluorinated peroxides, ionic MF_x) and extremely stable (CF_4, SF_6, covalent MF_x). Fluorinated steroids, other fluorinated drugs, and anesthetics have medical applications. The stability, lack of reactivity and, therefore, lack of toxicity of some fluorine compounds are also demonstrated by studies reporting survival of animals in an atmosphere of 80% SF_6 and 20% oxygen, and use of perfluorochemicals as short-term blood substitutes because of the ability to efficiently transport oxygen and carbon dioxide (see BLOOD, ARTIFICIAL; FLUORINE COMPOUNDS, ORGANIC). Fluorides including HF, BF_3, SbF_5, PF_5, and several complexes, eg, BF_4^-, PF_6^-, SbF_6^-, and AsF_6^-, are used in many applications in catalysis (qv).

History

The names fluorine and fluorospar are derived from the Latin *fluere* meaning flow or flux. In 1529 the use of fluorspar as a flux was described. In 1670 the etching of glass by acid-treated fluorspar was reported. Elemental fluorine was isolated by Moissan in 1886 (7).

Fluorspar has been used as a flux in the steel industry since the introduction of the open-hearth process. Historically, hydrogen fluoride was used in limited quantities for glass etching, polishing, scale removal, and small-volume production of fluorides. In the 1930s the first significant commercial HF production was applied toward the production of aluminum (see ALUMINUM AND ALUMINUM ALLOYS) and chlorofluorohydrocarbons for refrigerants (see REFRIGERATION AND REFRIGERANTS). During World War II HF was used in alkylation catalysis to produce aviation gasoline and in the manufacture of fluorine to produce volatile UF_6 for isotopic enrichment of the uranium essential for nuclear devices. The use of the very corrosive UF_6 also stimulated development of fluorinated organic compounds for lubricants and seals that are resistant to UF_6 (see LUBRICATION AND LUBRICANTS). Anhydrous HF is used both as a reactant and as a solvent (4) in the manufacture of inorganic fluorides.

Sources and Applications

The earth's crust consists of 0.09% fluorine. Among the elements fluorine ranks about thirteenth in terrestrial abundance.

The ores of most importance are fluorspar, CaF_2; fluorapatite, $Ca_5(PO_4)_3F$; and cryolite [*15096-52-3*], Na_3AlF_6. Fluorspar is the primary commercial source of fluorine. Twenty-six percent of the world's high quality deposits of fluorspar are in North America. Most of that is in Mexico. United States production in 1987–1991 was 314,500 metric tons, most of which occurred in the Illinois–Kentucky area. Imported fluorspar in 1990–1991 represented about 82% of U.S. consumption; 31% of U.S. fluorspar imports were from Mexico and 29% from China compared to 66% from Mexico in the 1973–1978 period. The majority of the fluorine in the earth's crust is in phosphate rock in the form of fluorapatite which has an average fluorine concentration of 3.5%. Recovery of these fluorine values as by-product fluorosilicic acid from phosphate production has grown steadily, partially because of environmental requirements (see PHOSPHORIC ACID AND THE PHOSPHATES).

Production of hydrogen fluoride from reaction of CaF_2 with sulfuric acid is the largest user of fluorspar and accounts for approximately 60–65% of total U.S. consumption. The principal uses of hydrogen fluoride are in the manufacture of aluminum fluoride and synthetic cryolite for the Hall aluminum process and fluoropolymers and chlorofluorocarbons that are used as refrigerants, solvents, aerosols (qv), and in plastics. Because of the concern that chlorofluorocarbons cause upper atmosphere ozone depletion, these compounds are being replaced by hydrochlorofluorocarbons and hydrofluorocarbons. The balance of hydrogen fluoride is used in applications such as stainless steel pickling, inorganic fluoride production, alkylation (qv), uranium enrichment, and fluorine production. Hydrogen fluoride is used to convert uranium oxide to UF_4 which then reacts with elemental fluorine to produce volatile UF_6. The UF_6 is then isotopically enriched by gaseous diffusion or gas centrifuge processes for nuclear applications.

The steel (qv) industry is also an extremely large user of fluorspar which is added to slag to make it more reactive. Smaller amounts are also used in the aluminum, ceramic, brick, cement, glass fiber, and foundry industries.

Synthesis

Most inorganic fluorides are prepared by the reaction of hydrofluoric acid with oxides, carbonates, hydroxides, chlorides, or metals. Routes starting with carbonate, hydroxide, or oxide are the most common and the choice is determined by the most economical starting material. In many cases, the water produced by the reaction cannot be removed without at least partial hydrolysis of the metal fluoride. This hydrolysis frequently can be reduced by dehydrating in a stream of hydrogen fluoride. If hydrolysis is unavoidable, reaction of anhydrous HF and the metal or the metal chloride may be required. The reaction of the metal and HF can be the most desirable if the metal is inexpensive relative to its salts and if the metal has an oxidation potential higher than hydrogen. If the metal is not finely divided, formation of a fluoride coating on the metal surface may occur. This may slow the reaction. Another route to metal fluorides is by reaction of the metal or its salts with elemental fluorine or with interhalogen fluorides. These reactions occur rapidly and frequently are violent. Because hydrofluoric acid is much less expensive than fluorine, it normally is used whenever possible. However, many of the compounds containing elements in the higher oxidation states can be achieved only by use of elemental fluorine.

Analysis and Characterization

The most popular device for fluoride analysis is the ion-selective electrode (see ELECTROANALYTICAL TECHNIQUES). Analysis using the electrode is rapid and this is especially useful for dilute solutions and water analysis. Because the electrode responds only to free fluoride ion, care must be taken to convert complexed fluoride ions to free fluoride to obtain the total fluoride value (8). The fluoride electrode also can be used as an end point detector in titration of fluoride using lanthanum nitrate [10099-59-9]. Often volumetric analysis by titration with thorium nitrate [13823-29-5] or lanthanum nitrate is the method of choice. The fluoride is preferably steam distilled from perchloric or sulfuric acid to prevent interference (9,10). Fusion with a sodium carbonate–sodium hydroxide mixture or sodium may be required if the samples are covalent or insoluble.

Because fluorine has a nuclear spin of one-half, a strong signal, and a large coupling constant, nmr spectroscopic analysis is an invaluable tool (see MAGNETIC SPIN RESONANCE). The use of gas chromatography on the many volatile fluorine compounds also has greatly enhanced separation and identification. However, nuclear magnetic resonance and infrared spectroscopy remain key techniques in the study of fluorine compounds (see INFRARED AND RAMAN SPECTROSCOPY). Raman spectroscopy and mass spectrometry (qv) are also useful tools.

Safety, Toxicity, and Handling

Hazards associated with fluorides are severe. Anhydrous or aqueous hydrogen fluoride is extremely corrosive to skin, eyes, mucous membranes, and lungs; it can cause permanent damage and even death. Detailed information about safety, toxicity, and handling can be obtained from the producers of hydrogen fluoride, eg, Elf Atochem North America, Inc., Du Pont, and AlliedSignal. Fluorides susceptible to hydrolysis can generate aqueous hydrogen fluoride. Ingestion of excess fluorides may cause poisoning or damage to bones and/or teeth. Fluorine-containing oxidizers can react with the body in addition to causing burns.

Hydrogen fluoride or compounds that can produce it and fluorine-containing oxidizers should be handled with adequate safety equipment and extreme care by well-trained personnel. Often the effect of skin exposure is not immediately evident, especially when dilute solutions are handled. Pain may develop several hours later.

Fluorides in small (1 ppm in water, 0.1% in dentifrices) quantities have been shown to provide dramatic reduction in dental decay. Fluorides also show promise for bone treatment and in pharmaceuticals (qv) (see also CHEMOTHERAPEUTICS, ANTICANCER; STEROIDS). However, larger quantities of fluorides can lead to dental fluorosis, bone fracture, and even death. The oral LD_{50} for free fluoride ion in rats appears to be 50 to 100 mg/kg body weight based on LD_{50} values for several fluorides.

Because hydrogen fluoride is extremely reactive, special materials are necessary for its handling and storage. Glass reacts with HF to produce SiF_4 which leads to pressure buildup and potential ruptures. Anhydrous hydrogen fluoride is produced and stored in mild steel equipment. Teflon or polyethylene are frequently used for aqueous solutions.

The OSHA permissible exposure limit (11) and the American Conference of Governmental Industrial Hygienists (ACGIH) established threshold limit value (TLV) (12) for fluorides is 2.5 mg of fluoride per cubic meter of air. This is the TLV–TWA concentration for a normal 8-h work day and a 40-h work week.

BIBLIOGRAPHY

"Fluorine Compounds, Inorganic," in *ECT* 1st ed., Vol. 6, pp. 667–668, by C. R. Hough, Polytechnic Institute of Brooklyn; in *ECT* 2nd ed., Vol. 9, pp. 527–529, by J. B. Beal, Jr., Ozark-Mahoning Co.; in *ECT* 3rd ed., Vol. 10, pp. 655–659, by C. B. Lindahl and D. T. Meshri, Ozark-Mahoning Co.

1. L. Pauling, *The Nature of the Chemical Bond*, 3rd ed., Cornell University Press, Ithaca, N.Y., 1960, pp. 82 and 90.
2. R. J. Gillespie, *Acc. Chem. Res.* **1**, 202 (1968).
3. E. Frmy, *Ann. Chim. Phys.* **47**, 5 (1856).
4. R. E. Banks, *J. Fluorine Chem.* **33**, 3–26 (1986).
5. O. Glemser, *J. Fluorine Chem.* **33**, 45–69 (1986).
6. J. M. Shreeve, *J. Fluorine Chem.* **33**, 179–193 (1986).
7. H. Moissan, *Compt. Rend.* **12**, 1543 (1886).
8. *Analytical Procedures for Fluoride Analyses*, Orion Research Inc., Boston, Mass., 1990–1991.
9. G. H. Cady, *Anal. Chem.* **48**, 655–660 (1976).
10. H. H. Willard and O. B. Winter, *Ind. Eng. Chem. Analyt. Edn.* **5**, 7–10 (1933).
11. *Code of Federal Regulations*, Title 29, Part 1910.1000, Washington, D.C.
12. *Threshold Limit Values for Chemical Substances and Physical Agents, 1992–1993*, The American Conference of Governmental Industrial Hygienists, Cincinnati, Ohio.

General References

Chemistry and characteristics

P. Tarrant, ed., *Fluorine Chemistry Reviews*, Vol. 6, Marcel Dekker Inc., New York, 1971; Vol. 1, 1967; Vol. 2, 1968; Vol. 3, 1969; Vol. 4, 1969; Vol. 5, 1973; Vol. 7, 1974; Vol. 8, 1977.

J. H. Simons, ed., *Fluorine Chemistry*, Academic Press, Inc., New York, Vol. 1, 1950; Vol. 2, 1954; Vol. 3, 1963; Vol. 4, 1965; Vol. 5, 1964.

M. Stacey and co-eds., *Advances in Fluorine Chemistry*, Butterworth Inc., Washington, D.C., Vol. 1, 1960; Vol. 2, 1961; Vol. 3, 1963; Vol. 4, 1965; Vol. 5, 1965; Vol. 6, 1970; Vol. 7, 1973.

G. Brauer, *Handbuch der Praparativen Anorganischen Chemie*, Ferdinand Enke, Stuttgart, Band 1, 1960; Band 2, 1962.

I. G. Ryss, *The Chemistry of Fluorine and Its Inorganic Compounds*, State Publishing House of Scientific, Technical, and Chemical Literature, Moscow, 1956; Eng. trans., AEC-tr-3927, Office of Technical Services, U.S. Dept. of Commerce, Washington D.C., 1960.

F. A. Cotton, ed., *Progress in Inorganic Chemistry*, Vol. 2, Interscience Publishers, Inc., New York, 1960.

Mellor's Comprehensive Treatise of Inorganic and Theoretical Chemistry, Vol. 2, Suppl. 1, Longmans Green & Co. Ltd., London, 1962.

Gmelin, *Handbuch der Anorganischen Chemie*, Fluorination System-Number 5, Verlag Chemie, GmbH, Weinheim, Germany, 1959.

M. F. A. Dove and A. F. Clifford, *Inorganic Chemistry in Liquid Hydrogen Fluoride*, Pergamon Press Ltd., Oxford, U.K., 1971.

E. Newbrun, ed., *Fluorides and Dental Caries*, Charles C Thomas Publishers, Springfield, Ill., 1975.

W. E. Jones and E. G. Skolnik, *Chem. Rev.* **76**, 563 (1976).

R. E. Banks, D. W. A. Sharp, and J. C. Tatlow, eds., *Fluorine, The First Hundred Years*, Elsevier, Science Publishing Co., Inc., New York, 1986, and *Journal of Fluorine Chemistry*, Vol. 33.

Production, consumption, uses, prices, and imports

Minerals Yearbooks Chapter on Fluorspar (annual comprehensive reports but delayed approximately two years).

Mineral Industry Surveys on Fluorspar, quarterly, and annual (less comprehensive but current).

U.S. Dept. of the Interior, Bureau of Mines, Washington, D.C. (periodic).

CHARLES B. LINDAHL
TARIQ MAHMOOD
Elf Atochem North America, Inc.

ALUMINUM

Both the binary and complex fluorides of aluminum have played a significant role in the aluminum industry. Aluminum trifluoride [*7784-18-1*], AlF_3, and its trihydrate [*15098-87-0*], $AlF_3 \cdot 3H_2O$, have thus far remained to be the only binary fluorides of industrial interest. The nonahydrate [*15098-89-2*], $AlF_3 \cdot 9H_2O$, and the monohydrate [*12252-28-7, 15621-55-3*], $AlF_3 \cdot H_2O$, are of only academic curiosity. The monofluoride [*13595-82-9*], AlF, and the difluoride [*13569-23-8*], AlF_2, have been observed as transient species at high temperatures.

Of the fluoroaluminates known, cryolite, ie, sodium hexafluoroaluminate [*15096-52-2*], Na_3AlF_6, has been an integral part of the process for production of aluminum. Recently, the mixtures of potassium tetrafluoroaluminate [*14484-69-6*], $KAlF_4$, and potassium hexafluoroaluminate [*13575-52-5*], K_3AlF_6, have been employed as brazing fluxes in the manufacture of aluminum parts.

Aluminum Monofluoride and Aluminum Difluoride

Significant vapor pressure of aluminum monofluoride [*13595-82-9*], AlF, has been observed when aluminum trifluoride [*7784-18-1*] is heated in the presence of reducing agents such as aluminum or magnesium metal, or is in contact with the cathode in the electrolysis of fused salt mixtures. AlF disproportionates into AlF_3 and aluminum at lower temperatures. The heat of formation at 25°C is −264 kJ/mol (−63.1 kcal/mol) and the free energy of formation is −290 kJ/mol (−69.3 kcal/mol) (1). Aluminum difluoride [*13569-23-8*] has been detected in the high temperature equilibrium between aluminum and its fluorides (2).

Aluminum Trifluoride

Aluminum trifluoride trihydrate [*15098-87-0*], $AlF_3 \cdot 3H_2O$, appears to exist in a soluble metastable α-form as well as a less soluble β-form (3). The α-form can be obtained only when the heat of the reaction between alumina and hydrofluoric acid is controlled and the temperature of the reaction is kept below 25°C. Upon warming the α-form changes into a irreversible β-form which is insoluble in water and is much more stable. The β-form is commercially available.

Aluminum trifluoride trihydrate is prepared by reacting alumina trihydrate and aqueous hydrofluoric acid. The concentration of acid can vary between 15 to 60% (4). In the beginning of the reaction, addition of $Al(OH)_3$ to hydrofluoric acid produces a clear solution which results from the formation of the soluble α-form of $AlF_3 \cdot 3H_2O$. As the addition of $Al(OH)_3$ is continued and the reaction temperature increases, irreversible change takes place and the α-form of $AlF_3 \cdot 3H_2O$ gets converted to the β-form and precipitation is observed. After all the alumina is added, the reaction mixture is continuously agitated for several hours at 90–95°C. After the precipitate settles down, the supernatant liquid is removed using rotary or table vacuum filters and the slurry is centrifuged. The cake is washed with cold water, dried, and calcined in rotating horizontal kilns (5), flash dryers, or fluid-bed calciners to produce anhydrous AlF_3 for aluminum reduction cells. This process is known as a wet process.

Aluminum trifluoride can also be advantageously made by a dry process in which dried $Al(OH)_3$ is treated at elevated temperatures with gaseous hydrogen fluoride. High temperature corrosion-resistant alloys, such as Monel, Inconel, and titanium are used in the construction of fluidized-bed reactors. In one instance, an Inconel reactor is divided into three superimposed compartments by two horizontal fluidizing grid sieve plates. Aluminum hydroxide is fed into the top zone where it is dried by the existing gases. The gases such as HF and SiF_4 are scrubbed from stack gases with water. These gases are recycled or used in the manufacture of cryolite [*15096-52-3*]. Solids are transported from top to bottom by downcomers while HF enters at the bottom zone getting preheated by heat exchange from the departing AlF_3. The bulk of the reaction occurs in the middle compartment which is maintained at 590°C.

The third process involves careful addition of aluminum hydroxide to fluorosilicic acid (6) which is generated by fertilizer and phosphoric acid-producing plants. The addition of $Al(OH)_3$ is critical. It must be added gradually and slowly so that the silica produced as by-product remains filterable and the $AlF_3 \cdot 3H_2O$ formed is in the soluble α-form. If the addition of $Al(OH)_3 \cdot 3H_2O$ is too slow, the α-form after some time changes into the insoluble β-form. Then separation of silica from insoluble β-$AlF_3 \cdot 3H_2O$ becomes difficult.

$$H_2SiF_6 + 2\,Al(OH)_3 \rightarrow 2\,AlF_3 \cdot 3H_2O + SiO_2 + H_2O$$

Environmentally sound phosphate fertilizer plants recover as much of the fluoride value as H_2SiF_6 as possible. Sales for production of $AlF_3 \cdot 3H_2O$ is one of the most important markets (see FERTILIZERS; PHOSPHORIC ACID AND THE PHOSPHATES).

Dehydration of $AlF_3 \cdot 3H_2O$ above 300°C leads to a partial pyrohydrolysis forming HF and Al_2O_3 which can be avoided by heating the trihydrate gradually to 200°C to remove 2.5 moles of water and then rapidly removing the remainder at 700°C. This latter procedure yields a product having less than 3.5% water content and Al_2O_3 content below 8% (7). This product is a typical material used in aluminum reduction cells. The presence of alumina does not interfere in the process of aluminum reduction because it replaces part of the alumina that is fed to the cells.

The principal producers of aluminum trifluoride in North America are Alcan, Alcoa, and AlliedSignal. It is also produced in other countries, eg, France, Mexico, Norway, Italy, Tunisia, and Japan. Total worldwide production of aluminum trifluoride in 1990 was 400,000 metric tons and the price was $1100/t. In 1993, because of excess recovery of fluorine values, use of energy efficient smelters, and the worldwide economic climate, the price was down to $750/t.

The principal use of AlF_3 is as a makeup ingredient in the molten cryolite, $Na_3AlF_6 \cdot Al_2O_3$, bath used in aluminum reduction cells in the Hall-Haroult process and in the electrolytic process for refining of aluminum metal in the Hoopes cell. A typical composition of the molten salt bath is 80–85% Na_3AlF_6, 5–7% AlF_3, 5–7% CaF_2, 2–6% Al_2O_3, and 0–7% LiF with an operating temperature of 950°C. Ideally fluorine is not consumed in the process, but substantial quantities of fluorine are absorbed by the cell lining and fluorine is lost to the atmosphere. Modern aluminum industry plants efficiently recycle the fluorine values.

Minor uses of aluminum fluoride include flux compositions for casting, welding (qv), brazing, and soldering (see SOLDERS AND BRAZING ALLOYS) (8,9); passivation of stainless steel (qv) surfaces (10); low melting glazes and enamels (see ENAMELS, PORCELAIN OR VITREOUS); and catalyst compositions as inhibitors in fermentation (qv) processes. Table 1 gives typical specifications for a commercial sample of AlF_3.

Other hydrates of aluminum trifluoride are the nonahydrate [15098-89-2], $AlF_3 \cdot 9H_2O$, which is stable only below 8°C, and aluminum trifluoride monohydrate [12252-28-7], [15621-55-3], $AlF_3 \cdot H_2O$, which occurs naturally as a rare mineral, fluellite found in Stenna-Gwyn Cornwall, U.K. (11).

Table 1. Specification for Commercial Aluminum Trifluoride

Parameter	Specification
assay as AlF_3, %	90–92
Al_2O_3, typical, %	8–9
SiO_2, max, %	0.1
iron as Fe_2O_3, %	0.1
sulfur as SO_2, %	0.32
bulk density, g/cm^3	
loose	1.3
packed	1.6
screen analysis, % retained	
105 μm (140 mesh)	20
74 μm (200 mesh)	60
44 μm (325 mesh)	90

Table 4. Physical Properties of Cryolite

Property	Value
mol wt	209.94
mp, °C	1012
transition temperature, °C	
monoclinic-to-rhombic	565
second-order	880
dimensions of unit cell, nm	
a	0.546
b	0.561
c	0.780
vapor pressure of liquid at 1012°C, Pa[a]	253
heat of fusion at 1012°C, kJ/mol[b]	107
heat of vaporization at 1012°C, kJ/mol[b]	225
heat of transition, kJ/mol[b]	
monoclinic-to-rhombic at 565°C	8.21
second-order at 880°C	0.4
heat capacity, J/(mol·K)[b]	
monoclinic crystal at 25°C	215
cubic crystal at 560°C	281
liquid at 1012°C	395
S, J/(mol·K)[b,c]	238
ΔH_f^0 at 25°C,[c] kJ/mol[b]	-3297
ΔG_f^0 at 25°C,[c] kJ/mol[b]	-3133
density, g/cm^3	
monoclinic crystal at 25°C	2.97
cubic crystal from x-ray	2.77
solid at 1012°C	2.62
liquid at 1012°C	2.087
hardness, Mohs'	2.5
refractive index	
α-fom	1.3385
β-fom	1.3389
τ-fom	1.3396
electrical conductivity, (Ω·cm)$^{-1}$	
solid at 400°C	4.0×10^{-6}
liquid at 1012°C	2.82
viscosity, liquid at 1012°C, mPa·s(=cP)[a]	6.7
surface tension, liquid in air, mN/m(=dyn/cm)	125
activity product constant in water at 25°C	1.46×10^{-34}
solubility in water, g/100 g	
at 25°C	0.0042
at 100°C	0.0135

[a] To convert Pa to mm Hg, multiply by 7.
[b] To convert J to cal, divide by 4.184.
[c] Monoclinic crystal.

The mineral cryolite is usually white, but may also be black, purple, or violet, and occasionally brownish or reddish. The lustre is vitreous to greasy, sometimes pearly, and the streak is white. The crystals are monoclinic, differing only slightly from orthorhombic symmetry, and have an axial angle of 90°11′. The space group is $P2_1/m$. The [001] and [110] axes are usually dominant, giving the crystals a cubic appearance. Twinning is ubiquitous, and because the lamellae tend to be perpendicular, cleavage appears to be cubic. The fracture of individual crystals, however, is uneven. Because its refractive indexes are close to that of water, powdered cryolite becomes nearly invisible when immersed in water, but because the optical dispersion is different for the two materials the suspension shows Christiansen colors.

Upon heating the crystallographic angles approach 90° and the transition to the cubic form at 565°C is accompanied by a small heat change. The transition also involves a substantial change in density as evidenced by a characteristic decrepitation (17). The second transformation occurs at 880°C as indicated by the slope of the heating curve. It is also accompanied by a sharp rise in electrical conductivity. The heat change is very small and the transitions with rising temperatures probably mark the onset of a lattice disorder. The more plastic character of the solid near the melting point seems to corroborate this view (18).

Liquid cryolite is an equilibrium mixture of the products of the dissociation:

$$Na_3AlF_6 \rightarrow 2\ NaF + NaAlF_4$$

The composition to the melting point is estimated to be 65% Na_3AlF_6, 14% NaF, and 21% $NaAlF_4$ [1382-15-3]. The ions Na^+ and F^- are the principal current carrying species in molten cryolite whereas the AlF^{4-} is less mobile. The structural evidences are provided by electrical conductivity, density, thermodynamic data, cryoscopic behavior, and the presence of $NaAlF_4$ in the equilibrium vapor (19,20).

Molten cryolite dissolves many salts and oxides, forming solutions of melting point lower than the components. Figure 1 combines the melting point diagrams for cryolite–AlF_3 and for cryolite–NaF. Cryolite systems are of great importance

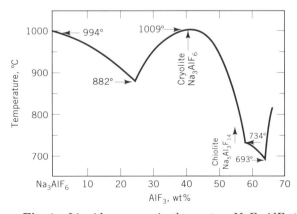

Fig. 1. Liquidus curves in the system NaF–AlF_3 (21).

High Purity Aluminum Trifluoride. High purity anhydrous aluminum trifluoride that is free from oxide impurities can be prepared by reaction of gaseous anhydrous HF and $AlCl_3$ at 100°C, gradually raising the temperature to 400°C. It can also be prepared by the action of elemental fluorine on metal/metal oxide and subsequent sublimation (12) or the decomposition of ammonium fluoroaluminate at 700°C.

Relatively smaller amounts of very high purity AlF_3 are used in ultra low loss optical fiber–fluoride glass compositions, the most common of which is ZBLAN containing zirconium, barium, lanthanum, aluminum, and sodium (see FIBER OPTICS). High purity AlF_3 is also used in the manufacture of aluminum silicate fiber and in ceramics for electrical resistors (see CERAMICS AS ELECTRICAL MATERIALS; REFRACTORY FIBERS).

Anhydrous aluminum trifluoride, AlF_3, is a white crystalline solid. Physical properties are listed in Table 2. Aluminum fluoride is sparingly soluble in water (0.4%) and insoluble in dilute mineral acids as well as organic acids at ambient temperatures, but when heated with concentrated sulfuric acid, HF is liberated, and with strong alkali solutions, aluminates are formed. AlF_3 is slowly attacked by fused alkalies with the formation of soluble metal fluorides and aluminate. A series of double salts with the fluorides of many metals and with ammonium ion can be made by precipitation or by solid-state reactions.

Health and Safety. Owing to very low solubility in water and body fluids, AlF_3 is relatively less toxic than many inorganic fluorides. The toxicity values are oral LD_{LO}, 600 mg/kg; subcutaneous, 3000 mg/kg. The ACGIH adopted (1992–1993) TLV for fluorides as F^- is TWA 2.5 mg/m^3. Pyrohydrolysis and strong acidic conditions can be a source of toxicity owing to liberated HF.

Table 2. Physical Properties of Anhydrous Aluminum Trifluoride

Property	Value
mol wt	83.977
mp, °C	1278[a]
transition point, °C	455
density, g/cm^3	3.10
dielectric constant	6
heat of transition at 455°C, kJ/mol[b]	0.677
heat of sublimation for crystals at 25°C, kJ/mol[b]	300
ΔH_f at 25°C, kJ/mol[b]	−1505
ΔG_f at 25°C, kJ/mol[b]	−1426
S at 25°C, J/(mol·K)[b]	66.23
C_p at 25°C J/(mol·K)[b]	
α-crystals	74.85
β-crystals	100.5

[a]Sublimes.
[b]To convert J to cal, divide by 4.184.

Fluoroaluminates

Several fluoroaluminates are known to exist but sodium hexafluoroaluminate [*15096-52-2*], Na_3AlF_6, has dominated industrial applications. More recently potassium tetrafluoroaluminate [*14484-69-6*], $KAlF_4$, has provided a noncorrosive and inexpensive flux in the manufacture of aluminum parts for various applications. The naturally occurring fluoroaluminates are listed in Table 3.

Table 3. Naturally Occurring Fluoroaluminates

Name	Cas Registry Number	Molecular formula
cryolite	[*15096-52-2*]	Na_3AlF_6
chiolite	[*1302-84-7*]	$Na_5Al_3F_{14}$
cryolithionate	[*15491-07-3*]	$Na_3Li_3(AlF_6)_2$
thomsenolite, hagemannite	[*16970-11-9*]	$NaCaAlF_6 \cdot H_2O$
ralstonite	[*12199-10-9*]	$Na_{2x}(Al_{2x},Na_x)(F,OH)_6 \cdot yH_2O$
prosopite	[*12420-95-0*]	$CaAl_2(F,OH)_8$
jarlite, *meta*-jarlite	[*12004-61-4*]	$NaSr_3Al_3F_{16}$
weberite	[*12423-93-7*]	Na_2MgAlF_7
gearksutite	[*12415-96-2*]	$CaAl(F,OH)_5 \cdot H_2O$
pachnolite	[*15489-46-0*]	$NaCaAlF_6 \cdot H_2O$

The common structural element in the crystal lattice of fluoroaluminates is the hexafluoroaluminate octahedron, AlF_6^{3-}. The differing structural features of the fluoroaluminates confer distinct physical properties to the species as compared to aluminum trifluoride. For example, in AlF_3 all corners are shared and the crystal becomes a giant molecule of very high melting point (13). In $KAlF_4$, all four equatorial atoms of each octahedron are shared and a layer lattice results. When the ratio of fluorine to aluminum is 6, as in cryolite, Na_3AlF_6, the AlF_6^{3-} ions are separate and bound in position by the balancing metal ions. Fluorine atoms may be shared between octahedrons. When opposite corners of each octahedron are shared with a corner of each neighboring octahedron, an infinite chain is formed as, for example, in Tl_2AlF_5 [*33897-68-6*]. More complex relations exist in chiolite, wherein one-third of the hexafluoroaluminate octahedra share four corners each and two-thirds share only two corners (14).

Cryolite. Cryolite constitutes an important raw material for aluminum manufacturing. The natural mineral is accurately depicted as $3NaF \cdot AlF_3$, but synthetic cryolite is often deficient in sodium fluoride. Physical properties are given in Table 4.

Cryolite derives its name from its resemblance to ice when immersed in water as a result of the closely matched refractive indexes. The only commercially viable source of cryolite deposits has been found in the south of Greenland at Ivigtut (15). Minor localities, not all authenticated, are in the Ilmen Mountains in the former USSR; Sallent, in the Pyrenees, Spain; and Pikes Peak, Colorado (16). For the most part the ore from Ivigtut is a coarse-grained aggregate carrying 10–30% of admixtures, including siderite, quartz, sphalerite, galena, chalcopyrite, and pyrite, in descending order of frequency.

in the Hall-Heroult electrolysis process for the manufacture of aluminum (see ALUMINUM AND ALUMINUM ALLOYS). Table 5 lists the additional examples of cryolite as a component in minimum melting compositions.

The vapor from molten cryolite is largely $NaAlF_4$, the vapor pressures of Na_3AlF_6, NaF, and $NaAlF_4$ near the melting point are about in the ratios 5:1:30. Therefore, the liquid tends to become depleted in AlF_3, and the composition of the aluminum cell electrolyte has to be regularly adjusted by the addition of AlF_3 (20,22).

In contact with moist air, molten cryolite loses HF and is depleted in AlF_3.

$$2\ Na_3AlF_6 + 3\ H_2O \rightarrow 6\ NaF + 6\ HF + Al_2O_3$$

The more electropositive metals react with cryolite, liberating aluminum or aluminum monofluoride (22,23). The reduction of cryolite by magnesium is a current method for removal of magnesium in the refining of aluminum. Upon contact with strong acids cryolite liberates hydrogen fluoride.

Synthetic Cryolite. As of this writing, the supply of cryolite is almost entirely met by synthetic material which possesses the same properties and composition with a minor difference in that it is deficient in NaF. Synthetic cryolite also commonly contains oxygen, hydroxyl group, and/or sulfate groups. The NaF

Table 5. Minimum Melting Compositions Containing Cryolite

Material composition	Added component Wt %	Melting point, °C
NaF	24.5	882
AlF_3	64	693
Al_2O_3	10.5	962
Li_3AlF_6	62	710
CaF_2	25.8	945
ZrO_2	14	969
MgO	7.5	902
CaO	11.3	896
ZnO	2.4	974
CdO	6.0	971
TiO_2	4.0	970
BaF_2	62.5	835
PbF_2	40	730
feldspar	70	830
NaF Al_2O_3	34.0 12.0	870
CaF_2 Al_2O_3	23.0 17.7	867
CaF_2 AlF_3	37.8 6.2	675
SiO_2 Al_2O_3	17 50	ca 800

deficiency does not interfere for most applications but the presence of moisture leads to the fluorine losses as HF on heating. Because synthetic cryolite is lighter than the natural mineral, losses by dusting are also higher.

There are several processes available for the manufacture of cryolite. The choice is mainly dictated by the cost and quality of the available sources of soda, alumina, and fluorine. Starting materials include sodium aluminate from Bayer's alumina process; hydrogen fluoride from kiln gases or aqueous hydrofluoric acid; sodium fluoride; ammonium bifluoride, fluorosilicic acid, fluoroboric acid, sodium fluosilicate, and aluminum fluorosilicate; aluminum oxide, aluminum sulfate, aluminum chloride, alumina hydrate; and sodium hydroxide, sodium carbonate, sodium chloride, and sodium aluminate.

The manufacture of cryolite is commonly integrated with the production of alumina hydrate and aluminum trifluoride. The intermediate stream of sodium aluminate from the Bayer alumina hydrate process can be used along with aqueous hydrofluoric acid, hydrogen fluoride kiln gases, or hydrogen fluoride-rich effluent from dry-process aluminum trifluoride manufacture.

$$NaAlO_2 + Na_2CO_3 + 6\ HF \rightarrow Na_3AlF_6 + 3\ H_2O + CO_2$$

The HF and Na_2CO_3 give a sodium fluoride solution. Bayer sodium aluminate solution is added in the stoichiometric ratio. Cryolite is precipitated at 30–70°C by bubbling CO_2, until the pH reaches 8.5–10.0. Seed crystals are desirable. The slurry is thickened and filtered, or settled and decanted, or centrifuged. The resulting product is calcined at 500–700°C. The weight ratio of fluorine to aluminum in the product should exceed 3.9. The calculated value is 4.2 (24). Cryolite can also be made by passing gaseous HF over briquettes of alumina hydrate, sodium chloride, and sodium carbonate at 400–700°C, followed by sintering at 720°C (25).

In addition, there are other methods of manufacture of cryolite from low fluorine value sources, eg, the effluent gases from phosphate plants or from low grade fluorspar. In the former case, making use of the fluorosilicic acid, the silica is separated by precipitation with ammonia, and the ammonium fluoride solution is added to a solution of sodium sulfate and aluminum sulfate at 60–90°C to precipitate cryolite (26,27):

$$12\ NH_4F + 3\ Na_2SO_4 + Al_2(SO_4)_3 \rightarrow 2\ Na_3AlF_6 + 6\ (NH_4)_2SO_4$$

The ammonia values can be recycled or sold for fertilizer use. The most important consideration in this process is the efficient elimination of the phosphorus from the product, because as little as 0.01% P_2O_5 in the electrolyte causes a 1–1.5% reduction in current efficiency for aluminum production (28).

Significant amounts of cryolite are also recovered from waste material in the manufacture of aluminum. The carbon lining of the electrolysis cells, which may contain 10–30% by weight of cryolite, is extracted with sodium hydroxide or sodium carbonate solution and the cryolite precipitated with carbon dioxide (28). Gases from operating cells containing HF, CO_2, and fluorine-containing dusts may be used for the carbonation (29).

The specifications for natural cryolite include 95% content of sodium aluminum fluorides as Na_3AlF_6, 4% of other fluorides calculated as CaF_2, and 88%

of the product passing through 44 μm sieve (325 mesh). Product for the ceramic industry contains a small amount of selected lump especially low in iron. The following is a typical analysis for commercial-grade cryolite: cryolite as Na_3AlF_6, 91%; fluorine, 48–52%; sodium, 31–34%; aluminum, 13–15%; alumina, 6.0%; silica (max), 0.70%; calcium fluoride, 0.04–0.06%; iron as Fe_2O_3, 0.10%; with moisture at 0.05–0.15%, bulk density at 1.4–1.5 g/cm^3, and screen analysis passing through 74 μm (200 mesh) at 65–75%.

In spite of the fact that cryolite is relatively less soluble, its fluoride toxicity by oral routes are reported to be about the same as for soluble fluorides: LD_{50} = 200 mg/kg; for NaF, 180 mg/kg; KF, 245 mg/kg (30). Apparently, stomach fluids are acid enough to bring the solubility of cryolite up to values comparable with other fluorides. Chronic exposure may eventually lead to symptoms of fluorosis. The toxicity to insects is in many cases high enough for control. Because of its variable composition, synthetic cryolite may show physiological activity greater than the natural mineral (31).

The effective dissolution of Al_2O_3 by molten cryolite to provide a conducting bath has spurred the need for its use in manufacture of aluminum. Additives enhance the physical and electrical properties of the electrolyte, for example the lowering of melting point by AlF_3 (Fig. 1). Figure 2 illustrates the effect of various additives on the electrical conductivity of liquid cryolite. AlF_3 has the adverse effect of decreasing the electrical conductivity. Calcium fluoride is better in this regard but again too much of it can lead to rise in density of the melt close to that of aluminum (ca 2.28 g/cm^3), inhibiting the separation of metal and electrolyte as indicated in Figure 3. Sodium fluoride has the disadvantage of reducing the current efficiency while increasing density and conductivity. Small amounts of lithium fluoride may also improve electrical conductivity. Compromises on all of these factors have led to the following composition of the electrolyte: 80–85% cryolite, 5–7% AlF_3, 5–7% CaF_2, 0–7% LiF, and 2–8% Al_2O_3.

Another use for cryolite is in the production of pure metal by electrolytic refining. A high density electrolyte capable of floating liquid aluminum is needed, and compositions are used containing cryolite with barium fluoride to raise the density, and aluminum fluoride to raise the current efficiency.

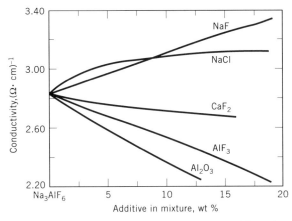

Fig. 2. Effect of additives on the electrical conductivity of liquid cryolite at 1009°C (32).

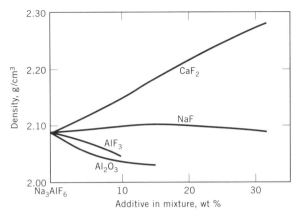

Fig. 3. Effect of additives on the density of liquid cryolite at 1009°C.

Other applications of cryolite include use in reworking of scrap aluminum as flux component to remove magnesium by electrochemical displacement; as a flux in aluminizing steel as well as in processing a variety of metals; in the compounding of welding-rod coatings; as a flux in glass manufacture owing to its ability to dissolve the oxides of aluminum, silicon, and calcium, and also because of the low melting compositions formed with the components; for lowering the surface tension in enamels and thereby improving spreading (33); as a filler for resin-bonded grinding wheels for longer wheel life, reducing metal buildup on the wheel, and faster and cooler grinding action; and in insecticide preparations making use of the fines residue from the refining operation of the cryolite.

Canada, the United States, and South America are the principal exporters of cryolite and Russia and Europe import cryolite. Primary producers in North America are Alcan, Alcoa, and Reynolds Aluminum. The 1993 price of recovered-grade cryolite, which has SO_4^{2-} as impurity, was $400/t, and of high purity cryolite, $800/t. There was a surplus of cryolite in 1992 in the United States and Canada.

Potassium Tetrafluoroaluminate. Potassium tetrafluoroaluminate, $KAlF_4$, is a more recent addition to the industrially important fluoroaluminates, mainly because of developments in the automotive industry involving attempts to replace the copper and solder employed in the manufacture of heat exchangers. The source mineral for aluminum radiator manufacture, bauxite, is highly abundant and also available in steady supply. Research and developmental work on the aluminum radiators started in the 1960s using chloride salt mixtures for brazing. The resulting products and the process itself could not compete with conventional radiators because these processes were comparatively uneconomical. This led to the development of an all fluoride-based flux which confers corrosion-resistant features to the product as well as to the process. Potassium tetrafluoroaluminate in mixtures with other fluoroaluminates, potassium hexafluoroaluminate [13775-52-5], K_3AlF_6, and potassium pentafluoroaluminate monohydrate [41627-26-3], $K_2AlF_5 \cdot H_2O$, has emerged as a highly efficient, noncorrosive, and nonhazardous flux for brazing aluminum parts of heat exchangers. Nocolok 100 Flux (Alcan Aluminum Corp.) developed by Alcan (Aluminum Co. of Canada) has been the first commercial product. Its use and mechanistic aspects of the associated brazing process have been well documented (33–37).

The important task performed by all brazing processes is the removal of oxide films lying on the surfaces of metals to be joined. The process should also permit wetting and flow of the molten filler metal at the brazing temperature (38). The fluxes employed should melt and become active for a successful brazing action. Thus if the flux melts at a temperature higher than that of the filler metal, it leads to the development of thick oxide films on the liquid filler metal inhibiting the flux action. The system $KF \cdot AlF_3$ (Fig. 4) (39) provides the most suitable flux for this application. The system presents a eutectic mixture of $KAlF_4$ and K_3AlF_6 which melts at $559 \pm 2°C$ (40). This is just below the eutectic temperature of the Al–Si filler metal, which is 577°C. The melting point of pure $KAlF_4$ is $574 \pm 1°C$ and that of K_3AlF_6 is 990°C (40).

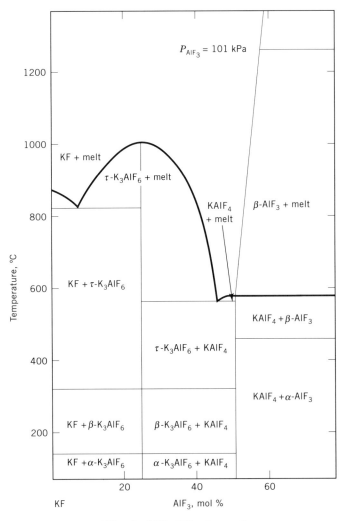

Fig. 4. $KF-AlF_3$ phase diagram.

Both $KAlF_4$ and K_3AlF_6 are white solids. The former is less soluble (0.22%) in water than the latter (1.4%). The generally cubic form of $KAlF_4$ inverts to the orthorhombic modification between -23 and 50°C. On heating the cubic form is stable to its congruent melting temperature. The materials are generally inert and infinitely stable under ambient conditions. At melting temperatures and more significantly at temperatures above 730°C they react with water releasing hydrogen fluoride (41). Dissolution in strong acids is also slow but is enhanced at higher temperatures leading to the evolution of HF. Several possible interactions of $KAlF_4$ and the metal oxides in the brazing processes have been proposed as part of the mechanism for the latter (34).

An early method of preparation of $KAlF_4$ (42) involved combining aqueous solutions of HF, AlF_3, and KHF_2 in stoichiometric proportions and evaporating the suspension to a dry mixture. The product was subsequently melted and recrystallized. Some of the other conventional technical methods comprise reacting hydrated alumina, hydrofluoric acid, and potassium hydroxide followed by separation of the product from the mother liquor; concentrating by evaporation, a suspension obtained by combining stoichiometric amounts of components; and melting together comminuted potassium fluoride and aluminum fluoride at 600°C and grinding the resulting solidified melt.

Several other proprietary methods have been reported, which in general have the aim of producing lower melting products thereby aiming more at the preparation of a eutectic mixtures of the fluoroaluminates as discussed in the beginning of this section. One process (42) describes the making of $KAlF_4$, melting below 575°C, by addition of potassium hydroxide to the aqueous solution of fluoroaluminum acid. The fluoroaluminum acid is prepared from a reaction of hydrofluoric acid and hydrated alumina. A fairly similar method has been reported in making a flux mixture comprising of K_2AlF_5 or $K_2AlF_5 \cdot H_2O$ and $KAlF_4$, wherein a potassium compound is added to the mixed aqueous fluoroaluminic acid ($HAlF_4$, H_2AlF_5, and H_2AlF_6) solution (43).

The toxicity of these fluoroaluminates is mainly as inorganic fluorides. The ACGIH adopted (1992–1993) values for fluorides as F^- is TLV 2.5 mg/m^3. The oral toxicity in laboratory animal tests is reported to be LD_{50} rat 2.15 mg/kg (41). Because of the fine nature of the products they can also be sources of chronic toxicity effects as dusts.

BIBLIOGRAPHY

"Aluminum Fluoride" under "Fluorine Compounds, Inorganic," in *ECT* 1st ed., Vol. 6, pp. 668–671 by R. G. Danehower, Pennsylvania Salt Manufacturing Co.; "Aluminum Fluorides" under "Fluorine Compounds, Inorganic," in *ECT* 2nd ed., Vol. 9, pp. 529–533, by J. F. Gall, Pennsalt Chemicals Corp.; "Fluoroaluminates" under "Fluorine Compounds, Inorganic," in *ECT* 1st ed., Vol. 6. pp. 671–675, by I. Mockrin, Pennsylvania Salt Manufacturing Co.; "Fluoroaluminates" under "Fluorine Compounds, Inorganic," in *ECT* 2nd ed., Vol. 9, pp. 534–548, by J. F. Gall, Pennsalt Chemicals Corp.; "Aluminum" under "Fluorine Compounds, Inorganic," in *ECT* 3rd ed., Vol. 10, pp. 660–675, by J. F. Gall, Philadelphia College of Textiles and Science.

1. *JANAF Thermochemical Tables*, 2nd ed., NSR DS-NBS 37, National Bureau of Standards, Washington, D.C., 1985.

2. T. C. Ehlert and J. L. Margrave, *J. Am. Chem. Soc.* **86**, 3901 (1964).
3. W. F. Ehret and F. J. Frere, *J. Am. Chem. Soc.* **67**, 64 (1945).
4. U.S. Pat. 2,958,575 (Nov. 1, 1960), D. R. Allen (to The Dow Chemical Co.).
5. J. K. Callaham, *Chem. Met. End.* **52**(3), 94 (1945).
6. F. Weinratter, *Chem. Eng.* **71**, 132 (Apr. 27, 1964).
7. J. K. Bradley, *Chem. Ind., London*, 1027 (1960).
8. Jpn. Kokai Tokkyo Koho 04 04,991 [92 04,991] (Jan. 9, 1992), T. Usui and S. Kagoshige (to Showa Aluminum Corp.).
9. Jpn. Kokai Tokkyo Koho 04 09,274 [92 09,274] (Jan. 14 1992), K. Toma and co-workers (to Mitsubishi Aluminum Co. Ltd.).
10. Jpn. Kokai Tokkyo Koho 03 215,656 [91 215,656] (Sept. 20, 1991), T. Omi and co-workers (to Hashimoto Industries Co. Ltd.).
11. J. D. Dana and co-workers, *The System of Mineralogy*, 7th ed., Vol. 2, John Wiley & Sons, Inc., New York, 1951, pp. 124–125.
12. U.S. Pat. 4,983,373 (Jan. 8, 1991), H. P. Withers Jr. and co-workers (to Air Products & Chemicals, Inc.).
13. P. J. Durrant and B. Durrant, *Introduction to Advanced Inorganic Chemistry*, John Wiley & Sons, Inc., New York, 1970, p. 570.
14. A. F. Wells, *Structural Inorganic Chemistry*, 4th ed., Clarendon Press, Oxford, U.K., 1975, pp. 388–390.
15. H. Pauly, *Met. Assoc. Acid. Magmat.* **I**, 393 (1974).
16. C. Palache and co-workers, in Ref. 4, pp. 110–113.
17. P. P. Fedotiev and V. Hyinskii, *Z. Anorg. Chem.* **80**, 113 (1913).
18. G. G. Landon and A. R. Ubbelohde, *Trans. Faraday Soc.* **52**, 647 (1955).
19. P. A. Foster, Jr. and W. B. Frank, *J. Electrochem. Soc.* **107**, 997 (1960).
20. L. M. Foster, *Ann. N.Y. Acad. Sci.* **79**, 919 (1960).
21. N. W. F. Philips and co-workers, *J. Electrochem. Soc.* **102**, 648–690 (1955).
22. K. Grjotheim and co-workers, *Light. Met.* **1**, 125 (1975).
23. M. Feinleib and B. Porter, *J. Electrochem. Soc.* **103**, 231 (1956); W. E. Haupin, *J. Electrochem. Soc.* **107**, 232 (1960).
24. U.S. Pat. 3,061,411 (Oct. 30, 1962), D.C. Gernes (to Kaiser Aluminum & Chemicals Corp.).
25. U.S. Pat. 3,104,156 (Sept. 17, 1963), P. Saccardo and F. Gozzo (to Sicedoison SpA).
26. G. Tarbutton and co-workers *Ind. Eng. Chem.* **50**, 1525 (1958).
27. U.S. Pat. 2,687,341 (Aug. 24, 1954), I. Mockrin (to Pennsylvania Salt Manufacturing Co.).
28. E. Elchardus, *Compt. Rend.* **206**, 1460 (1938).
29. U.S. Pat. 3,065,051 (Nov. 20, 1962), H. Mader (to Vereinigte Metallwerke Ranshofen-Berndorf A.G.).
30. *The Toxic Substances List*, 1974 ed., U.S. Dept. of Health, Education & Welfare, National Institute for Occupational Safety & Health, Rockville, Md., June 1974.
31. E. J. Largent, *J. Ind. Hyg. Toxicol.* **30**, 92 (1948).
32. J. D. Edwards and co-workers, *J. Electrochem. Soc.* **100**, 508 (1953); K. Matiasovsky and co-workers, *J. Electrochem. Soc.* **111**, 973 (1964).
33. R. Marker, *Glas Email Keramo Tech.* **4**, 117 (1957); **5**, 178 (1957).
34. Y. Ando and co-workers, *SAE Technical Paper Series, International Congress and Exposition*, paper no. 870180, Detroit, Mich., Feb. 23–27, 1987; D. J. Field and N. I. Steward, *ibid.*, paper no. 870186.
35. D. G. W. Claydon and A. Sugihara, in Ref. 34, paper no. 830021.
36. W. E. Cooke and H. Bowman, *Welding J.* (Oct. 1980).
37. W. E. Cooke and co-workers, *SAE Technical Paper Series, Congress and Exposition* paper no. 780300, Detroit, Mich., Feb. 27–Mar. 3, 1978.
38. J. R. Terril and co-workers, *Welding J.* **50**(12), 833–839 (1971).

39. B. Jensen, *Phase and Structure Determination of a New Complex Alkali Aluminum Fluoride*, Institute of Inorganic Chemistry, Norwegian Technical University, Trandheim, 1969.
40. B. Philips and co-workers, *J. Am. Ceram. Soc.* **49**(2), 631–634 (1966).
41. *Nocolok 100 Flux, Material Safety Data Sheet*, Alcan Aluminum Corp., Apr. 1986.
42. U.S. Pat. 4,428,920 (Jan. 31, 1984), H. Willenberg and co-workers (to Kali-Chemie Aktiengesellachaft).
43. U.S. Pat. 4,579,605 (Apr. 1, 1986), H. Kawase and co-workers (to Furukawa Aluminum Co., Ltd.).

DAYAL T. MESHRI
Advance Research Chemicals, Inc.

AMMONIUM

Two well-known salts of ammonia (qv) are the normal ammonium fluoride [*12125-01-8*], NH_4F, and ammonium bifluoride [*1341-49-7*], NH_4HF_2; the latter is sometimes named ammonium acid, or hydrogen, difluoride. Much of the commercial interest in the ammonium fluorides stems from their chemical reactivity as less hazardous substitutes for hydrofluoric acid.

Ammonium Fluoride

Ammonium fluoride is a white, deliquescent, crystalline salt. It tends to lose ammonia gas to revert to the more stable ammonium bifluoride. Its solubility in water is 45.3 g/100 g of H_2O at 25°C and its heat of formation is -466.9 kJ/mol (-116 kcal/mol). Ammonium fluoride is available principally as a laboratory reagent. If it is needed in large quantities, one mole of aqueous ammonia can be mixed with one mole of the more readily available ammonium bifluoride (1).

Ammonium Bifluoride

Properties. Ammonium bifluoride, NH_4HF_2, is a colorless, orthorhombic crystal (2). The compound is odorless; however, less than 1% excess HF can cause an acid odor. The salt has no tendency to form hydrates yet is hygroscopic if the ambient humidity is over 50%. A number of chemical and physical properties are listed in Table 1.

Corrosion. Ammonium bifluoride dissolves in aqueous solutions to yield the acidic bifluoride ion; the pH of a 5% solution is 3.5. In most cases, NH_4HF_2 solutions react readily with surface oxide coatings on metals; thus NH_4HF_2 is used in pickling solutions (see METAL SURFACE TREATMENTS). Many plastics, such as polyethylene, polypropylene, unplasticized PVC, and carbon brick, are resistant to attack by ammonium bifluoride.

Manufacture. Anhydrous ammonium bifluoride containing 0.1% H_2O and 93% NH_4HF_2 can be made by dehydrating ammonium fluoride solutions and

Table 1. Properties of Ammonium Bifluoride, NH_4HF_2

Property	Value	Reference
melting point, °C	126.1	3
boiling point, °C	239.5	3
index of refraction, n_D	1.390	3
solubility at 25°C, wt %		
water	41.5	3
90% ethanol	1.73	3
specific gravity	1.50	3
standard heat of formation, kJ/mol^a	−798.3	4
heat of fusion, kJ/mol^a	19.1	4
heat of vaporization, kJ/mol^a	65.3	4
heat of solution, kJ/mol^a	20.3	4
heat of dissociation,^{b} kJ/mol^a	141.4	4
heat capacity, C_p, J/(mol·K)^a at 25°C	106.7	4
vapor pressure,^{c} $\log P_{Pa} = a - bT^{-1}$		
153–207°C	$a = 11.72, b = 3370$	5
207–245°C	$a = 9.38, b = 2245$	5
oral LD (guinea pig), mg/kg	150	6

^{a}To convert kJ to kcal, divide by 4.184.
^{b}$NH_4HF_2 \rightarrow NH_3 + 2\,HF$.
^{c}To convert Pa to mm Hg, multiply by 7.5×10^{-3}.

by thermally decomposing the dry crystals (7). Commercial ammonium bifluoride, which usually contains 1% NH_4F, is made by gas-phase reaction of one mole of anhydrous ammonia and two moles of anhydrous hydrogen fluoride (8); the melt that forms is flaked on a cooled drum. The cost of the material in 1992 was $1.48/kg.

Production of bifluoride from fluoride by-products from the phosphate industry (9) has had little if any commercial significance.

Precautions in Handling. Ammonium bifluoride, like all soluble fluorides, is toxic if taken internally. Hydrofluoric acid burns may occur if the material comes in contact with moist skin. Ammonium bifluoride solutions should be thoroughly washed from the skin with mildly alkaline soap as soon as possible; however, if contact has been prolonged, the affected areas should be soaked with 0.13% solution of Zephiran chloride, or 0.2% Hyamine 1622 (Lonza, Inc.) or calcium gluconate, the treatment recommended for hydrofluoric acid burns. If any of these solutions come in contact with the eyes, they should be washed with water for at least 10 min and a physician should be consulted.

Applications. Ammonium bifluoride solubilizes silica and silicates by forming ammonium fluorosilicate [*16919-19-0*], $(NH_4)_2SiF_6$. Inhibited 15% hydrochloric acid containing about 2% ammonium bifluoride has been used to acidize oil wells in siliceous rocks to regenerate oil flow (10) (see PETROLEUM). Ammonium fluoride solution is made on-site near the well bore from ammonium bifluoride and ammonia and mixed with methyl formate to prevent rapid consumption of most of the HF (11). The use of ammonium bifluoride is important in locations where dissolved silicates foul boiler tubes with scale that cannot be removed using usual cleaning aids (12). Ammonium bifluoride is also used as an etching agent for silicon wafers.

Rapid frosting of glass is accomplished in a concentrated solution of ammonium bifluoride and hydrofluoric acid with nucleating agents that assure uniform frosts (13). A single dip in an aqueous solution of NH_4HF_2, HF, and sorbitol at <20°C for less than 60 s produces the low specular-reflecting finish on television face plates and on glass (qv) for picture framing (14). Treating glass, eg, often badly weathered window panes, with 2–5% solutions of ammonium bifluoride results in a polishing effect. Glass ampuls for parenteral solutions (15) and optical lenses (16) are best cleaned of adhering particulate matter in dilute ammonium bifluoride solutions.

Ammonium bifluoride is used as a sour or neutralizer for alkalies in commercial laundries and textile plants. Treatment also removes iron stain by forming colorless ammonium iron fluorides that are readily rinsed from the fabric (17).

Ammonium fluorides react with many metal oxides or carbonates at elevated temperatures to form double fluorides; eg,

$$2\ NH_4HF_2\ +\ BeO\ \xrightarrow{-H_2O}\ (NH_4)_2BeF_4\ \xrightarrow{\Delta}\ BeF_2\ +\ 2\ NH_3\ +\ 2\ HF$$

The double fluorides decompose at even higher temperatures to form the metal fluoride and volatile NH_3 and HF. This reaction produces pure salts less likely to be contaminated with oxyfluorides. Beryllium fluoride 7787-49-7], from which beryllium metal is made, is produced this way (18) (see BERYLLIUM AND BERYLLIUM ALLOYS). In pickling of stainless steel and titanium, NH_4HF_2 is used with high concentrations of nitric acid to avoid hydrogen embrittlement. Ammonium bifluoride is used in acid dips for steel (qv) prior to phosphating and galvanizing, and for activation of metals before nickel plating (19,20). Ammonium bifluoride also is used in aluminum anodizing formulations. Ammonium bifluoride is used in treatments to provide corrosion resistance on magnesium and its alloys (21). Such treatment provides an excellent base for painting and good abrasion resistance, heat resistance, and protection from atmospheric corrosion. A minor use for ammonium bifluoride is in the preservation of wood (qv) (22).

BIBLIOGRAPHY

"Ammonium" under "Fluorine Compounds, Inorganic," in *ECT* 1st ed., Vol. 6, p. 676, by G. C. Whitaker, The Harshaw Chemical Co.; "Ammonium Fluoride" under "Fluorine Compounds, Inorganic," in *ECT* 2nd ed., Vol. 9, pp. 548–549, by G. C. Whitaker, The Harshaw Chemical Co.; "Ammonium" under "Fluorine Compounds, Inorganic," in *ECT* 3rd ed., Vol. 10, pp. 675–678, by H. S. Halbedel and T. E. Nappier, The Harshaw Chemical Co.

1. Fr. Pat. 1,546,234 (Nov. 15, 1968), (to Farbenfabriken Bayer A.-G.).
2. O. Hassel and H. Luzanski, *Z. Kristallogr.* **83**, 448 (1932).
3. R. C. Weast, ed., *Handbook of Chemistry and Physics*, 59th ed., The Chemical Rubber Co., Cleveland, Ohio, 1978.
4. H. Schutza, M. Eucken, and W. Namesh, *Z. An. All. Chem.* **292**, 293 (1957).
5. L. N. Lazarev and B. V. Andronov, *J. Appl. Chem. USSR* **46**, 2087 (1973).

6. H. C. Hodge and F. A. Smith, in J. H. Simon, ed., *Fluorine Chemistry*, Vol. 4, Academic Press, Inc., New York, 1965, p. 192.
7. U.S. Pat. 3,310,369 (Mar. 21, 1967), J. A. Peterson (to Hooker Chemical Corp.).
8. U.S. Pat. 2,156,273 (Apr. 28, 1939), A. R. Bozarth (to Harshaw Chemical Co.).
9. U.S. Pat. 3,501,268 (Mar. 17, 1970), R. J. Laran, A. P. Giraitix, and P. Kobetz (to Ethyl Corp.).
10. H. K. van Poolen, *Oil Gas J.* **65**, 93 (Sept. 11, 1967).
11. U.S. Pat. 3,953,340 (Apr. 27, 1976), C. C. Templeton, E. H. Street, Jr., and E. A. Richardson (to Shell Oil Co.).
12. W. S. Midkiff and H. P. Foyt, *Mater. Perform.* **17**(2), 17 (1978).
13. *Glass Frosting and Polishing Technical Service Bulletin 667*, Harshaw Chemical Co., Solon, Ohio.
14. U.S. Pat. 3,373,130 (Mar. 19, 1968), E. E. Junge and J. Chabal (to PPG Industries).
15. A. L. Hinson, *Bull. Parenter. Drug. Assoc.* **25**, 266 (1971).
16. R. L. Parkes and M. R. Browne, *Appl. Opt.* **17**, 1845 (1978).
17. *Control of Souring Operations, Special Report #7*, American Institute of Laundering, Joliet, Ill.
18. Brit. Pat. 833,808 (Apr. 27, 1960), A. R. S. Gough and E. W. Bennet (to the United Kingdom Atomic Energy Commission).
19. U.S. Pat. 3,767,582 (Oct. 23, 1973), G. A. Miller (to Texas Instruments, Inc.).
20. U.S. Pat. 3,296,141 (Jan. 3, 1967), W. A. Lieb and E. Billow (to R. O. Hull Co.).
21. L. F. Spencer, *Met. Finish.* **68**(10), 52 (1970); H. K. DeLong, *Met. Prog.* **97**, 105 (June 1970); *Met Prog.* **98**, 43 (Mar. 1971); W. F. Higgins, *Light Met. Age* **17**(12), 8 (1959); A. E. Yaniv and H. Schick, *Plating* **55**, 1295 (1968).
22. E. Panck, *Am. Wood Preservers Assoc.* **59**, 189 (1963).

JOHN R. PAPCUN
Atotech

ANTIMONY

Antimony forms both a trifluoride and a pentafluoride. It also forms the very stable hexafluoroantimonate ion [*17111-95-4*], SbF_6^-, present in solution and a number of salts.

Antimony Trifluoride

Properties. Antimony trifluoride [*7783-56-4*], SbF_3, is a very hygroscopic, white, crystalline solid, mp = 292°C. It can be sublimed under vacuum. It is very soluble in water, hydrofluoric acid, and polar organic solvents such as alcohols and ketones. Its solubility in water is 384.7 g/100 g at 0°C, 492.4 g/100 g at 25°C, and 563.6 g/100 g at 30°C (1). The solubility at 25°C is 154 g/100 mL CH_3OH, 33 g/100 mL C_3H_7OH, and 55.3 g/100 mL acetone. It is practically insoluble in benzene, chlorobenzene, and heptane. The density of SbF_3 at 25°C is 4.385 g/cm^3. It does hydrolyze in water, but the rate of hydrolysis is very slow, much slower than $SbCl_3$.

Antimony trifluoride is a mild fluorinating reagent. However, it is much more effective in the Swarts reactions where its effectiveness as a fluorinating reagent is dramatically increased by addition of Cl_2, Br_2, or $SbCl_5$ to the reaction mixture (2). Antimony trifluoride can be used for the replacement of chlorine or bromine in halocarbons, hydrohalocarbons, and nonmetal and metal halides. Typical reactions can be summarized as follows:

$$3\ RCCl_3 + 2\ SbF_3 \rightarrow RCCl_2F + RCClF_2 + RCF_3 + 2\ SbCl_3$$

In aliphatic compounds, the ease of fluorination is of the order of $-CCl_3 > -CCl_2F > -CClF_2$. Other groups, eg, $-C=CCl_2$ and $-CHCl_2$, react, but not readily. Antimony trifluoride is not a suitable reagent for the replacement of hydrogen in organic compounds.

Inorganic compounds also can be fluorinated using SbF_3, eg,

$$PCl_3 + SbF_3 \xrightarrow{SbCl_5} PF_3 + SbCl_3 \cdot SbCl_5$$

$$SiHCl_3 + SbF_3 \xrightarrow{SbCl_5} SiHF_3 + SbCl_3 \cdot SbCl_5$$

In aqueous solutions SbF_3 reacts with many metal fluorides to form compounds such as $MSbF_4$ where M = Li [72121-39-2], Na [34109-83-6], K [15273-81-1], Cs [36195-09-0], and Tl [54189-44-5], and M_2SbF_5 where M = K [20645-41-4], Cs [40902-54-3], and NH_4 [32516-50-0]. In addition, triantimonate [65176-04-7], Na_3SbF_6, and MSb_4F_{13} where M = Tl [60719-48-4], Na [56094-73-6], K [56094-72-5], Rb [12776-50-0], Cs [12775-92-7], and NH_4 [52015-24-4] have been reported (3).

Preparation. Antimony trifluoride can be readily prepared by dissolving Sb_2O_3 in an excess of anhydrous hydrogen fluoride or in aqueous acid of 40% or higher strength hydrofluoric acid, followed by evaporation of the solution to dryness (4). It can also be prepared by thermal decomposition of the graphite intercalation compound with SbF_3Cl_2 (5), by heating ammonium hexafluorantimonate (6), and by the reaction of metal with anhydrous hydrogen fluoride in the presence of nitrile (7).

Uses. Early manufacturing processes for fluorocarbons and chlorofluorocarbons used SbF_3 on a large scale, but development of alternative routes to the Swarts reactions have greatly reduced usage. Its main use is in the manufacture of antimony pentafluoride. The market for SbF_3 in the United States is less than 5 t/yr. More recent uses of SbF_3 have been in the manufacture of fluoride glass and fluoride glass optical fiber preform (8), and fluoride optical fiber (9) in the preparation of transparent conductive films (10) (see FIBER OPTICS).

Antimony Pentafluoride

Properties. Antimony pentafluoride [7783-70-2], SbF_5, is a colorless, hygroscopic, very viscous liquid that fumes in air. Its viscosity at 20°C is 460 mPa·s(= cP) which is very close to the value for glycerol. The polymerization of high purity SbF$_5$ at ambient temperature can be prevented by addition of 1% anhydrous hydrogen fluoride, which can be removed by distillation prior to the use of SbF$_5$. The pure product melts at 7°C (11), boils at 142.7°C, and has a specific gravity (12) of 3.145 g/cm^3 at 15.5°C. The viscous, pure liquid can be handled briefly in glass if moisture and air are carefully excluded. However, it must never be stored in glass because any HF or moisture present leads to a dangerous reaction. Any moisture reacts with SbF$_5$ to produce HF which reacts with glass to produce SiF$_4$ and water which, in turn, reacts with SbF$_5$ to again produce HF. The reaction continues until the SiF$_4$ pressure ruptures the container. Commercial antimony pentafluoride is shipped in steel cylinders or polytetrafluoroethylene bottles (generally 1-kg or less). Nickel is rapidly attacked by a mixture of SbF$_5$ and HF, although there is little effect on mild steel or aluminum.

Preparation. Antimony pentafluoride can be prepared by direct fluorination of SbF$_3$ or antimony or by reaction of SbCl$_5$ with HF (13). The reaction of SbCl$_5$ with anhydrous hydrogen fluoride proceeds with the formation of intermediate products. These chlorofluoroantimonates can be prepared separately (14). Reaction of SbCl$_5$ with anhydrous hydrogen fluoride at -60°C produces SbCl$_4$F [*14913-58-7*], mp = 83°C, which polymerizes on sublimation to (SbCl$_4$F)$_4$. SbCl$_4$F can also be prepared by the reaction of SbCl$_5$ and anhydrous hydrogen fluoride in chlorofluorocarbons (15). Pure SbCl$_3$F$_2$ [*24626-20-6*], mp ~55°C, can be crystallized from a mixture of SbCl$_3$F$_2$ and SbCl$_2$F$_3$ [*7791-16-4*]. SbCl$_2$F$_3$, which is a thick liquid, can be prepared from the reaction of SbF$_3$ and Cl$_2$ at 135°C (16). SbClF$_4$ [*15588-48-4*] is obtained at 100°C by the reaction of SbCl$_5$ with a large excess of hydrogen fluoride. Other methods for preparation of antimony pentafluoride include reacting Sb with HF/F$_2$ (17) and fluorination of Sb using F$_2$ in a quartz tube (18).

Uses. Antimony pentafluoride is a moderate fluorinating reagent and a powerful oxidizer. It spontaneously inflames phosphorus and sodium but it is practically inert toward arsenic. Powdered antimony reduces SbF$_5$ to solid SbF$_5$·2SbF$_3$. SbF$_5$ reacts with water to form the solid antimony pentafluoride dihydrate [*65277-49-8*], SbF$_5$·2H$_2$O, which reacts violently with an additional amount of water to form a clear solution. Antimony pentafluoride undergoes very slow hydrolysis in the presence of a dilute NaOH solution to form Sb(OH)$_6^-$. Sulfur dioxide and nitrogen dioxide react with SbF$_5$ to form the adducts SbF$_5$·SO$_2$ [*19344-14-0*] and SbF$_5$·NO$_2$ [*72121-47-2*], respectively. These adducts decompose in water.

Antimony pentafluoride is used to saturate double bonds in straight-chain olefins, cycloolefins, aromatic rings (19–21), and in the fluorination of halocarbons and CrO$_2$Cl$_2$, MoCl$_5$, WCl$_6$, PCl$_3$, P$_4$O$_{10}$, SiCl$_4$, TiCl$_4$, and SiO$_2$.

Antimony pentafluoride forms intercalation compounds with graphite (22,23) and fluorinated graphite (24), CF$_x$, where x = 1.06, which have much higher conductivity than graphite and fluorinated graphite, respectively (25). These nonstoichiometric substances may have potential use as superconducting materials. When a mixture of O$_2$, F$_2$, and SbF$_5$ or NF$_3$, F$_2$, and SbF$_5$ is subjected to elevated temperature and pressure, it gives the dioxygenyl salt O$_2$SbF$_6$ [*51681-88-0*] (26) and the perfluoroammonium salt NF$_4$SbF$_6$ [*16871-76-4*] (27), respectively. The dioxygenyl salt is a solid that can oxidize xenon (28) and has been used for removal of xenon, radon, and radon daughter elements from contaminated atmospheres. SbF_5 has also been used in the conversion of methane to gasoline range hydrocarbons (29), in the syntheses of fluorocarboranes (30), in superacids (31), and in the preparation of stable carbocations (32).

Hexafluoroantimonates

Hexafluoroantimonic acid [72121-43-8], $HSbF_6 \cdot 6H_2O$, is prepared by dissolving freshly prepared hydrous antimony pentoxide in hydrofluoric acid or adding the stoichiometric amount of 70% HF to SbF_5. Both of these reactions are exothermic and must be carried out carefully.

The superacid systems $HSO_3F \cdot SbF_5$ [33843-68-4] and $HF \cdot SbF_5$ [16950-06-4] (fluoroantimonic acid) are used in radical polymerization (33) and in carbocation chemistry (34). Addition of SbF_5 drastically increases the acidities of HSO_3F and HF (35,36).

Anhydrous salts, $MSbF_6$, where M = H, NH_4, and alkali metal, and $M(SbF_6)_2$, where M is an alkaline-earth metal, can be prepared by the action of F_2 on MF or MF_2 and SbF_3 (37) by the oxidation of Sb(III) with H_2O_2 or alkali metal peroxide in HF (38), by the action of HF on a mixture of $SbCl_5$ and MF where M = NH_4, Li, Na, K, Ru, Cs, Ag, and Tl (39). These compounds can be used as photoinitiators for the production of polymers (40).

Environmental and Safety Aspects

OSHA has a TWA standard on a weight of Sb basis of 0.5 mg/m^3 for antimony in addition to a standard TWA of 2.5 mg/m^3 for fluoride. NIOSH has issued a criteria document on occupational exposure to inorganic fluorides. Antimony pentafluoride is considered by the EPA to be an extremely hazardous substance and releases of 0.45 kg or more reportable quantity (RQ) must be reported. Antimony trifluoride is on the CERCLA list and releasing of 450 kg or more RQ must be reported.

BIBLIOGRAPHY

"Antimony Compounds" under "Fluorine Compounds, Inorganic" in *ECT* 1st ed., Vol. 6, pp. 676–677, by F. D. Loomis and C. E. Inman, Pennsylvania Salt Manufacturing Co.; "Antimony" under "Fluorine Compounds, Inorganic" in *ECT* 2nd ed., Vol. 9, pp. 549–551, by W. E. White, Ozark-Mahoning Co.; in *ECT* 3rd ed., Vol. 10., pp. 679–681, by D. T. Meshri and C. B. Lindahl, Ozark-Mahoning Co., a subsidiary of the Pennwalt Corp.

1. A. Rosenheim and H. Grünbaum, *Z. Anorg. Chem.* **61**, 187 (1909).
2. F. Swarts, *Bull. Acad. Roy. Belg.* **24**, 309 (1892).
3. I. G. Ryss, *The Chemistry of Fluorine and Its Inorganic Compounds*, State Publishing House of Scientific, Technical, and Chemical Literature, Moscow, USSR, 1956; English trans., AEC-tr-3927, Office of Technical Services, U.S. Department of Commerce, Washington, D.C., 1960, pp. 283–295 (Part I).
4. Z. Xie, *Huaxue Shijie* **26**(5), 165–166 (1985).
5. H. Preiss, E. Alsdorf, and A. Lehman, *Carbon* **25**(6), 727–733 (1987).
6. Eur. Pat. 156,617 A2 (Oct. 2, 1985), M. Watanabe and S. Nishimura.

7. U.S. Pat. 4,0340,780 (July 5, 1977), J. A. Wojtowicz and D. F. Gavin.
8. Eur. Pat. 331,483 A2 (Sept. 6, 1989), K. Fujiura, Y. Ohishi, M. Fujiki, T. Kanamori, and S. Takahashi.
9. Jpn. Pat. 6011239 A2 (Jan. 21, 1983), (to Nippon Telegraph & Telephone Public Co.).
10. Jpn. Pat. 63314713 A2 (Dec. 22, 1988), N. Sonoda and N. Sato.
11. O. Ruff and co-workers, *Chem. Ber.* **42**, 4021 (1909).
12. R. C. Shair and W. F. Shurig, *Ind. Eng. Chem.* **43**, 1624 (1951).
13. O. Ruff and W. Plato, *Ber.* **37**, 673 (1904).
14. M. F. A. Dove and Md. O. Ali, *J. Inorg. Nucl. Chem.* 77. (1976 Suppl.).
15. E. Santacesaria and M. DiSerio, Jr., *J. Fluorine Chem.* **44**, 87–111 (1989).
16. A. L. Henne and P. Trott, *J. Am. Chem. Soc.* **69**, 1820 (1947).
17. Jpn. Pat. 03242326 A2 (Oct. 29, 1991), K. Kuge, S. Saito, A. Chuma, and S. Takenuki.
18. D. Ganter, A. Boles, and B. Erlec, *Vestn. Slov. Kem. Drus.* **30**(3), 289–93 (1983).
19. E. T. McBee, P. A. Wisemen, and G. B. Bachman, *Ind. Eng. Chem.* **39**, 415 (1947).
20. U.S. Pat. 2,488,216 (Nov. 15, 1949), E. T. McBee, V. V. Lindgren, and W. B. Ligett (to Purdue Research Foundation).
21. U.S. Pat. 2,533,217 (May 15, 1951), F. B. Stilmar (to the United States of America as represented by the Atomic Energy Commission).
22. J. M. Lalancette and J. Lafontaine, *J. Chem. Soc., Chem. Commun.*, 815 (1973).
23. A. A. Opalvskii, A. S. Nazarov, and A. A. Uminskii, *Zh. Neorg. Khim.* **19**, 1518 (1974); English ed., *Russ. J. Inorg. Chem.* **19**, 827 (1974).
24. L. B. Ebert, R. A. Muggins, and J. I. Brauman, *Mater. Res. Bull.* **2**, 615 (1976).
25. L. Vogel, *J. Mater. Sci.* **12**, 982 (1977).
26. J. B. Beal, Jr., C. Pupp, and W. E. White, *Inorg. Chem.* **8**, 828 (1969).
27. U.S. Pat. 3,708,570 (Jan. 2, 1973), W. E. Tolberg, R. S. Stringham, and R. T. Rewich (to Stanford Research Institute).
28. U.S. Pat. 3,829,551 (Aug. 13, 1974), L. Stein (to the United States Atomic Energy Commission).
29. U.S. Pat. 4,973,776A (Nov. 27, 1990), V. M. Allenger and R. N. Pandey.
30. V. N. Lebedev and co-workers *J. Organomet. Chem.* **385**(3), 307–308 (1990).
31. U.S. Pat. 4,369,107A (Jan. 18, 1983), S. C. Amendola.
32. G. A. Olah and co-workers, *J. Am. Chem. Soc.* **97**(19), 5477–5481 (1975).
33. K. K. Laali, E. Geleginter, and R. Filler, *J. Fluorine Chem.* **53**(1), 107–126 (1991).
34. G. A. Olah, A. Germain, and H. C. Lin, *J. Am. Chem. Soc.* **97**(19), 5481–5488 (1975).
35. R. J. Gillespie and T. E. Peel, *J. Am. Chem. Soc.* **95**, 5173 (1973).
36. R. J. Gillespie, in V. Gold, ed., *Proton Transfer Reactions*, Chapman and Hall, London, 1975, p. 27.
37. Jpn. Pat. 62027306 A2 (Feb. 5, 1987), Y. Mochida and co-workers.
38. Jpn. Pat. 62108730 A2 (May 20, 1987), Y. Mochida and co-workers.
39. Ger. Offen. DE 3432221 A1 (Mar. 13, 1986), A. Guenther.
40. U.S. Pat. 4,136,102 (Jan. 23, 1979), J. V. Crivello (to General Electric Co.).

TARIQ MAHMOOD
CHARLES B. LINDAHL
Elf Atochem North America, Inc.

ARSENIC

Arsenic forms the binary compounds arsenous trifluoride and arsenic pentafluoride, as well as a series of compounds and the acid of the very stable hexafluoroarsenate ion.

Great care should be exercised in the handling and use of all arsenic compounds (qv) because NIOSH has determined inorganic arsenic to be a carcinogen and OSHA considers inorganic arsenic to be a cancer hazard. The OSHA permissible exposure limit is 10 $\mu g/m^3$, averaged over any 8-h period. The OSHA action level is 5 $\mu g/m^3$, averaged over any 8-h period. The OSHA limits have the force of law and are much lower than the 0.2 mg/m^3 of ACGIH.

Arsenous Fluoride

Arsenous fluoride [7784-35-2], AsF_3, is a colorless liquid, mp = $-5.95°C$, bp = $57.13°C$ at 99 kPa (742.5 mm Hg) (1), and sp gr = 2.67, having a standard enthalpy of formation of -858.1 kJ/mol (-205.1 kcal/mol) (2). Arsenic(III) fluoride can be prepared by fluorination of arsenous oxide using sulfuric acid and calcium fluoride (3), or using hydrofluoric acid or fluorosulfuric acid; from thermal decomposition of $AsBr_4AsF_6$ (4); from the fluorination of gallium arsenide using F_2 or NF_3 (5); from As_2O_3, CaF_2, and concentrated H_2SO_4 (6); from disproportionation of graphite intercalated compounds of AsF_5 (7); from the reaction of arsenous trichloride with NaF at 300°C in the presence of $ZnCl_2$ or KCl (8), and from the fluorination of arsenous trichloride with antimony trifluoride or zinc fluoride (9).

It is used as a fluorinating reagent in semiconductor doping, to synthesize some hexafluoroarsenate compounds, and in the manufacture of graphite intercalated compounds (10) (see SEMICONDUCTORS). AsF_3 has been used to achieve >8% total area simulated air-mass 1 power conversion efficiencies in Si *p-n* junction solar cells (11) (see SOLAR ENERGY). It is commercially produced, but usage is estimated to be less than 100 kg/yr.

Arsenic Trifluoride Oxide

Arsenic trifluoride oxide [15120-14-6], $AsOF_3$, has been reported to be produced by the uv photolysis of O_3 or HOF in the presence of AsF_3 (12,13).

Arsenic Pentafluoride

Arsenic pentafluoride [7784-36-3], AsF_5, melts at $-79.8°C$ and boils at $-52.8°C$ (14). At the boiling point the liquid has a density of 2.33 g/mL. The standard enthalpy of formation is -1237 kJ/mol (-295.6 kcal/mol), and the average bond strength is 387 kJ/mol (92.4 kcal/mol), compared to 484.1 kJ/mol (115.7 kcal/mol) for AsF_3 (15).

Arsenic pentafluoride can be prepared by reaction of fluorine and arsenic trifluoride or arsenic; from the reaction of NF_3O and As (16); from the reaction of

$Ca(FSO_3)_2$ and H_3AsO_4 (17); or by reaction of alkali metal or alkaline-earth metal fluorides or fluorosulfonates with H_3AsO_4 or H_2AsO_3F (18).

It is used as a fluorinating reagent and in syntheses of some hexafluoroarsenate compounds. Arsenic pentafluoride is also used to dope semiconductors (19); to produce conductive polymers (20,21); and in conducting-oriented fibers (22). Arsenic pentafluoride has been found to react with graphite to form AsF_5–graphite intercalation compounds (23) having electrical conductivity as high as that of silver (24,25). Arsenic pentafluoride is produced commercially and usage is estimated to be less than 100 kg/yr.

Hexafluoroarsenic Acid and the Hexafluoroarsenates

The AsF_6^- ion is very stable toward hydrolysis in aqueous solution. It is not hydrolyzed by boiling a strongly basic solution almost to dryness (26), although it is hydrolyzed in sulfuric acid (27) or in boiling perchloric acid (26). The hydrolysis of AsF_6^- in concentrated sulfuric acid (27) and in base (28) at 193–222°C is first order in AsF_6^-. The hydrolysis of AsF_6^- in alkaline solution is slower than either PF_6^- or SbF_6^-.

Hexafluoroarsenic acid [17068-85-8] can be prepared by the reaction of arsenic acid with hydrofluoric acid or calcium fluorosulfate (29) and with alkali or alkaline-earth metal fluorides or fluorosulfonates (18). The hexafluoroarsenates can be prepared directly from arsenates and hydrofluoric acid, or by neutralization of $HAsF_6$. The reaction of 48% HF with potassium dihydrogen arsenate(V), KH_2AsO_4, gives potassium hydroxypentafluoroarsenate(V) [17068-84-7], $KAsF_5OH$, which hydrolyzes rapidly in water solution (26). Anhydrous HF reacts with KH_2AsO_4 or $KAsF_5OH$ to produce $KAsF_6$ [17029-22-0]. O_2AsF_6 [12370-43-3] can be prepared from the reaction of OF_2 and AsF_5 or a mixture of O_2, F_2, and AsF_5 (30). Reactions of $XeF(AsF_6)$ and water give $H_2OF(AsF_6)$ which reacts with SF_4 to produce $OSF_3(AsF_6)$ and with ClF_3 to produce $OClF_2AsF_6$ (31). The compound $SCl_3 \cdot AsF_6$ has also been reported (32).

Because of the special stability of the hexafluoroarsenate ion, there are a number of applications of hexafluoroarsenates. For example, onium hexafluoroarsenates (33) have been described as photoinitiators in the hardening of epoxy resins (qv). Lithium hexafluoroarsenate [29935-35-1] has been used as an electrolyte in lithium batteries (qv). Hexafluoroarsenates, especially alkali and alkaline-earth metal salts or substituted ammonium salts, have been reported (34) to be effective as herbicides (qv). Potassium hexafluoroarsenate [17029-22-0] has been reported (35) to be particularly effective against prickly pear. However, environmental and regulatory concerns have severely limited these applications.

BIBLIOGRAPHY

"Arsenic Compounds" in *ECT* 1st ed., Vol. 2, pp. 119–123, by I. E. Campbell, Battelle Memorial Institute; in *ECT* 2nd ed., Vol. 2, pp. 718–733, by G. O. Doak, L. D. Freedman, and G. G. Long, North Carolina State of the University of North Carolina at Raleigh;

"Arsenic" under "Fluorine Compounds, Inorganic" in *ECT* 3rd ed., Vol. 10, pp. 682–683, by C. B. Lindahl, Ozark-Mahoning Co., a subsidiary of the Pennwalt Corp.

1. H. Russell, Jr., R. E. Rundle, and D. M. Yost, *J. Am. Chem. Soc.* **63**, 2825 (1941).
2. A. A. Woolf, *J. Fluorine Chem.* **5**, 172 (1975).
3. O. Ruff, *Die Chemie des Fluors*, Springer-Verlag, Berlin, 1920, p. 27.
4. B. Ponsold, and H. Kath, *Z. Gesamte Hyg. Ihre Grenzgeb.* **37**(2), 58–63 (1991).
5. Eur. Pat. 333084 A2 (Sept. 20, 1989), I. Harada, Y. Yoda, N. Iwanaga, T. Nishitsuji, and A. Kikkawa.
6. Ger. Pat. DD248249 A3 (Aug. 5, 1987), P. Wolter, M. Schoenherr, D. Hass.
7. J. G. Hooley, *Ext. Abstr. Program*, 16th, Biennial Conference on Carbon, 1983, pp. 240–241.
8. U.S. Pat. 4,034,069 (July 5, 1977), D. M. Curtis.
9. F. Kober, *J. Fluorine Chem.* **2**(3), 247–256 (1973).
10. Y. Yacoby, *Synth. Met.*, **34**(1-3), 437–438 (1989).
11. E. J. Caine and E. J. Charlson, *J. Electron. Mater.* **13**(2), 341–372 (1984).
12. E. A. Evans, A. J. Downs, and C. J. Gardner, *J. Phys. Chem.* **93**(2), 598–608 (1989).
13. A. J. Downs, G. P. Gaskill, and S. B. Saville, *Inorg. Chem.* **21**(9), 3385–3393 (1982).
14. O. Ruff, A. Braida, O. Bretschneider, W. Menzel, and H. Plaut, *Z. Anorg. Allgem. Chem.* **206**, 59 (1932).
15. P. A. G. O'Hare and W. N. Hubbard, *J. Phys. Chem.* **69**, 4358 (1965).
16. O. D. Gupta, R. L. Kirchmeier, and J. M. Shreeve, *Inorg. Chem.* **29**(3), 573–574 (1990).
17. U.S. Pat. 3,875,292 (Apr. 1, 1975), R. A. Wiesboeck and J. D. Nickerson.
18. U.S. Pat. 3,769,387 (Oct. 30, 1973), R. A. Wiesboeck and J. D. Nickerson.
19. D. G. H. Ballard, A. Courtis, I. M. Shirley, and S. C. Taylor, *Air Force Off. Sci. Res.* (Technical Report), AFOSR-TR (U.S.), AFSOR-TR-87 1884, Biotechnol. Aided Synth. Aerosp. Compos. Resins 53-92, CA110(24):213428x.
20. M. Aldissi, *Polymer Prepr. (Am. Chem. Soc., Div. Polym. Chem.)* **26**(2), 269–270 (1985).
21. Jpn. Pat. 59 133,216 A2 (July 31, 1984) Showa (to Orient Watch Co.).
22. M. Stamm, *Mol. Cryst. Liquid Cryst.* **105**(1-4), 259–271 (1984).
23. L. Chun-Hsu, H. Selig. M. Rabinovitz, I. Agranat, and S. Sarig, *Inorg. Nucl. Chem. Lett.* **11**, 601 (1975).
24. E. R. Falardeau, G. M. T. Foley, C. Zeller, and F. L. Vogel, *Chem. Commun.*, 389 (1977).
25. G. M. T. Foley, C. Zeller, E. R. Falardeau, and F. L. Vogel, *Solid State Commun.* **24**, 371 (1977).
26. H. M. Dess and R. W. Parry, *J. Am. Chem. Soc.* **79**, 1589 (1957).
27. W. L. Lockhart, Jr., M. M. Jones, and D. O. Johnston, *J. Inorg. Nucl. Chem.* **31**, 407 (1969).
28. I. G. Ryss, V. B. Tul'chinskii and Y. A. Mazurov, *Izv. Sib. Otd. Akad. Nauk. SSSR Ser., Khim. Nauk*, 81 (1968).
29. U.S. Pat 3,875,292 (Apr. 1, 1975), R. A. Wiesboeck and J. D. Nickerson (to U.S. Steel Corp.).
30. J. B. Beal, Jr., P. Christian, and W. E. White, *Inorg. Chem.* **8**(4), 828–830 (1969).
31. R. Minkwitz and G. Nowicki, *Angew. Chem.* **102**(6), 692–693 (1990).
32. F. Claus and R. Minkwitz, *J. Fluorine Chem.* **19**(3-6), 243–252 (1982).
33. Ger. Offen. 2,618,871 (Nov. 11, 1976) and 2,518,652 (May 2, 1974), J. V. Crivello (to General Electric Co.).
34. Belg. Pat. 659,342 (Aug. 5, 1965), T. N. Russell (to Pennsalt Chemicals Corp.).
35. P. E. Buckley, J. D. Dodd, and W. H. Culver, *Proc. West. Soc. Weed Sci.* **22**, 17 (1968).

CHARLES B. LINDAHL
TARIQ MAHMOOD
Elf Atochem North America, Inc.

BARIUM

Barium Fluoride

Barium fluoride [7782-32-8], BaF_2, is a white crystal or powder. Under the microscope crystals may be clear and colorless. Reported melting points vary from 1290 (1) to 1355°C (2), including values of 1301 (3) and 1353°C (4). Differences may result from impurities, reaction with containers, or inaccurate temperature measurements. The heat of fusion is 28 kJ/mol (6.8 kcal/mol) (5), the boiling point 2260°C (6), and the density 4.9 g/cm³. The solubility in water is about 1.6 g/L at 25°C and 5.6 g/100 g (7) in anhydrous hydrogen fluoride. Several preparations for barium fluoride have been reported (8–10).

High purity BaF_2 can be prepared from the reaction of barium acetate and aqueous HF (11), by dissolving the impure material in 2-12N HCl and recrystallizing at $-40°C$ (12), by vacuum distillation of the metal fluoride impurities from a BaF_2 melt (13), by purification of the aqueous acetate solution by ion exchange followed by fluorination (14), by solvent extraction using dithiocarbamate and CCl_4 (15–17), and by solvent extraction using acetonitrile (18).

A typical analysis of the commercial product is 99% with a loss on ignition of 0.9%; sulfates as SO_4, 0.2%; hexafluorosilicate as SiF_4, 0.02%; heavy metals as lead, 0.02%; and iron, 0.005%.

Barium fluoride is used commercially in combination with other fluorides for arc welding (qv) electrode fluxes. However, this usage is limited because of the availability of the much less expensive naturally occurring calcium fluoride.

Other reported uses of barium fluoride include the manufacture of fluorophosphate glass (19); stable fluoride glass (20); fluoroaluminate glass (21); fluorozirconate glass (22); infrared transmitting glass (23); in oxidation-resistant ceramic coatings (24); in the manufacture of electric resistors (25,26); as a superconductor with copper oxide (27); and as a fluoride optical fiber (28) (see FIBER OPTICS; GLASS; SUPERCONDUCTING MATERIALS).

The toxicity of barium fluoride has received only little attention. A value for oral LD_{LO} of 350 mg/kg in guinea pigs has been reported (29). OSHA has a TWA standard on the basis of Ba of 0.5 mg/m³ for barium fluoride (29) in addition to a standard TWA on the basis of F of 2.5 mg/m³ (30). NIOSH has issued a criteria document (30) on occupational exposure to inorganic fluorides.

BIBLIOGRAPHY

"Barium Fluoride" under "Fluorine Compounds, Inorganic," in *ECT* 1st ed., Vol. 6, p. 677, by F. D. Loomis, Pennsylvania Salt Manufacturing Co.; in *ECT* 2nd ed., Vol. 9, p. 551, by W. E. White, Ozark-Mahoning Co.; "Barium" under "Fluorine Compounds, Inorganic," in *ECT* 3rd ed., Vol. 10, p. 684, by C. B. Lindahl, Ozark-Mahoning Co.

1. I. Barin and O. Knache, *Thermochemical Properties of Inorganic Substances*, Springer Verlag, Berlin, 1973.
2. H. Kojima, S. G. Whiteway, and C. R. Masson, *Can. J. Chem.* **46**, 2698 (1968).
3. I. Jackson, *Phys. Earth Planet. Inter.* **14**, 143 (1977).
4. B. Porter and E. A. Brown, *J. Am. Ceram. Soc.* **45**, 49 (1962).

5. G. Petit and A. Cremieo, *C. R. Acad. Sci.* **243**, 360 (1956).
6. O. Ruff and L. LeBoucher, *Z. Anorg. Chem.* **219**, 376 (1934).
7. A. W. Jache and G. H. Cady, *J. Phys. Chem.* **56**, 1106 (1952).
8. SU 1325018 Al (July 23, 1985), V. A. Bogomolov and co-workers.
9. SU 998352 A1 (Feb. 23, 1983), A. A. Luginina and co-workers.
10. A. A. Lugina and co-workers *Zh. Neorg. Khim 1981*, **26**(2), 332–336.
11. Jpn. Pat. 90-144378 (June 4, 1990), K. Kobayashi, K. Fujiura, and S. Takahashi.
12. EP 90-312689 (Nov. 21, 1990), J. A. Sommers, R. Ginther, and K. Ewing.
13. A. M. Garbar, A. N. Gulyaikin, G. L. Murskii, I. V. Filimonov, and M. F. Churbanov, *Vysokochist, Veshchestva* (6), 84–85 (1990).
14. A. M. Garbar, A. V. Loginov, G. L. Murskii, V. I. Rodchenkov, and V. G. Pimenov, *Vysokochist, Veshchestva* (3), 212–213 (1989).
15. Jpn. Pat. 01028203 A2 (Jan. 30, 1989), K. Kobayashi (to Heisei).
16. K. Kobayashi, *Mater. Sci. Forum*, **32–33**(5), 75–80 (1988).
17. DE 3813454 A1 (Nov. 3, 1988); Jpn. Pat. 87-100025 (Apr. 24, 1987), H. Yamashita and H. Kawamoto.
18. J. Guery and C. Jacoboni, in Ref. 16, pp. 31–35.
19. V. D. Khalilev, V. G. Cheichovskii, M. A. Amanikov, and Kh. V. Sabirov, *Fiz. Khim. Stekla* **17**(5), 740–743 (1991).
20. Y. Wang, *J. Non. Cryst. Solids* **142**(1–2), 185–188 (1992).
21. H. Hu, F. Lin, and J. Feng, *Guisudnyan Xuebao* **18**(6), 501–505 (1990).
22. M. N. Brekhovskikh, V. A. Fedorov, V. S. Shiryaev, and M. F. Churbanov, *Vysokochist Veshchestva* (1), 219–223 (1991).
23. A. Jha and J. M. Parker, *Phys. Chem. Glasses* **32**(1), 1–12 (1991).
24. EP 392822 A2 17 (Oct. 17, 1990), L. M. Niebylski.
25. Jpn. Pat. 63215556 A2 (Sept. 8, 1988), T. Honda, T. Yamada, K. Onigata, and S. Tosaka (to Showa).
26. Jpn. Pat. 63215553 A (Sept. 28, 1988), T. Honda, T. Yamada, K. Onigata, and S. Tosaka (to Showa).
27. S. R. Ovshinsky, R. T. Young, B. S. Chao, G. Fournier, and D. A. Pawlik, *Rev. Solid State Sci.* **1**(2), 207–219 (1987).
28. J. Chen and co-workers, *J. Non-Cryst. Solids* **140**(1–3), 293–296 (1992).
29. *Registry of Toxic Effects of Chemical Substances*, Vol. II, NIOSH, Washington, D.C., 1977, p. 141.
30. *Criteria for a Recommended Standard-Occupational Exposure to Inorganic Fluorides*, PB 246 692, NIOSH 76-103, U.S. Department of Health, Education, and Welfare, Washington, D.C., 1975.

Tariq Mahmood
Charles B. Lindahl
Elf Atochem North America, Inc.

BORON

Boron trifluoride, **60**
Fluoroboric acid and fluoroborates, **69**

BORON TRIFLUORIDE

Boron trifluoride [7637-07-2] (trifluoroborane), BF_3, was first reported in 1809 by Gay-Lussac and Thenard (1) who prepared it by the reaction of boric acid and fluorspar at dull red heat. It is a colorless gas when dry, but fumes in the presence of moisture yielding a dense white smoke of irritating, pungent odor. It is widely used as an acid catalyst (2) for many types of organic reactions, especially for the production of polymer and petroleum (qv) products. The gas was first produced commercially in 1936 by the Harshaw Chemical Co. (see also BORON COMPOUNDS).

The boron atom in boron trifluoride is hybridized to the sp^2 planar configuration and consequently is coordinatively unsaturated, ie, a Lewis acid. Its chemistry centers around satisfying this unsaturation by the formation with Lewis bases of adducts that are nearly tetrahedral (sp^3). The electrophilic properties

Table 1. Physical Properties of Boron Trifluoride

Property	Value	Reference
molecular weight	67.8062	5
melting point, °C	−128.37	6
boiling point, °C	−99.9	6
vapor pressure of liquid, kPa[a]		
at 145 K	8.43	
at 170 K	80.19	
at 220 K	1156	
at 260 K	4842	7
triple point at 8.34 kPa,[a] K	144.78	5
critical temperature, T_c, °C	−12.25 ± 0.03	7
critical pressure, P_c, kPa[a]	4984	7
density		
critical, d_c, g/cm^3	ca 0.591	8
gas at STP, g/L	3.07666	9
gas limiting, L_N, g/L	3.02662	9
liquid, for 148.9 to 170.8 K, g/cm^3	$1.699 - 0.00445\,(t + 125.0)$	10
enthalpy of fusion, $\Delta H_{144.45}$, kJ/mol[b]	4.2417	11
enthalpy of vaporization, $\Delta H_{154.5}$, kJ/mol[b]	18.46	12
entropy, $S_{298.15}$, J/(mol·K)[b]	254.3	
Gibbs free energy of formation, $\Delta G_{f298.15}$, kJ/mol[b]	−1119.0	13
enthalpy of formation, $\Delta H_{f298.15}$, kJ/mol[b]	−1135.6	13
infrared absorption frequencies, cm^{-1}		
v_1	888	
v_2	696.7	
v_3	1463.3	
v_4	480.7	

[a] To convert kPa to mm Hg, multiply by 7.5. [b] To convert J to cal, divide by 4.184.

(acid strengths) of the trihaloboranes have been found to increase in the order $BF_3 < BCl_3 < BBr_3 < BI_3$ (3,4).

Physical Properties. The physical properties are listed in Table 1. The molecule has a trigonal planar structure in which the F—B—F angle is 120° and the B—F bond distance is 0.1307 ± 0.0002 nm (13).

Nuclear magnetic resonance ^{11}B spectral studies of BF_3 have given a value of 9.4 ± 1.0 ppm for the chemical shift relative to $BF_3 \cdot O(C_2H_5)_2$ as the zero reference (14). Using methylcyclohexane as a solvent at 33.5°C and $BF_3 \cdot O(CH_2CH_3)_2$ as the internal standard, a value of 10.0 ± 0.1 ppm was obtained for the chemical shift (15). A value for the ^{19}F chemical shift of BF_3 in CCl_3F relative to CCl_3F is reported to be 127 ppm (16). The coupling constant $J_{^{11}B - ^{19}F}$ is reported to be 15 ± 2 Hz for BF_3 (17). Additional constants are available (3,18). See Table 2 for solubilities.

Aqueous mineral acids react with BF_3 to yield the hydrates of BF_3 or the hydroxyfluoroboric acids, fluoroboric acid, or boric acid. Solution in aqueous alkali gives the soluble salts of the hydroxyfluoroboric acids, fluoroboric acids, or boric acid. Boron trifluoride, slightly soluble in many organic solvents including saturated hydrocarbons (qv), halogenated hydrocarbons, and aromatic compounds, easily polymerizes unsaturated compounds such as butylenes (qv), styrene (qv), or vinyl esters, as well as easily cleaved cyclic molecules such as tetrahydrofuran (see FURAN DERIVATIVES). Other molecules containing electron-donating atoms such as O, S, N, P, etc, eg, alcohols, acids, amines, phosphines, and ethers, may dissolve BF_3 to produce soluble adducts.

Chemical Properties. In addition to the reactions listed in Table 3, boron trifluoride reacts with alkali or alkaline-earth metal oxides, as well as other inorganic alkaline materials, at 450°C to yield the trimer trifluoroboroxine [*13703-95-2*], $(BOF)_3$, MBF_4, and MF (29) where M is a univalent metal ion. The trimer is stable below −135°C but disproportionates to B_2O_3 and BF_3 at higher temperatures (30).

$$\begin{array}{c} F \\ | \\ B \\ O \diagup \diagdown O \\ | \quad\quad | \\ F-B \diagdown \diagup B-F \\ O \end{array}$$

The reaction of metal hydrides and BF_3 depends on the stoichiometry as well as the nature of the metal hydride. For example, LiH and $BF_3 \cdot O(C_2H_5)_2$ may form diborane (6) or lithium borohydride (31,32):

$$6\ LiH + 8\ BF_3 \cdot O(C_2H_5)_2 \rightarrow B_2H_6 + 6\ LiBF_4 + 8\ (C_2H_5)_2O$$
$$4\ LiH + 4\ BF_3 \cdot O(C_2H_5)_2 \rightarrow LiBH_4 + 3\ LiBF_4 + 4\ (C_2H_5)_2O$$

The first method is commonly used for preparing diborane.

Metal halides react with BF_3 (33) when heated to form BX_3 and the metal fluoride. For example,

$$AlBr_3 + BF_3 \rightarrow BBr_3 + AlF_3$$

Table 2. Solubilities of Boron Trifluoride

BF$_3$, g	Solvent, g	Temperature, °C	Product	CAS Registry Number	Reference
369.4	water,^{a} 100^b	6	BF$_3$·H$_2$O HBF$_3$(OH)	[15799-89-0] [16903-52-9]	19
2.06	sulfuric acid, conc, 100%	25			20
	nitric acid^a	20	HNO$_3$·2BF$_3$	[20660-63-3]	21
	orthophosphoric acid^a	25	H$_3$PO$_4$·BF$_3$	[13699-76-6]	22
2.18	hydrofluoric acid,^{c}	4.4			24
	hydrochloric acid, anhydrous (l)	24	miscible		25

^{a}Dissolves with reaction to form complexes and other species.
^{b}A higher dilution results in a mixture of H[BF$_2$(OH)$_2$], HBF$_4$, and H$_3$BO$_3$.
^{c}Equations for the solubility of BF$_3$ in liquid HF at 24, 49, and 90°C and up to 6.8 kPa (51 mm Hg) may be found in Reference 23.

Table 3. Reactions of Boron Trifluoride

Reactant	Temperature, °C	Products	Formula	Reference
sodium^a		boron, amorphous, sodium fluoride	NaF	26
magnesium, molten alloys	no reaction			
calcium	1600	calcium hexaboride	CaB$_6$	
aluminum	1200	aluminum boride (1:12), tetragonal boron	AlB$_{12}$	
	1650^b	β-rhombohedral boron		
titanium	1600	titanium boride	TiB$_2$	27
copper, mercury, chromium, iron	RT or below	no reaction^c		
sodium nitrate, sodium nitrite	180	sodium fluoroborate, boric oxide	NaBF$_4$	28

^{a}With incandescence.
^{b}Further reaction.
^{c}Even when subjected to pressure for a considerable length of time; also no reaction with red-hot iron.

The reaction of BF$_3$ with alkali halides yields the respective alkali fluoroborates (34):

$$3 \text{ KCl} + 4 \text{ BF}_3 \rightarrow 3 \text{ KBF}_4 + \text{BCl}_3$$

Alkyl and arylboranes are obtained (35) from BF$_3$ using the appropriate Grignard reagent, alkylaluminum halide, or zinc alkyl, using diethyl ether as the solvent (see also ORGANOMETALLICS):

$$\text{BF}_3 + 3 \text{ RMgX} \rightarrow \text{BR}_3 + 3 \text{ MgXF}$$

Tetraorganylborate complexes may be produced when tetrahydrofuran is the solvent (36).

Alkylfluoroboranes result from the reaction of the appropriate alkylborane and BF_3 under suitable conditions (37):

$$BR_3 + 2(C_2H_5)_2O \cdot BF_3 \rightarrow 3\ RBF_2 + 2\ (C_2H_5)_2O$$

Adducts of BF_3 and some organic compounds having labile hydrogen atoms in the vicinity of the atom bonding to the boron atom of BF_3 may form a derivative of BF_3 by splitting out HF. For example, β-diketones such as acetylacetone or benzoylacetone react with BF_3 in benzene (38):

$$BF_3 + CH_3COCH_2COCH_3 \rightarrow CH_3COCH=C(CH_3)OBF_2 + HF$$

In Group 14 (IV), carbon serves as a Lewis base in a few of its compounds. In general, saturated aliphatic and aromatic hydrocarbons are stable in the presence of BF_3, whereas unsaturated aliphatic hydrocarbons, such as propylene or acetylene, are polymerized. However, some hydrocarbons and their derivatives have been reported to form adducts with BF_3. Typical examples of adducts with unsaturated hydrocarbons are 1:1 adducts with tetracene and 3,4-benzopyrene (39), and 1:2 BF_3 adducts with α-carotene and lycopene (40).

In Group 15 (V), nitrogen compounds readily form molecular compounds with BF_3. Phosphorus compounds also form adducts with BF_3. Inorganic or organic compounds containing oxygen form many adducts with boron trifluoride, whereas sulfur and selenium have been reported to form only a few (41–43).

Boron trifluoride forms two hydrates, $BF_3 \cdot H_2O$ and boron trifluoride dihydrate [13319-75-0], $BF_3 \cdot 2H_2O$, (also $BF_3 \cdot D_2O$ [33598-66-2] and $BF_3 \cdot 2D_2O$ [33598-66-2]). According to reported nmr data (43,44), the dihydrate is ionic, $H_3O^+F_3BOH^-$. The trihydrate has also been reported (45). Acidities of BF_3–water systems have been determined (46). Equilibrium and hydrolysis of BF_3 in water have been studied (47–49).

Most of the coordination compounds formed by trifluoroborane are with oxygen-containing organic compounds (Table 4). Although the other boron halides frequently react to split out hydrogen halide, boron trifluoride usually forms stable molecular compounds. The reason is attributed to the back coordination of electrons from fluorine to boron forming a strong B—F bond which is 28% ionic (50).

It has been reported (51) that some adducts of alkyl ethers and/or alcohols are unstable and decompose at $-80°C$ to yield BF_3, H_2O, and the polyalkene. Adducts of BF_3 have been reported with hydrogen sulfide, sulfur dioxide, thionyl fluoride, and the sulfur analogues of many of the kind of oxygen-containing organic molecules cited in Table 4. The carbonyl oxygen or the carbonyl sulfur is the donor to BF_3 in 1:1 adducts such as $CH_3COOCH_3 \cdot BF_3$ [7611-14-5], $CH_3COSCH_3 \cdot BF_3$ [52913-04-9], and $CH_3CSOCH_3 \cdot BF_3$ [52912-98-8] (52).

Compounds containing fluorine and chlorine are also donors to BF_3. Aqueous fluoroboric acid and the tetrafluoroborates of metals, nonmetals, and organic radicals represent a large class of compounds in which the fluoride ion is coordinating with trifluoroborane. Representative examples of these compounds are given in Table 5. Coordination compounds of boron trifluoride with the chlorides of sodium,

Table 4. Boron Trifluoride Adducts with Oxygen-Containing Compounds

Donor	Adduct name	CAS Registry Number	Molecular formula
alcohols	ethanol trifluoroborane	[353-41-3]	$C_2H_5OH \cdot BF_3$
	bis(ethanol) trifluoroborane	[373-59-1]	$2C_2H_5OH \cdot BF_3$
	bis(2-chloroethanol) trifluoroborane	[72985-81-0]	$2ClCH_2CH_2OH \cdot BF_3$
	benzyl alcohol trifluoroborane	[456-31-5]	$C_6H_5CH_2OH \cdot BF_3$
acids	acetic acid trifluoroborane	[753-53-7]	$CH_3COOH \cdot BF_3$
	bis(acetic acid) trifluoroborane	[373-61-5]	$2CH_3COOH \cdot BF_3$
	stearic acid trifluoroborane	[60274-92-2]	$CH_3(CH_2)_{16}COOH \cdot BF_3$
	bis(phenol) trifluoroborane	[462-05-5]	$2C_6H_5OH \cdot BF_3$
ethers	diethyl ether trifluoroborane	[109-63-7]	$(C_2H_5)_2O \cdot BF_3$
	tetrahydrofuran trifluoroborane	[462-34-0]	$(CH_2)_4O \cdot BF_3$
	anisole trifluoroborane	[456-31-5]	$CH_3OC_6H_5 \cdot BF_3$
acid anhydride	acetic anhydride trifluoroborane	[591-00-4]	$(CH_3CO)_2O \cdot BF_3$
esters	ethyl formate trifluoroborane	[462-33-9]	$HCOOC_2H_5 \cdot BF_3$
	phenyl acetate trifluoroborane	[30884-81-6]	$CH_3COOC_6H_5 \cdot BF_3$
ketones	acetone trifluoroborane	[661-27-8]	$(CH_3)_2CO \cdot BF_3$
	benzophenone trifluoroborane	[322-21-4]	$(C_6H_5)_2CO \cdot BF_3$
	acetophenone trifluoroborane	[329-25-9]	$C_6H_5COCH_3 \cdot BF_3$
aldehydes	acetaldehyde trifluoroborane	[306-73-0]	$CH_3CHO \cdot BF_3$
	neopentanal trifluoroborane	[306-78-5]	$(CH_3)_3CCHO \cdot BF_3$
	benzaldehyde trifluoroborane	[456-30-4]	$C_6H_5CHO \cdot BF_3$

Table 5. Boron Trifluoride Adducts with Compounds Containing Chlorine and Fluorine

Name	CAS Registry Number	Molecular formula
potassium tetrafluoroborate	[14075-53-7]	KBF_4
hexamminenickel(II) tetrafluoroborate	[13877-20-8]	$[Ni(NH_3)_6](BF_4)_2$
nitrosyl tetrafluoroborate	[14635-75-7]	$NOBF_4$
acetylium tetrafluoroborate	[2261-02-1]	CH_3COBF_4
tetramethylammonium tetrafluoroborate	[661-36-9]	$(CH_3)_4NBF_4$
difluorobromine tetrafluoroborate	[14282-83-8]	BrF_2BF_4
anilinium tetrafluoroborate	[15603-97-1]	$C_6H_5NH_2HBF_4$

aluminum, iron, copper, zinc, tin, and lead have been indicated (53); they are probably chlorotrifluoroborates.

Trifluoroborane may form adducts with some of the transition elements. See Reference 54 for a detailed discussion of complexes of trifluoroborane with various Group 6–10 (VI, VII, and VIII) species.

Manufacture. Boron trifluoride is prepared by the reaction of a boron-containing material and a fluorine-containing substance in the presence of an acid. The traditional method used borax, fluorspar, and sulfuric acid.

In another process fluorosulfonic acid is treated with boric acid:

$$3 \ HSO_3F + H_3BO_3 \rightarrow BF_3 + 3 \ H_2SO_4$$

Numerous other reactions are available for the preparation of small quantities of boron trifluoride, some of which are of high purity (55).

Shipment and Handling. The gas is nonflammable and is shipped in DOT 3A and 3AA steel cylinders at a pressure of approximately 12,410 kPa (1800 psi). Boron trifluoride is classified as a poison gas, both domestically and internationally. Cylinders must have a poison gas diamond and an inhalation hazard warning label. Tube trailers carry both a poison gas placard and an inhalation hazard warning. Cylinders containing 27.2 kg and tube trailers containing 4.5–10 metric tons are available. If boron trifluoride is compressed using oil as a compressor lubricant, it must not be used with oxygen under pressure nor with gauges, valves, or lines that are to be used with oxygen.

Inasmuch as the gas hydrolyzes readily, all equipment should be purged repeatedly using inert dry gas before admitting boron trifluoride. Under anhydrous conditions, carbon steel equipment is satisfactory. Stainless steel and aluminum silicon bronze may also be used. Stainless steel tubing is recommended for both temporary and permanent connections.

In the presence of moisture, boron trifluoride may be handled in polytetrafluoroethylene (PTFE), polyethylene, Pyrex glass (limit to atmospheric pressure), or Hastelloy C containers. At 600°C, stainless steel (304 L) and Hastelloy N are attacked by BF_3; Hastelloy C is more resistant (56). Kel F and PTFE serve as satisfactory gasket and packing materials, whereas rubber, fiber, polymerizable materials, or organic oxygen- and nitrogen-containing compounds must be avoided. Because boron trifluoride is soluble in, and reacts with, many liquids, the gas must not be introduced into any liquid unless a vacuum break or similar safety device is employed.

Economic Aspects, Standards, and Analyses. The sole United States producer of boron trifluoride is AlliedSignal, Inc. The 1992 price of boron trifluoride was $9.59–12.46/kg, depending on purity and the quantity purchased.

Commercial boron trifluoride is usually approximately 99.5% pure. The common impurities are air, silicon tetrafluoride, and sulfur dioxide. An excellent procedure for sampling and making a complete analysis of gaseous boron trifluoride has been developed (57).

Health and Safety Factors. Boron trifluoride is primarily a pulmonary irritant. The toxicity of the gas to humans has not been reported (58), but laboratory tests on animals gave results ranging from an increased pneumonitis to death. The TLV is 1 ppm (59,60). Inhalation toxicity studies in rats have shown that exposure to BF_3 at 17 mg/m^3 resulted in renal toxicity, whereas exposure at 6 mg/m^3 did not result in a toxic response (61). Prolonged inhalation produced dental fluorosis (62). High concentrations burn the skin similarly to acids such as HBF_4 and, if the skin is subject to prolonged exposure, the treatment should be the same as for fluoride exposure and hypocalcemia. No chronic effects have been observed in workers exposed to small quantities of the gas at frequent intervals over a period of years.

Uses. Boron trifluoride is an excellent Lewis acid catalyst for numerous types of organic reactions. Its advantages are ease of handling as a gas and the absence of undesirable tarry by-products. As an electrophilic molecule, it is an excellent catalyst for Friedel-Crafts and many other types of reactions (63–65) (see FRIEDEL-CRAFTS REACTIONS).

BF$_3$·HF compositions have been reported to act as super acids in catalyzing condensation reactions (66). BF$_3$-catalyzed preparation of 1- or 2-naphthol is reported to be regioselective (67). Dehydration reactions may also be regioselective (68). Selected fluorinations may be catalyzed by BF$_3$ using HF as the fluoride source (69). BF$_3$ is widely used for the preparation of hydrocarbon resins (70), tall oil (qv) resins (71), and tackifier resins (72). Alpha olefin-based synthetic lubricants are commonly made using BF$_3$-based catalysts (73–75). BF$_3$ is widely used as a polymerization catalyst (76–78). A developing use for BF$_3$ is as an ion implant medium for semiconductor materials (79). BF$_3$ may be used as a chemical reagent for the manufacture of fluoroboro complexes (80), boron nitride [10043-11-5] (81), and boron trichloride [10294-34-5] (82). Carboxylic acids and esters may be prepared by reacting CO with olefins in the presence of BF$_3$-containing catalysts (83).

In addition, boron trifluoride and some of its adducts have widespread application as curing agents for epoxy resins (qv), and in preparing alcohol-soluble phenolic resins (qv) (41).

Boron trifluoride catalyst is used under a great variety of conditions either alone in the gas phase or in the presence of many types of promoters. Many boron trifluoride coordination compounds are also used.

Boron trifluoride catalyst may be recovered by distillation, chemical reactions, or a combination of these methods. Ammonia or amines are frequently added to the spent catalyst to form stable coordination compounds that can be separated from the reaction products. Subsequent treatment with sulfuric acid releases boron trifluoride. An organic compound may be added that forms an adduct more stable than that formed by the desired product and boron trifluoride. In another procedure, a fluoride is added to the reaction products to precipitate the boron trifluoride which is then released by heating. Selective solvents may also be employed in recovery procedures (see CATALYSTS, REGENERATION).

Boron trifluoride is also employed in nuclear technology by utilizing several nuclear characteristics of the boron atom. Of the two isotopes, ^{10}B and ^{11}B, only ^{10}B has a significant absorption cross section for thermal neutrons. It is used in ^{10}BF$_3$ as a neutron-absorbing medium in proportional neutron counters and for controlling nuclear reactors (qv). Some of the complexes of trifluoroborane have been used for the separation of the boron isotopes and the enrichment of ^{10}B as ^{10}BF$_3$ (84).

Boron trifluoride is used for the preparation of boranes (see BORON COMPOUNDS). Diborane is obtained from reaction with alkali metal hydrides; organoboranes are obtained with a suitable Grignard reagent.

Boron trifluoride has been used in mixtures to prepare boride surfaces on steel (qv) and other metals, and as a lubricant for casting steel (see LUBRICATION AND LUBRICANTS).

BIBLIOGRAPHY

"Boron Trifluroide" under "Flurorine Compounds, Inorganic" in *ECT* 1st ed., Vol. 6, pp. 678–684, by D. R. Martin, University of Illinois; "Boron Trifluoride" under "Boron" under "Fluorine Compounds, Inorganic" in *ECT* 2nd ed., Vol. 9, pp. 554–562, by D. R. Martin, The

Harshaw Chemical Co.; in *ECT* 3rd ed., Vol. 10, pp. 685–693 by D. R. Martin, University of Texas at Arlington.

1. J. L. Gay-Lussac and J. L. Thénard, *Rech. Phys.* **2**, 38 (1811); *Ann. Chim Phys.* **69**, 204 (1809).
2. J. A. Nieuwland, R. R. Vogt, and W. L. Foohey, *J. Am. Chem. Soc.* **52**, 1018 (1930).
3. H. C. Brown and R. R. Holmes, *J. Am. Chem. Soc.* **78**, 2173 (1956).
4. A. Oliva, *THEOCHEM 1991*, **82**(1–2), 75–84, 1991).
5. J. C. G. Calado and L. A. K. Staveley, *Trans. Faraday Soc.* **67**, 1261 (1971).
6. E. Pohland and W. Harlos, *Z. Anorg. Allgem. Chem.* **207**, 242 (1932).
7. H. S. Booth and J. M. Carter, *J. Phys. Chem.* **36**, 1359 (1932).
8. R. F. Smith, U.S. Atomic Energy Commission, *NAA-SR-5286*, 1960.
9. C. F. Rumold, PhD. dissertation, Case Western Reserve University, Cleveland, Ohio, 1931.
10. E. Wiberg and W. Mäthing, *Ber. Dtsch. Chem. Ges. B.* **70B**, 690 (1937).
11. A. Eucken and E. Schröder, *Z. Physik. Chem.* **341**, 307 (1938).
12. H. M. Spencer, *J. Chem. Phys.* **14**, 729 (1946).
13. D. R. Stull and H. Prophet, *Natl. Stand. Ref. Data Ser. Natl. Bur. Stand.* **37** (1971).
14. T. P. Onak and co-workers, *J. Phys. Chem.*, 63 (1959).
15. M. F. Lappert and co-workers, *J. Chem. Soc. A.*, 2426 (1971).
16. T. D. Coyle, S. L. Stafford, and F. G. Stone, *J. Chem. Soc.*, 3103 (1961).
17. T. D. Coyle and F. G. A. Stone, *J. Chem. Phys.* **32**, 1892 (1960); I. S. Jaworiwsky and co-workers, *Inorg. Chem.* **18**, 56 (1979).
18. *Gmelins Handbuch der Anorganischen Chemie*, Vol. 13, 8th ed., Verlag Chemie, GmbH, Weinheim/Bergstrasse, Germany, 1954, pp. 167–196.
19. S. Pawlenko, *Z. Anorg. Allegem. Chem.* **300**, 152 (1959).
20. N. N. Greenwood and A. Thompson, *J. Chem. Soc.*, 3643 (1959).
21. H. Gerding and co-workers, *Rec. Trav. Chim.* **71**, 501 (1952).
22. N. N. Greenwood and A. Thompson, *J. Chem. Soc.*, 3493 (1959).
23. R. J. Mikovsky, S. D. Levy, and A. L. Hensley, Jr., *J. Chem. Eng. Data* **6**, 603 (1961).
24. E. C. Hughes and S. M. Darling, *Ind. Eng. Chem.* **43**, 746 (1951).
25. H. S. Booth and D. R. Martin, *J. Am. Chem. Soc.* **64**, 2198 (1942).
26. K. L. Khachishvile and co-workers, *Zh. Neorg. Khim.* **6**, 1493 (1961).
27. P. Pichat, *C. R. Acad. Sci. Paris Ser. C* **265**, 385 (1967).
28. R. N. Scott and D. F. Shriver, *Inorg. Chem.* 5, 158 (1966).
29. P. Baumgarten and W. Bruns, *Ber. Dtsch. Chem. Ges. B.* **B72**, 1753 (1939); *Ibid.* **B74**, 1232 (1941).
30. H. D. Fishcher, W. J. Lehmann, and I. Shapiro, *J. Phys. Chem.* **65**, 1166 (1961).
31. H. I. Schlesinger and co-workers, *J. Am. Chem. Soc.* **75**, 195 (1953).
32. *Ibid.*, p. 199.
33. E. L. Gamble, *Inorg. Synth.* **3**, 27 (1950).
34. Brit. Pat. 226,490 (Dec. 20, 1923), A. F. Meyerhofer.
35. E. Krause and R. Nitsche, *Chem. Ber.* **54B**, 2784 (1921).
36. H. C. Brown and U. S. Racherla, *Organometallics 1986* **5**(2), 391–393 (1986).
37. B. M. Mikhailov and T. A. Schhegoleva, *J. Gen. Chem. U.S.S.R.* **29**, 3404 (1959).
38. G. T. Morgan and R. B. Tunstall, *J. Chem. Soc.* **125**, 1963 (1924).
39. W. I. Aalbersberg and co-workers, *J. Chem. Soc.*, 3055 (1959).
40. W. V. Bush and L. Zechmeister, *J. Am. Chem. Soc.* **80**, 2991 (1958).
41. H. S. Booth and D. R. Martin, *Boron Trifluoride and Its Derivatives*, John Wiley & Sons, Inc., New York, 1949.
42. P. Baumgarten and H. Henning, *Chem. Ber.* **72B**, 1743 (1939).
43. C. Gascard and G. Mascherpa, *J. Chim. Phys. Phys. Chim. Biol.* **70**, 1040 (1973).
44. R. J. Gillespie and J. L. Hartman, *Can. J. Chem.* **45**, 859 (1967).

45. H. S. Booth and D. R. Martin, *Boron Trifluoride and its Derivatives*, John Wiley & Sons, Inc., New York, 1948.
46. D. Farcasiu and A. Ghenciu, *J. Catal.* **134**(1), 126–133 (1992).
47. J. S. McGrath and co-workers, J.A.C.S., **66**, 126 (1944)
48. C. A. Wamser, *J. Am. Chem. Soc.* **73**, 409 (1951).
49. C. A. Wamser, *J. Am. Chem. Soc.* **70**, 1209 (1948)
50. V. I. Durkov and S. S. Batsanov, *Zh. Strukt. Khim.* **2**, 456 (1961).
51. E. F. Mooney and M. A. Qaseem, *Chem. Commun.*, 230 (1967).
52. M. J. Bula, J. S. Hartman, and C. V. Raman, *J. Chem. Soc. Dalton Trans.*, 725 (1974).
53. Brit. Pat 486,887 (June 13, 1938), (to E. I. du Pont de Nemours & Co., Inc.).
54. D. R. Martin and J. M. Canon, in G. A. Olah, ed., *Friedel-Crafts and Related Reactions*, Vol. 1, Wiley-Interscience, New York, 1963, pp. 399–567.
55. H. S. Booth and K. S. Wilson, *Inorg. Synth.* **1**, 21 (1939).
56. J. W. Koger, Oak Ridge National Laboratory, TM-4172, 1972; *Nucl. Sci. Abstr.* **28**, 11,211 (1973).
57. C. F. Swinehart, A. R. Bumblish, and H. F. Flisik, *Anal. Chem.* **19**, 28 (1947); *Ann. Proc.* 35-0049, internal document, Harshaw Chemical Co., Mar. 23, 1964.
58. K. H. Jacobson, R. A. Rhoden, and R. L. Roudabush, *HEW Pub. (NIOSH) Publ.* 77, (1976).
59. Code of Fed. Reg. 29, part. 1901, U.S. Govt. Printing Office, Washington, D.C., 1988.
60. A.C.G.I.H., *Threshold Limit Values for Chemical Substances*, 1989–1990.
61. G. M. Rusch and co-workers, *Toxicology and Applied Pharmacology* **83**, 69–78 (1986).
62. C. J. Spiegl, *Natl. Nucl. Energy Ser. Div. VI 1 (Book 4)*, 2291 (1953).
63. G. A. Olah, ed., in Ref. 54, pp. 228–235.
64. Ref. 41, Chapt. 6.
65. A V. Topchiev, S. V. Zavgorodnii, and Y. M. Paushkin, *Boron Fluoride and Its Compounds as Catalysts in Organic Chemistry*, Pergamon Press, New York, 1959.
66. Fr. Pat. 2,647,108 (Nov. 23, 1990), L. Gilbert and co-workers (to Rhône-Poulenc).
67. U.S. Pat. 4,419,528 (Dec. 6, 1983), G. A. Olah (to PCUK Ugine Kuhlman).
68. G. H. Posner and co-workers, *Tetrahedran Lett.* **32**(45) 6489–6492 (1991).
69. U.S. Pat. 4,962,244 (Oct. 9, 1990), M. Y. Elsheikh (to Atochem, N. Amer. Inc.).
70. U.S. Pat. 4,657,773 (Apr. 14, 1987), S. C. Durkee (to Hercules Inc.)
71. U.S. Pat. 4,657,706 (Apr. 14, 1987), S. C. Durkee (to Hercules Inc.)
72. U.S. Pat. 5,051,485 (Sept. 24, 1991), J. J. Schmid and J. W. Booth (to Arizona Chem.)
73. U.S. Pat. 4,434,309 (Feb. 28, 1984), J. M. Larkin and W. H. Brader (to Texaco Inc.)
74. U.S. Pat. 4,484,014 (Nov. 20, 1984), W. I. Nelson and co-workers (to Phillips Pet. Co.)
75. U.S. Pat. 4,935,570 (June. 19, 1990), M. B. Nelson and co-workers (to Ethyl Corp.)
76. U.S. Pat. 5,068,490 (Feb. 29, 1988), B. E. Eaton (to Amoco Corp.)
77. U.S. Pat. 5,071,812 (Mar. 31, 1989), D. R. Kelsey (to Shell Oil Co.)
78. M. C. Throckmorton, *J. Appl. Polym. Sci.* **42**(11), 3019–3024 (1991).
79. M. H. Juang and H. C. Cheng, *J. Appl. Phys.* **71**(3), 1265–1270 (1992).
80. B. K. Mohapatra and co-workers, *Indian. J. Chem., Sect. A* **30A**(11), 944–947 (1991).
81. W. Ahmed and co-workers, *J. Phys. IV*, **1**(C2) 119–126 (1991).
82. Jpn. Pat. 03,218,917[91,218,917] (Sept. 26, 1991), (to Hashimoto Chem. Ind. Co. Ltd.).
83. U.S. Pat. 5,034,368 (July 23, 1991), E. Drent (to Shell Int. Res. MIJ BV).
84. A. A. Palko and J. S. Drury, *J. Chem. Phys.* **47**, 2561 (1967).

FRANCIS EVANS
GANPAT MANI
AlliedSignal, Inc.

FLUOROBORIC ACID AND FLUOROBORATES

Fluoroboric Acid and the Fluoroborate Ion

Fluoroboric acid [16872-11-0], generally formulated as HBF_4, does not exist as a free, pure substance. The acid is stable only as a solvated ion pair, such as $H_3O^+BF_4^-$; the commercially available 48% HBF_4 solution approximates $H_3O^+ BF_4^- \cdot 4H_2O$. Other names used infrequently are hydrofluoroboric acid, hydroborofluoric acid, and tetrafluoroboric acid. Salts of the acid are named as fluoroborates or occasionally borofluorides. Fluoroboric acid and its salts were investigated as early as 1809 (1,2). The acid and many transition-metal salts are used in the electroplating (qv) and metal finishing industries. Some of the alkali metal fluoroborates are used in fluxes.

Properties. Fluoroboric acid is stable in concentrated solutions, and hydrolyzes slowly in aqueous solution to hydroxyfluoroborates. For the stability of the fluoroborate species, see Reference 3. The equilibrium quotients Q (4,5) in 1 molal NaCl at 25°C show the strong affinity of boron for fluoride:

$$B(OH)_3 + F^- \rightleftharpoons BF(OH)_3^- \qquad \log Q = -0.36 \pm 0.19$$
$$B(OH)_3 + 2F^- + H^+ \rightleftharpoons BF_2(OH)_2^- + H_2O \qquad \log Q = 7.06 \pm 0.02$$
$$B(OH)_3 + 3F^- + 2H^+ \rightleftharpoons BF_3OH^- + 2H_2O \qquad \log Q = 13.689 \pm 0.003$$
$$B(OH)_3 + 4F^- + 3H^+ \rightleftharpoons BF_4^- + 3H_2O \qquad \log Q = 19.0 \pm 0.1$$

The hydrolysis of BF_4^- occurs stepwise to BF_3OH^-, $BF_2(OH)_2^-$, and $BF(OH)_3^-$. By conductivity measurements the reaction of boric acid and HF was found to form $H[BF_3(OH)]$ [15433-40-6] rapidly; subsequently HBF_4 formed much more slowly from HBF_3OH. These studies demonstrate that BF_4^- is quite stable to hydrolysis yet is slow to form from BF_3OH^- and HF:

$$BF_4^- + H_2O \rightleftharpoons BF_3OH^- + HF$$

Kinetic results (5) and ^{19}F nmr experiments (6) illustrate clearly that the hydroxyfluoroborates are in rapid equilibrium and easily exchange fluoride.

Table 1 lists some of the physical properties of fluoroboric acid. It is a strong acid in water, equal to most mineral acids in strength and has a pK_{H_2O} of -4.9 as compared to -4.3 for nitric acid (9). The fluoroborate ion contains a nearly tetrahedral boron atom with almost equidistant B–F bonds in the solid state. Although lattice effects and hydrogen bonding distort the ion, the average B–F distance is 0.138 nm; the F–B–F angles are nearly the theoretical 109° (10,11). Raman spectra on molten, ie, liquid $NaBF_4$ agree with the symmetrical tetrahedral structure (12).

The fluoroborate ion has traditionally been referred to as a noncoordinating anion. It has shown little tendency to form a coordinate–covalent bond with transition metals as do nitrates and sulfates. A few exceptional cases have been reported (13) in which a coordinated BF_4^- was detected by infrared or visible spectroscopy.

Table 1. Physical Properties of Fluoroboric Acid

Property	Value	Reference
heat of formation, kJ/mol^a		
aqueous, 1 molal, at 25°C	−1527	
from boric oxide and HF (aq)	−123.34	2
BF_4^-, gas	−1765 ± 42	7
entropy of the BF_4^- ion, J/(mol·K)^a	167	
specific gravity		
48% soln	1.37	
42% soln	1.32	
30% soln	1.20	
surface tension, 48% soln at 25°C, mN/m(=dyn/cm)	65.3	
ir absorptions,^{b} cm^{-1}	ca 1100	8
	ca 530	

^{a}To convert J to cal, divide by 4.184.
^{b}Generally observed as strong absorptions.

Hydroxyfluoroborates are products of the reaction of BF_3 with water; $BF_3·2H_2O$ [13319-75-0] is actually $H_3O^+BF_3OH^-$. Salts such as sodium hydroxyfluoroborate [13876-97-6], $NaBF_3OH$, are made by neutralizing the acid. The BF_3OH^- anions are distorted tetrahedra (14). In the HBO_2–HF system, $HBO_2·2HF$ was found to be $HBF_2(OH)_2$, dihydroxyfluoroboric acid [17068-89-2] (15).

Manufacture, Shipping, and Waste Treatment. Fluoroboric acid (48%) is made commercially by direct reaction of 70% hydrofluoric acid and boric acid, H_3BO_3 (see BORON COMPOUNDS). The reaction is exothermic and must be controlled by cooling.

The commercial product is usually a 48–50% solution which contains up to a few percent excess boric acid to eliminate any HF fumes and to avoid HF burns. Reagent-grade solutions are usually 40%. A 61% solution can be made from metaboric acid, HBO_2, and 70% HF, and a lower grade by direct combination of fluorospar, CaF_2, sulfuric acid, and boric acid (16). The product contains a small amount of dissolved calcium sulfate. A silica-containing (0.11% SiO_2) fluoroboric acid is produced from inexpensive fluorosilicic acid (17). Boric acid is added to a 10% H_2SiF_6 solution and then concentrated in several steps to 45% HBF_4. Granular silicon dioxide must be filtered from the product.

Vessels and equipment must withstand the corrosive action of hydrofluoric acid. For a high quality product the preferred materials for handling HBF_4 solutions are polyethylene, polypropylene, or a resistant rubber such as neoprene (see ELASTOMERS, SYNTHETIC). Where metal must be used, ferrous alloys having high nickel and chromium content show good resistance to corrosion. Impregnated carbon (Carbate) or Teflon can be used in heat exchangers. Teflon-lined pumps and auxilliary equipment are also good choices. Working in glass equipment is not recommended for fluoroboric acid or any fluoroborate.

Fluoroboric acid and some fluoroborate solutions are shipped as corrosive material, generally in polyethylene-lined steel pails and drums or in rigid nonreturnable polyethylene containers. Acid spills should be neutralized with lime or soda ash.

Waste treatment of fluoroborate solutions includes a pretreatment with aluminum sulfate to facilitate hydrolysis, and final precipitation of fluoride with lime (18). The aluminum sulfate treatment can be avoided by hydrolyzing the fluoroborates at pH 2 in the presence of calcium chloride; at this pH, hydrolysis is most rapid at elevated temperature (19).

Economic Aspects. In the United States fluoroboric acid is manufactured by Atotech USA, Inc., General Chemical, C.P. Chemical Co., Fidelity Chemical Products, and Chemtech Harstan. Research quantities of reagent grade are made by Advance Research Chemical Co., Johnson-Mathey, and Ozark-Mahoning Co. The price for 48% fluoroboric acid in truckload quantities in 1993 was $2.13–2.25/kg (20).

Many specialty fluoroborates are available in research quantities from Advance Research Chemicals.

Analysis. Fluoroboric acid solutions and fluoroborates are analyzed gravimetrically using nitron or tetraphenylarsonsium chloride. A fluoroborate ion-selective electrode has been developed (21).

Toxicity. Fluoroborates are excreted mostly in the urine (22). Sodium fluoroborate is absorbed almost completely into the human bloodstream and over a 14-d experiment all of the $NaBF_4$ ingested was found in the urine. Although the fluoride ion is covalently bound to boron, the rate of absorption of the physiologically inert BF_4^- from the gastrointestinal tract of rats exceeds that of the physiologically active simple fluorides (23).

Uses. Printed circuit tin–lead plating is the main use of fluoroboric acid (24). However, the Alcoa Alzak process for electropolishing aluminum requires substantial quantities of fluoroboric acid. A 2.5% HBF_4 solution is used to produce a highly reflective surface (25). The high solubility of many metal oxides in HBF_4 is a decided advantage in metal finishing operations (see METAL SURFACE TREATMENTS). Before plating or other surface treatment, many metals are cleaned and pickled in fluoroboric acid solution; eg, continuous strip pickling of hot-rolled low carbon steel is feasible in HBF_4 solutions (26). Nontempered rolled steel requires 80°C for 60 s in HBF_4 130 g/L, whereas tempered rolled steel requires only 65°C for 60 s in 65 g/L. The spent pickling solution is recovered by electrodialysis.

Fluoroboric acid is used as a stripping solution for the removal of solder and plated metals from less active substrates. A number of fluoroborate plating baths (27) require pH adjustment with fluoroboric acid (see ELECTROPLATING).

A low grade fluoroboric acid (16) is used in the manufacture of cryolite (28) for the electrolytic production of aluminum:

$$4\ Na_2SO_4 \cdot NaF\ +\ 5\ HBF_4\ +\ 2\ Al_2O_3\ +\ 9\ H_2O \rightarrow 4\ Na_3AlF_6\ +\ 5\ H_3BO_3\ +\ 4\ H_2SO_4$$

The boric and sulfuric acids are recycled to a HBF_4 solution by reaction with CaF_2. As a strong acid, fluoroboric acid is frequently used as an acid catalyst, eg, in synthesizing mixed polyol esters (29). This process provides an inexpensive route to confectioner's hard-butter compositions which are substitutes for cocoa butter in chocolate candies (see CHOCOLATE AND COCOA). Epichlorohydrin is polymerized in the presence of HBF_4 for eventual conversion to polyglycidyl ethers (30) (see CHLOROHYDRINS). A more concentrated solution, 61–71% HBF_4, catalyzes the addition of CO and water to olefins under pressure to form neo acids (31) (see CARBOXYLIC ACIDS).

Main Group

Properties. A summary of the chemical and physical properties of alkali-metal and ammonium fluoroborates is given in Tables 2 and 3. Chemically these compounds differ from the transition-metal fluoroborates usually separating in anhydrous form. This group is very soluble in water, except for the K, Rb, and Cs salts which are only slightly soluble. Many of the soluble salts crystallize as hydrates.

Lithium fluoroborate crystallizes from aqueous solutions as $LiBF_4·3H_2O$ [39963-05-8] and $LiBF_4·H_2O$ [39963-03-6]. The heat of dehydration of the monohydrate at 91°C is 70.9 kJ/mol (16.95 kcal/mol); the melting point is 117°C (45). Magnesium, calcium, strontium, and barium fluoroborates crystallize as hydrates: $Mg(BF_4)_2·6H_2O$ [19585-07-0], $Ca(BF_4)_2·2H_2O$ [27860-81-7], $Sr(BF_4)_2·4H_2O$ [27902-05-2], and $Ba(BF_4)_2·2H_2O$ [72259-09-7], respectively. These hydrated fluoroborates can be dehydrated completely to the anhydrous salts, which show decreasing stabilities: Ba > Sr > Ca > Mg.

The anhydrous magnesium salt is least stable thermally. It forms MgF_2, which has the highest lattice energy. This has been confirmed by differential thermal analysis (dta) of the crystalline hydrates (46). Aluminum fluoroborate [14403-54-4], $Al(BF_4)_3·(H_2O)_n$, is soluble in strongly acid solutions and displays a tendency for fluoride exchange with BF_4^- to form aluminum fluorides. The aluminyl compound, $AlO^+BF_4^-$, is extremely hygroscopic and is prepared by the reaction of AlOCl, BF_3, and HF (47). Differential thermal analysis experiments show thermal decomposition beginning at 85°C, corresponding to removal of BF_3 and formation of AlOF.

Differential thermal analysis studies of ammonium fluoroborate showed the orthorhombic to cubic transition at 189 ± 5°C and BF_3 generation from 389 to 420°C (48). Sodium hydroxide reacts with NH_4BF_4 liberating ammonia and forming $NaBF_4$. When sodium fluoroborate was studied by infrared spectroscopy, sodium hydroxyfluoroborate, $NaBF_3OH$, was found to be present (49). Although pure sodium hydroxyfluoroborate is thermally unstable, decomposing to $Na_2B_2F_6O$ [18953-03-2] and H_2O, in a melt of $NaBF_4^-NaF$ no instability of the small amount of $NaBF_3OH$ present was detected. Fusion of $NaBF_4$ or KBF_4 with boric oxide generates BF_3 and complex borates such as KFB_4O_6 (50). Most fluoroborates decompose readily to give BF_3 when treated with sulfuric acid or when calcined (see Table 3 for dissociation pressure). Under strongly basic conditions the chemical equilibrium is shifted away from BF_4^- to borates and fluorides.

Manufacture. Fluoroborate salts are prepared commercially by several different combinations of boric acid and 70% hydrofluoric acid with oxides, hydroxides, carbonates, bicarbonates, fluorides, and bifluorides. Fluoroborate salts are substantially less corrosive than fluoroboric acid but the possible presence of HF or free fluorides cannot be overlooked. Glass vessels and equipment should not be used.

Table 2. General Properties of Metal Fluoroborates

Compound	CAS Registry Number	Molecular weight	Color	Physical form	Mp, °C	Density,[a] g/cm³	Solubility H₂O g/100 mL[b]	Solubility Other	References
$LiBF_4$	[14283-07-9]	93.74	white		406 dec		very soluble		2,32
$NaBF_4$	[13755-29-8]	109.79	white	orthorhombic <240°C $a = 0.68358, b = 0.62619, c = 0.67916$ nm noncubic >240°C		2.47 210[c]	108 (26°C)	sl alcohol	13,32,33
KBF_4	[14075-53-7]	125.92	colorless	rhombic <283°C $a = 0.7032, b = 0.8674, c = 0.5496$ nm cubic >283°C	530 dec	2.498	0.45 (20°C) 6.27 (100°C)	sl ethanol insol alkali	32–34
$RbBF_4$	[18909-68-7]	172.27		orthorhombic <245°C $a = 0.7296, b = 0.9108, c = 0.5636$ nm cubic >245°C	612 dec	2.820 10[c]	0.6 (17°C)		32–34
$CsBF_4$	[18909-69-8]	219.71	white	orthorhombic <140°C $a = 0.7647, b = 0.9675, c = 0.5885$ nm cubic >140°C	555 dec	3.20 30[c]	1.6 (17°C)		32–34
NH_4BF_4	[13826-83-0]	104.84	white	orthorhombic <205°C $a = 0.7278, b = 0.9072, c = 0.5678$ nm	487 dec	1.871[d]	3.09 (−1.0°C) 5.26 (−1.5°C) 10.85 (−2.7°C) 12.20 (0°C) 25 (16°C) 25.83 (25°C) 44.09 (50°C) 67.50 (75°C) 98.93 (100°C) 113.7 (108.5°C)	HF[e]	32,34,35 36
$NaBF_3OH$	[13876-97-6]			hexagonal $a = 0.8084, c = 0.7958$ nm		2.46			10

[a]Unless otherwise stated, at 20°C. [b]Temperature given in parentheses. [c]At 100°C. [d]At 15°C. [e]Value at 0°C is 19.89%.

Table 3. Thermodynamic Data^a for Metal Fluoroborates, kJ/mol^b

Compound	ΔH_{diss}	Lattice energy, $-U$	ΔH_{fus}	ΔH_f	Other	$\log P_{Pa} = -aT^{-1} + b$			References
						a	b^c	T, °C	
LiBF$_4$	15.9	699		−1838.4	$\Delta H^d = -89.54$	833	6.40	210–320	37–39
NaBF$_4$	69.83	657.3	13.6	−1843.5	$\Delta H^e = -134.1$	3650	8.75	400–700	33, 38–40
KBF$_4$	121	598	18.0	−1881.5	$\Delta H^f = -180.5$	6317	8.15	510–830	33, 38, 39, 41, 42
					$\Delta H_{sub} = 330$				
					$S = 130^g$				
					$C_p = 112.1^g$				
RbBF$_4$	112.8	577	19.6			5960	9.57	600–1000	33, 38, 41
CsBF$_4$	112.5	556	19.2			5880	9.47	610–1040	33, 38, 41
NaBF$_3$OH	77.0			−1754		4024	9.11	400–700	40
NH$_4$BF$_4$		607^h			$\Delta H_{sub} = 47.3$	2469	8.94		43, 44

$^a\Delta H_{diss}$ = heat of dissociation, ΔH_{fus} = heat of fusion, ΔH_f = heat of formation, ΔH_{sub} = sublimation. All thermodynamic data at 25°C, unless otherwise stated.
^{b}To convert J to cal, divide by 4.184.
^{c}To convert $\log P_{Pa}$ to $\log P_{mm\,Hg}$, subtract 2.12 from b.
^{d}LiF(s) + BF$_3$(g) → LiBF$_4$(s).
^{e}NaF(s) + BF$_3$(g) → NaBF$_4$(s).
^{f}KF(s) + BF$_3$(g) → KBF$_4$(s).
^{g}Units are in J/(mol·K).
^{h}At 260°C.

Sodium Fluoroborate. Sodium fluoroborate is prepared by the reaction of NaOH or Na_2CO_3 with fluoroboric acid (51), or by treatment of disodium hexafluorosilicate with boric acid.

Potassium Fluoroborate. Potassium fluoroborate is produced as a gelatinous precipitate by mixing fluoroboric acid and KOH or K_2CO_3. Alternatively, fluorosilicic acid is treated with H_3BO_3 in a 2:1 molar ratio to give HBF_3OH, which reacts with HF and KCl to yield 98% of KBF_4 in 98.5% purity (52). Commercial KBF_4 normally contains less than 1% KBF_3OH.

Ammonium and Lithium Fluoroborates. Ammonia reacts with fluoroboric acid to produce ammonium fluoroborate (53). An alternative method is the fusion of ammonium bifluoride and boric acid (54):

$$2\ NH_4HF_2 + H_3BO_3 \rightarrow NH_4BF_4 + 3\ H_2O + NH_3$$

The water and ammonia must be removed from the melt. Lithium hydroxide or carbonate react with HBF_4 to form $LiBF_4$.

Magnesium Fluoroborate. Treatment of magnesium metal, magnesium oxide, or magnesium carbonate with HBF_4 gives magnesium fluoroborate [14708-13-5]. The MgF_2 is filtered and the product is sold as a 30% solution.

Economic Aspects. In the United States the sodium, potassium, ammonium, and magnesium fluoroborates are sold by Advance Research Chemicals, Atotech USA, Inc., and General Chemical. The lithium compound is available from Advance Research Chemicals, Cyprus Foote Mineral, and FMC Lithium Corp. of America. Small amounts of other fluoroborates are sold by Alfa Inorganics, Inc. and Ozark-Mahoning Co. Prices in 1993 for truckload quantities were $NaBF_4$, $4.95–6.25/kg; KBF_4, $3.55/kg; and NH_4BF_4 $5.03–6.35/kg.

Uses. Alkali metal and ammonium fluoroborates are used mainly for the high temperature fluxing action required by the metals processing industries (see METAL SURFACE TREATMENTS; WELDING). The tendency toward BF_3 dissociation at elevated temperatures inhibits oxidation in magnesium casting and aluminum alloy heat treatment.

The molten salts quickly dissolve the metal oxides at high temperatures to form a clean metal surface. Other uses are as catalysts and in fire-retardant formulations (see FLAME RETARDANTS).

Potassium Fluoroborate. The addition of potassium fluoroborate to grinding wheel and disk formulations permits lower operating temperatures (55). Cooler action is desirable to reduce the burning of refractory materials such as titanium and stainless steels. Excellent results in grinding wheels are also obtained with $NaBF_4$ (56). A process for boriding steel surfaces using B_4C and KBF_4 as an activator improves the hardness of the base steel (57). Fluxes for aluminum bronze and silver soldering and brazing contain KBF_4 (58) (see SOLDERS). Fire retardance is imparted to acrylonitrile polymers by precipitating KBF_4 within the filaments during coagulation (59). In polyurethanes, KBF_4 and NH_4BF_4 reduce smoke and increase flame resistance (60). Both the potassium and ammonium salts improve insulating efficiency of intumescent coatings (61). The endothermic characteristics of these fillers (qv) (release of BF_3) counteract the exothermic nature of the intumescent agents (nitroaromatic amines) in the coating. The sodium and potassium salts are claimed to have a synergistic effect with polyhalogenated aromatics that improve flame-retardant properties of polyesters (62). Elemental boron is prepared by the Cooper electrolysis of a KBF_4 melt with B_2O_3 and KCl (63). The boron may be up to 99.5% purity and, if KBF_4 containing the ^{10}B isotope is used, the product is ^{10}B which is used in the nuclear energy field as a neutron absorber (see NUCLEAR REACTORS).

Sodium Fluoroborate. Sodium fluoroborate can be used in the transfer of boron to aluminum alloys but the efficiency is lower than for KBF_4 (64). Sodium fluoroborate in an etching solution with sulfamic acid and H_2O_2 aids in removing exposed lead in printed circuit manufacture (65). During the annealing of galvanized iron (galvannealing), the surface becomes oxidized. The resulting oxide coating, which causes difficulty in soldering, can be removed by aqueous $NaBF_4$ or NH_4BF_4 (66). Work at Oak Ridge National Lab (Tennessee) has shown that a $NaBF_4$, with 8 mol % NaF, salt mixture could be used as the coolant in the molten breeder reactor (67); in this molten salt at nearly 600°C the corrosion rate of Hastelloy N is about 8 μm/yr. Sodium fluoroborate acts as a catalyst for cross-linking cotton cellulose with formaldehyde (68); transesterification in the preparation of polycarbonates (69); and preparation of cyclic oligoethers from ethylene oxide (70). Sodium and lithium fluoroborates are effective flame retardants for cotton and rayon (71).

Ammonium Fluoroborate. Ammonium fluoroborate blends with antimony oxide give good results in flame-retarding polypropylene (72). The complete thermal vaporization makes ammonium fluoroborate an excellent gaseous flux for inert-atmosphere soldering (73). A soldering flux of zinc chloride and ammonium fluoroborate is used in joining dissimilar metals such as Al and Cu (74). Ammonium fluoroborate acts as a solid lubricant in cutting-oil emulsions for aluminum rolling and forming.

Lithium Fluoroborate. Lithium fluoroborate is used in a number of batteries (qv) as an electrolyte, for example in the lithium–sulfur battery (75).

Miscellaneous. Flame-resistant cross-linked polyethylene can be made with a number of fluoroborates and antimony oxide. This self-extinguishing material may contain the fluoroborates of NH_4^+, Na^+, K^+, Ca^{2+}, Mg^{2+}, Sr^{2+}, or Ba^{2+} in amounts of 4–20% (76). Magnesium fluoroborate catalyzes the epoxy treatment of cotton fabrics for permanent-press finishes (77) (see TEXTILES).

Transition-Metal and Other Heavy-Metal Fluoroborates

The physical and chemical properties are less well known for transition metals than for the alkali metal fluoroborates (Table 4). Most transition-metal fluoroborates are strongly hydrated coordination compounds and are difficult to dry without decomposition. Decomposition frequently occurs during the concentration of solutions for crystallization. The stability of the metal fluorides accentuates this problem. Loss of HF because of hydrolysis makes the reaction proceed even more rapidly. Even with low temperature vacuum drying to partially solve the decomposition, the dry salt readily absorbs water. The crystalline solids are generally soluble in water, alcohols, and ketones but only poorly soluble in hydrocarbons and halocarbons.

Table 4. Properties of Metal Fluoroborates[a]

Compound	CAS Registry Number	Color	Specific gravity	Solubility	Miscellaneous
$Mn(BF_4)_2 \cdot 6H_2O$	[26044-57-5]	pale pink	1.982	water, ethanol	
$Fe(BF_4)_2 \cdot 6H_2O$	[13877-16-2]	pale green	2.038	water, ethanol	
$Co(BF_4)_2 \cdot 6H_2O$	[15684-35-2]	red	2.081	water, alcohol	
$Ni(BF_4)_2 \cdot 6H_2O$	[14708-14-6]	green	2.136	water, alcohol	
$Cu(BF_4)_2 \cdot 6H_2O$	[72259-10-0]	blue	2.175	water, alcohol	
$AgBF_4 \cdot H_2O$	[72259-11-1]	colorless		water, less sol in alcohol, sol benzene, sol ether	dec 200°C, light sensitive
$Zn(BF_4)_2 \cdot 6H_2O$	[27860-83-9]	white	2.120	water, alcohol	dehydrates at 60°C
$Cd(BF_4)_2 \cdot 6H_2O$	[27860-84-0]	white	2.292	water, alcohol	
$In(BF_4)_3 \cdot xH_2O$	[27765-48-6]	colorless		water	
$TlBF_4 \cdot H_2O$	[72259-12-2]	colorless		water	orthorhombic, $a = 0.947$, $b = 0.581$, $c = 0.740$ nm, light sensitive
$Sn(BF_4)_2 \cdot xH_2O$	[72259-13-3]	white		water	$Sn(BF_4)_2 \cdot SnF_2 \cdot 5H_2O$ crystallizes from soln
$Pb(BF_4)_2 \cdot H_2O$	[26916-34-7]	colorless			

[a] Crystalline solids (2,78).

Differential thermal analysis in air on the crystalline hexahydrates of Zn, Cd, Fe, Co, and Ni fluoroborates show the loss of BF_3 and H_2O simultaneously at 195, 215, 180, 185, and 205°C, respectively (46,79). The dta curves also indicate initial melting at 107, 117, and 150°C for Zn, Cd, and Fe fluoroborates, respectively. The anhydrous metal fluoride and/or oxide is usually isolated. The copper salt also decomposes with liberation of BF_3 and H_2O (80).

The water of hydration of these complexes can be replaced with other coordinating solvents. For example, the ethanol and methanol solvates were made by dissolving the hydrates in triethyl and trimethyl orthoformate, respectively (81,82). The acetic acid solvates are made by treating the hydrates with acetic anhydride (83). Conductivity and visible spectra, where applicable, of the Co, Ni, Zn, and Cu fluoroborates in N,N-dimethylacetamide (L) showed that all metal ions were present as the ML_6^{2+} cations (84). Solvated fluoroborate complexes of Cr^{3+}, Fe^{2+}, Co^{2+}, Ni^{2+}, Cu^{2+}, Cu^+, and Zn^{2+} in diethyl ether, nitromethane, and benzene solutions have been prepared. Solutions of $Ti(BF_4)_3$, $V(BF_4)_3$, and $Fe(BF_4)_3$ could not be prepared probably because of formation of BF_3 and the metal fluoride (85). Ammonia easily replaces the coordinated water; the products are usually tetrammine or hexammine complexes (2) (see COORDINATION COMPOUNDS). The hexahydrate of $Ni(BF_4)_2$ was found to be stable from 25 to 100°C; solubility also was determined to 95°C (86). At 120°C the solid decomposed slowly to NiF_2 with loss of HF, H_3BO_3, and H_2O.

Manufacture. The transition- and heavy-metal fluoroborates can be made from the metal, metal oxide, hydroxide, or carbonate with fluoroboric acid. Because of the difficulty in isolating pure crystalline solids, these fluoroborates are usually available as 40–50% solutions, $M(BF_4)_x$. Most of the solutions contain about 1–2% excess fluoroboric acid to prevent precipitation of basic metal complexes. The solutions are usually sold in 19 and 57 L polyethylene containers.

In some cases, particularly with inactive metals, electrolytic cells are the primary method of manufacture of the fluoroborate solution. The manufacture of Sn, Pb, Cu, and Ni fluoroborates by electrolytic dissolution (87,88) is patented. A typical cell for continous production consists of a polyethylene-lined tank with tin anodes at the bottom and a mercury pool (in a porous basket) cathode near the top (88). Fluoroboric acid is added to the cell and electrolysis is begun. As tin fluoroborate is generated, differences in specific gravity cause the product to layer at the bottom of the cell. When the desired concentration is reached in this layer, the heavy solution is drawn from the bottom and fresh HBF_4 is added to the top of the cell continuously. The direct reaction of tin with HBF_4 is slow but can be accelerated by passing air or oxygen through the solution (89). The stannic fluoroborate is reduced by reaction with mossy tin under an inert atmosphere. In earlier procedures, HBF_4 reacted with hydrated stannous oxide.

Anhydrous silver fluoroborate [1404-20-2] is made by the addition of BF_3 gas to a suspension of AgF in ethylbenzene (90). An $AgBF_4 \cdot C_8H_{10}$ complex is precipitated with pentane and the complex is washed with pentane to give anhydrous $AgBF_4$.

Economic Aspects. Most fluoroborate solutions listed in Table 5 are manufactured by Atotech USA, Inc., General Chemical, Chemtec/Harstan, C.P. Chemical Co., and Fidelity Chemical Products. Prices are shown in Table 5.

Uses. Metal fluoroborate solutions are used primarily as plating solutions and as catalysts. The Sn, Cu, Zn, Ni, Pb, and Ag fluoroborates cure a wide range of epoxy resins at elevated or ambient room temperature (91,92). In the textile industry zinc fluoroborate is used extensively as the curing agent in applying resins for crease-resistant finishes (93). Emulsions of epoxy resins (94), polyoxymethylene compounds (95), or aziridinyl compounds (96) with $Zn(BF_4)_2$ and other additives are applied to the cloth. After the excess is removed, the cloth is dried and later cured at a higher temperature. Similarly treated acrylic textiles using epoxy resins take on an antistatic finish (97), or the acrylic textiles can be coated

Table 5. Commercial Metal Fluoroborate Solutions

Metal cation	CAS Registry Number	Formula	% Metal	Specific gravity, g/cm^3	1992 price, $/kg
antimony(II)	[14486-20-5]	Sb(BF$_4$)$_3$	12.8	1.42	
cadmium	[14886-19-2]	Cd(BF$_4$)$_2$	19.7	1.60	8.15
cobalt(II)	[26490-63-1]	Co(BF$_4$)$_2$	11.8	1.42	
copper(II)	[38465-60-0]	Cu(BF$_4$)$_2$	12.2	1.48	3.40–4.0
indium	[27765-48-6]	In(BF$_4$)$_3$	15.3	1.55	
iron(II)	[13877-16-2]	Fe(BF$_4$)$_2$	10.3	1.47	
lead(II)	[13814-96-5]	Pb(BF$_4$)$_2$	28.9	1.75	1.85–2.16
nickel(II)	[14708-14-6]	Ni(BF$_4$)$_2$	11.2	1.47	10.25–11.8
tin(II)	[13814-96-5]	Sn(BF$_4$)$_2$	20.2	1.61	6.38–6.6
zinc	[13826-88-5]	Zn(BF$_4$)$_2$	11.0	1.39	1.75–2.1

with 20% $Zn(BF_4)_2$ which results in up to 5.5% added solids for a fire-resistant finish (98).

The use of silver fluoroborate as a catalyst or reagent often depends on the precipitation of a silver halide. Thus the silver ion abstracts a Cl^- from a rhodium chloride complex, $((C_6H_5)_3As)_2(CO)RhCl$, yielding the cationic rhodium fluoroborate [*30935-54-7*] hydrogenation catalyst (99). The complexing tendency of olefins for $AgBF_4$ has led to the development of chemisorption methods for ethylene separation (100,101). Copper(I) fluoroborate [*14708-11-3*] also forms complexes with olefins; hydrocarbon separations are effected by similar means (102).

The manufacture of linear polyester is catalyzed by Cd, Sn (103), Pb, Zn, or Mn (104) fluoroborates. The Beckmann rearrangement of cyclohexanone oxime to caprolactam is catalyzed by $Ba(BF_4)_2$ [*13862-62-9*] or $Zn(BF_4)_2$ [*13826-88-5*] (105). The caprolactam is polymerized to polyamide fibers using $Mn(BF_4)_2$ [*30744-82-2*] catalyst (106). Nickel and cobalt fluoroborates appear to be good catalysts for the polymerization of conjugated dienes to *cis*-1,4-polydienes; the cis configuration is formed in up to 96% yields (107–109).

Electroplating. Metal fluoroborate electroplating (qv) baths (27,110,111) are employed where speed and quality of deposition are important. High current densities can be used for fast deposition and near 100% anode and cathode efficiencies can be expected. Because the salts are very soluble, highly concentrated solutions can be used without any crystallization. The high conductivity of these solutions reduces the power costs. The metal content of the bath is also easily maintained and the pH is adjusted with HBF_4 or aqueous ammonia. The disadvantages of using fluoroborate baths are treeing, lack of throwing power, and high initial cost. Treeing and throwing power can be controlled by additives; grain size of the deposits can also be changed. As of this writing, metals being plated from fluoroborate baths are Cd, Co, Cu, Fe, In, Ni, Pb, Sb, and Zn. Studies on Fe (112,113), Ni (113), and Co (113) fluoroborate baths describe the compositions and conditions of operation as well as the properties of the coatings. Iron foils electrodeposited from fluoroborate baths and properly annealed have exceptionally high tensile strength (113).

The Fe, Co, and Ni deposits are extremely fine grained at high current density and pH. Electroless nickel, cobalt, and nickel–cobalt alloy plating from fluoroborate-containing baths yields a deposit of superior corrosion resistance, low stress, and excellent hardenability (114). Lead is plated alone or in combination with tin, indium, and antimony (115). Sound insulators are made as lead–plastic laminates by electrolytically coating Pb from a fluoroborate bath to 0.5 mm on a copper-coated nylon or polypropylene film (116) (see INSULATION, ACOUSTIC). Steel plates can be simultaneously electrocoated with lead and poly(tetrafluoroethylene) (117). Solder is plated in solutions containing $Pb(BF_4)_2$ and $Sn(BF_4)_2$; thus the lustrous solder-plated object is coated with a Pb–Sn alloy (118).

BIBLIOGRAPHY

"Fluoroboric Acid" under "Fluorine Compounds, Inorganic," in *ECT* 1st ed., Vol. 6, pp. 684–688, by F. D. Loomis, Pennsylvania Salt Manufacturing Co.; "Fluoroboric Acid and Fluoroborates" under "Fluorine Compounds, Inorganic" in *ECT* 2nd ed., Vol. 9, pp. 562–572, by H. S. Halbedel, The Harshaw Chemical Co.; in *ECT* 3rd ed., Vol. 10, pp. 693–706, by H. S. Halbedel and T. E. Nappier, The Harshaw Chemical Co.

1. J. W. Mellor, *Comprehensive Treatise on Inorganic and Theoretical Chemistry*, Vol. 5, Longman, Green and Co., New York, 1929, pp. 123–129.
2. H. S. Booth and D. R. Martin, *Boron Trifluoride and Its Derivatives*, John Wiley & Sons, Inc., New York, 1949, pp. 87–165.
3. R. E. Mesmer, K. M. Palen, and C. F. Baes, *Inorg. Chem.* **12**(1), 89 (1973).
4. I. G. Ryss, *The Chemistry of Fluorine and Its Inorganic Compounds*, State Publishing House for Scientific, Technical, and Chemical Literature, Moscow, USSR, 1956; F. Haimson, English trans., *AEC-tr-3927*, U.S. Atomic Energy Commission, Washington, D.C., 1960, pp. 505–579.
5. C. H. Wamser, *J. Am. Chem. Soc.* **70**, 1209 (1948); **73**, 409 (1951).
6. R. E. Mesmer and A. C. Rutenberg, *Inorg. Chem.* **12**(3), 699 (1973).
7. R. D. Srinastava, M. O. Uy, and M. Faber, *J. Chem. Soc. Farad. Trans. 1* **70**, 1033 (1970).
8. H. Bonadeo and E. Silberman, *J. Mol. Spect.* **32**, 214 (1969).
9. J. Bessiere, *Anal. Chim. Acta* **52**(1), 55 (1970).
10. M. J. R. Clark, *Can. J. Chem.* **47**, 2579 (1969).
11. G. Brunton, *Acta Crystallogr. Sect. B* **24**, 1703 (1968).
12. A. S. Quist and co-workers, *J. Chem. Phys.* **54**, 4896 (1971); **55**, 2836 (1971).
13. M. R. Rosenthal, *J. Chem. Ed.* **50**(5), 331 (1973).
14. M. J. R. Clark and H. Linton, *Can. J. Chem.* **48**, 405 (1970).
15. I. Pawlenko, *Z. Anorg. Allgem. Chem.* **340**(3–4), 201 (1965).
16. H. W. Heiser, *Chem. Eng. Prog.* **45**(3), 169 (1949); U.S. Pats. 2,182,509–11 (Dec. 5, 1939), (to Alcoa).
17. U.S. Pat. 2,799,559 (July 16, 1957), T. J. Sullivan, C. H. Milligan, and J. A. Grady.
18. U.S. Pat. 3,959,132 (May 25, 1976), J. Singh (to Gilson Technical Services, Inc.).
19. U.S. Pat. 4,045,339 (Aug. 30, 1977), T. F. Korenowski, J. L. Penland, and C. J. Ritzert (to Dart Industries Inc.).
20. *Chemical Economics Handbook*, Stanford Research Institute, Menlo Park, Calif., 1975, p. 739.5030H.
21. D. C. Cornish and R. J. Simpson, *Meas. Contr.* **4**(11), 308 (1971).
22. E. J. Largent, "Metabolism of Inorganic Fluoride" in *Fluoridation as a Public Health Measure*, American Association for the Advancement of Science, Washington, D.C., 1954, pp. 49–78.
23. I. Zipkin and R. C. Likens, *Am. J. Physiol.* **191**, 549 (1957).
24. U.S. Pat. 3,888,778 (Mar. 13, 1973), M. Beckwith and G. F. Hau.
25. J. F. Jumer, *Met. Finish.* **56**(8), 44 (1958); **56**(9), 60 (1958).
26. R. M. Hudson, T. J. Butler, and C. J. Warning, *Met. Finish.* **74**(10), 37 (1976); U.S. Pat. 3,933,605 (Jan. 20, 1976), T. J. Butler, R. M. Hudson, and C. J. Warning (to U.S. Steel Corp.).
27. R. D. Mawiya and K. P. Joshi, *Indian Chem. J.* **6**(2), 19 (1971).
28. U.S. Pat. 2,925,325 (Feb. 16, 1960), J. Kamlet (to Reynolds Metals Co.).
29. U.S. Pat. 3,808,245 (Apr. 30, 1974), D. E. O'Connor and G. R. Wyness (to Procter & Gamble Co.).
30. U.S. Pat. 3,305,565 (Feb. 21, 1967), A. C. Mueller (to Shell Oil Co.).
31. U.S. Pat. 3,349,107 (Oct. 24, 1967), S. Pawlenko (to Schering Akliengessellshaft).
32. R. C. Weast, ed., *Handbook of Chemistry and Physics*, Vol. 59, The Chemical Rubber Co., Cleveland, Ohio, 1978.
33. A. S. Dworkin and M. A. Bredig, *J. Chem. Eng. Data* **15**, 505 (1970).
34. M. J. R. Clark and H. Lynton, *Can. J. Chem.* **47**, 2579 (1969).
35. V. S. Yatlov and E. N. Pinaevskays, *Zh. Obshch. Khim.* **15**, 269 (1945).

36. H. Boch, *Z. Naturforsch.* **17b**, 426 (1962).
37. L. J. Klinkenberg, doctoral thesis, Leiden, Germany, 1937.
38. T. C. Waddington, *Adv. Inorg. Chem. Radiochem.* **1**, 158 (1959).
39. P. Gross, C. Hayman, and H. A. Joel, *Trans. Faraday Soc.* **64**, 317 (1968).
40. L. J. Klinkenberg, *Rec. Trav. Chim.* **56**, 36 (1937).
41. J. H. de Boer and J. A. H. Van Liempt, *Rec. Trav. Chim.* **46**, 24 (1927).
42. *JANAF Thermochemical Tables*, Clearinghouse for Federal Scientific and Technical Information, U.S. Dept. of Commerce, Springfield, Va., Dec. 1963.
43. A. W. Laubengayer and G. F. Condike, *J. Am. Chem. Soc.* **70**, 2274 (1948).
44. A. P. Altschuller, *J. Am. Chem. Soc.* **77**, 6515 (1955).
45. V. N. Plakhotnik, V. B. Tul'chinski, and V. K. Steba, *Russ. J. Inorg. Chem.* **22**, 1398 (1977).
46. T. V. Ostrovskaya and S. A. Amirova, *Russ. J. Inorg. Chem.* **15**, 338 (1970).
47. A. V. Pankratov and co-workers, *Russ. J. Inorg. Chem.* **17**, 47 (1972).
48. R. T. Marano and J. L. McAtee, *Thermochimica Acta* **4**, 421 (1972).
49. J. B. Bates and co-workers, *J. Inorg. Nucl. Chem.* **34**, 2721 (1972).
50. L. Maya, *J. Am. Ceram. Soc.* **60**(7–8), 323 (1977).
51. V. Pecak, *Chem. Prum.* **23**(2), 71 (1973).
52. Ger. Pat. 2,320,360 (Nov. 7, 1974), H. K. Hellberg, J. Massonne, and O. Gaertner (to Kali-Chemie Fluor GmbH).
53. U.S. Pat. 2,799,556 (Feb. 1, 1954), T. J. Sullivan and C. G. Milligan (to American Agriculture Chemical Co.).
54. H. S. Booth and S. Rhemar, *Inorganic Synthesis*, Vol. 2, McGraw-Hill Book Co., New York, 1946, p. 23.
55. U.S. Pat. 3,541,739 (Nov. 24, 1970), J. P. Bryon and A. G. Rolfe (to English Abrasives Limited).
56. U.S. Pat. 3,963,458 (June 15, 1976), M. T. Gladstone and S. J. Supkis (to Norton Co.).
57. G. von Matuschka, *Kunstofftechnik* **11**(11), 304 (1972).
58. USSR Pat. 495,178 (Dec. 15, 1975), V. Boiko.
59. U.S. Pat. 3,376,253 (Apr. 2, 1968), E. V. Burnthall and J. J. Hirshfeld (to Monsanto Co.).
60. Ger. Pat. 2,121,821 (Dec. 2, 1971), K. C. Frisch (to Owens Corning Fiberglass Co.).
61. P. M. Sawko and S. R. Riccitiello, *Tech. Brief ARC-11043*, NASA-Ames Research Center, Moffett Field, Calif., July 1977.
62. U.S. Pat. 3,909,489 (Sept. 30, 1975), D. D. Callander (to Goodyear Tire and Rubber Co.).
63. U.S. Pats. 2,572,248-9 (Oct. 23, 1951), H. S. Cooper (to Walter M. Weil).
64. J. D. Donaldson, C. P. Squire, and F. E. Stokes, *J. Mater. Sci.* **13**, 421 (1978).
65. U.S. Pat. 3,305,416 (Feb. 21, 1967), G. J. Kahan and J. L. Mees (to International Business Machines Corp.).
66. U.S. Pat. 3,540,943 (Nov. 17, 1970), E. M. Grogan (to U.S. Steel Corp.).
67. W. R. Huntley and P. A. Gnadt, *Report ORNL-TM-3863*, Oak Ridge National Laboratory, Oak Ridge, Tenn., 1973.
68. L. Kravetz and G. R. Ferrante, *Text. Res. J.* **40**, 362 (1970).
69. Fr. Pat. 1,578,918 (Aug. 22, 1968), J. Borkowski.
70. J. Dale and K. Daasvet'n, *J. Chem. Soc. Chem. Commun.*, (8), 295 (1976).
71. M. A. Kasem and H. R. Richard, *Ind. Eng. Chem. Prod. Res. Dev.* **11**(2), 114 (1972).
72. *Technical Bulletin FR175*, Harshaw Chemical Co., Cleveland, Ohio, 1975.
73. U.S. Pat. 2,561,565 (July 24, 1951), A. P. Edson and I. L. Newell (to United Aircraft Corp.).
74. Br. Pat. 1,181,753 (Feb. 18, 1970), (to Aluminum Co. of America).
75. Ger. Pat. 2,334,660 (Jan. 23, 1975), H. Lauck.

76. U.S. Pat. 3,287,312 (Nov. 22, 1966), T. H. Ling (to Anaconda Wire and Cable Co.).
77. T. Hongu, *S. Gakkaishi* **26**(1), 38 (1970).
78. D. W. A. Sharp, in M. Stacy, J. C. Tatlow, and A. G. Sharpe, eds., *Advances in Fluorine Chemistry*, Vol. 1, Academic Press, Inc., New York, 1960, pp. 68–128.
79. T. V. Ostrovskaya, S. A. Amirova, and N. V. Startieva, *Russ. J. Inorg. Chem.* **12**, 1228 (1967).
80. R. T. Marano and J. L. McAtee, *Therm. Anal. Proc. Int. Conf. 3rd, 1971* **2**, 335 (1972).
81. A. D. Van Ingen Schenau, W. L. Groenveld, and J. Reedijk, *Recl. Trav. Chim. Pays-Bas* **9**, 88 (1972).
82. P. W. N. M. Van Leeuwen, *Recl. Trav. Chim. Pays-Bas* **86**, 247 (1967).
83. U.S. Pat. 3,672,759 (July 4, 1972), T. Yamawaki and co-workers.
84. E. Kamienska and I. Uruska, *Bull. Akad. Pol. Sci. Ser. Sci. Chim.* **21**, 587 (1973).
85. D. W. A. Sharp and co-workers, *Proc. Int. Conf. Coord. Chem. 8th, Vienna*, 322 (1964).
86. V. N. Plakhotnik and V. V. Varekh, *Izv. Vyssh. Uchebn. Zaved. Khim. Khim. Tekhnol.* **16**, 1619 (1973).
87. U.S. Pat. 3,795,595 (Mar. 5, 1974), H. P. Wilson (to Vulcan Materials Co.).
88. U.S. Pat. 3,300,397 (Jan. 24, 1967), G. Baltakmens and J. P. Tourish (to Allied Chemical Corp.).
89. U.S. Pat. 3,432,256 (Mar. 11, 1969), H. P. Wilson (to Vulcan Materials Co.).
90. S. Buffagni and I. M. Vezzosi, *Gazz. Chim. Ital.* **97**, 1258 (1967).
91. U.S. Pat. 4,092,296 (May 30, 1978), R. A. Skiff.
92. U.S. Pat. 3,432,440 (Mar. 11, 1969), D. A. Shimp, W. F. McWhorter, and N. G. Wolfe (to Celanese Coatings Co.).
93. *Technical Bulletin ZBF873*, Harshaw Chemical Co., Solon, Ohio.
94. A. Zemaitaitis and J. Zdanavicius, *Cellul. Chem. Technol.* **4**, 621 (1970).
95. U.S. Pat. 3,854,869 (Dec. 17, 1974), Y. Yanai (to Nisshin Spinning Co., Ltd.).
96. C. E. Morris and G. L. Drake, Jr., *Am. Dyestuff Rep.* **58**(4), 31 (1969).
97. Jpn. Pat. 71 11,080 (Mar. 20, 1971), S. Hiroaka and K. Mitsumura (to Mitsubishi Rayon Co., Ltd.).
98. U.S. Pat. 3,577,342 (May 4, 1971), L. I. Fidell (to American Cyanamid Co.).
99. U.S. Pat. 3,697,615 (Oct. 10, 1972), W. B. Hughes (to Phillips Petroleum Co.).
100. E. Rausz and S. Hulisz, *Chemik* **28**(7), 256 (1975).
101. H. W. Quinn and R. L. Van Gilder, *Can. J. Chem.* **48**, 2435 (1970).
102. U.S. Pat. 3,514,488 (May 26, 1970), C. E. Uebele, R. K. Grasselli, and W. C. Nixon (to Standard Oil Co. of Ohio).
103. Jpn. Pat. 714,030 (Dec. 3, 1971), Y. Fujita and T. Morimoto (to Mitsui Petrochemical Industries, Ltd.).
104. Jpn. Pat. 70 19,514 (July 3, 1970), I. Hiroi (to Toho Rayon Co., Ltd.).
105. Jpn. Pat. 76 04,163 (Jan. 14, 1976), J. Takeuchi, F. Iwata, and K. Kubo (to Ube Industries, Ltd.).
106. Jpn. Pat. 72 18,227 (May 26, 1972), S. Sugiura and co-workers (to Ube Industries, Ltd.).
107. Jpn. Pat. 73 06,185 (Feb. 23, 1973), T. Yamawaki, T. Suzuki, and S. Hino (to Mitsubishi Chemical Industries Co. Ltd.).
108. Fr. Pat. 2,039,808 (Jan. 15, 1971), (to Mitsubishi Chemical Industries, Co., Ltd.).
109. Jpn. Pat. 72 06,411 (May 4, 1972), T. Yamawaki and co-workers (to Mitsubishi Chemical Industries Co., Ltd.).
110. *Plating Processes*, Harshaw Chemical Co., Solon, Ohio, Mar. 1977.
111. Y. M. Faruq Marikan and K. I. Vasu, *Met. Finish.* **67**(8), 59 (1969).
112. F. Wild, *Electroplat. Met. Finish.* **13**, 331 (Sept. 1960).
113. E. M. Levy and G. J. Hutton, *Plating* **55**(2), 138 (1968).
114. U.S. Pat. 3,432,338 (Mar. 11, 1969), R. E. Sickles (to Diamond Shamrock Corp.).

115. N. J. Spiliotis, *Galvanotech. Oberflaechenschutz* **7**(8), 192 (1966).
116. Jpn. Pat. 76 02,633 (Jan. 10, 1976), J. Hara, R. Miyashata, and Y. Fukuoka (to Nippon Kayaku Co., Ltd.).
117. Ger. Pat. 2,146,908 (Mar. 23, 1972), K. Ishiguro and H. Shinohara (to Toyota Motor Co., Ltd.).
118. *Plating Processes, Tin-Lead Solder Alloy Fluoborate Plating Process for Printed Circuit Applications, HTPB5N 0272*, Harshaw Chemical Co., Solon, Ohio.

JOHN R. PAPCUN
Atotech

CALCIUM

Fluorine chemistry began with observations by Georgius Agricola as early as 1529 that fluorspar lowers the melting point of minerals and reduces the viscosity of slags. This property of fluxing (Latin *fluoere*, to flow) is the origin of the name fluorine. The term fluorspar correctly describes ores containing substantial amounts of the mineral fluorite [*14542-23-5*], CaF_2, but the word fluorspar is often used interchangeably with fluorite and calcium fluoride (see also CALCIUM COMPOUNDS).

Calcium Fluoride

Significant mining of fluorspar began in England about 1775 and in the United States after 1820. Substantial use of fluorspar began about 1880 in the basic open-hearth process for making steel (qv). Large increases in demand came with the need for fluorides in the aluminum industry, starting about 1900. A large fluorine chemicals industry based on hydrogen fluoride made from fluorspar followed in production of refrigerants (see REFRIGERATION AND REFRIGERANTS) (1930), alkylation (qv) catalysts for gasoline (1942), materials for nuclear energy (ca 1942), aerosol propellants (see AEROSOLS) (ca 1942), fluoroplastics (ca 1942), and fluorocarbons for soil-repellant surface treatments (early 1950s). Fluorspar is used directly in the manufacture and finishing of glass (qv), in ceramics (qv) and welding (qv) fluxes, and in the extraction and processing of nonferrous metals (see METALLURGY, EXTRACTIVE).

In the geochemistry of fluorine, the close match in the ionic radii of fluoride (0.136 nm), hydroxide (0.140 nm), and oxide ion (0.140 nm) allows a sequential replacement of oxygen by fluorine in a wide variety of minerals. This accounts for the wide dissemination of the element in nature. The ready formation of volatile silicon tetrafluoride, the pyrohydrolysis of fluorides to hydrogen fluoride, and the low solubility of calcium fluoride and of calcium fluorophosphates, have provided a geochemical cycle in which fluorine may be stripped from solution by limestone and by apatite to form the deposits of fluorspar and of phosphate rock (fluoro-apatite [*1306-01-0*]), approximately $CaF_2 \cdot 3Ca_3(PO_4)_2$ which are the world's main resources of fluorine (1).

On average, fluorine is about as abundant as chlorine in the accessible surface of the earth including oceans. The continental crust averages about 650 ppm fluorine. Igneous, metamorphic, and sedimentary rocks all show abundances in the range of 200 to 1000 ppm. As of 1993, fluorspar was still the principal source of fluorine for industry.

Fluorspar deposits are commonly epigenetic, ie, the elements moved from elsewhere into the country rock. For this reason, fluorine mineral deposits are closely associated with fault zones. In the United States, significant fluorspar deposits occur in the Appalachian Mountains and in the mountainous regions of the West, but the only reported commercial production in 1993 was from the faulted carbonate rocks of Illinois.

Worldwide, large deposits of fluorspar are found in China, Mongolia, France, Morocco, Mexico, Spain, South Africa, and countries of the former Soviet Union. The United States imports fluorspar from most of these countries (Table 1).

Properties. Some of the important physical properties of calcium fluoride are listed in Table 2. Pure calcium fluoride is without color. However, natural fluorite can vary from transparent and colorless to translucent and white, wine-yellow, green, greenish blue, violet-blue, and sometimes blue, deep purple, bluish black, and brown. These color variations are produced by impurities and by radiation damage (color centers). The color of fluorite is often lost upon heating, sometimes with luminescence. Mineral specimens are usually strongly fluorescent, and the mineral thus gives its name to this phenomenon. Specimens vary from well-formed crystals (optical grade) to massive or granular forms.

The crystal structure of fluorite gives its name to the fluorite crystal type. The lattice is face-centered cubic (fcc), where each calcium ion is surrounded by eight fluoride ions situated at the corners of a cube, and each fluoride ion lies within a tetrahedron defined by four calcium ions (3). The bonding is ionic. The unit cell (space group O_h^5) can be pictured as made up of eight small cubes, each containing a fluoride ion, and the eight forming a cube with a calcium ion on each corner and one in the center of each face (Fig. 1). The lattice constant is 0.54626 nm at 25°C (4). The habit is usually cubic, less frequently octahedral, rarely dodecahedral. Cleavage on the [111] planes is perfect. The crystals are brittle with flat-conchoidal or splintery fracture. Luster is vitreous, becoming dull in massive varieties.

Systems of metal oxides with calcium fluoride usually have a simple freezing point composition diagram, commonly exhibiting a eutectic point and no abnormal lowering of the melting point (5). When silicates are present, the systems become more complicated, and a striking decrease in the viscosity of the glassy melts is observed. The viscosity most likely decreases because of depolymerization of chains or networks of SiO_4 tetrahedra via the replacement of oxide ion by the singly charged fluoride ion which is close in both size and electronegativity to oxide ion (1). The benefits of calcium fluoride as a metallurgical flux result from both the freezing point depression and the decrease in slag viscosity.

Although stable at ambient temperature, calcium fluoride is slowly hydrolyzed by moist air at about 1200°C, presumably to CaO and HF. Calcium fluoride is not attacked by alkalies or by reactive fluorine compounds, but is decomposed by hot, high boiling acids, as in the reaction with concentrated sulfuric acid which is the process used to produce hydrogen fluoride. Calcium fluoride is slightly sol-

Table 1. U.S. Imports for Consumption of Fluorospar[a,b]

Country	1991[c] Quantity, t	1991[c] Value, $ × 10^6$	1992[d] Quantity, t	1992[d] Value, $ × 10^6$	1993[e] Quantity, t	1993[e] Value, $ × 10^6$	1993[e] Quantity, t	1993[e] Value, $ × 10^6$
			$CaF_2 > 97\%$					
Canada	18	9						
China	56,311	6,352	128,960	13,127	73,942	7,175	34,817	3,077
France	36	15	52	26				
Japan	4,627	416						
Mexico	52,475	6,570	34,058	4,228	8,917	1,096	7,289	884
Morocco	9,828	1,382						
Republic of South Africa	79,495	11,107	106,066	13,147	5,932	702	25,519	2,990
Spain	11,278	1,545						
Total	*214,350*	*27,387*	*267,328*	*30,528*	*88,791*	*8,973*	*77,791*	*7,895*
			$CaF_2 \leq 97\%$					
Canada	285	22						
China	11,051	909	59,130	4,619	17,968	1,270	3,937	335
Japan	4,674	387						
Mexico	33,970	3,449	54,405	5,069	822	55	9,117	712
Republic of South Africa	15,860	2,108	11,152	1,370				
Total	*65,840*	*6,875*	*124,687*	*11,058*	*18,790*	*1,325*	*13,054*	*1,047*

[a] Imports for consumption include imports of immediate entry plus warehouse withdrawals (2).
[b] Cost, insurance, freight (cif) at U.S. ports.   [c] 1991 Numbers represent the total of 2nd, 3rd, and 4th quarters only.
[d] 1992 Numbers represent the total of 1st, 2nd, 3rd, and 4th quarters.   [e] 1993 Numbers represent 1st, 2nd, and 3rd quarters, respectively.

Table 2. Physical Properties of Calcium Fluoride

Property	Value	Reference
formula weight	78.08	
composition, wt %		
Ca	51.33	
F	48.67	
melting point, °C	1402	6
boiling point, °C	2513	7
heat of fusion, kJ/mol[a]	23.0	8
heat of vaporization at bp, kJ/mol[a]	335	9
vapor pressure at 2100°C, Pa[b]	1013	9
heat capacity, C_p, kJ/(mol·K)[a]		
solid at 25°C	67.03	10
solid at mp	126	11
liquid at mp	100	11
entropy at 25°C, kJ/(mol·K)[a]	68.87	10
heat of formation, solid at 25°C, kJ/mol[a]	−1220	10
free energy of formation, solid at 25°C, kJ/mol[a]	−1167	10
thermal conductivity, crystal at 25°C, W/(m·K)	10.96	12
density, g/mL		
solid at 25°C	3.181	13
liquid at mp	2.52	14
thermal expansion, average 25 to 300°C, K^{-1}	22.3×10^{-6}	15
compressibility, at 25°C and 101.3 kPa (=1 atm)	1.22×10^{-8}	16
hardness		
Mohs' scale	4	
Knoop, 500-g load	158	17
solubility in water, g/L at 25°C	0.146	18
refractive index at 24°C, 589.3 nm	1.43382	19
dielectric constant at 30°C	6.64	20
electrical conductivity of solid, (Ω·cm)$^{-1}$		
at 20°C	1.3×10^{-18}	21
at 650°C	6×10^{-5}	22
at mp	3.45	23
optical transmission range, nm	150 to 8000	24

[a] To convert J to cal, divide by 4.184.
[b] To convert Pa to mm Hg, multiply by 7.5×10^{-3}.

uble in cold dilute acids, and somewhat more soluble in solutions of aluminum halides.

Preparation. CaF_2 is manufactured by the interaction of H_2SiF_6 with an aqueous carbonate suspension (25–28); by the reaction of $CaSO_4$ with NH_4F (29); by the reaction of HF with $CaCO_3$ in the presence of NH_4F (30); by reaction of $CaCO_3$ and NH_4F at 300–350°C followed by calcining at 700–800°C (31); by reaction of NH_4F and $CaCO_3$ (32–37); and from the thermal decomposition of calcium trifluoroacetate (38).

High purity CaF_2 is obtained from micro- and ultrafiltration (qv) of raw materials and then crystallization of CaF_2 (39) from the reaction of $CaCO_3$ and the

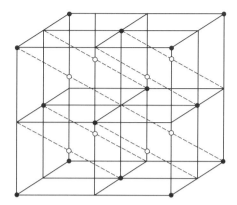

Fig. 1. Structure of fluorite where ● is Ca and ○ is F (3).

product of K_2SiF_6 and NH_3 (40). High purity CaF_2 having particles of 0.0005–0.5 mm is produced from the reaction of NaF, KF, or NH_4F and $CaCO_3$ with a particle size distribution of 0.01–0.05 nm (41). High purity CaF_2 is also prepared from the reaction of $Ca(NO_3)_2 \cdot 4H_2O$ and NH_4F or a mixture of NH_4HF_2 and NH_4F (42), and obtained by heating impure CaF_2 with 10–15% HCl at 95–100°C (43). Very pure calcium fluoride for the manufacture of special glasses is made by the reaction of hydrofluoric acid with precipitated calcium carbonate. Acicular (whisker-form) CaF_2 particles have been manufactured by continuous feeding of an aqueous $Ca(OH)_2$ solution into water containing CO_2 and subsequent reaction with HF (44). Coarse grain CaF_2 crystals can be prepared by several routes (45–58). CaF_2 can be is crystallized from a wastewater containing fluoride by adding $CaCl_2$ (59). Calcium fluoride can be produced from waste H_2SiF_6 from phosphate product operations and from treating fluoride solution from industrial wastewater with KOH and then with lime (60).

The large amount of fluorine values released from phosphate rock in the manufacture of fertilizers (qv) gives a strong impetus to develop fluorine chemicals production from this source (see PHOSPHORIC ACID AND THE PHOSPHATES). Additional incentive comes from the need to control the emission of fluorine-containing gases. Most of the fluorine values are scrubbed out as fluorosilicic acid, H_2SiF_6, which has limited usefulness. A procedure to convert fluorosilicic acid to calcium fluoride is available (61).

Mining. Underground mining procedures are used for deep fluorspar deposits, and open-pit mines are used for shallow deposits or where conditions do not support underground mining techniques (see MINERAL RECOVERY AND PROCESSING).

Fluorspar occurs in two distinct types of formation in the fluorspar district of southern Illinois and Kentucky; in vertical fissure veins and in horizontal bedded replacement deposits. A 61-m bed of sandstone and shale serves as a cap rock for ascending fluorine-containing solutions and gases. Mineralizing solutions come up the faults and form vein ore bodies where the larger faults are plugged by shale. Bedded deposits occur under the thick sandstone and shale roofs. Other elements of value associated with fluorspar ore bodies are zinc, lead, cadmium, silver, germanium, iron, and thorium. Ore has been mined as deep as 300 m in this district.

In 1993, Illinois was the only state reporting production of acid-grade fluorspar, typically 96.5–97.5% CaF_2, and accounted for 100% of all reported shipments. Ozark-Mahoning Co., a wholly owned subsidiary of Elf Atochem North America, Inc., operated three deep mines and a flotation mill in Hardin County, Illinois. A limited amount of metallurgical-grade gravel was produced by Hastie Mining, also located in Hardin County, Illinois (1).

Outside of the United States, there are six primary producers in China, France, Mexico, Morocco, South Africa, and Spain. Mines in Newfoundland, Canada, were closed in 1990. Both Mexico and South Africa have lost market share to China which has high grade, low cost fluorspar. China is expected to dominate world markets because reserves are vast and production cost is low. Table 3 (2) shows a list of world producers by country of fluorspar in the early 1990s.

Beneficiation. Most fluorspar ores as mined must be concentrated or beneficiated to remove waste. Metallurgical-grade fluorspar is sometimes produced by

Table 3. World Production of Fluorspar,[a] t

Country	1990	1991	1992
Argentina	20,000[b]	20,000	19,000
Brazil	70,383	90,000	81,000
Canada	25,000		
China	1,700,000	1,600,000	1,600,000
Czechoslovakia	46,966	40,000	40,000
France	201,000	200,000	160,000
Germany	155,300	135,000	60,000
India	21,700	22,000	25,000
Iran	4,767	6,000	12,000
Italy	122,503	100,000	80,000
Kenya	112,295	90,000	80,630
North Korea	40,000	41,000	41,000
Mexico	634,000	352,000	364,000
Mongolia	614,000	520,000	277,000
Morocco	86,500	86,000	80,000
Namibia	25,980	29,246	40,480
Pakistan	5,312	5,300	5,000
Romania	15,000	14,000	15,000
Republic of South Africa	311,032	270,340	258,105
Spain	100,000[b]	90,000	90,000
Thailand	94,757	100,000	52,000
Tunisia	40,974	40,000	15,000
Turkey	13,000	13,000	13,000
former USSR	380,000	350,000	300,000
United Kingdom	118,498	80,000	85,000
United States (shipments)[b]	63,500	58,000	51,000
Total	*5,024,626*	*4,354,036*	*3,844,315*

[a]Only countries producing more than 2000 metric tons are listed (2).
[b]Values are estimated.

hand sorting lumps of high grade ore. In most cases the ore is beneficiated by gravity concentration with fluorspar and the waste minerals, having specific gravity values of > 3 and < 2.8, respectively. In preconcentration of fluorspar, barite and valuable sulfide minerals are separated from waste as the higher density valuable minerals sink while the waste floats and is discarded. This preconcentration can enrich ores as low as 14% to a concentration of 40%. Multistage froth flotation (qv) is used to take this preconcentrate and produce acid and ceramic grades of fluorspar as well as zinc and lead sulfides. In this process air bubbles are forced through a suspension of pulverized ore which float the ore into a froth that is continuously skimmed off. After flotation the fluorspar products are filtered and dried in rotary kilns.

In steelmaking, the preferred form of fluorspar flux is a washed gravel, 0.6–5 cm in diameter, containing less than 3% water, and assaying 60 to 80% effective CaF_2 units. The higher CaF_2 ranges are hard to supply in large amounts from some sources of fluorspar. The use of fluorspar briquettes and pellets in the steel (qv) industry has declined but these are still preferred by some producers. The briquettes contain 25–90% CaF_2, are frequently made to customer specifications, and may include fluxing agents and recycled steel mill wastes. Binders used include molasses, lime, and sodium silicate (see METALLURGY, EXTRACTIVE).

Economic Aspects. Pertinent statistics on the U.S. production and consumption of fluorspar are given in Table 4. For many years the United States has relied on imports for more than 80% of fluorspar needs. The principal sources are Mexico, China, and the Republic of South Africa. Imports from Mexico have declined in part because Mexican export regulations favor domestic conversion of fluorspar to hydrogen fluoride for export to the United States.

There were very high U.S. exports in 1991. However, this number reflects material imported into U.S. foreign trade zones and then transhipped to other countries.

Consumer stocks (2) at the end of 1993 were 65,000 t. The National Defense Stockpile of fluorspar inventory, at year end 1992, contained 809,000 t of acid-grade material, 281,000 t of metallurgical-grade material, 816 t of nonstockpile, acid-grade material, and 105,938 t of nonstockpile, metallurgical-grade material.

Table 4. United States Fluorspar Production and Consumption, t

Parameter	1987	1988	1989	1990	1991	1992
production[a]	63,500	63,500	66,000	63,500	58,000	51,000
exports	2,595	3,136	5,134	14,921	73,943	13,646
imports for consumption	531,530	689,139	655,590	513,921	376,081[b]	407,169
consumption						
reported	542,830	651,055	641,882	564,545	483,589	48,544
apparent[c]	643,659	723,804	693,121	566,885	365,831	480,376
world production	4,600,441	5,086,376	5,529,184	5,024,626	4,354,036	3,846,443

[a]Values are estimated (2).
[b]At a value of $50,780,000 cif.
[c]U.S. primary and secondary production plus imports minus exports plus adjustments for government and industry stock changes.

The prices for acid-grade fluorspar dry basis from Mexico and the Republic of South Africa for 1993 were Mexican spar, fob Tampico, $100–$112/t; South Africa acid spar, fob Durban, $95–$105/t. Mid-1992 quotes for Chinese acid-grade fluorspar delivered to Louisiana Gulfport were $89–$97/t and to Northern Europe, $105–$110/t. The price reductions generated by the Chinese caused producers to reduce production as market share was lost. Table 5 (2) contains U.S. imports of fluorspar by country.

Table 5. U.S. Imports for Consumption of Fluorspar by Country, t[a]

Country	1990	1991	1992	1993[b]
	$CaF_2 > 97\%$			
Canada	22,137			
China	125,190	74,853	128,960	148,932
Mexico	142,964	70,322	34,248	22,216
Morocco	10,950	9,828		
Republic of South Africa	86,660	105,031	106,866	63,864
	$CaF_2 \leq 97\%$			
China	28,088	28,228	59,130	27,086
Mexico	71,006	49,541	54,405	19,272
Republic of South Africa	8,280	17,360		

[a]Countries not having 10,000 metric tons in any one year were omitted (2).
[b]Through September 1993.

Grades, Quality Control. Fluorspar is marketed in several grades: metallurgical fluorspar (metspar) is sold as gravel, lump, or briquettes. The minimum acceptable assay is 60% effective calcium fluoride. The effective value is determined by subtracting from the contained calcium fluoride 2.5% for every percent of SiO_2 found in the complete analysis apparently based on the following stoichiometry (1):

$$2\ CaF_2 + SiO_2 \rightarrow 2\ CaO + SiF_4$$

Ceramic-grade fluorspar and acid-grade fluorspar have the typical analyses shown in Table 6. Both types are usually finely ground, the bulk of the powder passing a 0.23 mm (65 mesh) screen, and 22 to 81% held on a 44 µm (325 mesh) screen. Optical-grade calcium fluoride, for special glasses and for growing single crystals, is supplied in purities up to 99.99% CaF_2. This grade is especially low in transition elements. For process control (qv) and product specification, fluorspar is commonly analyzed for fluorine, calcium, silica, carbonate, sulfide, iron, barium, and where significant, for metal values.

Analytical Methods. Fluorite is readily identified by its crystal shape, usually simple cubes or interpenetrating twins, by its prominent octahedral cleavage, its relative softness, and the production of hydrogen fluoride when treated with sulfuric acid, evidenced by etching of glass. The presence of fluorite in ore speci-

Table 6. Analyses of Ceramic- and Acid-Grade Fluorspar, wt %[a]

Assay	Ceramic	Acid
CaF_2	90.0–95.5	96.5–97.5
SiO_2	1.2–3.0	1.0
$CaCO_3$	1.5–3.4	1.0–1.5
MgO		0.15
B		0.02
Zn		0.02
Fe_2O_3	0.10	0.10
P_2O_5		0.03
$BaSO_4$		0.2–1.3
R_2O_3[b]	0.15–0.25	0.1–0.3

[a] Refs. 62 and 63.
[b] R_2O_3 is any trivalent metal oxide, eg, Al_2O_3.

mens, or when associated with other fluorine-containing minerals, may be determined by x-ray diffraction.

For many years fluorine has been determined by the Willard-Winters method in which finely ground ore, after removal of organic matter, is distilled with 72% perchloric acid in glass apparatus. The distillate, a dilute solution of fluorosilicic acid, is made alkaline to release fluoride ion, adjusted with monochloroacetic acid at pH 3.4, and titrated with thorium nitrate, using sodium alizarine sulfonate as indicator.

The direct determination of fluoride using ion-selective electrodes has allowed analysis of fluorspar without the tedious distillation step (see ELECTROANALYTICAL TECHNIQUES). The fluoride electrode uses a single crystal of lanthanum fluoride doped with europium in contact with the solution, and a high impedance detector similar to that used for glass electrode pH measurements. The finely ground fluorspar sample is dissolved in boiling aluminum chloride and hydrochloric acid solution, and the solution is buffered using ammonium sulfosalicylate. Ethylenediaminetetraacetic acid (EDTA) is added to displace the fluoride ion from complexes with positive ions (64–66). X-ray fluorescence is also used to analyze for fluoride content.

Fluorspar assay may be completed by fluoride determination alone, because the mineralogical grouping rarely includes fluorine minerals other than fluorite. Calcium can be determined as oxalate or by ion-selective electrodes (67). Silica can be determined in the residue from solution in perchloric acid–boric acid mixture by measuring the loss in weight on fuming off with hydrofluoric acid. Another method for determining silica in fluorspar is the ASTM Standard Test Method E463-72.

Carbonate is measured by evolution of carbon dioxide on treating the sample with sulfuric acid. The gas train should include a silver acetate absorber to remove hydrogen sulfide, a magnesium perchlorate drying unit, and a CO_2-absorption bulb. Sulfide is determined by distilling hydrogen sulfide from an acidified slurry of the sample into an ammoniacal cadmium chloride solution, and titrating the precipitated cadmium sulfide iodimetrically.

Health and Safety Factors. The low solubility of calcium fluoride reduces the potential problem of fluoride-related toxicity. Water saturated with calcium fluoride has a fluoride concentration of 8.1 ppm as compared to the recommended water fluoridation level of 1 ppm fluoride ion. However, because the solubility of calcium fluoride in stomach acid is higher, continued oral ingestion of calcium fluoride could produce symptoms of fluorosis. The adopted TWA limit for fluorides as F is 2.5 mg/m^3 (68,69).

A significant hazard results from contact of calcium fluoride with high concentrations of strong acids because of evolution of toxic concentrations of hydrogen fluoride. A less recognized danger is the production of hydrogen fluoride by pyrohydrolysis when calcium fluoride is exposed to water vapor at high temperature, as in a direct-fired kiln.

Beneficiation facilities require air and water pollution control systems, including efficient control of dust emissions, treatment of process water, and proper disposal of tailings (see AIR POLLUTION CONTROL METHODS). In handling finished fluorspar, operators must avoid breathing fluorspar dust and contacting fluorspar with acids. Proper disposal of spills and the use of respirators and other personnel protective equipment must be observed. Contact with fluorspar may irritate the skin and eyes.

Consumption and Uses. Acid-grade fluorspar, which is > 97% calcium fluoride, is used primarily in the production of hydrogen fluoride. Ceramic-grade fluorspar, containing 85 to 95% CaF_2 content, is used in the production of glass and enamel, to make welding rod coatings, and as a flux in the steel industry. Metallurgical-grade fluorspar, containing 60 to 85% or more CaF_2, is used primarily as a fluxing agent by the steel industry. Fluorspar is added to the slag to decrease its melting point and to increase its fluidity, thus increasing the chemical reactivity of the slag. Reducing the melting point of the slag brings lime and other fluxes into solution to allow the absorption of impurities.

Reported domestic consumption (2) by the HF industry in 1992 was 347,367 t. Reported consumption by the steel industry decreased in 1991 by about 52% from 1990 partly because of a 13% decrease in the production of steel.

In the ceramic industry, fluorspar is used as a flux and as an opacifier in the production of flint glass, white or opal glass, and enamels (see ENAMELS, PORCELAIN OR VITREOUS). Fluorspar is used in the manufacture of aluminum, brick, cement, and glass fibers, and is used by the foundry industry.

A small but artistically interesting use of fluorspar is in the production of vases, cups, and other ornamental objects popularly known as Blue John, after the Blue John Mine, Derbyshire, U.K. Optical quality fluorite, sometimes from natural crystals, but more often artificially grown, is important in use as infrared transmission windows and lenses (70) and optical components of high energy laser systems (see INFRARED AND RAMAN SPECTROSCOPY; LASERS; OPTICAL FILTERS) (71).

Calcium fluoride is also used in the thermal plasma process for purification of silicon (72); in the manufacture of fluoroaluminate glass (73,74); and fluorophosphate glasses (75); in the removal of lead in flotation waste (76); in the manufacture of clean steel without deoxidation with aluminum (77); in the electrochemical preparation of lanthanum (78); in the leaching process to treat domestic Mn-bearing silicate resources (79); as pigments for paper (qv) with titania (80);

in ir-transmitting window material (81); in the formation of passivated layers on superconductor oxides (82); in the manufacture of electrically resistive pastes (83) and electric resistors (84); in the refining of molten pig steel (85); in the recovery and purification of uranium (86); and as a solid lubricant for hot rolling (87). It is also used to strengthen cement (88,89) and for microcrystalline fluoride fibers useful for making seals for liquid propellant engines (90).

BIBLIOGRAPHY

"Calcium Fluoride" under "Fluorine Compounds, Inorganic," in *ECT* 1st ed., Vol. 6, pp. 689–692, by H. C. Miller, Pennsylvania Salt Manufacturing Co.; "Calcium Fluoride" under "Fluorine Compounds, Inorganic," in *ECT* 2nd ed., Vol. 9, pp. 573–582, by J. F. Gall, Pennsalt Chemicals Corp.; "Calcium" under "Fluorine Compounds, Inorganic," in *ECT* 3rd ed., Vol. 10, pp. 707–717, by J. F. Gall, Philadelphia College of Textiles and Science.

1. D. R. Shawe, ed., *Geology and Resources of Fluorine in the United States*, U.S. Geological Survey Professional Paper 933, Washington, D.C., 1976, pp. 1–5, 18, 19, 82–87.
2. M. M. Miller, *Fluorspar 1991 Annual Report*, U.S. Department of Interior, Bureau of Mines, Washington, D.C.
3. A. F. Wells, *Structural Inorganic Chemistry*, 3rd ed., Clardon Press, Oxford, U.K., 1962, p. 77.
4. H. E. Swanson and E. Tatge, *Natl. Bur. Stand. U.S. Circ. 539*, **1**, 69 (1953).
5. H. Krainer, *Radex Rundsch.*, 19 (1949).
6. B. Porter and E. A. Brown, *J. Am. Ceram. Soc.* **45**, 49 (1962).
7. D. A. Schulz and A. W. Searcy, *J. Phys. Chem.* **67**, 103 (1963).
8. G. Petit and A. Cremieu, *Compt. Rend.* **243**, 360 (1956).
9. O. Ruff and L. Leboucher, *Z. Anorg, Allg. Chem.* **219**, 376 (1934).
10. *National Bureau of Standards Technical Notes*, Washington, D.C., 1971, pp. 270–276.
11. B. F. Naylor, *J. Am. Chem. Soc.* **67**, 150 (1945).
12. K. A. McCarthy and S. S. Ballard, *J. Appl. Phys.* **36**, 1410 (1960).
13. Ref. 4, p. 69.
14. A. V. Grosse and C. S. Stokes, U.S. Department of Commerce, Office of Technical Service, PB Report 161460, Washington, D.C., 1960.
15. O. J. Whittemore, Jr. and N. N. Ault, *J. Ceram. Soc.* **39**, 443 (1956).
16. E. W. Washburn, ed., *International Critical Tables*, Vol. 3, McGraw-Hill Book Co., Inc., New York, 1929, p. 50.
17. S. S. Ballard, L. S. Combes, and K. A. McCarthy, *J. Opt. Soc. Am.* **42**, 684 (1952).
18. D. W. Brown and C. E. Roberson, *J. Res. U.S. Geol. Surv.* **5**, 509 (1977).
19. *Natl. Bur. Stand. U.S., Tech. News Bull.* **47**, 91 (1963).
20. J. L. Pauley and H. Chessin, *J. Am. Chem. Soc.* **76**, 3888 (1954).
21. E. W. Washburn, ed., *International Critical Tables*, Vol. 6, McGraw-Hill Book Co., Inc., New York, 1929, p. 154.
22. R. W. Ure, Jr., *J. Chem. Phys.* **26**, 1365 (1957).
23. T. Baak, *J. Chem. Phys.* **29**, 1195 (1958).
24. Data sheet, *IR Transmission Materials*, Barnes Engineering Co., Instrument Division, Stamford, Conn., 1992.
25. V. V. Babkin, V. V. Koryakov, T. A. Sokolova, and N. K. Petrova, *Khim. Prom-St (Moscow)* **3**, 1963–164.25, (1992).
26. H. Gabryel, L. Kacalski, and U. Glabisz, *Chem. Stosow* **33**(4), 673–678 (1989).
27. SU 1286520 A1, (Jan. 30, 1987), I. A. Elizarov and co-workers.
28. U.S. Pat. 4,264,563 (Apr. 28, 1981) S. K. Sikdar.

29. SU 1708762 A1 (Jan. 30, 1992), I. G. Saiko, A. A. Perebeinos, L. M. Pupyshevea, N. A. Orel.
30. SU 1699922 A1 (Dec. 23, 1991), M. E. Rakhimov, D. D. Ikrami, L. F. Mansurhodzhaeva, and Sh. A. Khalimov.
31. SU 998352 A1 (Feb. 23, 1983), A. A. Luginina, L. A. Ol'Khovaya, V. A. Reiterov, and D. D. Ikrami.
32. SU 802185 (Feb. 7, 1981), V. I. Rodin and co-workers.
33. A. A. Luginina and co-workers, *Zh. Neorg. Khim* **26**(2) 332–336 (1981).
34. V. V. Tumanov and co-workers, *Khim. Prom-St (Moscow)* (9), 668–671 (1989).
35. SU 83-3558912 (Mar. 2, 1983), M. I. Lyapunov, V. V. Tumonov, L. P. Belova, and G. H. Alekseeva.
36. V. V. Tumanov, L. P. Belova, and G. N. Alekseeva, *Prom-St (Moscow)* **9**, 551–553 (1983).
37. PL 104419 (Nov. 30, 1979), W. Augustyn, M. Dziegielewska, and A. Kossuth.
38. C. Russell, *J. Mater. Sci. Lett.* **11**(3), 152–154 (1992).
39. T. N. Naumova and co-workers, *Zh. Priki. Khim (Leningrad)* **64**(3), 480–484 (1991).
40. W. Augustyn, and co-workers, *Prezm. Chem.* **68**(4), 153–155 (1989).
41. PL 106787 (Jan. 31, 1980), W. Augustyn, M. Dziegielewska, and A. Kossuth.
42. RO 88593 B1 (Mar. 31, 1986), H. Glieb, E. Apostol, and C. Dan.
43. SU 983052 A1 (Dec. 23, 1982), V. K. Fomin, N. I. Varlamova, O. V. Leleedev, and A. P. Krasnov.
44. Jpn. Pat. 01083514 A2 (Mar. 29, 1989), Y. Oata, N. Goto, I. Motoyama, T. Iwashita, and K. Nomura.
45. PL 85616 (Sept. 15, 1976), W. Augustyn and co-workers.
46. U.S. Pat. 77,810,047 (June 27, 1977), W. C. Warneke.
47. U. Glabisz and co-workers, *Prezm. Chem.* **68**(1), 20, 29–30 (1989).
48. U.S. Pat. 685,100, AO (Aug. 1, 1986), A. B. Kreuzmann and D. A. Palmer.
49. V. S. Sakharov and co-workers, *Khim Prom-St (Moscow)* (1), 257 (1982).
50. SU 79-2829664 (July 30, 1978), G. A. Loptkina, V. I. Chernykh, and O. D. Fedorova.
51. SU 709537 (Jan. 15, 1980), G. A. Lopatking and V. I. Chernykh.
52. EP 210937 A1 (Feb. 4, 1987), L. Siegneurin.
53. M. S. Nesterova and T. Yu Magda, *Tekhnal. Obogashch. Polezn. Iskop. Sredni.* **A3** 3, 96–99 (1981).
54. R. V. Chernov and D. L. Dyubova, *Zh. Priki Khim. (Leningrad)* **56**(5), 1133–1135 (1983).
55. U. Glabisz, H. Gabryel, L. Kacalski, and B. Kic, *Pr. Nauk. Akad. Ekon. Im. Oskara Langego Wroclawiu* **338**, 165–169 (1986).
56. SU 1224263 A1 (Apr. 15, 1986), V. K. Fomin, N. I. Varlamova, V. P. Kozma, and M. N. Esin.
57. V. V. Pechkovskii, E. D. Dzyuba, and L. P. Valyu, *Zh. Priki (Leningrad)* **53**(5), 961–965 (1980).
58. A. A. Opalovskii and co-workers, *Zh. Neorg. Khim.* **20**(5), 1179–1183 (1975).
59. EP 476773 A1 (Mar. 25, 1992), J. Dijkhorst.
60. U.S. Pat. 82,406,420 A (Nov. 8, 1982), J. P. Harrison.
61. R. C. Kirby and A. S. Prokopovitsh, *Science* **191**, 717 (Feb. 1976).
62. P. L. Braekner, Allied Chemical, Industrial Chemicals Division, Morristown, N.J., private communication, Nov. 1978.
63. Data sheet, *Fluorspar*, Reynolds Chemicals, Richmond, Va., Mar. 1978.
64. *Analytical Methods Guide*, 6th ed., Orion Research, Inc., Cambridge, Mass., Aug. 1973, p. 18.
65. R. T. Oliver and A. G. Clayton, *Anal. Chim. Acta* **51**, 409 (1970).

66. R. B. Fisher, *J. Chem. Educ.* **51**, 387 (1974). (Note: reference to lithium fluoride on p. 389 is in error. It should read lanthanum fluoride.)
67. A. Julaicki and M. Trojanowicz, *Anal. Chim. Acta* **68**, 155 (1974).
68. *Criteria for a Recommended Standard—Occupational Exposure to Inorganic Fluorides*, NTIS Document PB-246692, National Institute for Occupational Safety and Health, Rockville, Md., 1975.
69. *1992–1993 Threshold Limit Values for Chemical Substances and Physical Agents*, American Conference of Governmental Industrial Hygienists, Cincinnati, Ohio, 1992.
70. Data sheet, *IR Transmission Materials*, Barnes Engineering Co., Instrument Division, Stamford, Conn., 1992.
71. C. B. Willingham and R. T. Newberg, *Exploratory Development of Fusion Cost Calcium Fluoride for 1.06 Micrometer Pulsed LaserOptics*, Progress Report, Sept. 25 to Dec. 25, 1976, NTIS Document C00-4029-2, Raytheon Co., for U.S. Energy Research and Development Administration, Jan. 1977.
72. P. Humbert and co-workers, *E. C. Photovoltair Sol. Energy Conference, Proc. Int. Conf. 1991*, 10th, pp. 261–266.
73. H. Hu, F. Lin, and Y. Yhan, *J. Feng. Mater. Scien. Forum*, 67–68; (Halide Glasses VI) 239–243 (1991).
74. H. Hu, and F. Lin, *J. Feng. Guisuanyan Xuebau* **18**(6), 501–505 (1990).
75. V. D. Khaliev and co-workers, *Fiz. Khim. Stekla* **17**(5), 740–743 (1991).
76. A. G. Nimchik and Kh. L. Usmanov, *N.A. Sirazhiddinov Uzb, Khim. Zh.* (6), 68–70 (1991).
77. Jpn. Pat. 03291324 A2 (Dec. 20, 1991), K. Masame and T. Matsuo (to Heisei).
78. FR 2661425 A1 (Oct. 31, 1991), Y. Bertaud and co-workers.
79. P. Comba, K. P. V. Lei, and T. G. Carnahan, *Bur. Mines Rep. Invest.*, RI **9372** 7 pp. (1991).
80. F1 83664 B (Apr. 30, 1991), T. Helttula and O. Jokinen.
81. Jpn. Pat. 03023251 A2 (Jan. 31, 1991), K. Shibata.
82. WO 9003265 A1 (Apr. 5, 1990), J. H. Weaver and co-workers; PCT Int. Appl. (1990), J. H. Weaver and co-workers.
83. Jpn. Pat. 63215553 A2 (Sept. 8, 1988), T. Honda, T. Yamada, K. Onigata, and S. Tosaka (to Showa).
84. Jpn. Pat. 63215547 A2 (Sept. 8, 1988) T. Honda, T. Yamada, K. Omigata, and S. Tosaka (to Showa); Jpn. Pat. 63215555 A2 (Sept. 8, 1988) (to Showa); Jpn. Pat. 63215557 A2 (Sept. 8, 1988) (to Showa); Jpn. Pat. 63215554 A2 (Sept. 8, 1988) (to Showa).
85. Belg. Pat. 905858 A1 (Apr. 1, 1987), M. Palchetti, S. Palella, and A. Crisafull.
86. U.S. Pat. 4,591,382 A (May 27, 1986), G. R. B. Elliott.
87. Czech. Pat. 221212 B (Jan. 15, 1986), J. Kotrbaty, J. Bar, J. Gocal, and T. Pazdiora.
88. I. Vulkova and co-workers, *Stroit. Mater. Silik, Prom-St.* **24**(12), 18–20 (1983).
89. Jpn. Pat. 53091932 (Aug. 12, 1978), H. Kitagawa (to Showa).
90. U.S. Pat. 3,832,451 (Aug. 27, 1974), E. F. Abrams and R. G. Shaver.

TARIQ MAHMOOD
CHARLES B. LINDAHL
Elf Atochem North America, Inc.

COBALT

Cobalt Difluoride

Cobalt difluoride [10026-17-2], CoF_2, is a pink solid having a magnetic moment of 4.266×10^{-23} J/T (4.6 Bohr magneton) (1) and closely resembling the ferrous (FeF_2) compounds. Physical properties are listed in Table 1. Cobalt(II) fluoride is highly stable. No decomposition or hydrolysis has been observed in samples stored in plastic containers for over three years.

CoF_2 is manufactured commercially by the action of aqueous or anhydrous hydrogen fluoride (see FLUORINE COMPOUNDS, INORGANIC–HYDROGEN) on cobalt carbonate (see COBALT COMPOUNDS) in a plastic, ie, polyethylene/polypropylene, Teflon, Kynar, rubber, or graphite-lined container to avoid metallic impurities. The partially hydrated mass is lavender pink in color. It is dried at 150–200°C and then pulverized to obtain the anhydrous salt. A very high (99.9%) purity CoF_2 having less than 0.05% moisture content has also been prepared by reaction of $CoCO_3$ and liquid hydrogen fluoride. This is a convenient synthetic route giving quantitative yields of the pure product. The reaction of $CoCl_2$ and anhydrous HF is no longer commercially practical because of environmental considerations. The various hydrates, eg, the cobalt(II) fluoride dihydrate [13455-27-1], $CoF_2 \cdot 2H_2O$, cobalt(II) fluoride trihydrate [13762-15-7], $CoF_2 \cdot 3H_2O$, and cobalt(II) fluoride tetrahydrate [13817-37-3], $CoF_2 \cdot 4H_2O$, have been obtained by the reaction of freshly prepared oxide, hydroxide, or carbonate of cobalt(II) and aqueous hydrogen fluoride (2).

Cobalt difluoride, used primarily for the manufacture of cobalt trifluoride, CoF_3, is available from Advance Research Chemicals, Inc., Aldrich Chemicals, and PCR in the United States, Fluorochem in the UK, and Schuhardt in Germany. The 1993 price varied from $60 to $200/kg depending on the quantity and the price of cobalt metal. CoF_2 is shipped as a corrosive and toxic material in DOT-approved containers.

Table 1. Physical Properties of the Cobalt Fluorides

Parameter	Cobalt difluoride[a]	Cobalt trifluoride
molecular weight	96.93	115.93
melting point, °C	1127	926
solubility, g/100 g[b]		
water	1.36	dec
anhydrous HF	0.036	
density, g/cm^3	4.43	3.88
ΔH_f, kJ/mol[c]	−672	−790
ΔG_f, kJ/mol[c]	−627	−719
S, J/(mol·K)[c]	82.4	95
C_p, J/(mol·K)[c]	68.9	92

[a]The bp of CoF_2 is 1739°C.
[b]CoF_2 is also soluble in mineral acids.
[c]To convert J to cal, divide by 4.184.

Cobalt Trifluoride

Cobalt(III) fluoride [10026-18-3] or cobalt trifluoride, CoF_3, is one of the most important fluorinating reagents. Physical properties may be found in Table 1. It is classified as a hard fluorinating reagent (3) and has been employed in a wide variety of organic and inorganic fluorination reactions. CoF_3, a light brown, very hygroscopic compound, is a powerful oxidizing agent and reacts violently with water evolving oxygen. It should be handled in a dry box or in a chemical hood and stored away from combustibles, moisture, and heat. The material should not be stored in plastic containers for more than two years. The crystals possess a hexagonal structure.

Cobalt trifluoride is readily prepared by reaction of fluorine (qv) and $CoCl_2$ at 250°C or CoF_2 at 150–180°C. Direct fluorination of CoF_2 leads to quantitative yields of 99.9% pure CoF_3 (4).

CoF_3 is used for the replacement of hydrogen with fluorine in halocarbons (5); for fluorination of xylylalkanes, used in vapor-phase soldering fluxes (6); formation of dibutyl decalins (7); fluorination of alkynes (8); synthesis of unsaturated or partially fluorinated compounds (9–11); and conversion of aromatic compounds to perfluorocyclic compounds (see FLUORINE COMPOUNDS, ORGANIC). CoF_3 rarely causes polymerization of hydrocarbons. CoF_3 is also used for the conversion of metal oxides to higher valency metal fluorides, eg, in the assay of uranium ore (12). It is also used in the manufacture of nitrogen fluoride, NF_3, from ammonia (13).

CoF_3 is available from Advance Research Chemicals, Inc., Aldrich Chemicals, Aesar, Johnson/Matthey, PCR, Pfaltz & Bauer, Noah Chemicals, and Strem Chemicals of the United States, Fluorochem of the UK, and Schuhardt of Germany. Demand for cobalt trifluoride varies from 100 to 1500 kg/yr and the 1993 price for smaller quantities ranged from $300 to $350/kg.

The ACGIH adopted TLV/TWA for 1992–1993 for fluorides as F^- is TWA 2.5 mg/m^3, and for cobalt as Co metal dust TWA 0.05 mg/m^3. Dust masks should be used while handling both the cobalt fluorides and all other cobalt compounds. CoF_3 is shipped as an oxidizer and a corrosive material.

BIBLIOGRAPHY

"Cobalt Compounds" under "Fluorine Compounds, Inorganic" in *ECT* 1st ed., Vol. 6, p. 693, by F. D. Loomis; "Cobalt" under "Fluorine Compounds, Inorganic" in *ECT* 2nd ed., Vol. 9, pp. 582–583, by W. E. White; in *ECT* 3rd ed., Vol. 10, pp. 717–718, by D. T. Meshri, Advance Research Chemicals Inc.

1. A. G. Sharp, *Quart. Rev. Chem. Soc.* **11**, 49 (1957).
2. I. G. Ryss, *The Chemistry of Fluorine and its Inorganic Compounds*, State Publishing House for Scientific and Chemical Literature, Moscow, 1956, Eng. Trans. ACE-Tr-3927, Vol. II, Office of Technical Services, U.S. Department of Commerce, Washington, D.C., 1960, pp. 659–665.
3. D. T. Meshri and W. E. White, "Fluorinating Reagents in Inorganic and Organic Chemistry," in the *Proceedings of the George H. Cady Symposium, Milwaukee, Wis.*, June 1970; M. Stacy and J. C. Tatlow, *Adv. Fluorine Chem.* **1**, 166 (1960).
4. E. A. Belmore, W. M. Ewalt, and B. H. Wojcik, *Ind. Eng. Chem.* **39**, 341 (1947).

5. R. D. Fowler and co-workers, *Ind. Eng. Chem.* **39**, 292 (1947).
6. Eur. Pat. Appl EP 281,784 (Sept. 14, 1988), W. Bailey and J. T. Lilack (to Air Products and Chemicals, Inc.).
7. U.S. Pat. 4,849,553 (July 18, 1989), W. T. Bailey, F. K. Schweighardt, and V. Ayala (to Air Products and Chemicals, Inc.).
8. Jpn. Kokai Tokkyo Koho JP 03 167,141 (July 19, 1991), H. Okajima and co-workers (to Kanto Denka Kaggo Co. Ltd.).
9. U.S. Pat. 2,670,387 (Feb. 23, 1954), H. B. Gottlich and J. D. Park (to E.I. du Pont de Nemours & Co., Inc.).
10. D. A. Rausch, R. A. Davis, and D. W. Osborn, *J. Org. Chem.* **28**, 494 (1963).
11. Ger. Pat. DD 287,478 (Feb. 28, 1991), W. Radeck and co-workers (to Akademie der Wissenschaften der DDR).
12. R. Hellman, Westinghouse Corp., Cincinnati, Ohio, private communication, Jan. 1989.
13. Jpn. Kokai Tokkyo Koho, JP 03,170,306 (July 23, 1991), S. Lizuka and co-workers (to Kanto Denka Kogyo Co. Ltd.).

DAYAL T. MESHRI
Advance Research Chemicals, Inc.

COPPER

Copper(II) Fluorides

Copper(II) forms several stable fluorides, eg, cupric fluoride [7789-19-7], CuF_2, copper(II) fluoride dihydrate [13454-88-1], $CuF_2 \cdot 2H_2O$, and copper hydroxyfluoride [13867-72-6], CuOHF, all of which are interconvertible. When CuF_2 is exposed to moisture, it readily forms the dihydrate, and when the latter is heated in the absence of HF, $CuOHF \cdot H_2O$ results. The colorless crystals of anhydrous CuF_2 are triclinic in structure and are moisture sensitive, turning blue when exposed to moist air. Physical properties of CuF_2 are listed in Table 1. CuF_2 reacts with ammonia to form $CuF_2 \cdot 5NH_3$.

Copper(I) fluoride is believed to be unstable (1) and no evidence for its existence has been found using mass spectrometry (2).

Manufacture. Several methods of synthesis for anhydrous CuF_2 have been reported, the most convenient and economical of which is the reaction of copper carbonate and anhydrous hydrogen fluoride to form the monohydrate, $CuF_2 \cdot H_2O$. Part of the water content from the monohydrate is removed by addition of excess HF. The excess HF is decanted and the remaining mass transferred to a Teflon-lined tray and dried under an atmosphere of hydrogen fluoride. The decanted material may also be dehydrated in a nickel or copper tray under an atmosphere of fluorine at 150–300°C. Both routes have successfully resulted in ultrapure (99.95%) white CuF_2 in good yields. The other method for the preparation of high purity anhydrous copper(II) fluoride is by the direct fluorination of commercially available CuOHF (3), or the action of a mixture of HF and BF_3 on $CuF_2 \cdot 2H_2O$ (4).

Uses. Copper(II) fluoride is used as a fluorinating reagent (5–7) in the fluorination of partially hydrogenated silanes; in superconductors (8–10); as a cathode material for high energy density primary and secondary batteries (qv)

Table 1. Physical Properties of CuF$_2$

Property	Value
molecular weight	101.54
melting point, °C	785 ± 10
boiling point, °C	1676
solubility, g/100 g	
water	4.75
anhydrous HF	0.01
aqueous 21.2% HF	12.1
density, g/cm^3	4.85
ΔH_f, kJ/mol[a]	−539
ΔG_f, kJ/mol[a]	−492
S, J/(mol·K)[a]	77.45
C_p, J/(mol·K)[a]	65.55

[a] To convert from J to cal, divide by 4.184.

(11–14); for the skeletal rearrangements of olefins (15); low temperature isomerization of pentane and hexane (16); as a selective herbicide (17); as a termite repellant (18); as a fungicide (19); in the manufacturing of conductive bicomponent fibers for electromagnetic shields (20); as a catalyst for the removal of nitrogen oxides from flue gases (21), and for the synthesis of heterocyclic tetraaromatics (22). The dihydrate is used in the casting of gray iron.

The high purity anhydrous copper(II) fluoride must be stored in a tightly closed or sealed container under an atmosphere of argon. The dihydrate may be stored in polyethylene-lined fiber drums. The ACGIH (1992–1993) adopted toxicity value for copper as Cu is 1 mg/m^3, and for fluorides a F$^-$, 2.5 mg/m^3.

In spite of the many applications for copper(II) fluoride, demand is restricted to 1 to 10 kg lots. It is available in the United States from Advance Research Chemicals, Aldrich Chemicals, Atomergic, Aesar, Johnson/Matthey, Cerac Corp., and PCR Corp. The 1993 price for the anhydrous copper(II) fluoride varied from $400 to $600/kg depending on the amount required. The dihydrate is available at $22/kg.

BIBLIOGRAPHY

"Copper Compounds" under "Fluorine Compounds, Inorganic" in *ECT* 1st ed., Vol. 6, p. 693, by F. D. Loomis; "Copper" under "Fluorine Compounds, Inorganic" in *ECT* 2nd ed., Vol. 9, pp. 583–584, by W. E. White; in *ECT* 3rd ed., Vol. 10, pp. 719–720, by D. T. Meshri, Advance Research Chemicals Inc.

1. I. G. Ryss, *Zh. Fiz. Khim.* **29**, 936 (1955).
2. R. K. Kent, J. D. McDonald, and J. Margrave, *J. Phys. Chem.* **70**, 874 (1966).
3. J. R. Lundquist, *Final Report Pacific Northwest Laboratories*, Seattle, Wash. NASA CR-72571, June 12, 1969; U.S. Pat. 3,607,015 (Sept. 21, 1971), J. R. Lundquist, R. Wash, and R. B. King (to NASA).
4. U.S. Pat. 2,782,099 (Feb. 19, 1957), D. A. McCaulay (to Standard Oil of Indiana).

5. Jpn. Kokai Tokkyo Koho, 02, 302,311 (Dec. 14, 1990), I. Harada, M. Aritsuka, and A. Yoshikawa (to Mitsui Tiatsu Chemicals).
6. B. Leng and J. H. Moss, *J. Flourine Chem.* **8**, 165 (1976).
7. J. H. Moss, R. Ottie, and J. B. Wilford, *J. Fluorine Chem.* **3** 317 (1973).
8. Jpn. Kokai Tokkyo Koho 01, 133,921 (Nov. 18, 1987), S. Aoki and co-workers (to Fujikura Ltd.).
9. Jpn. Kokai Tokkyo Koho, 63,313,426 (Dec. 21, 1988), Y. Tanaka, T. Shibata, and N. Uno (to Furukawa Electric Co. Ltd.).
10. Jpn. Kokai Tokkyo Koho, 63,288,943 (Nov. 25, 1988), T. Kyodo, S. Hirai, and K. Takahashi (to Sumitomo Electric Industries Ltd.).
11. Eur. Pat. 286,990 (Apr. 17, 1987), F. W. Dampier and R. M. Mank (to GTE Laboratories, Inc.).
12. Ger. Offen. 2,215,210 (Oct. 19, 1972), O. S. Savinovw (to Honeywell Inc.).
13. U.S. Pat. 3,953,232 (Apr. 27, 1976), W. L. Roth and G. C. Farrington (to General Electric Co.).
14. J. H. Kennedy and J. C. Hunter, *J. Electrochem. Soc.* **123**(1), 10 (1976).
15. U.S. Pat. 3,751,513 (Aug. 7, 1973), J. J. Tazuma (to Goodyear Tire & Rubber Co.).
16. Fr. Pat. 2,157,083 (July 6, 1973), T. Bernard (to Institut Francois du Petrole, des Carburnats et Lubricants).
17. I. G. Ryss, *The Chemistry of Fluorine and its Inorganic Compounds*, State Publishing House for Scientific and Chemical Literature, Moscow, 1956, Eng. Trans. ACE-Tr-3927, Vol. II, Office of Technical Services, U.S. Department of Commerce, Washington, D.C., 1960, p. 643.
18. G. N. Wolcott, *P. R. Agri. Exp. Stu. Bull.*, **73** (1947).
19. H. Martin, R. L. Wain, and E. H. Wilkinson, *Ann. Appl. Biol.* **29**, 412 (1942).
20. Jpn. Kokai Tokyo Koho, 01, 61,570 (Aug. 31, 1987), M. Oshida (to Tijin Ltd.).
21. Jpn. Kokai Tokkyo Koho, 63, 49,255 (Mar. 2, 1988), Y. Kawasaki (to Matsushita Electric Industrial Co. Ltd.).
22. T. Kaufmann and Z. R. Otter, *Angew. Chem.* **88**, 513 (1976).

DAYAL T. MESHRI
Advance Research Chemicals, Inc.

GERMANIUM

Germanium forms both a difluoride and a tetrafluoride. It also forms a stable hexafluorogermanate complex ion, GeF_6^{2-}, that is present in the aqueous acid and a number of salts.

Germanium Difluoride

Germanium difluoride [*13940-63-1*] is a white solid, mp 110°C, and d_{23} = 3.7 g/cm^3. This compound can be vacuum distilled. In a mass spectrometer, ions corresponding to $(GeF_2)_n^+$, where n = 1–4 have been observed at 361–403 K (1). At higher temperatures, GeF_2 disproportionates to GeF_4 (g), Ge (s), and GeF (g). Presumably the GeF [*39717-71-0*] formed is unstable and subsequently condenses and disproportionates. The initially reported (2) orange-red solid becomes red-brown going finally to the black of metallic Ge. The difluoride deliquesces (2) in moist air producing germanium(II) hydroxide (see GERMANIUM AND GERMANIUM COMPOUNDS). The difluoride is soluble in aqueous hydrofluoric acid and gives a solution having the reducing properties expected of divalent germanium. GeF_2 reacts with aqueous solutions of alkali metal fluorides to produce trifluorogermanites (3), eg, cesium trifluorogermanite [72121-41-6], $CsGeF_3$, and potassium trifluorogermanite [72121-42-7], $KGeF_3$. The GeF_3^- ion is oxidized to GeF either by oxygen in neutral solution or by its reduction of H^+ to hydrogen in hydrofluoric acid solution. Germanium difluoride is soluble in ethanol (2), forms a reversible complex with diethyl ether (2), and forms a dimethyl sulfoxide complex [72121-40-5], $GeF_2 \cdot OS(CH_3)_2$, which is decomposed at 240°C (3).

Germanium difluoride can be prepared by reduction (2,4) of GeF_4 by metallic germanium, by reaction (1) of stoichiometric amounts of Ge and HF in a sealed vessel at 225°C, by Ge powder and HgF_2 (5), and by GeS and PbF_2 (6). GeF_2 has been used in plasma chemical vapor deposition of amorphous film (see PLASMA TECHNOLOGY; THIN FILMS) (7).

Germanium Tetrafluoride

Germanium tetrafluoride [7783-58-6] is a gas having a garlic-like odor, a reported (8) triple point of −15°C and 404.1 kPa (4.0 atm), and a vapor pressure near 100 kPa (ca 1 atm) at −36.5°C. Germanium tetrafluoride fumes strongly in air and is hydrolyzed in solution to form GeF_6^{2-} ions. Germanium tetrafluoride can be prepared (8,9) by thermal decomposition of barium hexafluorogermanate [60897-63-4], $BaGeF_6$. Direct fluorination of germanium has been reported to give GeF_4 of higher purity (2). High purity GeF_4 is also manufactured by reaction of a finely powdered GeO_2 suspension in H_2SO_4 with UF_6 (10) or by the reaction of Ge metal or its oxide with F_2 or NF_3 (11). COF_2 has been used as a mild fluorinating agent to produce GeF_4 from GeO_2 (12). GeF_4 is used in ion implantation (qv) in semiconductor chips (see SEMICONDUCTORS) (13,14). Germanium tetrafluoride acts as a Lewis acid (15,16) to form complexes with many donor molecules. The tetrafluoride is commercially available.

Fluorogermanates

Fluorogermanic acid [16950-43-9] solutions, H_2GeF_6, are prepared by reaction of germanium dioxide and hydrofluoric acid or by hydrolysis of germanium tetrafluoride. Addition of potassium fluoride, barium chloride (9), or other salts results in hexafluorogermanates such as potassium hexafluorogermanate [7783-73-5], K_2GeF_6, or $BaGeF_6$, both of which are stable at temperatures up to 500°C where $BaGeF_6$ starts to decompose to GeF_4 and BaF_2.

Germanium tetrafluoride produces hydrogen fluoride in aqueous acidic solutions. Hydrogen fluoride is toxic and very corrosive. The OSHA permissible exposure limit (17) and the American Conference of Governmental Industrial Hygienists' (ACGIH) TLV for fluoride is 2.5 mg/m³ of air (18).

BIBLIOGRAPHY

"Germanium" under "Fluorine Compounds, Inorganic," in *ECT* 2nd ed., Vol. 9, pp. 584–585, by W. E. White, Ozark-Mahoning Co.; in *ECT* 3rd ed., Vol. 10, pp. 720–721, by C. B. Lindahl, Elf Atochem North America, Inc.

1. K. F. Zmbov, J. W. Hastie, R. Hauge, and J. L. Margrave, *Inorg. Chem.* **7**, 608 (1968).
2. N. Barlett and K. C. Yu. *Can. J. Chem.* **39**, 80 (1961).
3. E. L. Muetterties, *Inorg. Chem.* **1**, 342 (1962).
4. Jpn. Kokai Tokyo Koho 61111520 A2, (May 29, 1986), S. Ishihara, M. Hirooka, and S. Oono (to Showa).
5. P. Rivere, A. Castel, J. Stage, and C. Abdenhadheric, *Organometallics* **10**(5), 1227–1228 (1991).
6. S. M. Van der Kerk, *Polyhedron* **2**(6), 509–512 (1983).
7. EP 229707 A1 (July 22, 1987), S. Ishihara, M. Hirooka, J. Hanna, and I. Shimizu.
8. L. M. Dennis and A. W. Laubengayer, *Z. Phys. Chem.* **130**, 420 (1927).
9. C. J. Hoffman and J. S. Gutowsky, *Inorg. Syn.* **4**, 147 (1953).
10. Ger. Offen. DE 3841212 A1 (June 13, 1990), R. Doetzer.
11. EP 89-104364 (Mar. 11, 1989), I. Harada, Y. Yoda, N. Iwanaga, T. Nishitsuti, and A. Kikkawa.
12. S. P. Mallela, O. D. Gupta, and J. M. Shreeve, *Inorg. Chem.* **27**(1), 208–209 (1988).
13. A. Gottdang and co-workers, *Nucl. Instrum. Methods Phys. Res., Sect. B* **B55**(1–9), 310–313 (1991).
14. A. Ferreiro, J. DePontcharra, C. Jaussaud, and E. Lora-Tamayo, *Vacuum* **39**(7–8), 775–779 (1989).
15. E. L. Muetterties, *J. Am. Chem. Soc.* **82**, 1082 (1960).
16. R. C. Aggarwal and M. Onyszchuk, *J. Inorg. Nucl. Chem.* **30**, 3351 (1968).
17. *Code of Federal Regulation*, Title 29, Part 1910.1000, Washington, D.C., 1993.
18. *Threshold Limit Values for Chemical Substances and Physical Agents, 1992–1993*, The American Conference of Governmental Industrial Hygienists, Cincinnati, Ohio.

TARIQ MAHMOOD
CHARLES B. LINDAHL
Elf Atochem North America, Inc.

HALOGENS

The halogen fluorides are binary compounds of bromine, chlorine, and iodine with fluorine. Of the eight known compounds, only bromine trifluoride, chlorine trifluoride, and iodine pentafluoride have been of commercial importance. Properties and applications have been reviewed (1–7) as have the reactions with organic compounds (8). Reviews covering the methods of preparation, properties, and analytical chemistry of the halogen fluorides are also available (9).

The halogen fluorides are best prepared by the reaction of fluorine with the corresponding halogen. These compounds are powerful oxidizing agents; chlorine trifluoride approaches the reactivity of fluorine. In descending order of reactivity the halogen fluorides are chlorine pentafluoride [*13637-63-3*], ClF_5; chlorine trifluoride [*7790-91-2*], ClF_3; bromine pentafluoride [*7789-30-2*], BrF_5; iodine heptafluoride [*16921-96-3*], IF_7; chlorine monofluoride [*7790-91-2*], ClF; bromine trifluoride [*7787-71-5*], BrF_3; iodine pentafluoride [*7783-66-6*], IF_5; and bromine monofluoride [*13863-59-7*], BrF.

The halogen fluorides offer an advantage over fluorine in that the former can be stored as liquids in steel containers and, unlike fluorine, high pressure is not required. Bromine trifluoride is used as an oxidizing agent in cutting tools used in deep oil-well drilling, whereas chlorine trifluoride is used to convert uranium to UF_6 in nuclear fuel processing (see NUCLEAR REACTORS; PETROLEUM).

Except for iodine pentafluoride, the halogen fluorides have no commercial importance as fluorinating agents. Their extreme reactivity and the accompanying energy release of the reaction can be sufficient to disrupt C—C bonds and can result in explosive reactions or fires. In addition, both halogens are generally then introduced into organic compounds, giving rise to a complex mixture of products.

Physical Properties

The physical properties of the halogen fluorides are given in Table 1. Calculated thermodynamic properties can be found in Reference 24.

Bromine Monofluoride. Bromine monofluoride is red to red-brown (4) and is unstable, disproportionating rapidly into bromine and higher fluorides. Therefore, the measurement of its physical properties is difficult and the values reported in Table 1 are only approximate. The uv-absorption spectrum is available (25).

Bromine Trifluoride. Bromine trifluoride is a colorless liquid. The commercial grade is usually amber to red because of slight bromine contamination. The molecule has a distorted T structure (26). Infrared spectral data (26–30), the uv-absorption spectrum (31), and vapor pressure data (32) may be found in the literature.

Bromine Pentafluoride. Bromine pentafluoride is a colorless liquid having the molecular structure of a tetragonal pyramid (5). The index of refraction n_D is 1.3529 (33). Infrared spectra (13,34), the uv-absorption spectrum (35), and vapor pressure data (11) are all available.

Chlorine Monofluoride. Chlorine monofluoride is a colorless gas that condenses to a liquid with a slight yellow cast and freezes to a white solid. The infrared spectrum of gaseous chlorine monofluoride and the Raman spectrum of the liquid have been studied (36). The uv-absorption spectrum (37) and vapor pressure data are also available (11).

An equilibrium exists between chlorine trifluoride, chlorine monofluoride, and fluorine gas (38). The equilibrium constant may be expressed as

$$K_p = \frac{(P_{ClF})(P_{F_2})}{(P_{ClF_3})}$$

where P_X is the partial pressure of substance X. Values for K_p are

Temperature, °C	Pa (mm Hg)
250	30 (0.22)
300	240 (1.8)
350	1450 (10.9)

Table 1. Physical Properties of the Halogen Fluorides[a]

Property	BrF	BrF$_3$	BrF$_5$	ClF	ClF$_3$	ClF$_5$	IF$_5$	IF$_7$
boiling point, °C	20	125.7	40.9	−100.1	11.75	−13.1	102	5.5
melting point, °C	−33	8.8	−60.6	−155.6	−76.3	−103	8.5	4.5
liquid density at 25°C, g/mL		2.803	2.463[b]	1.620[b]	1.825[b]	1.790[c]	3.252	2.669
critical temperature, °C					154.5	142.6		
$-\Delta H_f(g)$ at 25°C, kJ/mol[d]	58.5	255.4	443.9	56.4	164.5	254.6	839.3	961.0
$-\Delta G_f(g)$ at 25°C, kJ/mol[d]	73.6	229.1	351.5	57.7	124.4	163.0	771.6	841.4
heat of vaporization, kJ/mol[d]		42.8	30.6	20.1	27.50	22.21	35.92	24.7
E_{diss}, kJ/mol[d]	254			253	105			122
	260				160			
heat of fusion, kJ/mol[d]		12.01	5.66		7.60		11.21	
specific heat, gas, J/(mol·K)[d]	32.9	66.5[e]		32.0	65.2		99.1	136.3
specific conductivity, liquid, at 25°C, W·cm		8.0×10^{-3}	9.1×10^{-8}	1.9×10^{-7f}	4.9×10^{-9}	1.25×10^{-9g}	5.4×10^{-6}	10^{-9}

[a]Compiled from References 8, 10–23.
[b]At boiling point.
[c]At 20°C.
[d]To convert J to cal, divide by 4.184.
[e]The specific heat of the liquid is 124.5 J/(mol·K) (29.8 cal/(mol·K)).
[f]At 145 K.
[g]At 256 K.

Chlorine Trifluoride. Chlorine trifluoride is a pale yellow liquid or a colorless gas. It freezes to a white solid and undergoes a transition in the solid state at −82.66°C (11). The infrared and Raman spectra have been studied (28,29,39,40) as has the uv absorption spectrum (41). Vapor pressure data are given in Reference 42. The viscosity of the liquid is 0.448 mPa·s(=cP) at 290 K, and the surface tension is 26.6 mN/m (=dyn/cm) at 273 K (43). The density of the solid is 2.530 g/cm^3 at 153 K (44). The vapors of chlorine trifluoride are nonideal and this has been attributed to the following equilibrium:

$$2\ ClF_3 \rightleftharpoons Cl_2F_6$$

The equilibrium constant at 24.2°C is 2.84 kPa (21.3 mm Hg) (11):

$$K_p = \frac{P_{Cl_2F_6}}{(P_{ClF_3})^2}$$

Chlorine Pentafluoride. Chlorine pentafluoride is a colorless gas at room temperature. The ir and Raman spectra of the liquid and gas phase have been studied (34,39). The uv absorption spectrum (45) and vapor pressure data may be found in the literature (18).

Iodine Pentafluoride. Iodine pentafluoride is a straw-colored liquid; the ir and Raman spectra of the gas phase have been studied (19,46,47); vapor pressure data are given in References 14 and 48.

Iodine Heptafluoride. Iodine heptafluoride is a colorless liquid; the ir and Raman spectra of the gas have been studied (47,49); vapor pressure data are available (19).

Chemical Properties

Reactions With Metals. All metals react to some extent with the halogen fluorides, although several react only superficially to form an adherent fluoride film of low permeability that serves as protection against further reaction. This protective capacity is lost at elevated temperatures, however. Hence, each metal has a temperature above which it continues to react. Mild steel reacts rapidly above 250°C. Copper and nickel lose the ability to resist reaction above 400 and 750°C, respectively.

Metals that form no protective fluoride film react readily with the halogen fluorides. Chlorine trifluoride reacts with Hg, As, Ca, Ti, Co, Pt, and Pb at elevated temperatures to give HgF_2, AsF_5, CaF_2, TiF_3, CoF_3, PtF_5, and PbF_{3-4}, respectively (50). Titanium alloys and molybdenum alloys and niobium metal react vigorously with ClF_3 (51). Molybdenum and tungsten react with BrF_3 to form the volatile MoF_6 and WF_6 (52,53). Chlorine trifluoride (54) converts Nb and Ta to pentafluorides, and Mo and Re to hexafluorides.

Uranium is converted by ClF_3, BrF_3, and BrF_5 to UF_6. The recovery of uranium from irradiated fuels has been the subject of numerous and extensive investigations sponsored by atomic energy agencies in a number of countries

(55–63). The fluorides of the nuclear fission products are nonvolatile; hence the volatile UF_6 can be removed by distillation (see NUCLEAR REACTORS; URANIUM AND URANIUM COMPOUNDS).

The rapid reaction of ClF_3 and BrF_3 with metals is the basis of the commercial use in cutting pipe in deep oil wells (64–68). In this application, the pipe is cut by the high temperature reaction of the halogen fluoride and the metal.

Reactions With Nonmetals. Few elements withstand the action of interhalogen compounds at elevated temperatures and many react violently at or below ambient temperatures. The oxidation of the element proceeds to its highest valence state, whereas the halogen other than fluorine is reduced either to the element or a lower valent interhalogen derivative. The oxidizing capacity of the interhalogens varies substantially from compound to compound. For example, chlorine trifluoride reacts vigorously with virtually every element at room temperature; on the other hand, iodine pentafluoride has a much milder oxidizing power. Thus chlorine trifluoride oxidizes xenon to xenon fluorides whereas iodine pentafluoride does not react (11). Furthermore, all stable halogen fluorides, except iodine pentafluoride, oxidize radon between -195 and $25°C$ (69) (see HELIUM-GROUP GASES, COMPOUNDS).

In general, reactions of halogens and halogen fluorides yield mixtures (4). Bromine pentafluoride reacts with iodine at ambient temperatures and with chlorine at 250–300°C, giving mixtures of interhalogen compounds. Bromine pentafluoride has been stored with bromine in steel cylinders at room temperature for extended periods of time without appreciable reaction. At elevated temperatures BrF_3 is formed. Bromine and chlorine trifluoride give bromine trifluoride (60–80% yields) and chlorine (70). Iodine and chlorine trifluoride produce IF_5 and ICl (70). Chlorine reacts with chlorine trifluoride to produce chlorine monofluoride. Bromine and iodine react with chlorine monofluoride to produce BrF_3 and IF_5, respectively. Bromine reacts with IF_5 on warming to give IBr and BrF_3 (4). Chlorine reacts with IF_7 to give ClF and interhalogens of iodine and chlorine (4).

Halogen fluorides react with sulfur, selenium, tellurium, phosphorus, silicon, and boron at room temperature to form the corresponding fluorides. Slight warming may be needed to initiate the reactions (4) which, once started, proceed rapidly to completion accompanied by heat and light. The lack of protective film formation allows complete reaction.

Reactions With Inorganic Compounds. In an investigation of the reactions of BrF_3 with oxides (71–73), little or no reaction was found with the oxides of Be, Mg, Ce, Ca, Fe, Zn, Zr, Cd, Sn, Hg, Th, and the rare earths, whereas the oxides of Mo and Re formed stable oxyfluorides. Manganese dioxide reacted incompletely but $KMnO_4$ released oxygen quantitatively. Complete replacement of oxygen took place with oxides of B, Ti, V, Cr, Cu, Ge, As, Se, Nb, Sb, Te, I, Ta, W, Tl, Pb, Bi, and U at 75°C.

Oxygen was partially replaced when P_2O_5, V_2O_5, and CrO_3 were dissolved in IF_5 to form POF_3, VOF_3, and CrO_2F_2. With WO_3 and MoO_3, $WO_3 \cdot IF_5$ and $2 MoO_3 \cdot 3IF_5$ complexes were formed (74). Reaction of excess IF_5 with $KMnO_4$ gives MnO_3F, IOF_3 [19058-78-7], and IO_2F [28633-62-7] (74).

Water reacts violently with all halogen fluorides. The hydrolysis process can be moderate by cooling or dilution. In addition to HF, the products may include oxygen, free halogens (except for fluorine), and oxyhalogen acids.

Fused silica and Pyrex glass (qv) are not significantly attacked by halogen fluorides up to 100°C if HF is absent.

Salts of halides other than fluorides react with halogen fluorides to produce the corresponding metal fluoride and release the free higher halogen. Filter paper moistened with KI solution darkens readily in the presence of ClF_3 and the bromine fluorides. This serves as a sensitive detector for leaks in equipment containing these halogen fluorides. If a metal exhibits more than one valence, reactions of halogen fluorides with halides (including fluorides) yields the fluoride in which the metal is usually at its highest valence. Chlorine trifluoride converts silver salts to AgF_2, cobalt compounds to CoF_3, and so on. Such reactions are useful in the preparation of fluorinated organic materials and also regenerate fixed-bed fluorinating agents without using fluorine itself (75). A comparison of the efficacy with which various halogen fluorides convert cobalt(II) chloride to cobalt(III) fluoride is indicated below (76).

fluorinating agent	ClF_3	BrF_5	BrF_3	IF_5
solid product	100% CoF_3	55% CoF_3	45% CoF_3	
		45% CoF_2	55% CoF_2	72% CoF_2

Reactions With Organic Compounds. Most organic compounds react vigorously exhibiting incandescence or even explosively with ClF_3 and BrF_3 (8,77,78). For this reason, only the less reactive iodine pentafluoride is used as a fluorinating agent to any extent. The reaction of iodine pentafluoride and various organic compounds is described in the literature (79–84).

Inert diluents in which the halogen fluorides are soluble, such as carbon tetrachloride and methylene dichloride (85), have been used for control in liquid-phase reactions. Anhydrous hydrogen fluoride is a good diluent, because it does not react with halogen fluorides but rather is miscible in all proportions with them (86). Control of vapor-phase reactions may be improved by diluting the halogen fluoride with an inert gas such as nitrogen or argon. However, any reaction of the halogen fluorides with an organic compound in either the gas or liquid phase should be approached with extreme caution.

Bromine trifluoride in bromine solution reacts smoothly with bromofluoroethanes to give a clean, progressive substitution of the bromine by fluorine with no replacement of the hydrogen (87). The relative ease of replacement of bromine in various groups is $CBr_3 > CBr_2F > CHBr_2 > CF_2Br > CHBrF > CH_2Br$.

The reactions with IF_5 are more amenable to control giving good yields of identifiable products and lower losses from oxidative fragmentation. The reaction of IF_5 and iodine with tetrafluoroethylene produces the telomer perfluoroethyl iodide [354-64-3] in yields that exceed 98% based on $CF_2=CF_2$ using SbF_5 as a catalyst (88).

$$2\ I_2 + IF_5 + 5\ CF_2=CF_2 \rightarrow 5\ CF_3CF_2I$$

Aryl and alkyl isothiocyanates are converted in good yields by IF_5 in pyridine to thiobis(N-trifluoromethylamines) (89,90):

$$RN{=}C{=}S + IF_5 \rightarrow RN\underset{\underset{CF_3}{|}}{\overset{\overset{CF_3}{|}}{S}}NR$$

Fluorination of aromatic isothiocyanates occurs much more readily than that of alkyl isothiocyanates. Alcohols treated with IF_5 in DMF give 30–70% yields of their respective formates.

Iodine pentafluoride fluorinates CCl_4 at room temperature to give $CClF_3$ and traces of CCl_2F_2 (91). It reacts with CHI_3 to yield CHF_3 and $CHIF_2$ (92) and with CI_4 to form C_2F_2 (92) and CIF_3 (93). With CBr_4 at 90°C, IF_5 forms 83% CBr_2F_2 and minor amounts of CBr_3F and $CBrF_3$.

Liquid Halogen Fluorides as Reaction Media. Bromine trifluoride and iodine pentafluoride are highly dimerized and behave as ionizing solvents:

$$(BrF_3)_2 \rightleftharpoons BrF_2^+ + BrF_4^-$$

$$(IF_5)_2 \rightleftharpoons IF_4^+ + IF_6^-$$

Antimony pentafluoride dissolves in each to form $BrF_2^+SbF_6^-$ and $IF_4^+SbF_6^-$ which act as acids. Potassium fluoride likewise forms $KBrF_4$ [15705-87-0] and KIF_6 [20916-97-6] which are both stable, white, crystalline solids (3,94,95). These compounds dissociate at 200°C to KF and the corresponding halogen fluoride. Other salts are formed similarly (71,95–99). Some of the acids and bases of these systems are listed in Table 2.

Table 2. Acids and Bases Derived from Halogen Fluorides

Name	CAS Registry Number	Formula	Reference
Acid			
difluorobromine hexafluoroantimonate	[19379-47-6]	BrF_2SbF_6	73
bis(difluorobromine) hexafluorostannate	[72229-86-8]	$(BrF_2)_2SnF_6$	73
difluorobromine hexafluoroniobate	[72229-87-9]	BrF_2NbF_6	100
difluorobromine hexafluorotantalate	[35967-87-4]	BrF_2TaF_6	100
difluorobromine hexafluorobismuthate	[36608-81-8]	BrF_2BiF_6	100
tetrafluoroiodine hexafluoroantimonate	[41646-48-4]	IF_4SbF_6	101
Base			
potassium hexafluorobromate	[32312-22-4]	$KBrF_6$	95
silver tetrafluorobromate	[35967-89-6]	$AgBrF_4$	95
barium tetrafluorobromate	[35967-90-9]	$Ba(BrF_4)_2$	95
potassium hexafluoroiodate	[20916-97-6]	KIF_6	94

The use of ClF_3 and BrF_3 as ionizing solvents has been studied (102,103). At 100°C and elevated pressures, significant yields of $KClF_4$ [19195-69-8], $CsClF_4$ [15321-04-7], $RbClF_4$ [15321-10-5], $KBrF_6$ [32312-22-4], $RbBrF_6$ [32312-22-4], and $CsBrF_6$ [26222-92-4] were obtained. Chlorine trifluoride showed no reaction with lithium fluoride or sodium fluoride.

Manufacture

Bromine Trifluoride. Bromine trifluoride is produced commercially by the reaction of fluorine with bromine in a continuous gas-phase process where the ratio of fluorine to bromine is maintained close to 3:1. It is also produced in a liquid-phase batch reaction where fluorine is added to liquid bromine at a temperature below the boiling point of bromine trifluoride.

Chlorine Trifluoride. Chlorine trifluoride is produced commercially by the continuous gas-phase reaction of fluorine and chlorine in a nickel reactor at ca 290°C. The ratio of fluorine to chlorine is maintained slightly in excess of 3:1 to promote conversion of the chlorine monofluoride to chlorine trifluoride. Sufficient time in the reactor must be provided to maintain high conversions to chlorine trifluoride. Temperature control is also critical because the equilibrium shift of chlorine trifluoride to chlorine monofluoride and fluorine is significant at elevated temperatures.

Iodine Pentafluoride. Iodine pentafluoride is produced by the reaction of iodine and fluorine. Because iodine has a high melting point, the reaction is either performed in a solvent or the reaction is maintained at a temperature where the iodine is liquid. In a continuous process using a solvent (104), ca 1% I_2 is dissolved in IF_5 and passed to a reactor where it is contacted with F_2 gas. The IF_5 is continuously discharged from the reactor where a small portion is taken off as product and the larger portion of the stream is recycled.

In another process (105), fluorine gas reacts under pressure with liquid I_2 held above its melting point (113°C) but below a temperature (150°C) that would result in the formation of significant amounts of IF_7. Fluorine is added continuously until all the iodine has been converted and yields of IF_5 in excess of 95% are reported. The reaction pressure is ca 300 kPa (3 atm) so that the IF_5 produced in the reaction is maintained as a liquid.

Economic Aspects

U.S. production of bromine trifluoride is several metric tons per year mostly used in oil-well cutting tools. Air Products and Chemicals, Inc. is the only U.S. producer. The 1992 price was ca $80/kg.

U.S. chlorine trifluoride production is several metric tons per year. Most of the product is used in nuclear fuel processing. A large production plant for chlorine trifluoride was operated in Germany during World War II with a reported capacity of 5 t/d (106,107). As of 1993, Air Products and Chemicals, Inc. was the only U.S. producer. The 1992 price was ca $100/kg.

United States production of iodine pentafluoride is several hundred metric tons per year. The two U.S. producers are Air Products and Chemicals, Inc. and AlliedSignal, Inc. The 1992 price was ca $50/kg.

Shipping, Specifications, and Analytical Methods

Bromine trifluoride is commercially available at a minimum purity of 98% (108). Free Br_2 is maintained at less than 2%. Other minor impurities are HF and BrF_5. Free Br_2 content estimates are based on color, with material containing less than 0.5% Br_2 having a straw color, and ca 2% Br_2 an amber-red color. Fluoride content can be obtained by controlled hydrolysis of a sample and standard analysis for fluorine content. Bromine trifluoride is too high boiling and reactive for gas chromatographic analysis. It is shipped as a liquid in steel cylinders in quantities of 91 kg or less. The cylinders are fitted with either a valve or plug to facilitate insertion of a dip tube. Bromine trifluoride is classified as an oxidizer and poison by DOT.

Chlorine trifluoride is commercially available at 99% minimum purity (108) and is shipped as a liquid under its own vapor pressure in steel cylinders in quantities of 82 kg per cylinder or less. Chlorine trifluoride is classified as an oxidizer and poison by DOT.

Iodine pentafluoride is commercially available at a minimum purity of 98% (108). Iodine heptafluoride is the principal impurity and maintained at less than 2%. Free I_2 and HF are minor impurities. Iodine pentafluoride is shipped as a liquid in steel cylinders in various quantities up to 1350 kg cylinders. It is classified as an oxidizer and poison by DOT.

Volatile impurities, eg, F_2, HF, ClF, and Cl_2, in halogen fluoride compounds are most easily determined by gas chromatography (109–111). The use of Ftoroplast adsorbents to determine certain volatile impurities to a detection limit of 0.01% has been described (112–114). Free halogen and halide concentrations can be determined by wet chemical analysis of hydrolyzed halogen fluoride compounds.

Handling

The halogen fluorides are highly reactive compounds and must be handled with extreme caution (115–120). The more reactive compounds, such as bromine trifluoride and chlorine trifluoride, are hypergolic oxidizers and react violently and sometimes explosively with many organic and inorganic materials at room temperature. At elevated temperatures, these cause immediate ignition of most organic substances and many metals.

Materials of Construction. Nickel, Monel, copper, mild steel, 304 stainless steel, and aluminum have been found to be suitable metals of construction for handling halogen fluorides (51). Silver solder is acceptable; lead solder is not recommended. Nickel and Monel are more suitable for elevated temperatures. Steel is not dependable above 150°C. Gaskets may be made of soft copper or calcium fluoride-impregnated polytetrafluoroethylene. Packing and gasketing should have smooth surfaces and the surfaces should be free from organic greases and embedded impurities, which may ignite in the presence of halogen fluorides.

Equipment should be carefully and completely degreased and passivated with low concentrations of fluorine or the gaseous halogen fluoride before use. Special care should be taken that valves are completely disassembled and each part carefully cleaned.

Disposal. Moderate amounts of chlorine trifluoride or other halogen fluorides may be destroyed by burning with a fuel such as natural gas, hydrogen, or propane. The resulting fumes may be vented to water or caustic scrubbers. Alternatively, they can be diluted with an inert gas and scrubbed in a caustic solution. Further information on disposal of halogen fluorides is available (115–118).

Toxicity

The time-weighted average (TWA) concentrations for 8-h exposure to bromine trifluoride, bromine pentafluoride, chlorine trifluoride, chlorine pentafluoride, and iodine pentafluoride have been established by ACGIH on a fluoride basis to be 2.5 mg/m^3. NIOSH reports (121) the following inhalation toxicity levels for chlorine trifluoride: LC$_{50}$ monkey, 230 ppm/h; LC$_{50}$ mouse, 178 ppm/h; for chlorine pentafluoride: LC$_{50}$ monkey, 173 ppm/h; mouse, 57 ppm/h.

No toxicity data have been reported on the other halogen fluorides, but all should be regarded as highly toxic and extremely irritating to all living tissue.

Uses

Chlorine trifluoride is utilized in the processing of nuclear fuels to convert uranium to gaseous uranium hexafluoride. Chlorine trifluoride has also been used as a low temperature etchant for single-crystalline silicon (122,123).

Bromine trifluoride and chlorine trifluoride are used in oil-well tubing cutters (65–68). Chemical cutter tools are commercially available for use in wells at any depth. The cutter consists of three tubular chambers with the top chamber carrying an explosive charge, the middle chamber containing the halogen fluoride, and the lower chamber containing a catalyst. At the extreme end is the cutter head which guides the halogen fluoride against the pipe to be cut. Arranged around the head is a row of evenly spaced orifices or nozzles. The cable used to lower the cutter into the hole also serves to carry an electric charge to set off the explosive. The force drives the chemical into the head where it jets out of the orifices under enormous pressure to impinge against the inner walls of the tube to be cut. The catalyst raises the temperature of the halogen fluoride to trigger a high speed reaction so that the tube is cut in a fraction of a second. The cut is clean and unflared. Tension on the pipe at the top of the well aids in completing the separation (see PETROLEUM).

Iodine pentafluoride is an easily storable liquid source of fluorine having little of the hazards associated with other fluorine sources. It is used as a selective fluorinating agent for organic compounds. For example, it adds iodine and fluorine to tetrafluoroethylene in a commercial process to produce a useful telomer (124).

BIBLIOGRAPHY

"Halogen Fluorides" under "Fluorine Compounds, Inorganic" in *ECT* 1st ed., Vol. 6, pp. 694–695, by H. S. Booth, Case Western Reserve University, and J. T. Pinkston, Harshaw Chemical Co.; in *ECT* 2nd ed., Vol. 9, pp. 585–598, by H. S. Halbedel, Harshaw Chemical Co.; "Halogens" in *ECT* 3rd ed., Vol. 10, pp. 722–733, by A. J. Woytek, Air Products and Chemicals, Inc.

1. H. S. Booth and J. T. Pinkston, *Chem. Rev.* **41**, 421 (1947).
2. H. J. Emeleus, in J. H. Simons, ed., *Fluorine Chemistry*, Vol. 2, Academic Press, Inc., New York, 1954, pp. 39–49.
3. A. G. Sharpe, *Q. Rev. Chem. Soc.* **4**, 115 (1950).
4. H. S. Booth and J. T. Pinkston, in J. H. Simons, ed., *Fluorine Chemistry*, Vol. 1, Academic Press, Inc., New York, 1950, pp. 189–224.
5. H. C. Clark, *Chem. Rev.* **58**, 869 (1958).
6. J. C. Bailer and co-workers, *Comprehensive Inorganic Chemistry*, Vol. 2, Pergamon Press, Compendium Publishers, Elmsford, N.Y., 1973, pp. 1054–1062.
7. F. A. Cotton and G. Wilkinson, *Advanced Inorganic Chemistry*, 5th ed., John Wiley & Sons, Inc., New York, 1988, pp. 572–574.
8. W. K. R. Musgrave, in M. Stacey, J. C. Tatlow, and A. G. Sharpe, eds., *Advances in Fluorine Chemistry*, Vol. 1, Academic Press, Inc., New York, 1960, pp. 1–28; L. S. Boguslavskaya and N. N. Chuvatkin, in L. German and S. Zemskov, eds., *New Fluorinating Agents in Organic Synthesis*, Springer-Verlag, Berlin, 1989, pp. 140–196.
9. K. R. Brower, *J. Org. Chem.* **52**, 798 (1987); N. S. Nikolaev and co-workers, *Khimiyn Galoidnykh*, Soldinerii Ftorn, Moscow, 1968 (U.S. translation NTIS no. *AD-702-974*).
10. J. A. Dean, ed., *Lange's Handbook of Chemistry*, 11th ed., McGraw-Hill Book Co., New York, 1973, p. 4:139.
11. L. Stein, in V. Gutmann, ed., *Halogen Chemistry*, Vol. 1, Academic Press, Inc., New York, 1967, p. 133.
12. D. Pilipovich and co-workers, *Inorg. Chem.* **6**, 1918 (1967).
13. *Mellor's Comprehensive Treatise on Inorganic and Theoretical Chemistry*, Suppl. 2, Part I, Longmans, Green and Co., London, 1956.
14. H. Selig, C. W. Williams, and G. J. Moody, *J. Phys. Chem.* **71**, 2739 (1967).
15. *National Bureau of Standards Technical Note 270-3*, U.S. Government Printing Office, Washington, D.C., 1968.
16. L. Stein, *J. Phys. Chem.* **66**, 288 (1962).
17. R. C. King and G. T. Armstrong, *J. Res. Natl. Bur. Stand.* **74A**, 769 (1970).
18. H. H. Rogers and co-workers, *J. Chem. Eng. Data* **13**, 307 (1968).
19. D. W. Osborne, F. Schreiner, and H. Selig, *J. Chem. Phys.* **54**, 3790 (1971).
20. O. Ruff and H. Krug, *Z. Anorg. Alleg. Chem.* **190**, 270 (1930).
21. R. C. King and G. T. Armstrong, *J. Res. Nat. Bur. Stand.* **74A**, 769 (1970).
22. A. A. Banks, H. J. Emeleus, and A. A. Woolf, *J. Chem. Soc.*, 2861 (1949).
23. M. T. Rogers, J. L. Speirs, and M. B. Panish, *J. Am. Chem. Soc.* **78**, 3288 (1956).
24. *JANAF Thermochemical Tables, NSRDA-NBS37*, 2nd ed., NBS, Washington D.C., 1971.
25. S. N. Buben and A. M. Chaikin, *Kinet. Katal.* **21**, 1591 (1980).
26. D. W. Magnuson, *J. Chem. Phys.* **27**(1), 233 (1957).
27. H. M. Haendler and co-workers, *J. Chem. Phys.* **22**, 1939 (1954).
28. H. Selig, H. H. Claassen, and J. H. Holloway, *J. Chem. Phys.* **52**, 3517 (1970).
29. R. A. Frez, R. L. Redington, and A. L. K. Aljiburty, *J. Chem. Phys.* **54**, 344 (1971).
30. Y. A. Rymarchuk and V. S. Ivanov, *Zh. Prikl. Spektrosk.* **22**, 950 (1975).
31. S. N. Buben and A. M. Chaikin, *Kinet. Katal.* **21**, 1591 (1980).

32. J. W. Grisard and G. D. Oliver, *The Vapor Pressure and Heat Vaporization of Bromine Trifluoride, K-25, Plant Report K-766*, U.C.C. Nuclear Co., Oak Ridge, Tenn., June 8, 1951.
33. R. D. Long, J. J. Martin, and R. C. Vogel, *Ind. Eng. Chem. Data Ser.* **3**, 28 (1958).
34. G. M. Begun, W. H. Fletcher, and D. F. Smith, *J. Chem. Phys.* **42**, 2236 (1965).
35. S. N. Buben and A. M. Chaikin, *Kinet. Katal.* **21**, 1591 (1980).
36. E. A. Jones, T. F. Parkinson, and T. G. Burke, *J. Chem. Phys.* **18**, 235 (1950).
37. Y. A. Rymarchuk and V. S. Ivanov, *Zh. Prikl. Spektrosk.* **22**, 950 (1975).
38. H. Schmitz and H. J. Schumacker, *Z. Naturforsch Teil 2*, 363 (1947).
39. E. A. Jones, T. F. Parkinson, and R. B. Murray, *J. Chem. Phys.* **17**, 501 (1949).
40. K. Schaefer and E. Wicke, *Z. Elektrochem.* **52**, 205 (1948).
41. Y. A. Rymarchuk and V. S. Ivanov, *Zh. Prikl. Spektrosk.* **22**, 950 (1975).
42. J. W. Grisard, H. A. Burnhardt, and G. D. Oliver, *J. Am. Chem. Soc.* **73**, 5725 (1951).
43. A. A. Banks, A. Davies, and A. J. Rudge, *J. Chem. Soc.*, 732 (1953).
44. R. D. Burbank and F. N. Bensey, *J. Chem. Phys.* **21**, 602 (1953).
45. Y. A. Rymarchuk and V. S. Ivanov, *Zh. Prikl. Spektrosk.* **22** (1975).
46. L. E. Alexander and I. R. Beattie, *J. Chem. Soc. A*, 3091 (1971).
47. H. H. Claassen, E. L. Gasner, and H. Selig, *J. Chem. Phys.* **49**, 1803 (1968).
48. C. J. Schack and co-workers, *J. Phys. Chem.* **72**, 4697 (1968).
49. R. K. Khanna, *J. Mol. Spectroscopy* **8**, 134 (1962).
50. W. Huckel, *Nachr. Akad. Wiss. Gottingen Math. Phys. Kl.*, 36 (1946).
51. J. C. Grigger and H. C. Miller, *Met. Protect.*, 33 (Sept. 1964).
52. B. Cox, D. W. A. Sharp, and A. G. Sharpe, *Proc. Chem. Soc.*, 1242 (1956).
53. N. S. Nikolaev and A. A. Opalovskii, *Zh. Neorgan. Khim.* **4**, 1174 (1959).
54. N. S. Nikolaev and E. G. Ippolitov, *Dokl. Akad. Nauk* **134**, 358 (1960).
55. H. A. Bernhardt, E. J. Barber, and R. A. Gustison, *Ind. Eng. Chem.* **51**, 179 (1959).
56. J. F. Ellis and L. H. Brooks, *U. K. Atomic Energy Authority* **8111-D** (1959).
57. R. A. Gustison and co-workers, in F. R. Bruce and co-eds., *Progress in Nuclear Chemistry, Series 3: Process Chemistry*, McGraw-Hill Book Co., Inc., New York, 1956, pp. 281–285.
58. U.S. Pat. 3,012,849 (Dec. 12, 1961), F. L. Horn (to U.S. Atomic Energy Commission).
59. W. J. Mechan and co-workers, *Chem. Eng. Prog.* **53**(2), 72-F-77F (1957).
60. L. Stein and R. Vogel, *Ind. Eng. Chem.* **48**, 418 (1956).
61. G. Strickland, F. L. Horn, and R. Johnson, *U.S. At. Energy Comm. BNL-471 (T-107)* (1957), *BNL-457 (1–21)* (1957).
62. U.S. Pat. 3,825,650 (July 23, 1974), R. A. Gustison and co-workers (to U.S. Atomic Energy Commission).
63. Ger. Pat. 2,209,628 (1971), P. Cousin and co-workers.
64. E. McGhee, *Oil Gas J.* **54**(14), 67 (1955).
65. Ger. Pat. 1,029,770 (May 14, 1958), W. G. Sweetman.
66. *Oil Forum*, 332 (Sept. 1955).
67. U.S. Pat. 4,315,797 (Feb. 16, 1982) J. M. Peppers (to Gearhart Industries, Inc.).
68. *Chemical Cutter Bulletin*, Pipe Recovery Systems, Houston, Tex., 1993.
69. L. Stein, *J. Am. Chem. Soc.* **91**, 5396 (1969).
70. R. Burnett and R. E. Banks, "Fluorine Chemistry," paper presented at *British Symposium*, Nov. 30, 1949.
71. H. J. Emeleus and A. A. Woolf, *J. Chem. Soc.*, 164 (1950).
72. H. R. Hoekstra and J. J. Katz, *Anal. Chem.* **25**, 1608 (1953).
73. A. A. Woolf and H. J. Emeleus, *J. Chem. Soc.*, 2865 (1949).
74. E. E. Aynsley, R. Nichols, and P. L. Robinson, *J. Chem. Soc.*, 623 (1953).
75. E. G. Rochow and I. Kukin, *J. Am. Chem. Soc.* **74**, 1615 (1952).

76. J. F. Gall and co-workers, "Interhalogen Compounds of Fluorine," papers presented at *Annual ACS Meeting*, New York, Sept. 1947.
77. S. Rosen and co-workers, *Accounts Chem. Res.* **21**, 307 (1988).
78. K. R. Brower, *J. Org. Chem.* **52**, 798 (1987).
79. H. J. Frohn and W. Pahlmann, *J. Fluorine Chem.* **24**, 219 (1984).
80. H. J. Frohn, *Chem. -Ztg.* **108**, 146 (1984).
81. H. J. Frohn and W. Puhlmann, *J. Fluorine Chem.* **26**, 243 (1984).
82. H. J. Frohn and W. Puhlmann, *J. Fluorine Chem.* **28**, 191 (1985).
83. H. J. Frohn and H. Maurer, *J. Fluorine Chem.* **34**, 73 (1986).
84. H. J. Frohn and H. Maurer, *J. Fluorine Chem.* **34**, 129 (1986).
85. U.S. Pat. 1,961,622 (June 5, 1934), H. S. Nutting and P. S. Petrie (to The Dow Chemical Co.).
86. U.S. Pat. 2,918,434 (Dec. 22, 1959), J. G. Gall and C. E. Inman (to Pennsalt Chemicals Corp.).
87. R. A. Davis and E. C. Larsen, *J. Org. Chem.* **32**, 3478 (1967).
88. U.S. Pat. 3,123,185 (May 5, 1964), R. E. Parsons (to E. I. du Pont de Nemours & Co., Inc.).
89. T. E. Stevens, *Tetrahedron Lett.*, (17), 16 (1959).
90. T. E. Stevens, *J. Org. Chem.* **26**, 3451 (1961).
91. O. Ruff and R. Keim, *Z. Anorg. Allg. Chem.* **201**, 245 (1931).
92. J. H. Simons, R. L. Bond, and R. E. McArthur, *J. Am. Chem. Soc.* **62**, 3477 (1940).
93. A. A. Banks and co-workers, *J. Chem. Soc.*, 2188 (1948).
94. H. J. Emeleus and A. G. Sharpe, *J. Chem. Soc.*, 2206 (1949).
95. A. G. Sharpe and H. J. Emeleus, *J. Chem. Soc.*, 2135 (1948).
96. A. A. Woolf and H. J. Emeleus, *J. Chem. Soc.*, 1050 (1950).
97. A. A. Woolf, *J. Chem. Soc.*, 1053 (1950).
98. A. G. Sharpe, *J. Chem. Soc.*, 2901 (1949).
99. A. G. Sharpe and A. A. Woolf, *J. Chem. Soc.*, 798 (1951).
100. V. Gutman and H. J. Emeleus, *J. Chem. Soc.*, 1046 (1950).
101. A. A. Woolf, *J. Chem. Soc.*, 3678 (1950).
102. U.S. Pat. 3,143,391 (Aug. 4, 1964), T. L. Hurley, R. O. MacLaren, and E. D. Whitney (to Olin Matheson Chemical Corp.).
103. E. D. Whitney and co-workers, *J. Am. Chem. Soc.* **86**, 2583 (1964).
104. U.S. Pat. 3,367,745 (Feb. 6, 1968), H. G. Tepp (to Allied Chemical Corp.).
105. U.S. Pat. 4,108,966 (Aug. 22, 1978), J. T. Lileck (to Air Products and Chemicals, Inc.).
106. *Chem. Ind.* **57**, 1084 (1945).
107. H. R. Neumark, *Trans. Electrochem. Soc.* **91**, 367 (1947).
108. *Inorganic Fluorine Compounds*, bulletin, Air Products and Chemicals, Inc., Allentown, Pa., 1974.
109. J. C. Million, C. W. Weber, and P. R. Kuehn, *Gas Chromatography of Some Corrosive Halogen-Containing Gases, Report No. K-1639*, Union Carbide Corp., Nuclear Division, New York, 1966.
110. J. F. Ellis and G. Iveson, *The Application of Gas-Liquid Chromatography to the Analysis of Volatile Halogen and Interhalogen Compounds, Gas Chromatography*, Butterworths, London, 1956 pp. 300–309.
111. U.S. Pat. 3,877,894 (Apr. 15 1975), L. G. Swope and E. A. Emory (to U.S. Atomic Energy Commission).
112. V. F. Sukhoverkhov and L. G. Podzolko, *Zh. Anal. Khim.* **38**, 715 (1983).
113. V. F. Sukhoverkhov and L. G. Podzolko, *Zh. Anal. Khim.* **45**, 1101 (1990).
114. V. F. Sukhoverkhov, L. G. Podzolko, and V. F. Garanin, *Zh. Anal. Khim.* **33**, 1360 (1978).

115. *Chlorine Trifluoride Handling Manual, AD266,121*, U.S. Department of Commerce, OTS, Washington, D.C., 1961.
116. R. L. Farrar, *Safe Handling of Chlorine Trifluoride*, report K-1416, U.C.C. Nuclear Company, Oak Ridge, Tenn., 1960.
117. J. M. Siegmund and co-workers, *NASA Document N62-14523, AD 281-818*, NASA, Washington, D.C., 1962.
118. "Chemical Rocket/Propellant Hazards," in *Liquid Propellant Handling, Storage and Transportation, CPLA Publication No. 194*, Vol. 3, CPLA, Silver Spring, Md., May 1970, Chapt. 8.
119. *Handling Hazardous Materials, NASA SP-5032*, NASA, Washington, D.C., 1965, Chapt. 4.
120. R. L. Farrar, Jr., and E. J. Barber, *Some Considerations in the Handling of Fluorine and the Chlorine Fluorides*, report K/ET-252, Oak Ridge Gaseous Diffusion Plant, Oak Ridge, Tenn., 1979.
121. *Registry of Toxic Effects of Chemical Substances*, 1985–1986 ed., U.S. Dept. of Health, Education, and Welfare, Washington, D.C., 1987.
122. Y. Saito, O. Yamaoka, and A. Yoshida, *J. Vac. Sci. Technol.* **B9**, 2503 (1991).
123. Y. Saito, M. Hirabaru, and A. Yoshida, *J. Vac. Sci. Technol.* **B10**, 175 (1992).
124. H. C. Fielding, in R. E. Banks ed., *Organofluorine Chemicals and their Industrial Applications*, Ellis Horwood Publishers, Chichester, UK, 1979.

General Reference

R. A. Rhein and M. H. Miles, *Bromine and Chlorine Fluorides: A Review*, Naval Weapons Center technical publication 6811, NWC, China Lake, Calif., 1988.

WEBB I. BAILEY
ANDREW J. WOYTEK
Air Products and Chemicals, Inc.

HYDROGEN

Hydrogen fluoride [7664-39-3], HF, is the most important manufactured fluorine compound. It is the largest in terms of volume, and serves as the raw material for most other fluorine-containing chemicals. It is available either in anhydrous form or as an aqueous solution (usually 70%). Anhydrous hydrogen fluoride is a colorless liquid or gas having a boiling point of 19.5°C. It is a corrosive, hazardous material, fuming strongly, which causes severe burns upon contact. Rigorous safety precautions are the standard throughout the industry, and in practice hydrogen fluoride can be handled quite safely.

Although it was known in the early nineteenth century, commercial use of hydrogen fluoride was limited. All early production was as aqueous solutions for uses such as glass etching, foundry scale removal, and production of chemicals such as sodium fluoride and sodium bifluoride. Some hydrogen fluoride was also produced for captive use in aluminum manufacture. Production of anhydrous hydrogen fluoride began in the early 1930s, but the demand at that time was limited to the small market for chlorofluorocarbons (see FLUORINE COMPOUNDS, ORGANIC).

World War II brought a revolution in the HF field. The need for high octane aviation fuels (see AVIATION AND OTHER GAS TURBINE FUELS), the birth of the nuclear industry requiring uranium hexafluoride (see NUCLEAR REACTORS; URANIUM AND URANIUM COMPOUNDS), and the rapid growth of the chlorofluorocarbon market all contributed to a steadily rising demand for hydrogen fluoride, especially in the anhydrous form. Whereas earlier anhydrous production had been solely via distillation of aqueous HF, technology emerged allowing direct production of anhydrous hydrogen fluoride.

Properties

Physical Properties. Physical properties of anhydrous hydrogen fluoride are summarized in Table 1. Figure 1 shows the vapor pressure and latent heat of vaporization. The specific gravity of the liquid decreases almost linearly from 1.1 at $-40°C$ to 0.84 at $80°C$ (4). The specific heat of anhydrous HF is shown in Figure 2 and the heat of solution in Figure 3.

Table 2 summarizes the properties of the hydrogen fluoride–water system. The freezing and boiling point curves of this system are shown in Figures 4 and 5, respectively. Figure 6 gives the partial pressures of HF and H_2O in aqueous HF solutions. The specific gravity of the solutions at various temperatures is shown in Figure 7. Specific conductivity of this system is given (27,28).

HF, wt %	Conductivity at 0°C, $(\Omega \cdot cm)^{-1}$
70	7.9×10^{-1}
80	7.1×10^{-1}
85	6.3×10^{-1}
90	4.9×10^{-1}
92.5	3.8×10^{-1}
95	2.5×10^{-1}
96	1.95×10^{-1}
97	1.04×10^{-1}
98	9.4×10^{-2}
99	5.6×10^{-2}
99.5	3.4×10^{-2}
99.75	1.8×10^{-2}
99.9	5.7×10^{-3}
99.95	2.8×10^{-3}
100	$<1.6 \times 10^{-6}$

Hydrogen fluoride is unique among the hydrogen halides in that it strongly associates to form polymers in both the liquid and gaseous states. At high temperatures or low partial pressures, HF gas exists as a monomer. At lower temperatures and higher partial pressures hydrogen bonding leads to the formation of chains of increasing length, and molecular weights of 80 and higher are observed. Electron diffraction study of the gas (29) has shown the hydrogen to fluorine distances to be about 0.10 nm and 0.155 nm, the F—H . . . F distance to be

Table 1. Properties of Anhydrous Hydrogen Fluoride

Property	Value	Reference
formula weight	20.006	
composition, wt %		
H	5.038	
F	94.96	
boiling point at 101.3 kPa,^{a} °C	19.54	1
critical pressure, MPa^a	6.48	2
critical temperature, °C	188.0	2
critical density, g/mL	0.29	2
critical compressibility factor	0.117	2
melting point, °C	−83.55	3
density, liquid, 25°C, g/mL	0.958	4
heat of vaporization, 101.3 kPa,^{a} kJ/mol^b	7.493	1
heat of fusion, −83.6°C, kJ/mol^b	3.931	5
heat capacity, constant pressure, liquid at 16°C, J/(mol·K)^b	50.6	5
heat of formation, ideal gas, 25°C, kJ/mol^b	−272.5	6
free energy of formation, ideal gas, 25°C, kJ/mol^b	−274.6	6
entropy, ideal gas, 25°C, J/(mol·K)^b	173.7	6
vapor pressure, 25°C, MPa^a	122.9	7
viscosity, liquid, 0°C, mPa·s(=cP)	0.256	8
surface tension, mN/m(=dyne/cm), 0°C	10.2	9
refractive index, liquid, 25°C, 589.3 nm	1.1574	10
molar refractivity, cm^3	2.13	10
dielectric constant, at 0°C	83.6	11
dipole moment, C·m^c	6.104×10^{-30}	12
thermal conductivity, at 25°C, J/(s·cm·°C)^b		
liquid	4.1×10^{-3}	13
vapor	2.1×10^{-4}	13
cryoscopic constant, K_f, mol/(kg·°C)	1.52	3
ebullioscopic constant, K_b, mol/(kg·°C)	1.9	14

^{a}To convert kPa to psi, multiply by 0.145.
^{b}To convert J to cal, divide by 4.184.
^{c}To convert C·m to debye, divide by 3.336×10^{-30}.

0.255 nm, and the polymer to have a linear zig-zag configuration. The angle H—F—H is reported to be about 120°. Monomeric HF has an H—F distance of 0.0917 nm. Cyclical polymers (possibly H_6F_6) also probably occur. In general, polymers of differing molecular weights are present in equilibrium at a given temperature and pressure, and an average molecular weight encompasses many different actual molecules. The apparent molecular weight of anhydrous HF vapor is shown in Figure 8.

This high degree of association results in highly nonideal physical properties. For example, heat effects resulting from vapor association may be significantly larger than the latent heat of vaporization (Fig. 9). Vapor heats of associ-

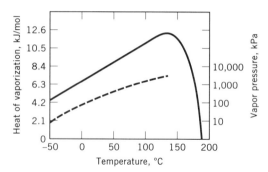

Fig. 1. (——) Latent heat of vaporization (1,7) and (— — —) vapor pressure (1,4,7,15) of anhydrous hydrogen fluoride. To convert kPa to psi, multiply by 0.145. To convert kJ to kcal, divide by 4.184.

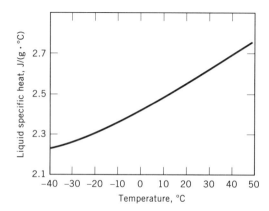

Fig. 2. Specific heat of liquid anhydrous hydrogen fluoride (5,16). To convert J to cal, divide by 4.184.

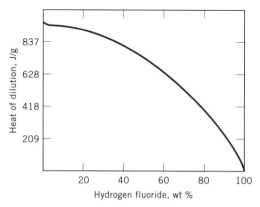

Fig. 3. Heat of solution per gram of anhydrous hydrogen fluoride in water when mixed to the final concentration shown in wt % of HF (16–18).

Table 2. Properties of 70% Aqueous Hydrogen Fluoride

Property	Value	Reference
boiling point at 101.3 kPa,[a] °C	66.4	19
freezing point, °C	−69	20
density, 0°C, g/mL	1.258	21
vapor pressure, 25°C, kPa[a]	20	22
viscosity, 25°C, mPa·s(=cP)	0.61	23
specific heat, 25°C, J/(g·°C)[b]	0.675	23

[a]To convert kPa to psi, multiply by 0.145.
[b]To convert J to cal, divide by 4.184.

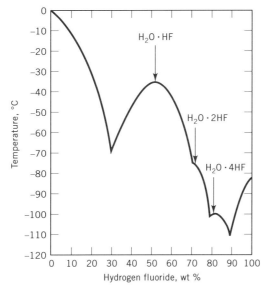

Fig. 4. Freezing point, HF–water system (24).

ation (ΔH_{assoc}) for HF to $(HF)_n$ per mole of $(HF)_n$ are as follows. To convert kJ to kcal, divide by 4.184.

n	$-\Delta H_{assoc}$, kJ
2	33.0
3	67.4
4	101.7
5	133.9
6	167.4
7	198.7
8	224.3

Chemical Properties. Hydrogen fluoride, characterized by its stability, has a dissociation energy of 560 kJ (134 kcal), which places HF among the most stable

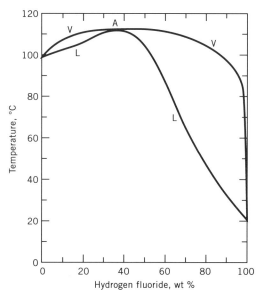

Fig. 5. Boiling point curve for the HF–water system, where A represents an azeotrope at 37.73 wt % HF, L is the liquid, and V the vapor (25,26).

diatomic molecules. Hydrogen fluoride is, however, highly reactive, and it has a special affinity for oxygen compounds, reacting with boric acid to form boron trifluoride and with sulfur trioxide and sulfuric acid to form fluorosulfonic acid. This last reaction demonstrates the dehydrating power of anhydrous hydrogen fluoride. HF belongs to the only class of compounds that readily react with silica and silicates, including glass (qv). With organic compounds, HF acts as a dehydrating agent, a fluorinating agent, a polymerizing agent, a catalyst for condensation reactions, and a hydrolysis catalyst. Hydrogen fluoride reacts with alcohols and unsaturated compounds to form fluorides and with alkylene oxides to give alkylene fluorohydrins.

The strong catalytic activity of anhydrous hydrogen fluoride results from the ability to donate a proton, as in the dimerization of isobutylene (see BUTYLENES):

$$CH_2=C(CH_3)_2 + HF \rightarrow (CH_3)_3C^+ + F^-$$

$$(CH_3)_3C^+ + CH_2=C(CH_3)_2 \rightarrow (CH_3)_3C-CH_2C^+(CH_3)_2 \rightarrow (CH_3)_3CCH=C(CH_3)_2 + H^+$$

Anhydrous hydrogen fluoride is an excellent solvent for ionic fluorides (Table 3). The soluble fluorides act as simple bases, becoming fully ionized and increasing the concentration of HF_2^-. For example,

$$HF + KF \rightarrow K^+ + HF_2^-$$

Because of the small size of the fluoride ion, F^- participates in coordination structures of high rank. Tantalum and niobium form stable hexafluorotantalate and hexafluoroniobate ions and hydrogen fluoride attacks these usually acid-

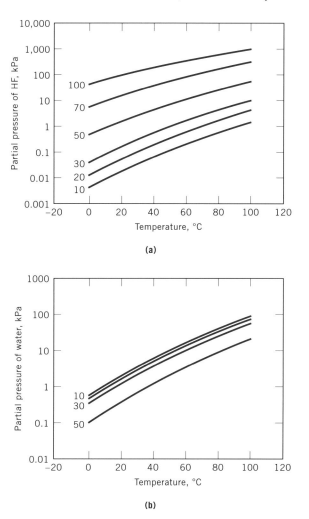

Fig. 6. Partial pressures over HF–water solutions where the numbers represent the quantity of HF in solution expressed as wt % (**a**) of HF and (**b**) of H$_2$O (22).

resistant metals. Hydrogen fluoride in water is a weak acid. Two dissociation constants are

$$K_1 = \frac{[H^+][F^-]}{[HF]} = 6.46 \times 10^{-4}\ M$$

$$K_2 = \frac{HF_2^-}{[HF][F^-]} = 5 \text{ to } 25\ M$$

Whereas hydrogen fluoride is a fairly weak acid as a solute, it is strongly acidic as a solvent. As the concentration of hydrogen fluoride increases in aqueous mixtures, the system becomes more acidic, with water acting as a very strong

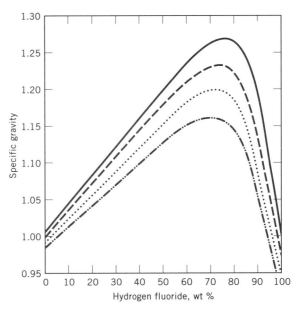

Fig. 7. Specific gravity of HF–water solutions where (——) represents 0°C, (– – –) 20°C, (····) 40°C, and (– ·· – ··) 60°C (21–23).

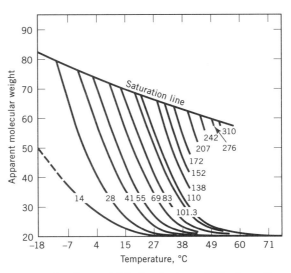

Fig. 8. Apparent vapor molecular weight of anhydrous HF where the numbers represent the partial pressure of HF in kPa. To convert kPa to psi, multiply by 0.145 (7,17,30–32).

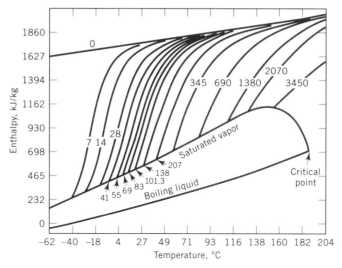

Fig. 9. Vapor-phase enthalpy of anhydrous HF where the numbers represent the partial pressure of HF in kPa (1,17,20,31,33). The critical point occurs at 188°C. To convert kPa to psi, multiply by 0.145. To convert kJ/kg to Btu/lb, multiply by 4.302×10^{-4}.

Table 3. Solubility of Metal Fluorides in Anhydrous Hydrogen Fluoride[a]

Fluoride	Temperature, °C	Solubility, g/100 g
LiF	12	10.3
NaF	11	30.1
KF	8	36.5
NH_4F	17	32.6
CaF_2	12	0.817
SrF_2	12	14.83
BaF_2	12	5.6
MgF_2	12	0.025
FeF_2	12	0.006
FeF_3	12	0.008
AlF_3	11	<0.002
SbF_5	25	miscible

[a]Ref. 8.

base. In dilute aqueous solution, an isolated hydrogen fluoride molecule donates a proton to an aggregate of water molecules and forms an aquated fluoride ion. When small amounts of water are present in the system, a proton is transferred to an isolated water molecule from polymeric hydrogen fluoride. The fluoride ion thus formed is part of a stable polymeric anionic complex. This difference in the solvation of the fluoride ion at the extremes of composition in the H_2O—HF system is probably the principal factor affecting the ease of proton transfer (14).

FLUORINE COMPOUNDS, INORGANIC (HYDROGEN)

For anhydrous hydrogen fluoride, the Hammett acidity function H_0 approaches -11. The high negative value of H_0 shows anhydrous hydrogen fluoride to be in the class of superacids. Addition of antimony pentafluoride to make a 3 M solution in anhydrous hydrogen fluoride raises the Hammett function to -15.2, nearly the strongest of all acids (34).

Manufacture

Raw Materials. Essentially all hydrogen fluoride manufactured worldwide is made from fluorspar and sulfuric acid, according to the reaction:

$$CaF_2(s) + H_2SO_4 \rightarrow CaSO_4(s) + 2\ HF(g)$$

Generally, yields on both fluorspar and sulfuric acid are greater than 90% in commercial plants.

Fluorspar. A typical acid-grade fluorspar analysis gives:

Component	Composition, wt %
calcium fluoride	97.0
silica	0.7
calcium carbonate	1.0
organic (as carbon)	0.1
sulfur	
total	0.02
sulfide	0.01
phosphorus pentoxide	0.02
chloride, total	0.02

Also present are 0.3 wt % mixed metal oxides (R_2O_3) and 5 ppm of arsenic. Impurities in fluorspar may affect yield, plant operability, or product quality.

Silica. Silica, which has the greatest impact on yield losses, reacts with HF and is discharged from the manufacturing process as H_2SiF_6. Yield losses can be calculated based on the chemical stoichiometry:

$$SiO_2 + 3\ H_2SO_4 + 3\ CaF_2 \rightarrow H_2SiF_6 + 2\ H_2O + 3\ CaSO_4$$

or

$$6\ HF + SiO_2 \rightarrow H_2SiF_6 + 2\ H_2O$$

Calcium fluoride loss is equal to 3.9% for each 1% silica; sulfuric acid loss is equal to 4.9% for each 1% silica.

Mixed-Metal Oxides. Generally, iron oxide is the principal component of mixed-metal oxides. These affect the sulfuric and oleum consumption in HF production.

$$R_2O_3 + 3\ H_2SO_4 \rightarrow R_2(SO_4)_3 + 3\ H_2O$$

Sulfuric acid loss is approximately 1.84% H_2SO_4 for each percentage of R_2O_3. Oleum consumption is increased to consume the water that is formed. The metal sulfates are more stable than metal fluorides under furnace conditions and are discharged from the process with the residue.

Calcium Carbonate. Calcium carbonate, like R_2O_3, affects sulfuric and oleum consumption in the HF process. Sulfuric acid loss is approximately 0.98% H_2SO_4 for each percentage of $CaCO_3$. The carbon dioxide evolved by the reaction increases the noncondensable gas flow, and because it carries HF, contributes to yield losses in the vent stream.

Magnesium Oxide. Magnesium oxide behaves in a similar manner to other metal oxides. However, most spars contain practically no magnesium oxide, so it does not affect yield loss or plant operation.

Organic Carbon. Organic materials interfere with plant operation because these compounds react with sulfuric acid under furnace conditions to form sulfur dioxide. There is a reducing atmosphere in the furnace which may reduce sulfur dioxide to elemental sulfur, which results in sulfur deposits in the gas handling system.

Total Sulfur and Sulfide Sulfur. Total sulfur is predominately in the form of metal sulfate, and because sulfates act as inerts, these materials have little impact on the process. Sulfide sulfur compounds, on the other hand, react and leave the furnace as a sulfur vapor, which may deposit in the gas handling system. A possible mechanism for this is the partial reaction of SO_2 to H_2S, followed by

$$2\ H_2S + SO_2 \rightarrow 3\ S + 2\ H_2O$$

Phosphorus Pentoxide. Phosphorus compounds form PF or POF compounds in the furnace. Some may be hydrolyzed to higher boiling forms in downstream process operation. Some of the phosphorus compounds do appear in the final product. This is objectionable to some users.

Chloride. Chloride is known to significantly increase the rate of corrosion in acidic fluoride media. The level of chloride that can be tolerated in the HF process before corrosion hinders plant operation is quite low.

Arsenic and Boron. Arsenic and boron form volatile fluorides which are difficult to separate from high purity HF. Special equipment and techniques must be used to remove the arsenic.

Sulfuric Acid. Generally, sulfuric acid of 93–99% is used. The sulfuric values may be fed to the plant as H_2SO_4, oleum (20% SO_3), or even SO_3 (see SULFURIC ACID AND SULFUR TRIOXIDE). Commonly, both H_2SO_4 and oleum are used. The split between the two is determined by water balance. All water entering the process or produced by side reactions reacts with the SO_3 component of the oleum:

$$H_2O + SO_3 \rightarrow H_2SO_4$$

The ratio of fluorspar to sulfuric acid fed depends on the relative cost of each raw material. As of this writing, fluorspar is more expensive than sulfuric acid; thus, most often a slight excess of sulfuric acid is desirable. Too much sulfuric acid, followed by hydrolysis of the SiF_4 in either the vapor or liquid phases:

$$3\ SiF_4(l) + 2\ H_2O \longrightarrow 2\ H_2SiF_6 + SiO_2$$

$$SiF_4(g) + 2\ H_2O \xrightarrow[>600°C]{} 4\ HF + SiO_2$$

Other technologies proceeding via intermediates such as NH_4F or KHF_2 are also possible. A more recently developed process (36) involves the reaction of the H_2SiF_6 and phosphate rock, producing a calcium silicon hexafluoride ($CaSiF_6$) intermediate that can be converted to CaF_2 and then to HF by reaction of H_2SO_4. All of the processes produce silica. The quality of the silica varies greatly, and its value as a coproduct has a significant impact on the processes' economics.

The future for these technologies is uncertain. Economic comparisons with fluorspar-based processes indicate that as long as fluorspar supplies remain abundant, there is little justification to proceed with such processes.

Specifications, Shipping, and Analysis. Hydrogen fluoride is shipped in bulk in tank cars (specification 112S400W) and tank trucks (specification MC312). A small volume of overseas business is shipped in ISO tanks. Bulk shipments are made of anhydrous HF as well as 70% aqueous solutions. A small amount of aqueous solution may be shipped as 50%. Cars and trucks used for anhydrous HF transport are of carbon steel construction. It is possible to ship 70% aqueous in steel from a corrosion standpoint; however, rubber lining is commonly used to eliminate iron pickup, which is detrimental to product quality in a number of applications. Hydrogen fluoride of less than 60% strength must always be shipped in lined containers.

Anhydrous hydrogen fluoride is also available in cylinders, and aqueous hydrogen fluoride, either 50% or 70%, is also shipped in polyethylene bottles and carboys. Typical product specifications and analysis methods are given in Table 4.

Table 4. Hydrogen Fluoride Product Specifications[a]

Component	Specification[b]	Analytical method
Anhydrous HF		
HF, wt %	99.95[c]	difference
nonvolatile acid, ppm	100	evaporation/titration
sulfur dioxide, ppm	50	iodimetry
water, ppm	200	conductivity
arsenic, ppm	25	colorimetry
fluosilicic acid, ppm	100	colorimetry
Aqueous HF		
HF, wt %	70–72	titration
nonvolatile acid, ppm	200	evaporation/titration
sulfur dioxide, ppm	100	iodimetry
arsenic, ppm	18	colorimetry
fluosilicic acid, ppm	100	colorimetry

[a]Ref. 37.
[b]Values are the maximum allowable unless otherwise stated.
[c]Value is the minimum allowable.

Materials of Construction. Acceptable materials of construction for hydrogen fluoride handling are a function of such variables as temperature, hydrogen fluoride strength, and method of use. As examples of the latter, corrosion is greater for higher velocities as well as for metals used in reboiler heat-transfer surfaces. Mild steel is generally used for most anhydrous hydrogen fluoride applications, up to 66°C. Steel, in contact with HF, forms a passive film of iron fluoride, which then protects the metal against further corrosion. Any physical or chemical action which disrupts this passive film can lead to substantial increases in corrosion rate. In steel service, hydrogen blistering, caused by accumulation of hydrogen released by corrosion at laminations and inclusions in the steel, may occur and must be evaluated during periodic inspections.

At higher temperatures, Monel, a nickel–copper alloy, is suitable, as is Hastelloy-C, a nickel molybdenum–chromium alloy.

Aqueous hydrogen fluoride of greater than 60% may be handled in steel up to 38°C, provided velocities are kept low (< 0.3 m/s) and iron pickup in the process stream is acceptable. Otherwise, rubber or polytetrafluoroethylene (PTFE) linings are used. For all applications, PTFE or PTFE-lined materials are suitable up to the maximum use temperature of 200°C. PTFE is also the material of choice for gasketing. Alloy 20 or Monel is typically used for valve and pump applications. Materials unacceptable for use in HF include cast iron, type 400 stainless steel, hardened steels, titanium, glass, and silicate ceramics.

Economic Factors

Production. Global hydrogen fluoride production capacity in 1992 was estimated to be 875,000 metric tons. An additional 204,000 metric tons was used captively for production of aluminum fluoride. Worldwide capacity is tabulated in Table 5 (38). Pricing for hydrogen fluoride in 1990 was about $1.52/kg (39).

North America accounts for about 38% of the worldwide hydrogen fluoride production and 52% of the captive aluminum fluoride production. Table 6 (38) summarizes North American capacity for hydrogen fluoride as well as this captive capacity for aluminum fluoride production. In North America, HF is produced in the United States, Canada, and Mexico, but represents a single market, as well over 90% of the consumption is in the United States.

North American HF production capacity has declined since the early 1980s and several smaller producers, such as Harshaw and Essex, have closed plants. Production is expected to continue to decline in the short term because of chlorofluorocarbon (CFC) cutbacks, but is expected to rebound later in the 1990s as replacement hydrochlorofluorocarbons are introduced to the marketplace.

At least in the short term, European production is expected to be impacted by two trends: the move away from planned economics in the East should lead to more rapid demand growth; and the phaseout of CFCs, including, in Europe, aerosols (qv), should lead to overcapacity in the West. This excess capacity in western Europe could be used to supply the East.

Asian production of hydrogen fluoride is concentrated in Japan. The Japanese are leaders in the production of high quality HF. Hashimoto has the capacity

Table 5. Worldwide Hydrofluoric Acid Capacity[a]

Country	Hydrogen fluoride, t/yr	
	Market	Captive[b]
North America[c]	330,000	106,000
Brazil	18,000	
Venezuela	5,000	
Germany	90,000	
the Netherlands	7,000	
United Kingdom	73,000	
Spain	28,000	
Italy	32,000	20,000
France	45,000	53,000
Greece	5,000	
CIS	100,000	
Czechoslovakia	8,000	
Norway		25,000
South Africa	3,000	
India	9,000	
Australia	9,000	
Japan	113,000	
Total	875,000	204,000

[a] Ref. 38.
[b] Anydrous HF for the production of aluminum fluoride.
[c] See also Table 6.

Table 6. Hydrofluoric Acid Capacity in North America[a]

Producer	Market, t/yr	AlF_3 production
Alcan		55,000
Alcoa		45,000
Allied-Signal	142,000	
Du Pont	68,000	
Atochem	22,000	
Quimica Fluor	68,000	
Fluorex	18,000	
Industrias Quimica de Mexico	6,000	16,000
Quimobasicos	6,000	
Total	330,000	116,000

[a] Ref. 38.

for 3000 t/yr of ultrahigh purity product. For the future, increased production in many of the developing Asian nations is likely.

Fluorspar Supply. Production costs of hydrogen fluoride are heavily dependent on raw materials, particularly fluorspar, and significant changes have occurred in this area. Identified world fluorspar resources amount to approximately 400×10^6 metric tons of fluorspar (40). Of these 400×10^6 t, however,

$$R_2O_3 + 3\ H_2SO_4 \rightarrow R_2(SO_4)_3 + 3\ H_2O$$

Sulfuric acid loss is approximately 1.84% H_2SO_4 for each percentage of R_2O_3. Oleum consumption is increased to consume the water that is formed. The metal sulfates are more stable than metal fluorides under furnace conditions and are discharged from the process with the residue.

Calcium Carbonate. Calcium carbonate, like R_2O_3, affects sulfuric and oleum consumption in the HF process. Sulfuric acid loss is approximately 0.98% H_2SO_4 for each percentage of $CaCO_3$. The carbon dioxide evolved by the reaction increases the noncondensable gas flow, and because it carries HF, contributes to yield losses in the vent stream.

Magnesium Oxide. Magnesium oxide behaves in a similar manner to other metal oxides. However, most spars contain practically no magnesium oxide, so it does not affect yield loss or plant operation.

Organic Carbon. Organic materials interfere with plant operation because these compounds react with sulfuric acid under furnace conditions to form sulfur dioxide. There is a reducing atmosphere in the furnace which may reduce sulfur dioxide to elemental sulfur, which results in sulfur deposits in the gas handling system.

Total Sulfur and Sulfide Sulfur. Total sulfur is predominately in the form of metal sulfate, and because sulfates act as inerts, these materials have little impact on the process. Sulfide sulfur compounds, on the other hand, react and leave the furnace as a sulfur vapor, which may deposit in the gas handling system. A possible mechanism for this is the partial reaction of SO_2 to H_2S, followed by

$$2\ H_2S + SO_2 \rightarrow 3\ S + 2\ H_2O$$

Phosphorus Pentoxide. Phosphorus compounds form PF or POF compounds in the furnace. Some may be hydrolyzed to higher boiling forms in downstream process operation. Some of the phosphorus compounds do appear in the final product. This is objectionable to some users.

Chloride. Chloride is known to significantly increase the rate of corrosion in acidic fluoride media. The level of chloride that can be tolerated in the HF process before corrosion hinders plant operation is quite low.

Arsenic and Boron. Arsenic and boron form volatile fluorides which are difficult to separate from high purity HF. Special equipment and techniques must be used to remove the arsenic.

Sulfuric Acid. Generally, sulfuric acid of 93–99% is used. The sulfuric values may be fed to the plant as H_2SO_4, oleum (20% SO_3), or even SO_3 (see SULFURIC ACID AND SULFUR TRIOXIDE). Commonly, both H_2SO_4 and oleum are used. The split between the two is determined by water balance. All water entering the process or produced by side reactions reacts with the SO_3 component of the oleum:

$$H_2O + SO_3 \rightarrow H_2SO_4$$

The ratio of fluorspar to sulfuric acid fed depends on the relative cost of each raw material. As of this writing, fluorspar is more expensive than sulfuric acid; thus, most often a slight excess of sulfuric acid is desirable. Too much sulfuric acid, however, yields a reaction mixture which becomes wet, sticky, corrosive, and hard to handle.

Technology. The key piece of equipment in a hydrogen fluoride manufacturing plant is the reaction furnace. The reaction between calcium fluoride and sulfuric acid is endothermic (1400 kJ/kg of HF) (334.6 kcal/kg), and for good yields, must be carried out at a temperature in the range of 200°C. Most industrial furnaces are horizontal rotating kilns, externally heated by, for example, circulating combustion gas in a jacket. Other heat sources are possible, eg, supplying the sulfuric acid value as SO_3 and steam (qv), which then react and condense, forming sulfuric acid and releasing heat.

Even at the small production rates involved with the earliest HF production, the fundamental technical problem of HF production was apparent. When finely ground spar and acid are mixed in the proper proportions, a thin, almost watery slurry is obtained. Little reaction occurs until the suspension is heated. Upon heating, the slurry thickens rapidly, passing into a sticky paste that can build up on the furnace walls, thus reducing heat transfer. Additionally, intimate mixing of spar and acid is required, and paste formation interferes with this. The fundamental problem in designing an HF furnace is thus to find a method to keep the heat-transfer surfaces clean enough to allow the reaction to proceed at a reasonable rate, and to keep the reaction mass from forming a sticky material.

Historically, internal scrapers or paddles were used in some designs, and loose rails were used in others to break up any caking material which formed. The nature of these designs mechanically limited the furnaces to relatively small sizes producing about 3000 t/yr. High maintenance costs were also involved.

In the 1960s the need for improved technology extendable to higher capacities became apparent, and several new approaches were commercialized. One example is the use of a heavy-duty mechanical mixer to partially react the spar and acid. This first reaction phase carries the reaction past the point where sticky material forms, enabling the reaction to be completed on a flowable solid material in a standard externally jacketed rotating kiln. A second technology employs an Archimedean screw fixed to the rotating shell, to bring sufficient quantities of hot, dry solid from the discharge end of the furnace to the feed end, such that any sticky material formed is absorbed by the dry solid. This method also serves to bring the reactants up to reaction temperature quickly, by contact with the hot recycled material (35). In both of these technologies, large (> 10,000 t/yr) furnace capacities are attainable, and the furnaces can be made largely from inexpensive carbon steel.

In all HF processes, the HF leaves the furnace as a gas, contaminated with small amounts of impurities such as water, sulfuric acid, SO_2, or SiF_4. Various manufacturers utilize different gas handling operations, which generally include scrubbing and cooling. Crude HF is condensed with refrigerant, and is further purified by distillation (qv). Plant vent gases are scrubbed with the incoming sulfuric acid stream to remove the bulk of the HF. The sulfuric acid is then fed to the furnace. Water or alkali scrubbers remove the remainder of the HF from the plant vent stream.

Some manufacturers recover by-products from the process. Fluosilicic acid [16961-83-4], which is used in water fluoridation, can easily be recovered from the plant vent gases, which contain SiF_4:

$$SiF_4 + 2\ HF \rightarrow H_2SiF_6\ (aq)$$

The calcium sulfate [7778-18-9] discharged from the furnace can also be recovered. This is less the practice in the United States where natural gypsum is plentiful and inexpensive than in Europe, where $CaSO_4$ recovery for use in cement (qv) and self-leveling floors is common. Some $CaSO_4$ is recovered in the United States, primarily for lower end uses such as road aggregate.

Figure 10 is a schematic of a typical HF process.

Alternative Processes. Because of the large quantity of phosphate rock reserves available worldwide, recovery of the fluoride values from this raw material source has frequently been studied. Strategies involve recovering the fluoride from wet-process phosphoric acid plants as fluosilicic acid [16961-83-4], H_2SiF_6, and then processing this acid to form hydrogen fluoride.

Numerous processes have been proposed, but none has been commercialized on a large scale (36). The overall reaction in such processes is

$$H_2SiF_6 + 2\ H_2O \rightarrow 6\ HF + SiO_2$$

However, this reaction does not take place in a single step, and multiple reactions must be used. One such route involves using sulfuric acid to decompose the H_2SiF_6:

$$H_2SiF_6 \xrightarrow{H_2SO_4} 2\ HF + SiF_4$$

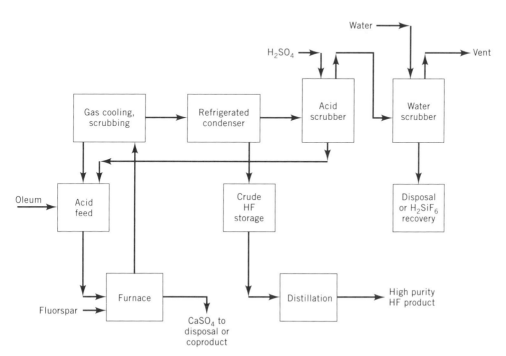

Fig. 10. Hydrogen fluoride manufacturing process.

followed by hydrolysis of the SiF_4 in either the vapor or liquid phases:

$$3\ SiF_4(l) + 2\ H_2O \longrightarrow 2\ H_2SiF_6 + SiO_2$$

$$SiF_4(g) + 2\ H_2O \xrightarrow{>600°C} 4\ HF + SiO_2$$

Other technologies proceeding via intermediates such as NH_4F or KHF_2 are also possible. A more recently developed process (36) involves the reaction of the H_2SiF_6 and phosphate rock, producing a calcium silicon hexafluoride ($CaSiF_6$) intermediate that can be converted to CaF_2 and then to HF by reaction of H_2SO_4. All of the processes produce silica. The quality of the silica varies greatly, and its value as a coproduct has a significant impact on the processes' economics.

The future for these technologies is uncertain. Economic comparisons with fluorspar-based processes indicate that as long as fluorspar supplies remain abundant, there is little justification to proceed with such processes.

Specifications, Shipping, and Analysis. Hydrogen fluoride is shipped in bulk in tank cars (specification 112S400W) and tank trucks (specification MC312). A small volume of overseas business is shipped in ISO tanks. Bulk shipments are made of anhydrous HF as well as 70% aqueous solutions. A small amount of aqueous solution may be shipped as 50%. Cars and trucks used for anhydrous HF transport are of carbon steel construction. It is possible to ship 70% aqueous in steel from a corrosion standpoint; however, rubber lining is commonly used to eliminate iron pickup, which is detrimental to product quality in a number of applications. Hydrogen fluoride of less than 60% strength must always be shipped in lined containers.

Anhydrous hydrogen fluoride is also available in cylinders, and aqueous hydrogen fluoride, either 50% or 70%, is also shipped in polyethylene bottles and carboys. Typical product specifications and analysis methods are given in Table 4.

Table 4. Hydrogen Fluoride Product Specifications[a]

Component	Specification[b]	Analytical method
Anhydrous HF		
HF, wt %	99.95[c]	difference
nonvolatile acid, ppm	100	evaporation/titration
sulfur dioxide, ppm	50	iodimetry
water, ppm	200	conductivity
arsenic, ppm	25	colorimetry
fluosilicic acid, ppm	100	colorimetry
Aqueous HF		
HF, wt %	70–72	titration
nonvolatile acid, ppm	200	evaporation/titration
sulfur dioxide, ppm	100	iodimetry
arsenic, ppm	18	colorimetry
fluosilicic acid, ppm	100	colorimetry

[a]Ref. 37.
[b]Values are the maximum allowable unless otherwise stated.
[c]Value is the minimum allowable.

only 243×10^6 t are considered reserves and an additional 93×10^6 t is considered reserve base, ie, recoverable at higher market prices.

Fluorspar is marketed in three grades: acid, ceramic, and metallurgical. Metallurgical grade is commonly sold as lump or gravel, and ceramic-grade as a dried flotation filter cake or as briquettes or pellets. Acid-grade is used for HF manufacture and is the purest form, having a minimum CaF_2 content of 97%.

Based on previous splits in milling operations, about a 60% yield or 146×10^6 t of acid-grade spar could be expected. At the production rates of the early 1990s, this would be a 24-yr supply. Additional supplies are expected to be brought into production, however, and no decline in available reserves is expected through the year 2000.

Most of the acid-grade spar used for HF production in the United States is imported. More than two-thirds of the fluorspar consumed in the United States goes into production of HF; nearly 30% is consumed as a flux in steelmaking; and the remainder is consumed in glass manufacture, enamels, welding rod coatings, and other end uses or products (see FLUORINE COMPOUNDS, INORGANIC–CALCIUM).

A most significant development has been the bringing on line of mining capacity in the People's Republic of China. China now produces more than 1×10^6 t/yr of fluorspar, making it the world's largest supplier. Kenyan production has also increased substantially.

Purity is expected to become a significant concern as reserves are depleted. Higher levels of impurities in the fluorspar may require modifications to HF production technology to produce high quality hydrogen fluoride. This has already happened regarding high arsenic levels in some Mexican fluorspar. Both Fluorex and Allied-Signal have installed facilities to remove arsenic from the HF process. In addition, the new Kenyan fluorspar production contains high levels of phosphate impurity which must be dealt with.

Uses

In the North American HF market, approximately 70% goes into the production of fluorocarbons, 4% to the nuclear industry, 5% to alkylation processes, 5% to steel pickling, and 16% to other markets (41). This does not include the HF going to aluminum fluoride, the majority of which is produced captively for this purpose.

Fluorocarbons. Fluorocarbons are a family of products that have properties which render them valuable as refrigerants, blowing agents, solvents, and sources of raw materials for production of fluoropolymer materials. Other specialty fluorine-containing, organic chemicals are also produced, some of which are used as anesthetics and fire extinguishants. Certain chlorofluorocarbon products are thought to be damaging to the ozone layer of the upper atmosphere and this market is expected to change to nonozone-damaging fluorocarbon products.

HF is used as a source of fluorine for production of all the various fluorocarbon products. HF reacts in the presence of a suitable catalyst and under the appropriate temperature and pressure conditions with various organic chemicals to yield a family of products. A by-product stream of hydrochloric acid may be co-produced.

Projection of HF requirements for this market segment is uncertain. The ultimate volume of this market segment is negatively impacted by replacement of fluorocarbons with nonfluorine-containing products for foam blowing and solvents. Sales of fluorocarbons would also be reduced by conservation, recovery, and recycle encouraged by high fluid cost, taxation, or regulation. The production of high growth fluorine-containing plastics would be unaffected.

It appears that the ultimate replacements for the high volume chlorofluorocarbon products are to be more highly fluorinated organic chemicals, thus requiring significantly higher volumes of HF in their manufacture.

Nuclear Industry. Technology in the manufacture of uranium reactor fuel for commercial electric power generation requires the uranium to be converted to gaseous uranium hexafluoride [7783-81-5], UF_6, so that enrichment may occur. UF_6, the only gaseous form of uranium, is the form used for the enrichment processes (see DIFFUSION SEPARATION METHODS). Following enrichment, reactor fuel elements are manufactured by converting to UO_2. Domestic U.S. nuclear power generation is not expected to grow. Most of the future demands for UF_6 are expected to be dictated by development of overseas reactor installations. Some other nuclear markets employing UF_6 involve weapons systems but usage and future growth is expected to be low.

To convert naturally occurring uranium oxide, yellow cake or U_3O_8, to the gaseous UF_6, hydrofluoric acid is first used to convert the U_3O_8 to UF_4. Further fluorination using fluorine (generated from more HF) is employed to convert the UF_4 to UF_6. The UF_6 is then processed at gaseous diffusion enrichment plants.

Alkylation. Petroleum and, to a lesser extent, detergent alkylation are processes which make use of the particular catalytic properties of anhydrous HF. Petroleum alkylation produces a very high octane gasoline blending component (C-7 or C-8 compounds) by condensation of C-3 or C-4 olefins obtained in the catalytic cracking process along with isobutane. Detergent alkylation generates a desirable biodegradable detergent intermediate. Although HF is used as a catalyst in these processes, the HF is slowly consumed because of side reactions, with impurities contained in the various feedstocks (qv). As of this writing there are 69 petroleum alkylation and two detergent alkylation units operating in North America.

In the petroleum alkylation process, liquid anhydrous HF is intimately contacted with isobutane [75-28-5] and mixed light olefins under pressure at an elevated (about 40°C) temperature to produce a branched-chain fuel having very high octane value. The mixture of HF and hydrocarbon is settled, the acid is recycled, and the alkylate is water-washed and dried. The HF catalyzes a broad range of desirable reactions in this process and although some HF is lost through reaction with impurities in the various feedstocks, regeneration of most of the HF is easily accomplished within the alkylation unit. The drawoff from this acid regeneration system is neutralized before disposal. This equates to the HF consumption of this process.

Increasing demand for higher octane, lead-free motor fuel having low volatility makes the alkylation process a proven way to maximize profitability. Alkylation using sulfuric acid as the catalyst competes with the HF process even though acid consumption and regeneration costs are much greater.

The choice between sulfuric and hydrofluoric acid-catalyzed processes for new alkylation capacity is influenced by proximity to sulfuric acid regeneration plants, energy costs, and the nature of the unit feed. HF produces a higher quality alkylate when the unit feed is rich in propenes or isobutene. Sulfuric acid is preferred when the feed is rich in pentenes or n-butene. Additionally, in some quarters the perception exists that the sulfuric acid alkylation process may be less hazardous and may present a lesser potential environmental threat. As a result, future installation of additional grass roots HF alkylation units in North America may be affected. Industry is working toward development of chemical additives which, when combined with HF, reduce the risks associated with an accidental release. The success of such work could have a significant impact on the alkylation market.

Chemicals. Both organic and inorganic fluorine-containing compounds, most of which have highly specialized and valuable properties, are produced from HF. Typically these fluorinated chemicals are relatively complex, sometimes difficult to manufacture, and of high value. These materials include products used as fabric and fiber treatments, herbicide and pharmaceutical intermediates, fluoroelastomers, and fluorinated inert liquids. Other products include BF_3, SF_6, and fluoborates.

Many different processes using HF as a reactant or source of fluorine are employed in the manufacture of fluorinated chemical derivatives. In many cases the chemistry employed is complex and in some cases proprietary. Electrochemical fluorination techniques and gaseous fluorine derived from HF are used in some of these applications.

Some of the chemical derivatives, especially those tied to agricultural uses, tend to experience some cyclical demand. However, because of the specialized nature of many of the fluorinated chemicals, these products are positioned in strong, high performance market areas having above average growth rates.

Aqueous HF. Aqueous solutions of hydrofluoric acid are used in stainless steel pickling (see METAL TREATMENTS), chemical milling, glass (qv) etching, exotic metals extraction, quartz purification, and a variety of other uses including metal coatings (qv) and other, small-volume, upgraded inorganic fluorine compounds. A substantial portion of aqueous HF is marketed through distributors as drummed or packaged product.

A small but significant use for aqueous HF is in the electronics industry (see ELECTRONIC MATERIALS). Aqueous HF (typically 49%) of extremely high purity is used as an etchant for silicon wafers (see ULTRAPURE MATERIALS).

Aluminum Industry. Large amounts of HF are consumed in the production of aluminum fluoride [7784-18-1], AlF_3, and cryolite [15096-52-3] (sodium aluminum fluoride), used by the aluminum industry. Both of these compounds are used in the fused alumina bath from which aluminum is produced by the electrolytic method.

Most AlF_3 and cryolite producers have their own HF production facilities. HF vapor is reacted with alumina trihydrate to form AlF_3 in a fluid-bed reactor. HF is reacted with sodium hydroxide to form sodium fluoride, which is then used to produce cryolite. Producers who manufacture these products solely for use in the aluminum industry do not generally install liquid HF storage and handling facilities, and do not participate in the merchant HF market.

Health, Safety, and Environmental Aspects

Although it is widely recognized as a hazardous substance, large volumes of HF are safely manufactured, shipped, and used, and have been for many years. Excellent manuals describing equipment and procedures for the safe handling of hydrogen fluoride are available from manufacturers (16,17,42).

Mild exposure to HF via inhalation can irritate the nose, throat, and respiratory system. The onset of symptoms may be delayed for several hours. Severe exposure via inhalation can cause nose and throat burns, lung inflammation, and pulmonary edema, and can also result in other systemic effects including hypocalcemia (depletion of body calcium levels), which if not promptly treated can be fatal. Permissible air concentrations are (42) OSHA PEL, 3 ppm (2.0 mg/m^3) as F; OSHA STEL, 6 ppm (5.2 mg/m^3) as F; and ACGIH TLV, 3 ppm (2.6 mg/m^3) as F. Ingestion can cause severe mouth, throat, and stomach burns, and may be fatal. Hypocalcemia is possible even if exposure consists of small amounts or dilute solutions of HF.

Both liquid HF and the vapor can cause severe skin burns which may not be immediately painful or visible. HF can penetrate skin and attack underlying tissues, and large (over 160 cm^2) burns may cause hypocalcemia and other systemic effects which may be fatal. Even very dilute solutions may cause burns. Both liquid and vapor can cause irritation to the eyes, corneal burns, and conjunctivitis.

Unlike other acid burns, HF burns always require specialized medical care. The fluoride ion is extremely mobile and easily penetrates deeply into the skin. Immediate first aid consists of flushing the affected area with copious quantities of water for at least 20 minutes. Subsequently the area is immersed in iced 0.13% benzalkonium chloride solutions or massaged with 2.5% calcium gluconate gel. For larger burns, subcutaneous injection of 5% calcium gluconate solution beneath the affected area may be required. Eye exposure requires flushing with water for at least 15 minutes, and subsequent treatment by an eye specialist. Exposure to HF vapor should be treated by moving the victim to fresh air, followed by artificial respiration if required, and administration of oxygen if the victim is having difficulty breathing. As in other cases of exposure, a qualified physician must be called, and the victim should be held under observation for at least 24 hours. First aid for swallowed HF consists of drinking large quantities of water; milk or several ounces of milk of magnesia may be given. Vomiting should not be induced.

Hydrogen fluoride is not a carcinogen. However, HF is highly reactive, and heat or toxic fumes may be evolved. Reaction with certain metals may generate flammable and potentially explosive hydrogen (qv) gas.

The hydrogen fluoride industry has undertaken a significant effort to investigate the behavior of HF releases so as better to define the risks associated with an accidental spill, and to design effective mitigation systems. A series of tests conducted in the Nevada desert in 1986 showed that spills of pressurized, superheated HF under certain conditions could form a heavier-than-air vapor cloud consisting of flashed, cold HF vapor and an entrained aerosol of HF droplets. The HF did not form liquid pools as expected, reducing the effectiveness of diking in mitigating the effect of a release (43). The effect of water sprays in mitigating an

HF release was studied in detail as one of several components of the Industry Cooperative Hydrogen Fluoride Mitigation and Ambient Impact Assessment Program (ICHMAP). Water spray curtains or water monitors were found to remove between 25 and 95% of HF released in field tests. The removal efficiency depended primarily on the ratio of water to HF volume. The higher removal efficiency was obtained at a 50:1 ratio with a single spray curtain (44).

BIBLIOGRAPHY

"Hydrogen Fluoride" under "Fluorine Compounds, Inorganic," in *ECT* 1st ed., Vol. 6, pp. 695–708, by A. S. Woodard, Pennsylvania Salt Manufacturing Co.; in *ECT* 2nd ed., Vol. 9, pp. 610–625, by J. F. Gall, Pennsalt Chemicals Corp.; "Hydrogen" under "Fluorine Compounds, Inorganic," in *ECT* 3rd ed., Vol. 10, pp. 733–753, by J. F. Gall, Philadelphia College of Textiles and Science.

1. C. E. Vanderzee and W. W. Rosenberg, *J. Chem. Thermodyn.* **2**, 461 (1970).
2. J. F. Mathews, *Chem. Rev.* **72**(1), 85, 97 (1972).
3. R. J. Gillespie and D. A. Humphreys, *J. Chem. Soc.* **92**, 2311 (1970).
4. E. U. Franck, and W. Spalthoff, *Z. Electrochem.* **61**, 348 (1957).
5. J. H. Hu, D. White, and H. L. Johnston, *J. Am. Chem. Soc.* **75**, 1232 (1953).
6. D. R. Stull and H. R. Prophet, *JANAF Thermochemical Tables*, 2nd ed., National Bureau of Standards, NSRDS-NBS 37, U.S. Government Printing Office, Washington, D.C., 1971.
7. R. L. Jarry and W. Davis, Jr., *J. Phys. Chem.* **57**, 600 (1953).
8. H. H. Hyman and J. J. Katz, in T. C. Woddington, ed., *Non-Aqueous Solvent Systems*, Academic Press, Inc., London, 1965, pp. 47–81.
9. J. H. Simons and J. W. Bouknight, *J. Am. Chem. Soc.* **54**, 129 (1932).
10. A. J. Perkins, *J. Phys. Chem.* **68**, 654 (1964).
11. K. Fredenhagen and J. Dahmlos, *Z. Anorg. Allg. Chem.* **178**, 272 (1929).
12. S. I. Chan, D. Ikenberry, and T. P. Das, *J. Chem. Phys.* **41**, 2107 (1964).
13. C. L. Yaws and L. S. Adler, *Chem. Eng.*, 119 (Oct. 28, 1974).
14. T. A. O'Donnell, in J. C. Bailar and co-workers, eds., *Comprehensive Inorganic Chemistry*, Vol. 2, Pergamon Press, Oxford, UK, 1973, pp. 1038–1054.
15. W. H. Claussen and J. H. Hildebrand, *J. Am. Chem. Soc.* **56**, 1820 (1934).
16. *Hydrofluoric Acid, Anhydrous—Technical, Properties, Uses, Storage, and Handling*, E. I. du Pont de Nemours & Co., Inc., Wilmington, Del., 1984.
17. *Hydrofluoric Acid*, Allied-Signal Corp., Morristown, N.J., 1978.
18. G. K. Johnson, P. N. Smith, and W. N. Hubbard, *J. Chem. Thermodyn.* **5**, 793 (1973).
19. K. Fredenhagen, *Z. Anorg. Allg. Chem.* **210**, 210 (1933).
20. E. U. Franck and F. Meyer, *Z. Electrochem.* **63**, 571 (1959).
21. Hodgman, C. D., *Handbook of Chemistry and Physics*, 33rd ed., Chemical Rubber Publishing Co., Boca Raton, Fla., 1951, p. 1677.
22. P. A. Munter, O. T. Aepli, and R. A. Kossatz, *Ind. Eng. Chem.* **41**, 1504 (1949).
23. Allied-Signal Corp. data.
24. G. H. Cady and J. H. Hildebrand, *J. Am. Chem. Soc.* **52**, 3843 (1930).
25. P. A. Munter, O. T. Aepli, and R. A. Kossatz, *Ind. Eng. Chem.* **39**, 427 (1947).
26. N. Miki, M. Maeno, K. Maruhashi, and T. Ohmi, *J. Electrochem Soc.* **137**(3), 787 (1990).
27. K. Fredenhagen, *Z. Phys. Chem.* **128**, 1 (1927).
28. K. Fredenhagen and M. Wellman, *Z. Phys. Chem.* **162**, 454 (1932).
29. S. H. Bauer, J. Y. Beach, and J. H. Simons, *J. Am. Chem. Soc.* **61**, 19 (1939).
30. W. Strohmeier and G. Brieglab, *Z. Electrochem.* **57**(8), 662 (1953).

31. W. Spalthoff and E. U. Franck, *Z. Electrochem* **61**(8), 993 (1957).
32. R. W. Long, J. H. Hildebrand, and W. E. Morrell *J. Am. Chem. Soc.* **65**, 182 (1943).
33. R. M. Yabroff, J. C. Smith, and E. H. Lightcap *J. Chem. Eng. Data* **9**(2), 178 (1964).
34. M. Kilpatrick and J. G. Jones, in J. J. Lagowski, ed., *The Chemistry of Nonaqueous Solvents*, Vol. 2, Academic Press, Inc., New York, 1967, pp. 43–49.
35. U.S. Pat. 3,718,736 (Feb. 27, 1973), W. E. Watson and R. P. Troeger, (to Allied Chemical Corp.).
36. *Chem. Eng.*, 27 (Mar. 1993).
37. *Aqueous and Anhydrous Hydrogen Fluoride product specifications*, Allied-Signal Inc., Morristown, N.J., 1991.
38. *Chem-Intell Database*, Reed Telepublishing, London, update Feb. 1990.
39. *Chem. Mark. Rep.*, 23 (Aug. 10, 1992).
40. *Annual Mineral Industry Surveys.* U.S. Bureau of Mines, Washington, D.C., 1990.
41. *Chem. Mark. Rep.*, 30 (July 29, 1991).
42. *Hydrofluoric Acid, Anhydrous*, Product Safety Data Sheet, Allied-Signal Inc., Morristown, N.J., 1991.
43. *Industry Cooperative Hydrogen Fluoride Mitigation and Ambient Impact Assessment Program, Summary Report*, National Technical Information Service, Aug. 1989.
44. *Effectiveness of Water Spray Mitigation Systems for Accidental Releases of Hydrogen Fluoride, Summary Report*, National Technical Information Service, June 1989.

ROBERT A. SMITH
AlliedSignal Inc.

IRON

Iron(II) Fluoride

Anhydrous iron(II) fluoride [7789-28-8], FeF_2, is a white solid. The off-white to buff-colored appearance of the material is attributed to the partial oxidation of Fe^{2+} to Fe^{3+}. FeF_2 is highly stable and does not decompose when heated in the presence of nitrogen. It is sparingly soluble in water but the solubility can be increased by the addition of aqueous HF or any strong acid. Physical properties are listed in Table 1. FeF_2 holds great promise in the field of advanced magnets known as the iron–boron–rare-earth-alloy sintered magnets (1).

Table 1. Physical Properties of Iron Fluorides

Property	FeF_2	FeF_3
mol wt	93.84	112.84
density, g/cm^3	4.09	3.87
mp, °C	1100	1000^a
bp, °C	1837	
C_p, J/(mol·K)^b	68.12	+91.0

^{a}Sublimes.
^{b}To convert J to cal, divide by 4.184.

FeF$_2$ was first prepared by the action of gaseous hydrogen fluoride over FeCl$_2$ in an iron boat (2). The reaction of anhydrous FeCl$_2$, FeCl$_2 \cdot$4H$_2$O, or FeSO$_4 \cdot$7H$_2$O and anhydrous HF in plastic reaction vessels such as vessels of polyethylene, polypropylene, or Teflon results in quantitative yields of very high purity FeF$_2$. The anhydrous salt has also been prepared from a solid-state reaction of a mixture of FeC$_2$O$_4$ and NH$_4$F (weight ratio 1:3) at 300°C and 13.3 Pa (0.1 torr) (3). Other methods of preparation are also available (4,5).

Colorless crystals of iron(II) fluoride tetrahydrate [*13940-89-1*], FeF$_2 \cdot$4H$_2$O, can be obtained by dissolving metallic iron or the anhydrous salt in hydrofluoric acid. The crystals of FeF$_2 \cdot$4H$_2$O are sparingly soluble in water and decompose to Fe$_2$O$_3$ when heated in air.

The only reported industrial application for FeF$_2$ is its use in rust removal solutions based on oxalic acid (6). The anhydrous salt is commercially available in 100 g to 5 kg lots from Advance Research Chemicals, Aldrich Chemicals, Cerac, Johnson/Matthey, PCR, and other suppliers in the United States. As of 1993, the prices varied between $500 to $700/kg.

Toxicity of iron(II) fluoride has not been determined. FeF$_2$ is shipped as a nonhazardous material in plastic containers. The ACGIH has adopted (1991–1992) a TWA value of 1 mg/m^3 for iron as Fe, and 2.5 mg/m^3 for fluorides as F$^-$.

Iron(III) Fluoride

Iron(III) fluoride [*7783-50-8*], FeF$_3$, is the most widely known fluoride of iron. It is light greenish (lime green) in color and the crystals have a rhombic structure. Physical properties are listed in Table 1.

Anhydrous FeF$_3$ is prepared by the action of liquid or gaseous hydrogen fluoride on anhydrous FeCl$_3$ (see IRON COMPOUNDS). FeF$_3$ is insoluble in alcohol, ether, and benzene, and sparingly soluble in anhydrous HF and water. The pH of a saturated solution in water varies between 3.5 and 4.0. Low pH indicates the presence of residual amounts of HF. The light gray color of the material is attributed to iron oxide or free iron impurities in the product.

The most important industrial application of the iron(III) fluoride is in the manufacture of Fe–Co–Nd magnets. Other significant uses are as a hydrocracking catalyst (7), as a catalyst for the preparation of perfluoroacyl fluorides (8), as a catalyst for hydrorefining of lubricating oils (9), as a fluorinating agent (10), for pin-hole prevention in cast iron (11), as a catalyst for preparation of xenon–fluorine compounds (12), burning rate control catalyst, as a catalyst for aromatization, dealkylation, and polymerization, and conversion of vinylidene chloride to the fluoride (13), and in the manufacturing of flame-retardant polymers (14). The industrial market for iron(III) fluoride varies from 2000 kg/yr to 30,000 kg/yr and 1993 prices ranged from $25 to $100/kg. FeF$_3$ is available from Advance Research Chemicals, Aldrich Chemicals, Morrita Chemicals of Japan, and also Russian and European producers.

Hydrated Salts and Other Compounds

Hydrated iron(III) fluoride [*15469-38-2*], FeF$_3 \cdot$3H$_2$O, is easily prepared from yellow Fe$_2$O$_3$ and hydrofluoric acid. Dehydration of FeF$_3 \cdot$3H$_2$O produces oxyfluorides of iron.

In the presence of excess HF, complex ions such as FeF_4^- and FeF_6^- are formed in solution. Neutralization using a base such as NaOH produces $NaFeF_4$ [15274-99-4] and Na_3FeF_6 [20955-11-7], respectively. The latter is used as a fluorinating agent (15).

A mixed valency pale yellow crystalline iron pentafluoride heptahydrate, $FeF_5 \cdot 7H_2O$, is prepared by dissolving iron powder in 40% HF in the presence of air (16). No applications have been reported for this material.

BIBLIOGRAPHY

"Iron Compounds" under "Fluorine Compounds, Inorganic," in *ECT* 1st ed., Vol. 6, p. 709, by F. D. Loomis; "Iron" under "Fluorine Compounds, Inorganic," in *ECT* 2nd ed., Vol. 9, pp. 625–626, by W. E. White; in *ECT* 3rd ed., Vol. 10, pp. 754–755, by D. T. Meshri, Advance Research Chemicals, Inc.

1. Jpn. Kokai Koho, 63,249,304 (Oct. 17, 1988), A. Kobayashi and T. Sato (to Hitachi Metals Ltd.).
2. C. Poulenc, *Compt. Rend. Hebd. Acad. Sci.* **115**, 942 (1980).
3. USSR Pat. 1,502,473 (Aug. 23, 1989), S. V. Petrov, N. I. Kuznetsova, D. D. Ikrami, S. Ganiev, and V. S. Sidorov.
4. G. Pourroy and P. Poix, *J. Fluorine Chem.* **42**(2), 257–263 (1989).
5. I. G. Ryss, *The Chemistry of Fluorine and its Inorganic Compounds*, State Publishing House for Scientific and Chemical Literature, Moscow, Russia, 1956; Eng. transl. ACE-Tr-3927, Vol. II, Office of Technical Services, U.S. Department of Commerce, Washington, D.C., 1960, p. 665.
6. U.S. Pat. 4,828,743 (May 7, 1989), S. Rahfield and B. Newman (to Boyle Midway Household Products Inc.).
7. U.S. Pat. 4,895,822 (Jan. 23, 1990), H. Okazaki, M. Adachi, and M. Ushio (to Nippon Oil Co. Ltd.).
8. Eur. Pat. Appl. 260,713 (Sept. 19, 1986), P. Cuzzato, A. Castellan, and A. Paquale (to Ausimont SPA).
9. Pol. Pat. 138,387 (Jan. 30, 1988), E. Zienkiewicz, J. Kudmierczyk, A. Kubacki, and K. Kowalczyk (to Gdanskse Zaklady Refineryine, Politechnika, Wroclawska).
10. S. Okazaki, *Nippon Kagaku Zasshi* **89**, 1054 (1968).
11. Jpn. Kokai, 75,17,173 (June 19, 1975), T. Kuska (to Hinoshita Rare Metal Institute).
12. B. Z. Slivnik, *Inorg. Nucl. Chem.*, 173 (1976).
13. U.S. Pat. 4,827,055 (Mar. 7, 1988), M. Elsheikh (to Pennwalt Corp.).
14. Ger. Offen. 2,531,816 (Feb. 12, 1976), E. Dorfman, R. R. Hindersim, and W. T. Schwatz (to Hooker Chemical Plastics Corp.).
15. B. Cornils, M. Rassch, and G. Shcieman, *Chem. Ztg. Chem. Appl.* **92**(5), 137 (1968).
16. K. J. Galagher and M. J. Ottaway, *J. Chem. Soc. Dalton*, 978 (1975).

DAYAL T. MESHRI
Advance Research Chemicals, Inc.

LEAD

Lead Difluoride

Lead difluoride [7783-46-2], PbF_2, has the highest melting and boiling points among all the dihalides of lead. Two colorless crystalline forms are known. The α-PbF_2 is orthorhombic in structure and is stable at ordinary temperatures. Upon heating to 200°C it transforms to the cubic β-form. Table 1 lists some of the physical properties of PbF_2.

PbF_2 is readily prepared by the action of hydrogen fluoride on lead hydroxide, lead carbonate, or α-lead oxide. It can also be obtained by precipitation from lead nitrate or lead acetate solutions using potassium fluoride, ammonium fluoride, or ammonium bifluoride.

PbF_2 exhibits very good electrical insulating properties and optical transparency. It is thus used in a variety of glass (qv) such as sealing glass (1), low melting glass (2), near infrared absorbing glass for fiber optics (qv) (3), weather-resistant glass (4,5), and glass for active optical fibers (6). It is also used in printing, photography (qv), brazing, scintillation counters (7), dielectric interference filters (see OPTICAL FILTERS) (8), as a mild fluorinating reagent, as a source material for PbF_4, and as an ingredient in lead–acid batteries (qv) (9).

High purity lead difluoride is available from Advance Research Chemicals, Aldrich Chemicals, Johnson/Matthey, Atomergic, Cerac, and other suppliers in the United States. The U.S. annual consumption varies between 500 to 2500 kg/yr. The 1993 price varied between $10–20/kg.

Table 1. Physical Properties of Lead Fluorides

Property	PbF_2	PbF_4
mol wt	245.19	283.2
density, g/cm^3	8.24	6.7
melting point, °C	855	600
boiling point, °C	1290	decomposes
solubility, g/100 g		
water	0.0641	a
anhydrous HF	2.628	
ΔH_f, kJ/mol^b	−677	
ΔG_f, kJ/mol^b	−631	
C_p, J/(mol·K)^b	72.3	
S, J/(mol·K)^b	−113	

^{a}Material hydrolyzes to PbO_2 and HF.
^{b}To convert J to cal, divide by 4.184.

Lead Tetrafluoride

Like all the lead tetrahalides, lead tetrafluoride [7783-59-7], PbF_4, is very reactive. It is relatively the most stable halide, however. PbF_4 is a white crystalline powder which is highly moisture sensitive, turning yellowish brown in moist air owing to hydrolysis. It should be handled in a dry box or under an atmosphere of dry nitrogen. Properties for PbF_4 are in Table 1.

PbF_4, produced by various routes including the *in situ* species, is a very effective fluorinating agent and also an oxidizing agent. It is classified as a hard fluorinating agent (10), replacing hydrogen with fluorine or adding fluorine to double bonds of both halogenated and hydrocarbon olefins to produce difluorocarbons (11,12).

$$CCl_2=CCl_2 + PbF_4 \rightarrow CCl_2FCCl_2F + PbF_2 \qquad (1)$$

$$CF_3CCl=CCl_2 + PbF_4 \rightarrow CF_3CClFCCl_2F + PbF_2 \qquad (2)$$

$$CHCl=CHCl + PbF_4 \rightarrow CHClFCHClF + PbF_2 \qquad (3)$$

It is also used in the preparation of biologically active steroids where the fluorine is added in a cis configuration to the double bond (13,14).

Lead fluorides are highly toxic and should be handled with great care. The ACGIH adopted toxicity value for lead compounds as Pb is TWA 0.15 mg/m^3 and for fluorides as F^- 2.5 mg/m^3. PbF_4 is prepared by the action of elemental fluorine on very dry PbF_2 at 280–300°C (15).

BIBLIOGRAPHY

"Lead Compounds" under "Fluorine Compounds, Inorganic" in *ECT* 1st ed., Vol. 6, pp. 709–710, by F. D. Loomis, Pennsylvania Salt Manufacturing Co.; "Lead" under "Fluorine Compounds, Inorganic" in *ECT* 2nd ed., Vol. 9, pp. 626–627 by W. E. White, Ozark-Mahoning Co.; in *ECT* 3rd ed., Vol. 10, pp. 756–757 by D. T. Meshri, Ozark Mahoning Co.

1. USSR Pat. 1,701,656 (Dec. 30, 1991), N. B. Knyazyan and co-workers.
2. V. A. Kulgin and K. Pan, *Otkrytiya, Izobret* **33**, 84(1991); *Chem. Abstr.* **116** 157385r; USSR Pat. 1,675,238 (Sept. 7, 1991).
3. Jpn. Kokai Tokkyo Koho, JP 03, 32,735 (Oct. 16, 1991), N. Matsui and W. Takahashi (to Toshiba Glass Co. Ltd.).
4. Jpn. Kokai Tokkyo Koho, JP 04 21,541 (Jan. 24, 1992), T. Osuga and T. Kawaguchi (to Asahi Glass Co. Ltd.).
5. Jpn. Kokai Tokkyo Koho, JP 03 40,935 (Feb. 21, 1991), T. Osuga and T. Kawaguchi (to Asahi Glass Co. Ltd.).
6. J. L. Adam and co-workers, *Proc. SPIE, Int. Soc. Opt. Eng.*, (*Glass Optoelectronics*), 1513 1991, pp. 150–157.
7. J. L. Pauley and M. K. Testerman, *J. Am. Chem. Soc.* **76**, 4220 (1954).
8. G. Honcia and K. Krebs, *Optik* **19**(3), 156 (1962).
9. Jpn. Kokai Tokkyo Koho JP 02,119,055 (May 7, 1990), M. Terada, S. Saito, and A. Miura (to Shin-Kobe Electric Machinery Co. Ltd.).
10. D. T. Meshri and W. E. White, *George H. Cady ACS Symposium*, Milwaukee, Wis., June 1970.
11. A. L. Henne and T. H. Newby, *J. Am. Chem. Soc.* **70**, 130 (1948).
12. A. L. Henne and T. P. Waalkes, *J. Am. Chem. Soc.* **67**, 1639 (1945).
13. A. Bowers and co-workers, *J. Am. Chem. Soc.* **84**, 1050 (1962).
14. Ger. Pat. 1,167,828 (Apr. 16, 1964), K. Bruckner and H. J. Mannhardt (to E. Merck AG).
15. J. Bornstein and L. Skarlas, *J. Am. Chem. Soc.* **90**, 5046 (1968).

DAYAL T. MESHRI
Advance Research Chemicals, Inc.

LITHIUM

Lithium Fluoride

Properties. Lithium fluoride [7789-24-4], LiF, is a white nonhygroscopic crystalline material that does not form a hydrate. The properties of lithium fluoride are similar to the alkaline-earth fluorides. The solubility in water is quite low and chemical reactivity is low, similar to that of calcium fluoride and magnesium fluoride. Several chemical and physical properties of lithium fluoride are listed in Table 1. At high temperatures, lithium fluoride hydrolyzes to hydrogen fluoride when heated in the presence of moisture. A bifluoride [12159-92-1], LiF·HF, which forms on reaction of LiF with hydrofluoric acid, is unstable to loss of HF in the solid form.

Table 1. Properties of Lithium Fluoride

Property	Value	Reference[a]
melting point, °C	848	1
boiling point, °C	1681	1
solubility, g/100 g solvent		
water, 25.4°C	0.133	2
water, 81.8°C	0.150	2
acetic acid, 25°C	0.084	3
acetic acid, 50°C	0.152	3
liquid HF, 12°C	10.3	4
tetrahydrofuran, 25°C	0.6	5
crystalline form	cubic (NaCl)	6
a_0, nm	0.401736	
density at 20°C, g/cm^3	2.635	6
index of refraction	1.3915	6
lattice energy, kJ/mol[b]	1020 ± 10	7
standard heat of formation, kJ/mol[b]	−613.0	1
standard entropy, J/(mol·K)[b]	35.9	1
heat capacity, C_p, J/(mol·K)[b]	42.01	1
heat of fusion, kJ/mol[b]	27.09	1
heat of vaporization, kJ/mol[b]	213	8
heat of hydration, kJ/mol[b]	1.023	8
debye temperature, °C	449	9

[a] Properties listed in this table can be supplemented by the comprehensive collections in References 10 and 11.
[b] To convert kJ to kcal, divide by 4.184.

Manufacture. Lithium fluoride is manufactured by the reaction of lithium carbonate or lithium hydroxide with dilute hydrofluoric acid. If the lithium carbonate is converted to the soluble bicarbonate, insolubles can be removed by filtration and a purer lithium fluoride can be made on addition of hydrofluoric acid (12). High purity material can also be made from other soluble lithium salts such as the chloride or nitrate with hydrofluoric acid or ammonium bifluoride (13).

Optical crystals of high purity lithium fluoride are grown by use of the Stockbarger process (10) in sizes to 25 cm dia × 25 cm high (14). Typical commercial material contains 99.2% LiF; typical impurities include Li_2CO_3 and Fe_2O_3 at <0.1% levels, and SO_4^{2-}, PO_4^{3-}, and heavy metals as Pb at <0.01% levels. The price during 1991 was $10.91/kg in truckload quantities. Annual production is probably less than 100 metric tons. Lithium fluoride toxicity relative to use in thermoluminescent dosimetry is discussed in Reference 15; 10 mg/d is proposed as the maximum permissible daily intake for the average human body. Ingestion of 200 mg/kg of body weight is lethal to guinea pigs (16).

Uses. Lithium fluoride is used primarily in the ceramic industry to reduce firing temperatures and improve resistance to thermal shock, abrasion, and acid attack (see CERAMICS). Another use of LiF is in flux compositions with other fluorides, chlorides, and borates for metal joining (17) (see SOLDERS).

Lithium fluoride is an essential component of the fluorine cell electrolyte; 1% LiF in the KF·2HF electrolyte improves the wettability of the carbon anodes and lowers the tendency of the cells to depolarize (18). Thermoluminescent radiation dosimeters used in personnel and environmental monitoring and in radiation therapy contain lithium fluoride powder, extruded ribbons, or rods (19).

Molten lithium fluoride is used in salt mixtures for an electrolyte in high temperature batteries (qv) (FLINAK) (20), and as a carrier in breeder reactors (FLIBE) (21) (see NUCLEAR REACTORS).

Large high purity crystals are cut into windows and refracting components for use in x-ray monochromators (14), and in the vacuum uv, uv, visible, and ir ranges.

BIBLIOGRAPHY

"Lithium Fluoride" in *ECT* 1st ed., under "Fluorine Compounds, Inorganic," Vol. 6, pp. 709–710, by F. D. Loomis, Pennsylvania Salt Manufacturing Co.; in *ECT* 2nd ed., Vol. 9, p. 627, by G. C. Whitaker, The Harshaw Chemical Co.; "Lithium" under "Fluorine Compounds, Inorganic," in *ECT* 3rd ed., Vol. 10, pp. 757–759, by H. S. Halbedel and T. E. Nappier, The Harshaw Chemical Co.

1. *JANAF Thermochemical Tables*, Clearinghouse for Federal Scientific and Technical Information, U.S. Dept. of Commerce, Springfield, Va., Dec. 1963.
2. C. B. Stubblefield and R. O. Bach, *J. Chem. Eng. Data* **17**, 491 (1972).
3. J. Emsley, *J. Chem. Soc. A*, 2511 (1971).
4. A. W. Jache and G. W. Cady, *J. Phys. Chem.* **56**, 1106 (1952).
5. Brit. Pat. 787,771 (Dec. 18, 1957), (to Metropolitan Vickers Electric Co.).
6. C. A. Hutchison and H. L. Johnson, *J. Am. Chem. Soc.* **62**, 3165 (1940).
7. D. F. C. Morris, *Acta Crystallogr.* **9**, 197 (1956).
8. M. C. Ball and A. A. Norbury, *Physical Data for Inorganic Chemists*, Longman, Inc., New York, 1974.
9. W. W. Scales, *Phys. Rev.* **112**, 49 (1958).
10. *Gmelins Handbuch der Anorganischen Chemie*, 8th ed., Deutsche Vol. 6, Verlag-Chemie, Weinheim/Bergstrasse, 1960, pp. 305–327.
11. J. W. Mellor, *Comprehensive Treatise on Inorganic and Theoretical Chemistry*, Vol. 2, Suppl. 2, Longman, Green and Co., New York, 1961, pp. 174–178.
12. D. C. Stockbarger, *Rev. Sci. Instrum.* **7**, 133 (1936).

13. U.S. Pat. 3,132,922 (May 12, 1954), R. D. Goodenough and T. G. Cook (to Dow Chemical Co.).
14. *Harshaw Optical Crystals*, Harshaw Chemical Co., Solon, Ohio, 1967, pp. 32–33.
15. N. C. Spoor, *Ann. Occup. Hyg.* **11**(1), 23 (1968).
16. H. C. Hodge and F. N. Smith, in J. H. Simon, ed., *Fluorine Chemistry*, Vol. 4, Academic Press, New York, 1965, p. 199.
17. U.S. Pat. 3,958,979 (May 25, 1976), A. R. Valdo (to Ethyl Corp.).
18. J. T. Pinkston, *Ind. Eng. Chem.* **39**, 255 (1947).
19. F. M. Cox, *Proc. 2nd Int. Conf. on Luminescence Dosimetry*, Oak Ridge National Laboratory, CONF 680920, Oak Ridge, Tenn., Sept. 1968; F. M. Cox, A. C. Lucas, and B. M. Kaspar, *Health Phys.* **30**, 135 (1976); J. F. Valley, C. Pache, and P. Lerch, *Helv. Phys. Acta* **49**(2), 171 (1976).
20. G. L. Green, J. B. Hunt, and R. A. Sutula, *U.S. Nat. Tech. Inform. Serv.*, A.D. Rep. 1973, No. 758001.
21. C. D. Scott and W. L. Carter, *AEC Accession No. 43422, Rept. No. ORNL-3791*, Oak Ridge National Laboratory, Oak Ridge, Tenn., 1966.

JOHN R. PAPCUN
Atotech

MAGNESIUM

Magnesium Fluoride

Properties. Magnesium fluoride [7783-40-6], MgF_2, is a fine white crystalline powder with low chemical reactivity. This relative inertness makes possible some of its uses, eg, stable permanent films to alter light transmission properties of optical and electronic materials. The reaction with sulfuric acid is so sluggish and incomplete that magnesium fluoride is not a suitable substitute for calcium fluoride in manufacturing hydrogen fluoride. Magnesium fluoride resists hydrolysis to hydrogen fluoride up to 750°C (1). Bimetallic fluorides, such as $KMgF_3$ [28042-61-7], are formed on fusion of MgF_2 alkali metal and ammonium fluorides (2). Chemical and physical properties are listed in Table 1. MgF_2 is birefringent and only mildly affected by high energy radiation, making possible optics for the uv region.

Manufacture. Magnesium fluoride is manufactured by the reaction of hydrofluoric acid and magnesium oxide or carbonate:

$$MgO + 2\ HF \rightarrow MgF_2 + H_2O$$
$$MgCO_3 + 2\ HF \rightarrow MgF_2 + CO_2 + H_2O$$

Formation of a gelatinous precipitate that is difficult to filter can be avoided by addition of magnesium oxide to the acid solution. In order to increase particle size it is often necessary to keep the solution hot for several hours; however, this problem is avoided by heating an intimate mixture of ammonium bifluoride with magnesium carbonate to 150–400°C (11). Particles of MgF_2 similar in size to those of the magnesium carbonate are obtained.

Table 1. Chemical and Physical Properties

Property	Value	Reference
melting point, °C	1263	3
boiling point, °C	2227	3
standard heat of formation, kJ/mol^a	−112.4	4
standard entropy of formation, J/(mol·K)^a	178	5
heat of fusion, kJ/mol^a	58.2	3
heat of vaporization, kJ/mol^a	264	5
lattice energy, kJ/mol^a	2920	5
heat capacity, 25°C, J/(mol·K)^a	61.59	3
free energy of solution, kJ/mol^a	40.2	5
density, g/cm^3	3.127	6
index of refraction^b		
n_o^{c}	1.37770	7
n_e	1.38950	7
crystalline form (sellaite)	tetragonal	6
a, nm	0.4623	
c, nm	0.3052	
solubility, g/100 g of solvent		
water, 25°C	0.013	8
hydrogen fluoride, 12°C	0.025	9
acetic acid, 25°C	0.681	10

^{a}To convert kJ to kcal, divide by 4.184.
^{b}o, ordinary; e, extraordinary.
^{c}At 589 nm.

$$\text{MgCO}_3 + \text{NH}_4\text{HF}_2 \xrightarrow{150-400°C} \text{MgF}_2 + \text{NH}_3 + \text{CO}_2 + \text{H}_2\text{O}$$

The same results are obtained by adding magnesium carbonate to an aqueous solution of ammonium bifluoride and ammonium hydroxide and warming to 60°C (12). The resulting precipitate is ammonium magnesium fluoride [35278-29-6] which settles rapidly.

$$\text{MgCO}_3 + 3\ \text{NH}_4\text{F (aq)} \longrightarrow \text{NH}_4\text{MgF}_3 + (\text{NH}_4)_2\text{CO}_3$$

$$\text{NH}_4\text{MgF}_3 \xrightarrow[4\ \text{h}]{620°C} \text{MgF}_2 + \text{NH}_3 + \text{HF}$$

Magnesium fluoride is a by-product of the manufacture of metallic beryllium and uranium. The beryllium or uranium fluorides are intimately mixed with magnesium metal in magnesium fluoride-lined crucibles. On heating, a Thermite-type reaction takes place to yield the desired metal and MgF$_2$ (13). Part of the magnesium fluoride produced in this reaction is then used as a lining for the crucibles used in the process.

$$\text{BeF}_2 + \text{Mg} \rightarrow \text{Be} + \text{MgF}_2$$

A commercial grade of magnesium fluoride containing approximately 96–98% MgF_2 is manufactured by Advance Research Chemicals and the Bicron Co. in the United States. Imported technical grades suitable for fluxes containing 94–96% magnesium fluoride are available from Atomergic Chemetals Co., Fine Chemical Co., and Magnesium Elektron.

Magnesium fluoride optical crystals are made by hot-pressing (14) high quality MgF_2 powder. The optical quality powder is made by the NH_4HF_2 method described (11) or by reaction of magnesium bicarbonate and hydrofluoric acid (15). Lead fluoride can also be used in purification of MgF_2 for optical crystals (16). Such optical crystals are manufactured by Bicron Co.

Toxicity. The lethal dose of MgF_2 to guinea pigs by ingestion is 1000 mg/kg (17).

Uses. Established uses of magnesium fluoride are as fluxes in magnesium metallurgy and in the ceramics industry. A proposed use is the extraction of aluminum from arc-furnace alloys with Fe, Si, Ti, and C (18). The molten alloy in reacting with magnesium fluoride volatilizes the aluminum and magnesium which are later separated above the melting point of MgF_2. A welding (qv) flux for aluminum (19) as well as fluxes for steel (20) contain MgF_2.

Optical windows of highly purified magnesium fluoride which transmit light from the vacuum ultraviolet (140 nm) into the infrared (7) are recommended for use as ultraviolet optical components for use in space exploration.

BIBLIOGRAPHY

"Magnesium Fluoride" under "Fluorine Compounds, Inorganic," in *ECT* 1st ed., Vol. 6, p. 709, by F. D. Loomis, Pennsylvania Salt and Manufacturing Co.; in *ECT* 2nd ed., Vol. 9, pp. 627–628 by G. C. Whitaker, The Harshaw Chemical Co.; "Magnesium" under "Fluorine Compounds, Inorganic," Vol. 10, pp. 760–762, by T. E. Nappier and H. S. Halbedel, The Harshaw Chemical Co.

1. D. R. Messier, *J. Am. Ceram. Soc.* **48**, 452, 459 (1965).
2. I. G. Ryss, *The Chemistry of Fluorine and Its Inorganic Compounds*, State Publishing House for Scientific and Technical Literature, Moscow, 1956; Engl. transl. by F. Haimson for the U.S. Atomic Energy Commission, *AEC-tr-3927*, Washington, D.C., 1960, p. 812.
3. *JANAF Thermochemical Tables*, Clearinghouse for Federal Scientific and Technical Information, U.S. Dept. of Commerce, Springfield, Va., 1966.
4. E. Ruelzitis, H. M. Fedar, and W. N. Hubbard, *J. Chem. Phys.* **68**, 2978 (1964).
5. M. C. Ball and A. A. Norbury, *Physical Data for Inorganic Chemicals*, Longman, Inc., New York, 1974.
6. C. Palache, H. Berman, and C. Frandel, *Danas System of Mineralogy*, 7th ed., Vol. 2, John Wiley & Sons, Inc., New York, 1951, p. 38.
7. Technical data, The Harshaw Chemical Co., Crystal and Electronics Dept., Solon, Ohio.
8. D. D. Ikrami, A. S. Paramzin, and A. Kubr, *Russ. J. Inorg. Chem.* **16**, 425 (1971).
9. A. W. Jache and G. W. Cady, *J. Phys. Chem.* **56**, 1106 (1952).
10. C. J. Emsley, *J. Chem. Soc. (A)*, 2511 (1971).
11. U.S. Pat. 3,357,788 (Dec. 12, 1967), J. F. Ross (to General Electric Co.).
12. U.S. Pat. 3,848,066 (Nov. 12, 1974), C. D. Vanderpool and M. B. MacInnis (to G.T.E. Sylvania).

13. H. E. Thayer, *Proc. of the 2nd U. M. Internat. Conference, Peaceful Uses of Atomic Energy, (Geneva)*, **4**, 22 (1958); W. E. Dennis and E. Proudfoot, *U.K. At. Energy R&D* **B**(C) TN-88 (1954).
14. U.S. Pat. 3,294,878 (Dec. 29, 1960), E. Carroll and co-workers (to Eastman Kodak Co.).
15. U.S. Pat. 3,920,802 (Nov. 18, 1975), R. H. Moss, C. F. Swinehart, and W. F. Spicuzza (to Kewanee Oil Co.).
16. U.S. Pat. 2,498,186 (Feb. 21, 1950), D. C. Stockbarger and A. A. Blanchard (to Research Corp.).
17. H. C. Hodges and F. A. Smith, in J. H. Simons, ed., *Fluorine Chemistry*, Vol. 4, Academic Press, Inc., New York, 1965, p. 199.
18. G. S. Layne and co-workers, *Light Met. Age* **30**(3,4), 8 (1972).
19. U.S. Pat. 2,552,104 (May 8, 1951), M. A. Miller and W. E. Haupin (to Alcoa).
20. Jpn. Pat. 7,556,339 (Sept. 19, 1973), T. Tanigaki, T. Koshio, and T. Enomoto (to Nippon Steel Corp.).

JOHN R. PAPCUN
Atotech

MERCURY

Mercury(I) Fluoride

Mercury(I) fluoride [13967-25-4], Hg_2F_2, also known as mercurous fluoride, is a light-sensitive golden yellow material decomposing in water at 15°C. Some of the physical properties are listed in Table 1. Hg_2F_2 resembles AgF in activity with the exception that the former does not form complex compounds or mixed halogen fluoride salts. Consequently, almost an equivalent amount of Hg_2F_2 is sufficient for halogen exchange reactions (1). A mixture of Hg_2F_2 and I_2 (1:1 molar ratio) is much more effective than Hg_2F_2 alone (2). Mercury(I) fluoride is classified as a soft fluorinating reagent (3). Reactions of Hg_2F_2 with monobromides and monoiodides produce fairly good yields, but in polybromides and polyiodides only one halogen is replaced at 120–140°C. Loss of hydrogen halide gives the corresponding olefins (4).

Table 1. Properties of Mercury Fluorides

Property	Hg_2F_2	HgF_2
mol wt	439.22	238.61
density, g/cm^3	8.73	8.95
melting point, °C	>570 (dec)	645
ΔH_f, kJ/mol[a]	−485	−405
ΔG_f, kJ/mol[a]	−469	−362
C_p, J/(mol·K)[a]	+100.4	+74.86
S, J/(mol·K)[a]	161	134.3

[a]To convert J to cal, divide by 4.184.

Several preparatory methods for the manufacture of Hg_2F_2 have been reported (5). Whereas no commercial applications for Hg_2F_2 have been reported, it is available from Advance Research Chemicals and Aldrich Chemicals in the United States. As of 1993, the U.S. market was a few kilograms per year at a price of $1500/kg.

Mercury(II) Fluoride

Mercury(II) fluoride [*7783-39-3*], HgF_2, also known as mercuric fluoride, is a white, hygroscopic solid which turns yellow instantly on exposure to moist air. It must be handled in a dry box or under an atmosphere of dry nitrogen. Some of its physical properties are listed in Table 1. Whereas HgF_2, classified as a moderate fluorinating reagent (3), is superior to both AgF and Hg_2F_2, it has been replaced by anhydrous potassium fluoride [*7789-23-3*], KF, owing to the toxicity of mercury and the disposal regulations issued by the EPA (see MERCURY COMPOUNDS). HgF_2 is an excellent reagent for the addition of fluorine to olefins (1).

Mercury(II) fluoride is easily prepared by passing pure elemental fluorine over predried $HgCl_2$ at 100–150°C until all the chloride ions have been replaced. It is also produced *in situ* by condensing anhydrous HF over HgO (6) or over $HgCl_2$ (10).

Mercury(II) fluoride has been used in the process for manufacture of fluoride glass (qv) for fiber optics (qv) applications (11) and in photochemical selective fluorination of organic substrates (12). It is available from Advance Research Chemicals, Aldrich Chemicals, Johnson/Matthey, Aesar, Cerac, Strem, and PCR in the United States. The 1993 annual consumption was less than 50 kg; the price was $800–1000/kg.

Mercury salts are highly toxic and must be handled carefully. It is necessary to consult the material safety data sheet prior to handling. Strict adherence to OSHA/EPA regulations is essential. The ACGIH adopted (1991–1992) TLV for mercury as inorganic compounds is TWA 0.1 mg/m^3 and for fluorides as F^- 2.5 mg/m^3.

BIBLIOGRAPHY

"Mercury Fluorides" under "Fluorine Compounds, Inorganic" in *ECT* 1st ed., Vol. 6, pp. 747–748, by E. T. McBee and O. R. Pierce, Purdue University; "Mercury" under "Fluorine Compounds, Inorganic" in *ECT* 2nd ed., Vol. 9, p. 628, by W. E. White, Ozark-Mahoning Co.; in *ECT* 3rd ed., Vol. 10, pp. 763–764 by D. T. Meshri, Ozark-Mahoning Co.

1. F. Swarts, *Bull. Acad. Roy. Belg.* **(5)**7, 438 (1921).
2. A. L. Henne and M. W. Renoll, *J. Am. Chem. Soc.* **60**, 1060 (1938).
3. D. T. Meshri and W. E. White, *George H. Cady ACS Symposium*, Milwaukee, Wis., June 1970.
4. R. N. Hazeldine and B. R. Steele, *J. Chem. Soc.*, 1199 (1953).
5. I. G. Ryss, *The Chemistry of Fluorine Compounds*, State Publishing House for Scientific and Chemical Literature, Moscow, Russia, 1956, English tranl. ACE-Tr-3927, Vol. II, Office of Technical Services, U.S. Dept. of Commerce, Washington D.C., 1960, pp. 634–635.

6. A. L. Henne and M. W. Renoll, *J. Am. Chem. Soc.* **60**, 1960 (1938).
7. A. L. Henne and T. Midgley, Jr., *J. Am. Chem. Soc.* **58**, 882 (1936).
8. A. L. Henne and M. W. Renoll, *J. Am. Chem. Soc.* **58**, 887 (1936).
9. J. B. Dicky and co-workers, *Ind. Eng. Chem.* **46**, 2213 (1954).
10. O. Ruff and co-workers, *Chem. Ber.* **51**, 1752 (1918).
11. Jpn. Kokai Tokkyo Koho, JP 63239,137 (Oct. 5, 1988), N. Mitachi, Y. Ooishi, and S. Sakaguchi (to Nippon Telegraph and Telephone Co.).
12. M. H. Habibi and T. E. Mallouk, *J. Fluorine Chem.* **51**(2), 291–294 (1991).

DAYAL T. MESHRI
Advance Research Chemicals, Inc.

MOLYBDENUM

Molybdenum Hexafluoride

Molybdenum hexafluoride [7783-77-9], MoF_6, is a volatile liquid at room temperature. It is very moisture sensitive, hydrolyzing immediately upon contact with water to produce HF and molybdenum oxyfluorides. MoF_6 should therefore be handled in a closed system or in a vacuum line located in a chemical hood. The crystals possess a body-centered cubic structure that changes to orthorhombic below −96°C (1,2). The known physical properties are listed in Table 1.

Molybdenum hexafluoride can be prepared by the action of elemental fluorine on hydrogen-reduced molybdenum powder (100–300 mesh (ca 149–46 μm)) at 200°C. The reaction starts at 150°C. Owing to the heat of reaction, the tem-

Table 1. Physical Properties of MoF_6

Property	Value
mol wt	209.93
melting point, °C	17.4
boiling point, °C	35.0
solubility, g/100 g	[a]
density, g/cm³	
liquid	2.544
solid	2.888
ΔH_f, kJ/mol[b]	
liquid	−1626
gas	−1558
ΔG_f, kJ/mol[b]	
liquid	−1511
gas	−1468
S, J/(mol·K)[b]	259.69

[a] Hydrolyzes in water.
[b] To convert J to cal, divide by 4.184.

perature of the reactor rises quickly but it can be controlled by increasing the flow rate of the carrier gas, argon, or reducing the flow of fluorine.

Molybdenum hexafluoride is used in the manufacture of thin films (qv) for large-scale integrated circuits (qv) commonly known as LSIC systems (3,4), in the manufacture of metallized ceramics (see METAL-MATRIX COMPOSITES) (5), and chemical vapor deposition of molybdenum and molybdenum–tungsten alloys (see MOLYBDENUM AND MOLYBDENUM ALLOYS) (6,7). The latter process involves the reduction of gaseous metal fluorides by hydrogen at elevated temperatures to produce metals or their alloys such as molybdenum–tungsten, molydenum–tungsten–rhenium, or molybdenum–rhenium alloys.

Molybdenum hexafluoride is classified as a corrosive and poison gas. It is stored and shipped in steel, stainless steel, or Monel cylinders approved by DOT. Electronic and semiconductor industries prefer the cylinders equipped with valves which have Compressed Gas Association (CGA) 330 outlets. This material is produced on pilot-plant scale and the U.S. annual consumption is less than 50 kg/yr. As of 1993, the price was $1500/kg. It is available from Advance Research Chemicals Inc., Aldrich Chemicals, Atomergic, Cerac, Johnson/Matthey, Pfaltz & Bauer, and Strem Chemicals.

Other Molybdenum Fluorides

Three other binary compounds of molybdenum and fluorine are known to exist: molybdenum trifluoride [*20193-58-2*], MoF_3, molybdenum tetrafluoride [*23412-45-5*], MoF_4, and molybdenum pentafluoride [*13819-84-6*], MoF_5. Also known are the two oxyfluorides, molybdenum dioxydifluoride [*13824-57-2*], MoO_2F_2, and molybdenum oxytetrafluoride [*14459-59-7*], $MoOF_4$. The use of these other compounds is limited to research applications.

BIBLIOGRAPHY

"Molybdenum" under "Fluorine Compounds, Inorganic" in *ECT* 3rd ed., Vol. 10, pp. 764–765, by A. J. Woytek, Air Products & Chemicals, Inc.

1. M. Carles, *L. Hexafluorure de Molybdene, MoF_6*, Commissariat a L'Energie Atomique, Pierelatte, France; *Report CEA-BIB-124, NASA Technical Translation TT-F-12, 702*, Sept. 1968, pp. 1–25.
2. S. Siegel and D. A. Northrup, *Inorg. Chem.* **5**, 2187 (1966).
3. Jpn. Kokai Tokkyo Koho JP 61 224,313 (Oct. 6, 1986), S. Tsujiku and co-workers.
4. Ger. Offen. 3,639,080 (May 21, 1987), Y. S. Liu and C. P. Yakmyshyn.
5. Eur. Pat. Appl. EP 443,277 (Aug. 28, 1991), P. Jalby and co-workers.
6. J. G. Donaldson and H. Kenworthy, *Electrodepostion Surf. Treat.* **2**, 435 (1973–1974).
7. A. M. Shroff and G. Delval, *High Temp. High Pressure* **3**, 695 (1971).

DAYAL T. MESHRI
Advance Research Chemicals Inc.

NICKEL

Nickel Fluoride Tetrahydrate

Nickel fluoride tetrahydrate [*13940-83-5*], $NiF_2 \cdot 4H_2O$, and its anhydrous counterpart, nickel fluoride [*10028-18-9*], NiF_2, are the only known stable binary compounds of nickel and fluorine. The former is a greenish light yellow crystal or powder prepared by the addition of nickel carbonate to 30–50% aqueous HF solution. The nickel fluoride formed first goes into solution and then precipitates out as the tetrahydrate as the concentration of nickel fluoride increases and that of HF decreases. When the addition of nickel is complete, the solution and the precipitates are dried at 75–100°C until all the water is expelled. The tetrahydrate has high solubility in aqueous HF, eg, 13.3 wt % in 30% HF. It is slightly soluble in water and insoluble in alcohol and ether.

Historically, the annual consumption of nickel fluoride was on the order of a few metric tons. Usage is dropping because nickel fluoride is listed in the EPA and TSCA's toxic substance inventory. Nickel fluoride tetrahydrate is packaged in 200–500-lb (90.7–227-kg) drums and the 1993 price was $22/kg. Small quantities for research and pilot-plant work are available from Advance Research Chemicals, Aldrich Chemicals, Johnson/Matthey, Pfaltz and Bauer, PCR, and Strem Chemicals of the United States, Fluorochem of the United Kingdom, and Morita of Japan.

Nickel Fluoride, Anhydrous

Anhydrous nickel fluoride, a light yellow colored powder, is prepared by the action of anhydrous HF on anhydrous $NiCl_2$, or nickel fluoride tetrahydrate at 300°C. It is also prepared by heating a mixture of NH_4HF_2 and $NiF_2 \cdot 4H_2O$. The other methods include the fluorination of metal salts using excess SF_4 (1) or using ClF_3 (2) at elevated temperatures, or the reaction of $NiCO_3$ and anhydrous HF at 250°C (3).

Nickel fluoride is used in marking ink compositions (see INKS), for fluorescent lamps (4) as a catalyst in transhalogenation of fluoroolefins (5), in the manufacture of varistors (6), as a catalyst for hydrofluorination (7), in the synthesis of XeF_6 (8), and in the preparation of high purity elemental fluorine for research (9) and for chemical lasers (qv) (10).

The 1993 price of high purity anhydrous nickel difluoride was $0.55/g in 100- or 250-g quantities. Small quantities are stored and shipped in polyethylene bottles, whereas large amounts are shipped in fiber board drums with polyethylene liners.

All nickel compounds are considered as suspected carcinogens and are listed in the EPA and TSCA's toxic substances inventory. LD_{50} (mice iv) for NiF_2 is 130 mg/kg (11–13).

Physical Properties. Anhydrous nickel fluoride has a mol wt of 96.71; mp, 1450°C; bp, 1740°C; solubility in water of 4.0 g/100 g; density, g/mL, of 4.72; ΔH_f

of -651.5 kJ/mol (-135.3 kcal/mol); ΔG_f of -604.2 kJ/mol (-144.4 kcal/mol); S of 73.6 J/(mol·K) (17.6 cal/(mol·K)); and C_p, 75.3 J/(mol·K) (18.0 cal/(°C·mol)).

Other Nickel Fluorides. Nickel trifluoride has been observed during the electrolysis of the NiF_2·HF system as a brownish solid capable of liberating iodine from KI solution and turning into yellow powder (14). It has also been observed during the fluorination of NiF_2 at 200°C. A black substance obtained by the addition of AsF_5 to K_2NiF_6·HF solution and decomposing to NiF_2 during the purification process is also believed to be impure NiF_3 (15).

Nickel Fluoride Complexes

Nickel tetrafluoroborate [*14708-14-6*], $Ni(BF_4)_2 \cdot xH_2O$, can be prepared by dissolving nickel carbonate in tetrafluoroboric acid [*16872-11-0*], HBF_4. Nickel tetrafluoroborate, commercially available as a hydrated solid, and also as a 50% solution, plays an important role in the electroplating (qv) and electronics industries. Its consumption is several hundred metric tons a year and its 1993 price was $4.25/kg. It is available from Advance Research Chemicals, Aldrich Chemicals, Aesar Chemicals, Johnson/Matthey, Harshaw M & T Chemicals, and from various other sources.

The complex hexafluoronickelates, M_2NiF_6 (M = Na [*21958-95-2*], K [*17218-47-2*], Rb [*17218-48-3*], Cs [*17218-49-4*]) and M_3NiF_6 (M = Na [*22707-99-9*], K [*14881-07-3*], Rb [*72151-96-3*], and Cs [*72138-72-8*]), are prepared by reaction of elemental fluorine, chlorine trifluoride, or xenon difluoride and a mixture of nickel fluoride and alkali metal fluorides or other metal halides (16,17). If the fluorination is carried out using mixed fluorides, a lower temperature can be used, yields are quantitative, and the final products are of high purity. Bis(tetrafluoroammonium) hexafluoronickelate [*63105-40-8*], $(NF_4)_2NiF_6$, prepared from Cs_2NiF_6 and NF_4SbF_6 by a metathesis in anhydrous HF, is also known (18).

These hexafluoronickelates can be used as fluorinating reagents (15), as a source of high purity elemental fluorine (9,10), and as high energy solid propellant oxidizers (see EXPLOSIVES AND PROPELLANTS) (18).

BIBLIOGRAPHY

"Nickel Fluoride" under "Fluorine Compounds, Inorganic," in *ECT* 2nd ed., Vol. 6, p. 710, by F. D. Loomis; "Nickel" under "Fluorine Compounds, Inorganic," in *ECT* 2nd ed., Vol. 9, pp. 628–629, by W. E. White; in *ECT* 3rd ed., Vol. 10, pp. 766–767, by D. T. Meshri, Advance Research Chemical, Inc.

1. USSR Pat. 495,279 (Dec. 15, 1975), A. P. Kostyuk and L. M. Yagupolskii (to Odessa Polytechnic Inst.).
2. Y. I. Nikonorou and co-workers, *Izv. Sib. Otd. Akad. Nauk. SSSR Ser. Khim. Nauk*, (3), 88 (1976).
3. U.S. Pat. 3,836,634 (Sept. 17, 1974), J. Soldick (to FMC Corp.).
4. Jpn. Kokai Tokkyo Koho 61,278,576 (Dec. 9, 1986), M. Nakono (to Hitachi Ltd.).

5. Eur. Pat. 203,807 (Dec. 3, 1986), F. J. Weigert (to E.I. du Pont de Nemours & Co. Inc.).
6. Jpn. Kokai, 75,139,395 (Nov. 7, 1975), M. Matsuura and co-workers (to Matsushita Electrical Industrial Co.).
7. USSR Pat. 466,202 (Apr. 5, 1975), Kh. U. Usmanov and co-workers (to Tashkent State University).
8. B. Zemva and J. Slivnik, *Vestn. Slov. Kem. Drus.* **19**(1–4), 43 (1972).
9. L. B. Asprey, *J. Fluorine Chem.* **7**(1–3), 359 (1976).
10. U.S. Pat. 4,711,680 (Dec. 8, 1987), K. O. Christe (to Rockwell International Corp.).
11. *Metal Toxicity in Mammals*, Vol. 2, Plenum Publishing Corp., New York, 1978, p. 293.
12. *The Merck Index*, 11th ed., Merck & Co., Inc., Rahway, N.J., 1989, p. 1028.
13. N. I. Sax, *Dangerous Properties of Industrial Materials*, 6th ed., Van Nostrand Reinhold Co., New York, 1984, p. 1993.
14. L. Stein and co-workers, *Inorg. Chem.* **8**, 247 (1969).
15. T. L. Court and M. F. A. Dove, *J. Chem. Soc. Chem. Commun.*, 726 (1971).
16. S. V. Zemskov and co-workers, *Izv. Sib. Otd. Akad. Nauk. S.S.S.R. Ser. Khim. Nauk*, (3), 83 (1976).
17. W. Klemm and E. Huss, *Z. Anorg. Chem.* **258**, 221 (1949).
18. U.S. Pat. 4,108,965 (Aug. 22, 1978), K. O. Christe (to Rockwell International Corp.).

DAYAL T. MESHRI
Advance Research Chemicals, Inc.

NITROGEN

Nitrogen has four binary fluorides: nitrogen trifluoride [7783-54-2], NF_3; tetrafluorohydrazine [10036-47-2], N_2F_4; difluorodiazine [10578-16-2], N_2F_2; and fluorine azide [14986-60-8], FN_3. There are numerous other nitrogen fluorine compounds, the most significant of which are the perfluoroammonium salts based on the NF_4^+ cation. Of all the nitrogen fluorine compounds, only NF_3 has been of commercial importance. Nitrogen trifluoride is used as an etchant gas in the electronics industry and as a fluorine source in high power chemical lasers.

Nitrogen Trifluoride

Physical Properties. Nitrogen trifluoride, NF_3, is a colorless gas, liquefying at 101.3 kPa (1 atm) and $-129.0°C$, and solidifying at $-206.8°C$ (1). High purity NF_3 has little odor, but material contaminated with traces of active fluorides may have a pungent, musty odor. NF_3, a pyramidal molecule with C_{3v} point group symmetry, has a structure similar to ammonia. The N–F bonds are 0.137 nm and the F–N–F bond makes a 102.1° angle. Selected physical properties of NF_3 are given in Table 1. An extensive tabulation and correlation of all the physical properties of NF_3 is available (2). The infrared (3), Raman (4), and ultraviolet spectra (5) of NF_3 have been investigated.

Chemical Properties. NF_3 can be a potent oxidizer, especially at elevated temperature. At temperatures up to ca 200°C, its reactivity is comparable to oxygen. At higher temperatures, the homolysis of the N–F bond into NF_2 and F free radicals becomes significant. The F radical reacts with organic compounds and

Table 1. Physical Properties of Nitrogen Trifluoride

Property	Value	Reference
boiling point, °C	−129.0	1
liquid density at −129°C, g/mL	1.533	6
heat of vaporization, kJ/mol[a]	11.59	1
triple point, °C, 0.263 Pa[b]	−206.8	7
heat of fusion, J/mol[a]	398	1
solid transition point, °C	−216.5	2
heat of transition, kJ/mol[a]	1.513	2
critical temperature, °C	−39.25	6
critical pressure, kPa[b]	4530 (44.7 atm)	6
critical volume, cm^3/mol	123.8	8
heat of formation, kJ/mol[a]	−131.5	9
heat capacity at 25°C, J/(mol·K)[a]	53.39	9
water solubility, 101.3 kPa,[b] 25°C	1.4×10^{-5} mol NF$_3$/mol H$_2$O	10
vapor pressure equation, P in kPa[c]	$\log P = 5.90445 - \dfrac{501.913}{T - 15.37}$	6

[a] To convert kJ to kcal, divide by 4.184.
[b] To convert kPa to mm Hg, multiply by 7.5.
[c] For P in mm Hg, $\log P = 6.77966 - \dfrac{501.913}{T - 15.37}$

certain metals, liberating heat and causing further dissociation of the NF$_3$. At temperatures above 400°C, the reactivity of NF$_3$ becomes more like that of fluorine. The thermal dissociation of NF$_3$ has been studied by a number of investigators (11–14) and was found to peak in the temperature range of 800 to 1200°C.

Nitrogen trifluoride acts primarily upon the elements as a fluorinating agent, but is not a very active one at lower temperatures. At elevated temperatures, NF$_3$ pyrolyzes with many of the elements to produce N$_2$F$_4$ and the corresponding fluoride. The element used in this reaction scavenges the fluorine radical, allowing the NF$_2$ radicals to combine. The pyrolysis of NF$_3$ over copper turnings produces N$_2$F$_4$ in a 62–71% yield at 375°C (15). Pyrolysis over carbon at 400–500°C is more favorable (16,17). This process was the basis for the commercial production of N$_2$F$_4$ in the early 1960s for use in rocketry.

Hydrogen and hydrides react with NF$_3$ with the rapid liberation of large amounts of heat. This is the basis for the use of NF$_3$ in high energy chemical lasers (qv). The flammability range of NF$_3$–H$_2$ mixtures is 9.4–95 mol % NF$_3$ (17), whereas the flammability range of NF$_3$–SiH$_4$ mixtures is even broader at 4.7–99.34 mol % NF$_3$ (18). Nitrogen trifluoride reacts with organic compounds but generally an elevated temperature is required to initiate the reaction. Under these conditions, the reaction often proceeds explosively and great care must be exercised when exposing NF$_3$ to organic compounds. Therefore, NF$_3$ has found little use as a fluorinating agent for organic compounds. The reactions of NF$_3$ with the elements, and various inorganic and organic substances, are summarized in Reference 2.

Although NF$_3$ is an amine, it exhibits virtually no basic properties and is not protonated by the HSO$_3$F–SbF$_5$–SO$_3$ superacid medium at 20°C (19). Commercial scrubbing systems for unwanted NF_3 are available (20) and work on the principle of pyrolysis of the NF_3 over reactive substrates at high temperatures.

NF_3 reacts with F_2 and certain Lewis acids under heat or uv light to form the corresponding NF_4^+ salts. For example, when NF_3, SbF_5, and F_2 are mixed in a 1:1:1.5 ratio in a Monel vessel and heated to 200°C for 50 hours, a 41% yield of NF_4SbF_6 [16871-76-4] is obtained (21). NF_4^+ salts had been thought to be too unstable to be synthesized (22). Preparation methods have been described which use other sources of energy to initiate the reaction. NF_4AsF_6 [16871-75-3] (23) and NF_4BF_4 [15640-93-4] (24) were prepared using a low temperature glow discharge. NF_4PF_6 [58702-88-8] and NF_4GeF_5 [58702-86-6] were prepared by uv photolysis (25). This series of compounds has been further extended by metathesis between NF_4SbF_6 and other salts in HF solution. For example, $(NF_4)_2TiF_6$ [61128-92-5] was prepared by the metathesis of Cs_2TiF_6 and NF_4SbF_6 in HF (26).

The salts rapidly hydrolyze to form NF_3 and O_2, and react with glass at temperatures above 85°C to form NF_3 and SiF_4 (21). The NF_4^+ salts are stable in dry atmospheres to 200°C, but rapidly decompose above 300°C to yield NF_3, F_2, and the corresponding Lewis acid. Therefore, these salts are solid sources of NF_3 and F_2, free of atmospheric contaminants and HF.

Manufacture and Economics. Nitrogen trifluoride can be formed from a wide variety of chemical reactions. Only two processes have been technically and economically feasible for large-scale production: the electrolysis of molten ammonium acid fluoride; and the direct fluorination of the ammonia in the presence of molten ammonium fluoride. In the electrolytic process, NF_3 is produced at the anode and H_2 is produced at the cathode. In a divided cell of 4 kA having nickel anodes, extensive dilution of the gas streams with N_2 was used to prevent explosive reactions between NF_3 and H_2 (17).

In the direct process, NF_3 is produced by the reaction of NH_3 and F_2 in the presence of molten ammonium acid fluoride (27). The process uses a specially designed reactor (28). Because H_2 is not generated in this process, the hazards associated with the reactions between NF_3 and H_2 are eliminated.

As a result of the development of electronic applications for NF_3, higher purities of NF_3 have been required, and considerable work has been done to improve the existing manufacturing and purification processes (29). N_2F_2 is removed by pyrolysis over heated metal (30) or metal fluoride (31). This purification step is carried out at temperatures between 200–300°C which is below the temperature at which NF_3 is converted to N_2F_4. Moisture, N_2O, and CO_2 are removed by adsorption on zeolites (29,32). The removal of CF_4 from NF_3, a particularly difficult separation owing to the similar physical and chemical properties of these two compounds, has been described (33,34).

Production of NF_3 is less than 100 t/yr in the United States. Air Products and Chemicals, Inc. is the only commercial producer in the United States. The 1992 price ranged from $400–$800/kg depending on the grade.

Specifications and Analysis. Nitrogen trifluoride is shipped as a high pressure gas at 10 MPa (1450 psig) and is available in tube trailers and cylinders. Table 2 shows NF_3 specifications for a commercial grade typically used in chemical laser applications, and two higher grades of NF_3 used by the electronics industry. Analysis of NF_3 for impurities can be performed on the gas chromatograph (35).

Table 2. Specifications of NF_3[a]

Impurity	Commercial, ppmv	VLSI, ppmv	Megaclass, ppmv
total fluorides as HF	3,900	1	1
CO_2	130	16	4
CO	330	25	0.5
CF_4	1,200	560	25
N_2	19,000	130	10
O_2 + Ar	22,000	100	6
SF_6	50	25	1
N_2O	500	16	2
H_2O	1	1	1

[a]Commercial grades offered by Air Products and Chemicals, Inc.

Active fluorides are determined by scrubbing with a basic solution and wet-chemical analysis for fluoride content (35).

Handling and Toxicity. Nitrogen trifluoride gas is noncorrosive to the common metals at temperatures below 70°C and can be used with steel, stainless steel, and nickel. The corrosion rate of NF_3 on these materials significantly increases with moisture or HF. Nitrogen trifluoride is compatible with the fluorinated polymers such as Teflon, Kel-F, and Viton at ambient conditions. Extensive data on the corrosion rates of gaseous and liquid NF_3 to a variety of metals and nonmetals are compiled in Reference 2. In systems handling high pressure NF_3, precautions should be taken to avoid any sudden heating of the NF_3 which can occur during adiabatic compression of the gas during the introduction of NF_3 rapidly from a high pressure point to a low pressure dead-end space (20).

Nitrogen trifluoride is a toxic substance and is most hazardous by inhalation. NF_3 induces the production of methemoglobin which reduces the level of oxygen transfer to the body tissues. At the cessation of NF_3 exposure methemoglobin reverts back to hemoglobin. The OSHA permissible exposure limits is set as a TLV–TWA of 29 mg/kg or 10 ppm (36). Because NF_3 has very little odor, it cannot be detected by its odor at concentrations within the TLV. Therefore, adequate personnel protection or monitoring must be provided when handling NF_3. Commercially available monitors detect NF_3 by either pyrolysis to HF or NO_2 followed by electrochemical quantification, or by infrared absorption (20). All of the monitors can detect at least 1 ppm or 10 times below the TLV. The pyrolysis–electrolytic monitors are generally less expensive, but other halogen-containing compounds can interfere with the NF_3 detection.

The inhalation toxicity of NF_3 on animals has been studied extensively (37–40). These studies provide the basis of emergency exposure limits (EEL) that have been proposed for NF_3. The NAS–NRC Committee on Toxicology recommends that the EEL for NF_3 be 10 min at 2250 ppm, 30 min at 750 ppm, and 60 min at 375 ppm. Gaseous NF_3 is considered to be innocuous to the skin and a minor irritant to the eyes and mucous membranes. NF_3 does give a weakly positive metabolically activated Ames test but only at concentrations greater than 2% or 10 times the 10 minute EEL.

Environmental impact studies on NF_3 have been performed. Although undiluted NF_3 inhibits seed growth, no effect on plant growth was observed when exposed to 6,000 ppm·min of NF_3 and only minor effects were observed at the 60,000 ppm·min exposure level (41). Exposure of microbial populations to 25% NF_3 in air for seven hours showed normal growth. NF_3 is not an ozone-depleting gas (20).

Uses. The principal use of NF_3 is as a fluorine source in the electronics industry. The use of NF_3 as a dry chemical etchant has been reviewed (20,42–44). The advantages of using NF_3 as an etchant over traditional carbon-based etchants include high etch rates, high selectivities for nitride-over-oxide etching and single-crystal silicon over thermally grown oxide, and the production of only volatile reaction products resulting in an etch with no polymer or fluoride residues. *In situ* plasma or thermal cleaning of chemical vapor deposition (CVD) reactors is also a use of NF_3. Residual coatings are deposited on the internal surfaces of CVD reactors during deposition processes. A plasma of NF_3 can remove these deposits as volatile fluorides in minutes at the process temperature eliminating the need to remove the internal CVD reactor components to be cleaned by acid tank immersion.

Another use of NF_3 is as a fluorine source for the hydrogen and deuterium fluoride (HF/DF) high energy chemical lasers (qv). The HF/DF lasers are the most promising of the chemical lasers under development because a substantial fraction (ca 25%) of the energy of the reaction between H_2/D_2 and F_2 can be released as laser radiation (45,46). The use of NF_3 is preferred to F_2 because of its comparative ease of handling at ambient temperatures. Storage and handling of NF_3 do not require the precautions necessary for the large-scale use of fluorine.

BIBLIOGRAPHY

"Nitrogen Trifluoride" under "Fluorine Compounds, Inorganic" in *ECT* 1st ed., Vol. 6, p. 710; "Nitrogen" under "Fluorine Compounds, Inorganic" in *ECT* 2nd ed., Vol. 9, pp. 629–630 by W. E. White, Ozark-Mahoning Co.; in *ECT* 3rd ed., Vol. 10, pp. 768–772, by A. J. Woytek, Air Products and Chemicals, Inc.

1. L. Pierce and E. L. Pace, *J. Chem. Phys.* **23**, 551 (1955).
2. R. E. Anderson, E. M. Vander Wall, and R. K. Schaplowsky, "Nitrogen Trifluoride," *USAF Propellant Handbook*, AFRPL-TR-77-71, Contract F04611-76-C-0058, Aerojet Liquid Rocket Co., Sacramento, Calif., 1977.
3. P. N. Schatz and I. W. Levin, *J. Chem. Phys.* **29**, 475 (1958).
4. M. Gilbert, P. Nectoux, and M. Drifford, *J. Chem. Phys.* **68**, 679 (1978).
5. V. Legasov and co-workers, *J. Fluorine Chem.* **11**(2), 109 (1978).
6. R. L. Jarry and H. C. Miller, *J. Phys. Chem.* **60**, 1412 (1956).
7. *DIPPR on-line database on STN*, American Institute of Chemical Engineers, New York, 1992.
8. D. N. Seshadri, D. S. Viswanath, and N. R. Kuloor, *Indian J. Technol.* **8**(5), 153 (1970).
9. D. Wagman and co-workers, *J. Phys. Chem. Ref. Data*, Suppl. No. 2 (1982).
10. C. R. S. Dean, A. Finch, and P. J. Gardner, *J. Chem. Soc., Dalton Trans.* **23**, 2722 (1973).
11. K. O. McFadden and E. Tschuikow, *J. Phys. Chem.* **77**, 1475 (1973).
12. G. L. Schott, L. S. Blair, and J. D. Morgan, Jr., *J. Phys. Chem.* **77**, 2823 (1973).

13. E. A. Karko and co-workers, *J. Chem. Phys.* **63**, 3596 (1975).
14. P. J. Evans and E. Tschuikow-Roux, *J. Chem. Phys.* **65**, 4202 (1976).
15. C. B. Colburn and A. Kennedy, *J. Am. Chem. Soc.* **80**, 5004 (1958).
16. *Chem. Eng. News* **38**, 85 (Sept. 19, 1960).
17. U.S. Pat. 3,235,474 (Feb. 15, 1966), J. F. Tompkins (to Air Products and Chemicals, Inc.).
18. Y. Urano, K. Tokuhashi, S. Kondo, S. Horiguchi, and M. Iwsaka, *20th Safety Engineering Symposium*, Tokyo, Japan, June 1990.
19. R. J. Gillespie and G. P. Pez, *Inorg. Chem.* **8**, 1233 (1969).
20. *Nitrogen Trifluoride: Safety, Applications, and Technical Data Manual*, Air Products and Chemicals, Inc., Allentown, Pa., 1992.
21. W. E. Tolberg and co-workers, *Inorg. Chem.* **6**, 1156 (1967).
22. W. C. Price, T. R. Passmore, and D. M. Roessler, *Discussions Faraday Soc.* **35**, 201 (1963).
23. J. P. Guertin, K. O. Christe, and A. E. Pavlath, *Inorg. Chem.* **5**, 1921 (1966).
24. S. M. Sinel'nikov and V. Ya. Rosolovskii, *Dokl. Akad. Nauk SSSR* **194**, 1341 (1970).
25. K. O. Christe, C. J. Schack, and R. D. Wilson, *Inorg. Chem,* **15**, 1275 (1976).
26. K. O. Christe and C. J. Schack, *Inorg. Chem.* **16**, 353 (1977).
27. U.S. Pat. 4,091,081 (May 23, 1978), A. J. Woytek and J. T. Lileck (to Air Products and Chemicals, Inc.).
28. *Chem. Eng.* **84**(26), 116 (1977).
29. A. J. Woytek and P. B. Henderson, in *Proceedings of the Institute of Environmental Sciences 37th Annual Meeting*, IES, Mount Prospect, Ill., 1990, p. 570.
30. U.S. Pat. 4,193,976 (Mar. 18, 1980), J. T. Lileck, J. Papinsick, and E. J. Steigerwalt (to Air Products and Chemicals, Inc.).
31. U.S. Pat. 4,948,571 (Dec. 3, 1988), I. Harada, H. Hokonohara, and T. Yamaguchi (to Mitsui Toatsu Chemicals, Inc.).
32. Eur. Pat. Appl. 366 078 A2, (May 2, 1990), M. Aritsuka and N. Iwanaga (to Mitsui Toatsu Chemicals, Inc.).
33. U.S. Pat. 5,069,690 (Dec. 3, 1991), P. B. Henderson, C. G. Coe, D. E. Fowler, and M. S. Benson (to Air Products and Chemicals, Inc.).
34. U.S. Pat. 5,069,887 (Dec. 3, 1991), T. Suenaga, T. Fujii, and Y. Kobayashi (to Central Glass).
35. L. A. Dee, *Analysis of Nitrogen Trifluoride*, AFRPL-TR-76-20 (AD-A022887), Air Force Rocket Propulsion Laboratory, Edward Air Force Base, Calif., Apr. 1976.
36. J. O. Accrocco and M. Cinquanti, eds., *Right-To-Know Pocket Guide for Laboratory Employees*, Genium, Schenectady, N.Y., 1990.
37. E. H. Vernot and C. C. Haun, *Acute Toxicology and Proposed Emergency Exposure Limits of Nitrogen Trifluoride*, AMRL-TR-69-130-Paper No. 13 (AD 710062), Contract F33615-70-V-1046, Syste-Med Corp., Dayton, Ohio, Dec. 1969. Reprinted from the *Proceedings of the 5th Annual Conference on Atmospheric Contamination in Conhned Spaces, Sept. 16–18, 1969*, AMRL-TR-69-130 (AD 709994), Dayton, Ohio, Dec. 1969, pp. 165–171.
38. F. N. Dost and co-workers, "Metabolism and Pharmacology of Inorganic and Fluorine Containing Compounds," *Final Report (July 1, 1964–June 30, 1967)*, AMRL-TR-67-224 (AD 681-161), Contract AF 33(615)-1799, Oregon State University, Corvallis, Oreg., Aug. 1968.
39. F. N. Dost, D. J. Reed, and C. H. Wang, *Toxicol. Appl. Pharmacol.* **17**, 585 (1970).
40. G. L. Coppoc and S. J. Leger, "Effect of Nitrogen Trifluoride on Plasma Concentrations of Lactate, Methemoglobin, and Selected Enzymes," *ApriWune 1968*, SAM-TR-70-42 (AD 711044), School of Aerospace Medicine, Brooks Air Force Base, Texas, July 1970.

41. D. J. Reed, F. N. Dost, and C. H. Want, "Inorganic Fluoride Propellant Oxidizers, Vol. 1: Their Effects Upon Seed Germination and Plant Growth," *Interim Report (May 15, 1964–May 15, 1966)*, AMRLTR-66-187, Vol. I (AD 667-556), Contract AF 33(615)-1767, Oregon State University, Corvallis, Oreg., Nov. 1967.
42. A. J. Woytek, J. T. Lileck, and J. A. Barkanic, *Solid State Technol.* **27**, 109 (1979).
43. J. A. Barkanic and co-workers, *Solid State Technol.* **32**, 172 (1984).
44. J. M. Parks, R. J. Jaccodine, J. G. Langan, and M. A. George, in *Proceedings of the 8th Symposium on Plasma Processing*, Electrochemical Society, Montreal, Canada, May 6, 1990, p. 701.
45. E. R. Schulman, W. G. Burwell, and R. A. Meinzer, "Design and Operation of Medium Power CW HF/DF Chemical Lasers," *AIAA 7th Fluid and Plasma Dynamics Conference*, Palo Alto, Calif., AIAA Paper No. 74-546, June 1974.
46. P. J. Klass, *Aviat. Week Space Technol.* **107**, 34 (Aug. 1976).

General Reference

C. J. Hoffman and R. G. Neville, *Chem. Reo.* **62**, 1 (1962); A. W. Jache and W. E. Shite, in C. A. Hunpel, ed., *Encyclopedia of Electrochemistry*, Reinhold Publishing Corp., New York, 1964, pp. 856–859; R. E. Anderson, E. M. Vander Wall, and R. K. Schaplowsky, "Nitrogen Trifluoride," *USAF Propellant Handbook*, AFRPL-TR-77-71, Contract F0461 1-76-C-0058, Aerojet Liquid Rocket Co., Sacramento, Calif., 1977.

PHILIP B. HENDERSON
ANDREW J. WOYTEK
Air Products and Chemicals, Inc.

OXYGEN

Oxygen Difluoride

Oxygen difluoride [7783-41-7], OF_2, is the most stable binary compound of oxygen and fluorine. Under ambient conditions, it is a colorless gas that condenses to a pale yellow liquid at $-145°C$ (1) and freezes at $-224°C$ (2). Oxygen difluoride is a powerful oxidizer that has attracted considerable attention as an ingredient of high energy rocket propellant systems (see EXPLOSIVES AND PROPELLANTS). Several comprehensive reviews of the physical and chemical properties of OF_2 (3–5) and its handling (6) are available.

Physical Properties. An extensive tabulation of the physical properties of OF_2 is available (4). Selected data are mp $-224°C$ (2); bp, $-145°C$ (1); critical temperature $-58°C$ (7); density of liquid, in g/mL from -145 to $-153°C$, t in K, $d = 2.190 - 0.00523\,t$ (8); heat of formation 31.8 kJ/mol (7.6 kcal/mol) (9); and heat of vaporization 11.1 kJ/mol (2.65 kcal/mol) (10).

Spectroscopic investigations have shown that OF_2 is bent and has equivalent O—F bonds. The O—F distance is 0.139–0.141 nm and the FOF angle is 103–104° (11–13). Measurements of the dipole moment have yielded values of $0.6–1.3 \times 10^{-30}$ C·m (0.18–0.40 D) (12,14,15). The ir (16–18), uv (19), mass (20), and nmr (21) spectra of OF_2 have been reported.

Chemical Properties. The kinetics of decomposition of OF_2 by pyrolysis in a shock tube are different, as a result of surface effects, from those obtained by conventional decomposition studies. Dry OF_2 is stable up to 250°C (22).

Reactions with Metals. Many common metals react with OF_2, but the reaction stops after a passive metal fluoride coating is formed (3,4).

Reactions with Nonmetallic Elements and Inorganic Compounds. Mixtures of OF_2 with carbon, CO, CH_4, H_2, or H_2O vapor explode when ignited with an electrical shock. Elemental B, Si, P, As, Sb, S, Se, and Te react vigorously on slight warming to produce fluorides and oxyfluorides. Oxides such as CrO_3, WO_3, As_2O_3, and CaO react with OF_2 to form fluorides. The corresponding chlorides react with OF_2 to form the respective fluorides and liberate free chlorine in the process (3,4).

In aqueous solution, OF_2 oxidizes HCl, HBr, and HI (and their salts), liberating the free halogens. Oxygen difluoride reacts slowly with water and a dilute aqueous base to form oxygen and fluorine. The rate of this hydrolysis reaction has been determined (23).

Nitric oxide and OF_2 inflame on contact; emission and absorption spectra of the flame have been studied (24). Oxygen difluoride oxidizes SO_2 to SO_3, but under the influence of uv irradiation it forms sulfuryl fluoride [2699-79-8], SO_2F_2, and pyrosulfuryl fluoride [37240-33-8], $S_2O_5F_2$ (25). Photolysis of SO_3–OF_2 mixtures yields the peroxy compound FSO_2OOF [13997-94-9] (25,26).

Oxygen Difluoride as a Source of the OF Radical. The existence of the ·OF radical [12061-70-0] was first reported in 1934 (27). This work was later refuted (28). The ·OF radical was produced by photolysis of OF_2 in a nitrogen or argon matrix at 4 K. The existence of the ·OF species was deduced from a study of the kinetics of decomposition of OF_2 and the kinetics of the photochemical reaction (25,26):

$$OF_2 + SO_3 \xrightarrow[350 \text{ nm}]{h\nu} FSO_2OOF$$

The existence of the ·OF radical was further established by use of ^{17}O-labeled compounds and ^{17}O nmr studies to verify the mechanism (29):

$$OF_2 + h\nu \rightarrow F\cdot + \cdot OF$$
$$F\cdot + SO_3 \rightarrow FSO_3\cdot$$
$$FSO_3\cdot + \cdot OF \rightarrow FSO_2OOF$$

The ·OF radical has also been detected by CO_2 laser magnetic resonance (30). The O—F bond length is 0.135789 nm.

Carbonyl fluoride, COF_2, and oxygen difluoride react in the presence of cesium fluoride catalyst to give bis(trifluoromethyl)trioxide [1718-18-9], CF_3OOOCF_3 (31). CF_3OOF has been isolated from the reaction in the presence of excess OF_2 (32).

Reactions with Organic Compounds. Tetrafluoroethylene and OF_2 react spontaneously to form C_2F_6 and COF_2. Ethylene and OF_2 may react explosively, but under controlled conditions monofluoroethane and 1,2-difluoroethane can be recovered (33). Benzene is oxidized to quinone and hydroquinone by OF_2. Methanol and ethanol are oxidized at room temperature (4). Organic amines are extensively degraded by OF_2 at room temperature, but primary aliphatic amines in a fluorocarbon solvent at $-42°C$ are smoothly oxidized to the corresponding nitroso compounds (34).

The reaction of OF_2 and various unsaturated fluorocarbons has been examined (35,36) and it is claimed that OF_2 can be used to chain-extend fluoropolyenes, convert functional perfluorovinyl groups to acyl fluorides and/or epoxide groups, and act as a monomer for an addition-type copolymerization with diolefins.

Preparation. The synthesis of OF_2 was first achieved by the electrolysis of molten KHF_2 in the presence of water (37). The electrolysis of aqueous HF in the presence of O_2 and O_3 was also found to produce OF_2 (38–40).

The most satisfactory method of OF_2 generation is probably the fluorination of aqueous NaOH (3,22,41–45):

$$2 F_2 + 2 NaOH \rightarrow OF_2 + 2 NaF + H_2O$$

Yields of greater than 60% are obtained (46). This method has been used for the commercial production of OF_2 (8). The NaOH concentration, however, must be kept low to avoid the loss of product by a secondary reaction:

$$OF_2 + 2 OH^- \rightarrow O_2 + F^- + H_2O$$

An economic study of the preparation of OF_2 is available (47).

Analytical Procedures. Oxygen difluoride may be determined conveniently by quantitative application of ir, nmr, and mass spectroscopy. Purity may also be assessed by vapor pressure measurements. Wet-chemical analyses can be conducted either by digestion with excess NaOH, followed by measurement of the excess base (2) and the fluoride ion (48,49), or by reaction with acidified KI solution, followed by measurement of the liberated I_2 (4).

Handling and Safety Factors. Oxygen difluoride can be handled easily and safely in glass and in common metals such as stainless steel, copper, aluminum, Monel, and nickel, from cryogenic temperatures to 200°C (4). At higher temperatures only nickel and Monel are recommended. The compatibility of OF_2 with process equipment depends largely on the cleanliness of the equipment; contaminants such as dirt, moisture, oil, grease, scale slag, and pipe dope must be avoided. Equipment should be passivated with elemental fluorine before contact with OF_2.

Oxygen difluoride must be regarded as a highly poisonous gas, somewhat more toxic than fluorine. It has a foul odor with a limit of detectability of 0.1–0.5 ppm. Repeated exposure of rats to 0.5 ppm OF_2 produced death; repeated exposure to 0.1 ppm, however, caused no discernible effects.

Dioxygen Difluoride

Dioxygen difluoride [7783-44-0], O_2F_2, prepared by passing a 1:1 mixture of O_2 and F_2 through a high voltage electric discharge tube cooled by liquid nitrogen, has also been prepared by uv irradiation of O_2 and F_2 (50,51) and by radiolysis of liquid mixtures of O_2 and F_2 at 77 K using 3 MeV bremsstrahlung (52). Heating an O_2/F_2 mixture to 700°C in stainless steel tubes followed by rapid cooling produces O_2F_2 (53). This compound is also obtained in high yield by subjecting a flowing gas mixture of F_2 to microwave, then downstream and outside of the region of discharge, introducing molecular oxygen (54).

Physical Properties. Because O_2F_2 is unstable, it is difficult to purify. Consequently, some of the reported physical properties are open to question. Selected data are density, in g/mL, from -87 to $-156°C$, $d = 2.074 - 0.00291\,t$ (50); heat of formation 19.8 kJ/mol (4.73 kcal/mol) (55); and heat of vaporization 19.2 kJ/mol (4.58 kcal/mol) at $-57°C$ (55).

The structure of O_2F_2 is that of a nonlinear FOOF chain, having the following molecular constants (56,57): O—O distance, 0.122 nm; OOF angle, 109°30′; dihedral angle, 87°30′; dipole moment, 4.8×10^{-30} C·m (1.44 D). Additional physical and spectral data are summarized in References 4 and 58.

Chemical Properties. The bond distance of O—O is relatively short (121.7 ± 0.3 pm) and that of O—F is relatively long (157.5 ± 0.3 pm) (56). The weakest bond in O_2F_2 is thus the O—F bond and the mechanisms of reaction of O_2F_2 can probably be explained by the formation of F· and ·OOF and not two ·OF radicals. The ·OOF radical [15499-23-7] is a feasible intermediate as it has been shown to exist at low temperatures (56,59–61). If O_2F_2 is allowed to react quickly with other compounds, simple fluorination usually results. The controlled reactions of O_2F_2, however, yield products that appear to be formed via an ·OOF intermediate.

Simple Fluorination Reactions. Some examples (62) of O_2F_2 acting mainly as a fluorinating agent are

$$Xe \xrightarrow{O_2F_2} XeF_4$$

$$ClF_3 \xrightarrow{O_2F_2} ClF_5$$

$$Ag + ClF_5 \xrightarrow{O_2F_2} AgF_3$$

$$PuF_4 \xrightarrow{O_2F_2} PuF_6$$

Reactions Involving an ·OOF Intermediate. In controlled reactions of O_2F_2 and various compounds, ^{17}O tracer studies and other techniques have shown that the first step in the reaction appears to be

$$FOOF \rightarrow \cdot OOF + F\cdot$$

For example:

$$SO_2 + O_2F_2 \rightarrow FSO_2OOF$$

where the proposed mechanism (63) is

162 FLUORINE COMPOUNDS, INORGANIC (OXYGEN)

$$SO_2 + F\cdot \rightarrow FSO_2\cdot$$

$$FSO_2\cdot + \cdot OOF \rightarrow FSO_2OOF$$

Also:

$$2\,CF_3CF{=}CF_2 + 2\,O_2F_2 \rightarrow CF_3CF(OOF)CF_3 + CF_3CF_2CF_2OOF$$

in which the proposed mechanism (64) involves the transfer of an OOF group.

The formation of a new class of compounds, dioxygenyls, containing O_2^+, is also thought to take place via an ·OOF intermediate (65).

$$O_2F_2 \rightarrow \cdot OOF + F\cdot$$

$$\cdot O_2F + BF_3 \rightarrow O_2^+ BF_4^-$$

A number of fluorides have been shown to form O_2^+ compounds upon reaction with O_2F_2.

Uses

Oxygen difluoride is mainly a laboratory chemical. It has been suggested as an oxidizer for rocket applications and has been used for small tests in this area.

Dioxygen difluoride has found some application in the conversion of uranium oxides to UF_6 (66), in fluorination of actinide fluorides and oxyfluorides to AcF_6 (67), and in the recovery of actinides from nuclear wastes (68) (see ACTINIDES AND TRANSACTINIDES; NUCLEAR REACTION, WASTE MANAGEMENT).

Higher Oxygen Fluorides

Several higher oxygen fluorides, O_3F_2 [16829-28-0] (50,69), O_4F_2 [12020-93-8] (70), O_5F_2 [12191-79-6] (71), and O_6F_2 [12191-80-9] (71), and radicals such as ·O_3F (72,73) have been reported. Only ·OF, OF_2, O_2F_2, ·OOF, and O_4F_2, however, have been satisfactorily characterized. From cryogenic mass spectroscopy, it appears that O_3F_2 consists of loosely bonded ·O_2F and ·OF radicals (74). The ^{19}F nmr spectrum of O_3F_2 suggests an O_3F_2 model consisting of O_2F_2 and interstitial oxygen (75). However, ^{19}F and ^{17}O nmr (7,76), and other studies have shown that O_3F_2, as reported in the literature, is actually a mixture of O_4F_2 and O_2F_2.

Little is known about O_4F_2. It has been reported to behave similarly to O_2F_2 in that it can act as a fluorinating agent or a source of the ·OOF radical. In fact, it appears to be a better source of the ·OOF radical than O_2F_2 in its reactions with SO_2 and BF_3.

BIBLIOGRAPHY

"Oxygen Compounds" under "Fluorine Compounds, Inorganic," in *ECT* 1st ed., Vol. 6, pp. 710–711; "Oxygen" under "Fluorine Compounds, Inorganic," in *ECT* 2nd ed., Vol. 9, pp. 631–635, W. B. Fox and R. B. Jackson, Allied Chemical Corp.; in *ECT* 3rd ed., Vol. 10, pp. 773–778, by I. J. Solomon, IIT Research Institute.

1. J. Schnitzlstein and co-workers, *J. Phys. Chem.* **56**, 233 (1952).
2. O. Ruff and K. Clusius, *Z. Anorg. Allgem. Chem.* **190**, 267 (1930).
3. H. R. Leech, in Mellor, ed., *Comprehensive Treatise on Inorganic and Theoretical Chemistry*, Suppl. II, Part I, Longmans, Green & Co., Inc., New York, 1956, pp. 186–193.
4. A. G. Streng, *Chem. Rev.* **63**, 607 (1963).
5. R. B. Jackson, *Oxygen Difluoride Handling Manual*, Report No. NASA-CR-72401, Allied Chemical Corp., Morristown, N.J., Dec. 1970.
6. R. F. Muraca, J. Neff, and J. S. Whittick, *Physical Properties of Liquid Oxygen Difluoride and Liquid Diborane—A Critical Review*, Report No. NASA-CR-88519, SRI-951581-4, Jet Propulsion Lab., Calif. Inst. of Tech., Pasadena, Stanford Research Inst., Menlo Park, Calif., July 1967.
7. R. Anderson and co-workers, *J. Phys. Chem.* **56**, 473 (1952).
8. *Oxygen Difluoride*, Product Data Sheet, General Chemical Division, Allied Chemical Corp., Morristown, N.J.
9. W. Evans, T. Munson, and D. Wagman, *J. Res. Natl. Bur. Std.* **55**, 147 (1955).
10. O. Ruff and W. Menzel, *Z. Anorg. Chem.* **198**, 39 (1931).
11. A. Hilton and co-workers, *J. Chem. Phys.* **56**, 473 (1952).
12. L. Pierce, R. Jackson, and N. Dicianni, *J. Chem. Phys.* **35**, 2240 (1961).
13. J. Ibers and V. Schomaker, *J. Phys. Chem.* **57**, 699 (1953).
14. J. Bransford, A. Kunkel, and A. Jache, *J. Inorg. Nucl. Chem.* **14**, 159 (1960).
15. R. Dodd and R. Little, *Nature* **188**, 737 (1960).
16. H. Bernstein and J. Powling, *J. Chem. Phys.* **18**, 685 (1960).
17. E. Jones and co-workers, *J. Chem. Phys.* **19**, 337 (1951).
18. A. Nielsen, *J. Chem. Phys.* **19**, 379 (1951).
19. A. Glissman and H. Schumacher, *Z. Physik. Chem.* **324**, 328 (1934).
20. V. Dibeler, R. Reese, and J. Franklin, *J. Chem. Phys.* **27**, 1296 (1957).
21. H. Agahigian, A. Gray, and G. Vickers, *Can. J. Chem.* **40**, 157 (1962).
22. G. Brauer, *Handbuch der Preparativen Anorganischen Chemie*, Ferdinand Enke, Stuttgart, 1954.
23. S. N. Misra and G. H. Cady, *Kinetics of Hydrolysis of Oxygen Difluoride*, Report No. TR-70, University of Washington Department of Chemistry, Seattle, Jan. 1972.
24. P. Goodfriend and H. Woods, *J. Chem. Phys.* **39**, 2379 (1963).
25. G. Franz and F. Neumayr, *Inorg. Chem.* **3**, 921 (1964).
26. R. Gath and co-workers, *Angew. Chem.* **75**, 137 (1963).
27. O. Ruff and W. Z. Menzel, *Z. Anorg. Allg. Chem.* **217**, 85 (1934).
28. P. Frisch and H. J. Schumacher, *Z. Anorg. Allg. Chem.* **229**, 423 (1936); (Leipzig) **B34**, 322 (1936); **B37**, 18 (1937).
29. I. J. Solomon, A. J. Kacmarek, and J. Raney, *J. Phys. Chem.* **72**, 2262 (1968).
30. A. R. W. McKellar, *Can. J. Phys.* **57**, 2106 (1979).
31. L. R. Anderson and W. B. Fox, *J. Am. Chem. Soc.* **89**, 431B (1967).
32. I. J. Solomon and co-workers, *Inorg. Chem.* **11**, 195 (1972).
33. R. Rhein and G. Cady, *Inorg. Chem.* **3**, 1644 (1964).
34. R. Merritt and J. Ruff, *J. Am. Chem. Soc.* **86**, 1342 (1964).
35. M. S. Toy, *Utilization of Oxygen Difluoride for Syntheses of Fluoropolymers*, Report No. Patent-3,931,132, Pat. Appl.-45,549, NASA, Pasadena Office, Calif., Jan. 1976.
36. M. Dos Santos Afonso, E. Castellano, and H. J. Schumacher, *An. Asoc. Quim. Argent.* **74**, 465 (1986).
37. P. Lebeau and A. Damiens, *Compt. Rend.* **185**, 652 (1927).
38. A. Englebrecht and E. Nachbaur, *Monatsh. Chem.* **90**, 367 (1959).
39. J. A. Donohue, T. D. Nevitt, and A. Zletz, *Adv. Chem. Ser.* **54**, 192 (1966).

40. D. Hass and P. Wolter, *Z. Anorg. Allg. Chem.* **463**, 91 (1980).
41. G. Rohrbach and G. H. Cady, *J. Am. Chem. Soc.* **69**, 677 (1947).
42. D. Yost, *Inorg. Synth.* **1**, 109 (1939).
43. W. Koblitz and H. Schumacher, *Z. Physik. Chem.* **B25**, 283 (1934).
44. P. Lebeau and A. Damiens, *Compt. Rend.* **188**, 1253 (1929).
45. A. Borning and K. E. Pullen, *Inorg. Chem.* **8**, 1791 (1969).
46. G. H. Cady, *J. Am. Chem. Soc.* **57**, 246 (1935).
47. F. L. Hyman and J. F. Tompkins, *An Economic Study of Oxygen Difluoride*, Final Report No. NASA-CR-117317, Air Products and Chemicals, Inc., Allentown, Pa., June 1970.
48. H. Willard and C. Horton, *Anal. Chem.* **22**, 1190 (1950).
49. H. Willard and C. Horton, *Anal. Chem.* **24**, 862 (1952).
50. S. Aoyama and S. Sakuraba, *J. Chem. Soc. Japan* **59**, 1321 (1938).
51. A. Kirshenbaum, A. Grosse, and J. Astor, *J. Am. Chem. Soc.* **81**, 6398 (1959).
52. C. D. Wagner and co-workers, *J. Am. Chem. Soc.* **91**, 4702 (1969).
53. T. R. Mills, *J. Fluorine Chem.* **52**, 267 (1991).
54. U.S. Pat. Appl. 6,696,548 (Jan. 1986), W. H. Beattie (to U.S. Dept. of Energy).
55. A. Streng, *J. Am. Chem. Soc.* **85**, 1380 (1963).
56. R. Jackson, *J. Chem. Soc.*, 4585 (1962).
57. L. Hedberg and co-workers, *Inorg. Chem.* **27**, 232 (1988).
58. K. C. Kim and G. M. Campbell, *J. Mol. Struct.* **129**, 263 (1985).
59. R. W. Fessenden and R. H. Schuler, *J. Chem. Phys.* **44**, 434 (1966).
60. A. Arkell, *J. Am. Chem. Soc.* **87**, 4057 (1965).
61. R. D. Sprately, J. J. Turner, and G. C. Pimentel, *J. Chem. Phys.* **44**, 2063 (1966).
62. J. B. Nielsen and co-workers, *Inorg. Chem.* **29**, 1779 (1990); S. A. Kinkead, L. B. Asprey, and P. G. Eller, *J. Fluorine Chem.* **29**, 459 (1985); Yu. M. Kiselev and co-workers, *Zh. Neorg. Khim.* **33**, 1252 (1988); J. G. Malm, P. G. Eller, and L. B. Asprey, *J. Am. Chem. Soc.* **106**, 2726 (1984).
63. I. J. Solomon, A. J. Kacmarek, and J. M. McDonough, *Chem. Eng. Data* **13**, 529 (1968).
64. I. J. Solomon, A. J. Kacmarek, and J. Raney, *Inorg. Chem.* **7**, 1221 (1968).
65. I. J. Solomon and co-workers, *J. Am. Chem. Soc.* **90**, 6557 (1968).
66. L. B. Asprey, S. A. Kinkead, and P. G. Eller, *Nucl. Technol.* **73**, 69 (1986).
67. U.S. Pat. Appl. 6,636,656 (Oct. 1985), P. G. Eller, J. G. Malm, and R. A. Penneman (to U.S. Dept. of Energy).
68. U.S. Pat. Appl. 6,649,626 (Oct. 1985), L. B. Asprey and P. G. Eller (to U.S. Dept. of Energy).
69. J. N. Keith and co-workers, *Inorg. Chem.* **7**, 320 (1968).
70. A. D. Kirshenbaum and A. V. Grosse, *J. Am. Chem. Soc.* **81**, 1277 (1959).
71. A. V. Grosse, A. G. Streng, and A. D. Kirshenbaum, *J. Am. Chem. Soc.* **83**, 1004 (1961).
72. A. G. Streng and A. V. Grosse, *J. Am. Chem. Soc.* **88**, 169 (1966).
73. A. D. Kirshenbaum and A. V. Grosse, *Production, Isolation, and Identification of the ·OF, ·O_2F, and ·O_3F Radicals*, Research Institute, Temple University, Philadelphia, Pa., June 1964.
74. T. J. Malone and H. A. McGee, *J. Phys. Chem.* **71**, 3060 (1967).
75. J. W. Nebgen, F. I. Metz, and W. B. Rose, *J. Am. Chem. Soc.* **89**, 3118 (1967).
76. I. J. Solomon and co-workers, *J. Am. Chem. Soc.* **89**, 2015 (1967).

I. J. Solomon
IIT Research Institute

Jean'ne M. Shreeve
University of Idaho

PHOSPHORUS

The majority of the fluorine in the earth's crust is present in the form of the phosphorus fluoride fluoroapatite [1306-05-4], $Ca_5(PO_4)_3F$. Phosphate rock deposits contain an average concentration of 3.5 wt % fluorine. During phosphate processing these fluorine values are partially recovered as by-product fluorosilicic acid. The amount of fluorosilicic acid recovered has grown steadily, in part because of environmental requirements (see PHOSPHORIC ACID AND THE PHOSPHATES).

The compounds phosphorus trifluoride [7783-55-3], PF_3; phosphorus pentafluoride [7647-19-0], PF_5; phosphorus oxyfluoride [13478-20-1], POF_3; and phosphorus thiofluoride [2404-52-6], PSF_3, were prepared prior to 1900. The most widely studied of these are PF_5 and PF_3. Physical properties are given in Table 1. The mixed chlorofluorides PCl_xF_y where $x + y = 3$ and 5 have also been studied. Diphosphorus tetrafluoride [13537-32-1], P_2F_4, was first reported in 1966 (1–3).

Table 1. Properties of Phosphorus Fluorides[a]

Property	PF_5	POF_3	PF_3	PSF_3
melting point, °C	−91.6	−39.1	−151.5	0.15
boiling point, °C	−84.8	−39.7	−101.8	−0.5
density, liquid, at bp, g/mL			1.6	
critical temperature, °C	>25	73.3	−2.05	−0.73
critical pressure, MPa[b]		4.23	4.33	
heat of fusion, kJ/mol[c]	12.1	14.9		
heat of vaporization, kJ/mol[c]	16.7	23.2[d]	16.5	
heat of formation, $-\Delta H_f$, kJ/mol[c]	1210[e]		946[f]	

[a]Refs. 4–6 unless otherwise noted.
[b]To convert MPa to atm, divide by 0.101.
[c]To convert kJ to kcal, divide by 4.184.
[d]Ref. 7.
[e]Ref. 8.
[f]Ref. 9.

Phosphorus Pentafluoride

Phosphorus pentafluoride was first prepared in 1876 through fluorination of phosphorus pentachloride using arsenic trifluoride (1). Other routes to PF_5 have included fluorination of PCl_5 by HF, AgF, benzoyl fluoride, SbF_3, PbF_2, or CaF_2 (10). It can also be made by the reaction of PF_3 and fluorine, chlorine (11), or chlorine in contact with calcium fluoride (12); by the reaction of FSO_3H on fluoride and phosphate-containing rocks (13); by the reaction of SF_6 and PF_3 at high (~400°C) temperature (14); VF_5 and POF_3 (15), and Ca_3P_2 and NF_3 (16); by the reaction of alkali or alkaline-earth metal fluorides or fluorosulfonate fluorides with P_2O_5 or H_3PO_4 at 180–200°C (17); by the reaction of POF_3 and HF at 60–80°C (18); $HPF_6 \cdot XH_2O$ and SO_3 or HSO_3F (19); by reaction of $POF_3 \cdot SO_3$ adduct with HF (20); and by the reaction of carbonyl fluoride with POF_3 (21). From 1968–1973 a series of patents (22–25) reported the production of PF_5, PF_3, POF_3, and the hexa-, di-, and monofluorophosphoric acids by fluorination of phosphoric and phosphorus acids using calcium fluorosulfate fluoride, $CaFSO_3F$, produced by reaction of CaF_2 and SO_3 (22–25). Based on these patents a pilot plant was established to produce fluorine–phosphorus chemicals, but the activity was not commercially successful and was terminated in 1972.

Phosphorus pentafluoride is a colorless gas which fumes in contact with moist air and reacts immediately with water to hydrolyze, first to POF_3 and then to the fluorophosphoric acids. Although PF_5 is probably stable in completely dry glass, glass (qv) is not recommended for storage as even a trace of moisture generates HF which regenerates moisture converting the glass to SiF_4 and PF_5 to POF_3.

Phosphorus pentafluoride behaves as a Lewis acid showing electron-accepting properties. It forms complexes, generally in a ratio of 1:1 with Lewis bases, with amines, ethers, nitriles, sulfoxides, and other bases. These complexes are frequently less stable than the similar BF_3 complexes, probably owing to stearic factors. Because it is a strong acceptor, PF_5 is an excellent catalyst especially in ionic polymerizations. Phosphorus pentafluoride is also used as a source of phosphorus for ion implantation (qv) in semiconductors (qv) (26).

Phosphorus Trifluoride

Phosphorus trifluoride was prepared by Moissan in 1884 by reaction of copper phosphide and lead fluoride [7783-55-3]. It is usually prepared by fluorination of PCl_3 with CaF_2 (11), AsF_3, SbF_3, AgF, PbF_2, ZnF_2, or NaF; reaction of fluorosulfonate, $CaF(FSO_3)$, using molten H_3PO_3 (27); by the reaction of phosphorus oxide and F_2 or NF_3 gas (28) or reaction of PH_3 and NF_3 (29); by the reaction of KHF_2 and PCl_3 or PBr_3 (30). PF_3 can be purified for semiconductor devices by contacting it with nickel silicide (31) or with Cu arsenide, phosphide, or silicide (32). Laboratory-scale syntheses have been published (33,34) and PF_3 is commercially available.

Phosphorus trifluoride is an almost odorless gas that does not fume in air and reacts slowly with water but rapidly with base. It may be very toxic, and great care should be taken in handling it. It reacts with ferrohemoglobin to form an unstable complex (35). Because the action may be similar to carbon monoxide poisoning, oxygen may be helpful in treatment. Phosphorus trifluoride acts as a Lewis base and forms many complexes by donating the lone electron pair on the phosphorus atom. Complexes similar to those of CO (see CARBONYLS) are $Ni(PF_3)_4$ [13859-65-9] (36), $Fe(PF_3)_5$ [13815-34-4] (37), $(PF_3)_2PtCl_2$ [15977-33-0] (38), $Mo(PF_3)_6$ [15339-46-5] (39), and $Cr(PF_3)_6$ [26117-61-3] (39). Although AsF_3 and SbF_3 have acceptor properties and function as Lewis acids, no evidence has been found for molecular complex formation by PF_3 as a Lewis acid nor for stable fluorophosphites (PF_4^-).

Phosphorus Oxyfluoride

Phosphorus oxyfluoride is a colorless gas which is susceptible to hydrolysis. It can be formed by the reaction of PF_5 with water, and it can undergo further hydrolysis to form a mixture of fluorophosphoric acids. It reacts with HF to form PF_5. It can be prepared by fluorination of phosphorus oxytrichloride using HF, AsF_3, or SbF_3. It can also be prepared by the reaction of calcium phosphate and ammonium fluoride (40), by the oxidization of PF_3 with NO_2Cl (41) and NOCl (42); in the presence of ozone (43); by the thermal decomposition of strontium fluorophosphate hydrate (44); by thermal decomposition of $CaPO_3F \cdot 2H_2O$ (45); and reaction of SiF_4 and P_2O_5 (46).

Phosphorus Thiofluoride

Phosphorus thiofluoride can be prepared at a low temperature by uv radiation of OCS and PF_3 (47); by the reaction of PF_5 and $(C_2H_5)_4NSH$ in acetonitrile (48); by the reaction of PF_3 and SF_6 at elevated temperature (49), or with H_2S (50); by the reaction of $PSCl_3$ and NaF (51); and by the high temperature reaction of PF_3 and S (52).

Fluorophosphoric Acids and the Fluorophosphates

The three primary fluorophosphoric acids, monofluorophosphoric acid [*13537-32-1*], H_2O_3PF (**1**), difluorophosphoric acid [*13779-41-4*], HO_2PF_2 (**2**), and hexafluorophosphoric acid [*16940-811*], HPF_6 (**3**), were discovered in 1927 (53). In 1963 *sym*-difluorodiphosphoric acid [*44801-72-1*] (**4**) was reported (54). The three primary acids can be prepared by reaction of phosphoric acid or phosphoric anhydride (55) using varying amounts of HF or phosphorus oxyfluoride (23–26) and HF or water, or both. The reaction of anhydrous hydrogen fluoride and phosphoric anhydride is extremely violent. These acids are in equilibrium with each other, HF, and phosphoric acid. The interrelationships of the acids are

$$\underset{(1)}{\overset{HO}{\underset{HO}{\diagdown}}P{\rightarrow}O} \underset{H_2O}{\overset{HF}{\rightleftharpoons}} \underset{(2)}{\overset{HO}{\underset{F}{\diagdown}}P{\rightarrow}O} \underset{H_2O}{\overset{HF}{\rightleftharpoons}} \underset{}{\overset{HO}{\underset{F}{\diagdown}}P{\rightarrow}O} \underset{H_2O}{\overset{HF}{\rightleftharpoons}} \underset{(3)}{\left[\overset{FF}{\underset{FF}{\diagdown\!\diagup}}P\diagup\!\!\!\diagdown\right]^-} H^+$$

$$\underset{}{\overset{F}{\underset{F}{\diagdown}}P{\rightarrow}O} \underset{H_2O}{\overset{HF}{\rightleftharpoons}} \underset{}{\overset{FF}{\underset{FF}{\diagdown\!\diagup}}P} \overset{HF}{\longrightarrow} (3)$$

$$\left[\underset{HOF}{\overset{O\uparrow}{\diagdown\!P\!\diagup}}\right]_2 O$$

(**4**)

All three fluorophosphoric acids are commercially available. The mono- and difluoro acids can be made as anhydrous or hydrated liquids. Commercial hexafluorophosphoric acid is an aqueous solution. Anhydrous hexafluorophosphoric acid may be prepared at reduced temperatures and pressures but it dissociates rapidly into PF_5 and HF at 25°C (56). When diluted with water all the fluorophosphoric acids hydrolyze producing orthophosphoric acid. The hexafluoro acid is the most stable of the three fluorophosphoric acids.

A number of salts of the monofluoro- and hexafluorophosphoric acids are known and some are commercially important. The salts of difluorophosphoric acid are typically less stable toward hydrolysis and are less well characterized. Sodium monofluorophosphate [7631-97-2], the most widely used dentifrice additive for the reduction of tooth decay, is best known (see DENTIFRICES). Several hexafluorophosphates can be prepared by neutralization of the appropriate base using hexafluorophosphoric acid. The monofluorophosphates are usually prepared by other methods (57) because neutralization of the acid usually results in extensive hydrolysis.

Because HF is present in the aqueous acids and can be generated by hydrolysis of the anhydrous acids, glass should be avoided in handling or processing. Teflon or other plastics can generally be used for laboratory work although there is evidence for migration of the acids through both thin plastic and plastic bottle closures. The acids are generally shipped and stored in United Nations (UN) 6HA1 heavy plastic drums with steel overpacks. Aluminum is also satisfactory for storage and use of concentrated solutions of the difluoro acid and hexafluoro acid.

Experimentation with test animals and laboratory and plant experience indicate that the fluorophosphoric acids are less toxic and dangerous than hydrogen fluoride (58). However, they contain, or can hydrolyze to, hydrofluoric acid and must be treated with the same care as hydrofluoric acid. Rubber gloves and face shields are essential for all work with these acids, and full rubber dress is necessary for handling larger quantities. The fumes from these acids contain HF.

Monofluorophosphoric Acid. Monofluorophosphoric acid (**1**) is a colorless, nonvolatile, viscous liquid having practically no odor. On cooling it does not crystallize but sets to a rigid glass at $-78°C$. It has a density of $d_{25} = 1.818$ g/mL. Little decomposition occurs up to 185°C under vacuum but it cannot be distilled. An aqueous solution shows the normal behavior of a dibasic acid; the first neutralization point in 0.05 N solution is at pH 3.5 and the second at pH 8.5. Conductance measurements, however, indicate H_2PO_3F behaves as a monobasic acid in aqueous solution (59). The permanent end point shows the stability of PO_3F^{2-} ions at this pH. Slow hydrolysis occurs at low pH to give orthophosphoric and hydrofluoric acids. These kinetics have been studied (60,61). Equilibrium concentrations of H_2PO_3F at varying acid strengths have been determined by nmr studies (62). The anhydrous acid causes rapid swelling of cellulose, and acts as a polymerization catalyst. It has no oxidizing character.

Monofluorophosphoric acid is one of the hydrolysis products of POF_3. It is the primary product of reaction between theoretical amounts of phosphoric anhydride and aqueous hydrogen fluoride. However, the product of this reaction usually contains up to 20% each of orthophosphoric and difluorophosphoric acids. Reaction of P_2O_5 and 40% aqueous hydrogen fluoride gives more complete conversion to monofluorophosphoric acid. Although difluorophosphoric acid is formed initially, it hydrolyzes to give an aqueous solution of primarily monofluorophosphoric acid, some orthophosphoric acid, and HF.

Difluorophosphoric Acid. Difluorophosphoric acid (**2**) is a mobile, colorless liquid. It fumes on contact with air, probably owing to HF aerosol formation. The mp of anhydrous difluorophosphoric acid has been reported to be $-96.5 \pm 1°C$ (63) and $-91.3 \pm 1°C$ (64). The density at 25°C is 1.583 g/mL. It partially decomposes on heating above 80–100°C. An extrapolated normal boiling point is 116°C (63) although it boils at 107–111°C (64,65) with decomposition. Vapor pressure data are available (65,66), however, the data above ca 80°C may be unreliable.

A freshly made solution behaves as a strong monobasic acid. Neutralized solutions slowly become acidic because of hydrolysis to monofluorophosphoric acid and hydrofluoric acid. The anhydrous acid undergoes slow decomposition on distillation at atmospheric pressure, reacts with alcohols to give monofluorophosphoric acid esters, and is an alkylation (qv) and a polymerization catalyst.

The commercially available difluorophosphoric acid has a formula approximating 2 $HPO_2F_2 \cdot H_2O$ and contains some monofluorophosphoric acid (**1**) and some hexafluorophosphoric acid (**3**) as well as HF. It is primarily used in catalytic applications.

Hexafluorophosphoric Acid. Hexafluorophosphoric acid (**3**) is present under ambient conditions only as an aqueous solution because the anhydrous acid dissociates rapidly to HF and PF at 25°C (56). The commercially available HPF_6 is approximately 60% HPF_6 based on PF_5^- analysis with HF, HPO_2F_2, HPO_3F, and H_3PO_4 in equilibrium equivalent to about 11% additional HPF_6. The acid is a colorless liquid which fumes considerably owing to formation of an HF aerosol. Frequently, the commercially available acid has a dark honey color which is thought to be reduced phosphate species. This color can be removed by oxidation with a small amount of nitric acid. When the hexafluorophosphoric acid is diluted, it slowly hydrolyzes to the other fluorophosphoric acids and finally phosphoric acid. In concentrated solutions, the hexafluorophosphoric acid establishes equilibrium with its hydrolysis products in relatively low concentration. Hexafluorophosphoric acid hexahydrate [40209-76-5], $HPF_6 \cdot 6H_2O$, mp ca 31.5°C, also forms (66). This compound has been isolated in good yield when a concentrated acid solution is cooled rapidly to prevent renewed equilibration. The hexahydrate forms hard, coarse crystals having a cubic structure (67) which are very hygroscopic. However, upon melting, an equilibrium is again established. Liquid $HPF_6 \cdot (C_2H_5)_2O$ [4590-57-2], boiling at 114°C and melting at $-35°C$, has been reported (68). Hexafluorophosphoric acid is used catalytically, as a fluorinating reagent, as a fungicide, and extensively in preparation of numerous hexafluorophosphates (see FUNGICIDES, AGRICULTURAL).

The PF_6 ion can be determined by precipitation with nitron or tetraphenylarsonium chloride (69).

Monofluorophosphates. Monofluorophosphates are probably the best characterized series of fluoroxy salts. The PO_3F^{2-} ion is stable in neutral or slightly alkaline solution. The alkali metal and ammonium monofluorophosphates are soluble in water but the alkaline-earth salts are only slightly soluble, eg, $CaPO_3F$ is not water-soluble and precipitates as the dihydrate.

Monofluorophosphates of ammonium, lithium, sodium, potassium, silver, calcium, strontium, barium, mercury, lead, and benzidine have been described (70) as have the nickel, cobalt, and zinc salts (71), and the cadmium, manganese, chromium, and iron monofluorophosphates (72). Many of the monofluorophosphates are similar to the corresponding sulfates (73).

The monofluorophosphates can be prepared by neutralization of monofluorophosphoric acid (1). Sodium monofluorophosphate [7631-97-2] is prepared commercially (57) by fusion of sodium fluoride and sodium metaphosphate, and the potassium monofluorophosphate [14104-28-0] can be prepared similarly. Insoluble monofluorophosphates can be readily prepared from reaction of nitrate or chloride solutions with sodium monofluorophosphate. Some salts are prepared by metathetical reactions between silver monofluorophosphate [66904-72-1] and metal chlorides.

Molten alkali metal monofluorophosphates are reactive and corrosive, hydrolyzing to generate HF and reacting with many metals and ceramics. They readily dissolve metal oxides and are effective metal surface cleaners and fluxes (see METAL SURFACE TREATMENTS). They also have bactericidal and fungicidal properties (74). However, the main commercial application among monofluorophosphates is of sodium monofluorophosphate in dentifrices.

Sodium monofluorophosphate, mp 625°C, is soluble in water to the extent of 42 g/100 g solution. The pH of a 2% solution is between 6.5 and 8.0. Dilute solutions are stable indefinitely in the absence of acid or cations that form insoluble fluorides.

Sodium monofluorophosphate is used in most dentifrices at a concentration of 0.76 wt % which produces the desired fluoride level of 1000 ppm although one extra strength dentifrice has 1.14 wt % and 1500 ppm F. Although the mechanism of its efficacy in reducing dental decay is not completely understood (75), it almost certainly reacts with the apatite of the tooth converting it to fluoroapatite which is less soluble in mouth acids (see DENTIFRICES).

The *United States Pharmacopeia* (76) specifications for sodium monofluorophosphate require a minimum of 12.1% fluoride as PO_3F^{2-} (theoretical 13.2%) and a maximum of 1.2% fluoride ion reflecting unreacted sodium fluoride. Analysis for PO_3F^{2-} is by difference between total fluoride in the product less fluoride ion as determined by a specific ion electrode. The oral LD_{50} of sodium monofluorophosphate in rats is 888 mg/kg.

Sodium monofluorophosphate decahydrate [7727-73-3], $Na_2PO_3F \cdot 10H_2O$, melts at 9°C. Commercial sodium monofluorophosphate can be purified using this salt. The commercial salt is dissolved in about 10% more than the theoretical water and filtered. The filtrate is cooled to 0°C and allowed to stand. If crystals do not form, the solution may be seeded with $Na_2PO_3F \cdot 10H_2O$ or $Na_2SO_4 \cdot 10H_2O$. The resulting crystals are filtered cold and washed with small quantities of ice water. Attempts to remove the water of hydration by heating lead to hydrolysis of the sodium monofluorophosphate. The water can be extracted by multiple extractions with ethyl alcohol (77).

Calcium, strontium, and barium monofluorophosphates can be precipitated from aqueous solutions using sodium monofluorophosphate. The salts obtained are $CaPO_3F \cdot 2H_2O$ [37809-19-1], $SrPO_3F \cdot H_2O$ [72152-36-4], and $BaPO_3F \cdot xH_2O$ [58882-62-5] where $x < 1$. The solubility of $CaPO_3F \cdot 2H_2O$ is 0.417 g/100 mL solution at 27°C (78). It can be partially dehydrated to the hemihydrate [72152-38-6] but further attempts cause hydrolysis to CaF_2 and phosphate. Heating $SrPO_3F \cdot H_2O$ to 450°C gives anhydrous $SrPO_3F$ [66546-46-1] but $BaPO_3F \cdot xH_2O$ retains water even at 500°C (79).

Difluorophosphates. Difluorophosphates have limited applications largely because of hydrolytic instability of the $PO_3F_2^-$ ion. The ammonium salt can be prepared from ammonium fluoride and phosphoric anhydride.

$$3\ NH_4F + P_2O_5 \rightarrow NH_4PO_2F_2 + (NH_4)_2PO_3F$$

The $NH_4PO_2F_2$ can be extracted from the solid reaction product with boiling methanol (80). Alkali metal difluorophosphates are prepared from the hexafluorophosphates by one of the following fusion reactions (81):

$$KPF_6 + 2\ KPO_3 \rightarrow 3\ KPO_2F_2$$
$$3\ KPF_6 + 2\ B_2O_3 \rightarrow 3\ KPO_2F_2 + 4\ BF_3$$

Even though the $PO_2F_2^-$ is considered to be hydrolytically unstable, hydrolysis is slow in a neutral solution. However, in a solution initially 0.1 N in NaOH, at 70°C, $NaPO_2F_2$ is quantitatively hydrolyzed to give the PO_3F^{2-} and F^- ions within 10 min (82).

A number of organic nitrogen-containing basic compounds give insoluble difluorophosphates. Among these is nitron which can be used for the gravimetric determination of $PO_2F_2^-$ (83). Potassium and other metal difluorophosphates have been reported as stabilizing agents in chloroethylene polymers (84).

Hexafluorophosphates. There is a great deal of interest in the hexafluorophosphate anion [1691-18-8], mostly as organic hexafluorophosphates for catalysis in photopolymerization. A number of the compounds are diazonium compounds (see PHOTOREACTIVE POLYMERS).

The hexafluorophosphates are among the most stable halogen complexes known. The highly symmetrical PF_6^- ion is stable to boiling aqueous alkali and is decomposed only slowly in acidic solutions at ambient temperatures. It does hydrolyze rapidly at elevated temperatures in acid (85). The stability of the PF_6^- ion can be compared to the isoelectronic SF_6. The hexafluorophosphates can be decomposed yielding PF_5 although in many cases only at temperatures where the PF_5 reacts with the metal containers. Benzenediazonium hexafluorophosphate can be decomposed to PF_5, N_2, and fluorobenzene at 120°C (86) and is a convenient source for laboratory amounts of PF_5 as well as a frequently used catalyst.

Many of the organic and inorganic hexafluorophosphates can be prepared by reaction of hexafluorophosphoric acid and the appropriate base. Another method involves reaction of the appropriate chloride and PCl_5 with anhydrous HF (87).

Potassium hexafluorophosphate [17084-13-8], KPF_6 and ammonium hexafluorophosphate [16941-11-0], NH_4PF_6, are the most readily available PF_6^- salts. The KPF_6 salt melts with slow decomposition to PF_5 and KF at ca 565°C. The density of KPF_6 is 2.55 g/mL, and its solubility in water is 3.56 g/100 g solution at 0°C, 8.35 g/100 g solution at 25°C and 38.3 g/100 g solution at 100°C (88). A solution of KPF_6 is neutral and stable against hydrolysis unless the pH is reduced to about 3. The salt has a minimum intraperitoneal lethal dose of 1120 mg/kg for female albino mice (89). Potassium hexafluorophosphate is a soluble neutral salt which can be used in syntheses of other PF_6^- salts as a substitute for HPF_6 (3). Ammonium hexafluorophosphate has a water solubility of 60.4 g/100 g solution at 25°C (90).

Anhydrous silver hexafluorophosphate [26042-63-7], $AgPF_6$, as well as other silver fluorosalts, is unusual in that it is soluble in benzene, toluene, and *m*-xylene and forms 1:2 molecular crystalline complexes with these solvents (91). Olefins form complexes with $AgPF_6$ and this characteristic has been used in the separation of olefins from paraffins (92). $AgPF_6$ also is used as a catalyst. Lithium hexafluorophosphate [21324-40-3], $LiPF_6$, as well as KPF_6 and other PF_6^- salts, is used as electrolytes in lithium anode batteries (qv).

Substituted ammonium hexafluorophosphates are of decreasing water solubility corresponding to the greater number of hydrogens on the ammonium group replaced by the organic radicals. Unusual thermal stability is found in the quaternary compounds when the R groups are the low alkyl radicals. For example, tetramethylammonium hexafluorophosphate [558-32-7], $(CH_3)_4NPF_6$, is stable up to about 400°C.

Fluorophosphate Esters

The esters of monofluorophosphoric acid are of great interest because of their cholinesterase inhibiting activity which causes them to be highly toxic nerve gases and also gives them medical activity (see ENZYME INHIBITORS). The most studied is the bis(1-methylethyl)ester of phosphorofluoridic acid also known as diisopropyl phosphorofluoridate [55-91-4], DFP (5), and as the ophthalmic ointment or solution Isoflurophate USP. It is used as a parasympathomimetic agent, and as a miotic in glaucoma and convergent strabismus. Developed during World War II as a nerve gas, (93) it is prepared by reaction of PCl_3 and isopropanol, followed by chlorination and conversion to the desired product using NaF (94).

$$(CH_3)_2CHO \diagdown P=O \diagup F \quad\quad CH_3 \diagdown P=O \diagup F$$
$$(CH_3)_2CHO \quad\quad\quad (CH_3)_2CHO$$

(5)   (6)

In the nerve gas known as Sarin [107-44-8] or GB (6), one of the isopropoxy groups of DFP is replaced by a methyl group. Sarin is more toxic than DFP; oral LD_{50} in rats are 550 µg/kg and 6 mg/kg (95), respectively (see CHEMICALS IN WAR).

The esters of monochloro- and dichlorophosphoric acids having polyfluoroalkyl groups Rf, eg, Rf = $(CF_3)_2CH_3C-$, $CF_3(CH_3)_2C-$, CF_3CH_2-, $CH_2-(CF_2CH_2\!\!-\!\!)_{\overline{n}}$, n = 2, 4, have been synthesized by the reaction of PCl_3 and the lithium salt of the respective polyfluoroalkyl alcohols followed by oxidation with

N_2O_4 (96,97). Toxicity data of these compounds are not available. The esters have been hydrolyzed to acid phosphates. These acids have potential as fuel cell electrolytes (see FUEL CELLS).

$$RfOLi + PCl_3 \longrightarrow RfOPCl_2 + LiCl$$
$$2\ RfOPCl_2 + N_2O_4 \longrightarrow 2\ RfOP(O)Cl_2 + 2\ NO$$
$$RfP(O)Cl_2 \xrightarrow[2\ HCl]{H_2O} RfOP(O)OH_2$$

The perfluoroalkylphosphorus(V) acids and bis(perfluoroalkyl)phosphorus(V) acids, $RfPO(OH)_2$ and $(Rf)_2P(O)(OH)$, where $Rf = CF_3, C_2F_5, C_3F_7$, and C_4F_9 have also been reported (98).

BIBLIOGRAPHY

"Phosphorus Compounds" under "Fluorine Compounds, Inorganic" in *ECT* 1st ed., Vol. 6, pp. 711–721, by W. E. White, Ozark-Mahoning Co.; "Phosphorus" under "Fluorine Compounds, Inorganic" in *ECT* 2nd ed., Vol. 9, pp. 635–649, by W. E. White and C. Pupp, Ozark-Mahoning Co., in *ECT* 3rd ed., Vol. 10, pp. 779–788, by C. B. Lindahl, Elf Atochen North America, Inc.

1. R. W. Rudolph, R. C. Taylor, and R. W. Parry, *J. Am. Chem. Soc.* **88**, 3729 (1966).
2. M. Lustig, J. K. Ruff, and C. B. Colburn, *J. Am. Chem. Soc.* **88**, 3875 (1966).
3. L. B. Centofanti and R. W. Rudolph, *Inorg. Syn.* **12**, 281 (1970).
4. S. Johnson, Ph.D. dissertation, Purdue University, Lafayette, Ind., 1953.
5. T. D. Farr, *Phosphorus—Properties of the Element and Some of its Compounds*, Chemical Engineering Report No. 8, Tennessee Valley Authority, Part XI, U.S. Government Printing Office, Washington, D.C., 1950.
6. *Handbook Volume*: P: Mvol. C, 1965, p. 589.
7. G. Tarbutton, E. P. Egan, Jr., and S. G. Frary, *J. Am. Chem. Soc.* **63**, 1783 (1941).
8. C. J. Hoffman, *Phosphorus—Fluorine Oxidizers*, PF-150613-1, Part 7, Propulsion Chemistry Part II, Lockheed Aircraft Corp., Missiles and Space Div., Burbank, Calif., 1959.
9. H. C. Duus and D. P. Mykytiuk, *J. Chem. Eng. Data* **9**, 585 (1964).
10. E. L. Muetterties and co-workers, *J. Inorg. Nucl. Chem.* **16**, 52 (1960).
11. Brit. Pat. 822,539 (Oct. 28, 1959), K. C. Brinker (to E. I. du Pont de Nemours & Co., Inc.).
12. U.S. Pat. 2,810,629 (Oct. 22, 1957), E. L. Muetterties (to E. I. du Pont de Nemours & Co., Inc.).
13. Fr. Pat. 2,476,054, A1, (Aug. 21, 1981), W. C. Cannon and N. R. Hall.
14. A. P. Hagen and D. L. Terrell, *Inorg. Chem.* **20**(4), 1325–1326 (1981).
15. B. R. Fowler and K. C. Moss, *J. Fluorine Chem.* **15**(1), 67–73 (1980).
16. A. Tasaka and O. Gleuser, *Dashisha Daigaku, Rikogaku Kenkyu Hokoku*, **14**(3), 175–196 (1973).
17. U.S. Pat. 3,769,387 (Oct. 30, 1973), R. A. Wiesboeck and J. D. Nickerson.
18. Fr. Pat. 2,082,502 (Jan. 14, 1972).
19. Ger. Offen. DE 2013858. R. A. Wiesboeck.
20. U.S. Pat. 3,584,999 (Jan. 15, 1971), R. A. Wiesboeck.
21. S. P. Mallela, O. D. Gupta, and J. M. Shreeve, *Inorg. Chem.* **27**(1), 208–209 (1988).

22. U.S. Pat. 342,019 (Sept. 17, 1968), H. L. Bowkley and R. B. Thurman (to Armour Agricultural Chemical Co.).
23. U.S. Pat. 3,428,422 (Feb. 18, 1969), R. A. Wiesboeck (to USS Agri-Chemicals, Inc.).
24. U.S. Pat. 3,634,034 (Jan. 11, 1972), J. D. Nickerson and R. A. Wiesboeck (to United States Steel Corp.).
25. U.S. Pat. 3,728,435 (Apr. 17, 1973), R. A. Wiesboeck (to United States Steel Corp.).
26. R. G. Wilson and D. M. Jamba, *Appl. Phys. Lett.* **22**, 176 (1973).
27. V. S. Zuev, L. D. Mikheev, and I. V. Pogorelskii, *Kvantovanya Elektron (Moscow)* (2), 394–400 (1974).
28. Eur. Pat. 333,084 Aw (Sept. 20, 1989), I. Harada and co-workers.
29. J. K. McDonald and R. W. Jones, *Proceedings of the SPIE-International Society of Opt. Engineering, 1986*, pp. 99–104, and 669.
30. R. G. Kalbandkeru and co-workers, *Indian J. Chem.* **23**(A)(12), 990–991 (1984).
31. Jpn. Pat. 03,164,429 A2 (July 16, 1991), K. Kitahara, T. Shimada, and K. Iwata (to Heisei).
32. Jpn. Pat. 03,178,313 A2 (Aug. 2, 1991), K. Kitahara, T. Shimada, and K. Iwata (to Heisei).
33. C. J. Hoffman, *Inorg. Syn.* **4**, 149 (1953).
34. A. A. Williams, *Inorg. Syn.* **5**, 95 (1957).
35. G. Wilkinson, *Nature (London)* **168**, 514 (1951).
36. G. Wilkinson, *J. Am. Chem. Soc.* **73**, 5501 (1951).
37. R. J. Clark, *Inorg. Chem.* **3**, 1395 (1964).
38. J. Chatt and A. A. Williams, *J. Chem. Soc.*, 3061 (1951).
39. Th. Kruck, *Z. Naturforsch* **196**, 164 (1964).
40. Zh. K. Dzhanmuldaeva and co-workers, *Aktual. Vap. Poluch., Fosfora. Soedin. Ego OSU*, 99–101 (1990).
41. S. H. P. Kuma and K. D. Padma, *J. Fluorine Chem.* **49**(3), 301–311 (1990).
42. G. R. Kalbandkeri and co-workers, *Indian J. Chem. Sect. A* **20**(A)(12), 83–84 (1981).
43. B. W. Moores and L. Andrews, *J. Phys. Chem.* **93**(5), 1902–1907 (1989).
44. H. D. Menz and co-workers, *Z. Anorg. Allg. Chem.*, 540–541, 191–197 (1986).
45. K. Heide, H. D. Menz, and C. Schmidt, *K. Kolditz*, **32**(8), 520 (1985).
46. S. B. Suresh and K. D. Padma; *J. Fluorine Chem.* **24**(4), 399–407 (1984).
47. M. Hawkins, M. J. Almond, and A. J. Downs, *J. Phy. Chem.* **89**(15), 3325–3334 (1985).
48. L. Kolditz and co-workers, *Dokl. Akad. Nauk. SSSR* **267**(6), 1392–1395 (1982).
49. A. P. Hagen and D. L. Terrl, *Inorg. Chem.* **20**(4), 1325–1326 (1981).
50. A. P. Hagen and B. W. Callaway, *Inorg. Chem.* **17**(3), 554–555 (1978).
51. K. D. Padma, S. K. Vijayalashmi, and A. R. Vasudevamurthy, *J. Fluorine Chem.* **8**(6), 461–465 (1976).
52. A. P. Hagen and E. A. Elphingston, *Inorg. Chem.* **12**(2), 478–480 (1973).
53. W. Lange, *Chem. Ber.* **60**, 962 (1927).
54. K. B. Boerner, C. Stoelzer, and A. Simon, *Ber.* **96**, 1328 (1963).
55. L. C. Mosier and W. E. White, *Ind. Eng. Chem.* **43**, 246 (1951).
56. U.S. Pat. 2,718,456 (Sept. 20, 1955), A. J. Mulder and W. C. B. Smithuysen (to Shell Development Co.).
57. U.S. Pat. 2,481,807 (Sept. 13, 1949), C. O. Anderson (to Ozark-Mahoning Co.).
58. J. M. Godwin and W. E. White, *Chem. Eng. News* **28**, 2721 (1950).
59. Y.-T. Chen, T.-C Li, and K.-C. Yin, *Sci Sinica (Peking)* **13**, 1719 (1952).
60. I. G. Ryss and V. B. Tul'chimskii, *Dokl. Akad, Nauk SSSR* **142**, 141 (1962).
61. L. N. Devonshire and H. H. Rowley, *Inorg. Chem.* **1**, 680 (1962).
62. D. P. Ames and co-workers, *J. Am. Chem. Soc.* **81**, 6350 (1959).
63. W. Lange and R. Livingston, *J. Am. Chem. Soc.* **72**, 1280 (1950).

64. A. S. Lenskii, A. D. Shaposhnikova, and A. S. Allilueva, *Zh. Prikl. Khim. (Leningrad)* **35**, 760 (1962).
65. G. Tarbutton, E. P. Egan, Jr., and S. G. Frary, *J. Am. Chem. Soc.* **63**, 1782 (1941).
66. U.S. Pat. 2,488,298 (Nov. 15, 1949), W. Lange and R. Livingston (to Ozark-Mahoning Co.).
67. H. Bode and G. Teufer, *Acta Cryst.* **8**, 611 (1955).
68. Ger. Pat. 812,247 (Aug. 27, 1951), H. Jonas (to Farbenfabriken Bayer A.G.).
69. H. E. Affsprung and U. S. Archer, *Anal. Chem.* **35**, 1912 (1963).
70. W. Lange, *Chem. Ber.* **62**, 793 (1929).
71. H. C. Goswami, *J. Indian Chem. Soc.* **14**, 660 (1937).
72. E. B. Singh and P. C. Sinha, *J. Indian Chem. Soc.* **41**, 407 (1964).
73. I. V. Mardirosova and co-workers, *Izv. Akad, Nauk SSSR Neorg. Mater.* **9**, 970 (1973).
74. J. M. Godwin, Masters Thesis, University of Tulsa, Tulsa, Oklahoma, 1950.
75. E. J. Duff, *Caries Res.* **7**, 79 (1973).
76. *The United States Pharmacopeia XIX-National Formulary XIV*, 4th Suppl., United States Pharmacopeial Convention, Inc., Rockville, Md, Jan. 31, 1978, USP XX-NF XV, 1980.
77. G. F. Hill and L. F. Audrieth, *Inorg. Syn.* **3**, 108 (1950).
78. H. H. Rowley and J. E. Stuckey, *J. Am. Chem. Soc.* **78**, 4262 (1956).
79. Vu Quang Kinh, *Bull Soc. Chim. Fr.*, 1466 (1962).
80. W. Lange, in W. C. Fernelius, ed., *Inorganic Syntheses*, Vol. II, McGraw-Hill Book Co., Inc., New York, 1946, pp. 155–158.
81. Ger. Pat. 813–848 (Sept. 17, 1951), H. Jonas (to Farbenfabriken Bayer A.G.).
82. I. G. Ryss and V. B. Tul'chimskii, *Zh. Neorgna. Khim.* **7**, 1313 (1962); *Ibid.*, **9**, 831 (1964).
83. W. Lange, *Chem. Ber.* **62**, 786 (1929).
84. U.S. Pat. 2,846,412 (Aug. 5, 1958), C. B. Havens (to The Dow Chemical Co.).
85. W. Lange and E. Mueller, *Chem. Ber.* **63**, 1058 (1930).
86. R. Schmutzler, in M. Stacey, J. D. Tatlow, and A. G. Sharpe, eds., *Advances in Fluorine Chemistry*, Vol. 5, Butterworth & Co., Inc., Washington, D.C., 1965, pp. 31–287.
87. M. M. Woyski, in L. F. Audrieth, ed., *Inorganic Syntheses*, Vol. III, McGraw-Hill Book Co., Inc., New York, 1950, pp. 111–117.
88. J. N. Sarmousakis and M. J. D. Law, *J. Am. Chem. Soc.* **77**, 6518 (1955).
89. F. A. Smith and co-workers, *Toxicol. Appl. Pharmacol.* **2**, 54 (1960).
90. A. V. Nikolaev and co-workers, *Izv. Sib. otd. Akad. Nauk SSSR Ser. Khim. Nauk*, 48 (1976).
91. D. W. A. Sharp and A. G. Sharpe, *J. Chem. Soc.*, 1855 (1956).
92. U.S. Pat. 3,189,658 (June 15, 1965), H. W. Quinn (to The Dow Chemical Co.).
93. *Chemical Warfare Service TDMR 832*, Edgewood Arsenal, Apr. 1944.
94. U.S. Pat. 2,409,039 (Oct. 8, 1946), E. E. Hardy and G. M. Kosolapoff (to Monsanto Chemical Co.).
95. *Registry of Toxic Effects of Chemical Substances*, 1977 ed., Vol. II, NIOSH, U.S. Department of Health, Education, and Welfare, Washington, D.C., pp. 667, 686.
96. T. Mahmood and J. M. Shreeve, *Inorg. Chem.* **25**, 3830–3837 (1986).
97. Ref. 96, pp. 4081–4084.
98. Ref. 96, pp. 3128–3131.

General References

R. Schmutzler, in M. Stacey, J. D. Tatlow, and A. G. Sharpe, eds., *Advances in Fluorine Chemistry*, Vol. 5, Butterworth & Co., Inc., Washington, D.C., 1965, pp. 31–287.
G. I. Drozd, *Usp. Khim.* **39**, 3 (1970).

W. Lange, in J. H. Simons, ed., *Fluorine Chemistry*, Vol. I, Academic Press, Inc., New York, 1950, pp. 125–188.

E. L. Muetterties and co-workers, *J. Inorg., Nucl. Chem.* **16**, 52 (1960).

S. Johnson, "Some Chemical and Physical Properties of Phosphorus Pentafluoride," Ph.D. dissertation, Purdue University, Lafayette, Ind., 1953.

CHARLES B. LINDAHL
TARIQ MAHMOOD
Elf Atochen North America, Inc.

POTASSIUM

The two stable salts of potassium and fluorine of commercial significance are the normal fluoride [7789-23-3], KF, and potassium bifluoride [7789-29-9], KHF_2.

Potassium Fluoride

Properties. Anhydrous potassium fluoride [7789-23-3] is a white hygroscopic salt that forms two hydrates, $KF \cdot 2H_2O$ [13455-21-5] and $KF \cdot 4H_2O$ [34341-58-7]. The tetrahydrate exists at temperatures below 17.7°C. The dihydrate is stable at room temperature and starts to lose water above 40°C. Temperatures on the order of 250–300°C are required to remove the last few percent of water in a reasonable period of time. Potassium fluoride does not pyrohydrolyze at temperatures as high as 1000°C (1). Chemical and physical properties of KF are summarized in Table 1.

Halogen exchange with KF is not successful in acetic acid (10). Hydrogen bonding of the acid hydrogen with the fluoride ion was postulated to cause acetate substitution for the halide; however, the products of dissolved KF in acetic acid are potassium acetate and potassium bifluoride (11). Thus KF acts as a base rather than as a fluorinating agent in acetic acid.

Manufacture. Commercial KF is manufactured from potassium hydroxide and hydrofluoric acid followed by drying in a spray dryer or flaking from a heated drum. The KF assay is typically 97–99%; impurities are $KF \cdot 2H_2O$ and either potassium carbonate or potassium bifluoride. The 1992 price of the anhydrous salt was $4.68/kg and that of the reagent-grade dihydrate ranged from $8–$14/kg. Potassium fluoride can be purified by passing anhydrous HF through the melt (12).

Toxicology. By ingestion, the lethal dose of potassium fluoride in guinea pigs is 250 mg/kg body weight. The LD_{50} orally for rats is 245 mg/kg body weight (13). Ingestion of potassium fluoride may cause vomiting, abdominal pains, and diarrhea.

Uses. Potassium fluoride is used in the manufacture of silver solder fluxes and in fluxes for various metallurgical operations (see SOLDERS AND BRAZING ALLOYS). In tin deposition from halogen plating baths, KF is used to complex tetravalent tin to form K_2SnF_6 [16893-93-9] which may be filtered from the so-

Table 1. Physical and Chemical Properties of Potassium Fluoride

Property	Value	Reference
melting point, °C	857	2
boiling point, °C	1505	2
specific gravity at 25°C	2.48	2
solubility, g/100 g		
in H_2O, 25°C	49.6	3
in liquid HF, 8°C	36.5	4
in acetic acid, 25°C	28	5
in methanol, 25°C	2.3	6
standard heat of formation, kJ/mol[a]		
KF	−567.4	2
$KF \cdot 2H_2O$	−1159	7
heat of fusion, kJ/mol[a]	28.2	2
heat of vaporization, kJ/mol[a]	173	7
heat of dehydration, $KF \cdot 2H_2O$, kJ/mol[a]	58.2	8
entropy at 25°C, J/(mol·K)[a]	66.6	2
lattice energy, kJ/mol[a]	813.8	9
specific heat, J/(kg·K)[a]		
at 0°C	833	2
at 50°C	854	2

[a]To convert kJ to kcal, divide by 4.184.

lution (see ELECTROPLATING) (14). Highly purified KF is formed into single crystals by the Stockbarger process (15). The principal use of the crystals is in the studies of fundamental properties and defects in alkali halide crystals used in introducing fluorine in organic synthesis.

For many types of replacement of halogens by fluorine in organic compounds, KF is the most frequently used fluoride. As a fluorinating agent, KF must be as anhydrous as possible and very finely divided. The potassium chloride or bromide by-products from reactions with organic chlorides or bromides deposit on the potassium fluoride crystal surfaces, significantly retarding the reaction. Polar solvents such as acetonitrile, dimethyl sulfoxide, or formamide, and rapid stirring are useful in overcoming this drawback. Ball-milling the reaction mixture is also helpful in speeding up the reaction. This fluorination process is used commercially in the manufacture of sodium fluoroacetate, a useful rat poison sold as "1080" (see FLUORINE COMPOUNDS, ORGANIC–FLUORINATED ACETIC ACID). An alkyl haloacetate is converted to an alkyl fluoroacetate which is then hydrolyzed using sodium hydroxide to the final product. Similarly, fluoroacetamide, a systemic insecticide, is made from chloroacetamide. Organic fluorides (16) that may be prepared from corresponding bromides or chlorides using potassium fluoride include monofluoroalkanes; α-fluoroesters, -ethers, and -alcohols; acyl fluorides; sulfonyl fluorides; and 1-fluoro-2,4-dinitrobenzene. Potassium fluoride behaves as a base when used as a catalyst for reactions such as dehydrohalogenation (17), Michael addition (18), and the Knoevenagel reaction (19). Polyurethane foams can be formed when KF is used as a catalyst for the reaction of adiponitrile carbonate and a polyester polyol (20) (see URETHANE POLYMERS).

Potassium Bifluoride

Properties. Other names for potassium bifluoride are potassium hydrogen difluoride and potassium acid fluoride. This white crystalline salt is a soft, waxy solid. The crystal forms of potassium bifluoride are tetragonal and cubic (21). The bifluoride ion in KHF_2 averages 0.2292 nm between fluoride ions in the F–H–F group (22). At elevated temperatures, potassium bifluoride exhibits an appreciable vapor pressure of HF. At 440°C, KHF_2 is decomposed completely to KF and HF; this decomposition is a convenient means of obtaining very pure HF. Other chemical and physical properties are summarized in Table 2.

Manufacture. Potassium bifluoride is produced from potassium hydroxide or potassium carbonate and hydrofluoric acid. The concentrated solution is cooled and allowed to crystallize. The crystals are separated centrifugally and dried. The commercial product consists typically of 99.7% KHF_2 and 0.2% KF. Potassium bifluoride is available in the United States in 180-kg drums at $4.04/kg (1992).

Toxicology and Handling. The lethal dose by ingestion in guinea pigs is 150 mg/kg body weight (13). The TLV for KHF_2 is 2.5 mg/m^3 (25). Potassium bifluoride crystals may break down to a fine white powder that is readily airborne. In this form, the salt is quite irritating to the nasal passages, eyes, and skin. Therefore, the hands and eyes should be protected and acid dust masks should be worn while handling, as an acid fluoride KHF_2 can cause superficial hydrofluoric acid-type burns. Areas of skin that have been in contact with potassium bifluoride should be washed as soon as possible with mildly alkaline soaps or borax-containing hand cleaners. If there has been contact with the eyes, they should be washed well with water and a physician should be consulted.

Uses. A primary use for potassium bifluoride is in the electrolyte for cells in fluorine manufacture (26). Sufficient hydrogen fluoride is dissolved with KHF_2 to bring the total HF content up to 40–42 wt %. This mixture approximates the formula $KF \cdot 2HF$; it is molten at 90°C, the operating temperature of the cell. Fluxes for a wide variety of metal joining applications utilize KHF_2, usually in combination with potassium pentaborate, boric acid, and other fluorides and chlo-

Table 2. Properties of Potassium Bifluoride, KHF_2

Property	Value	Reference
melting point, °C	238.8	23
specific gravity	2.37	24
solubility in H_2O, 20°C, g/100 g	39.2	
crystal transition temperature, °C	196.7	23
standard heat of formation, kJ/mol[a]	−920.4	7
heat of fusion, kJ/mol[a]	6.6	23
heat of dissociation to KF + HF, 226.8°C, kJ/mol[a]	77.5	24
entropy, 25°C, J/(mol·K)[a]	104.3	7
heat capacity, C_p, 25°C, J/(mol·K)[a]	76.86	7
lattice energy, kJ/mol[a]	641.8	24

[a]To convert kJ to kcal, divide by 4.184.

rides such as $ZnCl_2$ (27). Solutions of KHF_2 and citric acid are used to etch aluminum prior to coating with an acrylic copolymer emulsion; this opaque white coating resists chipping and is useful for automotive, architectural, and decorative applications (28). Tetrahydrofuran is polymerized to poly(tetramethylene glycol) with fuming sulfuric acid and potassium bifluoride (29).

BIBLIOGRAPHY

"Potassium Compounds" under "Fluorine Compounds, Inorganic," in *ECT* 1st ed., Vol. 6, pp. 721–722, by D. C. Whitaker, The Harshaw Chemical Co.; "Potassium" under "Fluorine Compounds, Inorganic," in *ECT* 2nd ed., Vol. 9, pp. 649–650, by D. C. Whitaker, The Harshaw Chemical Co.; in *ECT* 3rd ed., Vol. 10, pp. 789–792, by H. S. Halbedel and T. E. Nappier, The Harshaw Chemical Co.

1. D. L. Deadman, J. S. Machin, and A. W. Allen, *J. Am. Ceram. Soc.* **44**(3), 105 (1961).
2. *JANAF Thermochemical Tables*, Clearinghouse for Federal, Scientific, and Technical Information, U.S. Dept. of Commerce, Springfield, Va., 1964.
3. J. H. Simons, *Fluorine Chemistry*, Vol. 1, Academic Press, Inc., New York, 1950, p. 28.
4. A. W. Jache and G. W. Cady, *J. Phys. Chem.* **56**, 1106 (1952).
5. J. Emsley, *J. Chem. Soc. A*, 2511 (1971).
6. R. E. Harner, J. B. Sydnor, and E. S. Gilbreath, *J. Chem. Eng. Data* **8**, 411 (1963).
7. F. D. Rossini and co-workers, *NBS, Circ. 500*, U.S. Government Printing House, Washington, D.C., 1952.
8. J. Bell, *J. Chem. Soc.*, 72 (1940).
9. H. Vaino and M. Kanko, *Ann. Univ. Turku. Ser. AI*, (40), 3 (1960).
10. J. H. Clark and J. Emsley, *J. Chem. Soc. Dalton Trans.*, 2129 (1975).
11. V. Kazakov and V. G. Kharchuk, *Zh. Obshch. Khim.* **45**, 2744 (1975).
12. H. Kojima, S. G. Whiteway, and C. R. Masson, *Can. J. Chem.* **46**, 2968 (1968).
13. H. C. Hodge and F. A. Smith, in J. H. Simons, ed., *Fluorine Chemistry*, Vol. 4, Academic Press, Inc., New York, 1965, p. 200.
14. I. Rajagonal and K. S. Rajams, *Met. Finish.* **76**(4), 43 (1978).
15. D. C. Stockbarger, *J. Opt. Soc. Am.* **14**, 448 (1927).
16. A. R. Basbour, L. F. Belf, and M. W. Bruxton, in M. Stacy and co-eds., *Advances in Fluorine Chemistry*, Vol. 3, Butterworth, Washington, D.C., 1963, pp. 181–250.
17. F. Naso and L. Ronzini, *J. Chem. Soc. Perkin Trans. 1*, 340 (1974); J. H. Clark and J. M. Miller, *J. Am. Chem. Soc.* **99**, 498 (1977); J. H. Clark, J. Emsley, and O. P. A. Hoyta, *J. Chem. Soc. Perkin Trans. 1*, 1091 (1977).
18. I. Belski, *Chem. Commun.*, 237 (1977).
19. L. Rand, J. V. Swisher, and C. J. Cromin, *J. Org. Chem.* **27**, 3505 (1962); L. Rand, D. Haidukewych, and R. J. Dohinski, *J. Org. Chem.* **31**, 1272 (1966).
20. U.S. Pat. 3,766,147 (July 31, 1972), L. G. Walgemuth (to Atlantic Richfield Co.).
21. R. Kruh, K. Fuwa, and T. E. McEver, *J. Am. Ceram. Soc.* **78**, 4526 (1956).
22. H. L. Carrell and J. Donohue, *Isr. J. Chem.* **10**(2), 195 (1972).
23. M. L. Davis and E. F. Westrum, Jr., *J. Phys. Chem.* **65**, 338 (1961).
24. T. C. Waddington, *Trans. Faraday Soc.* **54**, 25 (1958).
25. *Proceedings of American Conference Governing Individual Hygiene*, Cincinnati, Ohio, 1977.
26. S. P. Vavalides and co-workers, *Ind. Eng. Chem.* **50**(2), 178 (1958).

27. U.S. Pat. 2,829,078 (Apr. 1, 1958), H. B. Aull and A. S. Cross; U.S. Pat. 3,958,979 (May 29, 1976), A. R. Valdo (to Ethyl Corp.); Jpn. Kokai 75 113,449 (Sept. 5, 1975), K. Motoyoshi, M. Kume, and Y. Amano (to Sumitomo Electric Industries).
28. U.S. Pat. 3,849,208 (Aug. 13, 1973), M. N. Marosi (to Convertex, Ltd.).
29. Jpn. Kokai 73 01,100 (Jan. 9, 1973), K. Matsuzawa, Y. Suzuki, and K. Ohya (to Mitsubishi Chemical Industries Co.).

JOHN R. PAPCUN
Atotech

RHENIUM

Rhenium Hexafluoride

Rhenium hexafluoride [10049-17-9], ReF_6, is a pale yellow solid at 0°C, but a liquid at ambient temperature. In the presence of moisture it hydrolyzes rapidly forming HF, ReO_2, and $HReO_4$ (see RHENIUM AND RHENIUM COMPOUNDS). It is not safe to store ReF_6 in a glass trap or glass-lined container. Leaks in the system can initiate hydrolysis and produce HF. The pressure buildup causes the system to burst and an explosion may result.

Properties. Some physical properties of ReF_6 are mol wt, 300.19; mp, 18.5°C; bp, 33.7°C; solubility in HF, 52.5 g/100 g; specific gravity, 3.58; and vapor pressure at 20.3°C, 61 kPa (458 mm Hg). The transition point has been reported as -3.45 (1) and -1.9°C (2). The compound can be handled in dry metal vacuum lines made of copper, nickel, stainless steel, or Monel. It forms a passive fluoride film on the surface which protects these metals from further corrosion. Reaction with nitric oxide yields nitrosonium hexafluororhenate [60447-76-9], $NOReF_6$ (3), and with potassium fluoride yields potassium octafluororhenate [57300-90-0], K_2ReF_8 (4). Reaction with alkali metal iodides dissolved in SO_2 results in the reduction to rhenium(IV) complex salts, M_2ReF_6 (M = Na [12021-61-3], K [16962-12-2], Rb [16962-13-3], and Cs [16962-14-4]) (5).

Rhenium hexafluoride is readily prepared by the direct interaction of purified elemental fluorine over hydrogen-reduced, 300 mesh (ca 48 μm) rhenium powder at 120°C. The reaction is exothermic and temperature rises rapidly. Failure to control the temperature may result in the formation of rhenium heptafluoride. The latter could be reduced to rhenium hexafluoride by heating with rhenium metal at 400°C.

Rhenium hexafluoride is used for the deposition of rhenium metal films for electronic, semiconductor, laser parts (6–8), and in chemical vapor deposition (CVD) processes which involve the reduction of ReF_6 by hydrogen at elevated (550–750°C) temperatures and reduced (< 101.3 kPa (1 atm)) pressures (9,10).

Rhenium hexafluoride is a costly (ca $3000/kg) material and is often used as a small percentage composite with tungsten or molybdenum. The addition of rhenium to tungsten metal improves the ductility and high temperature properties of metal films or parts (11). Tungsten–rhenium alloys produced by CVD processes exhibit higher superconducting transition temperatures than those alloys produced by arc-melt processes (12).

Rhenium hexafluoride (99.5% pure) is commercially available from Advance Research Chemicals, Atomergic, Atochem, Spectra Gases, and Matheson Gas of the United States, Fluorochem of the United Kingdom, and other sources. The 1993 price for small quantities varied from $3000 to $3500/kg. Larger quantities were available at $2000 to $2500/kg depending on the price of rhenium metal. U.S. production is less than 100 kg/yr. Because of its high irritating and corrosive nature it is classified as corrosive, poisonous liquid and shipped in steel, stainless steel, or Monel cylinders. Upon exposure to air it hydrolyzes producing HF fumes that are corrosive to the lower respiratory tract, skin, and eyes. Prolonged exposure to fumes may cause pulmonary edema. ACGIH (1992–1993) adopted TLV for fluorides as F^- is 2.5 mg/m^3; therefore great care should be taken while handling ReF_6. Personnel working with this material should use vacuum lines or closed systems located in a chemical hood. All precautions must be taken to avoid breathing of vapors or contact with skin.

Other Rhenium Fluoride Compounds

Rhenium heptafluoride [17029-21-9], ReF_7, is obtained by the direct interaction of elemental fluorine with hydrogen-reduced rhenium powder at 400°C and slightly over atmospheric pressure of fluorine. It is a pale yellow solid, mol wt 319.19; mp, 48.3°C; and bp, 73.7°C.

Rhenium pentafluoride [30937-52-1], ReF_5, is obtained along with rhenium tetrafluoride [15192-42-4], ReF_4, when reduction of ReF_6 is carried out with metal carbonyls (qv). ReF_5 is a greenish yellow solid with mp 48°C. Its ready thermal decomposition and magnetic properties suggest that it may be $ReF_4^+ ReF_6^-$ (13). ReF_4, best prepared by the reduction of rhenium hexafluoride with hydrogen at 200°C, is a pale blue solid melting at 124.5°C and boiling at 795°C.

Rhenium also forms several important oxyfluorides: rhenium oxytetrafluoride [17026-29-8], $ReOF_4$; rhenium oxypentafluoride [23377-53-9], $ReOF_5$; rhenium dioxytrifluoride [57246-89-6], ReO_2F_3; and perrhenyl fluoride [25813-73-4], ReO_3F. All are solids at room temperature. Properties are summarized in Table 1.

Table 1. Rhenium Oxyfluorides[a]

Compound	$ReOF_5$	ReO_2F_3	ReO_3F	$ReOF_4$	$ReOF_3$
preparative route	$ReO_2 + F_2$	$ReO_2 + F_2$	$KReO_4 + IF_5$	$ReF_6 + M(CO)_x$	$ReF_6 + M(CO)_x$
color	cream	pale yellow	yellow	blue	black
mp, °C	40.8[b]	90	71	107.8[b]	> 200

[a]Refs. 14–16.
[b]Transition point.

BIBLIOGRAPHY

"Rhenium" under "Fluorine Compounds, Inorganic" in *ECT*, 3rd ed., Vol. 10, pp. 792–794, by A. J. Woytek, Air Products and Chemicals, Inc.

1. G. H. Cady and C. B. Hargreaves, *J. Chem. Soc.*, 1563–1568 (1961).
2. J. G. Malm and H. Selig, *J. Inorg. Nucl. Chem.* **20**, 189 (1961).
3. N. Bartlett, S. P. Beaton, and N. K. Jha, *Chem. Commun.*, 168 (1966).
4. E. G. Ippolitono, *J. Inorg. Chem., USSR* **7**, 485 (1962).
5. R. D. Peacock, *J. Chem. Soc.*, 467 (1957).
6. Jpn. Kokai Tokkyo Koho 03,293,725 (Dec. 25, 1991), T. Tsutsumi (to Mitsubishi Electric Corp.).
7. PCT Int. Appl. 9,201,310 (Jan. 23, 1992), T. Ohmi.
8. Jpn. Kokai Tokkyo Koho 03,36,734 (Feb. 18, 1991), T. Ooba (to Fujitsu Ltd.).
9. U.S. Pat. 3,565,676 (Feb. 23, 1971), R. A. Holzl (to Fanstell Metallurgical Corp.).
10. F. W. Hoertel and J. G. Donaldson, *J. Electrodep. Surf. Treat.* **2**, 343 (1974).
11. W. L. Roberts, *High Temp. Mater. Pap. Plansee Semin. 6th*, 880–884 (1969).
12. D. S. Easton and co-workers, *Philos Mag.* **30**, 1117 (1974).
13. F. A. Cotton and G. Wilkinson, *Advanced Inorganic Chemistry, A Comprehensive Text*, 3rd ed., Interscience Publishers, Division of John Wiley & Sons, New York, 1972, p. 977.
14. O. Ruff and W. Kwansik, *Z. Anorg. Chem.* **219**, 65 (1934).
15. E. E. Aynsley and M. L. Hair, *J. Chem. Soc.*, 3747 (1958).
16. A. Englebrecht and A. V. Grosse, *J. Am. Chem. Soc.* **76**, 2042 (1954).

DAYAL T. MESHRI
Advance Research Chemicals, Inc.

SILVER

Silver Subfluoride

Pure silver subfluoride [1302-01-8], Ag_2F, is a greenish shiny crystalline material, or yellowish green solid if contaminated with AgF. It decomposes in water but is stable in alcohol and saturated solutions of AgF. Ag_2F disproportionates to Ag and AgF when heated above 100°C (1,2).

Silver subfluoride is prepared by heating a concentrated solution of silver fluoride with metallic silver powder. It may also be obtained by electrolysis of a solution containing AgF, aqueous HF, and NH_4F at 50°C in a platinum dish using a silver electrode. Another process for the manufacture of Ag_2F involves heating a mixture of a saturated solution of AgF and fine Ag powder. This process yields a yellowish green solid mass. Silver subfluoride is a reagent of laboratory curiosity. No commercial applications are known.

Silver Fluoride

Anhydrous silver fluoride [7775-41-9], AgF, is a golden yellow solid in its pure form and is classified as a soft fluorinating agent (3). Several solid phases of the solvated species are reported for AgF in the system AgF–HF–H$_2$O (4), eg, AgF·2H$_2$O [72214-21-2], AgF·4H$_2$O [2242-42-6], 3AgF·2HF [72318-57-1], AgF·2HF [12444-84-7], AgF·3HF [12444-85-8], AgF·5HF [12444-86-9], and AgF·7HF·2H$_2$O [72318-56-0]. In addition, AgF·3HF is formed in the absence of water at 0°C. When this last is warmed to 25°C under an atmosphere of dry nitrogen it dissociates into AgF·HF or AgHF$_2$ [12249-52-4] and HF.

Preparation. Silver fluoride can be prepared by dissolving Ag$_2$O or Ag$_2$CO$_3$ in anhydrous hydrogen fluoride or aqueous hydrofluoric acid, evaporating to dryness, and then treating with methanol or ether.

Properties. Silver fluoride is light sensitive and has a specific gravity of 5.852. It melts at 435°C into a black liquid which boils at 1150°C. Unlike the other halides, it is extremely soluble (182 g/100 g) in water and in anhydrous hydrogen fluoride (83.2 g/100 g at 11.9°C). It is only slightly soluble in absolute methanol (1.5 g/100 mL). Its heat of formation, ΔH_f, is -204.6 kJ/mol (-48.9 kcal/mol); heat capacity, C_p, is 51.92 J/(mol·K) (12.4 cal/(mol·K)); and entropy, S, 83.7 J/(mol·K) (20.0 cal/(mol·K)).

Silver fluoride forms explosive adducts with ammonia (qv) (5,6), and therefore all of the reactions involving liquid or gaseous ammonia should be carried out with extreme precautions.

Uses. Silver fluoride has found many laboratory and special industrial applications. It is used as a soft (mild) fluorinating agent for selective fluorination (7–17), as a cathode material in batteries (qv) (18), and as an antimicrobial agent (19). Silver fluoride is commercially available from Advance Research Chemicals, Inc., Aldrich Chemicals, Cerac Corp., Johnson/Matthey, PCR, Atochem, and other sources in the United States. The U.S. price of silver fluoride in 1993 was $1000–$1400/kg and the total U.S. consumption was less than 200 kg/yr.

Silver Difluoride

Silver difluoride [7783-95-1], AgF$_2$, is a black crystalline powder. It has been classified as a hard fluorinating agent (3) which liberates iodine from KI solutions and ozone from dilute aqueous acid solutions on heating. It spontaneously oxidizes xenon gas to Xe(II) in anhydrous hydrogen fluoride solutions (20).

AgF$_2$ is prepared by the action of elemental fluorine on AgF or AgCl at 200°C. Both processes result in quantitative yields. Silver difluoride should be stored in Teflon, passivated metal containers, or in sealed quartz tubes.

Properties. Silver difluoride melts at 690°C, boils at 700°C, and has a specific gravity of 4.57. It decomposes in contact with water. Silver difluoride may react violently with organic compounds, quite often after an initial induction period. Provisions must be made to dissipate the heat of the reaction. Small-scale experiments must be run prior to attempting large-scale reactions.

Uses. AgF$_2$ is a powerful fluorinating agent and is used for substitution of hydrogen by fluorine in hydrohalocarbons (21), preparation of perfluorocompounds (21–23), purification of perfluoromorpholines containing partially fluorinated hydrocarbons (24), fluorination of compounds containing triple bonds (25), addition of fluorine to unsaturated halocarbons (25), and conversion of carbon monoxide into carbonyl fluoride.

Silver difluoride, commercially available from the same sources as those of AgF, had a 1993 price between $1000–$1400/kg. In spite of the technical success in laboratory experiments, silver fluorides have found limited use on a large scale mainly because of the high cost of the reagents. Demand for silver difluoride is less than 100 kg/yr.

Silver Trifluoride

The existence of diamagnetic salts of AgF_4^- was first reported in 1957 (26), but little was known about their properties. In 1988 (27) it was claimed that AgF_3 was prepared by a reaction of Ag metal and O_2F_2 in ClF_5. Silver trifluoride [91899-63-7], AgF_3, has since been prepared (28) from anhydrous HF solutions of AgF_4^- salts by addition of BF_3, PF_5, or AsF_5.

If excess AsF_5 is added, silver(III) is reduced and the $AgFAsF_6$ salt is produced.

$$AgF_3 + AsF_5 \rightarrow AgFAsF_6 + 1/2\ F_2$$

The red precipitates of AgF_3 are diamagnetic and isostructural with AuF_3. Silver trifluoride is a powerful oxidizing agent and thermodynamically unstable. Its powerful oxidizing properties result from the tight binding of its valence shell d-orbital electrons. No commercial source is available.

Silver Fluorocomplexes

The silver fluorocomplexes, ie, silver hexafluoroantimonate [26042-64-8], $AgSbF_6$; silver hexafluorophosphate [26042-63-7], $AgPF_6$; silver tetrafluoroborate [14104-20-2], $AgBF_4$; and other salts such as silver trifluoromethane sulfonate [2923-28-6], CF_3SO_3Ag, and silver trifluoroacetate [2966-50-9], CF_3COOAg, play an important role in the synthesis of organic compounds and have gained potential industrial importance.

These compounds perform a dual function in synthesis procedures. The introduction of a complex anion assists in the stabilization of the desired product and the generation of unique intermediates by chloride displacement, eg, silver hexafluorophosphate, $AgPF_6$, forms adducts with neutral diamagnetic organometallics which can act as controlled sources of highly reactive cations (29). Silver hexafluoroantimonate, $AgSbF_6$, is an electrophilic bromination catalyst (30) and is also used in promoting chlorination of reactive alkanes (31). Silver trifluoromethane sulfonate, CF_3SO_3Ag, is an excellent precursor to a number of derivatives useful as alkylating agents for aromatic compounds (32).

Silver fluorocomplexes are also used in the separation of olefin–paraffin mixtures (33), nitration (qv) of aromatic compounds (34), in the synthesis of o-bridged bicyclics (35), pyrroles (36), cyclo-addition of vinylbromides to olefins (37), and in the generation of thiobenzoyl cations (38).

These complex salts are very hygroscopic and light sensitive. They are slightly soluble in anhydrous hydrogen fluoride, very soluble in water, and soluble in organic solvents such as acetonitrile, benzene, toluene, and m-xylene. Except for the melting point of CF_3COOAg (257–260°C), not many other physical properties are known. Most of the salts decompose at higher temperatures.

These salts are corrosive and are to be considered toxic because of the presence of Ag^+ ions. The American Conference of Government Industrial Hygienists (ACGIH) (1992–1993) has adopted TWA values of 0.01 mg/m^3 for silver metal and 0.01 mg/m^3 for soluble silver salts. TWA for fluorides as F^- ions is 2.5 mg/m^3. The MSDS should be consulted prior to use. Skin contact and inhalation should be avoided.

These salts are commercially available. Worldwide consumption of fluorocomplex salts varies between 100 to 300 kg/yr. The most popular salt is $AgBF_4$. Prices vary between $1000 and $1400/kg.

BIBLIOGRAPHY

"Silver Compounds" under "Fluorine Compounds, Inorganic," in *ECT* 1st ed., Vol. 6, pp. 730–731, by F. D. Loomis, Pennsylvania Salt Manufacturing Co.; "Silver" under "Fluorine Compounds, Inorganic," in *ECT* 2nd ed., Vol. 9, pp. 661–662, by W. E. White, Ozark-Mahoning Co.; in *ECT* 3rd ed., Vol. 10, pp. 795–797, by D. T. Meshri, Ozark-Mahoning Co.

1. R. Scholder and K. Traulsen, *Z. Anorg. Allg. Chem.* **197**, 57 (1931).
2. X. L. Wang, *J. Phys. Soc. Jpn.* **60**(4), 1398–1405 (1991).
3. D. T. Meshri and W. E. White, *George H. Cady ACS Symposium*, Milwaukee, Wis., June 1970.
4. H. J. Thomas and A. W. Jache, *J. Inorg. Nucl. Chem.* **13**, 54 (1960).
5. L. J. Olner and M. Dervin, *Compt. Rend.* **175**, 1085 (1922).
6. W. Blitz and E. Rahlfs, *Z. Anorg. Allg. Chem.* **166**, 351 (1927).
7. D. T. Meshri and W. T. Miller, *158th ACS National Meeting*, New York, Sept. 1969, abstract 14.
8. W. T. Miller, R. A. Snider, and D. T. Meshri, *4th Winter Fluorine Conference*, Daytona Beach, Fla., Feb. 1979.
9. E. D. Bergmann and I. Shalhok, *J. Chem. Soc.*, 1418 (1959).
10. H. J. Emeleus and D. E. McDuffe, *J. Chem. Soc.*, 2597 (1961).
11. J. H. Simons, D. F. Herman, and W. H. Pearson, *J. Am. Chem. Soc.* **68**, 1672 (1946).
12. A. D. Britt and W. B. Moniz, *J. Am. Chem. Soc.* **91**, 6204 (1969).
13. A. Takashi, D. G. Cork, M. Fujta, K. Takahide, and T. Tatsuno, *Chem. Lett.* **11**, 187–188 (1988).
14. B. C. Jago and J. Gittins, *Am. Mineral* **74**(7–8), 936–937 (1989).
15. L. F. Chen, J. Mohthasham, and G. L. Gard, *J. Fluorine Chem.* **49**(3), 331–347 (1990).
16. C. M. Wang and T. E. Mallouk, *J. Am. Chem. Soc.* **112**(5), 2016–2018 (1990).
17. Jpn. Kokai Tokkyo Koho 02,169,523 (June 29, 1990), I. Yuji, A. Nakahara, and J. Nakajima.
18. Jpn. Kokai 75,131,034 (Oct. 16, 1961), T. M. Saaki, (to Japan Storage Battery Co. Ltd.).
19. *Chem. Week* **52** (Jan. 14, 1961).
20. B. Zemva, and co-workers, *J. Am. Chem. Soc.* **112**(12), 4849–4849 (1990).
21. W. B. Burford, III and co-workers, *Ind. Eng. Chem.* **39**, 379 (1947).
22. E. T. McBee and co-workers, in Ref. 21, p. 310.
23. Jpn. Kokai Tokkyo Koho 03,167,141 (July 19, 1991), H. Okajima, T. Hiroshi, I. Fuyuhiko, S. Masamichi, and S. Shiro (to Kanto Denka Kogyo Co. Ltd.).
24. Ger. Pat. 287,477 (Feb. 28, 1991), W. Radeck, S. Ruediger, A. V. Dimitrov, and H. Stewig (to Akademie der Wissenschaften der DDR).

25. D. A. Rausch, R. A. Davis, and D. W. Osborne, *J. Org. Chem.* **28**, 494 (1963).
26. R. Hoppe, *Z. Anorg. Allg. Chem.* **292**, 28 (1957).
27. Yu. M. Kiselev, A. I. Popov, A. A. Timakov, and K. V. Bukharin, *Zh. Neorg. Khim.* **33**(5), 1252–1256 (1988).
28. B. Zemva, and co-workers, *J. Am. Chem. Soc.* **113**, 4192–4198 (1991).
29. N. G. Connelly, A. R. Lucy, and A. M. R. Galas, *Chem. Commun.*, 43 (1981).
30. G. A. Olah and P. Schilling, *J. Am. Chem. Soc.* **95**, 7680 (1973).
31. G. A. Olah, R. Renner, P. Schilling, and Y. K. Mo, *J. Am. Chem. Soc.* **95**, 7686 (1973).
32. B. L. Booth, R. N. Hazeldine, and K. Laak, *J. Chem. Soc., Perkin Trans.*, 2887 (1980).
33. U.S. Pat. 3,189,658 (June 15, 1965), H. W. Quinn (to The Dow Chemical Co.).
34. G. A. Olah and S. Kahn, *J. Am. Chem. Soc.* **83**, 4564 (1961).
35. J. Mann and A. A. Usman, *Chem. Commun.*, 119 (1980).
36. J. E. Baeckvall and J. E. Nystroem, *Chem. Commun.*, 59 (1981).
37. M. Hanock, I. Harder, and K. R. Balinger, *Tetrahedron Lett.* **22**, 553 (1981).
38. G. A. Olah, G. K. S. Prakash, and T. Nakajima, *Angew. Chem., Int. Ed. Engl.* **19**, 812 (1980).

DAYAL T. MESHRI
Advance Research Chemicals, Inc.

SODIUM

Sodium has two fluorides, sodium fluoride [7722-88-5] and sodium bifluoride [1333-83-1].

Sodium Fluoride

Sodium fluoride, NaF, is a white, free-flowing crystalline powder, mp 992°C, bp 1704°C, with a solubility of 4.2 g/100 g water at 10°C, and 4.95 g/100 g water at 93.3°C. The purity of the commercial material is about 98%.

Sodium fluoride is normally manufactured by the reaction of hydrofluoric acid and soda ash (sodium carbonate), or caustic soda (sodium hydroxide). Control of pH is essential and proper agitation necessary to obtain the desired crystal size. The crystals are centrifuged, dried, sized, and packaged. Reactors are usually constructed of carbon brick and lead-lined steel, with process lines of stainless, plastic or plastic-lined steel; diaphragm, plug cock, or butterfly valves are preferred.

The salt is packaged in 45-kg multiwall bags or fiber drums of 45, 170, or 181 kg. It is available in both powdered and granular forms with densities of 1.04 and 1.44 g/cm^3 (65 and 90 lb/ft^3), respectively. Only the powdered grade is authorized by and registered with the EPA for use in pesticide formulations, with the further proviso that it must be tinted blue or green, or otherwise discolored. The word poison appears on all labels together with first-aid information.

Both sodium fluoride and sodium bifluoride are poisonous if taken internally. Dust inhalation and skin or eye contact may cause irritation of the skin, eyes, or respiratory tract, and should be avoided by the use of proper protective equipment (1).

Fluoridation of potable water supplies for the prevention of dental caries is one of the principal uses for sodium fluoride (see WATER, MUNICIPAL WATER TREATMENT). Use rate for this application is on the order of 0.7 to 1.0 mg/L of water as fluoride or 1.5 to 2.2 mg/L as NaF (2). NaF is also applied topically to teeth as a 2% solution (see DENTIFRICES). Other uses are as a flux for deoxidizing (degassing) rimmed steel (qv), and in the resmelting of aluminum. NaF is also used in the manufacture of vitreous enamels, in pickling stainless steel, in wood preservation compounds, casein glues, in the manufacture of coated papers, in heat-treating salts, and as a component of laundry sours.

Sodium Bifluoride

Sodium bifluoride (sodium acid fluoride, sodium hydrogen fluoride), $NaHF_2$ or $NaF \cdot HF$, is a white, free-flowing fine granular material. Its solubility in water is 3.7 g/100 g solution at 20°C, and 16.4 g/100 g at 80°C. It decomposes at temperatures above 160°C to give sodium fluoride and hydrogen fluoride. Commercial material is ca 99% pure. To prevent the formation of irritating dust, wetted products, containing 85–90% $NaHF_2$ and 10–15% water, are also in use.

The same reactants are used for manufacture as for sodium fluoride. An excess of acid is required to crystallize the bifluoride. The crystals are dewatered, dried, sized, and packaged. Cooling of the reaction is necessary to avoid overheating and decomposition. Reactors and auxiliary equipment are the same as for sodium fluoride.

The dried salt is shipped in 45-kg multiwall bags and in 57-, 170-, and 180-kg fiber drums. Densities range from ca 0.70 g/cm^3 (44 lb/ft^3) to 1.2 g/cm^3 (75 lb/ft^3) for crystalline material.

Sodium bifluoride, by itself or in conjunction with other materials, is a good laundry sour because, in the concentrations used, it does not create a pH below 4.0 and thus causes no damage to textile fibers, although it removes iron stains. Leather (qv) bleaching and cleaning of stone and brick building faces are other uses for this material (3).

BIBLIOGRAPHY

"Sodium Fluoride" under "Fluorine Compounds, Inorganic," in *ECT* 1st ed., Vol. 6, p. 731, by F. D. Loomis, Pennsylvania Salt Manufacturing Co.; "Sodium Bifluoride" under "Fluorine Compounds, Inorganic," in *ECT*, 1st ed., Vol. 6, pp. 731–732, by J. E. Dodgen, Pennsylvania Salt Manufacturing Co.; "Sodium Fluoride" and "Sodium Bifluoride," under "Fluorine Compounds, Inorganic," in *ECT* 2nd ed., Vol. 9, pp. 662–663, by J. Griswold, Allied Chemical Corp.; "Sodium" under "Fluorine Compounds, Inorganic" in *ECT* 3rd ed., Vol. 10, pp. 797–798, by K. Wachter, Olin Corp.

1. N. I. Sax, *Dangerous Properties of Industrial Materials*, 4th ed., Van Nostrand Reinhold, New York, 1975.

2. *Fluoridation Engineering Manual*, EPA, Office of Water Programs, Washington, D.C., 1972.
3. M. Windholz, ed., *Merck Index*, 9th ed., Merck & Co., Inc., Rahway, N.J. 1976.

WERNER H. MUELLER
Hoechst-Celanese Corporation

SULFUR

Sulfur fluorides, **188**
Fluorosulfuric acid, **202**

SULFUR FLUORIDES

The known binary compounds of sulfur and fluorine range in character from ephemeral to rock-like and provide excellent examples of the influence of electronic and structural factors on chemical reactivity. These marked differences are also reflected in the diversified technological utility.

Sulfur Hexafluoride

Sulfur hexafluoride [2551-62-4], SF_6, molecular weight 146.07, is a colorless, odorless, tasteless gas. It is not flammable and not particularly reactive. Its high chemical stability and excellent electrical characteristics have led to widespread use in various kinds of electrical and electronic equipment such as circuit breakers, capacitors, transformers, microwave components, etc (see ELECTRONIC MATERIALS). Other properties of the gas have led to limited usage in a variety of unique applications ranging from medical applications to space research.

Sulfur hexafluoride was first prepared in 1902 (1). The discovery in 1937 that its dielectric strength is much higher than that of air (2) led to its use as an insulating material for cables, capacitors (3), and transformers (4) (see INSULATION, ELECTRIC). Sulfur hexafluoride has been commercially available as AccuDri, SF_6 (AlliedSignal Inc.) since 1948. It is also produced by Air Products and Chemicals in the United States and by others in Germany, Italy, Japan, and Russia.

Properties. Sulfur hexafluoride is a good dielectric because a high gas density can be maintained at low temperatures. Properties are given in Table 1.

The vapor pressure of the liquid for the range −50 to 45.6°C and a standard % deviation of ±0.18 is calculated as (10):

Table 1. Physical Properties[a] of Sulfur Hexafluoride

Property	Value	References
sublimation point, °C	−63.9	5
heat of sublimation, kJ/mol[b]	23.59	6
triple point, °C	−50.52	7
pressure at triple point, kPa[c]	225.31	7
critical temperature, °C	45.55	8, 9
critical pressure, MPa[d]	3.759	9, 10
critical density, g/cm^3	0.737	10–12
density, g/cm^3		
solid at −195.2°C	2.863	13
liquid	1.336	10, 14
gas	6.0886 × 10^{-3}	15
vapor pressure of saturated liquid, MPa[d]	2.3676	15
heat of formation, kJ/mol[b]	−1221.66	
free energy of formation, kJ/mol[b]	−1117.73	16
heat of vaporization, kJ/mol[b]	9.6419	15
entropy, kJ/(mol·K)[b]	291.874	16
heat capacity, J/(mol·K)[b]		
liquid at −43°C	119.5	17
gas	97.234	15
surface tension at −20°C, mN/m(=dyn/cm)	8.02	18
viscosity, mPa·s(=cP)		
liquid	0.277	14
gas	0.01576	17
thermal conductivity, W/(m·K)		
liquid	0.0583	19
gas	0.01415	20
sound velocity, gas, m/s	136	14
refractive index, n_D	1.000783	21
dielectric constant		
liquid	1.81	22
gas	1.00204	23
loss tangent (liquid)	0.001	22

[a] All data refer to 25°C and 101.3 kPa (1 atm), unless otherwise stated.
[b] To convert J to cal, divide by 4.184.
[c] To convert kPa to mm Hg, multiply by 7.5.
[d] To convert MPa to atm, divide by 0.101.

$$\log P_{kPa} = 0.87652594 - 816.48995/T + 0.029287342T$$
$$- 0.40107549 \times 10^{-4} T^2$$
$$+ 0.7142667 \frac{(319.802 - T)}{T} \log(319.802 - T)$$

where T is in Kelvin. Equations for the calculation of sublimation pressure are available (5,24). Heats of vaporization, calculated from the Clausius-Clapeyron equation (15), are

Temperature, °C	ΔH_{vap}, kJ/mol (kcal/mol)	
45.6	0	
40	5.608	(1.340)
20	9.807	(2.344)
0	12.23	(2.923)
−20	13.58	(3.246)
−40	14.94	(3.571)

Chemical Properties. With few exceptions, SF_6 is chemically inert at ambient temperature and atmospheric pressure. Thermodynamically SF_6 is unstable and should react with many materials, including water, but these reactions are kinetically impeded by the fluorine shielding the sulfur. Sulfur hexafluoride does not react with alkali hydroxides, ammonia, or strong acids.

At elevated temperatures SF_6 forms the respective fluorides and sulfides with many metals (25). In quartz, it starts to decompose at 500°C (1); in copper or stainless steel, it is less stable (26). The stability of SF_6 at 200 and 250°C in the presence of aluminum, copper, silicon steel, and mild steel is shown in Table 2 (14). Careful exclusion of moisture from the system improves the stability of sulfur hexafluoride in the presence of most materials.

Sulfur hexafluoride is more stable in arcs (27) than fluorocarbons such as C_2F_6, or refrigerants such as CCl_2F_2, but less stable than CF_4, BCl_3, or SiF_4. Exposed to 1000°C temperatures, SF_6 decomposes to SOF_2 and SF_4 to the extent of 10 mol %. In spite of its decomposition, the dielectric strength of SF_6 remains the same.

The main products of SF_6 arc decomposition in the presence of air are SOF_2, SF_4, and SOF_4 plus metal fluorides and sulfides (28).

Electrical Properties. The electrical properties of SF_6 stem primarily from its effectiveness as an electron scavenger. To accomplish electrical breakdown in a dielectric gas, primary electrons must gain sufficient energy to generate appreciable numbers of secondary electrons on molecular impact. Sulfur hexafluoride interferes with this process by capturing the primary electrons, resulting in the formation of SF_6^- or SF_5^- ions and F atoms (29):

Table 2. Stability of Sulfur Hexafluoride in Various Materials of Construction[a]

Material	Decomposition, %/yr	
	200°C	250°C
aluminum		0.006
copper	0.18	1.4
silicon steel	0.005	0.01[b]
mild steel	0.2	ca 2

[a]Ref. 14.
[b]Estimated value.

$$SF_6 \rightarrow (SF_6^-)^* \rightarrow SF_6^-$$
$$(SF_6^-)^* \rightarrow SF_5^- + F$$

where $(SF_6^-)^*$ represents an activated complex. This complex is stable against autodissociation during tens of microseconds and thus can be stabilized by collisions.

Although production of SF_6^- is the primary process, formation of SF_5^- ions increases with temperature, and at 200°C the $SF_5^-:SF_6^-$ ratio is 1:25 (30). In addition to high dielectric strength, SF_6 can rapidly interrupt heavy currents at high voltages. At 550 or 756 kV, circuit breaker ratings can be 38,000 and 5,000 MVA, respectively (31). Both SF_6 and C_2F_6 have a more rapid arc recovery than nitrogen (32); six S—F bonds, mean energy 3.4–3.8 eV, are available for fast energy absorption. Both SF_6 and SF_4 have high stability in arcs (27).

Paschen's Rule and Breakdown Voltage. As pressure decreases to vacuum conditions, the breakdown voltage (BDV) first decreases, then increases, resulting in a minimum as shown in Figure 1. Table 3 gives BDV data for SF_6 and other

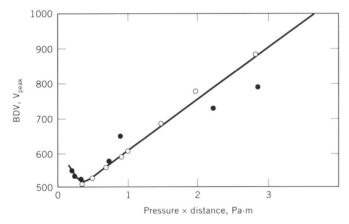

Fig. 1. Paschen's rule plot for SF_6 in uniform electric fields, 60 Hz or dc at 25°C, pressures ≤ 200 kPa, and a gap ≤ 0.3 mm. BDV = breakdown voltage. Compiled from References 33 (○) and 34 (●). To convert Pa to μm Hg, multiply by 7.50.

Table 3. Sulfur Hexafluoride Compared with other Dielectrics[a,b]

Compound	Molecular weight	Bp, °C	Relative BDV[c]
SF_6	146.05	−64 (sub)	2.7
CF_2Cl_2	120.9	−29	2.8
CF_3Cl	104.5	−81	1.44
CF_4	88.0	−128	1.14
N_2	28	−194	1.0

[a]Ref. 35.
[b]Conditions: 60 Hz, 0.5 cm gap, 5 cm spheres, at 101.3 kPa (1 atm) and ca 25°C.
[c]BDV = breakdown voltage.

dielectrics. For optimum utility of a dielectric, a compromise is needed between low boiling point and high BDV. At 300–400 kPa (3–4 atm), the BDV of SF_6 gas has been shown to be equivalent to that of transformer oil under uniform field conditions (35). The BDV of SF_6 deviates from Paschen's rule as pressure increases. This rule is obeyed well only in the case of uniform fields at pressures up to ca 400 kPa (4 atm).

The description of SF_6 electrical properties needed for practical applications is more complex than knowledge of BDV. Corona-onset voltages (COV) must be considered, particularly for the more usual nonuniform fields. Figure 2 illustrates this for a point-to-plane electrode configuration, where extensive prebreakdown corona occurs before sparkover (36).

The theory and application of SF_6 BDV and COV have been studied in both uniform and nonuniform electric fields (37). The ionization potentials of SF_6 and electron attachment coefficients are the basis for one set of correlation equations. A critical field exists at 89 kV/(cm·kPa) above which coronas can appear. Relative field uniformity is characterized in terms of electrode radii of curvature. Peak voltages up to 100 kV can be sustained. A second BDV analysis (38) also uses electrode radii of curvature in rod-plane data at 60 Hz, and can be used to correlate results up to 150 kV. With d-c voltages (39), a similarity rule can be used to treat BDV in fields up to 500 kV/cm at pressures of 101–709 kPa (1–7 atm). It relates field strength, SF_6 pressure, and electrode radii to coaxial electrodes having 2.5-cm gaps. At elevated pressures and large electrode areas, a fall-off from this rule appears. The BDV properties of liquid SF_6 are described in the literature (40–41).

High Frequency Dielectric Strength. Dielectric strength at high frequency is important in microwave power uses such as radar (see MICROWAVE TECHNOLOGY). Because SF_6 has zero dipole moment, its dielectric strength is substantially constant as frequency increases. At 1.2 MHz, SF_6 has been shown to have a dielectric strength of 2.3–2.5 relative to N_2 (42). At 3 GHz, SF_6 has

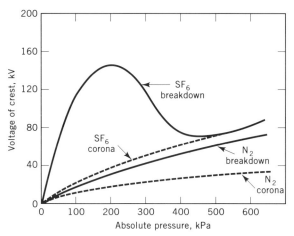

Fig. 2. (——) Sparkover and (- - -) corona onset voltages for SF_6 and N_2 (36). To convert kPa to atm, divide by 101.

about 10 times the power-carrying capacity of air (43), whereas at 9.375 GHz in a waveguide assembly the power-carrying capacity was 7.5 times that of air (42).

Particle Contamination. In assembling large, high voltage equipment such as coaxial lines, contamination by metal particles may occur which may decrease the dielectric strength under various conditions by 5 to 10-fold (44–45). Metal needles are the worst contaminants and electrostatic traps or adhesive areas have been designed to cope with them (46).

In some cases particles have been added to electrical systems to improve heat removal, for example with an SF_6-fluidized particulate bed to be used in transformers (47). This process appears feasible, using polytetrafluoroethylene (PTFE) particles of low dielectric constant. For a successful application, practical problems such as fluidizing narrow gaps must be solved.

Manufacture and Quality Control. Sulfur hexafluoride is manufactured by combining sulfur vapor and pure elemental fluorine (48,49). It is then given a preliminary scrubbing with caustic. Any disulfur decafluoride, S_2F_{10}, formed is decomposed by heating the product to 400°C, to give SF_4 and SF_6. The SF_4 and any remaining lower fluorides of sulfur are removed by a second caustic scrubber. The gas is then dried in a sulfuric acid tower, distilled, and packaged. A routine manufacturing quality-control test establishes the absence of toxic impurities. In this test, mice are exposed to an atmosphere of 80% sulfur hexafluoride and 20% oxygen for 16–20 h and must show no visible effects from the gas.

Economic Aspects and Shipping. Consumption of SF_6 has increased gradually as dielectric uses have broadened. The estimated worldwide annual consumption for 1992 was about 6000 metric tons. The 1992 U.S. price was ca $8–15/kg.

Sulfur hexafluoride is packaged as a liquefied gas in DOT 3AA 2015 steel cylinders containing 52 kg. Larger quantities are available in tube trailers containing ca 11,000 kg.

Specifications and Analytical Methods. Sulfur hexafluoride is made to rigid specifications. Per ASTM D2472-81 (reapproved 1985) (50), the only permissible impurities are traces of air, carbon tetrafluoride (0.05 wt % max), and water (9 ppm by wt max; dew point $-45°C$ max).

Sulfur hexafluoride may be analyzed chromatographically using a molecular sieve or a Porapak QS column. Using an electron-capture detector, a sensitivity of 10^{-3} to 10^{-4} ppb is possible (51–53).

Health and Safety Factors. Sulfur hexafluoride is a nonflammable, relatively unreactive gas that has been described as physiologically inert (54). The current OSHA standard maximum allowable concentration for human exposure in air is 6000 mg/m^3 (1000 ppm) TWA (55). The Underwriters Laboratories classification is Toxicity Group VI. It should be noted, however, that breakdown products of SF_6, produced by electrical decomposition of the gas, are toxic. If SF_6 is exposed to electrical arcing, provision should be made to absorb the toxic components by passing the gas over activated alumina, soda-lime, or molecular sieves (qv) (56).

Because of concerns about the production of the highly toxic S_2F_{10} in arced SF_6 gas, an electrical industry-supported research program was begun in 1992. An interim report indicated that S_2F_{10} may be present in SF_6 gas exposed to electrical discharge, but as of this writing, this could not be precisely quantified

(57). The effects of SF_6 on the environment and the Greenhouse effect have been discussed (58).

Applications. For use as a gaseous dielectric, other specific properties are needed in addition to high breakdown strength, and a compromise must be made between electrical and mechanical requirements. Desirable properties include low toxicity, thermal stability toward materials of construction, good heat transfer, and absence of electrically conducting carbon. Sulfur hexafluoride has a good balance of properties (see Table 1), good dew points, and chemical attributes.

Reviews of gas-phase kinetics (59) and ionization energies (60) have also listed some of the advantages SF_6 enjoys in service as a gaseous dielectric.

Circuit Breakers. Current interruption is essential in high voltage equipment when overloads or other emergencies occur. Circuit breakers consist of relays having contacts equipped with SF_6 jets and high voltage transformers holding fault-sensing coils that can activate a relay contact for each phase of current (61). At 60 Hz, a 765 kV root-mean-square (rms) breaker can have a rating of 50,000 MVA (31). High (1.7 MPa (17 atm)) pressure SF_6 jets extinguish the arcs generated on opening the contacts. Designs of circuit breakers called puffer breakers have encouraged the design of smaller devices, which has extended the use of these types of units (62,63). Sulfur hexafluoride can also act as the insulation for the sensing transformer and supply pressure to activate the relay contacts. At higher pressures and winter temperatures, heaters serve to prevent SF_6 condensation.

High Voltage Coaxial Lines. Sulfur hexafluoride is the main insulation for power transmission in high voltage coaxial lines which can move power above ground, underground, or underwater (64–66). Voltage ratings range between 65 and 500 kV and higher. Above ground, powers of 300 MVA have been transported. Generally each phase needs one coaxial line, but three-phase lines in one envelope have been developed (64). Lines now handle a-c power, and d-c lines have been tested. To reduce costs, SF_6–N_2 mixtures can be used (67). A long coaxial line having a dielectric constant very near unity is possible before reactive compensation is needed. Usually lines of only a few hundred meters are used, especially where land values are high or where passage under multilane highways or other power lines is necessary.

Mini-Substations. Development of SF_6-insulated lines and circuit breakers made possible development of a compact electric substation that requires one-tenth the land area of conventional designs (64,68). Other advantages of the mini-station are freedom from effects of weather and pollution, and reduced risk of vandalism (69). Substation units in which SF_6 insulates 400 and 500 kV d-c equipment have been developed (70).

Transformers. Units insulated with SF_6 are used mainly in circuit breakers. Some SF_6 power transformers have been designed, but the temperature limitation of ca 200°C restricts this use. The problem in high power units is heat transfer through the gas; this can be alleviated with a fluidized particulate bed (47). Although this approach appears feasible, it has yet to be commercialized.

Other Electrical Uses. Using SF_6 insulation, waveguides can transport 7 to 10-fold more microwave power, which results in doubling radar ranges (21,42,43). Voltage ratings in Van de Graaf generators and linear accelerators are also increased by replacing N_2 or air with SF_6 (21). Incorporation of SF_6 in polyethylene-insulated cables increases the starting voltage for tree-formation breakdown processes in the polymer (71,72).

Nonelectrical Uses. Because of its inertness under normal conditions, SF_6 has been used as a tracer for a variety of studies such as air flow patterns (73), underground pipe leak detection (74), and dispersion of air pollutants (75). It has also been proposed as a refrigerant, either alone (76) or with $CHClF_2$ (77) or CHF_3 as an azeotrope (78). Owing to its low sound velocity, it can improve the performance of loudspeakers at lower pitch registers (79).

At elevated temperatures or under laser radiation, SF_6 becomes a source of fluorine atoms. In the operation of a chemical laser, it abstracts hydrogen from suitable molecules (80–81). SF_6 is also being used in etching of semiconductor surfaces (82,83). Mixed with air or CO_2 in the amount of a few tenths of a percent (0.22 vol %), SF_6 protects molten magnesium during its casting process, resulting in reduced slag and an improved metal surface (84,85). It has also been found useful in removing hydrogen and other gases from aluminum melts (86).

Additional uses include (*1*) filling the enclosed space in double-pane window units to reduce noise and heat transmission (87,88); (*2*) acting as a source of power from the reaction with lithium to produce heat (89,90); and (*3*) pressurizing recreation ball packages such as tennis balls to give improved shelf life (91).

Sulfur Tetrafluoride

Sulfur tetrafluoride [7783-60-0], SF_4, molecular weight 108.06, is a highly reactive colorless gas that fumes in moist air and has an irritating odor that resembles sulfur dioxide. Interest in this compound as a fluorinating agent was spurred by its unique ability to replace oxygen in compounds containing carbonyl groups. It was first reported in 1929 (92).

Physical Properties. Sulfur tetrafluoride has the structure of a distorted trigonal bipyramid, the sulfur having hybrid sp^3d orbitals and an unshared electron pair (93). The FSF bond angles have been found to be 101° and 187°, and the bond distances 0.1646 and 0.1545 nm (94).

Selected physical properties are given in Table 4. The nmr data (97) and ir and Raman spectra (98) have also been determined. Thermodynamic functions have been calculated from spectral data (99).

Chemical Properties. Sulfur tetrafluoride reacts rapidly with water to give hydrofluoric acid and thionyl fluoride [7783-42-8]:

$$SF_4 + H_2O \rightarrow SOF_2 + 2\ HF$$

With alcohols, mixtures of alkyl fluorides and alkyl ethers are obtained (100). Alcohols bearing electron-withdrawing groups can be converted to the corresponding fluorides in high yield (101). Sulfur tetrafluoride replaces the carbonyl oxygen with fluorine (100,102).

$$\underset{R'}{\overset{R}{>}}C=O + SF_4 \rightarrow \underset{R'}{\overset{R}{>}}CF_2 + SOF_2$$

Table 4. Physical Properties of Sulfur Tetrafluoride

Property	Value	Reference
molecular weight	108.055	
melting point, °C	−121.0	95
boiling point, °C	−38	96
critical temperature, °C	90.9	96
surface tension at −73°C, mN/m(=dyn/cm)	257	95
density, liquid, at −73°C, g/mL	1.9190	95
vapor pressure at 25°C, MPa^a	2.0219	95
heat of vaporization, kJ/mol^b	26.4	95
heat of formation at 25°C, kJ/mol^b	−781.1	16
free energy of formation at 25°C, kJ/mol^b	−740.4	16
entropy at 25°C, J/(mol·K)^b	300.7	16
dipole moment, C·m^c	2.11×10^{-30}	94

^{a}To convert MPa to atm, divide by 0.101.
^{b}To convert J to cal, divide by 4.184.
^{c}To convert C·m to debye, divide by 3.3366×10^{-30}.

Sulfur tetrafluoride reacts with most inorganic oxides and sulfides to give the corresponding fluorides (103):

$$SF_4 + SnS_2 \rightarrow SnF_4 + 3\,S$$
$$5\,SF_4 + I_2O_5 \rightarrow 2\,IF_5 + 5\,SOF_2$$
$$3\,SF_4 + UO_3 \rightarrow UF_6 + 3\,SOF_2$$

Extensive reviews of SF_4 in organic fluorination are available (104,105).

Preparation. In the laboratory, sulfur tetrafluoride is made by combining SCl_2 and NaF suspended in acetonitrile at ca 77°C (106). For commercial production, SF_4 is made by direct combination of sulfur with elemental fluorine (107). Commercial applications of SF_4 are limited. It is available from Air Products and Chemicals.

Toxicity. Sulfur tetrafluoride has an inhalation toxicity comparable to phosgene. The current OSHA standard maximum allowable concentration for human exposure in air is 0.4 mg/m^3 (TWA) (54). On exposure to moisture, eg, on the surface of skin, sulfur tetrafluoride liberates hydrofluoric acid and care must be taken to avoid burns. One case of accidental exposure of electrical workers to decomposed SF_6 gas containing SF_4 has been cited (108).

Other Sulfur Fluorides

Although eight other binary sulfur fluorides have been synthesized and characterized, proof of the existence of several members of this group was dependent on modern instrumental methods of analysis because of extreme instability. SF_5 and S_2F_{10} are stable, however, the latter is noted for its extreme toxicity. All sulfur fluorides other than SF_6 must be considered extremely toxic.

As a group, these materials have no technological utility because of instability, toxicity, and difficulty of preparation. An excellent review of many of these compounds is available (109).

Sulfur Pentafluoride. Sulfur pentafluoride [10546-01-7] is thought to be formed during the electrical breakdown of SF_6 and also to be present in plasma reactions involving SF_6. A number of theoretical studies have been reported (110–113).

Disulfur Decafluoride. Disulfur decafluoride [5714-22-7], S_2F_{10}, is an extremely toxic, colorless, volatile liquid (114). Electron diffraction studies show the molecule to be composed of two octahedral SF_5 groups joined by a sulfur–sulfur bond. The S—F bond distance is 0.156 nm, almost identical to that of SF_6, and the S—S distance is 0.221 nm (115). Table 5 summarizes the known physical properties.

Disulfur decafluoride does not react rapidly with water, mercury, copper, or platinum at ambient temperatures. There is evidence that it slowly decomposes on various surfaces in the presence of water when stored in the vapor state (118). It is decomposed by molten KOH to give a mixture of potassium compounds of sulfur and fluorine. The gas reacts vigorously with many other metals and silica at red heat (114). At ca 156°C it combines with Cl_2 or Br_2 to form SF_5Cl or SF_5Br (119,120). At ca 200°C, S_2F_{10} is almost completely thermally decomposed into the hexa- and tetrafluoride (121).

In the laboratory, S_2F_{10} is prepared by the photochemical reduction of SF_5Cl in the presence of hydrogen (122).

$$2\ SF_5Cl + H_2 \xrightarrow{h\nu} S_2F_{10} + 2\ HCl$$

The OSHA standard maximum allowable concentration for human exposure in air is 0.10 mg/m³ (TWA) (55). No commercial uses for this compound have developed.

Table 5. Physical Properties of S_2F_{10}

Property	Value	Reference
molecular weight	254.13	
melting point, °C	−55	116
boiling point,[a] °C	28.7	114
critical temperature,[b] °C	165	14
density, liquid, at 25°C, g/cm³	2.08	114
surface tension at 25°C, mN/m(= dyn/cm)	13.9	114
heat of vaporization, kJ/mol[c]	29.18	114
heat of formation[b,d] at 25°C, kJ/mol[c]	−2.08	117
entropy[b,d] at 25°C, J/(mol·K)[c]	444.93	117
Trouton constant	3.0	114
dielectric constant at 35°C	2.042	114

[a]Vapor pressure calculated.
[b]Estimated.
[c]To convert J to cal, divide by 4.184.
[d]Ideal gas.

Because of the extreme toxicity of this material and the possibility it could be present in failed circuit breakers containing decomposed SF_6, several studies have been carried out to develop analytical methods and investigate possible ways to minimize environmental exposure. One method capable of determining S_2F_{10} in the ppb range has been reviewed (123).

Studies on the kinetics of formation of S_2F_{10} and reviews of applicable literature have been reported (124–126). Other work has concentrated on the use of cell culture evaluation methods for assessing cytotoxic activity of SF_6 decomposition products (127,128). Several laboratories seek to provide methods for accurately determining S_2F_{10} in operating electrical units (57).

Thiothionyl Fluoride and Difluorodisulfane. Thiothionyl fluoride [1686-09-9], S=SF_2, and difluorodisulfane [13709-35-8], FSSF, are isomeric compounds which may be prepared as a mixture by the action of various metal fluorides on sulfur vapor or S_2Cl_2 vapor. Chemically, the two isomers are very similar and extremely reactive. However, in the absence of catalytic agents and other reactive species, FSSF is stable for days at ordinary temperatures and S=SF_2 may be heated to 250°C without significant decomposition (129). Physical properties of the two isomers are given in Table 6. The microwave spectrum of S=SF_2 has been reported (130).

Difluoromonosulfane and Difluorodisulfane Difluoride. Difluoromonosulfane [13814-25-0] (sulfur difluoride), SF_2, and its dimer, disulfane tetrafluoride [27245-05-2], SF_3SF, are both extremely unstable compounds which have only a fleeting existence except under rigorously controlled laboratory conditions. These compounds may be prepared by passing SCl_2 vapor over HgF_2 at 150°C (131). Electronic and nmr examinations of SF_2 have been reported (132,133).

Other Fluorosulfanes. Difluorotrisulfane [31517-17-6], FSSSF, and difluorotetrasulfane [31517-18-7], FSSSSF, have been identified as the constituents of the yellow oil obtained when sulfur vapor reacts with AgF. Their existence was demonstrated by nmr and mass spectroscopy (134,135).

Table 6. Physical Properties of S_2F_2 Isomers[a]

	Value	
Property	FSSF	SSF_2
molecular weight	102.13	102.13
melting point, °C	−133	−164.6
boiling point, °C	15	−10.6
heat of vaporization, kJ/mol[b]	24.80	22.72
entropy of vaporization, J/(mol·K)[b]	86.67	78.08
heat capacity at 25°C, J/(mol·K)[b]	64.0	41.50
entropy[c] at 25°C, J/(mol·K)[b]	299	290.77
heat content at 25°C, kJ/mol[b]	14.017	13.342
free energy, kJ/(mol·K)[b]	72.48	73.26

[a]Ref. 129.
[b]To convert J to cal, divide by 4.184.
[c]Ideal gas.

BIBLIOGRAPHY

"Sulfur Compounds" under "Fluorine Compounds, Inorganic" in *ECT* 1st ed., Vol. 6, pp. 732–734, by H. C. Miller, Pennsylvania Salt Manufacturing Co.; "Sulfur Hexafluoride" in *ECT* 1st ed., Suppl. 2, pp. 793–802, by W. Mears, Allied Chemical & Dye Corp.; "Sulfur" under "Fluorine Compounds, Inorganic" in *ECT* 2nd ed., Vol. 9, pp. 664–676, by J. A. Brown, Allied Chemical Corp.; "Sulfur Fluorides" under "Fluorine Compounds, Inorganic–Sulfur" in *ECT* 3rd ed., Vol. 10, pp. 799–811, by R. E. Eibeck and W. Mears, Allied Chemical Corp.

1. H. Moissan and P. Lebeau, *Ann. Chim. Phys.* **26**, 145 (1902).
2. E. E. Charlton and F. S. Cooper, *Gen. Electr. Rev.* **40**, 438 (1937).
3. B. M. Hokhberg and co-workers, *J. Tech. Phys. (USSR)* **12**, 3 (1942).
4. Brit. Pat. 532,670 (Jan. 29, 1941), (to The British Thompson-Houston Co., Ltd.).
5. W. E. Schumb and E. L. Gamble, *J. Am. Chem. Soc.* **52**, 4302 (1930).
6. D. M. Yost and H. Russell, Jr., *Systematic Organic Chemistry*, Prentice-Hall, Inc., Englewood Cliffs, N.J., 1944, pp. 297–309.
7. V. P. Borisoylekskii, L. I. Strokovskii, and I. S. Zhizuleva *Zh. Fiz. Khim.* **48**, 119 (1974).
8. L. A. Makarevich, E. S. Sokolova, and G. A. Sorena, *Zh. Fiz. Khim.* **42**, 22 (1968).
9. K. E. MacCormack and W. G. Schneider, *Can. J. Chem.* **29**, 699 (1951).
10. W. H. Mears, E. Rosenthal, and J. V. Sinka, *J. Phys. Chem.* **73**, 2254 (1969).
11. L. A. Makarevich and O. N. Sokolova, *Zh. Fiz. Khim.* **47**, 763 (1973).
12. D. Balzarini and P. Palffy, *Can. J. Phys.* **52**, 2007 (1974).
13. T. G. Pearson and R. L. Robinson, *J. Chem. Soc.*, 1427 (1933).
14. Unpublished data, Allied Chemical Corp., Morristown, N.J., 1993.
15. E. Rosenthal, *Sulfur Hexafluoride—Thermodynamic Properties*, unpublished study, Specialty Chemicals Div., Allied Chemical Corp., Morristown, N.J., 1969.
16. *JANAF Thermochemical Tables, NSRDS-NBS-37*, 2nd ed., National Bureau of Standards, Washington, D.C., June 1971.
17. A. Eucken and E. Schroder, *Z. Phys. Chem.* **B41**, 307 (1938).
18. J. Neudorffer, *Ann. Chim. (Paris)* **8**, 501 (1953).
19. P. Grassman and W. Tauscher, *Allied Chemical Research Contract*, Institute of Heat and Engineering, EIDG Technical University, Zurich, Switzerland, 1967.
20. W. A. Tauscher, *Kaltetech.-Klimatisierin* **24**, 67 (1972).
21. *Technical Bulletin SFBR-1 (Sulfur Hexafluoride)*, Specialty Chemicals Division, Allied Chemical Corp., Morristown, N.J., 1973.
22. D. Berg, *J. Chem. Phys.* **31**, 572 (1959).
23. H. E. Watson, G. C. Rao, and K. L. Ramaswamy, *Proc. R. Soc. Ser. A* **132**, 569 (1931).
24. B. Genot, *J. Chim. Phys. Physiochem. Biol.* **68**, 111 (1971).
25. F. A. Cotton and G. Wilkinson, *Advanced Inorganic Chemistry*, 3rd ed., Wiley-Interscience, New York, 1962, p. 419.
26. D. K. Padma and A. R. Vasuderamurthy, *J. Fluor. Chem.* **5**, 181 (1975).
27. J. P. Manion, J. A. Philosophos, and M. B. Robinson, *IEEE Trans., Trans. Electr. Insul.* **E1-2**(1), 1 (Apr. 1967).
28. C. Boudene and co-workers, *Rev. Gen. Electr.* (Special No.), 45 (June 1974).
29. R. L. Champion, *Gaseous Dielectric 6*, Proceedings of the 6th International Symposium, 1990, (Pub. 1991), pp. 1–8.
30. F. C. Fehsenfeld, *J. Chem. Phys.* **53**, 2000 (1970).
31. R. N. Yerkley and C. F. Cromer, *IEEE Trans., Power Appar. Syst.* **89**(8), 2065 (1970).
32. M. Hudis, *CP74-090-7, IEEE Power Engineering Society, Winter Meeting, New York*, Jan. 27–Feb. 1, 1974.

33. G. Luxa and co-workers, *Item D. Paschen Curve for SF_6, 1975 CIGRE International Conference on Large High Tension Electrical Systems*.
34. S. Schreier, *IEEE Trans., Power Appar. Syst.* **83**, 468 (1964).
35. P. R. Howard, *Proc. Inst. Elect. Eng.* **104**(A), 123 (1957).
36. C. N. Works and T. W. Dakin, *Trans. AIEE* **72**(1), 682 (1953).
37. T. Nitta and Y. Shibuya, *IEEE Trans., Power Appar. Syst.* **90**, 1965 (1971).
38. A. A. Azer and P. P. Comsa, *IEEE Trans. Electr. Insul.* **8**(4), 136 (1973).
39. I. M. Bortnik and C. M. Cooke, *IEEE Trans., Power Appar. Syst.* **91**, 2196 (1972).
40. C. N. Works, T. W. Dakin, and R. W. Rogers, *N.A.S.N.R.C. Publ.* (1080), 69 (1963).
41. Y. V. Torshin, Conference Paper 118, *Third International Conference on Gas Discharges*, IEE, London, Sept. 9–12, 1974.
42. T. Anderson and co-workers, *AIEE Conf. Paper 52-82*, 1957.
43. J. W. Sutherland, *Electron. Eng.*, 538 (1955).
44. C. M. Cook, R. E. Wootton, and A. H. Cookson, *IEEE Trans., Power Appar. Syst.* **96**(3), 768 (1977).
45. A. H. Cookson, O. Farish, and G. M. Sommerman, *IEEE Trans., Power Appar. Syst.* **9**(4), 1329 (1972).
46. J. G. Trump, *IEEE Trans. Nucl. Sci.* **14**, 113 (1962).
47. *Gas Insulated Fluidized Bed Transformer, Final Report EL-302*, Project 479-1, Electric Power Research Inst. (EPRI), Buffalo, N.Y., May 1977.
48. U.S. Pat. 3,336,111 (Aug. 15, 1967), W. E. Watson, H. G. Tepp, and M. H. Cohen (to Allied Chemical Corp.)
49. E. P. 87338 (Aug. 31, 1983) M. Jaccaud and A. J. F. Ducouret (to PCUK-Ugine Kuhlmann-Atochem).
50. *ASTM D2472-81* (reapproved 1985), American Society of Testing and Materials, Philadelphia, Pa., 1981.
51. J. E. Lovelock and S. B. Lipsky, *J. Am. Chem. Soc.* **82**, 860 (1960).
52. *Ibid.*, 431 (1960).
53. P. G. Simonds and co-workers, *Anal. Chem.* **44**, 860 (1972).
54. D. Lester and L. A. Greenberg, *Arch. Ind. Hyg. Occup. Med.* **2**, 348 (1950).
55. *Threshold Limit Values for Chemical Substances and Physical Agents*, American Conference of Governmental Industrial Hygienists, Cincinnati, Ohio, 1990–1991.
56. W. C. Schumb, J. G. Trump, and G. L. Priest, *Ind. Eng. Chem.* **41**, 1348 (1949).
57. D. R. James, Technical Note No. 1, *Cooperative Research and Development Agreement (CRADA), Investigation of S_2F_{10} Production and Mitigation in Compressed SF_6-Insulated Power System*, Oak Ridge National Laboratory, Oak Ridge, Tenn., Dec. 28, 1992.
58. L. Niemayer, F. Y. Chu, *IEEE Trans. Elec. Insul.* **27**(1), 184–187 (Feb. 1992).
59. J. T. Herron, *Int. J. Chem. Kinet.* **19**(2) 129-42 (1987).
60. M. L. Lanferd and co-workers, *Int. J. Mass. Spectrum Ion Processes*, **98**(2) 147-53 (1990).
61. T. Ushio, I. Shimura, and G. Tominago, *IEEE Trans., Power Appar. Syst.* **89**, 2615 (1970).
62. U.S. Pat. 5,059,753 (Oct. 22, 1991), S. R. Hamm (to Cooper Industries).
63. U.S. Pat. 4,752,860 (June 21, 1988), A. Giboulet and P. Romanet (to Merlin-Gerin SA).
64. S. D. Barrett, *The A to Z of SF_6, Special Report Electric Light and Power*, TID ed., Dec. 1972.
65. *Electr. World Trans. Dist.* **54**, (Feb. 15, 1972).
66. B. O. Pedersen, H. C. Doepken, and D. C. Bolin, *IEEE Trans., Power Appar. Syst.* **90**, 2631 (1971).

67. R. Nakata and co-workers, *An Underground High Voltage Direct Current Transmission Line, IEEE Underground Distribution and Transmission Conference, Dallas, Tex., Apr. 1–5, 1974.*
68. *Electr. World* **61**, (Feb. 23, 1976).
69. *IEEE Trans. Power Appar. Syst.* **94**, (1975).
70. E. E. Fischer, A. Glassanos, and N. G. Hingorani, *Electr. World Trans. Dist.* **40**, (Feb. 1, 1977).
71. T. Kojima and co-workers, *Showa Wire Cable Rev.* **221**, 11 (1972).
72. U.S. Pat. 4,783,576 (Nov. 8, 1988), W. G. Lawson and D. A. Silver (to Pirelli Cable Co.)
73. P. J. Drivas and F. H. Shair, *Atmos. Environ.* **8**, 1155 (1974).
74. U.S. Pat. 5,046,353 (Sept. 10, 1991), G. M. Thompson (to Tracer Research Corp.)
75. B. K. Lamb, D. E. Stock, HTD (*Am. Soc. Mech. Eng.*) **152** (*Mixed Convert Enveron Flows*) 55–59, (1990).
76. R. Plank, *Kaltetechnik* **8**, 302 (1956).
77. U.S. Pat. 3,642,639 (Feb. 15, 1972), K. P. Murphy and R. F. Stahl (to Allied Chemical Corp.).
78. U.S. Pat. 3,719,603 (Mar. 6, 1973), R. F. Stahl (to Allied Chemical Corp.).
79. U.S. Pat. 2,797,766 (July 2, 1957), H. W. Sullivan (to D. Bogen and Co.).
80. D. N. Kaye, *New Sci.* **14**, 65 (1971).
81. D. J. Spenser and co-workers, *Int. J. Chem. Kinet.* **1**, 493 (1969).
82. U.S. Pat. 4,680,087 (July 14, 1987), S. M. Bobbio (to AlliedSignal).
83. U.S. Pat. 4,980,018 (Dec. 25, 1990), Mu Xiao-Chun and Multani Jagin (to Intel Corp.).
84. J. W. Fruehling and J. D. Hanawalt, *Am. Foundry Soc. Trans.* **16**, 159 (1969).
85. O. Schlem, *Giesserei* **19**, 558 (1971).
86. U.S. Pat. 4,959,010 (Sept. 25, 1990), R. R. Corns and co-workers (to AGA AB).
87. U.S. Pat. 4,800,693 (Jan. 31, 1989), I. Fasth and J. Karlsen (to Barrier HB).
88. Fr. Pat. 2,529,609 (Jan. 6, 1984), M. Rehfeld (to Saint-Gobain Vitrage).
89. U.S. Pat. 4,959,566 (Sept. 25, 1990) Dobran Flavio
90. S. H. Chan and co-workers, *23rd Symposium on Internal Combustion. Proceedings, 1990* (Pub. 1991), pp. 1139–1146.
91. U.S. Pat. 4,358,111 (Nov. 11, 1982), J. J. Oransky and co-workers (to Air Products and Chemicals).
92. J. Fischer and W. Jaenckner, *Z. Angew. Chem.* **42**, 810 (1929).
93. F. H. Cotton, J. W. George, and J. S. Waugh, *J. Chem. Phys.* **28**, 994 (1958).
94. W. M. Tolles and W. D. Gwinn, *J. Chem. Phys.* **36**, 1119 (1962).
95. E. Brown and P. L. Robinson, *J. Chem. Soc.*, 3147 (1955).
96. *Sulfur Tetrafluoride, Tech. Bulletin, 2B*, E. I. du Pont de Nemours & Co., Inc., Wilmington, Del., 1946.
97. J. Bacon and R. J. Gillespie, *Can. J. Chem.* **41**, 1016 (1963).
98. K. O. Christe and co-workers, *Spectrochim. Acta* **32A**, 1141 (1976).
99. M. Radharkrishnan, *Z. Naturforsch.* **18a**, 103 (1963).
100. W. R. Hasek, W. C. Smith, and V. A. Engelhardt, *J. Am. Chem. Soc.* **82**, 543 (1960).
101. U.S. Pat. 2,980,740 (Apr. 18, 1961), W. R. Hasek and A. C. Haven, Jr. (to E. I. du Pont de Nemours & Co., Inc.).
102. U.S. Pat. 2,859,245 (Nov. 4, 1958), W. C. Smith (to E. I. du Pont de Nemours & Co., Inc.).
103. A. L. Oppegard and co-workers, *J. Am. Chem. Soc.* **82**, 3835 (1960).
104. C. L. J. Wang, *Org. React. (N.Y.)* **34**, 319–400 (1985).
105. W. Dmowski, *J. Fluorine Chem.* **32**(3) 255–282 (1986).
106. F. S. Fawcett and C. W. Tullock in J. Kleinberg, ed., *Inorganic Syntheses*, Vol. 7, McGraw-Hill Book Co., Inc., New York, 1963, pp. 119–124.

107. U.S. Pat. 3,399,036 (Aug. 27, 1968), S. Kleinberg and J. F. Tompkins, Jr. (to Air Products and Chemicals, Inc.).
108. A. Kraut, R. Lilis, *Br. J. Ind. Med.* **47**(12) 829–832 (1990).
109. F. Seel in H. J. Emeleus and G. G. Sharpe, eds., *Advances in Inorganic and Radio Chemistry*, Vol. 16, Academic Press, Inc., New York, 1974, pp. 297–333.
110. J. T. Herron, *Proceedings of the 5th International Symposium on Gaseous Dielectricics*, 1987, pp. 199–204.
111. M. Ticky and co-workers, *Int. J. Mass. Spectrum Ion Processes*, **76**(3) 231–235 (1987).
112. W. L. Sieck and P. J. Ausloos, *J. Chem. Phys.* **93**(11) 8374-8 (1990).
113. I. C. Plant and K. R. Ryan, *Plasma Chem. Plasma Process.*, **6**(3) 247–258 (1986).
114. K. B. Denbigh and R. W. Gray, *J. Chem. Soc.*, 1346 (1934).
115. R. B. Harvey and S. H. Bauer, *J. Am. Chem. Soc.* **75**, 2840 (1953).
116. N. R. S. Hollies and R. L. McIntosh, *Can. J. Chem.* **29**, 494 (1951).
117. G. Pass, *J. Appl. Chem.* **19**, 77 (1969).
118. J. K. Olthoff and co-workers, *Conference Record of the IEEE International Symposium on Electrical Insulation*, 1990, pp. 248–252.
119. B. Cohen and A. G. MacDiarmid, *Inorg. Chem.* **4**, 1782 (1965).
120. T. A. Kovacina, A. D. Berry, and W. B. Fox., *J. Fluor. Chem.* **7**, 430 (1976).
121. W. R. Trost and R. L. McIntosh, *Can. J. Chem.* **29**, 508 (1951).
122. H. L. Roberts, *J. Chem. Soc.*, 3183 (1962).
123. J. K. Olthoff and co-workers, *Anal. Chem.* **63**(7) 726 (1991).
124. F. Y. Chu and co-workers, *Conference Record of the IEEE International Symposium on Electrical Insulation*, 1988, pp. 131–134.
125. J. T. Herron, *Int. J. Chem. Kinet.* **19**(2) 129–142 (1987).
126. J. T. Herron, *J. Phys. Chem. Ref. Data* **16**(1) 1–6 (1987).
127. G. D. Griffin and co-workers, *Toxicol. Environ. Chem.* **9**(2) 139–166 (1984).
128. G. D. Griffin, *IEE Proc. Part A*, **137**(4) 221–227 (1990).
129. F. Seel, *Chimia* **22**, 79 (1968).
130. R. W. Davis, *J. Mol. Spectrosc* **116**(2) 371–383 (1986).
131. F. Seel, H. Heinrich, and W. Gombler, *Chimia* **23**, 73 (1969).
132. W. Gombler and co-workers, *Inorg. Chem.* **29**(14) 2697-8 (1990).
133. R. J. Glinski and co-workers, *J. Chem. Phys. Chem.* **94**(16) 6196–6201 (1990).
134. F. Seel and co-workers, *Z. Anorg. Allgem. Chem.* **380**, 262 (1971).
135. F. Seel, R. Budenz, and D. Werner, *Ber.* **97**, 1369 (1964).

FRANCIS E. EVANS
GANPAT MANI
AlliedSignal Inc.

FLUOROSULFURIC ACID

Fluorosulfuric acid [7789-21-1], HSO_3F, is a colorless-to-light yellow liquid that fumes strongly in moist air and has a sharp odor. It may be regarded as a mixed anhydride of sulfuric and hydrofluoric acids. Fluorosulfuric acid was first identified and characterized in 1892 (1). It is a strong acid and is employed as a catalyst and chemical reagent in a number of chemical processes, such as alkylation (qv), acylation, polymerization, sulfonation, isomerization, and production of organic fluorosulfates (see FRIEDEL-CRAFTS REACTIONS).

Properties. Selected physical properties of fluorosulfuric acid are shown in Table 1. Fluorosulfuric acid is soluble in acetic acid, ethyl acetate, nitrobenzene,

Table 1. Physical and Chemical Constants of Fluorosulfuric Acid

Property	Value[a]	References
molecular weight	100.07	
boiling point, °C	162.7	1, 2
freezing point, °C	−88.98	3
density, g/mL	1.726	2
viscosity, mPa·s(=cP)	1.56	2
dielectric constant	ca 120	2
specific conductance, $(\Omega \cdot m)^{-1}$	1.085×10^{-6}	2
heat of formation,[b] ΔH_f, kJ/mol[c]	792.45	4

[a] All values at 25°C.
[b] From SO_3 and HF.
[c] To convert kJ to kcal, divide by 4.184.

and diethyl ether, and insoluble in carbon disulfide, carbon tetrachloride, chloroform, and tetrachloroethane. Many inorganic and organic materials dissolve in fluorosulfuric acid; the physical and chemical properties of such solutions have been extensively investigated (5–8). The structure of fluorosulfuric acid has been determined (9), and the ir, Raman, and nmr spectra have been reported (10). The solution of antimony pentafluoride [7783-70-2] in fluorosulfuric acid results in a superacid possessing protonating power orders of magnitude greater than 100% sulfuric acid. Extensive studies on the properties of superacid compositions containing fluorosulfuric acid have been published (11,12).

Fluorosulfuric acid is stable to heat up to decomposition at about 900°C (13), where vapor-phase dissociation into hydrogen fluoride and sulfur trioxide probably occurs. Reviews of the chemistry and properties of fluorosulfuric acid have been published (14–16).

Reactions. The reaction of fluorosulfuric acid and water is violent and exothermic; it proceeds as follows:

Fast hydrolysis $HSO_3F + H_2O \rightleftharpoons H_2SO_4 + HF$

Ionization $HSO_3F + H_2O \rightleftharpoons H_3O^+ + SO_3F^-$

Slow hydrolysis $SO_3F^- + H_2O \rightleftharpoons HSO_4^- + HF$

The extent of the initial hydrolysis depends on temperature and how the water is added. Hydrolysis is reduced at slower addition rates and lower temperatures. The hydrolysis subsequent to the initial fast reaction is slow, presumably because part of the acid is converted to fluorosulfate ions which hydrolyze slowly even at elevated temperatures. The hydrolysis in basic solution has also been studied (17). Under controlled conditions, hydrates of HSO_3F containing one, two, and four molecules of water have been observed (18,19).

The pure acid does not react in the cold with sulfur, selenium, tellurium, carbon, silver, copper, zinc, iron, chromium, or manganese, but slowly dissolves mercury and tin (20). At higher temperatures, lead, mercury, tin, and sulfur react rapidly, eg:

$$S + 2\ HSO_3F \rightarrow 3\ SO_2 + 2\ HF$$

Precipitated (hydrated) silica reacts vigorously with fluorosulfuric acid to give silicon tetrafluoride [7783-61-1] (21), but glass (qv) is not attacked in the absence of moisture (20). Alkali and alkaline-earth metal chlorides are readily converted to fluorosulfates by treatment with fluorosulfuric acid (7,13,22,23).

Electrolysis of fluorosulfuric acid produces either $S_2O_6F_2$ [13709-32-5] (24) or SO_2F_2 [13036-75-4] plus OF_2 (25), depending on specific conditions. Various reactions of fluorosulfuric acid with inorganic compounds are shown in Table 2, and with organic compounds in Table 3.

Other studies which have been reported describe unusual chemistry such as $HSO_3F-Nb(SO_3F)_5$ systems (42). Also the unique properties of fluorosulfuric acid have been found to provide unusual solvent systems, which can vary properties such as acidity, heats of solution, enthalpy, and heats of neutralization (43).

Table 2. Reactions of Fluorosulfuric Acid and Inorganic Compounds

Reactant	Product Name	Product Formula	References
nitrogen oxides	nitrosyl fluorosulfate, nitryl fluorosulfate	FSO_3NO, FSO_3NO_2	26, 27
H_3BO_3	boron trifluoride	BF_3	28
$KClO_4$	perchloryl fluoride	ClO_3F	29
$KMnO_4$	manganese(VII) fluoride trioxide	MnO_3F	30
As_2O_3	arsenic trifluoride	AsF_3	31
As_2O_5	arsenic pentafluoride	AsF_5	31
CrO_3	difluorodioxochromium(VI)	CrO_2F_2	32
P_4O_{10}	phosphoryl fluoride	POF_3	31
H_2S in ethanol	monothiosulfuric acid	$H_2S_2O_3$	33

Table 3. Reactions of Fluorosulfuric Acid and Organic Compounds

Reactant	Product Name	Product Formula	References
benzene[a]	benzenesulfonic acid	$C_6H_5SO_3H$	20, 34
	diphenyl sulfone	$(C_6H_5)_2SO_2$	34
	benzenesulfonyl fluoride	$C_6H_5SO_2F$	35
aliphatic amines	amidosulfuric acids	R_2NSO_3H or $RNHSO_3H$	36
aromatic compound	arylsulfonyl fluorides[b]	$ArSO_2F$	
carboxylic acids	acid fluorides	$RCOF$	37
alcohol or alkene	alkyl fluorosulfates	$ROSO_2F$	20, 38, 39
perhaloolefins or perhaloalkyl iodides	perhaloalkyl fluorosulfates	R_fOSO_2F	40, 41

[a] Product is dependent on reaction conditions and proportions of reagents.
[b] Ar represents an aryl group.

Fluorosulfuric acid may be used to prepare diazonium fluorosulfates, $ArN_2^+ SO_3F^-$ (44), which decompose on heating to give aryl (Ar) fluorosulfates (36,45). Aryl fluorosulfates are also obtained from arylsulfonyl chlorides and fluorosulfuric acid (35). Alkyl and other organofluorosulfates form during electrolysis of fluorosulfuric acid in the presence of organic species (46,47).

Preparation and Manufacture. Fluorosulfuric acid, first prepared by combining anhydrous HF and cooled, anhydrous SO_3 in a platinum container (1), has also been prepared from ionic fluorides or fluorosulfates and sulfuric acid (20,48). The reaction of chlorosulfuric acid (qv) with ionic fluorides also gives fluorosulfuric acid (49).

Commercially, fluorosulfuric acid is made by processes utilizing the product as a solvent. Solutions of HF and SO_3 in fluorosulfuric acid are mixed in stoichiometric quantities, or SO_3 and HF are separately introduced into a stream of fluorosulfuric acid to produce essentially pure HSO_3F. Some of the product is then recycled (50,51).

Fluorosulfuric acid can be very corrosive. A study of the corrosive properties of fluorosulfuric acid during preparation and use showed carbon steel to be acceptable up to 40°C, stainless steel up to 80°C, and aluminum alloys up to 130°C (52).

Economic Aspects. U.S. manufacturers of fluorosulfuric acid are AlliedSignal and Du Pont. These companies have a combined annual capacity estimated at 20,000 metric tons, most of which is used internally although some merchant sales exist. Fluorosulfuric acid is shipped in tank cars.

Specifications and Analysis. Commercial fluorosulfuric acid contains approximately 98% HSO_3F and approximately 1% H_2SO_4 and lesser amounts of sulfur trioxide and dioxide. No free HF is present.

The free sulfur trioxide can be titrated with water; the end point is determined conductimetrically. The sulfuric acid content is determined from the specific conductivity of the liquid at the point in the titration where no free SO_3 or excess water is present. If the presence of HF is suspected, a known amount of SO_3 is added to the acid and the excess SO_3 is determined as above. The content of another common impurity, SO_2, may be determined iodometrically in a dilute, aqueous solution.

Health and Safety Factors. Fluorosulfuric acid is a strong acid capable of causing severe burns similar to those experienced with sulfuric and hydrofluoric acids. In addition, the fumes of fluorosulfuric acid are extremely irritating, and breathing of the fumes is to be avoided. Precautions and first aid measures generally observed in handling strong sulfuric acid and hydrofluoric acid are applicable to fluorosulfuric acid. Small containers of fluorosulfuric acid should be well cooled before opening and precautions taken to relieve any gas pressure that may have developed. In the laboratory, fluorosulfuric acid may be handled in glass if water is not present; otherwise, containers of inert polymers or platinum should be used. For larger-scale equipment, iron or carbon steel (not stainless steel) may be used. Material safety data sheets and other literature from manufacturers describe additional precautions in handling large quantities of fluorosulfuric acid.

Uses. Fluorosulfuric acid serves as catalyst in the alkylation (qv) of branched-chain paraffins (53–58) and aromatic compounds (59), and in the polymerization of monoolefins (60) and rosin (61). Addition of strong Lewis acids, such as SbF_5, TaF_5, and NbF_5, to fluorosulfuric acid markedly increases the system acidity and catalytic activity (62–69). Other examples which show the marked catalytic effect of fluorosulfuric acid alone or in systems including SbF_5, etc, are the synthesis of methyl *tert*-butyl ether (70), the stereospecific formation of 2-naphthol (71), formation of aromatic aldehydes using CO (72), and polymerization of tetrahydrofuran (73,74) (see CATALYSIS).

As a reagent, fluorosulfuric acid has been employed in the preparation of boron trifluoride (28), silicon tetrafluoride (75,76), alkyl fluorosulfates (20,38), arenesulfonyl fluorides (35), acyl fluorides (77), sulfamic acid (78), and diazonium fluorosulfates (44). Among its other uses are the removal of small amounts of organic fluorides from petroleum alkylate made by the hydrogen fluoride process (79), the removal of HF from process exhaust gases (80), the removal of HF from F_2 (81), and as a constituent of baths for electropolishing metals (82,83) and glass polishing (84).

Derivatives. The nonmetallic inorganic derivatives of fluorosulfuric acid are generally made indirectly, although complex fluorosulfates of the Group 15 (V) elements and of xenon can be made directly (85,86), as can the NO^+ and NO_2^+ salts (26,27).

Peroxydisulfuryl difluoride [*13709-32-5*], FSO_2OOSO_2F, prepared from fluorine and SO_3 (87), is a ready source of fluorosulfate radicals, $FSO_2O\cdot$, (88) which react with many substances to form stable fluorosulfates (89,90). By using the route

$$X_2 + n\, S_2O_6F_2 \rightarrow 2\, X(OSO_2F)_n$$

where $n = 1$ or 3, compounds of the type $FOSO_2F$ [*13536-85-1*] (91), $BrOSO_2F$ [*13997-93-8*], and $I(OSO_2F)_3$ [*13709-37-0*] (92) have been prepared. Fluorosulfates of most metallic elements have been prepared but none have any commercial significance. The physical properties of some fluorosulfates are summarized in Table 4.

Table 4. Physical Properties of Some Fluorosulfates

Salt	CAS Registry Number	Appearance	Mp, °C	Solubility Water[a]	Other solvents[a]
NH_4SO_3F	[*13446-08-7*]	long colorless needles	245	s	sl s ethanol; v s methanol
$LiSO_3F$	[*13453-75-3*]	white powder	360	v s	v s ethanol, ether, acetone, amyl alcohol, ethyl acetate; i ligroin
$LiSO_3F\cdot 3H_2O$		long shiny needles	60–61		
$NaSO_3F$	[*14483-63-7*]	shiny leaflets, hygroscopic		s	s ethanol, acetone; i ether
KSO_3F	[*13455-22-6*]	short white prisms	311	6.9[b]	sl s methanol
$RbSO_3F$	[*15587-05-0*]	colorless needles	304	s	sl s methanol
$CsSO_3F$	[*13530-70-6*]	colorless rhombic	292	2.23[b]	

[a] i = insoluble; sl s = slightly soluble; s = soluble; v s = very soluble.
[b] In g/100 mL water.

Ammonium fluorosulfate is produced from ammonium fluoride by reaction with sulfur trioxide, oleum, or potassium pyrosulfate, $K_2S_2O_7$ (48). Solutions of ammonium fluorosulfate show little evidence of hydrolysis and the salt may be recrystallized from hot water. Ammonium fluorosulfate absorbs anhydrous ammonia to form a series of liquid amines that contain 2.5–6 moles of ammonia per mole of salt (77).

Sodium fluorosulfate may be prepared by the action of fluorosulfuric acid on powdered, ignited sodium chloride (13) or of sulfur trioxide on sodium fluoride (48). In general, the alkali metal fluorosulfates may be prepared from the ammonium salt by evaporating a solution containing that salt and an alkali metal hydroxide (77). The solubilities of some Group 1 and 2 fluorosulfates in fluorosulfuric acid have been determined (93).

BIBLIOGRAPHY

"Fluosulfonic Acid" under "Fluorine Compounds, Inorganic" in *ECT* 1st ed., Vol. 6, pp. 734–738, by W. S. W. McCarter, Pennsylvania Salt Manufacturing Co.; "Fluorosulfuric Acid" under "Fluorine Compounds, Inorganic" in *ECT* 2nd ed., Vol. 9, pp. 676–681, by R. E. Eibeck, Allied Chemical Corp.; "Fluorosulfuric Acid" under "Fluorine Compounds, Inorganic–Sulfur" in *ECT* 3rd ed., Vol. 10, pp. 812–817, by R. E. Eibeck, Allied Chemical Corp.

1. T. E. Thorpe and W. Kirman, *J. Chem. Soc.*, 921 (1892).
2. J. Barr, R. J. Gillespie, and R. C. Thompson, *Inorg. Chem.* **3**, 1149 (1964).
3. R. J. Gillespie, J. B. Milne, and R. C. Thompson, *Inorg. Chem.* **5**, 468 (1966).
4. G. W. Richards and A. A. Woolf, *J. Chem. Soc. A*, 1118 (1967).
5. A. A. Woolf, *J. Chem. Soc.*, 2840 (1954); *ibid.*, 433 (1955).
6. R. J. Gillespie and co-workers, *Can. J. Chem.* **40**, 675 (1962); **41**, 148, 2642 (1963); **42**, 502, 1433 (1964).
7. R. J. Gillespie and co-workers, *Inorg. Chem.* **3**, 1149 (1964).
8. R. J. Gillespie and co-workers, *Inorg. Chem.* **4**, 1641 (1965); **8**, 63 (1969).
9. K. Bartmann and D. Mootz, *Acta Crystallogr., Sect. C* **C46**(2), 319–320 (1990).
10. R. J. Gillespie and E. A. Robinson, *Can. J. Chem.* **40**, 644, 675 (1962); R. Savoie and P. A. Giguere, *Can. J. Chem.* **42**, 277 (1964).
11. B. Carre and J. Devynck, *Anal. Chim. Acta* **159**, 149–158 (1984).
12. V. Gold and co-workers, *J. Chem. Soc., Perkin Trans.* **2**(6) 859–864 (1985).
13. O. Ruff, *Chem. Ber.* **47**, 646 (1914).
14. R. J. Gillespie, *Accounts. Chem. Res.* **1**(7), 202 (1968).
15. R. C. Thompson in G. Nickless, ed., *Inorganic Sulphur Chemistry*, Elsevier, Amsterdam, the Netherlands, 1968, pp. 587–606.
16. A. W. Jache in H. J. Emeleus and A. G. Sharpe, eds., *Advances in Organic Chemistry and Radiochemistry*, Vol. 16, Academic Press, Inc., New York, 1974, pp. 177–200.
17. I. G. Ryss and A. Drabkina, *Kinet. Katal.* **7**, 319 (1966).
18. R. C. Paul, K. K. Paul, and K. C. Malhotra, *Inorg. Nucl. Chem. Lett.* **5**, 689 (1969).
19. D. Mootz, K. Bartmann, *Z. Anorg. Allg. Chem.* **592**, 171–178 (1991).
20. J. Meyer and G. Schramm, *Z. Anorg. Allg. Chem.* **206**, 24 (1932).
21. L. J. Belf, *Chem. Ind. (London)*, 1296 (1955).
22. P. Bernard, Y. Parent, and P. Vast, *C. R. Acad. Sci. Ser. C* **269**, 767 (1969).
23. E. Kemnitz and D. Hass, *Z. Chem.* **30**(7), 264–265 (1990).
24. J. M. Shreeve and G. H. Cady, *J. Am. Chem. Soc.* **83**, 4521 (1961).
25. H. Schmidt and H. D. Schmidt, *Z. Anorg. Allg. Chem.* **279**, 289 (1955).

26. D. R. Goddard, E. D. Hughes, and C. K. Ingold, *J. Chem. Soc.*, 2559 (1950).
27. W. Lange, *Chem. Ber.* **60B**, 967 (1927).
28. U.S. Pat. 2,416,133 (Feb. 18, 1947), D. Young and J. Pearson (to Allied Chemical Corp.).
29. G. Barth-Wehrenalp, *J. Inorg. Nucl. Chem.* **2**, 266 (1956).
30. A. Engelbrecht and A. V. Grosse, *J. Am. Chem. Soc.* **76**, 2042 (1954).
31. E. Hayek, A. Aignesberger, and A. Engelbrecht, *Monatch. Chem.* **86**, 470735 (1955).
32. A. Engelbrecht, *Angew. Chem. Int. Ed. Engl.* **4**, 641 (1965).
33. M. Schmidt and G. Talsky, *Z. Anorg. Allg. Chem.* **303**, 210 (1960).
34. J. H. Simons, H. J. Passino, and S. Archer, *J. Am. Chem. Soc.* **63**, 608 (1941).
35. W. Steinkopf and co-workers, *J. Prakt. Chem.* **117**, 1 (1927).
36. Ger. Pat. 532,394 (Aug. 8, 1930), W. Lange.
37. W. Traube and A. Krahmer, *Chem. Ber.* **B52**, 1293 (1919).
38. Ger. Pats. 342,898 (Oct. 25, 1921), 346,245 (Dec. 27, 1921), W. Traube.
39. G. Olah, J. Nishimura, and Y. Mo., *Synthesis* **4**(11), 661 (1973).
40. U.S. Pats 3,254,107 (May 31, 1966), 3,255,228 and 3,255,229 (June 7, 1966), M. Hauptschein and M. Braid (to Pennsalt Chemicals Corp.); 3,083,220 (Mar. 26, 1963), E. L. Edens (to E. I. du Pont de Nemours & Co., Inc.); 2,878,156 (Mar. 17, 1959), R. A. Davis (to The Dow Chemical Co.); 2,628,927 (Feb. 17, 1953), J. D. Calfee and P. A. Florio (to Allied Chemical Corp).
41. M. Hauptschein and M. Braid, *J. Am. Chem. Soc.* **83**, 2502 (1961).
42. W. V. Cicha and F. Aubke, *J. Am. Chem. Soc.* **111**(12), 4328–4331 (1989).
43. R. C. Paul, K. S. Dhindsa, *Proc. Indian Natl. Sci. Acad. Part A*, **47**(3), 357–372 (1981).
44. U.S. Pat. 1,847,513 (Mar. 1, 1932), W. Hentrich, M. Hardtmann, and H. Ossenbeck (to General Aniline Works).
45. W. Lange and E. Müller, *Chem. Ber.* **B63**, 2653 (1930).
46. J. P. Coleman and D. Pletcher, *Tetrahedron Lett.* (2), 147 (1974).
47. D. Pletcher and C. Smith, *Chem. Ind. (London)* **8**, 371 (1976).
48. W. Traube, *Chem. Ber.* **46**, 2525 (1913).
49. U.S. Pat. 2,312,413 (Mar. 2, 1943), R. K. Iler (to E. I. du Pont de Nemours & Co., Inc.).
50. U.S. Pats. 2,430,963 (Nov. 18, 1947), R. Stephenson and W. Watson (to Allied Chem. Corp.); U.S. Pat. 3,957,959 (May 18, 1976), R. Wheatley, D. Treadway, and R. Toennies (to E. I. du Pont de Nemours & CO., Inc.).
51. Jpn. Pat. 55126509 (Sept. 30, 1980), (to Akita Chem Co.).
52. V. I. D. Daritskii and co-workers, *Khim. Prom. (Moscow)*, (3) 183–184 (1991).
53. Br. Pat. 537,589 (June 27, 1941), (to Standard Oil Development Co.).
54. U.S. Pat. 2,313,103 (Mar. 9, 1943), C. L. Thomas (to Universal Oil Products Co.).
55. U.S. Pat. 3,778,489 (Dec. 11, 1973), P. T. Parker and I. Mayer (to Esso Research and Engineering Co.).
56. U.S. Pat. 3,922,319 (Nov. 25, 1975), J. W. Brockington (to Texaco, Inc.).
57. U.S. Pat. 3,928,487 (Dec. 23, 1975) D. A. McCauley (to Standard Oil Co.).
58. U.S. Pat. 4,008,178 (Feb. 15, 1977), J. W. Brockington (to Texaco, Inc.).
59. U.S. Pat. 2,428,279 (Sept. 3, 1947), V. N. Ipatieff and C. B. Linn (to Universal Oil Products Co.).
60. U.S. Pat. 2,421,946 (June 10, 1947), V. N. Ipatieff and C. B. Linn (to Universal Oil Products Co.).
61. U.S. Pat. 2,419,185 (Apr. 15, 1947), C. A. Braidwood and A. G. Hovey (to Reichold Chemicals).
62. U.S. Pat. 3,594,445 (July 20, 1971) P. T. Parker (to Esso Research and Engineering Co.).
63. U.S. Pat. 3,636,129 (Jan. 18, 1972), P. T. Parker and C. N. Kimerlin, Jr. (to Esso Research and Engineering Co.).
64. U.S. Pat. 3,678,120 (July 18, 1972), H. S. Bloch (to Universal Oil Products Co.).

65. U.S. Pat. 3,708,553 (Jan. 2, 1973), G. A. Olah (to Esso Research and Engineering Co.).
66. D. T. Roberts, Jr. and L. E. Calihan, *J. Macromol. Sci. Chem.* **7**, 1629 (1973).
67. U.S. Pat. 3,819,743 (June 25, 1974), D. A. McCauley (to Standard Oil Co.).
68. U.S. Pat. 3,925,495 (Dec. 9, 1975), P. G. Rodewald (to Mobil Oil Corp.).
69. U.S. Pat. 3,984,352 (Oct. 5, 1976), P. G. Rodewald (to Mobil Oil Corp.).
70. U.S. Pat. 5,081,318 (Mar. 4, 1991), J. F. Knifton (to Texaco).
71. G. A. Olah and co-workers, *J. Org. Chem.* **56**(21) 6148–6151 (1991).
72. Jpn. Pat. 01075442 A2 (Mar. 22, 1989), Y. Sama (to Agency of Industrial Science and Technology).
73. U.S. Pat. 4,544,774 (Oct. 1, 1985), R. Pick (to Du Pont).
74. U.S. Pat. 4,569,990 (Feb. 11, 1986), W. W. Kasper and co-workers (to Du Pont).
75. Can. Pat. 448,662 (May 25, 1948), A. C. Hopkins, Jr., R. M. Stephenson, and W. E. Watson (to Allied Chemical Corp.).
76. Brit. Pat. 755,692 (Aug. 22, 1956), A. J. Edwards (to National Smelting Co., Ltd.).
77. W. Traube, J. Horenz, and F. Wunderlich, *Chem. Ber.* **B52**, 1272 (1919).
78. W. Traube and E. Brehmer, *Chem. Ber.* **B52**, 1284 (1919).
79. U.S. Pat. 2,428,753 (Oct. 7, 1947), C. B. Linn (to Universal Oil Product Co.).
80. U.S. Pat. 2,434,040 (Jan. 6, 1948), B. F. Hartman (to Socony-Vacuum Oil Co.).
81. Brit. Pat. 824,427 (Dec. 2, 1959), H. R. Leech and W. H. Wilson (to Imperial Chemical Industries, Ltd.).
82. C. B. F. Young and K. R. Hesse, *Met. Finish.* **45**(2), 63, 84 (1947); **45**(3), 64 (1947).
83. Ger. Pat. DE3438433 A1 (May 15, 1985), K. Tajiri, H. Nomura (to Mitsubishi Heavy Ind. Ltd.).
84. A. Kaiser, *Glastech Ber.* **62**(4) 127–134 (1989).
85. R. C. Paul and co-workers, *J. Inorg. Nucl. Chem.* **34**, 2535 (1972).
86. D. D. Des Marteau and M. Eisenberg, *Inorg. Chem.* **11**, 2641 (1972).
87. F. B. Dudley and G. H. Cady, *J. Am. Chem. Soc.* **79**, 513 (1957).
88. *Ibid.* **85**, 3375 (1963).
89. F. Aubke and D. D. Des Marteau, *Fluorine Chem. Rev.* **8**, 74 (1977).
90. R. A. DeMarco and J. M. Shreeve, *Adv. Inorg. Chem. Radiochem.* **16**, 115 (1974).
91. J. E. Roberts and G. H. Cady, *J. Am. Chem. Soc.* **81**, 4166 (1959).
92. *Ibid.* **82**, 352 (1960).
93. R. Seeley and A. W. Jache, *J. Fluorine Chem.* **2**(3), 225 (1973).

FRANCIS E. EVANS
GANPAT MANI
AlliedSignal Inc.

TANTALUM

Tantalum Pentafluoride

Tantalum pentafluoride [7783-71-3], TaF_5, a white solid with a reported mp of 97°C and a bp of 229°C (1), is the only known binary fluoride. The vapor pressure of TaF_5 in kPa is given by the equation log P_{kPa} = 7.649 − 2834/T over the temperature range of 80–230°C and the heat of vaporization is 54.4 kJ/mol (13 kcal/mol) (1).

There are a number of methods of preparation for TaF_5. For example, tantalum pentafluoride has been produced by the reaction of F_2 or ClF_3 and Ta metal (2,3), by contacting Ta_2O_5 with excess HF in the presence of a dehydrating agent (4), by the reaction of Ta-containing ores and $HF–H_2SO_4$ followed by extraction with an organic solvent (5,6), by reaction of Ta_2O_5 and COF_2 (7), by heating ammonium hexafluorotantalate (8), by contacting fluorotantalic acid with a dehydrating agent containing C–Cl or C–Br bonds (9) and by halogen exchange of $TaCl_5$ with HF (10).

TaF_5 has been characterized by ir, Raman, x-ray diffraction, and mass spectrometry (3,11,12). TaF_5 has been used as a superacid catalyst for the conversion of CH_4 to gasoline-range hydrocarbons (qv) (12); in the manufacture of fluoride glass and fluoride glass optical fiber preforms (13), and incorporated in semiconductor devices (14). TaF_5 is also a catalyst for the liquid-phase addition of HF to polychlorinated ethenes (15). The chemistry of TaF_5 has been reviewed (1,16–19). Total commercial production for TaF_5 is thought to be no more than a few hundred kilograms annually.

BIBLIOGRAPHY

"Tantalum" under "Fluorine Compounds, Inorganic," in *ECT* 2nd ed., Vol. 9, pp. 681, W. E. White, Ozark-Mahoning Co.; in *ECT* 3rd ed., Vol. 10, p. 818, by A. J. Woytek, Air Products & Chemicals, Inc.

1. J. H. Canterford and R. Cotton, *Halides of the Second and Third Row Transition Metals*, John Wiley & Sons, Inc., New York, 1968.
2. J. K. Gibson, *J. Fluorine Chem.* **55**(3), 299–311 (1991).
3. B. Frlec, *Vestu. Slov. Kem. Drus* **16**(1–4), 47–50 (1969).
4. U.S. Pat. 5,091,168 A (Feb. 25, 1992), M. J. Nappa and J. Mario (to E. I. du Pont de Nemours and Co., Inc.).
5. Jpn. Kokai Tokkyo Koho 63236716 A2 (Oct. 3, 1988), M. Watanabe, M. Nanjo, and Y. Nishimura (to Solex Research Corp of Japan).
6. Jpn. Pat. 63147827 A2 (June 20, 1988), M. Watanabe, M. Nanjo, and Y. Nishimura (to Solex Research Corp. of Japan).
7. S. P. Mallela, O. D. Gupta, and J. M. Shreeve, *Inorg. Chem.* **27**(1), 208–209 (1988).
8. Eur. Pat. 85-301897 (Mar. 1985), M. Watanabe and S. Nishimura.
9. U.S. Pat. 77,864,687 (Dec. 27, 1977), C. J. Kim and D. Farcasiu.
10. S. Ruff, *Z. Anorg. Allgem. Chem.* **72**, 329 (1911).
11. A. I. Popov, V. F. Sukhoverkhov, and N. A. Chumae-Vskii, *Zh. Neorg. Khim.* **35**(5), 1111–1122 (1990).
12. I. R. Beattie, K. M. S. Livingston, G. A. Ozin, and D. J. Reynolds, *J. Chem. Soc. A.*, (6), 958–965 (1969).

13. U.S. Pat. 4,973,776 A (Nov. 27, 1990), V. M. Allenger and R. N. Pandey.
14. Eur. Pat. 331,483 A2, (Sept. 6, 1989), K. Fujiura, Y. Ohishi, M. Fujiki, T. Kanamori, and S. Takahashi.
15. A. E. Feiring, *J. Fluorine Chem.* **14**, 7 (1979).
16. Eur. Pat. 89-104364 (Mar. 11, 1989), I. Haroda and co-workers.
17. F. Fairbrother, in V. Gutmann, ed., *Halogen Chemistry*, Vol. 3, Academic Press, Inc., New York, 1966, p. 123.
18. F. Fairbrother, *The Chemistry of Niobium and Tantalum*, Elsevier Scientific Publishing Co., London, 1967.
19. D. Brown, in J. C. Barter, ed., *Comprehensive Inorganic Chemistry*, Vol. 3, Pergamon Press, Elmsford, N.Y., 1973, p. 565.

TARIQ MAHMOOD
CHARLES B. LINDAHL
Elf Atochem North America, Inc.

TIN

The main binary tin fluorides are stannous fluoride and stannic fluoride. Because the stannous ion, Sn^{2+}, is readily oxidized to the stannic ion, Sn^{4+}, most reported tin and fluorine complexes are of tin(IV) and fluorostannates. Stannous fluoroborates have also been reported.

Stannous Fluoride

Stannous fluoride [7783-47-3], SnF_2, is a white crystalline salt that has mp 215°C (1), bp 850°C, and is readily soluble in water and hydrogen fluoride. At 20°C stannous fluoride dissolves in water to a concentration of 30–39%; in anhydrous hydrogen fluoride to 72–82% (2–4).

The pH of a freshly prepared 0.4% solution of stannous fluoride is between 2.8 and 3.5. Initially clear aqueous solutions become cloudy on standing owing to hydrolysis and oxidation. The insoluble residue is a mixture containing stannous and stannic species, fluoride, oxide, oxyfluorides, and hydrates.

Stannous fluoride probably was first prepared by Scheele in 1771 and was described by Gay-Lussac and Thenard in 1809. Commercial production of stannous fluoride is by the reaction of stannous oxide and aqueous hydrofluoric acid, or metallic tin and anhydrous hydrogen fluoride (5,6). SnF_2 is also produced by the reaction of tin metal, HF, and a halogen in the presence of a nitrile (7).

Stannous fluoride is used widely in dentifrices (qv) and other dental preparations because of its anticaries effect (8). The chemistry (9) involved in cavity prevention is thought to be reaction of stannous fluoride and the hydroxyapatite, $Ca_5(PO_4)_3OH$, of the tooth to form the more insoluble fluoroapatite, $Ca_5(PO_4)_3F$. More concentrated solutions of stannous fluoride react with hydroxyapatite to produce $Sn_3F_3PO_4$ [12592-27-7] (10). The role of SnF_2 in reducing acidogenicity of dental plaque *in vivo* has also been studied (11). On heating stannous fluoride under nitrogen with stannic fluoride, Sn_7F_{16}, Sn_3F_8, Sn_2F_6, and $Sn_{10}F_{34}$ are formed (12).

Other uses of SnF_2 are in the synthesis of fluorophosphate glasses having low melting temperatures (13–15), in formation of transparent film (16), and in the preparation of optically active alcohols (17).

Fluorostannites and Fluorostannates

Complexes of the type $MSnF_3$, where M is NH_4 [15660-29-4], Na [13782-22-4], K [13782-23-5], and Cs [13782-25-7], have been crystallized from aqueous solutions (18–20). Solutions of these salts deposit tin(II) oxide crystals indicating hydrolysis but not oxidation. From molten mixtures of SnF_2 and NaF, RbF, and CsF, both the $MSnF_3$ (M = Na [13782-22-4], K [13782-23-5], Rb [13782-24-6], and Cs [13782-25-7]) and the fluorostannate salts, $MSnF_5$ (M = Na [58179-42-3], K [58179-40-1], Rb [72264-75-6], and Cs [72264-76-7]) have been obtained (21). Complexes of the type $Cd(H_2O)_6 \cdot (SnF_3)_2$ [125445-76-3], $Zn(H_2O)_6 \cdot SnF_3$ [125445-75-2] (22), $SnCl_3 \cdot SnF_3$ [108632-61-7], $N_2H_6 \cdot (SnF_3)_2$ [99625-93-1] (23), $N_2H_5 \cdot SnF_3$ [73953-53-4] (24), $Ca(SnF_3)_2$ [69244-56-0] (25), $Ni(SnF_3)_2$ [26442-44-4] (26), $Co(SnF_3)_2$ [26442-43-3] (26) have also been reported.

Stannic Fluoride

Stannic fluoride [7783-62-2], SnF_4, is a white solid that sublimes at 705°C and hydrolyzes in water to form insoluble stannic acid. It can be prepared by reaction of fluorine and probably ClF_3 or BrF_3 with virtually any tin(II) or tin(IV) compound, eg, Sn, SnO, SnO_2, SnS, SnS_2 (27), and $SnCl_2$ (28). Reaction of $SnCl_4$ and HF (29) forms $SnCl_4 \cdot SnF_4$ which can be decomposed by heating to 750°C where pure SnF_4 sublimes. Stannic fluoride forms numerous complexes as a Lewis acid. The other methods of preparation for stannic fluoride include the oxidation of SnF_2 by a halogen in acetonitrile (30,31); the reaction of NF_3O and Sn (32); and the reaction of COF_2 and SnO_2 (33). Stannic fluoride is used in the manufacture of glass (qv) (34).

Stannous Fluoroborate

Stannous fluoroborate [13814-97-6], $Sn(BF_4)_2$, is prepared in electrochemical cells using tin and fluoroboric acid (35,36), by reaction of 80% HF and H_3BO_3 followed by reaction with $Sn(OH)_2$ (37); and from the reaction of mossy tin and 30–70% HBF_4 (38). The main use of stannous fluoroborate is in electroplating (qv) (39).

Hexafluorostannates

The hexafluorostannate anion [21340-04-5], SnF_6^{2-}, forms readily and is stable over a wide pH range. Numerous hexafluorostannates have been prepared by dissolving stannates in excess hydrofluoric acid, dissolving stannic acid in excess HF and neutralizing, or by reaction of salts and SnF_4. Many of these stable and generally water-soluble hexafluorostannates were prepared as early as 1857. Spectral studies of the SnF_6^{2-} anion have been reported (40). Some of the newer hexafluorostannates are K·NaSnF$_6$ [112813-21-5] (41), CsNa·SnF$_6$ [112813-23-7], Rb·NaSnF$_6$ [112813-22-6] (42), and N_2H_6·SnF$_6$ [128493-43-6] (43).

Safety, Handling, and Toxicity

Stannous fluoride is used in dentifrices and dental preparations. The OSHA permissible exposure limit (44) and ACGIH (45) established TLV for fluoride is 2.5 mg/m^3 of air.

BIBLIOGRAPHY

"Tin Fluoride" under "Tin Compounds," in *ECT* 1st ed., Vol. 4, p. 160, by H. Richter, Metal & Thermit Corp.; "Tin" under "Fluorine Compounds, Inorganic" in *ECT* 2nd ed., Vol. 9, pp. 682–683, by W. E. White, Ozark-Mahoning Co.; in *ECT* 3rd ed., Vol. 10, pp. 819–820, by C. B. Lindahl and D. T. Meshri, Ozark-Mahoning Co.

1. J. J. Dudash and A. W. Searcy, *High Temp. Sci.* **1**, 287 (1969).
2. W. H. Nebergall, J. C. Muhler, and H. G. Day, *J. Am. Chem. Soc.* **74**, 1604 (1952).
3. J. B. Bearl, Jr., Ph.D. dissertation, Texas A&M University, College Station, 1963.
4. J. E. Gilliland, M.S. thesis, Oklahoma State University, Stillwater, 1960.
5. U.S. Pats. 2,924,508 (Feb. 9, 1960); 2,955,914 (Oct. 11, 1960); 3,097,063 (July 9, 1963), J. E. Gilliland, R. Ray, and W. E. White (to Ozark-Mahoning Co.).
6. I. V. Murin, S. V. Chernov, and M. Yu. Vlasov, *Zh. Prikl. Khim. (Leningrad)* **58**(10), 2340–2342 (1985).
7. U.S. Pat. 4,034,070 (July 21, 1975) J. A. Wojtowicz and D. F. Gavin.
8. J. C. Muhler and co-workers, *J. Am. Dent. Assoc.* **50**, 163 (1955).
9. E. J. Duff, *Caries Res.* **7**, 79 (1973).
10. S. H. Y. Wei, *J. Dent. Res.* **53**, 57 (1974).
11. J. E. Ellingsen, B. Svatun, and G. Roella; *Acta Odontol. Scand.* **38**(4), 219–222 (1980).
12. R. Sabtier, A. M. Hebrard, J. D. Cousseinsv, and C. R. Hebd, *Seances Acad. Sci. Ser. C* **279**(26), 1121–1123 (1974).
13. J. Leissner, K. Sebastian, H. Roggendorf, and H. Schmidt, *Mater. Sci. Forum* **67–68**, 137–142 (1991).
14. C. M. Shaw and J. E. Shelby, *Phys. Chem. Glasses* **29**(2), 49–53 (1988).
15. N. Sakamoto and K. Morinaga; *Sogo Rikogaku Hokoku (Kyushu Daigaku Daigakuin)* **12**(3), 283–289 (1990).
16. Jpn. Pat. 633147B A2 (Dec. 22, 1988) N. Sonoda and N. Sato (to Showa).
17. Jpn. Pat. 62158222 A2 (July 14, 1987) M. Mukoyama, N. Minowa, T. Oriyama, and K. Narasaka (to Showa).
18. E. L. Muetterties, *Inorg. Chem.* **I**, 342 (1962).
19. W. B. Schaap, J. A. Davis, and W. N. Nebergall, *J. Am. Chem. Soc.* **76**, 5226 (1954).
20. J. D. Donaldson and J. D. O'Donoghue, *J. Chem. Soc.*, 271 (1964).
21. J. D. Donaldson, J. D. O'Donoghue, and R. Oteng, *J. Chem. Soc.*, 3876 (1965).
22. Yu. V. Kokunov, and co-workers, *Dokl. Akad. Nauk. SSSR* **307**(5), 1126–1130 (1989).
23. V. Kancic and co-workers, *Acta. Crystallogr., Sect. C: Crystal Struct. Commun.* **C44**(8), 1329–1331 (1988).

24. W. Granier and M. Lopez, *Calorim. Anal. Therm.* **16**, 358–371 (1985).
25. F. Babcock, K. David, C. Jeana, and T. H. Jordan, *J. Dent. Res.* **57**(9–10), 933–938 (1978).
26. J. D. Donaldson and R. Oteng, *J. Chem. Soc. A*, (18), 2696–2699 (1969).
27. H. M. Haendler and co-workers, *J. Am. Chem. Soc.* **76**, 2179 (1954).
28. A. A. Woolf and H. J. Emeleus, *J. Chem. Soc.*, 2864 (1949).
29. H. J. Emeleus, in J. H. Simons, ed., *Fluorine Chemistry*, Vol. 1, Academic Press, New York, 1950, pp. 1–76.
30. D. Tudela and F. Rey, *Z. Anorg. Allg. Chem.* **575**, 202–208 (1989).
31. O. D. Gupta, R. L. Kirchmeier, and J. M. Shreeve, *Inorg. Chem.* **29**(3), 573–574 (1990).
32. S. P. Mallela, O. D. Gupta, and J. M. Shreeve, *Inorg. Chem.* **27**(1), 208–209 (1988).
33. EP 156617 A (Oct. 22, 1985) M. Watanabe and S. Nishimura.
34. G. D. Lukiyanchuk, V. K. Gonsharuk, E. V. Merkulov, and T. I. Usol'tseva, *Fiz. Khim. Stekla*, **18**(2), 141–145 (1992).
35. DD 293,609 A5 (Sept. 5, 1991), D. Ohms and co-workers.
36. C. J. Chen and C. C. Wan, *Electrochim. Acta* **30**(10), 1307–1312 (1985).
37. RO 67814 (Jan. 15, 1980), V. Grigore and N. Cretu.
38. U.S. Pat. 3,432,256 (Mar. 11, 1969), H. P. Wilson.
39. EP 45,471 A1 (Feb. 10, 1982), H. Willenberg, W. Becher, and K. H. Hellberg.
40. H. Kreigsmann and G. Kessler, *Z. Anorg. Allegm. Chem.* **318**, 266–276 (1962).
41. A. V. Gerasimenko, S. B. Antokhina, and V. I. Sergienko, *Koor. Khim.* **18**(2), 129–132 (1992).
42. V. I. Sergienko, V. Ya. Kavun, and L. N. Ignat'eva, *Zh. Neorg. Khim.* **36**(5), 1265–1268 (1991).
43. A. Rahten, D. Gantar, and I. Leban, *J. Fluorine Chem.* **46**(3), 521–528 (1990).
44. *Code of Federal Regulations*, Title 29, Part 1910.1000, Washington, D.C.
45. *Threshold Limit Values for Chemical Substances and Physical Agents, 1992–1993*, The American Conference of Governmental Industrial Hygienists, Cincinnati, Ohio.

CHARLES B. LINDAHL
TARIQ MAHMOOD
Elf Atochem North America, Inc.

TITANIUM

Titanium(III) Fluoride

Titanium trifluoride [13470-08-1], TiF_3, is a blue crystalline solid that undergoes oxidation to TiO_2 upon heating in air at 100°C (see TITANIUM COMPOUNDS). In the absence of air, disproportionation occurs above 950°C to give TiF_4 and titanium metal. TiF_3 decomposes at 1200°C, has a density of 2.98 g/cm^3, and is insoluble in water but soluble in acids and alkalies. The magnetic moment is 16.2 × 10^{-24} J/T (1.75 µB).

Titanium trifluoride is prepared by dissolving titanium metal in hydrofluoric acid (1,2) or by passing anhydrous hydrogen fluoride over titanium trihydrate at 700°C or over heated titanium powder (3). Reaction of titanium trichloride and anhydrous hydrogen fluoride at room temperature yields a crude product that can be purified by sublimation under high vacuum at 930–950°C.

Titanium trifluoride can be stored in tightly closed polyethylene containers for several years. Shipping regulations classify the material as a corrosive solid and it should be handled in a fully ventilated area or in a chemical hood. The ACGIH adopted toxicity values (1992–1993) for TiF_3 is as TWA for fluorides as F^- 2.5 mg/m^3.

This material is available from Advance Research Chemicals, Inc., Aldrich Chemical Company, Inc., Aesar, Johnson/Matthey, Cerac, PCR, and Pfaltz & Bauer in the United States, Fluorochem of the United Kingdom, and Schuchardt of Germany. Its 1993 price was approximately $500/kg. No commercial applications have been reported.

Titanium(IV) Fluoride

Titanium tetrafluoride [7783-63-3], TiF_4, has potential for use in dental hygiene products. It is used in infrared transmitting halide glass.

TiF_4 is a colorless, very hygroscopic solid and is classified as a soft fluorinating reagent (4), fluorinating chlorosilanes to fluorosilanes at 100°C. It also forms adducts, some of them quite stable, with ammonia, pyridine, and ethanol. TiF_4 sublimes at 285.5°C, and melts at temperatures >400°C. It is soluble in water, alcohol, and pyridine, hydrolyzing in the former, and has a density of 2.79 g/mL.

Titanium tetrafluoride may be prepared by the action of elemental fluorine on titanium metal at 250°C (5) or on TiO_2 at 350°C. The most economical and convenient method is the action of liquid anhydrous HF on commercially available titanium tetrachloride in Teflon or Kynar containers. Polyethylene reacts with $TiCl_4$ and turns dark upon prolonged exposure. The excess of HF used is boiled off to remove residual chloride present in the intermediates.

Titanium(IV) fluoride dihydrate [60927-06-2], $TiF_4 \cdot 2H_2O$, crystals can be prepared by the action of aqueous HF on titanium metal. The solution is carefully evaporated to obtain the crystals. Neutral solutions when heated slowly hydrolyze and form titanium(IV) oxyfluoride [13537-16-1], $TiOF_2$ (6). Upon dissolution in hydrogen fluoride, TiF_4 forms hexafluorotitanic acid [17439-11-1], H_2TiF_6.

The most promising application of titanium tetrafluoride is for use in topical applications for prevention of dental caries (7–13). It is being evaluated and compared to NaF, MFP, and SnF_2 used in these applications. The other use is in mixed optical halide glass (14–16), and in the preparation of fluorotitanates (17–19).

Total consumption of TiF_4 in both the United States and Europe is less than 500 kg/yr. TiF_4 is available from Advance Research Chemicals, Inc., Aldrich, Aesar, Johnson/Matthey, Cerac, PCR, and Pfaltz & Bauer of the United States, Fluorochem of the United Kingdom, and Schuchardt of Germany. Its 1993 price varied between $300 to $400/kg.

Fluorotitanates

Hexafluoroanions of Group 4 (IVB) are octahedral crystals that are quite stable in acidic media. Solutions having pH > 4 tend to hydrolyze forming the metal dioxides. All three hexafluoroacids are known, ie, hexafluorotitanic acid, hexafluorozirconic acid [12021-95-3], H_2ZrF_6, and hexafluorohafnic acid [12021-47-5], H_2HfF_6. These acids exist only in aqueous media in the presence of excess hydrofluoric acid. Alkali, alkaline-earth, and other metal salts of these acids, M_2XF_6, where X = Ti and M = Li [19193-50-1], Na [17116-13-1], K [16919-27-0], Rb [16962-41-7], Cs [16919-28-1], NH_4 [16962-40-6], and Tl [26460-00-4], have been isolated as stable solids at ambient temperatures (20). Maximum concentration of hexafluorotitanic acid is found to be 63% in the presence of 0.5% excess HF. Its salts, NH_4^+, Li^+, and Na^+ are quite soluble in water, whereas those of K^+, Rb^+, and Cs^+ are only slightly soluble.

Fluorotitanic acid is used as a metal surface cleaning agent, as a catalyst, and as an aluminum finishing solvent (see METAL SURFACE TREATMENTS). Fluorotitanates are used in abrasive grinding wheels and for incorporating titanium into aluminum alloys (see ABRASIVES; ALUMINUM AND ALUMINUM ALLOYS).

Although titanium compounds are considered to be physiologically inert (21), fluorides in general are considered as toxic above 3 ppm level and extreme care should be taken in handling large amounts of titanium salts as well as hexafluorotitanic acid. The ACGIH adopted (1992–1993) toxicity limits are as TWA for fluorides as F^- 2.5 mg/m^3.

The total U.S. consumption of H_2TiF_6 is 20 t/yr. The 1993 price varied between $2.80 to $7.50/kg depending on quantity and specifications. It is packaged in DOT approved polyethylene-lined drums and the salts in polyethylene-lined fiber board drums.

BIBLIOGRAPHY

"Titanium Fluorides" under "Titanium Compounds," in *ECT* 1st ed., Vol. 14, p. 217, by L. R. Blair, H. H. Beecham, and W. K. Nelson; "Titanium" under "Fluorine Compounds, Inorganic," in *ECT* 2nd ed., Vol. 9, pp. 683–684, by W. E. White; in *ECT* 3rd ed., Vol. 10, pp. 821–822, by D. T. Meshri, Ozark-Mahoning Co.

1. M. E. Straumenis and J. I. Ballas, *Z. Anorg. Chem.* **278**, 33 (1955).
2. P. H. Woods and L. D. Cockrell, *J. Am. Chem. Soc.* **80**, 1534 (1958).
3. P. Ehrlich and G. Pietzka, *Z. Anorg. Chem.* **275**, 121 (1954).
4. D. T. Meshri and W. E. White, "Fluorinating Reagents in Inorganic and Organic Chemistry" in the *Proceedings of George H. Cady Symposium*, Milwaukee, Wis., June 1970.
5. H. M. Haendler, *J. Am. Chem. Soc.* **76**, 2177 (1954).
6. K. S. Vorres and F. B. Dutton, *J. Am. Chem. Soc.* **77**, 2019 (1955).
7. B. Regolati and co-workers, *Helv. Odontol. Acta* **18**(2), 92 (1974).
8. A. S. Mundorff, M. F. Little, and B. G. Bibby, *J. Dent. Res.* **51**, 1567 (1972).
9. A. J. Reed and B. G. Bibby, *J. Dent. Res.* **55**, 357 (1976).
10. L. Skartveit, K. A. Selvig, S. Myklebust, and A. B. Tveit, *Acta Odontol. Scand.* **48**(3), 169–174 (1990).
11. L. Skartveit, A. B. Tveit, B. Klinge, B. Toetdal, and K. A. Selvig, *Acta Odontol. Scand.* **47**(2), 65–68 (1989).
12. L. Skartveit, A. B. Tveit, B. Toetdal, and K. A. Selvig, *Acta Odontol. Scand.* **47**(1), 25–30 (1989).
13. A. B. Tveit, K. Bjorn, B. Toetdal, and K. A. Selvig, *Scand. J. Dent. Res.* **96**(6), 536–540 (1988).

14. A. Jha and J. M. Parker, *Phys. Chem. Glasses.* **32**(1), 1–2 (1991).
15. B. Boulard and C. Jacoboni, *Mater. Res. Bull.* **25**(5), 671–677 (1990).
16. Eur. Pat. EP 331,483 (Sept. 6, 1989), K. Fujiura and co-workers (to Nippon Telegraph & Telephone Corp.).
17. Pol. Pat. PL 153,066 (May 11, 1988), L. Stoch, S. Mocydlarz, M. Laczka, and I. Waclawska (to Akademia Gorniczo Hutnicza).
18. Pol. Pat. PL 153,702 (May 31, 1991), I. Kustra, A. Chajduga, J. Konczal, and M. Jarzynowski (to Instytut Chemï Nieorganicznej).
19. B. N. Chernyshov and co-workers, *Zh. Neorg. Khim.* **34**(9), 2179–2186 (1989).
20. B. Cox and A. G. Sharpe, *J. Chem. Soc.*, 1783 (1953).
21. N. I. Sax, *Dangerous Properties of Industrial Materials*, 6th ed., Van Nostrand Reinhold Co., New York, 1984, p. 2585.

DAYAL T. MESHRI
Advance Research Chemicals, Inc.

TUNGSTEN

Tungsten has three readily prepared binary fluorides, tungsten hexafluoride [7783-82-6], tungsten pentafluoride [19357-83-6], and tungsten tetrafluoride [13766-47-7] (1,2). The three lower oxidation state tungsten binary fluorides have been observed only in high energy systems (3). Several complex oxyfluorides are known including WOF_4 [13520-79-1] and WO_2F_2 [14118-73-1] (4). Only tungsten hexafluoride is made commercially. Tungsten hexafluoride is used as a tungsten source in chemical vapor deposition (CVD) for very large-scale integration (VSLI) devices.

Tungsten Hexafluoride

Physical Properties. Tungsten(VI) fluoride [7783-82-6], WF_6, is a colorless gas that condenses at ca 100 kPa (1 atm) and 17.1°C to a water-white liquid that may be colored owing to metallic impurities. Below 2°C it forms a white solid. Tungsten hexafluoride has a symmetrical octahedral structure at near room temperature and a phase of lower symmetry below −8.5°C (5). The Raman and uv spectra (6), as well as the ir spectrum (7) have been studied. The physical properties of tungsten hexafluoride are given in Table 1.

Chemical Properties. Tungsten hexafluoride is readily hydrolyzed by water to give tungsten trioxide and hydrogen fluoride. It is a strong fluorinating agent and reacts with many metals at room temperature. Tungsten hexafluoride reacts with the alkali fluorides KF, RbF, and CsF to form the complex salts K_2WF_8 [57300-87-5], Rb_2WF_8 [57300-88-6], and Cs_2WF_8 [57300-89-7], respectively (10). The alkali iodides and WF_6 react in sulfur dioxide to form the tungsten(V) compounds $NaWF_6$ [55822-76-9], KWF_6 [34629-85-1], $RbWF_6$ [53639-97-7], and $CsWF_6$ [19175-38-3] (11). Tungsten hexafluoride reacts with hydrogen and hy-

Table 1. Physical Properties of Tungsten Hexafluoride

Property	Value	References
boiling point, °C	17.2	8
triple point, °C, 55.1 kPa^a	2.0	9
liquid density at 15°C, g/mL	3.441	8
transition point, °C, 32.0 kPa^a	−8.2	9
heat of vaporization, kJ/mol^b	26.5	9
heat of fusion, kJ/mol^b	1.76	9
heat of transition, kJ/mol^b	5.86	9
heat of sublimation, kJ/mol^b		
above transition	32.4	9
below transition	38.3	9
entropy of vaporization, J/(mol·K)^b	91.2	9
entropy of fusion, J/(mol·K)^b	6.07	9
entropy of transition, J/(mol·K)^b	22.1	9
specific heat at 25°C, J/(mol·K)^b	118.92	9
vapor pressure equation	$\log P_{kPa} = a - b/T$	
liquid		9
T, °C	2.0 to 17.1	
a	6.760^c	
b	1380.5	
solid		9
T, °C	−8.2 to 2.0	
a	7.883^d	
b	1689.9	
T, °C	−60 to −8.2	
a	9.076^e	
b	2006.0	

^{a}To convert kPa to mm Hg, multiply by 7.5.
^{b}To convert kJ to kcal, divide by 4.184.
^{c}For P in mm Hg, $a = 7.635$.
^{d}For P in mm Hg, $a = 8.758$.
^{e}For P in mm Hg, $a = 9.951$.

drogen-containing reducing agents at elevated temperature to form tungsten metal and hydrogen fluoride. This reaction is the basis of the primary use of tungsten hexafluoride in CVD (see THIN FILMS). The CVD chemistry of WF$_6$ has been reviewed (12,13). Reduction with hydrogen generally requires temperatures of 450–750°C and pressures of < 100 kPa (14.5 psi) (14). Other gaseous reductants include GeH$_4$ (15), SiH$_2$F$_2$ (16), SiH$_4$ (17), and diethyl silane (18).

Manufacture and Economics. Tungsten hexafluoride is produced commercially by the reaction of tungsten powder and gaseous fluorine at a temperature in excess of 350°C (19). Tungsten hexafluoride is the principal product of the reaction, and there are no by-products when high purity tungsten powder and fluorine are used. U.S. production is several metric tons per year. Essentially all of the product is used in CVD. Air Products and Chemicals, Inc. (Allentown, Pennsylvania) and Bandgap Technology Corp. (Broomfield, Colorado) are the only U.S. producers. The 1992 price ranges from $300–$850/kg, depending on the purity.

Because of the development of electronic applications for WF_6, higher purities of WF_6 have been required, and considerable work has been done to improve the existing manufacturing and purification processes (20). Most metal contaminants and gaseous impurities are removed from WF_6 by distillation. HF, which has a similar vapor pressure to WF_6, must be removed by adsorption (see ELECTRONIC MATERIALS; ULTRAPURE MATERIALS).

Specifications. The use of tungsten hexafluoride in CVD applications in the manufacture of high density silicon chips requires a high purity product, essentially free of all metallic contaminants. Several grades of WF_6 are available. Table 2 shows the specifications for three grades of WF_6.

Tungsten hexafluoride is shipped as a liquid under its own vapor pressure in nickel or steel cylinders in quantities of 45 kilograms per cylinder or less; however, it has been shown that the purity of WF_6 packaged in steel cylinders can degrade over time (21). It is classified as a corrosive liquid by the DOT.

Handling and Toxicity. Tungsten hexafluoride is irritating and corrosive to the upper and lower airways, eyes, and skin. It is extremely corrosive to the skin, producing burns typical of hydrofluoric acid. The OSHA permissible exposure limits is set as a time-weighted average of 2.5 mg/kg or 0.2 ppm (22).

Monel and nickel are the preferred materials of construction for cylinders and delivery systems; however, copper, brass, steel, and stainless steel can be used at room temperature, providing that these metals are cleaned, dried, and passivated with a fluoride film prior to use. Studies have shown that fluorine passivation of stainless steel and subsequent formation of an iron fluoride layer prior to WF_6 exposure prevents reaction between the WF_6 and the stainless steel surface (23).

Uses. The primary use of WF_6 is for blanket and selective deposition of tungsten and tungsten silicide films in the manufacture of VLSI electronic devices. The important aspects of this application have been reviewed (24). Addi-

Table 2. Specifications on Three Grades[a] of WF_6

Impurity	Electronic	VLSI	Megaclass
HF, ppmv	150	10	1
CO_2, CF_4, SF_6, SiF_4, ppmv each	10	0.5	0.5
N_2, ppmv	15	1	0.5
O_2 + Ar, ppmv	10	0.5	0.5
CO, ppmv	[b]	1	1
total metals, ppb[c]	1000	1000	1000
Cr, Fe, K, Na, ppb[c] each	10	10	10
U, ppb[c]	0.1	0.05	0.05
Th, ppb[c]	0.1	0.1	0.1

[a]Commercial grades offered by Air Products and Chemicals, Inc.
[b]No specifications given.
[c]By weight.

tionally, several conferences have been devoted to CVD using WF_6 (13,20). Non-electronic applications of tungsten hexafluoride include the CVD of tungsten to form hard tungsten carbide coatings on steel (25) and to fabricate solid tungsten pieces such as tubing or crucibles. Composite coatings of tungsten and rhenium are produced by the simultaneous chemical vapor deposition from these hexafluorides (26) and the addition of rhenium improves the ductility and high temperature properties of the deposit.

BIBLIOGRAPHY

"Tungsten" under "Fluorine Compounds, Inorganic" in *ECT* 3rd ed., Vol. 10, pp. 823–825, by A. J. Woytek, Air Products and Chemicals, Inc.

1. N. N. Greenwood and A. Earnshaw, *Chemistry of the Elements*, Pergamon Press, Elmsford, N.Y., 1984, pp. 1187–1191.
2. J. Schröeder and F. J. Grewe, *Angew. Chem., Int. Ed. Engl.* **7**, 132 (1968).
3. A. Bensaoula, E. Grossman, and A. Ignatiev, *J. Appl. Phys.* **62**, 4587 (1987).
4. J. C. Bailar and co-eds., *Comprehensive Inorganic Chemistry*, Vol. 3, Pergamon Press, Compendium Publishers, Elmsford, N.Y., 1973, pp. 749–763.
5. S. Siegal and D. A. Northrup, *Inorg. Chem.* **5**, 2187 (1966).
6. K. N. Tanner and A. B. F. Duncan, *J. Am. Chem. Soc.* **73**, 1164 (1951).
7. J. Gaunt, *Trans. Faraday Soc.* **49**, 1122 (1953).
8. J. A. Dean, ed., *Lange's Handbook of Chemistry*, 13th ed., McGraw-Hill Book Co., New York, 1985, pp. 4–125.
9. G. H. Cady and G. B. Hargreaves, *J. Chem. Soc.*, 1563 (1961).
10. B. Cox, D. W. Sharp, and A. G. Sharpe, *J. Chem. Soc.*, 1242 (1956).
11. G. B. Hargreaves and R. D. Peacock, *J. Chem. Soc.*, 4212 (1957).
12. M. L. Yu, K. Y. Ahn, and R. V. Joshi, *IBM J. Res. Dev.* **34**(6), 875 (1990).
13. V. V. S. Rana, R. V. Joshi, and I. Ohdomari, eds., *Advanced Metallization for ULSI Applications* and references therein, Materials Research Society, Pittsburgh, Pa., 1992.
14. U.S. Pat. 3,565,676 (Feb. 23, 1971), R. A. Holzl (to Fansteel Metallurgical Corp.).
15. C. A. Van der Jeugd, G. J. Leusink, G. C. A. M. Janssen, and S. Radelaar, *Appl. Phys. Lett.* **57**, 354 (1990).
16. Jpn. Pat. J63250463 (Oct. 18, 1988) T. Kusumoto and co-workers (to ULVAC Corp.).
17. H. L. Park and co-workers, *J. Electrochem Soc.* **137**, 3213 (1990).
18. D. A. Roberts and co-workers, in Ref. 13, p. 127.
19. H. F. Priest, *Inorg. Synth.* **3**, 181 (1950).
20. *Forum on Process Gases, TechWeek/East*, Semiconductor Equipment and Materials International, Cambridge, Mass., Sept. 29, 1992.
21. M. A. George and D. Garg, in *Proceedings of 1990 Microcontamination Conference*, Canon Communications, Santa Monica, Calif., 1990.
22. *Documentation of the Threshold Limit Values and Biological Exposure Indices*, 5th ed., American Conference of Governmental Industrial Hygienists, Inc., Cincinnati, Ohio, 1986.
23. D. A. Bohling and M. George, *Semicond. Int.* **14**, 104 (1991).
24. J. E. J. Schmitz, *Chemical Vapor Deposition of Tungsten and Tungsten Silicides for VLSI/ULSI Applications*, Noyes Publications, Park Ridge, N.J., 1992.
25. N. J. Archer, *Proceedings Conference Chemical Vapor Deposition, 5th International Conference*, Electrochemical Society, Princeton, N.J., 1975.

26. J. L. Federer and A. C. Schaffhauser, *9th Thermionic Conversion Specialist Conference*, Miami Beach, Fla., 1970, pp. 74–81.

General References

J. H. Simons, ed., *Fluorine Chemistry*, Vol. 5, Academic Press, Inc., New York, 1964.
J. W. Mellor, *A Comprehensive Treatise on Inorganic and Theoretical Chemistry*, Vol. 11, John Wiley & Sons, Inc., New York, 1962.

PHILIP B. HENDERSON
ANDREW J. WOYTEK
Air Products and Chemicals, Inc.

FLUORINE COMPOUNDS, INORGANIC (URANIUM). See
FLUORINE; URANIUM AND URANIUM COMPOUNDS.

ZINC

Zinc Fluoride

Anhydrous zinc fluoride [7783-49-5], ZnF_2, melts at 872–910°C, has a solubility of only 0.024 g/100 g anhydrous HF at 14.2°C (1), and can be prepared by slowly drying zinc fluoride tetrahydrate [13986-18-0], $ZnF_2 \cdot 4H_2O$, in a current of anhydrous hydrogen fluoride to minimize hydrolysis and formation of the oxide. There is x-ray evidence for dihydrate formation during dehydration of the tetrahydrate (2). Anhydrous zinc fluoride can also be prepared from the reaction of Zn metal powder and pyridinium poly(hydrogen fluoride) at ambient temperature (3); by treating zinc hydroxycarbonate with NH_4F followed by thermal decomposition (4); by the reaction of NF_3O (5) or NH_4F and ZnO (6,7); by the thermal decomposition of $(NH_4)_2ZnF_4$ (8); by the reaction of SOF_2 and Zn (9); by the reaction of Zn and HF in the presence of acetonitrile (10); by the reaction of SF_6 and Zn (11); by the reaction of PF_3 and ZnO (12); and by the reaction of ZnO and hydrogen fluoride (13). Zinc fluoride of ca 96% purity is commercially produced for use as a flux in metallurgy (qv). Production is only on a small scale.

Zinc fluoride has been used as a mild fluorinating reagent in replacement of chlorine in halogenated hydrocarbons (14,15). It is also used as a catalyst in several applications including cyclization processes (15). High purity ZnF_2 is used in the synthesis of fluorophosphate glass (16,17), fluoride glass (18,19), high conducting oxyfluoride glass (20), as fluoride glass films (21), in the manufacture of fluoride glass optical fibers (22), and in the preparation of optical transmitting glass (23) (see GLASS; FIBER OPTICS).

The only reported toxicity data on zinc fluoride in the NIOSH RTECS file is a LD_{LO} of 280 mg/kg for subcutaneous administration in frogs. OSHA has a standard time-weighted average (TWA) of 2.5 mg/m^3 based on fluoride. NIOSH has issued a criteria document (24) on occupational exposure to inorganic fluorides.

Zinc Fluoride Tetrahydrate. Zinc fluoride tetrahydrate [13986-18-0] is prepared by reaction of ZnO and aqueous HF. $ZnF_2 \cdot 4H_2O$ has a water solubility of about 1.6 g/100 mL solution at 25°C. Addition of HF increases the solubility to 11.8 g/100 mL in a 29% HF solution. The tetrahydrate loses water at temperatures above 75°C.

Fluorozincates. Fluorozincates of the formula $MZnF_3$, where M = Na [18251-84-8], K [13827-02-6], Rb [29987-38-0], Cs [29507-53-7], NH$_4$ [14972-88-4], Ag [28667-89-2], N$_2$H$_5$ [63439-12-3], and Li [106207-44-7] (25–28); as well as M_2ZnF_4 where M = K [37732-22-2], Rb [35944-46-8], Cs [72161-48-9], and Li [155007-51-9]; Ba [13825-40-6], Sr [15154-47-9], and Ca [15246-41-0] (25–31), have been reported. Potassium fluorozincate [13827-02-6], $KZnF_3$, and sodium fluorozincate [18251-84-8], $NaZnF_3$, are used as catalysts in alginate dental impression materials (see DENTAL MATERIALS) (32).

BIBLIOGRAPHY

"Zinc Fluoride" under "Fluorine Compounds, Inorganic," in *ECT* 1st ed., Vol. 6, p. 738, by F. D. Loomis, Pennsylvania Salt Manufacturing Co.; "Zinc" under "Fluorine Compounds, Inorganic," in *ECT* 2nd ed., Vol. 9, pp. 684–685, by W. E. White, Ozark-Mahoning Co.; in *ECT* 3rd ed., Vol. 10, p. 826, by C. B. Lindahl, Elf Atochem North America, Inc.

1. A. W. Jache and G. H. Cady, *J. Phys. Chem.* **56**, 1106 (1952).
2. E. A. Secco and R. R. Martin, *Can. J. Chem.* **43**, 175 (1965).
3. K. R. Muddukrishna, R. N. Singh, and D. K. Padma, *J. Fluorine Chem.* **57**(1–3), 155–158 (1992).
4. USSR Pat. 1,590,433 (Sept. 7, 1990), R. Okhunov., N. N. Levina, and D. D. Ikrami.
5. O. D. Gupta, R. L. Kirchmeier, and J. M. Shreeve, *Inorg. Chem.* **29**(3), 573–574 (1990).
6. G. Pourroy, and P. Poix, *J. Fluorine Chem.* **42**(2), 257–263 (1989).
7. G. A. Lopatkina and co-workers, *Khim. Prom-st. (Moscow)*, (11), 846–847 (1978).
8. Eur. Pat. 156,617 A2 (Oct. 2, 1985), M. Watanabe and S. Nishimura.
9. D. K. Padma and co-workers, *J. Inorg. Nucl. Chem.* **43**(12), 3099–3101 (1981).
10. U.S. Pat. 597,546 (July 21, 1975), J. A. Wojtowicz and D. F. Gavin.
11. A. A. Opalovskii and co-workers, *Izv. Sib. Otd. Akad. Nauk SSSR, Ser. Khim. Nauk.*, (6), 83–86 (1974).
12. M. Chaigneau and M. Santarromana, *C. R. Acad. Sci., Ser. C* **278**(25), 1453–1455 (1974).
13. USSR Pat. 265,091 (Mar. 9, 1970), G. A. Lopatkina, T. N. Kolosova, and O. S. Suslova.
14. A. Sekiya and N. Ishikawa, *Bull. Chem. Soc. Jpn.* **51**, 1267 (1978).
15. U.S. Pat. 3,728,405 (Sept. 14, 1970), J. Allan (to E. I. du Pont de Nemours & Co., Inc.).
16. J. Leissner and co-workers, *Mater. Sci. Forum*, 67–68, 137–142 (1991).
17. M. Matecki and M. Poulain, *J. Non-Cryst. Solids* **56**(1–3) (1983).
18. Y. Wang, *J. Non-Cryst. Solids* **142**(1–2), 185–188 (1992).
19. K. Zhang, *J. Chem.*, (2), 136–140 (1990).
20. K. Hirao, A. Tsujimura, and N. Soga, *Zairyo* **39**(438), 283–286 (1990).
21. B. Boulard and C. Jacoboni, *Mater. Res. Bull.* **25**(5), 671–677 (1990).
22. Eur. Pat. 331,483 (Sept. 6, 1989), K. Fujiura and co-workers.
23. M. Poulain and Y. Messaddeq, *Mater. Sci. Forum*, 32–33, 131–136 (1988).

24. *Criteria for a Recommended Standard-Occupational Exposure to Inorganic Fluorides, PB 246 692, NIOSH 76-103*, U.S. Dept. of Health, Education, and Welfare, Washington, D.C., 1975.
25. O. Schmitz-Dumont and A. Bornefeld, *Z. Anorg. Allg. Chem.* **287**, 120 (1956).
26. J. Portier, A. Tressaud, and J. L. Dupin, *C. R. Acad. Sci., Ser. C* **270**(2), 216–218 (1970).
27. J. Slivnik and co-workers, *Vestn. Slov. Kem. Drus.* **26**(1), 19–26 (1979).
28. S. H. Pulcinelli and co-workers, *Rev. Chem. Miner* **23**(2), 238–249 (1986).
29. P. A. Rodnyi, M. A. Terekhin, and E. N. Melchakov, *J. Lumin.* **47**(6), 281–284 (1991).
30. H. G. Von Schnering, D. Vu, and K. Peters, *Z. Kristallogr.* **165**(1–4), 305–308 (1983).
31. Yu. Wan-Lun, and Z. Min-Guang, *J. Phys. Chem.* **17**(20), L525–L527 (1984).
32. U.S. Pat. 2,769,717 (Nov. 6, 1956), J. Cresson (to L. D. Caulk Co.).

CHARLES B. LINDAHL
TARIQ MAHMOOD
Elf Atochem North America, Inc.

ZIRCONIUM

Three binary zirconium fluorides ZrF_2, ZrF_3, and ZrF_4, are known to exist. The most important compounds industrially are zirconium tetrafluoride, ZrF_4, and fluorozirconic acid [*12021-95-3*], H_2ZrF_6, and its salts (see ZIRCONIUM AND ZIRCONIUM COMPOUNDS).

Zirconium Difluoride

Zirconium difluoride [*7783-49-5*], ZrF_2, has been isolated in rare gas–solid matrices (1). ZrF_2, prepared by Knudsen cell techniques, is not commercially available.

Zirconium Trifluoride

Zirconium trifluoride [*13814-22-7*], ZrF_3, was first prepared by the fluorination of ZrH_2 using a mixture of H_2 and anhydrous HF at 750°C (2). It can also be prepared by the electrolysis of Zr metal in KF–NaF melts (3). Zirconium trifluoride is stable at ambient temperatures but decomposes at 300°C. It is slightly soluble in hot water and readily soluble in inorganic acids. This compound is of academic interest rather than of any industrial importance.

Zirconium Tetrafluoride

Zirconium tetrafluoride [*7783-64-4*], ZrF_4, is one of the many important inorganic fluorides that have played a role in the development of heavy-metal fluoride glass (HMFG) technology (see GLASS). Table 1 summarizes some of the physical properties of zirconium tetrafluoride. Zirconium tetrafluoride monohydrate [*14956-*

Table 1. Properties of ZrF_4

Property	Value
mol wt	167.21
sublimation point, °C	912
specific gravity	4.54
solubility in water at 25°C	1.388
ΔH_f, kJ/mol^a	−1911
ΔG_f, kJ/mol^a	−1807
S, J/(mol·K)^a	104.6
C_p, J/(mol·K)^a	103.7

^{a}To convert J to cal, divide by 4.184.

11-3], $ZrF_4 \cdot H_2O$, and the trihydrate [14517-16-9], $ZrF_4 \cdot 3H_2O$, also exist. The hydrated forms can be prepared by dissolving zirconium hydroxy carbonate, commonly known as basic zirconium carbonate, in aqueous hydrogen fluoride and evaporating the solution to dryness. This produces the trihydrate $ZrF_4 \cdot 3H_2O$. The monohydrate, $ZrF_4 \cdot H_2O$, is prepared by removing two moles of water from the trihydrate under dynamic vacuum at 70°C. The monohydrate can also be prepared by dissolving the anhydrous salt in aqueous hydrofluoric acid and evaporating the solution to dryness.

The anhydrous salt is prepared by several methods, eg, by reacting $ZrCl_4$ with liquid anhydrous HF. It is necessary to use an excess of HF which also acts as a wetting agent. The reaction is instantaneous and is carried out in a polyethylene jar or carboy. When the evolution of HCl ceases, the material is transferred to a tray and dried under an atmosphere of nitrogen. By proper selection of equipment, purification of raw material, and drying conditions, materials of spectrographic purity can be produced (4).

Other methods of preparation of anhydrous ZrF_4 include the decomposition of $(NH_4)_3ZrF_7$ [17250-81-6] at 297°C (5). NH_4F sublimes and leaves the flow reactor.

The principal application of ZrF_4 has been in the manufacture of HMFGs of which the most widely investigated is the system composed of Zr, Ba, La, Al, and Na, also popularly known as the ZBLAN glasses. This system has revolutionized the optics industry because of the significantly superior qualities of these glasses over conventional silica glasses. The theoretical transmission losses for fluoride glass fibers are calculated to be 0.001 dB/km at 3.2 μm and 0.005 dB/km at 3.5 μm (6) (see FIBER OPTICS). Transmission loss rates have been observed as low as 0.65 dB/km (4). Fluoride glasses, unlike the silica glasses, are expected to transmit light in the range of 2–4 micrometers or in the infrared range. ZBLAN fibers transmit light at a 5 to 10 times lower optical loss than that of silica fibers, and can be used for long distance data transmission, for use in mid-ir and multispectral optical components, ir domes, laser windows, laser hosts, for medical applications such as surgery and cauterization, and for nuclear radiation resistant transmitting devices (6–16).

Other applications of zirconium tetrafluoride are in molten salt reactor experiments; as a catalyst for the fluorination of chloroacetone to chlorofluoroacetone (17,18); as a catalyst for olefin polymerization (19); as a catalyst for the conversion of a mixture of formaldehyde, acetaldehyde, and ammonia (in the ratio of 1:1:3:3) to pyridine (20); as an inhibitor for the combustion of NH_4ClO_4 (21); in rechargeable electrochemical cells (22); and in dental applications (23) (see DENTAL MATERIALS).

High purity ZrF_4 is available in the United States from Advance Research Chemicals, Inc., Air Products and Chemicals, Inc., Johnson–Matthey/AESAR group, Aldrich Chemical, and EM Industries, Inc. Ultrahigh purity (99.999%) material is available only from Air Products and Chemicals, Inc. of Allentown, Pennsylvania. The price varies depending on the purity of the material from $25/kg (99%) through $250/kg (99.99%), to $1500–2500/kg (99.999%). Consumption of ZrF_4 in the United States is less than 5000 kg/yr.

Fluorozirconic Acid and Fluorozirconates

Hexafluorozirconic acid [12021-95-3], H_2ZrF_6, is formed by dissolving freshly prepared oxide, fluoride, or carbonate of zirconium in aqueous HF. This acid is produced commercially in a concentration range of 10 to 47%. The acid can be stored at ambient temperatures in polyethylene or Teflon containers without decomposition for at least two years. By neutralization of the acid, several stable fluorozirconates can be produced, such as $(NH_4)_2ZrF_6$ [16919-31-6], Na_2ZrF_6 [16925-26-1], K_2ZrF_6 [16923-06-8], Rb_2ZrF_6 [16923-95-8], and Cs_2ZrF_6, $CaZrF_6$ [30868-51-0]. The acid is regulated as a corrosive liquid NOS for shipping purposes (UN ID 1760) and packaging requires a corrosive label. The toxicity of zirconium compounds in general is represented by the TWA 5 mg/m^3 as Zr. The TWA for fluorides is 2.5 mg/m^3.

Hexafluorozirconic acid is used in metal finishing and cleaning of metal surfaces, whereas the fluorozirconates are used in the manufacture of abrasive grinding wheels, in aluminum metallurgy, ceramics industry, glass manufacturing, in electrolytic cells, in the preparation of fluxes, and as a fire retardant (see ABRASIVES; METAL SURFACE TREATMENTS).

High purity hexafluorozirconic acid and its salts are produced by Advance Research Chemicals of the United States, and Akita and Moritta of Japan. The technical-grade green-colored material is supplied by Cabot Corp. of the United States. In 1993, the U.S. market for fluorozirconic acid was about 250,000 kg/yr; the world market was less than 500,000 kg/yr. A principal part of this production is consumed by the wool, garment, and upholstery industries. The 1993 price varied between $2.4 to $6.6/kg depending on the quality and quantity required. Potassium fluorozirconate [16923-95-8], K_2ZrF_6, is commercially important; the world market is about 750,000 kg/yr. The most important application is as a fire-retardant material in the wool (qv) industry, for the manufacture of garments, upholstery for aeroplane industry, and children's clothes (see FLAME RETARDANTS). The 1993 unit price was between $5.0 and $6.6/kg.

BIBLIOGRAPHY

"Zirconium Fluorides" under "Zirconium Compounds" in *ECT* 1st ed., Vol. 15, pp. 297–298, by W. B. Blumenthal, National Lead Co.; "Zirconium" under "Fluorine Compounds, Inorganic", *ECT* 2nd ed., Vol. 9, pp. 685–686, by W. E. White, Ozark-Mahoning Co.; in *ECT* 3rd ed., Vol. 10, pp. 827–828, by D. T. Meshri, Ozark-Mahoning Co.

1. R. H. Hange, J. L. Margrave, and J. W. Hastie, *High Temp. Sci.* **5**(2), 89 (1973).
2. P. Ehrlich, F. Ploeger, and E. Kotch, *Z. Anorg. Allg. Chem.* **333**, 209 (1964).
3. Y. V. Baimakov, *Freiberg. Forschbungsh. B.* **118**, 43 (1967).
4. H. P. Withers Jr., V. A. Monk, and G. A. Cooper, *Proceedings of the SPIE International Society of Optical Engineers*, Vol. 1048, 1989, pp. 72–77; U.S. Pat. 4,983,372 (Jan. 8, 1991), H. P. Withers, Jr. and V. A. Monk (to Air Products & Chemicals, Inc.).
5. H. M. Haendler, C. M. Wheeler, Jr., and D. W. Robinson, *J. Am. Chem. Soc.* **74**, 2352 (1952).
6. K. Ohsawa and T. Shibata, *J. Lightwave Technol.* **LT-2**(5), 602–606 (1984).
7. J. M. Jewell and I. D. Aggarwal, *Proceedings of the SPIE International Society of Optical Engineers*, Vol. 1327, 1990, pp. 190–197.
8. G. C. Devyatykh and M. F. Churbanov, *Z. Anorg. Allg. Chem.* **576**, 25–32 (1989).
9. R. Mossadegh, P. M. Kutty, N. J. Garrito, and D. C. Tran, Proceedings of the SPIE International Society of Optical Engineers, Vol. 1112, 1989, pp. 40–46.
10. L. J. Moore, D. R. MacFarlane, and P. J. Newman, *J. Non-Cryst. Solids* **140**(1–3), 159–165 (1992).
11. L. A. Bursill, J. Peng, and J. R. Sellar, *Mater. Sci. Forum* **14**(1), 41–59 (1990).
12. R. N. Schwartz, M. Robinson, and G. L. Tangonan, *Mater. Sci. Forum*, **19–20** 275–285 (1987).
13. H. W. Schneider, A. Schoberth, and A. Standt, *Glasstech. Ber.* **60**(6), 205–210 (1987).
14. Y. Dai, T. Kawaguchi, K. Suzuki, S. Suzuki, and K. Yamamoto, *J. Non-Cryst. Solids* **142**(1–2), 159–164 (1992).
15. W. A. Sibley, *Mater. Sci. Forum* **6**, 611–616 (1985).
16. T. Iqbal, M. Shahriari, G. Merberg, and G. H. Sigel, *J. Mater. Res.* **6**(2), 401–406 (1991).
17. U.S. Pat. 2,807,646 (Sept. 24, 1957), C. B. Millen and C. Woolf (to Allied Chemicals and Dye Corp.).
18. U.S. Pat. 2,805,121 (Sept. 3, 1957), C. Woolf (to Allied Chemicals and Dye Corp.).
19. U.S. Pat. 3,165,504 (Jan. 12, 1965), J. P. Hogan (to Philips Petroleum Co.).
20. Jpn. Kokai 7,663,176 (June 1, 1976), W. Yasuo and co-workers (to Koci Chemical Co., Ltd.).
21. A. P. GalzKova, *Dokl. Akad. Nauk. USSR* **213**, 622 (1973).
22. U.S. Pat. 3,725,128 (Nov. 26, 1974), S. Senderoff (to Union Carbide).
23. R. M. Shrestha, S. A. Mundorff, and B. G. Bibby, *J. Dent. Res.* **51**, 1561 (1972).

DAYAL T. MESHRI
Advance Research Chemicals, Inc.

FLUORINE COMPOUNDS, ORGANIC

Introduction, 227
Direct fluorination, 242
Fluorinated aliphatic compounds, 259
Fluoroethanols, 281
Fluoroethers and fluoroamines, 285
Perfluoroepoxides, 294
Fluorinated acetic acids, 304
Fluorinated higher carboxylic acids, 311
Perfluoroalkanesulfonic acids, 318
Fluorinated aromatic compounds, 325
Polytetrafluoroethylene, 381
Perfluorinated ethylene–propylene copolymers, 404
Tetrafluoroethylene–ethylene copolymers, 417
Tetrafluoroethylene–perfluorovinyl ether copolymers, 431
Poly(vinyl fluoride), 443
Poly(vinylidene fluoride), 454
Polychlorotrifluoroethylene, 472
Bromotrifluoroethylene, 479
Poly(fluorosilicones), 482
Fluorocarbon elastomers, 490

INTRODUCTION

Organic fluorine compounds were first prepared in the latter part of the nineteenth century. Pioneer work by the Belgian chemist, F. Swarts, led to observations that antimony(III) fluoride reacts with organic compounds having activated carbon–chlorine bonds to form the corresponding carbon–fluorine bonds. Preparation of fluorinated compounds was facilitated by fluorinations with antimony(III) fluoride containing antimony(V) halides as a reaction catalyst.

It was the 1930s before the direction of organic fluorine chemistry turned commercial (1). Facing the problem of replacing methyl chloride and ammonia in household refrigerators, researchers found that dichlorodifluoromethane (CFC-12) was the best alternative as a safe, stable gas whose liquefied state had low compressibility. In addition, it was not flammable. General Motors and E.I. du Pont de Nemours and Co. jointly led the early application of chlorofluorocarbons (CFCs) as refrigerants. Later, other diverse CFC applications were commercialized as cleaning agents, as blowing agents to make foam products, and as sources of monomers for many fluoropolymers.

Another impetus to expansion of this field was the advent of World War II and the development of the atomic bomb. The desired isotope of uranium, ^{235}U, in the form of UF_6 was prepared by a gaseous diffusion separation process of the mixed isotopes (see FLUORINE). UF_6 is extremely reactive and required contact with inert organic materials as process seals and greases. The wartime Manhattan Project successfully developed a family of stable materials for UF_6 service.

These early materials later evolved into the current fluorochemical and fluoropolymer materials industry. A detailed description of the fluorine research performed on the Manhattan Project has been published (2).

Concern arose during the 1970s about cumulative CFC emissions into the atmosphere with progressive depletion of the stratospheric ozone layer by Cl atoms and led to the formation and global support of a multinational forum, called the Montreal Protocol on Substances That Deplete the Ozone Layer. As a result, CFC production has been dramatically decreased and will likely be totally phased out before the year 2000 (3) (see FLUORINATED ALIPHATICS). If hydrogen atoms are introduced into the CFC structure to lower the chlorine content, the resulting hydrochlorofluorocarbon (HCFC) is more susceptible to degradation in the lower atmosphere before it can reach the stratosphere. However when a hydrogen atom is introduced into a one-carbon compound, the boiling point is lowered and may be too low for the same CFC application. Therefore two-carbon compounds bearing some hydrogen are more attractive substitutes than the one-carbon modified CFCs. As hydrogen content increases, there is a counter effect of increasing flammability, which in turn limits some HCFC applications.

Physical Properties

Substitution of fluorine for hydrogen in an organic compound has a profound influence on the compound's chemical and physical properties. Several factors that are characteristic of fluorine and that underlie the observed effects are the large electronegativity of fluorine, its small size, the low degree of polarizability of the carbon–fluorine bond and the weak intermolecular forces. These effects are illustrated by the comparisons of properties of fluorocarbons to chlorocarbons and hydrocarbons in Tables 1 and 2.

The replacement of chlorine by fluorine results in a nearly constant boiling point (bp) drop of approximately 50°C for every chlorine atom that is replaced (see Table 1). In Table 2, a similar boiling point effect with hydrocarbons is apparent, even though the molecular weight of the fluorocarbon is much higher than the corresponding hydrocarbon analogue. An analogous drop in the corresponding fluorocarbon freezing point results in a widened liquid range for applications like lubricating fluids and greases. One other significant property difference, attributed to weak intermolecular forces, can be found in the very low surface tensions of fluorocarbons as compared to hydrocarbons and water (Table 3).

Table 1. Boiling Points of Halomethanes

Chloro-hydrocarbon	Bp, °C	Fluoro-hydrocarbon	CAS Registry Number	Bp, °C	Difference per F atom, °C
CH_3Cl	−24	CH_3F	[593-53-3]	−78	54
CH_2Cl_2	40	CH_2F_2	[75-10-5]	−52	46
$CHCl_3$	61	CHF_3	[75-46-7]	−83	48
CCl_4	77	CF_4	[75-73-0]	−128	51

Table 2. Boiling Points of Hydrocarbons and Fluorocarbons

Hydrocarbon	Bp, °C	Fluorocarbon	CAS Registry Number	Bp, °C
CH_4	−161	CF_4	[75-73-0]	−128
C_2H_6	−88	C_2F_6	[76-16-4]	−78
C_3H_8	−45	C_3F_8	[76-19-7]	−38
C_4H_{10}	0.6	C_4F_{10}	[355-25-9]	−5
C_7H_{16}	98	C_7F_{16}	[335-57-9]	82
C_6H_6	80	C_6F_6	[392-56-3]	80
C_6H_{12} (cyclic)	81	C_6F_{12} (cyclic)	[355-68-0]	52

Table 3. Surface Tensions of Selected Fluids

Compound	CAS Registry Number	Surface tension, 20 °C, mN/m (= dyn/cm)
perfluoroheptane	[335-57-9]	13.6
perfluoromethylcyclohexane	[355-02-2]	15.4
perfluoro-1,4-dimethylcyclohexane	[374-77-6]	16.3
octane	[111-65-9]	21.8
benzene	[71-43-2]	28.9
methyl bromide	[74-83-9]	41.5
water	[7732-18-5]	72.8

The low surface tension of highly fluorinated organic compounds is commercially important for their application in surfactants, antisoiling textile treatments, lubricants, and specialty wetting agents.

In contrast, the viscosities of fluorocarbons are higher than those of the corresponding hydrocarbons. This can be explained by the greater stiffness of the fluorocarbon chain arising from the large repulsive forces between molecules, and from the greater density imparted by the more massive fluorine atoms (vs hydrogen). The fluorocarbon viscosity drops rapidly with increasing temperature and is accompanied by a simultaneous large decrease in density.

The refractive indexes and dielectric constants for the fluorocarbons are both lower than that for the corresponding hydrocarbon analogue.

Preparation

There are many known ways to introduce fluorine into organic compounds, but hydrogen fluoride [7664-39-3], HF, is considered to be the most economical source of fluorine for many commercial applications.

Halogen Exchange. The exchange of another halogen atom in an organic compound for a fluorine atom is the most widely used method of fluorination. The relative ease of replacement follows the general order I > Br > Cl. Commonly used fluorinating agents are the fluorides of the alkali metals (especially KF), antimony, and mercury.

Antimony trifluoride [7783-56-4], SbF_3, can be used in the following preparations (4):

$$RCOCl \xrightarrow{SbF_3} RCOF \qquad (1)$$

$$RSO_2Cl \xrightarrow{SbF_3} RSO_2F \qquad (2)$$

$$\diagdown\!\!\!\!\!C\!=\!C(R)\!-\!CCl_3 \xrightarrow{SbF_3} \diagdown\!\!\!\!\!C\!=\!C(R)\!-\!CF_3 \qquad (3)$$

$$RCCl_2R \xrightarrow{SbF_3} RCF_2R \qquad (4)$$

$$CHCl_2OCH_2Cl \xrightarrow{SbF_3} CHF_2OCH_2F \qquad (5)$$

The limitations of this reagent are several. It cannot be used to replace a single unactivated halogen atom with the exception of the chloromethyl ether (eq. 5) to form difluoromethyl fluoromethyl ether [461-63-2]. It also cannot be used to replace a halogen attached to a carbon–carbon double bond. Fluorination of functional group compounds, eg, esters, sulfides, ketones, acids, and aldehydes, produces decomposition products caused by scission of the carbon chains.

The effectiveness of antimony fluoride is increased if it is used in conjunction with chlorine or with antimony pentachloride. The formation of either $SbCl_2F_3$ or a complex of SbF_3 and $SbCl_5$ probably accounts for the increased activity (4).

Antimony pentafluoride [7783-70-2], SbF_5, is a highly active fluorinating agent and is generally used to fluorinate only completely halogenated compounds, since those containing hydrogen as well as halogen undergo decomposition. In the case of halogenated olefins or aromatic compounds (5), addition as well as substitution occurs. Thus hexachlorobenzene forms a fluorochlorocyclohexene [27458-17-9]:

$$C_6Cl_6 \xrightarrow{SbF_5} C_6Cl_2F_8 \qquad (6)$$

Potassium fluoride [7789-23-3], KF, is the most frequently used of the alkali metal fluorides, although reactivity of the alkali fluorides is in the order CsF > RbF > KF > NaF > LiF (6). The preference for KF is based on cost and availability traded off against relative reactivity. In its anhydrous form it can be used to convert alkyl halides and sulfonyl halides to the fluorides. The versatility makes it suitable for halogen exchange in various functional organic compounds like alcohols, acids and esters (7). For example, 2,2-difluoroethanol [359-13-7] can be made as shown in equation 9 and methyl difluoroacetate [433-53-4] as in equation 10.

$$RX \xrightarrow{KF} RF \qquad (7)$$

$$RSO_2Cl \xrightarrow{KF} RSO_2F \qquad (8)$$

$$CHCl_2CH_2OH \xrightarrow{KF} CHF_2CH_2OH \qquad (9)$$

$$CHCl_2COOCH_3 \xrightarrow{KF} CHF_2COOCH_3 \qquad (10)$$

The preparation of fluoroaromatics by the reaction of KF with perhaloaromatics, primarily hexachlorobenzene, has received considerable attention. Two methods were developed and include either the use of an aprotic, polar solvent, such as N-methylpyrrolidinone (8), or no solvent (9). These methods plus findings that various fluoroaryl derivatives are effective fungicides (10) prompted development of a commercial process for the production of polyfluorobenzenes (11). The process uses a mixture of sodium and potassium fluorides or potassium fluoride alone in aprotic, polar solvents such as dimethyl sulfoxide or sulfolane.

Mercuric fluoride [7783-39-3], HgF_2, is used in the following conversions where X = Br, I (4,12):

$$RX \xrightarrow{HgF_2} RF \qquad (11)$$

Because alkyl chlorides react very slowly, chloroform or methylene chloride can be used as solvents in the above reaction. Polyhalides are known to react to form polyfluoride products. Oxygen-containing compounds, eg, esters, ethers, and alcohols, appear to inhibit this fluorination.

Mercurous fluoride [13967-25-4], Hg_2F_2, is less effective than HgF_2. The addition of chlorine or iodine to the reagent increases its reactivity owing to the formation of a complex between HgF_2 and HgX_2 (4,12).

Hydrogen fluoride, HF, when used alone is a comparatively ineffective exchange agent and replaces only active halogens (13), eg, acyl fluorides from acyl chlorides and benzotrifluoride [98-08-8] from benzotrichloride (eq. 12).

$$C_6H_5CCl_3 \xrightarrow{HF} C_6H_5CF_3 \qquad (12)$$

When used with antimony pentachloride, the reactivity of HF is comparable to $SbCl_2F_3$ alone. Therefore a continuous fluorination exchange process is possible where antimony is the fluoride carrier from HF to the organic fluoride.

$$SbCl_5 + 3\,HF \rightarrow SbCl_2F_3 + 3\,HCl \qquad (13)$$

$$SbCl_2F_3 + CCl_4 \rightarrow SbCl_4F + CCl_2F_2 \qquad (14)$$

$$SbCl_4F + 2\,HF \rightarrow SbCl_2F_3 + 2\,HCl \qquad (15)$$

Since antimony halides serve as fluorine carriers, the actual fluorination agent is HF. This process is the principal one used in the production of CFC-11, CFC-12, and HCFC-22 as well as many others. The application is well suited to the fluorination of one-carbon through three-carbon containing organic halides. In practice, the HF and organic halide enter by separate feeds into the process vessel. Under pressure, they react in the liquid phase with heating to form HCl and the organic fluoride. By suitable control of the feed ratios, temperature, pressure, and residence time, the degree of fluorination can be controlled. As the reaction progresses, vapors of HCl and the organic fluoride are continuously vented off for separation and recovery. Antimony salts stay in the process vessel and are periodically reactivated by treatment with chlorine to form the desired antimony(V) halide salts prior to reuse.

Heterogeneous vapor-phase fluorination of a chlorocarbon or chlorohydrocarbon with HF over a supported metal catalyst is an alternative to the liquid phase process. Salts of chromium, nickel, cobalt or iron on an AlF_3 support are considered viable catalysts in pellet or fluidized powder form. This process can be used to manufacture CFC-11 and CFC-12, but is hampered by the formation of over-fluorinated by-products with little to no commercial value. The most effective application for vapor-phase fluorination is where all the halogens are to be replaced by fluorine, as in manufacture of 3,3,3-trifluoropropene [677-21-4] (14) for use in polyfluorosilicones.

$$CCl_3CH_2CH_2Cl \xrightarrow[\text{excess}]{\text{HF}} CF_3CH=CH_2 \quad (16)$$

Another use of hydrogen fluoride, although not in halogen exchange, is the reaction with ethylenes or acetylenes to form the addition products, 1,1-difluoroethane [75-37-6] and vinyl fluoride [75-02-5]:

$$HC{\equiv}CH + HF \rightarrow CH_3CHF_2 + CH_2{=}CHF \quad (17)$$

Reaction conditions must be controlled since HF is also an excellent polymerization catalyst. Controlled reaction conditions can alternatively lead to vinyl fluoride or to HFC-152a (CH_3CHF_2). The latter can be thermally cracked to form vinyl fluoride.

Sulfur tetrafluoride [7783-60-0], SF_4, replaces halogen in haloalkanes, haloalkenes, and aryl chlorides, but is only effective (even at elevated temperatures) in the presence of a Lewis acid catalyst. The reagent is most often used in the replacement of carbonyl oxygen with fluorine (15,16). Aldehydes and ketones react readily, particularly if no alpha-hydrogen atoms are present (eg, benzal fluoride [455-31-2] from benzaldehyde), but acids, esters, acid chlorides, and anhydrides are very sluggish. However, these reactions can be catalyzed by Lewis acids (HF, BF_3, etc).

$$C_6H_5CHO + SF_4 \rightarrow C_6H_5CHF_2 \quad (18)$$

$$C_6H_5COOH + SF_4 \rightarrow C_6H_5CF_3 \quad (19)$$

Halogen Fluorides. These include compounds such as IF_3, IF_5, ClF, etc, of which only a few, ClF, ClF_3, BrF_3, and IF_5, are used to some extent. They act both as halogen exchange agents and, in the case of the monofluorides, as addition agents to unsaturated bonds (17).

Replacement of Hydrogen. Three methods of substitution of a hydrogen atom by fluorine are (*1*) reaction of a C–H bond with elemental fluorine (direct fluorination, (*2*) reaction of a C–H bond with a high valence state metal fluoride like AgF_2 or CoF_3, and (*3*) electrochemical fluorination in which the reaction occurs at the anode of a cell containing a source of fluoride, usually HF.

Direct Fluorination. The principal disadvantage of the use of elemental fluorine as a fluorinating agent is the high heat of reaction. A considerable degree of carbon–carbon bond scission can occur as well as polymer formation. In order to prevent these complications, fluorine is diluted with nitrogen and the reaction zone is constructed such that good heat conductivity is possible. Low temperatures are favored to achieve maximum selectivity and yield (18). Fluorine is also effective for the replacement of residual hydrogen in a highly fluorinated organic molecule to produce the corresponding fluorocarbon.

The fluorination reaction is best described as a radical-chain process involving fluorine atoms (19) and hydrogen abstraction as the initiation step. If the molecule contains unsaturation, addition of fluorine also takes place (17). Complete fluorination of complex molecules can be conducted using this method (see FLUORINE COMPOUNDS, ORGANIC–DIRECT FLUORINATION).

Reaction with a Metal Fluoride. A second technique for hydrogen substitution is the reaction of a higher valence metal fluoride with a hydrocarbon to form a fluorocarbon:

$$2\ CoF_2 + F_2 \rightarrow 2\ CoF_3 \tag{20}$$

$$2\ CoF_3 + RH \rightarrow RF + 2\ CoF_2 + HF \tag{21}$$

The principal advantage to this method is that the heat evolved for each carbon–fluorine bond formed, 192.5 kJ/mol (46 kcal/mol), is much less than that obtained in direct fluorination, 435.3 kJ/mol (104 kcal/mol). The reaction yields are therefore much higher and less carbon–carbon bond scisson occurs. Only two metal fluorides are of practical use, AgF_2 and CoF_3.

The reactivity of the metal fluoride appears to be associated with the oxidation potential of the metal. For example, AgF replaces halogen in organic compounds, whereas AgF_2 replaces hydrogen.

The reaction is conducted by passing fluorine through a bed of AgCl or $CoCl_2$ at an elevated temperature to form the higher valence state fluorides. The organic reactant is then passed through the bed to realize a semicontinuous fluorination process. In general, the method is used for the preparation of fluorocarbons since any unsaturation or functionality in the reactant is removed. The process can also be used to fluorinate polychlorohydrocarbons, whereby replacement of both chlorine and hydrogen occurs (2).

Electrochemical Fluorination. The electrochemical fluorination (ECF) of highly fluorinated organic compounds (20) involves the electrolysis of an organic reactant in liquid anhydrous HF at a voltage below that for liberation of fluorine. The reaction is limited by temperature (usually done at 0°C) and by the solubility of the reactant in HF. Electrical conductivity is required for current to flow and the reaction to proceed. Current density is 10–20 mA/cm^2. Fluorination takes place at the nickel anode by a free-radical process not involving the intermediate formation of elemental fluorine. Hydrogen is liberated at the cathode. The method is used to fluorinate acyl halides, sulfonyl halides, ethers, carboxylic acids, and amines. The product is a fluorocarbon having no residual hydrogen. Olefins and carbocyclics, as well as heterocyclic compounds, become saturated. Side reactions, resulting in reduced yields, include cleavage of carbon–carbon bonds and polymer formation. The electrochemical yields decrease with increasing number of carbon atoms in the structure.

ECF is successfully used on a commercial scale to produce certain perfluoroacyl fluorides, perfluoroalkylsulfonyl fluorides, perfluoroalkyl ethers, and perfluoroalkylamines. The perfluoroacyl fluorides and perfluoroalkylsulfonyl fluorides can be hydrolyzed to form the corresponding acid and acid derivatives. Examples include perfluorooctanoyl fluoride [335-66-0], perfluorooctanoic acid [335-67-1], perfluorooctanesulfonyl fluoride [307-35-7], perfluorooctanesulfonic acid [763-23-1], and tris(perfluoro-n-butyl)amine [311-89-7].

$$C_7H_{15}COCl \xrightarrow[HF]{ECF} C_7F_{15}COF \rightarrow C_7F_{15}COOH \qquad (22)$$

$$C_8H_{17}SO_2Cl \xrightarrow[HF]{ECF} C_8F_{17}SO_2F \rightarrow C_8F_{17}SO_3H \qquad (23)$$

$$(C_4H_9)_3N \xrightarrow[HF]{ECF} (C_4F_9)_3N \qquad (24)$$

Telomer Formation. Fluorinated compounds with active C–Br or C–I bonds can add to fluoroolefins to form addition products in high yield. The olefin most often used is tetrafluoroethylene [116-14-3]. Telomerization involves reaction of a telogen, or addition agent like $CBrF_3$, CF_3I [2314-97-8], or C_2F_5I [354-64-3], with the olefin to form longer chain addition products called telomers. The reaction is initiated by thermolysis, photolysis, peroxides, and other free-radical initiators, certain metal complexes, and various redox chemicals. By control of the stoichiometry and reaction conditions, a simple addition product or telomers with high fluorine content can be formed. The yield is higher than that seen with ECF production methods. The route does suffer from a distribution of adducts formed vs formation of one specific reaction product. The usual adducts are perfluoroalkyl iodides having up to 14 carbon atoms in the alkyl chain.

$$C_2F_5I + x\ CF_2{=}CF_2 \rightarrow C_2F_5(CF_2CF_2)_xI$$

Often used as mixtures, the telomers are subsequently converted to commercial surfactants and stain-resistant fiber finishes through functionalizing steps using standard chemical reactions of the C–I bond.

Aromatic Ring Fluorination. The formation of an aryl diazonium fluoride salt, followed by decomposition, is a classical reaction (the Schiemann reaction) for aryl fluoride preparation (21). This method has been adapted to the production-scale manufacture of fluorobenzene [462-06-6] (22) where HF is the source of the fluoride.

$$C_6H_5NH_2 \xrightarrow[HF]{NaNO_2} C_6H_5N_2{}^+F^- \xrightarrow{-N_2} C_6H_5F \qquad (25)$$

Chemical Properties and Applications

Substitution of fluorine into an organic molecule results in enhanced chemical stability. The resulting chemical reactivity of adjacent functional groups is drastically altered due to the large inductive effect of fluorine. These effects become more pronounced as the degree of fluorine substitution is increased, especially on the same carbon atom. This effect demonstrates a maximum in fluorocarbons and their derivatives.

Fluorinated Alkanes. As the fluorine content increases, the chemical reactivity decreases until complete fluorination is achieved, after which they are inert to most chemical attack, including the highly reactive element fluorine. Their lack of reactivity leads to their use in certain commercial applications where stability is valued when in contact with highly reactive chemicals.

Fluorinated Olefins. In electrophilic addition reactions, the reactivity of the unsaturated linkage is reduced by the inductive effect of fluorine. Nucleophilic additions are enhanced by this same effect. Amines, phenols, alcohols, and many other nucleophiles, including fluoride ion, add to the carbon–carbon double bond of highly fluorinated olefins (23). Free-radical addition of halogen halides proceeds easily using either peroxide or thermal initiation. Some halides, especially those derived from iodine, eg, ICl, react by an ionic mechanism. Fluorinated olefins also undergo free-radical polymerization producing a wide range of valuable fluoropolymers.

Certain CFCs are used as raw materials to manufacture key fluorinated olefins to support polymer applications. Thermolysis of HCFC-22 affords tetrafluoroethylene and hexafluoropropylene [116-15-4] under separate processing conditions. Dechlorination of CFC-113 forms chlorotrifluoroethylene [79-38-9]. Vinylidene fluoride [75-38-7] is produced by the thermal cracking of HCFC-142b.

Fluorinated Aromatic Hydrocarbons. Many aromatic fluorocarbon derivatives, eg, hexafluorobenzene, pentafluorotoluene [771-56-2], and perfluoronaphthalene [313-72-4] are examples of compounds that readily undergo nucleophilic ring substitution reactions with loss of one or more fluorine substituents. This is in sharp contrast to the fluorine substituents in perfluoroalkanes. Perfluoroalkyl substitution on the aromatic ring has a strong inductive effect, making the ring more susceptible to nucleophilic attack. Fluorine hyperconjugative effects are considered to be unimportant (24) in these reactivity patterns, leaving inductive effects as the primary factor to describe the substituent effect.

Fluorinated Heterocyclic Compounds. Heterocyclic compounds containing the CF_3 group are prepared by methods similar to those used in the fluorination of aliphatic compounds. The direct action of fluorine on uracil yields the cancer chemotherapy agent, 5-fluorouracil [51-21-8], as one special example of a selective fluorination on a commercial scale (25).

Fluorinated Acids. This class of compounds is characterized by the strength of the fluorocarbon acids, eg, CF_3COOH, approaching that of mineral acids. This property results from the strong inductive effect of fluorine and is markedly less when the fluorocarbon group is moved away from the carbonyl group. Generally, their reactions are similar to organic acids and they find applications, particularly trifluoroacetic acid [76-05-1] and its anhydride [407-25-0], as promotors in the preparation of esters and ketones and in nitration reactions.

Fluorinated Biologically Active Compounds. Many biologically active compounds are prepared from fluorobenzene, difluorobenzene, benzotrifluoride and fluorinated steroids. The preparation of fluorinated compounds for use in medicine has increased rapidly (26,27) since the 1950s. The strong interest in such substances is based on the following considerations: (1) fluorine most closely resembles bioactive hydrogen analogues with respect to steric requirements at the receptor sites; (2) fluorine alters electronic effects, owing to its high electronegativity; (3) fluorine imparts improved oxidative and thermal stability to the parent molecule; and (4) fluorine imparts lipid solubility, thereby increasing the *in vivo* absorption and transport rates in membranes.

Many fluorinated, biologically active agents have been developed and successfully used in the treatment of diseases. The biological property of fluorinated organics has been further extended to applications in the agrochemical and pest management fields.

Analgesics. Four examples of antiinflammatory agents are Sulindac [38194-50-2], based on a monofluoro indene derivative; diflunisal [22494-42-4], based on a substituted difluorobenzene; and dexamethasone [50-02-2] and fluocinonide [356-12-7], based on monofluorinated and difluorinated steroids, respectively.

Antiviral Agents. Trifluridine [70-00-8] is a trifluoromethyl substituted heterocyclic antiviral agent.

Appetite Depressants. Fenfluramine hydrochloride [404-82-0] is an anorexiant based on a meta-substituted benzotrifluoride.

Tranquilizers. Fluphenazine hydrochloride [146-56-5], trifluoperazine hydrochloride [440-17-5], and triflupromazine [146-54-3] are all trifluoromethyl substituted phenothiazine chemicals useful in the management of psychotic disorders.

Diuretics. The diuretic and antihypertensive agent bendroflumethiazide [73-48-3] is a benzotrifluoride-based pharmaceutical.

Inhalation Anesthetics. Examples of highly fluorinated halocarbons and ethers are halothane [151-67-7], fluroxene [406-90-6], enflurane [13838-16-9], methoxyflurane [76-38-0], sevoflurane [28523-86-6], desflurane [57041-67-5], isoflurane [26675-46-7], and pure enantiomers (28) of isoflurane as potentially more effective anesthetic compounds than the racemic mixture. Isoflurane is the leading inhalation anesthetic marketed in North America since its introduction in 1981. Desflurane is being newly introduced into the United States market. Enflurane is largely used in Europe while sevoflurane is used in Japan and Korea. Methoxyflurane is used in veterinary applications only.

Herbicides. Fluometuron [2164-17-2] is a fluorophenyl-substituted urea effective against grassy and broadleaf weeds in bean, grain, fruit, and cotton crops. Trifluralin [1582-09-8] is a trifluoromethyl-substituted dinitroaniline used as preemergence control against weeds in cotton and soybean crops. Profluralin [26399-36-0] and benfluralin [1861-40-1] are structural analogues of trifluralin. Fluorodifen [15457-05-3] is a trifluoromethyl-substituted diphenyl ether used for weed control in bean and rice crops. Fluroxypyr [81406-37-3] is a fluoropyridine compound used on cereals for post-emergent control of broadleaf weeds.

Insecticides. Diflubenzuron [35367-38-5] is a difluorobenzoyl urea. It inhibits insect chitin formation during larval molting. Application is with management of fruit, bean, and cotton crops. Perfluoroalkylsulfonamides of carbon chain length equal to 6 or 8 show good control against fire ants.

Fungicides. Flusilazole [85509-19-9] is a newer broad-spectrum foliar fungicide containing two fluorophenyl substituents with application on cereal, fruit, and vegetable crops. Flutriafol [76674-21-0] is another fluorobenzene derivative useful on small grain cereal diseases.

Economic Aspects

The CFC commodity application is undergoing significant change due to environmental pressures. Development of acceptable, alternative fluorinated compounds is extremely expensive. As the largest global supplier, Du Pont plans to spend $1 billion by the mid-1990s to develop CFC alternatives. The five-year toxicity testing program alone has been estimated at up to $5 million for each candidate compound tested.

The HCFC and HFC refrigeration alternatives are estimated to be two to five times higher in price, and some of the viable alternatives demonstrate a lower heat-transfer efficiency than the current CFCs. Total production will continue to drop due to conservation in use and elimination of emissive uses along with substitutions in refrigeration applications. The global CFC market is estimated at $4 billion with one-half of that being in the United States. The largest supplier is Du Pont with 25% of the global market share.

The 1990 United States production for CFCs, also generically called fluorocarbons in some sources, was 417,009 metric tons (29). Global production is approximately three times this value, with roughly one-third of the total produced in the North American area, one-third in Europe, and the remaining one-third being spread over all remaining global areas. The peak production year was in 1988 with a total of 624,550 t of CFCs. Production of CFC-11 and CFC-12 has been dropping due to environmental pressures while HCFC-22 has been growing modestly to pick up some of the refrigeration growth not addressed by CFC-11 or CFC-12 use. The continued refrigeration use of HCFC-22 is based on its very low relative ozone depletion potential. Some of the HCFC-22 production growth is attributable to growing captive conversion to tetrafluoroethylene for polymer manufacture at the same HCFC-22 production sites. Global poly(tetrafluoroethylene) demand has been currently growing at 4–5% per year, which in turn grows HCFC-22 demand.

United States CFC production is spread over different sites with a variety of companies. Allied-Signal Inc. has 143,180 t of capacity for CFC-11, CFC-12, HCFC-22, CFC-113, CFC-114, and HCFC-141b at its three sites in Baton Rouge, La., Danville, Ill., and El Segundo, Calif.; Ausimont USA, Inc. has 11,360 t of capacity for HCFC-141b and HCFC-142b at their Thorofare, N.J. site, where vinylidene fluoride is produced. Both HCFC-141b and HCFC-142b can be used as alternative feedstocks for vinylidene fluoride manufacture. Elf Atochem North America (former Pennwalt Corp. merged with Elf Atochem) can produce 72,730 t of combined CFC-11, CFC-12, HCFC-22, HFC 143a, HCFC-141b, and HCFC-142b at Calvert City, Ky., and another 36,360 t of combined CFC-11, CFC-12, and HCFC-22 capacity at their Wichita, Kans. site. Du Pont has an estimated 318,180 t of combined capacity for CFC-11, CFC-12, CFC-13, CFC-14, HCFC-22, HFC-23, CFC-113, CFC-114, CFC-115, CFC-116, HFC 125, HFC-134a, HFC 143a, and HFC-152a at the Antioch, Calif., Corpus Christi, Tex., Deepwater, N.J., Louisville, Ky., and Montague, Mich. sites. La Roche Chemicals Inc. has a plant in Gramercy, La. with 36,360 t per year capacity for the combined CFC-11, CFC-12, and HCFC-22. Great Lakes Chemical Co. has a Halon production facility at their El Dorado, Ark. site with unspecified capacity.

In Western Europe, the CFC producers are equally varied. The following is a partial list of the larger companies with total CFC production capacity (10^3 t) at all sites shown in parentheses: Atochem SA (148.5, France and Spain), Hoescht AG (102.0, Germany), Kali-Chemie AG (66.0, Germany and Spain), Montefluos SpA (100.0, Italy), and ICI Chemicals and Polymers Ltd. (> 113.6, United Kingdom). These producers account for over 80% of the Western European CFC production.

In Japan, the primary suppliers are Asahi Glass Co., Ltd. (60,000 t), Daikin Industries, Ltd. (115,000 t), and Du Pont-Mitsui Fluorochemicals Co. (64,000 t). Together these three producers account for over 90% of the Japanese CFC production.

Investments in plants to produce refrigeration alternatives have been announced by a variety of companies. ICI has an HFC-134a plant built in St. Gabriel, La. and is scheduled for 1993 startup. Their HFC-134a plant in Runcorn, UK is operational; another one is planned for the Mihara, Japan site. ICI's new HFC-32 pilot plant at Widnes, UK will produce this low temperature refrigerant beginning in 1992. Mixtures of HFC-32 and HFC-134a are also being pursued as a refrigerant. Initial HFC-32 supplies will support toxicology testing required prior to commercialization. ICI has also announced it will close its last Halon-1211 production facility in Manchester, UK by the end of 1993. In 1990, Du Pont built a commercial HFC-134a plant in Corpus Christi, Tex. The Du Pont HFC-134a facilities at Chiba, Japan, and Dordrecht, the Netherlands are targeted for 1995 completion. Another Du Pont plant in Maitland, Canada commercially manufactures HCFC-123, and a conversion of their Montague, Michigan plant is targeted at manufacture of HCFC-141b. Du Pont reports their global HFC 134a capacity at 55,000 t per year. Exact capacity of these facilities has not been confirmed but each is assumed to be well over 5000 t per year.

The Minnesota Mining and Manufacturing Co., or 3M, manufactures specialty perfluorochemicals using mainly electrochemical fluorination methods at their St. Paul, Minn., Decatur, Ala., and Cordova, Ill. sites. Their capacity is not reported, but is estimated at over 5000 t as fluorinated inert fluids, surfactants, and fire extinguishment chemicals. Asahi, Du Pont, and Hoescht all use fluoro-olefin telomerization technology at a variety of their sites to manufacture a line of perfluorinated specialty chemicals for stain-resistant treatment and surfactant applications. Globally, these telomer-based fluorochemicals are estimated to be over 5000 t per year with Du Pont having one-third of the total. Daikin and 3M recently formed a U.S. joint venture called MDA Manufacturing. This venture will build a plant at the Decatur, Ala. site for production of about 4550 t of HCFC-22 and other unspecified fluorochemical intermediates. The HCFC-22 production will be dedicated to the manufacture of tetrafluoroethylene for Daikin use at their own proposed Decatur plant and of hexafluoropropylene to be used by 3M at its existing Decatur plant.

Aromatic fluorine compounds are varied in kind and in volume. Mallinckrodt Specialty Chemicals division of the Imcera Group claim their continuous process capacity for fluorobenzene is 1200 t per year from their St. Louis, Mo. site. EniChem reports a 910 t per year capacity for fluorinated aromatics at their Trissino, Italy site using a novel continuous diazotization process coupled with electrofluorination technology. Hoescht AG has announced plans to double its unspecified capacity at Griesheim, Germany for fluoroaromatics. ICI in the United Kingdom and Du Pont at their Deepwater, N.J. facility also have fluoroaromatic capabilities. Other smaller suppliers also manufacture fluorinated aromatic compounds for specialty applications, but their capacities are again unreported. Fluoroaromatics are basic intermediate building blocks leading toward the more advanced aromatic fluorine intermediates (AFI) including fluorinated aniline, quinoline, biphenyl, and phenol compounds. AFI uses are in surfactants, pharmaceuticals, agrochemicals, electronics, and biomedical applications. The worldwide AFI demand is estimated at 4000 t growing to 10,000 t by 1994. The 1994 market value is projected at about $400 million (30).

Safe Handling Aspects

The safety of fluorine compounds is possibly as varied as the numbers of compounds known which bear fluorine substituents. Most compounds bearing the C–F bond are synthetic and therefore not normally encountered in nature. Aerosol or vapor inhalation is the most likely route of exposure where adverse health effects may occur. As such all new fluorine compounds should be handled with caution as one would with any potentially hazardous substance until full toxicological properties are known. Existing fluorine compounds cover the range from biologically inert materials like fluorocarbon fluids suitable for potential blood substitutes (31) (see BLOOD, ARTIFICIAL) through biologically active materials like the very highly toxic octafluoroisobutylene [382-21-8]. The toxicity of one chemical vs another chemical is not predictable based on the number or the site of fluorine substituents. The main commercial fluorinated compounds, like the CFCs, exhibit a very low order of toxicity (32). The potential cardiotoxicity from inhalation of bronchiodilator aerosols using CFCs as propellants is well documented in the medical literature.

Many new fluorinated drug and agrochemical agents were discovered based initially on the properties that a C–F bond imparts to a molecule. Its size similarity to the C–H bond analogue allows entry into a binding site for a subsequent biological effect. The more stable linkage allows for longer term effects with slower metabolism and excretion. Some compounds possess toxic effects attributable to fluoride ion toxicity and irritation of the respiratory system may be a common response, but many others demonstrate a more complex biological behavior. Recent long-term inhalation studies with HCFC-123 as a CFC alternative have now shown nonmalignant tumors in male rats (33). This behavior is characteristic of this specific compound, and while unexpected, it is not representative of the family of fluorinated two-carbon compounds as a whole. Many of the other members of this family show excellent safety from inhalation toxicity testing. Fluorinated inhalation anesthetics require inhalation efficacy and safety acquired only through trial and error testing. Much of the work on anesthetics heralds back to the early refrigerant development work and the safety studies that were conducted to, in turn, identify this important property (34).

BIBLIOGRAPHY

"Nomenclature, Physical Properties, Reactions and Methods of Fluorination" under "Fluorine Compounds, Organic," in *ECT* 1st ed, Vol. 6, pp. 735–751, by E. T. McBee, O. R. Pierce, and W. F. Edgell, Purdue University; "Introduction" in *ECT* 2nd ed., under "Fluorine Compounds, Organic," Vol. 9, pp. 686–704, by E. T. McBee, C. J. Norton, and T. Hodgins, Purdue University; in *ECT* 3rd ed., Vol. 10, pp. 829–839, by O. R. Pierce, Dow Corning Corp.

1. U.S. Pat. 1,833,847 (Nov. 24, 1931), T. Midgley, Jr., A. L. Henne, and R. McNary (to Frigidaire Corp.) and related patents.
2. C. Slesser and S. R. Schram, *Preparation, Properties and Technology of Fluorine and Organic Fluorine Compounds*, McGraw-Hill Book Co., Inc., New York, 1951.
3. P. S. Zurer, *Chem. Eng. News*, 7–13 (July 24, 1989); 7–13 (June 22, 1992).
4. A. L. Henne, in R. Adams and co-workers, eds., *Organic Reactions*, Vol. II, John Wiley & Sons, Inc, New York, 1944, pp. 49–93.
5. E. T. McBee, P. A. Wiseman, and G. B. Bachman, *Ind. Eng. Chem.* **39**, 415–417 (1947).
6. N. N. Vorozhtsov and G. G. Yacobson, *Khim. Nauka i Prom.* **3**, 403 (1958).
7. E. Gryszkiewicz-Trochimowski, A. Sporzynski, and J. Wnuk, *Rec. Trav. Chim.* **66**, 413–418 (1947).
8. J. T. Maynard, *J. Org. Chem.* **28**, 112–115 (1963).
9. N. N. Vorozhtsov, V. E. Platonov, and G. G. Yakobson, *Izvest. Akad. Nauk SSSR, Ser. Khim.*, (8), 1524 (1963).
10. G. C. Finger, F. H. Reed, and L. R. Tehon, *Ill. State Geol. Surv. Circ.* **199**, 1–15 (1955).
11. W. Prescott, *Chem. Ind. (London)*, (2), 56–63 (1978).
12. W. Bockemuller, in *Newer Methods of Preparative Organic Chemistry*, rev. ed., Interscience Publishers, Inc., New York, 1948, pp. 229–248.
13. K. Wiechert, in Ref. 12, pp. 315–368.
14. U.S. Pat. 4,465,786 (Aug. 14, 1984), M. F. Zimmer, W. E. Smith, and D. F. Malpass (to General Electric Co.); U.S. Pat. 4,798,818 (Jan. 17, 1989), W. X. Bajzer and co-workers (to Dow Corning Corp.).
15. G. W. Parshall, *J. Org. Chem.* **27**, 4649–4651 (1962).
16. W. C. Smith, *Angew Chem. Int. Ed. Engl.* **1**, 467–475 (1962).
17. W. K. R. Musgrave, in M. Stacey and co-workers, eds., *Advances in Fluorine Chemistry*, Vol. 1, Butterworth & Co. Ltd., London, 1960, pp. 1–28.
18. W. E. Jones and E. G. Skolnik, *Chem. Reviews* **76**, 563–592 (1976); R. J. Lagow and J. L. Margrave, in S. J. Lippard, ed., *Progress in Inorganic Chemistry*, Vol. 26, John Wiley & Sons, Inc., New York, 1979, pp. 161–210.
19. S. T. Purrington, B. S. Kagen, and T. B. Patrick, *Chem. Reviews* **86**, 997–1018 (1986).
20. J. H. Simons and co-workers, *J. Electrochem. Soc.* **95**, 47–67 (1949); J. Burdon and J. C. Tatlow, in Ref. 17, pp. 129–165; S. Nagase, in P. Tarrant, ed., *Fluorine Chemistry Reviews*, Vol. 1, Marcel Dekker, Inc., New York, 1967, pp. 77–106.
21. A. Roe, in R. Adams and co-workers, eds., *Organic Reactions*, Vol. V, John Wiley & Sons, Inc., New York, 1949, pp. 193–228.
22. U.S. Pat. 4,822,927 (Apr. 18, 1989), N. J. Stepaniuk and B. J. Lamb (to Mallinckrodt, Inc.).
23. R. D. Chambers and R. H. Mobbs, in M. Stacey and co-workers, eds., *Advances in Fluorine Chemistry*, Vol. 4, Butterworth & Co., Ltd., London, 1965, pp. 50–112.
24. D. Holtz, *Chem. Reviews* **71**, 139–145 (1971).
25. U.S. Pat. 4,082,752 (Apr. 4, 1978), S. Misaki and T. Takahara (to Daikin Kogyo Co. Ltd.); U.S. Pat. 4,029,661 (June 14, 1977), R. Anderson, P. Schuman, and G. Westmoreland (to PCR Inc.); U.S. Pat. 3,954,758 (May 4, 1976), P. Schuman and co-workers, (to PCR Inc.).

26. R. Filler, in R. E. Banks, ed., *Organofluorine Chemicals and Their Industrial Applications*, Ellis Horwood Ltd., Chichester, UK, 1979.
27. *AMA Drug Evaluation*, 3rd ed., Publishing Sciences Group, Inc., Littleton, Mass., 1977.
28. C. G. Huang, in *Abstracts of the 203rd National Meeting of the American Chemical Society*, abstract #FLUO-018, San Francisco, Apr. 5–10, 1992.
29. United States International Trade Commission (USITC), *Synthetic Organic Chemicals–United States Production and Sales, 1990*, publication #2470, Washington, D.C., Dec. 1991; R. F. Bradley, A. Leder, and Y. Sakuma, *Fluorocarbons*, in *Chemical Economics Handbook*, SRI International, Menlo Park, Calif., 1990, sections 543.7000–543.7003, plus 1992 supplemental data.
30. *Chem. Eng. News*, 13 (Jan. 4, 1988); 27 (Nov. 27, 1989).
31. J. G. Riess and M. LeBlanc, in K. C. Lowe, ed., *Blood Substitutes: Preparation, Physiology and Medical Applications*, Ellis Horwood, Ltd., Chichester, UK, 1988, pp. 94–129.
32. G. D. Clayton and F. E. Clayton, eds., *Patty's Industrial Hygiene and Toxicology*, Vol. 2B, 3rd rev. ed., John Wiley & Sons, Inc., New York, 1981.
33. *Chem. Eng. News*, 26 (July 8, 1991); P. Zurer, *Chem. Eng. News*, 21 (July 22, 1991).
34. E. R. Larsen, in P. Tarrant, ed., *Fluorine Chemistry Reviews*, Vol. 3, Marcel Dekker, Inc., 1969, pp. 1–44.

General References

R. E. Banks, *Preparation, Properties and Industrial Applications of Organofluorine Compounds*, Ellis Horwood Ltd., Chichester, UK 1982.

R. E. Banks and M. G. Barbour, *Fluorocarbon and Related Chemistry*, Vols. 1–3, The Chemical Society, Burlington House, London, 1971–1976.

R. E. Banks, ed., *Organofluorine Chemicals and Their Industrial Applications*, Ellis Horwood, Ltd., Chichester, UK, 1979.

R. D. Chambers, *Fluorine in Organic Chemistry*, John Wiley & Sons, Inc., New York, 1973.

R. Filler and Y. Kobayashi, *Biochemical Aspects of Fluorine Chemistry*, Kodansha Scientific Books, Tokyo, and Elsevier Biomedical Press, Amsterdam, 1982.

M. Hudlicky, *Chemistry of Organic Fluorine Compounds*, 2nd ed., Ellis Horwood, Ltd., Chichester, UK, 1976.

I. L. Knunyants and G. G. Yakobson, *Synthesis of Fluoroorganic Compounds*, Springer-Verlag, New York, 1985.

J. F. Liebman, A. Greenberg, and W. R. Dolbier, Jr., eds., *Fluorine-Containing Molecules: Structure, Reactivity, Synthesis and Applications*, VCH Publishers, Inc., New York, 1988.

A. M. Lovelace, D. A. Rausch, and W. Postelnek, *Aliphatic Fluorine Compounds*, Reinhold Publishing Corp., New York, 1954.

G. A. Olah, R. D. Chambers, and G. K. S. Prakash, eds., *Synthetic Fluorine Chemistry*, John Wiley & Sons, Inc., New York, 1992.

A. E. Pavlath and J. E. Leffler, *Aromatic Fluorine Compounds*, Reinhold Publishing Corp., New York, 1962.

G. Siegemund and co-workers, *Fluorine Compounds, Organic*, in W. Gerhartz and co-workers, eds., *Ullmann's Encyclopedia of Industrial Chemistry*, 5th rev. ed., Vol. A11, VCH Publishers, New York, 1988.

J. H. Simons, ed., *Fluorine Chemistry*, Vols. 1–5, Academic Press, Inc., New York, 1950–1964.

W. A. Sheppard and C. M. Sharts, *Organic Fluorine Chemistry*, Benjamin, Inc., New York, 1969.

P. Tarrant, ed., *Fluorine Chemistry Reviews*, Vols. 1–8, Marcel Dekker, Inc., New York, 1967–1977.

J. C. Tatlow and co-workers, eds., *Advances in Fluorine Chemistry*, Vols. 1–7, W. A. Benjamin, Inc., New York, 1960–1973.

J. T. Welch and S. Eswarakrishnan, *Fluorine in Bioorganic Chemistry*, John Wiley & Sons, Inc., New York, 1991.

J. T. Welch, *Selective Fluorination in Organic and Bioorganic Chemistry*, ACS Symposium Series #456, American Chemical Society, Washington, D.C., 1991.

WILLIAM X. BAJZER
YUNG K. KIM
Dow Corning Corporation

DIRECT FLUORINATION

Organic compounds containing fluorine are well known for their special properties, especially their inertness, very low boiling points, comparatively high melting points, and high thermal stability. Binary compounds of carbon and fluorine, ie, fluorocarbons, can be traced back to Moissan who, in 1886, discovered and isolated fluorine (1). The simplest solid fluorocarbon (2) is poly(fluoromethylidyne) $(CF)_n$, CFX (MarChem), which has a layered structure (Fig. 1). It can be synthesized by the reaction of fluorine with graphite at 300–600°C and fluorine pressures up to 101 kPa (1 atm) (3,4). Other solids (C_4F, C_2F, etc) have also been reported from the reaction of F_2 with graphite (5,6). This material is a grayish to white powder of variable stoichiometry and stable up to 600°C in air. Fluorinated graphite has a very low coefficient of friction, similar to Teflon, and is an excellent lubricant that can be used up to 600°C, surpassing Teflon which softens and decomposes above 350°C (see FLUORINE COMPOUNDS, ORGANIC–POLYTETRAFLUOROETHYLENE). When $(CF_x)_n$ decomposes above 650°C, it yields mainly car-

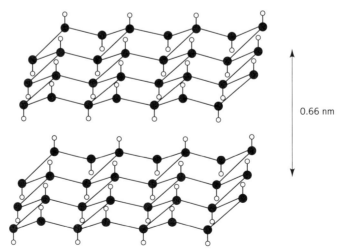

Fig. 1. Structure of poly(carbon monofluoride). ●, carbon; ○, fluorine. The interstitial space is 0.66 nm.

bon black (finely divided soot) and the inert gas carbon tetrafluoride (7). Thus it can be used safely at high temperatures without risk of poisonous gases being produced on decomposition (2,8).

Initial attempts at reactions between fluorine and hydrocarbons were described as similar to combustion and the reaction products contained mostly carbon tetrafluoride and hydrogen fluoride:

$$C_xH_y + \left(\frac{4x + y}{2}\right) F_2 \rightarrow x\, CF_4 + y\, HF + \text{energy}$$

This reaction has often reached explosive proportions in the laboratory. Several methods were devised for controlling it between 1940 and 1965. For fluorination of hydrocarbons of low (1–6 carbon atoms) molecular weight at room temperature or below by these methods, yields as high as 80% of perfluorinated products were reported together with partially fluorinated species (9–11). However, fluorination reactions in that era involving elemental fluorine with complex hydrocarbons at elevated temperatures led to appreciable cleavage of the carbon–carbon bonds and the yields invariably were only a few percent.

Before the LaMar process was developed in 1969, the use of direct fluorination was usually considered the classical method of fluorination (12,13) and other approaches were regarded as modern methods. Now only telomerization reactions using tetrafluorethylene and reactions in hydrogen fluoride-based electrochemical cells are more widely used than direct fluorination on a commercial scale; however, this may change in the future. Direct fluorination not only gives higher yields in most cases but preparation in this manner is applicable to a wider range of organofluorine compounds and classes of compounds inaccessible by these more established technologies. Many compounds are uniquely prepared in the laboratory by direct fluorination, and ton quantities of various fluorocarbon materials are available from 3M Co. manufactured by new direct fluorination technology.

Metal Fluoride Method

Before the LaMar process was developed, it was generally believed that because the reaction of elemental fluorine with hydrocarbons is highly exothermic, the carbon–carbon bonds are first ruptured, creating free radicals and eventually leading to polymerization. Therefore in most instances the expected product was a complex mixture of polymers and degradation products, including pure carbon and various fragments of the original hydrocarbon material.

In the 1940s researchers developed diverse fluorination methods that avoid these problems by using certain metal fluorides, inorganic fluorides, or halogen fluorides and electrochemical or indirect methods using fluorides as reagents. Hydrogen fluoride has been successfully used as a fluorinating agent and is, in fact, still widely used industrially.

Fluorination of organic compounds using high valency metallic fluorides (14) may be represented as follows:

Exchange of halogen with fluorine of the metal fluorides, MF_n:

$$\mathrm{\underset{/}{\overset{\backslash}{-}}C{-}X + M^+F^- \rightarrow \underset{/}{\overset{\backslash}{-}}C{-}F + M^+X^-} \tag{1}$$

where X = Cl, Br, or I and M = K, Sb, $AgHg_2$, or Hg

Replacement of hydrogen with the fluorine of metal fluorides:

$$\mathrm{\underset{/}{\overset{\backslash}{-}}C{-}H + 2\,MF_n \rightarrow \underset{/}{\overset{\backslash}{-}}C{-}F + HF + 2\,MF_{n-1}} \tag{2}$$

Addition to double bonds:

$$\mathrm{\underset{/}{\overset{\backslash}{}}C{=}C\underset{\backslash}{\overset{/}{}} + 2\,MF_n \rightarrow F{-}\underset{/}{\overset{\backslash}{}}C{-}C\underset{\backslash}{\overset{/}{}}{-}F + 2\,MF_{n-1}} \tag{3}$$

High valency metallic fluorides are very reactive compounds and most decompose in water. They include CoF_3, AgF_2, MnF_3, CeF_4, PbF_4, and possibly BiF_5 and UF_6. As shown in equations 1–3, at elevated temperatures, usually 100–400°C, they react with organic compounds producing the fluoro derivatives and the corresponding lower fluorides such as CoF_2 and AgF. Using cobalt trifluoride [10026-18-3] at 200–400°C, all hydrogen atoms of a hydrocarbon can be replaced by fluorine if the product is thermally stable. Similar addition of fluorine to unsaturated linkages and to aromatic nuclei takes place.

The requirement that organic compounds be vaporized at temperatures averaging 280°C across the bed of cobalt trifluoride or silver difluoride causes serious limitations to the broad applicability of the synthesis of organofluorine compounds using metal fluoride technology. There are at least two companies, Imperial Smelting, Ltd. of Britain and Air Products and Chemicals of Allentown, Pennsylvania, still active in this field; the number of organic compounds that can be prepared effectively with this technique numbers approximately 100. Fused-ring aromatic compounds are the most able to survive these harsh fluorination conditions. Fluorination of polymers is almost impossible using this method because few polymers are easily volatilized. The most effective of these processes utilizing metal fluorides, the cobalt trifluoride technology, is much less flexible than the 3M electrochemical cell (Simons' cell) or the Du Pont tetrafluoroethylene telomerization technology. In general, cobalt trifluoride fluorination has been more successful in the vapor phase than the liquid phase; yields tend to be low with extensive thermal degradation occurring at high temperatures.

Hydrogen Fluoride Electrochemical Cell Methods

Direct fluorination using hydrogen fluoride electrochemical cell methods is mechanistically similar in some regards to direct fluorination with F_2. This method uses an electrolytically activated fluoride ion produced by a Simons' designed hydrogen fluoride electrochemical cell as its primary means of fluorination. The Simons' electrochemical cell fluorination technology is practiced widely by Minnesota Mining & Manufacturing Co. (3M) of St. Paul, Minnesota. In this method, organic precursors are dissolved in liquid hydrogen fluoride and a voltage slightly under the voltage required for generation of elemental fluorine is applied across carbon electrodes. This technique, invented by J. H. Simons (15–17), has been a successful source of organofluorine compounds, functional fluids, and low molecular weight perfluorocarbon acids and diacids. There are also a number of companies in Japan and Europe that now use electrochemical fluorination for production of fluorocarbons.

The principal disadvantage to electrochemical fluorination is the requirement that the organic material be at least somewhat soluble in the polar liquid hydrogen fluoride. Therefore 3M product lines are generally based on perfluoro amines and functionalized materials such as carboxylic acids or sulfonic acids which are soluble in hydrogen fluoride (see FLUORINE COMPOUNDS, ORGANIC–FLUORINATED HIGHER CARBOXYLIC ACIDS; FLUOROETHERS AND FLUORAMINES; PERFLUOROALKANESULFONIC ACIDS). Even so this technology is not capable of producing high molecular weight functional products beyond C-8 or C-10 in yields which make the process an economically viable technique. 3M uses the electrochemical cell technology to produce a well-known line of fluids known as Fluorinerts, largely based on perfluoro amines.

Other limitations of electrochemical fluorination are that compounds such as ethers and esters are decomposed by hydrogen fluoride and cannot be effectively processed. Branching and cross-linking often take place as a side reaction in the electrochemical fluorination process. The reaction is also somewhat slow because the organic reactant materials have to diffuse within 0.3 nm of the surface of the electrode and remain there long enough to have all hydrogen replaced with fluorine. The activated fluoride is only active within 0.3 nm of the surface of the electrode.

Fluorocarbons produced by electrochemical fluorination often have small quantities (1–5%) of up to 20 by-products produced by rearrangement. Rearrangement is not characteristic of modern direct fluorination technology using elemental fluorine. By-product formation is a particular disadvantage for applications such as production of fluorocarbon oxygen carriers, fluorocarbon blood, and other biomedical fluorocarbon products where high purity materials are required. Single compound materials are essential for advantageous consideration by the U.S. Food and Drug Administration (FDA) because each often requires individual FDA approval before the mixture receives approval.

Direct Fluorination Using Elemental Fluorine

Kinetic as well as thermodynamic problems are encountered in fluorination. The rate of reaction must be decelerated so that the energy liberated may be absorbed or carried away without degrading the molecular structure. The most recent advances in direct fluorination are the LaMar process (18–20) and the Exfluor process (21–24), which is practiced commercially by 3M.

Thermochemistry. Thermodynamic considerations are of utmost importance in fluorinations. Table 1 is based on JANAF data (25) for CH_4, which indicate an average carbon–hydrogen bond strength of 410.0 kJ/mol (98 kcal/mol) based on the atomization energy of CH_4.

Table 1. Thermodynamic Data[a] for Fluorination of CH_4, kJ/mol[b]

Step	Reaction	ΔH_{25}	ΔH_{325}	ΔG_{25}	ΔG_{325}
initiation					
1a	$F_2 \rightarrow 2\,F\cdot$	157.7	161.0	123.6	87.4
1b	$F_2 + RH \rightarrow R\cdot + HF + F\cdot$	16.3	21.3	−24.4	−79.1
propagation					
2a	$RH + F\cdot \rightarrow R\cdot + HF$	−141.4	−139.7	−151.2	−156.9
2b	$R\cdot + F_2 \rightarrow RF + F\cdot$	−289.1	−290.8	−284.9	−268.4
termination					
3a	$R\cdot + F\cdot \rightarrow RF$	−446.8	−451.8	−407.9	−356.0
3b	$R\cdot + R\cdot \rightarrow R-R$	−350.6	−347.5	−294.1	−240.6
overall reaction	$RH + F_2 \rightarrow RF + HF$	−430.5	−430.5	−432.6	−430.9

[a] Based on JANAF table data (25).
[b] To convert J to cal, divide by 4.184.

The limiting parameter to be considered in attempting to develop a satisfactory method for controlling reactions of elemental fluorine is the weakest bond in the reactant compound. For hydrocarbons the average carbon–carbon single-bond strength is 351.5–368.2 kJ/mol (84–88 kcal/mol). The overall reaction for the replacement of hydrogen by fluorine is exothermic enough [$\Delta G_{25} = -432.6$ kJ/mol (−103.4 kcal/mol)] for a fracture of carbon–carbon bonds if it were to occur via a concerted mechanism or on several adjacent carbon atoms simultaneously. This energy must be dissipated so as to avoid the fragmentation of the molecular skeleton. The comparison of 359.8 kJ/mol (86 kcal/mol) vs 430.9 kJ/mol (103 kcal/mol) has been cited in many previous discussions as an obvious basis to predict the failure of direct fluorination methods. For rapid reaction rates, which were employed in most previous experiments, this is a valid argument.

It can be seen from Table 1 that there are no individual steps that are exothermic enough to break carbon–carbon bonds except the termination of step 3a of −407.9 kJ/mol (−97.5 kcal/mol). Consequently, procedures or conditions that reduce the atomic fluorine concentration or decrease the mobility of hydrocarbon radical intermediates, and/or keep them in the solid state during reaction, are desirable. It is necessary to reduce the reaction rate to the extent that these hydrocarbon radical intermediates have longer lifetimes permitting the advantages of fluorination in individual steps to be achieved experimentally. It has been demonstrated by electron paramagnetic resonance (epr) methods (26) that, with high fluorine dilution, various radicals do indeed have appreciable lifetimes.

The two possible initiations for the free-radical reaction are step 1b or the combination of steps 1a and 2a from Table 1. The role of the initiation step 1b in the reaction scheme is an important consideration in minimizing the concentration of atomic fluorine (27). As indicated in Table 1, this process is spontaneous at room temperature [$\Delta G_{25} = -24.4$ kJ/mol (−5.84 kcal/mol)] although the enthalpy is slightly positive. The validity of this step has not yet been conclusively established by spectroscopic methods which makes it an unsolved problem of prime importance. Furthermore, the fact that fluorine reacts at a significant rate with some hydrocarbons in the dark at temperatures below −78°C indicates that step 1b is important and may have little or no activation energy at RT. At extremely low temperatures (ca 10 K) there is no reaction between gaseous fluorine and CH_4 or C_2H_6 (28).

A simple equilibrium calculation reveals that, at 25°C and atmospheric pressure, fluorine is less than 1% dissociated, whereas at 325°C an estimated 4.6% dissociation of molecular fluorine is calculated. Obviously, less than 1% of the collisions occurring at RT would result in reaction if step 1a were the only important initiation step. At 325°C the fluorine atom initiation step should become more important. From the viewpoint of energy control, as shown in Table 1, it would be advantageous to have step 1b predominate over step 2a and promote attack by molecular rather than atomic fluorine. Ambient or lower temperatures keep the atomic fluorine concentration low.

In the addition of fluorine to double bonds, the energetic situation is less severe, ie, the addition of fluorine to double bonds is only 251.4–292.9 kJ/mol (60–70 kcal/mol) exothermic per carbon–carbon bond. This energy is not sufficient to fracture the carbon skeleton if care is taken to keep addition from occurring on several adjacent carbon atoms simultaneously. Here, as in the case of hydrogen removal, the individual steps are less exothermic than the overall reaction. It has been established experimentally that less fragmentation occurs and, correspondingly, a higher yield is obtained with most conventional fluorination processes when an unsaturated rather than a saturated hydrocarbon is the starting material. This is owing to the greater exothermicity of the reaction with hydrogen, ie, 434.7 kJ/mol (103.9 kcal/mol) per saturated carbon atom as compared with 207–289.9 kJ/mol (50–70 kcal/mol) per unsaturated carbon atom. In the case of addition of fluorine to double bonds, the initiation step (eq. 4) is probably exothermic by 20.7–190.4 kJ/mol (5–46 kcal/mol) and thus plays an important role.

$$R_2C{=}CR_2 + F_2 \rightarrow R_2\dot{C}{-}CFR_2 + F\cdot \qquad (4)$$

A second possibility is that a concerted mechanism (eq. 5) which is exothermic by 207.0–283.2 kJ/mol (50–58.4 kcal/mol) per carbon atom, is important.

$$R_2C{=}CR_2 + F_2 \rightarrow R_2CF{-}CFR_2 \qquad (5)$$

Steric Factors. Initially, most of the collisions of fluorine molecules with saturated or aromatic hydrocarbons occur at a hydrogen site or at a π-bond (unsaturated) site. When collision occurs at the π-bond, the double bond disappears but the single bond remains because the energy released in initiation (eq. 4) is insufficient to fracture the carbon–carbon single bond. Once carbon–fluorine bonds have begun to form on the carbon skeleton of either an unsaturated or alkane system, the carbon skeleton is somewhat sterically protected by the sheath of fluorine atoms. Figure 2, which shows the crowded helical arrangement of fluorine around the carbon backbone of polytetrafluoroethylene (PTFE), is an example of an extreme case of steric protection of carbon–carbon bonds (29).

The nonbonding electron clouds of the attached fluorine atoms tend to repel the oncoming fluorine molecules as they approach the carbon skeleton. This reduces the number of effective collisions, making it possible to increase the total

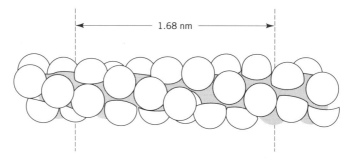

Fig. 2. The steric protection of the carbon backbone by fluorine of a polytetrafluoroethylene chain. The helical configuration with a repeat distance of 1.68 nm results from the steric crowding of adjacent fluorine.

number of collisions and still not accelerate the reaction rate as the reaction proceeds toward completion. This protective sheath of fluorine atoms provides the inertness of Teflon and other fluorocarbons. It also explains the fact that greater success in direct fluorination processes has been reported when the hydrocarbon to be fluorinated had already been partially fluorinated by some other process or was prechlorinated, ie, the protective sheath of halogens reduced the number of reactive collisions and allowed reactions to occur without excessive cleavage of carbon–carbon bonds or runaway exothermic processes.

Kinetic Control. In direct fluorination processes, concentration, time, and temperature can be controlled. In most previous work, the fluorine was diluted with an inert gas such as nitrogen, helium, or even carbon dioxide. However, the concentration of fluorine in the reactor was kept at a constant level, usually 10% or greater, by premixing the inert gas with fluorine in the desired proportion and then introducing this mixture into the reactor. The rate of reaction between a hydrocarbon compound and a 10% fluorine mixture is relatively high, and this very exothermic process can lead to fragmentation and, in some cases, to combustion. The initial stages of reaction are most critical; nearly all the fragmentation occurs at this time. An initial concentration of 10% fluorine or more is, for most compounds, much too high for nondestructive fluorination.

Molecular relaxation processes such as vibrational or rotational relaxations or thermal conduction make it possible to dissipate the energy released during fluorination. Such relaxation processes can minimize the chances that the energy required to break the weakest bond is appropriately localized if the reaction sites are widely distributed over the system. Therefore, in the initial stages of fluorination, it is necessary to reduce the probability of adjacent reaction sites simultaneously occurring in the same molecule, or in adjacent molecules in a crystal, by diluting the reactants and relying on relaxation processes to distribute the energy over the entire system and thus avoid fragmentation.

Reactant molecules are able to withstand more fluorine collisions, as they become more highly fluorinated, without decomposition because some sites are sterically protected, ie, collisions at carbon–fluorine sites are obviously nonreactive. The fluorine concentration may therefore be increased as the reaction proceeds to obtain a practical reaction rate. Actual dilution schemes to achieve successful fluorination must be individually tailored for specific reaction systems and may, in some cases, include a stepwise procedure. In Figure 3, the horizontal line at about 10 kPa (10%) represents the 1940s approach to direct fluorination. High initial concentrations result in extensive fragmentation. The curved lines that asymptotically approach 101.3 kPa (1 atm) of fluorine pressure or 50.6 kPa (0.5 atm) of fluorine pressure, etc, represent the controlled approach of the LaMar process (18–20).

To achieve the very low initial fluorine concentration in the LaMar fluorination process initially a helium or nitrogen atmosphere is used in the reactor and fluorine is bled slowly into the system. If pure fluorine is used as the incoming gas, a concentration of fluorine may be approached asymptotically over any time period (Fig. 3). It is possible to approach asymptotically any fluorine partial pressure in this manner. The very low initial concentrations of fluorine in the system greatly decreases the probability of simultaneous fluorine collisions on the same molecules or on adjacent reaction sites.

Thus, for a successful fluorination process involving elemental fluorine, the number of collisions must be drastically reduced in the initial stages; the rate of fluorination must be slow enough to allow relaxation processes to occur and a heat sink must be provided to remove the reaction heat. Most direct fluorination reactions with organic compounds are performed at or near room temperature unless reaction rates are so fast that excessive fragmentation, charring, or decomposition occurs and a much lower temperature is desirable.

Low temperature fluorination techniques ($-78°C$) are promising for the preparation of complex fluorinated molecules, especially where functional groups are present (30), eg, fluorination of hexamethylethane to perfluorohexamethylethane [39902-62-0], of norbornane to perfluoro- (C_7F_{12}) and 1-hydro undecafluoronorbornane [4934-61-6], C_7HF_{11}, and of adamantane to 1-hydropentadecafluoroadamantane [54767-15-6].

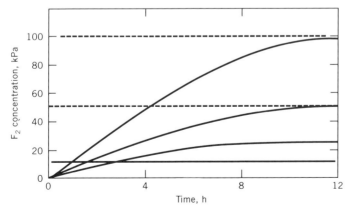

Fig. 3. Fluorine dilution scheme direct fluorination. 1 atm = 101.3 kPa.

Experimental Techniques

In early reaction systems (9,10,31,32) the vaporized hydrocarbon was combined with nitrogen in a reactor and mixed with a nitrogen–fluorine mixture from a preheated source. The jet reactor (11) for low molecular weight fluorocarbons was an important improvement. The process takes place at around 200–300°C, and fluorination is carried out in the vapor state.

At 200–300°C many compounds, both organic and inorganic, are marginally stable, and certainly not in their lowest vibrational states. They may even undergo some pyrolytic decomposition. Thus the addition of extra energy produced by the interaction of fluorine with these compounds is likely to produce substantial fragmentation. Vaporization of high molecular weight hydrocarbons is rather difficult, and their fluorination was not successful in the early jet fluorination studies.

The typical fluorination apparatus used in the LaMar process for these reactions is simple in design (Fig. 4) (33). It is essential that the materials of construction are resistant to fluorine (34). The presence of even traces of oxygen or moisture can have a deleterious effect and, therefore, extreme precautions must be taken to eliminate these contaminants.

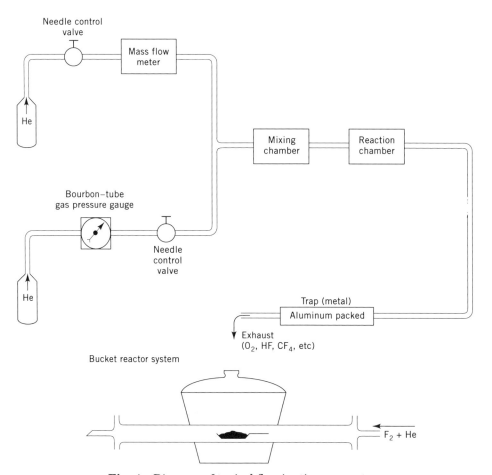

Fig. 4. Diagram of typical fluorination apparatus.

The connections are conveniently made of 0.635 cm OD copper tubing. When the fluorocarbon produced in the reaction is volatile, a cold finger-type trap can be placed between the reaction chamber and the trap to catch volatile products. The temperature of the trap must be high enough to pass unreacted or excess fluorine and nitrogen but cold enough to condense the reaction products. Before the reaction is started, the whole system is purged with helium or nitrogen for ca 30 minutes and then 0.5–2 mL of fluorine and 50–100 mL of nitrogen per minute are passed through the system. Solid material has to be ground to a very fine powder (37 mm (~100 mesh)) to achieve complete fluorination.

A special cryogenic reactor (35) in which the reactions of fluorine with liquid and gaseous samples can be controlled at very low temperatures is shown in Figure 5. Reactants are volatilized into the reaction zone of the cryogenic reactor from the heated oil evaporator prior to initiation of the reaction. The main reaction chamber is a nickel tube, 2.54 cm in diameter, packed with copper turnings. The compartments (10.1 × 10.1 × 20.2 cm) are constructed of stainless steel and insulated with urethane foam and act as heat sinks. All connections are made of 0.635 cm copper or aluminum tubing. A sodium fluoride trap is used to remove the hydrogen fluoride from the reaction products. By cooling or warming the compartments, they can be used to create a temperature gradient along the reaction tube. Because the products are highly fluorinated, they are usually volatile and tend to move through the reactor tube rapidly, depending on the temperature gradient. This provides a continually renewed surface of reactant at the optimum temperature for fluorination. Fluorinated copper turnings effectively increase the surface area of the compound exposed to fluorine. The individual zones of the reactor may be cooled with various solvent–solid carbon dioxide or with solvent–liquid nitrogen slushes. Preferably, the temperature is precisely regulated with

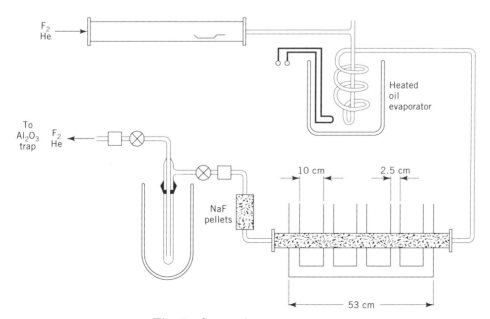

Fig. 5. Cryogenic reactor system.

an automatic liquid nitrogen temperature controller. In addition to the four-zone reactor shown in Figure 5, a multizone reactor can also be used; an eight-zone reactor has been found to be particularly efficient (36). Internal Freon cooling is effective for controlling the temperatures of the various compartments (37).

Oxygen or moisture has to be excluded because the presence of oxygen leads to cross-linking, presumably with epoxy bridges, to carbonyl groups, which give acid fluorides, and to peroxides (38). Cross-linking obviously decreases the yield of pure perfluorocarbon. It can be detected by infrared absorption in the 1600–2000 cm^{-1} region and by noting the polymeric nature of the products, ie, high melting points, low vapor pressures, etc. On the other hand, oxyfluorination is a technique offering unique possibilities for the functionalization of fluorocarbons and for the preparation of functional fluorocarbon membranes (39).

Aerosol-Based Direct Fluorination. A technology that works on liter and half-liter quantities has been introduced (40–42). This new aerosol technique, which functions on principles similar to LaMar direct fluorination (Fig. 5), uses fine aerosol particle surfaces rather than copper filings to maintain a high surface area for direct fluorination. The aerosol direct fluorination technique has been shown to be effective for the synthesis of bicyclic perfluorocarbon such as perfluoroadamantane, perfluoroketones, perfluoroethers, and highly branched perfluorocarbons.

Modern Direct Fluorination. Direct fluorination technology has been scaled up at Exfluor Research Corp. of Austin, Texas (21–24). The synthesis of perfluoroethers by this method is licensed to 3M Co. where it is practiced on a commercial (multiton) scale. Using direct fluorination it is possible to produce almost any desired fluorocarbon structure for which there is a hydrocarbon or organic structural precursor (22–24,43). There are two basic approaches to controlling direct fluorination: the LaMar method where the rate of fluorine addition is the limiting factor, and the Lagow-Exfluor method in which the rate of addition of the hydrocarbon is the limiting factor. A highly effective solvent fluorination technology with rapid heat transfer has been developed on this principle by Exfluor Research Corp. Multikilogram quantities of new fluorocarbons are produced in yields ranging from 95 to 99% with this technology (21). This technique is capable of producing very high molecular weight perfluoro acids and diacids that are precursors for new fluorocarbon copolymers. This is accomplished rapidly and on a commercial scale with essentially no branching or rearrangement.

Applications

In 1954 the surface fluorination of polyethylene sheets by using a solid CO_2 cooled heat sink was patented (44). Later patents covered the fluorination of PVC (45) and polyethylene bottles (46). Studies of surface fluorination of polymer films have been reported (47). The fluorination of polyethylene powder was described (48) as a fiery intense reaction, which was finally controlled by dilution with an inert gas at reduced pressures. Direct fluorination of polymers was achieved in 1970 (8,49). More recently, surface fluorinations of poly(vinyl fluoride), polycarbonates, polystyrene, and poly(methyl methacrylate), and the surface fluorination of containers have been described (50,51). Partially fluorinated poly(ethylene terephthalate) and polyamides such as nylon have excellent soil release properties as well as high wettability (52,53). The most advanced direct fluorination technology in the area of single-compound synthesis and synthesis of high performance fluids is currently practiced by 3M Co. of St. Paul, Minnesota, and by Exfluor Research Corp. of Austin, Texas.

The following companies manufacture organic fluorine compounds by direct fluorination techniques: 3M; Exfluor; Air Products and Chemicals, Inc., Allentown, Pennsylvania; MarChem, Inc., Houston, Texas; Ozark-Mahoning, Inc., Tulsa, Oklahoma; and PCR, Inc., Gainesville, Florida.

Simple and Complex Organic Molecules. Using modern direct fluorination technology, the synthesis of even the most complex perfluorocarbon structures from hydrocarbon precursors is now possible. For example, syntheses of the first perfluoro crown ethers, perfluoro 18-crown-6, perfluoro 15-crown-5, and perfluoro 12-crown-4 (54) have been reported. Perfluoro crown ethers (54,55) are becoming important as the molecules of choice for many ^{19}F-nmr imaging applications (56) in humans and are particularly effective in brain and spinal diagnostics when

administered to the cerebrospinal fluid compartment. Synthesis scale-up of perfluoro 15-crown-5 (54,55) and plans for commercialization are underway while research is being conducted on other biological applications of these new compounds (57). In collaboration with Air Products, excellent brain imaging scans have been obtained by infusing the perfluoro 15-crown-5 in spinal fluids. Toxicology reports on these are very favorable with essentially no toxic effects physiologically in several different animals.

Perfluoro crown ethers from the hydrocarbon dibenzo crown ethers have also been synthesized (58) and the first perfluorocryptand molecule [2.2.2] has been reported (59). The perfluorocryptand is a stable, inert, high boiling clear oil.

Hydrocarbon crown ethers coordinate cations; however, both the perfluoro crown ethers and the perfluorocryptands coordinate anions. For example, perfluoro crown ethers and perfluorocryptands tenaciously encapsulate O_2^- and F^- (60,61).

The first several perfluoro spiro compounds have also been synthesized (62). An example of this technology is the synthesis and crystal structure of perfluoro-1,4,9,12-tetraoxadispiro[4.2.4.2]tetradecane.

Many novel small molecule perfluoropolyethers have been made using direct fluorination technology. For example, even branched ethers such as perfluoro(pentaerythritol tetramethyl ether) can be prepared:

$$C(CH_2-O-CH_3)_4 \xrightarrow{F_2} C(CF_2-O-CF_3)_4$$

Very good low temperature fluids are obtained by direct fluorination of trialkylorthoformates (63):

$$HC(OCH_2CH_3)_3 \xrightarrow{F_2/He} FC(OCF_2CF_3)_3$$

Perfluoropolyethers emerged on the market in the early 1970s; however, for the next 15 years there were only two basic structures known. The first perfluoropolyether was the homopolymer of hexafluoropropylene oxide produced by Du Pont having the structure

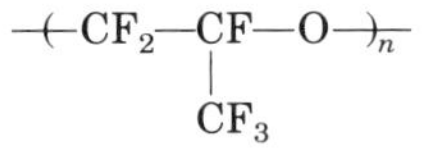

$$-(CF_2-CF(CF_3)-O)_n-$$

Du Pont called this new lubricant material Krytox (64,65) and initially it had such extraordinary properties that it sold for $200/kg ($187/kg ca 1993). Krytox was and is used in most of the vacuum pumps and diffusion oil pumps for the microelectronics industry in this country and in Japan because it produces no hydrocarbon (or fluorocarbon) vapor contamination. It has also found important applications in the lubrication of computer tapes and in other data processing applications as well as military and space applications.

Materials similar in high temperature properties to the Du Pont material with better low temperature properties have been synthesized using direct fluorination. The first was produced by reaction of fluorine with inexpensive hydrocarbon polyethers such as poly(ethylene oxide). In the simplest case, poly(ethylene oxide) is converted to the perfluoroethylene oxide polymer:

$$HO(CH_2CH_2O)_nH \xrightarrow{F_2/He} R_f(OCF_2CF_2)_nOR_f$$

This simple reaction chemistry was first reported in 1978 (66).

Other interesting perfluoro ether structures can be obtained by copolymerization of hexafluoroacetone with ethylene oxide, propylene oxide, and trimethylene oxide with subsequent fluorination to yield the following structures (67):

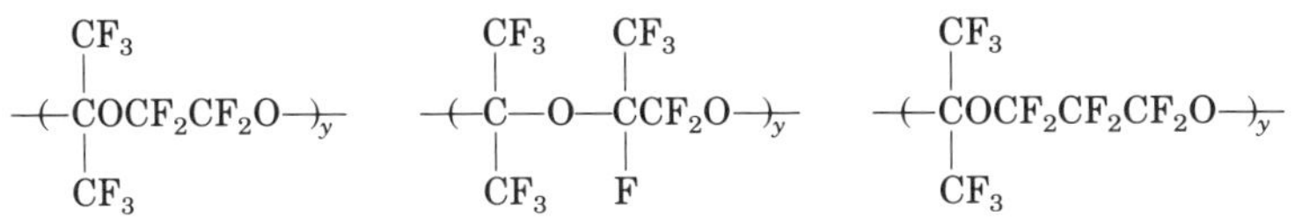

$$-(C(CF_3)_2OCF_2CF_2O)_y- \quad -(C(CF_3)(F)-O-C(CF_3)CF_2O)_y- \quad -(C(CF_3)_2OCF_2CF_2CF_2O)_y-$$

Two of the perfluoropolyether fluid structures yet to be commercialized are interesting. The first structure is a strictly alternating copolymer of ethylene oxide and methylene oxide, which has the longest liquid range of any molecule containing carbon (40). The second structure is the perfluoromethylene oxide polyether which has low temperature liquid properties down to $-120°C$:

$$-(-CF_2O-CF_2CF_2O-)_n- \qquad -(-CF_2O-CF_2-O-)_n-$$

Other perfluoropolyether structures that have been synthesized are (24,43)

$$-(-CF_2CFO-)_n- \qquad -(-CF_2CFO-)_n-$$
$$\quad\;\;| \qquad\qquad\qquad\;\; |$$
$$\;\;CF_2CF_3 \qquad\qquad\;\; CF_2Cl$$

Hydrocarbon Polymers. It is difficult to produce perfluorocarbon polymers by the usual methods. Many monomers, such as hexafluoropropylene, polymerize only slowly because of the steric hindrance of fluorine. Furthermore, some monomers are not very stable and are difficult to synthesize. Direct fluorination can be used for the direct synthesis of fluorocarbon polymers (68–70) and for producing fluorocarbon coatings on the surfaces of hydrocarbon polymers (8,29,44–47,49,68–71).

Thus fluorocarbon polymers can be produced with chemical compositions similar to polytetrafluoroethylene by the direct reaction of fluorine with polyethylene and the perfluoro analogues of polypropylene and polystyrene can be prepared. These fluorocarbon polymers differ from the more familiar linear structures because carbon–carbon cross-linking occurs to a significant extent during fluorination. Most of these fluoropolymers are white solids with high thermal stability; some are stable in air as much as 200°C above the ignition temperatures of their corresponding hydrocarbon precursors. Hydrocarbon polymers such as polyethylene and the new surfaces formed by direct fluorination have been studied by esca which shows that the surface is truly converted to a fluorocarbon polymer (72). Most of these fluorocarbon surfaces are inert and many of them have good lubricant properties. The fluorination of PVC has been followed with esca and $(CF_x)_n$ polymer was also identified (73).

Surface Fluorination of Polymers. Fluorocarbon-coated objects have many practical applications because the chemically adherent surface provides increased thermal stability, resistance to oxidation and corrosive chemicals and solvents, decreased coefficient of friction and thus decreased wear, and decreased permeability to gas flow. Unusual surface effects can be obtained by fluorinating the polymer surfaces only partially (74).

Natural and Synthetic Rubber. Fluorination of natural or synthetic rubber creates a fluorocarbon coating (29,75,76) which is very smooth and water repellent (see WATERPROOFING). Rubber articles such as surgical gloves, O-rings, gaskets, and windshield wiper blades can be fluorinated on the surface while the interior retains the elastic, flexible properties of the natural rubber. Fluorinated O-rings can be used without extra lubricant in corrosive atmospheres since the fluorocarbon is unreactive. In food-processing equipment, grease or lubricants are eliminated and do not contaminate the food products. Fluorinated O-rings have smooth surfaces, very low frictional coefficients, and enhanced thermal stabilities. Fluorinated windshield wiper blades have a very low coefficient of friction, run smoother with less squeak, their surface is more resistant to the sun's uv radiation and attack by ozone, and they require less electrical energy for operation.

Many applications of this technique are apparent in medicine, such as surgical rubber gloves, rubber sheets, drain tubes, catheters, etc. Since talcum or other lubricating powder often used with surgical gloves can cause allergic reactions, thin powderless gloves are desirable. Teflon or silicone films have proved unsuccessful for surgical gloves. However, an excellent direct fluorination process has been developed (76) whereby the inside surface is fluorinated under expanded conditions at elevated temperatures. A very smooth surface is obtained and powder is not required. At the same time, the outside surface remains rough so that surgical instruments can be held firmly without slippage. The tactile sensitivity of the tips of the fingers seems to be increased, and the problem of powder forming lumps or a mud-like slush inside the glove is eliminated.

Blow-Molded Containers. A surface-fluorination process (Airopak) has been developed by Air Products & Chemicals for the blow-molding industry to produce solvent-resistant polyolefin containers. In this application, the air that is normally used to blow-mold containers is replaced by a low concentration of fluorine in nitrogen. Airopak containers produced by this process show outstanding resistance to nonpolar solvents (50,51,77,78) and such blow-molding fluorination procedures have been widely used for the last 10 years by Ford Motor Co. and many European auto manufacturers to produce low cost–high performance gas tanks for cars and trucks. A similar technology is now practiced by Fluoroseal, Inc. (Houston, Texas). This group has generated a successful product line based on post-treating containers and other objects with elemental fluorine.

BIBLIOGRAPHY

"Direct Fluorination" under "Fluorine Compounds, Organic" in *ECT* 3rd ed., Vol. 10, pp. 840–855, by J. L. Margrave. R. H. Hauge, and R. B. Badachhape, Rice University, and R. J. Lagow, University of Texas.

1. H. Moissan, *Le Fluor et ses Composes*, Steinbeil, Paris, 1900.
2. P. Kamarchik and J. L. Margrave, *Acc. Chem. Res.* **11**, 296 (1978); U.S. Pat. 3,519,657 (July 7, 1970), G. A. Olah (to Dow Chemical Co.,); L. B. Ebert, J. I. Brauman, and R. A. Huggins, *J. Am. Chem. Soc.* **96**, 7841 (1974).
3. O. Ruff and O. Bretschneider, *Z. Anorg. Allgem. Chem.* **217**, 1 (1937); W. Rudorff and G. Rudorff, *Z. Anorg. Allgem. Chem.* **253**, 281 (1947); W. Rudorff and G. Rudorff, *Chem. Ber.* **80**, 413 (1947); W. Rudorff and K. Brodersen, *Z. Naturforsch* **12b**, 575 (1957); W. Rudorff, *Adv. Inorg. Chem. Radiochem.* **1**, 230 (1959).
4. A. K. Kuriakose and J. L. Margrave, *J. Phys. Chem.* **69**, 2772 (1965); U.S. Pat. 3,674,432 (July 4, 1972), J. L. Margrave, R. J. Lagow, and co-workers (to R. I. Patents, Inc.); J. L. Margrave, R. J. Lagow, and co-workers, *J. Am. Chem. Soc.* **96**, 1268, 2628 (1974); R. B. Badachhape, V. K. Mahajan, and J. L. Margrave, *Inorg. Nucl. Chem. Lett.* **10**, 1103 (1974).
5. N. Watanabe and M. Ishii, *J. Electrochem. Soc. Jpn.* **29**, 364 (1961); N. Watanabe, Y. Koyama, and S. Yoshizawa, *J. Electrochem. Soc. Jpn.* **31**, 756 (1963); N. Watanabe and K. Kumon, *J. Electrochem. Soc. Jpn.* **35**, 19 (1967); H. Imoto and N. Watanabe,

Bull. Chem. Soc. Jpn. **49**, 1736 (1976); M. Takashima and N. Watanabe, *Nippon Kagaku Kaishi*, 1222 (1976); N. Watanabe, Y. Kita, and T. Kawaguchi, *Nippon Kagaku Kaishi*, 191 (1977).

6. Y. Kita, N. Watanabe, and Y. Fujii, *J. Am. Chem. Soc.* **101**, 3823 (1979); R. J. Lagow and co-workers, *Inorg. Metalorg. Chem.* **2**, 145 (1972).
7. P. Kamarchik and J. L. Margrave, *J. Therm. Anal.* **11**, 259 (1977).
8. J. L. Margrave, R. J. Lagow, and co-workers, *IR-100 Award for CFX, a Lubricant Powder*, Rice University, Houston, Tex., 1970; *Ind. Res.* **12**, 47 (1970).
9. G. H. Cady and co-workers, *Ind. Eng. Chem.* **39**, 290 (1947).
10. R. N. Haszeldine and F. Smith, *J. Chem. Soc.*, 2689, 2787 (1950).
11. L. A. Bigelow, in J. H. Simons, ed., *Fluorine Chemistry*, Vol. 1, Academic Press, Inc., New York, 1970, p. 373; E. A. Tyczkowski and L. A. Bigelow, *J. Am. Chem. Soc.* **77**, 3007 (1955).
12. C. M. Sharts, *J. Chem. Ed.* **45**, 3 (1968).
13. R. E. Banks, *Fluorocarbons and Their Derivatives*, Oldbourne Press, London, 1964, p. 87.
14. W. A. Shepherd and C. M. Sharts, *Organic Fluorine Chemistry*, W. A. Benjamin, New York, 1969; M. Hudlicky, *Chemistry of Organic Fluorine Compounds*, 2nd ed., Halsted Press, a division of John Wiley & Sons, Inc., New York, 1976; M. Stacey and J. C. Tatlow, in *Advances in Fluorine Chemistry*, Vol. 1, Butterworths Scientific Publications, London, 1960, pp. 166–198.
15. J. H. Simons, *Trans. Electrochem. Soc.* **95**, 47 (1949).
16. J. H. Simons, *Fluorine Chemistry*, Vol. 1, Academic Press, Inc., New York, 1950, p. 401.
17. U.S. Pat. 2,490,099 (Dec. 6, 1949), J. H. Simons (to Minnesota Mining & Manufacturing Co.).
18. R. J. Lagow and J. L. Margrave, *Proc. Natl. Acad. Sci.* **67**(4)8A (1970).
19. R. J. Lagow, Ph.D. dissertation, Rice University, Houston, Tex., 1970.
20. B. Fegley, *MIT Technol. Eng. News*, 13 (Apr. 1973).
21. U.S. Pat. 5,093,432 (Mar. 3, 1992), T. R. Bierschenk, R. J. Lagow, T. J. Juhlke, and H. Kawa (to Exfluor Research Corporation).
22. U.S. Pat. 4,760,198 (July 26, 1988), T. R. Bierschenk, T. J. Juhlke, and R. J. Lagow (to Exfluor Research Corp.).
23. U. S. Pat. 4,827,042 (May 2, 1989), T. R. Bierschenk, T. J. Juhlke, and R. J. Lagow (to Exfluor Research Corp.).
24. U.S. Pat. Appl. Ser. No. 07/982,030 (Nov. 24, 1992), T. R. Bierschenk, R. J. Lagow, T. J. Juhlke, and H. Kawa (to Exfluor Research Corp.).
25. D. R. Stull and co-eds, *JANAF Thermochemical Tables*, 2nd ed., *NSRDS-NBS37*, U. S. Government Printing Office, Washington, D.C., June 1971, and subsequent revisions.
26. R. E. Florin and L. A. Wall, "Electron Spin Resonance Studies on Fluorination of Polymers," abstract no. 8, Fluorine Chemistry Div., *165th American Chemical Society Meeting*, Dallas, Tex., Apr. 10, 1973.
27. W. T. Miller, S. D. Koch, and F. W. McLafferty, *J. Am. Chem. Soc.* **78**, 4992 (1956).
28. R. H. Hauge, J. Wang, and J. L. Margrave, paper presented at the *First Winter Fluorine Conference*, St. Petersburg, Fla., Jan. 1972.
29. C. W. Bunn and E. R. Howell, *Nature* **174**, 549 (1954).
30. N. J. Maraschin and R. J. Lagow, *J. Am. Chem. Soc.* **94**, 8601 (1972).
31. O. Ruff, *Die Chemie des Fluors*, Springer, Berlin, 1920.
32. W. Bockemuller, *Organische Fluorverbindungen*, F. Enke, Stuttgart, 1936.
33. U.S. Pat. 3,758,450 (Sept. 11, 1973), R. J. Lagow and J. L. Margrave (to R. I. Patents, Inc.); U.S. Pat. 3,775,489 (Nov. 27, 1973), R. J. Lagow and J. L. Margrave (to R. I. Patents, Inc.)

34. H. F. Priest and A. V. Grosse, *Ind. Eng. Chem.* **39**, 279 (1947); R. Landau and R. Rosen, *Ind. Eng. Chem.* **39**, 281 (1947).
35. N. J. Maraschin and co-workers, *J. Am. Chem. Soc.* **97**, 513 (1975).
36. U.S. Pat. 3,904,501 (Sept. 9, 1975), R. J. Lagow and co-workers (to Massachusetts Institute of Technology).
37. U.S. Pat. 4,281,119 (July 28, 1981), R. J. Lagow and co-workers (to Massachusetts Institute of Technology).
38. U.S. Pat. 3,480,667 (Nov. 25, 1969), W. R. Siegart and W. D. Blackley.
39. S. Inoue, J. L. Adcock, and R. J. Lagow, *J. Am. Chem. Soc.* **100**, 1948 (1978).
40. J. L. Adcock, K. Horita, and E. B. Renh, *J. Am. Chem. Soc.* **103**, 6932 (1981).
41. J. L. Adcock and M. L. Robin, *J. Org. Chem.* **49**, 1442 (1984).
42. J. L. Adcock, *J. Fluorine Chem.* **33**, 327 (1986).
43. U.S. Pat. 4,931,199 (June 5, 1990), T. R. Bierschenk, R. J. Lagow, T. J. Juhlke, and H. Kawa (to Exfluor Research Corp.).
44. Brit. Pat. 710,523 (June 16, 1954), A. J. Rudge (to Exfluor Research Corp.).
45. U.S. Pat. 2,497,046 (Feb. 7, 1950), E. L. Kopra (to American Cyanamid Co.).
46. U.S. Pat. 2,811,468 (Oct. 29, 1957), S. P. Joffre (to Shulton, Inc.).
47. H. Schonhorn and R. Hansen, *J. Appl. Polym. Sci.* **12**, 1231 (1968).
48. M. Okade and K. Makuuchi, *Ind. Eng. Chem. Prod. Res. Dev.* **8**, 334 (1969).
49. *Chem. Eng. News* **48**, 40 (Jan. 12, 1970); *Chemistry* **43**(4), 30 (1970).
50. U.S. Pat. 3,862,284 (Jan. 21, 1975), D. D. Dixon, D. G. Manly, and G. W. Recktenwald (to Air Products and Chemicals, Inc.).
51. *Ind. Res. Dev.* **12**, 102 (1978).
52. U.S. Pats. 3,988,491 (Oct. 26, 1976), and 4,020,223 (Apr. 26, 1977), D. D. Dixson and L. J. Hayes (to Air Products and Chemicals, Inc.).
53. L. J. Hayes, *J. Fluorine Chem.* **8**, 69 (1976).
54. W. H. Lin, W. I. Bailey, Jr., and R. J. Lagow, *J. Chem. Soc., Chem. Commun.*, 1550 (1985).
55. U.S. Pat. 4,570,005 (Feb. 11, 1986), W. H. Lin and R. J. Lagow (to University of Texas System).
56. U.S. Pat. 4,838,274 (June 13, 1989), F. K. Schweighardt and J. A. Rubertone (to Air Products and Chemicals, Inc.).
57. T. Y. Lin, L. C. Clark, Jr., and R. J. Lagow, to be published, 1993.
58. T. Y. Lin and R. J. Lagow, *J. Chem. Soc., Chem. Commun.*, 12 (1991).
59. W. D. Clark and R. J. Lagow, *J. Org. Chem.* **55**, 5933 (1990).
60. J. Brodbelt, R. J. Lagow, and co-workers, *J. Am. Chem. Soc.* **113**, 5913 (1991).
61. J. Brodbelt, R. J. Lagow, and co-workers, *J. Chem. Soc., Chem. Commun.*, 1705 (1991).
62. T. Y. Lin and R. J. Lagow, in print, 1994.
63. R. J. Lagow, T. E. Mlsna and co-workers, *Eur. J. Solid State and Inorg. Chem.* **29**, 907 (1992).
64. J. T. Hill, *J. Macromol. Sci., Chem.* **8**, 499 (1974).
65. H. S. Eleuterio, *J. Macromol. Sci., Chem.* **6**, 1027 (1972).
66. G. E. Gerhardt and R. J. Lagow, *J. Org. Chem.* **43**, 4505 (1978).
67. D. F. Persico and R. J. Lagow, *Macromolecules* **18**, 1383 (1985).
68. J. L. Margrave and R. J. Lagow, *J. Polym. Sci. Polym. Lett. Ed.* **12**, 177 (1974).
69. A. J. Otsuka and R. J. Lagow, *J. Fluorine Chem.* **4**, 371 (1974).
70. R. J. Lagow, H. Kawa and co-workers, *J. Polym. Sci., Polym. Lett. Ed.* **28**, 297 (1990).
71. U.S. Pat. 3,647,613 (Mar. 7, 1972), J. L. Scotland (to British Resin Products Ltd.).
72. D. T. Clark and co-workers, *J. Polym. Sci. Polym. Chem. Ed.* **13**, 857 (1975); D. T. Clark and co-workers, in L. H. Lee, ed., *Advances in Friction Wear*, Vol. 5A, Plenum Press, New York, 1975, p. 373.
73. G. Parks, Ph.D. dissertation, Rice University, Houston, Tex., 1976.

74. J. Pederson, M.A. thesis, Rice University, Houston, Tex., 1979.
75. Can. Pat. 1,002,689 (Dec. 28, 1976), R. J. Lagow and J. L. Margrave (to DAMW Associates); Brit. Pat. 1,440,605 (Oct. 20, 1976), R. J. Lagow and J. L. Margrave (to DAMW Associates).
76. U.S. Pat. 3,992,221 (Nov. 16, 1976), R. B. Badachhape, C. Homsy, and J. L. Margrave (to Vitek Inc. and MarChem, Inc.).
77. A. J. Woytek and J. F. Gentilecore, "A New Blow Molding Process to Resist Solvent Permeation of Polyolefin Containers," paper no. 13, presented at *Advances in Blow Molding Conference*, Rubber and Plastics Institute, London, Dec. 6, 1977.
78. J. F. Gentilecore, M. A. Triolo, and A. J. Woytek, *Plast. Eng.* **34**, 23 (1978).

RICHARD J. LAGOW
University of Texas at Austin

FLUORINATED ALIPHATIC COMPOUNDS

The hydrogen atoms in alkanes can be partially or completely replaced by fluorine. Partially fluorinated alkanes are commonly called hydrofluorocarbons (HFCs) and the fully fluorinated derivatives are perfluorocarbons (PFCs). Alkanes whose hydrogens are replaced by both fluorine and chlorine are designated chlorofluorocarbons (CFCs), or hydrochlorofluorocarbons (HCFCs) if the replacement is incomplete. Similar designations are used for other halogenated fluorocarbons. Fluorinated aliphatics are further identified by a series of numbers related to the formula of the compound. In this numbering system for methane and ethane derivatives, the first digit on the right is the number of fluorine atoms in the compound, and the second digit from the right is one more than the number of hydrogen atoms. The third digit from the right is one less than the number of carbon atoms, but when this digit is zero, it is omitted. The remaining available positions in the compound are taken by chlorine atoms unless specified otherwise. For example, CCl_3F, $CHClF_2$, CF_3CHF_2, and CF_3CF_3 are designated CFC-11, HCFC-22, HFC-125, and PFC-116, respectively. When bromine is present, the same rules apply except that the letter B is used, followed by a number that indicates the number of chlorine atoms replaced by bromine. For example, CF_3Br and $CHBrF_2$ are coded BFC-13B1 and HBFC-22B1, respectively. In the fire extinguishing trade, the brominated derivatives are usually called Halons and have a different numbering system in which the digits from right to left are respectively the number of bromine, chlorine, fluorine, and carbon atoms. Any remaining available positions are hydrogen atoms. The above BFC and HBFC become H-1301 and H-1201 in this system. This numbering system has been extended systematically to both acyclic and cyclic compounds with more than two carbon atoms, but the code for distinguishing various isomers becomes rather complex (1).

Perfluorocarbons and Hydrofluorocarbons

Properties. Aliphatic PFCs have an unusual combination of physical properties relative to their hydrocarbon counterparts (2–5). The volatilities of PFCs are much higher than expected based on their molecular weights. For example, tetrafluoromethane, mol wt 88, boils at −128°C, whereas n-hexane, mol wt 86, boils at +69°C. Perfluorocarbons containing up to four carbon atoms boil somewhat higher than the corresponding hydrocarbons; the reverse is true of PFCs with more carbon atoms. Liquid PFCs are two to three times as dense as hydrocarbons with the same carbon skeleton, and aliphatic PFCs have among the lowest dielectric constants, refractive indexes, and surface tensions of any liquids at room temperature. The compressibilities and absolute viscosities of PFCs are considerably higher than those of hydrocarbons. Aliphatic PFCs are poor solvents for all materials except for those with low cohesive energies, such as gases and other PFCs. They are practically insoluble in water and only slightly soluble in hydrocarbons.

The extremely nonpolar character of PFCs and very low forces of attraction between PFC molecules account for their special properties. Perfluorocarbons boil only slightly higher than noble gases of similar molecular weight, and their solvent properties are much more like those of argon and krypton than hydrocarbons (2). The physical properties of some PFCs are listed in Table 1.

The physical properties of hydrofluorocarbons reflect their polar character, and possibly the importance of intermolecular hydrogen bonding (3). Hydrofluorocarbons often boil higher than either their PFC or hydrocarbon counterparts. For example, 1-$C_6H_{13}F$ boils at 91.5°C compared with 58°C for n-C_6F_{14} and 69°C for n-C_6H_{14}. Within the series of fluorinated methanes, the boiling point reaches a maximum for CH_2F_2, which contains an equal number of hydrogen and fluorine atoms for maximum hydrogen bonding. The methane boiling points, however, also parallel their dipole moments, which reflect relative polar character: CH_3F ($\mu = 1.85$D), CH_2F_2 (1.97D), CHF_3 (1.65D), CF_4 (0.0D) (1D = 3.336 × 10^{-30} C·m).

Hydrofluorocarbons invariably have higher refractive indexes, dielectric constants, and surface tensions, but lower densities than their PFC counterparts. The physical properties of some HFCs are listed in Table 2. Because of their very strong carbon–fluorine and carbon–carbon bonds (10), the chemical and thermal stability of PFCs is considerably higher in general than that of the corresponding hydrocarbons (3,11). Perfluorocarbons normally are significantly less reactive than hydrocarbons toward all chemical reagents except alkali metals. Molten alkali metals or alkali–metal hydrocarbon complexes degrade most fluorocarbons, and this reaction is used for their chemical analysis (3). Perfluorocarbons are not affected by acids or oxidizing agents and are not hydrolyzed below 500°C. Carbon tetrafluoride decomposes only slowly at carbon arc temperatures, and it does not react with Cu, Ni, W, or Mo at 900°C. Perfluorocarbons of higher molecular weight are less thermally stable, but temperatures approaching 1000°C are still required to decompose C_2F_6 or n-C_3F_8, and most PFCs are stable below 300°C. Partially fluorinated hydrocarbons are less stable and more reactive, especially when only one fluorine atom is present. Hydrogen fluoride can be eliminated by chemical or thermal action from hydrofluorocarbons.

Manufacture. The direct fluorination of hydrocarbons with elemental fluorine is extremely exothermic and difficult to control. Special methods including metal packing techniques, jet reactors, and high dilution have been developed to control the reaction, but currently they have limited industrial importance (12). Poly(carbon monofluoride), $(CF)_x$, is one product that is made commercially by direct fluorination (of graphite) (13). The disadvantages of direct fluorination have

Table 1. Physical Properties of Aliphatic Perfluorocarbons (PFCs)[a]

PFC number	Formula	CAS Registry Number	Molecular weight	Boiling point, °C	Melting point, °C	Liquid density, g/mL at °C	Liquid refractive index, n_D at °C	Critical temp, °C	Critical pressure, MPa[b]
14	CF_4	[75-73-0]	88.01	−128.1	−183.6	1.613_{-130}	1.151_{-73}	−45.6	3.74
116	CF_3CF_3	[76-16-4]	138.02	−78.2	−100.6	1.600_{-80}	1.206_{-73}	19.7	2.99
218	$CF_3CF_2CF_3$	[76-19-7]	188.03	−36.7	−183	1.350_{20}		71.9	2.68
31-10	$CF_3(CF_2)_2CF_3$	[355-25-9]	238.04	−2.2	−128	1.543_{20}		113.2	2.32
C-318	cyclo-C_4F_8	[115-25-3]	200.04	−5.9	−41.4	1.500_{25}	1.217_{25}	115.2	2.78
41-12	$CF_3(CF_2)_3CF_3$	[678-26-2]	288.05	29.2	−126	1.620_{20}	1.242_{15}	149	2.04
51-14	$CF_3(CF_2)_4CF_3$	[355-42-0]	338.07	58	−86	1.680_{25}	1.251_{22}	174.5	1.90
61-16	$CF_3(CF_2)_5CF_3$	[335-57-9]	388.08	82.5	−51	1.733_{20}	1.262_{20}	201.6	1.62
PP3[c]	cyclo-$C_6F_{10}(CF_3)_2$[d]	[335-27-3]	400.09	102	−70	1.828_{25}	1.290_{25}	241.5	1.88
PP6[c]	cyclo-$C_{10}F_{18}$[e]	[306-94-5]	462.11	142	[f]	1.917_{25}	1.313_{25}	292.0	1.75
PP9[c]	cyclo-$C_{10}F_{17}(CF_3)$[g]	[306-92-3]	512.12	160	−70	1.972_{25}	1.320_{25}	313.4	1.66
PP11[c]	cyclo-$C_{14}F_{24}$[h]	[306-91-2]	624.15	215	−20	2.03_{25}	1.335_{25}	377[i]	1.46[i]

[a] Refs. 6–8.
[b] To convert MPa to psi, multiply by 145.
[c] Flutec number (trademark of Rhône-Poulenc, Inc., RTZ Chemicals, ISC Division).
[d] Perfluoro-1,3-dimethylcyclohexane.
[e] Perfluorodecalin, cis/trans mixture.
[f] −11.2 to 18.0°C, depending on cis/trans ratio.
[g] Perfluoro-2-methyldecalin.
[h] Perfluorotetradecahydrophenanthrene.
[i] Estimated values.

Table 2. Physical Properties of Aliphatic Hydrofluorocarbons (HFCs)[a]

HFC number	Formula	CAS Registry Number	Molecular weight	Boiling point, °C	Melting point, °C	Liquid density, g/mL at °C	Liquid refractive index, n_D at °C	Critical temp, °C	Critical pressure, MPa[b]
23	CHF_3	[75-46-7]	70.01	−82.2	−155.2	1.442_{-80}	1.215_{-73}	25.7	4.83
32	CH_2F_2	[75-10-5]	52.02	−51.6	−136	1.200_{-50}	1.190_{20}		5.86
41	CH_3F	[593-53-3]	34.03	−78.3	−141.8	0.884_{-80}	1.1727_{20}	44.6	3.52
125	CHF_2CF_3	[354-33-6]	120.02	−48.5	−103		1.5012_{19}	72.4	
134	CHF_2CHF_2	[359-35-3]	102.03	−19.7	−89		1.250_{20}		
134a	CH_2FCF_3	[811-97-2]	102.03	−26.5	−101	1.21_{25}		101.1	4.14
143	CHF_2CH_2F	[430-66-0]	84.04	5.0	−84			71.2	
143a	CH_3CF_3	[420-46-2]	84.04	−47.4	−111.3	1.176_{-50}	1.22_{25}	73.1	3.76
152	CH_2FCH_2F	[624-72-6]	66.05	30.7	−117	0.913_{19}	1.28_{25}	107.5	
152a	CH_3CHF_2	[75-37-6]	66.05	−25.8	−143.2	1.023_{-30}	1.3011_{-72}	113.5	4.50
161	CH_3CH_2F	[353-36-6]	48.06	−37.4	−129.5	0.818_{-37}	1.3033_{-37}	102.2	4.72
227ea	CF_3CFHCF_3[c]	[431-89-0]	170.03	−18	−82	1.407_{25}		101.7	2.91
245ca	$CHF_2CF_2CH_2F$	[679-86-7]	134.05	26			1.30_{15}		
245cb	$CF_3CF_2CH_3$	[1814-88-6]	134.05	−18				106.9	
254fb	$CH_2FCH_2CF_3$	[460-36-6]	116.06	29.4		1.2584_{25}	1.2765_{25}		
272ca	$CH_3CF_2CH_3$	[420-45-1]	80.08	−0.4	−104.8	0.9205_{20}	1.2904_{20}		
272fa	$CH_2FCH_2CH_2F$	[462-39-5]	80.08	41.6		1.0057_{25}	1.3190_{26}		
281ea	CH_3CHFCH_3	[420-26-8]	62.09	−10	−133.4	0.7238_{-20}	1.3075_{-10}		
281fa	$CH_2FCH_2CH_3$	[460-13-9]	62.09	−2.5	−159	0.7956_{20}	1.3115_{20}		

[a] Refs. 5, 7–9.
[b] To convert MPa to psi, multiply by 145.
[c] Unpublished data, Great Lakes Chemical Co.

been overcome by the use of fluorine carriers, in particular, high valence metal fluorides such as cobalt trifluoride, CoF_3, or potassium tetrafluorocobaltate, $KCoF_4$. These reagents replace hydrogen and halogen atoms by fluorine and add fluorine to double bonds and aromatic systems (12).

$$2\ CoF_3 + RH \xrightarrow{F_2} RF + HF + 2\ CoF_2$$

Cobalt trifluoride is generated *in situ* by passing fluorine over cobalt difluoride contained in a horizontal, mechanically agitated steel reactor. The compound to be fluorinated is passed through the reactor at 150–300°C as a vapor in a stream of nitrogen. After the reaction is completed, the CoF_3 is regenerated by adding fluorine. Advances in process control technology have allowed the process to be run continuously by simultaneously introducing the fluorine and hydrocarbon into the cobalt fluoride bed (6). Principally, cyclic and higher molecular weight acyclic fluorocarbons are prepared by this method.

Fluorocarbons are made commercially also by the electrolysis of hydrocarbons in anhydrous hydrogen fluoride (Simons process) (14). Nickel anodes and nickel or steel cathodes are used. Special porous anodes improve the yields. This method is limited to starting materials that are appreciably soluble in hydrogen fluoride, and is most useful for manufacturing perfluoroalkyl carboxylic and sulfonic acids, and tertiary amines. For volatile materials with little solubility in hydrofluoric acid, a complementary method that uses porous carbon anodes and HF·2KF electrolyte (Phillips process) is useful (14).

Hydrofluorocarbons are also prepared from acetylene or olefins and hydrogen fluoride (3), or from chlorocarbons and anhydrous hydrogen fluoride in the presence of various catalysts (3,15). A commercial synthesis of 1,1-difluoroethane, a CFC alternative and an intermediate to vinyl fluoride, is conducted in the vapor phase over an aluminum fluoride catalyst.

$$HC{\equiv}CH + 2\ HF \xrightarrow{catalyst} CH_3CHF_2$$
$$CH_3CHF_2 + heat \longrightarrow CH_2{=}CHF + HF$$

Perfluorocyclobutane is prepared by the thermal cyclodimerization of tetrafluoroethylene [116-14-3].

Health and Safety Factors. Completely fluorinated alkanes are essentially nontoxic (16). Rats exposed for four hours to 80% perfluorocyclobutane and 20% oxygen showed only slight effects on respiration, but no pathological changes in organs. However, some fluorochemicals, especially functionalized derivatives and fluoroolefins, can be lethal. Monofluoroacetic acid and perfluoroisobutylene [382-21-8] are notoriously toxic (16).

Uses. The chemical inertness, thermal stability, low toxicity, and nonflammability of PFCs coupled with their unusual physical properties suggest many useful applications. However, the high cost of raw materials and manufacture has limited commercial production to a few, small-volume products. Carbon tetrafluoride and hexafluoroethane are used for plasma, ion-beam, or sputter etching of semiconductor devices (17) (see ION IMPLANTATION). Hexafluoroethane and octafluoropropane have some applications as dielectric gases, and perfluorocyclobutane is used in minor amounts as a dielectric fluid. Perfluoro-1,3-dimethylcyclohexane is used as an inert, immersion coolant for electronic equipment, and perfluoro-2-methyldecalin is used for pin-hole leak testing of encapsulated electronic devices (6,18). Perfluoroperhydrophenanthrene has several diverse applications, ranging from a vapor-phase soldering agent for fabrication of printed circuits to a substitute for internal eye fluid in remedial eye surgery (6,19).

Medical applications of PFC emulsions for organ perfusion and intravenous uses have received much attention in recent years. The first commercial blood substitute (Fluosol DA 20%, trademark of the Green Cross Corp.) employed perfluorodecalin, and improved, second generation products based on this PFC, or perfluorooctylbromide, are now under development (20,21). The relatively high oxygen dissolving capability of PFCs underlies these applications (see BLOOD, ARTIFICIAL).

Poly(carbon monofluoride) is used as a high temperature lubricant and as a cathode material in high energy lithium batteries (13), but the fluorocarbons of greatest commercial interest and volume are the high molecular weight fluoroplastics and elastomers derived from tetrafluoroethylene, hexafluoropropylene, vinylidene fluoride, and vinyl fluoride (22). Poly(tetrafluoroethylene) [9002-84-0], Teflon (trademark of E. I. du Pont de Nemours & Co.), and its copolymer [25067-11-2] with hexafluoropropylene, Teflon FEP, are fluorinated plastics, particularly notable for their outstanding chemical and thermal stability, electrical inertness, and nonflammability (23). Teflon has service temperatures in the range of −196 to 260°C. The copolymers [9011-17-0] of hexafluoropropylene and vinylidene fluoride, and the terpolymers [25190-89-0] of hexafluoropropylene, tetrafluoroethylene, and vinylidene fluoride (Viton fluoroelastomers, trademark of E. I. du Pont de Nemours & Co.) are rubbers with excellent thermal, chemical, and oxidative stability (24). They remain useful elastomers for indefinite periods of continuous exposure in air up to about 230°C (see ELASTOMERS, SYNTHETIC–FLUOROCARBON ELASTOMERS).

Chlorofluorocarbons and Hydrochlorofluorocarbons

Properties. The physical properties of aliphatic fluorine compounds containing chlorine are similar to those of the PFCs or HFCs (3,5). They usually have high densities and low boiling points, viscosities, and surface tensions. The irregularity in the boiling points of the fluorinated methanes, however, does not appear in the chlorofluorocarbons. Their boiling points consistently increase with the number of chlorines present. The properties of some CFCs and HCFCs are shown in Tables 3 and 4.

Although the CFCs and HCFCs are not as stable as the PFCs, they still can be rather stable compounds (3,11). Dichlorodifluoromethane, CCl_2F_2, is stable at 500°C in quartz; CCl_3F and $CHClF_2$ begin to decompose at 450 and 290°C, re-

Table 3. Physical Properties of Aliphatic Chlorofluorocarbons (CFCs)[a]

CFC number	Formula	CAS Registry Number	Molecular weight	Boiling point, °C	Melting point, °C	Liquid density, g/mL at °C	Liquid refractive index, n_D at °C	Critical temp, °C	Critical pressure, MPa[b]
11	CCl_3F	[75-69-4]	137.36	23.8	−111	1.476_{25}	1.374_{25}	198.0	4.41
12	CCl_2F_2	[75-71-8]	120.91	−29.8	−158	1.311_{25}	1.287_{25}	112.0	4.11
13	$CClF_3$	[75-72-9]	104.46	−81.4	−181	1.298_{-30}	1.199_{-73}	28.9	3.87
111	CCl_3CCl_2F	[354-56-3]	220.29	137	100	1.740_{25}			
112	CCl_2FCCl_2F	[76-12-0]	203.82	92.8	26	1.634_{30}	1.413_{25}	278	3.44
112a	CCl_3CClF_2	[76-11-9]	203.82	91.5	40.6	1.649_{20}			
113	CCl_2FCClF_2	[76-13-1]	187.38	47.6	−35	1.565_{25}	1.354_{25}	214.1	3.41
113a	CCl_3CF_3	[354-58-5]	187.38	45.8	14.2	1.579_{20}	1.361_{20}		
114	$CClF_2CClF_2$	[76-14-2]	170.92	3.8	−94	1.456_{25}	1.288_{25}	145.7	3.26
114a	CCl_2FCF_3	[374-07-2]	170.92	3.6	−94	1.455_{25}	1.309_0	145.6	3.29
115	$CClF_2CF_3$	[76-15-3]	154.47	−39.1	−106	1.291_{25}	1.214_{25}	80.0	3.12

[a]Refs. 5, 7–9.
[b]To convert MPa to psi, multiply by 145.

Table 4. Physical Properties of Aliphatic Hydrochlorofluorocarbons (HCFCs)[a]

HCFC number	Formula	CAS Registry Number	Molecular weight	Boiling point, °C	Melting point, °C	Liquid density, g/mL at °C	Liquid refractive index, n_D at °C	Critical temp, °C	Critical pressure, MPa[b]
21	$CHCl_2F$	[75-43-4]	102.92	8.92	−135	1.366_{25}	1.354_{25}	178.5	5.17
22	$CHClF_2$	[75-45-6]	86.47	−40.75	−160	1.194_{25}	1.256_{25}	96.0	4.97
31	CH_2ClF	[593-70-4]	68.48	−9.1	−133	1.271_{20}			
121	$CHCl_2CCl_2F$	[354-14-3]	185.84	116.6	−82.6	1.622_{20}	1.4463_{20}		
121a	$CHClFCCl_3$	[354-11-0]	185.84	116.5	−95.4	1.625_{20}	1.4525_{20}		
122	$CClF_2CHCl_2$	[354-21-2]	169.39	71.9	−140	1.5447_{25}	1.3889_{20}		
122a	$CHClFCCl_2F$	[354-15-4]	169.39	72.5		1.5587_{20}	1.3942_{20}		
122b	CHF_2CCl_3	[354-12-1]	169.39	73		1.566_{20}	1.3979_{20}		
123	$CHCl_2CF_3$	[306-83-2]	152.93	28.7	−107	1.475_{15}	1.3332_{15}	185	3.79
123a	$CHClFCClF_2$	[354-23-4]	152.93	28.2	−78	1.498_{10}	1.327_{20}		
124	$CHFClCF_3$	[2837-89-0]	136.48	−12	−199	1.364_{25}		122.2	3.57
124a	CHF_2CClF_2	[354-25-6]	136.48	−10.2	−117	1.379_{20}			
131	$CHCl_2CHClF$	[359-28-4]	151.40	102.5		1.5497_{17}	1.4390_{20}		
131a	CH_2ClCCl_2F	[811-95-0]	151.40	88	−140.7	1.4227_{20}	1.4248_{20}		
132	$CHFClCHFCl$	[431-06-1]	134.94	59	−155	1.46_{20}	1.391_{20}		
132a	CHF_2CHCl_2	[471-43-2]	134.94	60		1.4945_{17}	$1.3830_{16.4}$		
132b	$CH_2ClCClF_2$	[1649-08-7]	134.94	46.8	−101	1.416_{20}	1.362_{20}	222	
133	$CHClFCHF_2$	[431-07-2]	118.49	17.2		1.365_{10}			
133a	CH_2ClCF_3	[75-88-7]	118.49	6.1	−105.5	1.389_0	1.309_0	153.0	
141	$CH_2ClCHClF$	[430-57-9]	116.95	75.7	−60	1.3814_{20}	1.4113_{20}		
141b	CCl_2FCH_3	[1717-00-6]	116.95	32	−103.5	1.2500_{10}	1.3600_{10}	210.3	4.640
142	CHF_2CH_2Cl	[338-65-8]	100.50	35.1		1.312_{15}	1.3528_{15}		
142a	$CHClFCH_2F$	[338-64-7]	100.50	35			1.3416_{20}		
142b	CH_3CClF_2	[75-68-3]	100.50	−9.2	−130.8	1.113_{25}	1.3752_{20}	137.1	4.12
151	CH_2FCH_2Cl	[762-50-5]	82.50	53.2	<−50	1.1675_{25}	1.326_{20}	237.6	
225ca	$CF_3CF_2CHCl_2$[c]	[422-56-0]	202.94	51.1	−94	1.55_{25}	1.3262_{25}		
225cb	$CClF_2CF_2CHClF$[c]	[507-55-1]	202.94	56.1	−97	1.56_{25}			

[a]Refs. 5, 7–9.
[b]To convert MPa to psi, multiply by 145.
[c]Physical data from Ref. 25.

spectively (7). The pyrolysis of $CHClF_2$ at 650–700°C in metal tubes is the basis of a commercial synthesis of tetrafluoroethylene:

$$2\ CHClF_2 + \text{heat} \rightarrow CF_2{=}CF_2 + 2\ HCl$$

The chlorofluorocarbons react with molten alkali metals and CCl_2F_2 reacts vigorously with molten aluminum, but with most metals they do not react below 200°C. An exception is the dechlorination of chlorofluorocarbons with two or more carbon atoms in the presence of Zn, Mg, or Al in polar solvents. A commercial synthesis of chlorotrifluoroethylene [79-38-9] employs this reaction:

$$CClF_2CCl_2F + Zn \xrightarrow{\text{alcohol}} CF_2{=}CClF + ZnCl_2$$

Most chlorofluorocarbons are hydrolytically stable, CCl_2F_2 being considerably more stable than either CCl_3F or $CHCl_2F$. Chlorofluoromethanes and ethanes disproportionate in the presence of aluminum chloride. For example, CCl_3F and CCl_2F_2 give $CClF_3$ and CCl_4; $CHClF_2$ disproportionates to CHF_3 and $CHCl_3$. The carbon–chlorine bond in most chlorofluorocarbons can be homolytically cleaved under photolytic conditions (185–225 nm) to give chlorine radicals. This photochemical decomposition is the basis of the prediction that chlorofluorocarbons that reach the upper atmosphere deplete the earth's ozone shield.

Manufacture. The most important commercial method for manufacturing CFCs and HCFCs is the successive replacement of chlorine by fluorine using hydrogen fluoride (3,15). The traditional, liquid-phase process uses antimony pentafluoride or a mixture of antimony trifluoride and chlorine as catalysts. Continuous vapor-phase processes that employ gaseous hydrogen fluoride in the presence of heterogenous chromium, iron, or fluorinated alumina catalysts also are widely used. Carbon tetrachloride, chloroform, and hexachloroethane (or tetrachloroethylene plus chlorine) are commonly used starting materials for one- and two-carbon chlorofluorocarbons. The extent of chlorine exchange can be controlled by varying the hydrogen fluoride concentration, the contact time, or the reaction temperature.

$$CHCl_3 \xrightarrow{HF} CHCl_2F + CHClF_2 + CHF_3$$

$$CCl_4 \xrightarrow{HF} CCl_3F + CCl_2F_2 + CClF_3$$

$$CCl_3CCl_3 \xrightarrow{HF} CCl_2FCCl_3 + CCl_2FCCl_2F + CClF_2CCl_2F + CClF_2CClF_2$$

$$CH_3CCl_3 \xrightarrow{HF} CH_3CCl_2F + CH_3CClF_2 + CH_3CF_3$$

The direct chlorination of hydrofluorocarbons and fluoroolefins has also been used commercially, eg, in the preparations of CH_3CClF_2 from CH_3CHF_2 and $CClF_2CClF_2$ from tetrafluoroethylene.

Economic Aspects. The estimated worldwide production of important industrial CFCs is shown in Table 5. Trichlorofluoromethane, dichlorodifluoromethane, and trichlorotrifluoroethane account for over 95% of the total produc-

Table 5. Worldwide Production of Important Chlorofluorocarbons,[a] 10^3 t

Product	1986 Market	1991 Market	Percentage of 1986 market
CFC-11	415	263	63%
CFC-12	441	259	59%
CFC-113	241	143	59%
CFC-114	18	5	30%
CFC-115	13	11	85%

[a]Ref. 26.

tion. Between 1986 and 1991 the production of CFCs has decreased dramatically due to global adherence to the provisions of the Montreal Protocol and eventually will be phased out entirely. Estimates of the distribution by use in 1986 and subsequent reductions in use are shown in Table 6.

In 1990, approximately 115,000 t of CFCs were used as propellants, which represents a 58% decrease from the 1986 level. Most of this market segment likely will move to using hydrocarbon propellants, but the optimal choice of alternative will depend on the particular application. No suitable replacement for CFC-12 in pharmaceutical metered dose inhalers has been found. This single application represents a 6000 t/yr market. It also has been difficult to replace CFC-12 as the inert propellant for the potent sterilant ethylene oxide, which is a market for approximately 20,000 t per year.

Worldwide use of CFCs for refrigeration, air conditioning, and heat pumps totaled 260,000 t in 1991 (see REFRIGERATION) which is a dropoff of only 7% from 1986 levels.

Cleaning agent and solvent use has decreased 41% from 1986 levels, but in 1990, over 178,000 t of CFC-113 were still used in electronics, metal, precision, and dry cleaning (see SOLVENTS, INDUSTRIAL).

The use of CFCs as foam blowing agents has decreased 35% from 1986 levels. Polyurethanes, phenolics, extruded polystyrenes, and polyolefins are blown with CFCs, and in 1990 the building and appliance insulation markets represented about 88% of the 174,000 t of CFCs used in foams (see FOAMED PLASTICS).

Health and Safety Factors. The toxicity of aliphatic CFCs and HCFCs generally decreases as the number of fluorine atoms increases (16), as shown in

Table 6. Worldwide Estimates of CFC Use by Industry, 1991 vs 1986[a]

Application	1986 Total uses, %	Reduction since 1986, %
propellants	28	58
refrigerants	23	7
cleaning	21	41
foam blowing agents	26	35
other uses	2	

[a]Ref. 26.

Table 7, but there are exceptions as in the case of 141b vs 142b. Also, derivatives like HCFC-132b can have low acute but high chronic toxicities (29).

Chlorofluorocarbons and Stratospheric Ozone Destruction. In 1971, it was shown (30) that CFCs were accumulating in the atmosphere, and three years later a relationship between CFCs and stratospheric ozone destruction, wherein the longlived CFCs that migrated to the upper stratosphere were being photolyzed by the intense uv radiation from the sun to form chlorine atoms, was proposed

Table 7. Toxicity of Selected Halocarbons[a]

Compound	Formula	TLV,[b] ppm	ALC,[c] ppm
BFC-12B2	CBr_2F_2	100	7,102[d]
BCFC-12B1	$CBrClF_2$		131,000
BFC-13B1	$CBrF_3$	1000	800,000
HBCC-30B1	CH_2BrCl	200	28,800[e]
HBFC-22B1	$CHBrF_2$		108,000
CC-10	CCl_4	10	8,000
CFC-11	CCl_3F	1000	26,200
CFC-12	CCl_2F_2	1000	800,000
CFC-13	$CClF_3$	1000	>600,000[f]
HCC-20	$CHCl_3$	10[g]	8,861
HCFC-21	$CHCl_2F$	10	49,900
HCFC-22	$CHClF_2$	1000	220,000
HFC-23	CHF_3	1000[h]	>663,000
HCC-30	CH_2Cl_2	100	22,669[i]
CFC-113	CCl_2FCF_2Cl	1000	52,500
CFC-114	CF_2ClCF_2Cl	1000	720,000[i]
CFC-115	CF_3CF_2Cl	1000	>800,000
HCFC-123	CF_3CHCl_2	10[h]	32,000
HCFC-124	CF_3CHFCl	500[h]	>230,000
HFC-125	CF_3CF_2H	1000[h]	>709,000
HFC-134a	CF_3CH_2F	1000[h]	567,000
HCFC-141b	CH_3CFCl_2	500[h]	61,647
HCFC-142b	CH_3CF_2Cl	1000[h]	128,000
HFC-152a	CH_3CHF_2	1000[h]	383,000
HCFC-225ca	$CF_3CF_2CHCl_2$		31,000[j]
HCFC-225cb	CF_2ClCF_2CHFCl		31,000[j]
HFC-227ea	CF_3CHFCF_3		>800,000[k]
PFC-31-10	$CF_3CF_2CF_2CF_3$		>800,000[l]

[a] Refs. 7 and 27.
[b] Except for CO_2, no compound has a higher TLV than 1000 ppm.
[c] Approximate lethal concentration, inhalation by rats, 4 h exposure unless noted otherwise.
[d] 15 min exposure, mouse.
[e] 15 min exposure.
[f] 2 h exposure.
[g] Suspected carcinogen in humans.
[h] Du Pont Allowable Exposure Limit.
[i] 30 min exposure, mouse.
[j] Ref. 28.
[k] Unpublished results, Great Lakes Chemical Co.
[l] Unpublished results, 3M, Inc.

(31). These chlorine atoms then participated in an ozone destruction cycle, and it has been estimated that one chlorine atom destroys 10,000 ozone molecules before getting trapped as inactive HCl (32). Bromine also participates in a destruction cycle, but it is about 10 times more efficient than chlorine in destroying ozone (33).

The possibility that CFCs and Halons can deplete the earth's ozone layer has had a significant impact on the fluorochemicals industry. Also, CFCs have been cited as contributors to global warming owing to their absorption of infrared irradiation and long atmospheric lifetimes (34). Because aerosol products containing CFC-11 and CFC-12 had accounted for the biggest release of CFCs, their manufacture for this use was banned in 1978. The discovery of the Antarctic ozone hole in 1985 (35) has prompted the world community to take much more extensive action toward halting the release of CFCs.

The Montreal Protocol. In response to the growing scientific consensus that CFCs and Halons would eventually deplete the ozone layer, the United Nations Environmental Programme (UNEP) began negotiations in 1981 aimed at protecting the ozone layer. In March 1985, the Vienna Convention for the Protection of the Ozone Layer was convened and provided a framework for international cooperation in research, environmental monitoring, and information exchange. In September of 1987, the Montreal Protocol on Substances that Deplete the Ozone Layer was signed by 24 nations and took force on January 1, 1989. This treaty called for (*1*) limiting production of specified CFCs, including 11, 12, 113, 114, and 115, to 50% of 1986 levels by 1998, (*2*) freezing production of specified Halons 1211, 1301, and 2402 at 1986 levels starting in 1992, and (*3*) convening the signatories yearly to reevaluate and update the Protocol articles in light of recent developments. By 1988 the Ozone Trends Panel issued a report based on new scientific evidence that conclusively linked CFCs to ozone depletion in the stratosphere. In June of 1990, the parties to the Montreal Protocol met in London and amended the Protocol to strengthen the controls on ozone depleting chemicals, expand the list of chemicals to include carbon tetrachloride and 1,1,1-trichloroethane, and specify stepped-up timetables for total phaseout of ozone depleting chemicals by the year 2000.

In April of 1991, the U.S. National Aeronautics and Space Administration concluded that ozone depletion was occurring even faster than had been estimated, and at the third meeting of the parties to the Montreal Protocol in June of 1991, an earlier phaseout of controlled substances was proposed. An assessment of the technical and economic consequences of a 1997 phaseout is currently underway, and further acceleration of the phaseout schedule to as soon as 1995 seems likely. Many countries already have unilaterally banned or curbed the use of controlled substances well ahead of the Montreal Protocol timetable. As of early July 1992, there were 81 parties to the Protocol.

Chlorofluorocarbon Alternatives

Properties. The ideal substitute should have identical or better performance properties than the CFC it replaces. The ideal CFC substitute must not harm the ozone layer, and must have a short atmospheric lifetime to ensure a low greenhouse warming potential (GWP). It also must be nontoxic, nonflammable, thermally and chemically stable under normal use conditions, and manufacturable at a reasonable price. The chemical industry has found substitutes that match many but not all of these criteria.

The general strategy has been to incorporate at least one hydrogen atom in the proposed CFC substitute's structure which provides a means for its destruction via hydrogen atom abstraction by tropospheric hydroxyl radicals. The haloalkylradicals thus formed are then rapidly degraded to acids and CO_2, which are both removed from the atmosphere by natural processes. Since fluorine does not participate in the ozone destruction cycle, a substitute composed of only hydrogen, fluorine, and carbon would be ideal, but HFCs for every application have not yet been identified. In some applications, HCFCs and even PFCs have been suggested as transitional replacements to accelerate phaseout of the much more harmful CFCs. Trade-offs will likely be required in most applications, and the alternatives that have been identified for the various markets are listed in Table 8.

The physical and environmental properties of the leading commercial CFCs and their proposed substitutes are compared in Tables 9 and 10. The HCFCs have relatively small but non-zero ozone depletion potentials (ODP) and low global warming potentials (GWP). Recent results indicate even these values may be 15% too high (37). The HFCs have zero ODPs and low-to-moderate GWPs. Perfluorocarbons also have zero ODPs, but very large GWPs.

Manufacture. The manufacture of CFC alternatives is a far more complex challenge than production of the CFCs themselves (38). The very design feature

Table 8. Alternatives to CFCs

CFC	Application	Near-term substitute	Long-term substitute
CFC-11	blowing agents and refrigerants	HCFC-123 HCFC-22 HCFC-141b HCFC-142b	HFCs HFC-152a blends
CFC-12	refrigerants	HFC-134a HCFC-22	HFC-134a HFC-152a blends
CFC-113	cleaning agents	blends/azeotropes HCFC-225ca/cb	HFCs
CFC-114	blowing agents and refrigerants	HCFC-124 HCFC-142b blends/azeotropes	HFCs
CFC-115	refrigerants	HFC-125 blends/azeotropes	HFC-125
H-1301	fire extinguishant	HFC-23 HFC-125 PFC-31-10	HFC-23 HFC-227ea
H-1211	fire extinguishant	HCFC-123 HBFC-22B1 HBFC-124B1 PFC-51-14	HFCs

Table 9. Physical Property Comparisons of CFCs and Their HCFC or HFC Substitutes

Property	CFC CFC-11	Substitutes HCFC-123	HCFC-22	HCFC-141b	HCFC-142b	CFC CFC-113	Substitutes HCFC-225ca[a]	HCFC-225cb[a]
molecular weight	137.37	152.9	86.47	116.95	100.47	187.38	202.94	202.94
boiling point, °C	23.8	27.9	−40.7	32	−9.8	47.6	51.1	56.1
freezing point, °C	−111.1	−107	−157.4	−103.5	−130.8	−35.0	−94	−97
critical temperature, °C	198.0	185	96.0	210	137.1	214.2		
critical pressure, kPa[b]	4408	3789	4977	4641	4123	3415		
viscosity,[c] gas, mPa·s(=cP)	0.0105_{24}	0.0136_{60}	0.0122_0	0.0129_{60}	0.01099_{25}	0.0108_{49}		
viscosity,[c] liquid, mPa·s(=cP)	0.43_{20}	0.449_{25}		0.409_{25}	$0.453_{-20.9}$	$0.497_{48.9}$	0.58_{25}	0.60_{25}
surface tension,[c] mN/m(=dyn/cm)	18_{25}		8_{25}			17.3_{25}	15.8_{25}	16.7_{25}
heat capacity,[c] kJ/(kg·K)[d]								
liquid	0.870_{25}	1.0174_{25}	1.0962_{-40}	1.1556_{25}	1.2979_{25}	0.912_{25}		
vapor at 101.3 kPa[b]	0.565_{25}	0.7201_{25}		0.7913_{25}	0.8792_{25}	0.674_{60}		
latent heat of vaporization, kJ/kg[d]	180.3	174.17	233.84	223.15	223.15	146.73		
solubility in water at 101.3 kPa,[b] wt %	0.11_{25}	0.39	0.30	0.021	0.14	0.017		
atmospheric lifetime, years[e]	55	1.71	15.8	10.8	22	110	2.8	8.0
ozone depletion potential (ODP)[e]	1.00	0.02	0.055	0.11	0.065	1.07	0.025	0.033
global warming potential (GWP)[e]	1.00	0.02	0.39	0.14	0.44	1.80		

[a]Ref. 25.
[b]To convert kPa to atm, divide by 101.3.
[c]At the temperature (°C) indicated by subscripts.
[d]To convert J to cal, divide by 4.184.
[e]Ref. 36.

Table 10. Environmental Property Comparisons of CFCs and Substitutes

Property	CFC CFC-12	Substitute HFC-134a	CFC CFC-114	Substitutes HCFC-124	Substitutes HCFC-142b	CFC CFC-115	Substitute HFC-125
viscosity,[a] gas, mPa·s (=cP)	$0.0117_{4.4}$	0.0152_{60}	0.0118_{25}	0.0138_{60}	0.01099_{25}	0.0125_{25}	0.015_{25}
viscosity,[a] liquid, mPa·s (=cP)	0.398	0.205_{25}	0.485_{0}	0.0314_{25}	$0.453_{-20.9}$	0.193_{25}	0.104_{25}
heat capacity,[a] kJ/(kg·K)[b]							
liquid	0.971_{25}	1.428_{25}	1.016_{25}	1.130_{25}	1.298_{25}	1.192_{25}	1.260_{25}
vapor at 101.3 kPa[c]	0.607_{25}	0.854_{25}	0.711_{25}	0.741_{25}	0.879_{25}	0.686_{25}	0.707_{25}
latent heat of vaporization, kJ/kg[b]	165.1	219.8	136.0	167.9	223.2	126.0	159.0
solubility in water[a] at 101.3 kPa,[c] wt %	0.028_{25}	0.15_{25}	0.013_{25}	1.71_{24}	0.14	0.006_{25}	0.09_{25}
atmospheric lifetime, years[d]	116	15.6	220	7.0	22.4	550	40.5
ozone depletion potential (ODP)[d]	1.00	0.00	0.8	0.022	0.065	0.52	0.00
global warming potential (GWP)[d]	2.90	0.29	4.1	0.11	0.44	6.1	0.86

[a] At the temperature (°C) indicated by subscripts.
[b] To convert J to cal, divide by 4.184.
[c] To convert kPa to atm, divide by 101.3.
[d] Ref. 36.

which makes the alternatives tropospherically labile, the hydrogen atom substituent, also significantly complicates their manufacture because of potential by-product formation or catalyst inactivation. At least a dozen different routes to HFC-134a have been identified, but a simple, single-step process is very unlikely (39). A two-step process that has been commercialized first involves reaction of trichloroethylene with HF in the vapor or liquid phase to form HCFC-133a (40,41), which is then separated and reacts again with HF to form HFC-134a.

$$CCl_2{=}CHCl + 3\ HF \rightarrow CF_3CH_2Cl + 2\ HCl$$

$$CF_3CH_2Cl + \text{excess HF} \rightarrow CF_3CH_2F + HCl$$

The HCFC-123 alternative to CFC-11 is made by the fluorination of tetrachloroethylene with either liquid or gaseous HF (41). Further reaction of HCFC-123 with HF provides the HCFC-124 and HFC-125 alternatives.

$$CCl_2{=}CCl_2 + 3\ HF \longrightarrow CF_3CHCl_2 + 2\ HCl$$

$$CF_3CHCl_2 \xrightarrow{HF} CF_3CHClF + CF_3CHF_2$$

The HCFC-225 isomers designed to replace CFC-113 are manufactured by Lewis acid promoted addition of HCFC-21 to tetrafluoroethylene (25,42).

$$CHCl_2F + CF_2{=}CF_2 \xrightarrow{AlX_3} CF_3CF_2CHCl_2 + CClF_2CF_2CHClF$$

Since HFC-134a likely will be the single largest volume CFC alternative produced, many manufacturers around the world are in the process of or have plans to commercialize it, each under their own trade name. United States and foreign trademarks and manufacturers of CFC alternatives are listed in Table 11.

Economic Aspects. Manufacturing facilities for CFC alternatives are just now coming on line. The size of the markets for the alternatives is estimated to be quite large (several thousand t/yr), but it will not be as large as the prior markets for CFCs themselves. This is largely because of the higher cost of the alternatives, typically 3–5 times that of the incumbents. Low value-in-use applications which cannot support the cost of the alternatives will disappear or will switch to not-in-kind alternatives such as hydrocarbons for foam blowing.

Health and Safety Factors. The toxicity of CFC alternatives is the subject of intense study. Fifteen fluorocarbon producers have formed the Program for Alternative Fluorocarbon Toxicity testing (PAFT) to share the costs associated with determining safe operating and handling procedures for the proposed CFC alternatives. Long-term chronic toxicity studies are still underway and results to date generally look encouraging, although prolonged exposure to HCFC-123 produced benign tumors in rats (43). The approximate lethal concentrations (ALC) are shown in Table 7.

Hydrofluorocarbons generally are less toxic than HCFCs, with the notable exception of HFC-152, CH_2FCH_2F, which apparently can be metabolically converted to monofluoroacetic acid and is therefore quite toxic (44).

Table 11. Trademarks and Manufacturers of CFC Alternatives

Country and trademark	Manufacturer
France	
Forane	Elf Atochem
Belgium	
Solkane	Solvay
Germany	
Frigen	Farbwerke Hoechst
Fridohna	Hüls
Italy	
Algogrene	Montedison
Japan	
Asahiflon	Asahi Glass
Daiflon	Daikin Kogyo
Suva	Du Pont Mitsui Fluorochemicals
United Kingdom	
Klea	Imperial Chemical Industries
Isceon	Rhône-Poulenc
United States	
Genetron	AlliedSignal Corp.
Suva	Du Pont
FM	Great Lakes Chemical Co.
3M Brand	Minnesota Mining and Manufacturing Co.

Fluorocarbons Containing Other Halogens

Properties. The physical and chemical properties of bromo- and iodofluorocarbons are similar to those of the chlorofluorocarbons except for higher densities and generally decreased stability. The stability of these compounds decreases as the ratio of bromine or iodine to fluorine increases. The reactivity of carbon–halogen bonds toward exchange by fluorine or homolytic cleavage increases in the order C–Cl, C–Br, C–I. Iodofluorocarbons and most bromofluorocarbons readily lose iodine or bromine radicals under photolytic, thermal, or radical initiation to give the corresponding carbon-centered radical. The physical properties of several examples are shown in Tables 12 and 13.

Manufacture. Brominated fluoromethanes are prepared industrially by the halogen exchange of tetrabromomethane or by the bromination of CH_2F_2 or CHF_3 at elevated temperatures (3). Other bromo- or iodofluorocarbons can be prepared by halogenating suitable fluorocarbons, including fluoroolefins, or by halogen exchange of perfluoroiodocarbons (47).

$$CF_3CH_2Cl + Br_2 \xrightarrow{500°C} CF_3CHBrCl$$

$$CI_4 + IF_5 \longrightarrow CF_3I$$

$$CF_2{=}CF_2 + I_2 \longrightarrow ICF_2CF_2I$$

$$5\ CF_2{=}CF_2 + IF_5 + 2\ I_2 \longrightarrow 5\ CF_3CF_2I$$

$$CF_3(CF_2)_6CF_2I + Br_2 \xrightarrow{100-290°C} CF_3(CF_2)_6CF_2Br$$

Table 12. Physical Properties of Aliphatic Hydrobromofluorocarbons (HBFCs) and Hydroiodofluorocarbons (HIFCs)[a]

B(I)FC number	Formula	CAS Registry Number	Molecular weight	Boiling point, °C	Melting point, °C	Liquid density, g/mL at °C	Refractive index, n_D at °C
11B1	$CBrCl_2F$	[353-58-2]	181.82	52	−106	1.9317_{20}	1.4304_{20}
11B2	CBr_2ClF	[353-55-9]	226.28	80		2.3172_{20}	1.4750_{20}
11B3	CBr_3F	[353-54-8]	270.74	106	−75	2.765_{20}	1.5256_{20}
12B1	$CBrClF_2$	[353-59-3]	165.37	−3.9	−161	1.850_{15}	
12B2	CBr_2F_2	[75-61-6]	209.83	24.5	−110	2.306_{15}	
13B1	$CBrF_3$	[75-63-8]	148.92	−57.8	−168	1.538_{25}	1.238_{25}
21B1	$CHBrClF$	[593-98-6]	147.38	−45	36.1	1.977_0	1.4144_{25}
21B2	$CHBr_2F$	[1868-53-7]	191.84	64.9	26.5	2.421_{20}	1.4685_{20}
22B1	$CHBrF_2$	[1511-62-2]	130.92	−15.5	−145	1.825_{20}	
113aB1	$CBrCl_2CF_3$	[354-50-7]	231.83	69.2		1.950_{20}	1.3977_{20}
114B2	$CBrF_2CBrF_2$	[124-73-2]	259.85	47.3	−110	2.163_{25}	1.367_{25}
115B1	$CBrF_2CF_3$	[354-55-2]	198.92	−21		1.810_0	$1.2966_{29.8}$
123B1	$CHBrClCF_3$	[151-67-7]	197.39	50.2		1.860_{20}	1.3700_{20}
123aB1a	$CHClFCBrF_2$	[354-06-3]	197.39	52.5		1.864_{25}	1.3685_{25}
124B1	CF_3CHFBr[b]	[124-72-1]	180.94	8.6	−80	1.85_{20}	
124aB1	CF_2BrCF_2H	[354-07-4]	180.94	10.8		1.900_{15}	1.321_{15}
132bB1a	$CH_2BrCClF_2$	[421-01-2]	179.40	68.4	−76	1.830_{20}	1.4018_{20}
133aB1	CF_3CH_2Br	[421-06-7]	162.94	26	−94	1.7881_{20}	1.3331_{20}
142B1	CH_2BrCHF_2	[359-07-9]	144.95	57.3	−75	$1.824_{18.5}$	$1.3940_{10.5}$
11I1	CCl_2FI	[420-48-4]	228.82	90.0	−107	2.313_{20}	1.510_{20}
13I1	CF_3I	[2314-97-8]	195.91	−22.5		2.361_{-32}	1.379_{-42}
21I2	$CHFI_2$	[1493-01-2]	285.83	100.3	−34.5	3.197_{22}	
22I1	CHF_2I	[1493-03-4]	177.92	21.6	−122	3.238_{-19}	
31I1	CH_2FI	[373-53-5]	159.93	53.4		2.366_{20}	1.491_{20}
113I1	$CClFICClF_2$	[354-61-0]	278.83	99		2.196_{25}	1.447_{25}
133aI1	CH_2ICF_3	[353-83-3]	209.94	55		2.142_{25}	1.3981_{25}

[a]Refs. 8 and 9. [b]Ref. 45.

Table 13. Physical Properties of Telomer Iodides[a]

Formula	CAS Registry Number	Molecular weight	Boiling point, °C	Melting point, °C	Liquid density, g/mL at °C	Refractive index, n_D at °C
CF_3CF_2I	[354-64-3]	245.92	13		2.072_{28}	$1.3378_{0.5}$
$CF_3(CF_2)_3I$	[423-39-2]	345.91	67		2.0424_{25}	1.3258_{25}
$CF_3(CF_2)_5I$	[355-43-1]	445.92	118	−46	$2.028_{26.4}$	1.3220_{20}
$CF_3(CF_2)_7I$	[507-63-1]	545.96	163	20.8	2.008_{25}	
$CF_3(CF_2)_9I$	[423-62-1]	645.98	195	65.5	1.9400_{70}	$1.3350_{25.5}$
$CF_3(CF_2)_2I$	[754-34-7]	295.93	41.2	−95.0	2.0026_{20}	1.3281_{20}
$CF_3(CF_2)_4I$	[638-79-9]	395.94	94.4	−50.0	$2.0349_{27.8}$	$1.3389_{0.5}$
$CF_3(CF_2)_6I$	[335-58-0]	495.96	137.5			1.3230_{30}
$CF_3(CF_2)_8I$	[558-97-4]	595.97	181			
$I(CF_2)_2I$	[354-65-4]	353.82	112		2.629_{25}	1.4895_{25}
$I(CF_2)_4I$	[375-50-8]	453.82	150	−9.0	2.4739_{27}	$1.4273_{26.2}$

[a]Refs. 9 and 46.

The higher molecular weight perfluoroalkyl iodides are prepared by telomerization of tetrafluoroethylene with lower molecular weight perfluoroalkyl iodides (46,48).

$$CF_3CF_2I + m\ CF_2{=}CF_2 \xrightarrow{\text{cat. } SbF_5/IF_5} CF_3CF_2(CF_2CF_2)_mI$$

$$ICF_2CF_2I + n\ CF_2{=}CF_2 + \text{heat} \longrightarrow ICF_2CF_2(CF_2CF_2)_nI$$

Health and Safety Factors. Fluorocarbons containing bromine or iodine are more toxic than the corresponding chloro compounds. When the ratio of the fluorine to other halogens is high, the toxicity can be quite low, especially for bromofluorocarbons. Perfluoro-1-bromooctane [423-55-2] has an LD_{50} of greater than 64 mL/kg when administered into the gastrointestinal tract, and has little effect when instilled into the lungs (49). Other examples are included in Table 7.

Uses. The most important industrial products of this class have been the fire-extinguishing agents $CBrClF_2$ and $CBrF_3$. The latter is considerably more effective than CO_2, and is nontoxic as well as its decomposition products (50). It is used in commercial aircraft for in-flight engine fires and in portable fire extinguishers for both military and civilian markets. Both of these Halons have very high ozone depletion potentials, and their production is scheduled to be completely phased out by the year 1994 (51). Suitable replacements are being sought, and the properties of some leading candidates are compared in Table 14.

Halothane, $CF_3CHClBr$, was a widely used anesthetic which has the advantages of nonflammability, high anesthetizing power, and general lack of postnarcotic effects, but it has lost its leading market share to the fluorinated ethers enflurane, $CHClFCF_2OCHF_2$, and isofluorane, $CF_3CHClOCHF_2$ (52,53) (see ANESTHETICS).

Perfluorooctyl bromide [423-55-2], which has one of the highest oxygen-dissolving capabilities among fluorinated liquids and is readily emulsified, shows great promise in various medical applications such as tissue oxygenation, chemotherapy, and radiographic imaging (20,21,49).

Table 14. Physical Property Comparisons of BFCs and Their Substitutes

Property	BFC H-1301	Substitutes HFC-23	Substitutes HFC-227ea	Substitutes PFC-31-10	BFC H-1211	HCFC-123	Substitutes HBFC-22B1	Substitutes PFC-51-14
molecular weight	148.91	70.01	170.03	238.03	165.4	152.9	130.92	338.0
boiling point, °C	−57.8	−82.0	−16.4	−2.0	−4	27.9	−15.5	56
freezing point, °C	−168.0	−155.2	−131	−128.2	−160.5	−107	−145	−90
critical temperature, °C	67.0	25.7	101.7	113.2	153.8	185	138.8	178.0
critical pressure, kPa[a]	3964	4810	2909	2323	4104	3789	5129	1835
critical density, kg/cm³	745	525	621	630	713			
viscosity,[b] gas, mPa·s	0.0016_{25}	0.0144_{25}	0.0132_{25}		0.0138_{25}	0.0136_{60}	0.0153_{25}	
viscosity,[b] liquid, mPa·s	0.32_{-40}	0.368_{-80}	0.184_{25}	0.607_{25}	0.292_{25}	0.449_{25}	0.269_{25}	0.700_{25}
heat capacity, kJ/(kg·K)[c]								
liquid	0.828_0	1.211_{-100}	1.102_{25}	1.045_{25}	0.742_{25}	1.017	0.814_{25}	
vapor at 101.3 kPa[d]	0.474_{25}	0.736_{25}	0.777_{25}	0.805_{25}	0.474_{25}	0.720	0.478_{25}	
latent heat of vaporization, kJ/kg[c]	118.7	251.2_{-100}	132.7	96.3	149.35	174.17	172.08	88.4
relative dielectric strength, $N_2 = 1$	1.83	1.04		5.25				
solubility in water[b] at 101.3 kPa, wt %	0.03_{25}	0.10_{25}		0.001_{25}		0.39_{25}		0.001_{25}
atmospheric lifetime, years[e]	67	310[f]		>500	19	1.71	5.6	>500
ozone depletion potential (ODP)[e]	16.0	0.00	0.00	0.00	4.0	0.02	1.40	0.00
global warming potential (GWP)[e]	1.6	8	>8	>8		0.02		>8

[a] To convert kPa to atm, divide by 101.3. [b] At the temperature (°C) indicated by subscripts. [c] To convert J to cal, divide by 4.184. [d] 101.3 kPa = 1 atm. [e] Ref. 36. [f] Estimated value.

The use of α,ω-diiodoperfluoroalkanes as chain-transfer agents in the manufacture of fluoroelastomers (54) is the only direct commercial application of iodofluorocarbons, although several telomer iodides, such as $CF_3(CF_2)_7I$ [507-63-1] and $CF_3(CF_2)_7CH_2CH_2I$ [2043-53-0], are intermediates in the manufacture of oil and water repellants, surfactants, and fire-extinguishing foams (55). The alcohols derived from perfluoroalkane carboxylic acids, eg, $CF_3(CF_2)_nCH_2OH$, or from the telomer iodides, eg, $CF_3(CF_2)_nCH_2CH_2OH$, are used to prepare fluorinated acrylate esters. Aqueous dispersions of the fluoroacrylate polymers are used as soil, water, and oil repellents for fabrics (56) (see WATERPROOFING). They are sold under the Scotchgard (3M) and Zepel or Zonyl (Du Pont) trade names.

BIBLIOGRAPHY

"Aliphatic Fluorinated Hydrocarbons" under "Fluorine Compounds, Organic" in *ECT* 1st ed., Vol. 6, pp. 752–757, by A. F. Benning, E. I. du Pont de Nemours & Co., Inc., and L. J. Hals and W. H. Pearlson, Minnesota Mining & Manufacturing Co.; "Fluorinated Hydrocarbons" under "Fluorine Compounds, Organic" in *ECT* 2nd ed., Vol. 9, pp. 739–751, by R. C. Downing, E. I. du Pont de Nemours & Co., Inc.; "Fluorinated Aliphatic Compounds" under "Fluorine Compounds, Organic" in *ECT* 3rd ed., Vol. 10, pp. 856–870, by B. E. Smart, E. I. du Pont de Nemours & Co., Inc.

1. *Refrig. Eng.* **65**, 49 (1957), ASRE Standard 34.
2. A. Maciejewski, *J. Photochem. Photobio., A: Chemistry* **51**, 87 (1990).
3. M. Hudlicky, *Chemistry of Organic Fluorine Compounds*, 2nd ed., Ellis Horwood Ltd., Chichester, UK, 1976.
4. H. G. Bryce, in J. H. Simons, ed., *Fluorine Chemistry*, Vol. 5, Academic Press, Inc., New York, 1965, pp. 297–492.
5. T. M. Reed, in Ref. 4, pp. 133–236.
6. B. D. Joyner, *J. Fluorine Chem.* **33**, 337 (1986).
7. *Freon Fluorocarbons, B-2*, E. I. du Pont de Nemours & Co., Inc., Wilmington, Del., 1969.
8. *Selected Values of Properties of Chemical Compounds*, Thermodynamics Research Center Data Project, Texas Engineering Experimental Sta., Texas A&M University, College Station, 1977.
9. A. M. Lovelace, D. A. Rausch, and W. Postelnek, *Aliphatic Fluorine Compounds*, Reinhold, Inc., New York, 1958.
10. B. E. Smart, in J. F. Liebman and A. Greenberg, eds., *Molecular Structure and Energetics*, VCH Publishers, Inc., Deerfield Beach, Fl., 1986, pp. 141–191.
11. R. E. Banks, *Fluorocarbons and Their Derivatives*, Macdonald, Ltd., London, 1970.
12. G. G. Furin, *Sov. Sci. Rev. B. Chem.* **16**, 1 (1991).
13. N. Watanabe, T. Nakajima, and H. Touhara, *Graphite Fluorides*, Elsevier, Ltd., Oxford, 1988.
14. W. V. Childs and co-workers, in C. H. Lund and M. M. Baizer, eds., *Organic Electrochemistry*, 3rd ed., Marcel Dekker, Inc., New York, 1991, Chap. 6, p. 1103.
15. A. K. Barbour, L. J. Belf, and M. W. Buxton, *Adv. Fluorine Chem.* **3**, 181 (1963).
16. J. W. Clayton, Jr., *Fluorine Chem. Rev.* **1**, 197 (1967).
17. C. M. Melliar-Smith and C. J. Mogab, in J. L. Vossen and W. Kern, eds., *Thin Film Processes*, Academic Press, Inc., New York, 1978.
18. D. S. L. Slinn and S. W. Green, in R. E. Banks, ed., *Preparation, Properties, and Industrial Applications of Organofluorine Compounds*, Ellis Horwood, Chichester, UK, 1982, Chap. 2, pp. 45–82.

19. *Flutec*, Rhône-Poulenc, Inc., RTZ Chemicals, ISC Division, Princeton, N.J., 1989.
20. K. C. Lowe, *Adv. Mater.*, 87 (1991).
21. T. M. S. Chang and R. P. Geyer, eds., *Blood Substitutes*, Marcel Dekker, Inc., New York, 1989.
22. L. A. Wall, ed., *Fluoropolymers*, John Wiley & Sons, Inc., New York, 1972.
23. S. V. Gangal, in J. I. Kroschwitz, ed., *Encyclopedia of Polymer Science and Engineering*, Vol. 16, John Wiley & Sons, Inc., New York, 1989, pp. 577–642.
24. A. L. Logothetis, *Prog. Polym. Sci.* **14**, 251 (1989).
25. M. Yamabe, "HCFC-225s as CFC-113 Substitutes" in *Symposium on Progress on the Development and Use of Chlorofluorocarbon (CFC) Alternatives, 200th ACS National Meeting*, Abstract No. 22, Washington, D.C., Aug. 28, 1990.
26. United Nations Environment Programme (UNEP), *Report of the Technology and Economic Assessment Panel*, Dec. 1991.
27. *Threshold Limit Values for Chemical Substances and Physical Agents and Biological Exposure Indices*, American Conference of Governmental Industrial Hygienists, Cincinnati, Ohio, 1991.
28. S. R. Frame, M. C. Carakostas, and D. B. Warheit, *Fundam. Appl. Toxicol.* **18**, 590 (1992).
29. *Du Pont Toxicology Information System*, Vol. 1.02, Du Pont Co., 1988.
30. J. E. Lovelock, *Nature* **230**, 379 (1971).
31. F. S. Rowland and M. J. Molina, *Nature* **249**, 810 (1974).
32. L. T. Molina and M. J. Molina, *J. Phys. Chem.* **91**, 433, (1987).
33. M. B. McElroy and co-workers, *Nature* **321**, 759 (1986).
34. A. Lacis and co-workers, *Geophys. Res. Lett.* **8**, 1035 (1981).
35. J. C. Farman, B. G. Gardiner, and J. D. Shanklin, *Nature* **315**, 207 (1985).
36. *Scientific Assessment of Ozone Depletion: 1991*, Report No. 25, World Meteorological Organization, Global Ozone Research and Monitoring Project, Geneva, 1991.
37. R. K. Talukar and co-workers, *Science* **257**, 227 (1992).
38. L. E. Manzer, *Science* **249**, 31 (1990).
39. L. E. Manzer, *Catalysis Today*, **13**, 13 (1992).
40. E. Chynowyth, *European Chem. News*, 8 (Apr. 17, 1991); 21 (July 15, 1991).
41. G. W. Parshall and S. D. Ittel, *Homogeneous Catalysis*, 2nd ed., John Wiley & Sons, Inc., New York, 1992, pp. 305–308.
42. Eur. Pat. Appl. EP 456,841 (Nov. 21, 1991), K. Ohnishi and co-workers (to Asahi Glass).
43. H. J. Trochimowicz, *Toxicol. Lett.* **68**, 25 (1993).
44. P. H. Lieder and D. A. Keller, *Chem. Eng. News* **70**, 2 (1992).
45. C. N. Fletcher, P. Jones, and M. Winterton, "Clean Agent Fire Extinguishant: Break-Down Products," *1990 International Conference on CFC and Halon Alternatives*, Baltimore, Md., 1990.
46. U. S. Pat. 3,234,294 (Feb. 8, 1966) and 3,132,185 (May 5, 1964), R. E. Parsons (to Du Pont).
47. Ger. Offen. DE 4,116,361 (Jan. 2, 1992), B. Felix and H. Katezenberger (to Hoechst AG).
48. C. D. Bedford and K. Baum, *J. Org. Chem.* **45**, 347 (1980).
49. D. M. Long and co-workers, in R. Filler, ed., *Biochemistry Involving Carbon–Fluoride Bonds*, American Chemical Society, Washington, D.C., 1976, pp. 171–189.
50. *Du Pont Freon FE 1301 Fire Extinguishing Agent*, E. I. du Pont de Nemours & Co., Inc., Wilmington, Del., 1969.
51. *Handbook for the Montreal Protocol on Substances that Deplete the Ozone Layer*, 3rd ed., Ozone Secretariat, United Nations Environmental Program, Nairobi, 1993.
52. D. Noble and L. Martin, *Anaesthesia* **45**, 339 (1990).
53. J. Tarpley and P. Lawler, *Anaesthesia* **44**, 596 (1989).

54. M. Oka and M. Tatemoto, *Contemporary Topics in Polymer Science*, Vol. 4, Plenum Press, New York, 1984, p. 763.
55. H. C. Fielding, in R. E. Banks, ed., *Organofluorine Chemicals and their Industrial Applications*, Ellis Horwood Ltd., Chichester, U.K., 1979, pp. 214–234.
56. M. J. Owen, in J. I. Kroschwitz, ed., *Encyclopedia of Polymer Science and Engineering*, Vol. 14, John Wiley & Sons, Inc., New York, 1988, pp. 411–421.

BRUCE E. SMART
RICHARD E. FERNANDEZ
E. I. du Pont de Nemours & Co., Inc.

FLUOROETHANOLS

Ethanol may be readily fluorinated at C-2. Replacement of H by F at C-1 would lead to unstable compounds which readily form carbonyls by loss of HF.

Monofluoro Derivative

2-Fluoroethanol [371-62-0] (ethylene fluorohydrin, β-fluoroethyl alcohol), FCH_2CH_2OH, is a colorless liquid with an alcohol-like odor; mp, $-26.45°C$; bp, $103.55°C$; d_4, 1.1297; n_D^{18}, 1.13647; heat of combustion, -1214.0 kJ/mol (-290.16 kcal/mol) (1,2). It is miscible with water, stable to distillation, and low in flammability. It is the least acidic of the fluoroethanols, although more acidic than ordinary alcohols with a pK_a value of 14.42 ± 0.04 in aqueous solution (3). Its most notable difference from the other fluoroethanols is its extreme toxicity (4–12). In mice an LD_{50} of 10 mg/kg has been measured (10–12). The toxicity is due to its facile oxidation in animals to derivatives of fluoroacetic acid [144-49-0], a known inhibitor of the tricarboxylic acid cycle of respiration. No effective antidote to this poisoning is known, although ethanol appears to diminish the lethal effects of 2-fluoroethanol in rats and monkeys (9).

In its chemical reactions, 2-fluoroethanol behaves like a typical alcohol. Oxidation (12) yields fluoroacetaldehyde [1544-46-3] or fluoroacetic acid; reaction with phosphorus tribromide (12) gives 1-bromo-2-fluoroethane [762-49-2]; addition to olefins results in ethers (13); and additions to isocyanates give carbamates (14). The alcohol can be prepared in 50% yield by the reaction of potassium fluoride with 2-chloroethanol at 175°C in high boiling glycol solvents (15). Alternatively, the addition of hydrogen fluoride to ethylene oxide (16) or the fluorination of 2-bromoethyl or 2-chloroethyl acetate with silver, potassium, or mercuric fluoride followed by hydrolysis gives the alcohol (1,2). 2-Fluoroethanol is not currently produced in commercial quantities, although (in 1992) it was available in research quantities for ca \$3/g.

Because of its high toxicity, special procedures should be followed by users of 2-fluoroethanol. Suggested precautions include working with it in sealed reactors at subatmospheric pressure and careful monitoring to ensure that contamination of the surroundings is minimized (17). Another potential hazard is the formation of the alcohol as a minor by-product in reactions such as those involving boron trifluoride and ethylene oxide (18). Despite these problems, several potential uses for the alcohol and its derivatives have been reported. The alcohol has been used to control rodent populations (19) and, when labeled with ^{18}F, as a radiodiagnostic agent (20). Various derivatives have shown promise as herbicides or as agents to control mites and other plant pests (14,21–24).

Difluoro Derivative

2,2-Difluoroethanol [359-13-7], F$_2$CHCH$_2$OH, is a colorless liquid with an alcohol-like odor; mp, 28.2°C, bp, 96°C; d_1^{17}, 1.3084; n_D^{17}, 1.3320; heat of combustion, −1026 kJ/mol (−245.3 kcal/mol). It is stable to distillation and miscible with water and many organic solvents. As expected, its acidity lies between that of 2-fluoroethanol and 2,2,2-trifluoroethanol both in the gas phase (25) and in 50% aqueous ethanol solution (26), where its K_a of 1.0×10^{-12} is about 4.8 times smaller than that of trifluoroethanol.

2,2-Difluoroethanol is prepared by the mercuric oxide catalyzed hydrolysis of 2-bromo-1,1-difluoroethane with carboxylic acid esters and alkali metal hydroxides in water (27). Its chemical reactions are similar to those of most alcohols. It can be oxidized to difluoroacetic acid [381-73-7] (28); it forms alkoxides with alkali and alkaline-earth metals (29); with alkoxides of other alcohols it forms mixed ethers such as 2,2-difluoroethyl methyl ether [461-57-4], bp 47°C, or 2,2-difluoroethyl ethyl ether [82907-09-3], bp 66°C (29). 2,2-Difluoroethyl difluoromethyl ether [32778-16-8], made from the alcohol and chlorodifluoromethane in aqueous base, has been investigated as an inhalation anesthetic (30,31) as have several ethers made by addition of the alcohol to various fluoroalkenes (32,33). Methacrylate esters of the alcohol are useful as a sheathing material for polymers in optical applications (34). The alcohol has also been reported to be useful as a working fluid in heat pumps (35). The alcohol is available in research quantities for ca $6/g (1992).

Trifluoroethanol

2,2,2-Trifluoroethanol [75-89-8], CF$_3$CH$_2$OH, is a colorless liquid with an ethanol-like odor; mp, −45°C; bp, 73.6°C; d_4^{25}, 1.3823; n_D^{20}, 1.2907; flash point (open cup), 41°C; flash point (closed cup), 33°C; no fire point (36); heat of combustion, −886.6 kJ/mol (−211.9 kcal/mol) (36); and dielectric constant (25°C), 26.14 (37). Many other physical and thermodynamic properties of the alcohol and its solutions have been published (36,38–41). It is the most acidic fluoroethanol with an ionization constant of 4.3×10^{-13} (42). It is stable to distillation and miscible with water and many organic solvents. It has the unusual property of dissolving most polyamides, both nylons (43) and polypeptides (44), at room temperature. Because of its excellent combination of physical and thermodynamic properties, 2,2,2-trifluoroethanol–water mixtures (also known as fluorinols) have application as working fluids in Rankine-cycle engines for recovering energy from waste heat sources (36,45,46). Its high ionizing power and low specific conductance make the alcohol useful as a solvent for ionic reactions and conductometric titrations (47), and basic research into solvolysis mechanisms (48–50).

Chemically, 2,2,2-trifluoroethanol behaves as a typical alcohol. It can be converted to trifluoroacetaldehyde [75-90-1] or trifluoroacetic acid [76-05-1] by various oxidizing agents such as aqueous chlorine solutions (51) or oxygen in the presence of a vanadium pentoxide catalyst (52). Under basic conditions, it adds to tetrafluoroethylene and acetylene to give, respectively, 1,1,2,2-tetrafluoroethyl 2′,2′,2′-trifluoroethyl ether [406-78-0] (53) and 2,2,2-trifluoroethyl vinyl ether [406-90-6] which was used as the inhalation anesthetic Fluroxene. Its alkoxides react with bromoethane to give trifluoroethyl ethyl ether [461-24-5], bp 50.3°C. Similarly prepared is bis(trifluoroethyl) ether used as the convulsant drug Flurothyl as a substitute for electric shock therapy. As the trichlorosulfonate ester, trifluoroethanol is used to introduce the trifluoroethyl group into the anxiolytic drug Halazepam [23092-17-3] (54). 2,2,2-Trifluoroethanol is also the starting material for the anesthetic Isoflurane (1-chloro-2,2,2-trifluoroethyl difluoromethyl ether [26675-46-7]) (55,56) and Desflurane (2-difluoromethoxy-1,1,1,2-tetrafluoroethane [57041-67-5]) (57).

Trifluoroethanol was first prepared by the catalytic reduction of trifluoroacetic anhydride [407-25-0] (58). Other methods include the catalytic hydrogenation of trifluoroacetamide [354-38-1] (59), the lithium aluminum hydride reduction of trifluoroacetyl chloride [354-32-5] (60) or of trifluoroacetic acid or its esters (61,62), and the acetolysis of 2-chloro-1,1,1-trifluoroethane [75-88-7] followed by hydrolysis (60). More recently, the hydrogenation of 2,2,2-trifluoroethyl trifluoroacetate [407-38-5] over a copper(II) oxide catalyst has been reported to give the alcohol in 95% yield (63).

The largest producer of trifluoroethanol is Halocarbon Products Corp. Other producers include Japan Halon and Rhône-Poulenc. Commercial quantities sell for approximately $20/kg (1992).

Toxicity studies on trifluoroethanol show acute oral LD_{50}, 240 mg/kg; acute dermal LD_{50}, 1680 mg/kg; and acute inhalation $L(ct)_{50}$, 4600 ppm·h. Long-term subchronic inhalation exposure to 50–150 ppm of the alcohol has caused testicular depression in male rats, but no effects were noted at the 10 ppm level (32). Although the significance of the latter observations for human safety is unknown, it is recommended that continuous exposure to greater than 5 ppm or skin contact with it be avoided.

BIBLIOGRAPHY

"Fluoroethanols" under "Fluorine Compounds, Organic," in *ECT* 1st ed., Vol. 6, pp. 760–762, by J. F. Nobis, Xavier University; in *ECT* 2nd ed., Vol. 9, pp. 751–752, by L. L. Ferstandig, Halocarbon Products Corp.; in *ECT* 3rd ed., Vol. 10, pp. 871–874, by G. Astrologes, Halocarbon Products Corp.

1. F. Swarts, *Rec. Trav. Chim.* **33**, 252 (1914).
2. F. Swarts, *J. Chem. Soc.* **106**, 475 (1914).
3. W. L. Mock and J. Z. Zhang, *Tetrahedron Lett.*, 5687 (1990).
4. E. Gryszkiewicz-Trochimowski, *Rec. Trav. Chim.* **66**, 427 (1947).
5. H. McCombie and B. C. Saunders, *Nature* **158**, 382 (1946).

6. E. V. Avdeeva and N. M. Dukel'skaya, *Vestn. Mosk. Univ. Ser. VI* **21**(4), 49 (1966).
7. F. R. Johannsen and C. O. Knowles, *Comp. Gen. Pharmacol.* **5**(1), 101 (1974).
8. E. O. Dillingham and co-workers, *J. Pharm. Sci.* **62**, 22 (1973).
9. D. I. Peterson, J. E. Peterson, and M. G. Hardinge, *J. Pharm. Pharmacol.* **20**, 465 (1968).
10. F. L. M. Pattison, *Toxic Aliphatic Fluorine Compounds*, Elsevier Publishing Co., New York, 1959, p. 65.
11. F. L. M. Pattison and co-workers, *J. Org. Chem.* **21**, 739 (1956).
12. B. C. Saunders, G. J. Stacey, and I. G. E. Wilding, *J. Chem. Soc.*, 773 (1949).
13. R. J. Koshar, T. C. Simmons, and F. W. Hoffmann, *J. Am. Chem. Soc.* **79**, 1741 (1957).
14. East Ger. Pat. 111,149 (Feb. 5, 1975), H. G. Werchan and co-workers.
15. F. W. Hoffmann, *J. Am. Chem. Soc.* **70**, 2596 (1948).
16. I. L. Knunyants, O. V. Kil'disheva, and I. P. Petrov, *J. Gen. Chem. USSR, Eng. Transl.* **19**, 95 (1949).
17. G. DiDrusco and F. Smai, *Quad. Ing. Chim. Ital.* **9**(11), 156 (1973).
18. C. T. Bedford, D. Blair, and D. E. Stevenson, *Nature* **267**, 335 (1977).
19. A. I. Kryl'tsov and co-workers, *Tr. Kaz. Nauch. Issled Inst. Zashch. Rast.* **11**, 171 (1972).
20. G. D. Robinson, Jr., *Radiopharm. Label Compounds, Proc. Symp.* **1**, 423 (1973).
21. U.S. Pat. 4,022,609 (May 10, 1977), D. E. Hardies and J. K. Rinehart (to PPG Industries, Inc.).
22. U.S. Pat. 4,960,884 (Oct. 2, 1990), D. M. Roush and co-workers (to FMC Corp.).
23. U.S. Pat. 3,852,464 (Dec. 3, 1974), D. E. Hardies and J. K. Rinehart (to PPG Industries, Inc.).
24. Fr. Pat. 1,604,978 (July 30, 1971), G. Rossi, G. Michieli, and P. Paolucci (Montecatini Edison S.p.A.).
25. J. H. J. Dawson and K. R. Jenning, *Int. J. Mass Spectrom. Ion Phys.* **25**(1), 47 (1977).
26. R. N. Haszeldine, *J. Chem. Soc.*, 1757 (1953).
27. Jpn. Kokai 62 273,925 (Nov. 28, 1987), T. Komatsu and Y. Asai (to Asahi Chemical Industry Co., Ltd.).
28. F. Swarts, *Chem. Zentr.* **II**, 709 (1903).
29. F. Swarts, *Bull. Soc. Chim. Belg.* **11**, 731 (1902).
30. U.S. Pat. 3,769,433 (Oct. 30, 1973), R. C. Terrell (to Airco, Inc.).
31. U.S. Pat. 3,896,178 (July 22, 1975), R. C. Terrell (to Airco, Inc.).
32. U.S. Pat. 3,746,769 (July 17, 1973), R. C. Terrell (to Airco, Inc.).
33. U.S. Pat. 3,862,240 (Jan. 21, 1975), R. C. Terrell (to Airco, Inc.).
34. Jpn. Kokai 63 066154 (Mar. 24, 1988), T. Ide and T. Komatsu (to Asahi Chemical Industry Co., Ltd.).
35. Jpn. Kokai 62 013481 (Jan. 22, 1987), M. Sagami and H. Matsuo (to Asahi Glass Co.).
36. *Trifluoroethanol Brochure*, Halocarbon Products Corp., Hackensack, N.J., 1979.
37. J. M. Mukherjee and E. Grunwald, *J. Phys. Chem.* **62**, 1311 (1958).
38. J. Murto and E. Heino, *Suom. Kemistil.* **39**, 263 (1966).
39. C. H. Rochester and J. R. Symonds, *J. Fluorine Chem.* **4**, 141 (1974).
40. A. Kivenen, J. Murto, and M. Lehtonen, *Suom. Kemistil.* **B 41**, 359 (1968).
41. C. H. Rochester and J. R. Symonds, *J. Chem. Soc. Faraday Trans. I* **69**, 1274 (1973).
42. P. Ballinger and F. A. Long, *J. Am. Chem. Soc.* **81**, 1050 (1959).
43. Ger. Pat. 1,017,782 (Oct. 17, 1957), P. Schlack (to Farbwerke Hoechst AG).
44. M. Goodman, I. G. Rosen, and M. Safdy, *Biopolymers* **2**, 503,519,537 (1964).
45. U.S. Pat. 3,722,211 (Mar. 27, 1973), R. C. Conner and L. L. Ferstandig (to Halocarbon Products Corp.).
46. D. G. Shepherd, *Hydrocarbon Proc.*, 141 (Dec. 1977).
47. N. Paetzold, *J. Polym. Sci. Part B* **1**, 269 (1963).
48. V. J. Shiner, Jr. and co-workers, *J. Am. Chem. Soc.* **91**, 4838 (1969).

49. D. A. daRoza, L. J. Andrews, and R. M. Keefer, *J. Am. Chem. Soc.* **95**, 7003 (1973).
50. D. S. Noyce, R. L. Castenson, and D. A. Meyers, *J. Org. Chem.* **37**, 4222 (1972).
51. U.S. Pat. 3,088,896 (May 7, 1963), M. Braid (to Pennsalt Chemicals Corp.).
52. U.S. Pat. 3,038,936 (June 12, 1962), M. Braid (to Pennsalt Chemicals Corp.).
53. A. L. Henne and M. A. Smook, *J. Am. Chem. Soc.* **72**, 4378 (1950).
54. M. Steinman and co-workers, *J. Med. Chem.* **16**, 1354 (1973).
55. U.S. Pat. 3,535,425 (Oct. 20, 1970), R. C. Terrell (to Air Reduction Co., Inc.).
56. U.S. Pat. 3,637,477 (Jan. 25, 1972), L. S. Croix (to Air Reduction Co., Inc.).
57. U.S. Pat. 4,762,856 (Aug. 9, 1988), R. C. Terrell (to BOC, Inc.).
58. F. Swarts, *Compt. Rend.* **197**, 1201 (1933).
59. H. Gilman and R. G. Jones, *J. Am. Chem. Soc.* **70**, 1281 (1948).
60. A. L. Henne, R. M. Alm, and M. Smook, *J. Am. Chem. Soc.* **70**, 1968 (1948).
61. K. N. Campbell, J. O. Knobloch, and B. K. Campbell, *J. Am. Chem. Soc.* **72**, 4380 (1950).
62. D. R. Husted and A. H. Ahlbrecht, *J. Am. Chem. Soc.* **74**, 5422 (1952).
63. U.S. Pat. 4,072,726 (Feb. 7, 1978), H. R. Nychka and co-workers (to Allied Chemical Corp.).

ARTHUR J. ELLIOTT
Halocarbon Products Corporation

FLUOROETHERS AND FLUOROAMINES

Perfluoroaliphatic ethers and perfluorotertiary amines together with the perfluoroalkanes and cycloalkanes comprise a class of extremely unreactive materials known in the industry as inert fluids. These fluids are colorless, odorless, essentially nontoxic, nonflammable, dense, and extremely nonpolar. In the electronics industry, the lower molecular weight compounds find application in the areas of heat transfer, testing, and vapor-phase soldering. Higher molecular weight polymers and oligomers are used in a variety of applications, including hazardous-duty vacuum pump fluids, specialty greases, and various specialty cosmetics and lubricants.

Many perfluoroaliphatic ethers and tertiary amines have been prepared by electrochemical fluorination (1–6), direct fluorination using elemental fluorine (7–9), or, in a few cases, by fluorination using cobalt trifluoride (10). Examples of lower molecular weight materials are shown in Table 1. In addition to these, there are three commercial classes of perfluoropolyethers prepared by anionic polymerization of hexafluoropropene oxide [*428-59-1*] (11,12), photooxidation of hexafluoropropene [*116-15-4*] or tetrafluoroethene [*116-14-3*] (13,14), or by anionic ring-opening polymerization of tetrafluorooxetane [*765-63-9*] followed by direct fluorination (15).

Physical Properties

Perfluorinated compounds boil at much lower temperatures and have lower heats of vaporization than the corresponding hydrocarbon analogues even though they have considerably higher molecular weights. This holds true not only for the per-

Table 1. Physical Properties of Some Perfluorinated Liquids

Name	CAS Registry Number	Molecular formula	Bp, °C	d_4^{25}	n_D^{24}	Pour point,[a] °C
perfluoro-4-methylmorpholine	[382-28-5]	$CF_3N(CF_2)_2O(CF_2)_2$	51	1.70	1.267	−80
perfluoro-2-ethyltetrahydrofuran	[356-48-9]	$C_6F_{12}O$	56	1.69	1.263	−136
perfluorohexane	[355-42-0]	C_6F_{14}	58	1.68	1.252	−74
Galden HT 70[b]			70	1.73		−110
perfluorotriethylamine	[359-70-6]	$(C_2F_5)_3N$	71	1.73	1.262	−110
perfluoro-4-ethylmorpholine	[55716-11-5]	$C_2F_5N(CF_2)_2O(CF_2)_2$	72	1.74	1.273	−73
perfluoro-4-isopropylmorpholine	[1600-71-1]	$(CF_3)_2CFN(CF_2)_2O(CF_2)_2$	95	1.79	1.283	−112
perfluorobutyl ether	[308-48-5]	$(C_4F_9)_2O$	102	1.71	1.261	−100
Fluorinert FC-75	[11072-16-5]	$C_8F_{16}O$[c] (cyclic)	103	1.76	1.276	−93
perfluorooctane	[307-34-6]	C_8F_{18}	103	1.77	1.272	−42
perfluorononane	[375-96-2]	C_9F_{20}	123	1.80	1.276	−16
perfluorotripropylamine	[338-83-0]	$(C_3F_7)_3N$	130	1.82	1.279	−52
perfluorobis(2-butoxyethoxy)methane	[130085-23-3]	$(C_4F_9OC_2F_4O)_2CF_2$	178	1.76		< −110
perfluorotributylamine	[311-89-7]	$(C_4F_9)_3N$	178	1.86	1.291	−50
perfluoro(diethylamino)ethyl ether	[108709-75-7]	$[(C_2F_5)_2NC_2F_4]_2O$	178			−80
perfluorohexyl ether	[424-20-4]	$(C_6F_{13})_2O$	181	1.81	1.278	−90
K7 fluid	[59884-34-3]	$C_3F_7O[CF(CF_3)CF_2O]_5C_2F_5$	250	1.82		−80

[a] ASTM D97.
[b] A mixture of perfluorinated polyethers marketed by Montefluos.
[c] $C_8F_{16}O$ represents a mixture of isomers of cyclic perfluoroaliphatic ethers, primarily perfluoro-2-butyltetrahydrofuran [335-36-4].

fluoroalkanes and cycloalkanes but for the perfluorinated ethers and tertiary amines as well. The latter compounds have boiling points very close to the perfluoroalkanes having the same number of carbon atoms; the heteroatoms contribute little polarity to the molecules. Catenary oxygen and nitrogen atoms have marked effects on the freezing points and on the viscosity at low temperatures. This is illustrated by the C_{12}-perfluorohexyl ether and the C_{12}-perfluorotributylamine in Table 1 that freeze at -90 and $-50°C$, respectively, in contrast to the 32°C melting point of perfluorodecane. Note also the extremely low pour point of the Du Pont K7 fluid despite its high molecular weight. This effect is believed to be due to the increased flexibility that catenary oxygen or nitrogen atoms contribute to the perfluorinated chain (16) and is also observed for the higher molecular weight perfluoropolyethers that have relatively low pour points at relatively high molecular weights.

Many of the unusual properties of the perfluorinated inert fluids are the result of the extremely low intermolecular interactions. This is manifested in, for example, the very low surface tensions of the perfluorinated materials (on the order of 9–19 mN/m = dyn/cm) at 25°C which enables these liquids to wet any surface including polytetrafluoroethene. Their refractive indexes are lower than those of any other organic liquids, as are their acoustic velocities. They have isothermal compressibilities almost twice as high as water. Densities range from 1.7 to 1.9 g/cm^3 (17).

The absolute viscosities of the perfluorinated inert liquids are higher than the analogous hydrocarbons but the kinematic viscosities are lower due to the higher density of the perfluorinated compounds. The viscosity index, ie, the change in viscosity with temperature, is generally higher for the perfluorinated liquids than for hydrocarbons.

Thermal Stabilities. The perfluoroethers have thermal stabilities comparable to those of the perfluoroalkanes. Typically, although this depends somewhat on structure, they do not undergo significant decomposition until about 400°C. Perfluorotertiary amines are less stable thermally and begin to decompose at temperatures of about 250°C (18). Generally this slight instability is not a significant problem during use.

Electrical Properties. The low polarizability of perfluorinated liquids makes them excellent insulators. Their dielectric strengths are about 40 kV (ASTM D877); dissipation factors are about 0.0001 at 1 MHz; dielectric constants are about 1.8; volume resistivities are about 1×10^{15} ohm·cm (ASTM D257) (17).

Chemical Properties

The inert character of the perfluoroethers and tertiary amines is demonstrated by their lack of basicity or reactivity as compared with their hydrocarbon analogues. Both classes of compounds are nonflammable. The perfluorotertiary amines do not form salts with any protic acid nor do they form complexes with boron trifluoride. Neither class reacts with most oxidizing or reducing agents nor with strong acids or bases. As with the perfluoroalkanes, perfluoroethers and tertiary amines may, under some conditions, react violently with fused alkali metals. In contrast to the perfluoroalkanes, both classes of compounds react with aluminum chloride or bromide at elevated temperatures (ca 90–200°C) (19). The —OCF$_2$O— linkage is especially vulnerable to this attack (20,21).

Solvent Properties. In comparison to the more familiar hydrocarbon systems, the solvent properties of the perfluorinated inert liquids are also unusual due to their nonpolar nature and low intermolecular forces. They are generally very poor solvents for most organic compounds. Water and hydrocarbon alcohols are nearly completely insoluble in them. Lower aliphatic hydrocarbons, lower molecular weight ethers, and some highly chlorinated solvents such as carbon tetrachloride are relatively soluble. As the molecular weight of the perfluorinated compound increases, the hydrocarbon solubility decreases. Partially fluorinated compounds such as benzotrifluoride or bis(trifluoromethyl)benzene are soluble in perfluorinated liquids. Aliphatic highly fluorinated compounds containing hydrogen such as C$_7$F$_{15}$H are also miscible with perfluorinated liquids. The chlorofluorocarbons (CFCs) such as CFC-113, CF$_2$ClCFCl$_2$, or the hydrochlorofluorocarbons (HCFCs) such as HCFC-123, CF$_3$CHCl$_2$, HCFC-141b, or CH$_3$CFCl$_2$, are miscible in all proportions with the perfluoro compounds at 25°C. Typical solubility data are shown in Table 2.

The solubilities of gases such as oxygen, nitrogen, and carbon dioxide are generally high as shown in Table 3 (22). The oxygen and carbon dioxide solubilities of related compounds has led to their use as specialized synthetic blood substitutes.

These solubility relationships are consistent with the predictions based on the Hildebrand solubility parameter (23). For perfluorinated liquids, the solubility

Table 2. Solubility Relationships for C$_8$F$_{16}$O, C$_5$F$_{11}$NO, and (C$_4$F$_9$)$_3$N[a]

	Solubility at 20°C, mL/100 mL					
Solvent	C$_8$F$_{16}$O in solvent	Solvent in C$_8$F$_{16}$O	C$_5$F$_{11}$NO in solvent	Solvent in C$_5$F$_{11}$NO	(C$_4$F$_9$)$_3$N in solvent	Solvent in (C$_4$F$_9$)$_3$N
acetone	4.8	1.2	2.9	1.6	0.9	0.6
benzene	2.6	3.8	6.0	2.7	0.3	0.2
benzotrifluoride	miscible	miscible	miscible	miscible	miscible	miscible
benzyl alcohol	0.2	0.4	0.1	0.1	insoluble	insoluble
2-butanone	2.2	1.9	5.1	2.6	0.3	0.7
carbon tetrachloride	20.2	36.5	48.0	58.9	2.4	15.0
chlorobenzene	1.8	3.2	1.6	2.3	0.3	0.4
chloroform	7.7	4.5	13.2	8.1	1.2	5.4
cyclohexane	7.7	8.4	8.9	6.4	1.8	2.2
ethyl ether	miscible	miscible	miscible	miscible	miscible	5.0
ethyl acetate	7.5	6.5	8.8	5.5	2.2	2.4
heptane	25.5	11.6	25.1	10.1	6.4	3.4
isopropyl alcohol	4.1	1.3	4.7	0.7	insoluble	insoluble
methanol	1.0	0.1	1.6	0.6	insoluble	insoluble
petroleum ether	miscible	miscible	miscible	miscible	33.2	7.0
toluene	2.9	4.1	3.1	3.6	0.4	2.0
turpentine	5.3	1.0	6.4	0.4	0.9	insoluble
xylene	3.0	3.0	3.1	2.6	0.2	1.0
water	insoluble	insoluble	insoluble	insoluble	insoluble	insoluble

[a]C$_8$F$_{16}$O denotes a mixture of cyclic perfluoroaliphatic ethers, primarily perfluoro-2-butyltetrahydrofuran. C$_5$F$_{11}$NO denotes perfluoro-4-methylmorpholine.

parameters are on the order of 10–12 $J^{1/2}/cm^{3/2}$ (5–6 $cal^{1/2}/cm^{3/2}$) which are the lowest known values for liquids.

Table 3. Solubility of Gases in Perfluorinated Liquids[a]

Gas	Solubility	
	mL gas/100 mL $C_8F_{16}O$ (cyclic)	mL gas/100 mL $(C_4F_9)_3N$
oxygen	48.8	38.9
nitrogen	33.4	28.4
carbon dioxide	192.0	152.0
air	40.5	33.1

[a]At 25°C and 101 kPa (1 atm).

Methods of Preparation

Electrochemical Fluorination. In the Simons electrochemical fluorination (ECF) process the organic reactant is dissolved in anhydrous hydrogen fluoride and fluorinated at the anode, usually nickel, of an electrochemical cell. This process has been reviewed (6). Essentially all hydrogen atoms are substituted by fluorine atoms; carbon–carbon multiple bonds are saturated. The product phase is heavier than the HF phase and insoluble in it and is recovered by phase separation.

$$(C_4H_9)_3N + 27\ HF \xrightarrow{ECF} (C_4F_9)_3N + 27\ H_2$$

With an amine reactant, it has been shown that roughly one-third of the current passed makes liquid product, one-third gas, and one-third goes to HF-soluble polyfluorinated products (24). The ether perfluoro(2-butyltetrahydrofuran) [335-36-4] is made from a cyclization process during the ECF of perfluorooctanoyl chloride; other cyclic ethers have been prepared from certain ester reactants by a similar cyclization (25). Perfluoroaminoethers have been prepared by ECF (26).

Electrochemical fluorination leads to fragmentation, coupling, and rearrangement reactions as well as giving the perfluorinated product. In addition, small amounts of hydrogen can be retained in the crude product. The products are purified by treatment with base to remove the hydrogen-containing species and subsequently distilled.

Direct Fluorination. This is a more recently developed method for the synthesis of perfluorinated compounds. In this process, fluorine gas is passed through a solution or suspension of the reactant in a nonreactive solvent such as trichlorotrifluoroethane (CFC-113). Sodium fluoride may also be present in the reaction medium to remove the coproduct hydrogen fluoride. There has been enormous interest in this area since the early 1980s resulting in numerous journal publications and patents (7–9) (see FLUORINE COMPOUNDS, ORGANIC–DIRECT FLUORINATION). Direct fluorination is especially useful for the preparation of perfluoroethers.

Multiple ether oxygen atoms can be present in the molecule. Cleavage and coupling reactions occur with direct fluorination although to a lesser extent than with ECF. This allows the direct fluorination of acid-sensitive materials, such as the formal shown below, which would not survive ECF (8).

$$(C_4H_9OC_2H_4O)_2CH_2 + F_2 \rightarrow (C_4F_9OC_2F_4O)_2CF_2$$

As opposed to ECF, direct fluorination affords a much lower degree of isomerization so that the carbon skeleton of the reactant remains intact in the perfluorinated product. Direct fluorination is also complementary to ECF in the significantly higher yields observed for the direct fluorination of ethers. As with ECF the products are purified by treatment with base and subsequent distillation.

Polymerization. The higher molecular weight perfluoropolyethers are prepared by distinctly different technology. The anionic polymerization of hexafluoropropene oxide is carried out using cesium or potassium fluoride as catalyst in a polar aprotic solvent such as diglyme (11). This leads to repeating units (n can vary widely) of the perfluoroisopropoxy group in the oligomeric chain which is terminated with an acyl fluoride.

$$CF_3CF\underset{O}{-\!\!-\!\!-}CF_2 \xrightarrow{F^-} F[CF(CF_3)CF_2O]_nCF(CF_3)COF$$

This reactive end group must be removed to render the final compound inert. There are several methods to do this. Deacylation using antimony pentafluoride or aluminum fluoride as catalyst leads to the stable OC_2F_5 end group. The acyl fluoride may be converted to the corresponding acid salt by reaction with a base such as potassium hydroxide followed by thermal decarboxylation. This yields the hydrogen-containing $OCFHCF_3$ end group. The acid (27), or at higher temperatures the acyl fluoride, may be directly replaced by fluorine with either elemental fluorine or cobalt trifluoride. The lower molecular weight members of this family have been marketed by Du Pont as their K-Series fluids.

The photooxidation of hexafluoropropene or tetrafluoroethene or mixtures thereof leads to perfluoro polyether peroxides of varying molecular weights (14). The peroxidic links are subsequently decomposed by thermal treatment, the presence of base is optional, to give lower molecular weight fragments containing a variety of end groups. These include the CF_3O and the fluoroformyl FCO_2 groups. The polymers are then stabilized by removal of the functional group via direct fluorination. The final products of the photooxidation of hexafluoropropene have the following structure where the ratio q/p can vary between 0 and 0.1:

$$CF_3O[(CF(CF_3)CF_2O)_p(CF_2O)_q]\,CF_3$$

The lower molecular weight fractions from this process have been marketed by Montefluos under their trade name Galden.

The final products for the photooxidation of tetrafluoroethene have the following structure where the ratio m/n is between 0.6–1.5 and is typically about

0.8 (14). The average molecular weight of these polyethers is between ca 1,000 and 40,000.

$$CF_3O(C_2F_4O)_m(CF_2O)_nCF_3$$

Perfluoropolyethers with the linear perfluoropropoxy repeat unit have been commercialized (28). They are prepared by the anionic oligomerization of tetrafluorooxetane followed by direct fluorination to remove the acyl fluoride end group as well as to fluorinate the remaining CH_2 groups; n can vary widely.

$$\underset{\underset{CH_2-O}{||}}{CF_2-CF_2} + F^- \rightarrow F(CH_2CF_2CF_2O)_nCH_2CF_2COF \xrightarrow{F_2} C_3F_7O(C_3F_6O)_nC_2F_5$$

All three processes give perfluoropolyethers with a broad distribution of molecular weights. They are typically separated into fractions by vacuum distillation.

Economic Aspects

Information on the production levels of the perfluoroethers and perfluorotertiary amines is not disclosed, but the products are available commercially and are marketed, for instance, as part of the Fluorinert Electronic Liquids family by 3M Co. (17). These liquids have boiling points of 30–215°C with molecular weights of about 300–800. They range in price from $26–88/kg. Perfluoropropene oxide polyethers are marketed by Du Pont with the trade name Krytox (29). The linear perfluoropropene oxide polyethers are marketed by Daikin under the trade name Demnum (28). The perfluoropolyethers derived from photooxidation are marketed by Montefluos under the trade name of Fomblin (30). These three classes of polyethers are priced from about $100–150/kg.

Environmental, Health, and Safety Factors

Over the years animal studies have repeatedly shown that perfluorinated inert fluids are nonirritating to the eyes and skin and practically nontoxic by ingestion, inhalation, or intraperitoneal injection (17,22). Thermal degradation can produce toxic decomposition products including perfluoroisobutene which has a reported LC_{50} of 0.5 ppm (6 hr exposure in rats) (31). This decomposition generally requires temperatures above 200°C.

Perfluorinated ethers and perfluorinated tertiary amines do not contribute to the formation of ground level ozone and are exempt from VOC regulations (32). The commercial compounds discussed above have an ozone depletion potential of zero because they do not contain either chlorine or bromine which take part in catalytic cycles that destroy stratospheric ozone (33).

Uses

The unique combination of properties of the perfluorinated fluids makes them useful in a variety of applications in the electronics industry (34). The lower molecular weight materials are used in three principal areas in this industry: direct contact cooling of electronic components, testing, and reflow-soldering. Recently perfluorinated liquids have been used as the total immersion coolant for a new generation supercomputer. This fluid is used to cool the power supplies, memory boards, logic circuits, and main processors. Vapor-phase reflow-soldering has been described (35). Testing applications include liquid burn-in testing, gross and fine leak testing, and electrical environmental testing. In the early 1990s, these perfluorinated fluids have found use in some specialized applications as replacements for the chlorofluorocarbons (CFCs) which have come under tight regulation. Perfluorinated liquids have replaced CFCs in several applications including coolant for ion implanters, secondary blanket in vapor-phase soldering, solvent for magnetic media lubricants, and coolant for large rectifiers. They have also found use in blowing agent systems for polyurethane foams (36) and in the removal of water from precision parts that have been cleaned using an aqueous rinse (37). They are under investigation as potential replacements for the Halon fire-extinguishing agents (qv) (38).

The higher molecular weight perfluoropolyethers are useful as specialty lubricants. They provide good lubrication under boundary conditions in systems in which the mechanical parts are exposed to high temperatures or aggressive chemical environments. They are typically used as the working fluid in hazardous duty vacuum pumps used in plasma etching. Specialty greases, used in high temperature environments in which a hydrocarbon-based grease fails, have also been formulated by blending perfluoropolyethers with fluorinated polymers. Additionally, perfluoropolyethers are used as lubricants for magnetic media, lubricant and sealing agent for oxygen service, inert hydraulic fluids, etc (14). They have also found application in cosmetics (39) and as a protective coating for outdoor stone art and masonry (40).

Perfluorinated compounds are also potentially useful as inert reaction media, particularly when one of the reactants is gaseous. The high solubility of oxygen and carbon dioxide in perfluorinated liquids has allowed their use as blood substitutes (41) and as oxygenation media for biotechnology (42). One product, Fluosol DA (43) (Green Cross Corp.), has been commercialized, and there is an abundant patent art in this area (see BLOOD, ARTIFICIAL).

BIBLIOGRAPHY

"Fluoro Ethers and Amines" under "Fluorine Compounds, Organic" in *ECT* 1st ed., Vol. 6, pp. 762–763, by L. J. Hals and W. H. Pearlson, Minnesota Mining & Manufacturing Co.; in *ECT* 2nd ed., Vol. 9, pp. 753–754, by L. J. Hals and W. H. Pearlson, Minnesota Mining & Manufacturing Co.; in *ECT* 3rd ed., Vol. 10, pp. 874–881, by R. D. Danielson, Minnesota Mining & Manufacturing Co.

1. J. H. Simons and co-workers, *J. Electrochem. Soc.* **95**, 47 (1949).
2. J. H. Simons, ed., *Fluorine Chemistry*, Vol. 1, Academic Press, Inc., New York, 1950, pp. 414–420.

3. S. Nagase, in P. Tarrant, ed., *Fluorine Chemistry Reviews*, Vol. 1, Marcel Dekker, Inc., New York, 1967, pp. 77–106.
4. T. Abe and S. Nagase, in R. E. Banks, ed., *Preparation, Properties and Industrial Applications of Organofluorine Compounds*, Ellis Horwood, Chichester, U.K., 1982, p. 19.
5. I. N. Rozhkov, in M. M. Baizer and H. Lund, eds., *Organic Electrochemistry*, 2nd ed., Marcel Dekker, Inc., New York, 1983, p. 805.
6. W. V. Childs and co-workers, in H. Lund and M. M. Baizer, eds., *Organic Electrochemistry*, 3rd ed., Marcel Dekker, Inc., New York, 1991, p. 1103.
7. U.S. Pat. 5,093,432 (Mar. 3, 1992), T. R. Bierschenk and co-workers (to Exfluor Research Corp.).
8. World Pat. 90,06296 (June 14, 1990), M. G. Costello and G. G. I. Moore (to Minnesota Mining & Mfg. Co.).
9. R. J. Lagow, *Prog. Inorg. Chem.* **26**, 161–210 (1979).
10. R. D. Chambers and co-workers, *J. Fluorine Chem.* **29**, 323 (1985).
11. J. T. Hill, *J. Macromol. Sci.-Chem.* **A8**(3), 499 (1974); P. Tarrant and co-workers, in P. Tarrant, ed., *Fluorine Chemistry Reviews*, Vol. 5, Marcel Dekker, Inc., New York, 1971, pp. 77–113.
12. H. S. Eleuterio, *J. Macromol. Sci.-Chem.* **A6**(6), 1027 (1972).
13. D. Sianesi and co-workers, *Chim. Ind. (Milan)* **55**(2), 208 (1973).
14. G. Caporiccio, in R. E. Banks, D. W. A. Sharp, and J. C. Tatlow, eds., *Fluorine: The First Hundred Years (1886–1986)*, Elsevier Sequoia, New York, 1986, pp. 314–320.
15. Y. Ohsaka, *J. Jpn. Pet. Inst.* **8**(9), 2 (1985).
16. U.S. Pat. 3,810,874 (May 14, 1974), R. A. Mitsch and J. L. Zollinger (to Minnesota Mining & Mfg. Co.).
17. *Fluorinert Liquids*, technical notebook, 3M Co., St. Paul, Minn., 1987.
18. H. G. Bryce, in J. H. Simons, ed., *Fluorine Chemistry*, Vol. 5, Academic Press, Inc., New York, 1964.
19. G. V. D. Tiers, *J. Am. Chem. Soc.* **77**, 4837, 6703, 6704 (1955).
20. P. Kasai and P. Wheeler, *Appl. Surface Sci.* **52**, 91 (1991).
21. P. Kasai, W. Tang, and P. Wheeler, *Appl. Surface Sci.* **51**, 201 (1991).
22. J. W. Sargent and R. J. Seffl, *Fed. Proc.* **29**, 1699 (1970).
23. J. H. Hildebrand, J. M. Prausnitz, and R. L. Scott, *Regular and Related Solutions*, Van Nostrand Reinhold, New York, 1970, p. 207.
24. A. Dimitrov, St. Rudiger, and M. Bartoszek, *J. Fluorine Chem.* **47**, 23 (1990).
25. T. Abe and co-workers, *J. Fluorine Chem.* **12**, 359 (1978).
26. G. G. I. Moore and co-workers, *J. Fluorine Chem.* **32**, 41 (1986).
27. U.S. Pat. 4,847,427 (July 11, 1989), M. J. Nappa (to E. I. du Pont de Nemours & Co., Inc.).
28. *Demnum, Demnum Grease Technical Bulletin*, Daikin Industries, Ltd., Osaka, Japan, Jan. 1987.
29. *Krytox, Technical Bulletins No. E74823*, E. I. du Pont de Nemours & Co., Inc., Wilmington, Del., Aug. 1985.
30. *Fomblin y Fluorinated Fluids*, Montedison SpA, Milan, Italy, 1971.
31. J. W. Clayton, Jr., in Ref. 3, pp. 225–232.
32. *Fed. Reg.* **57**(22), 3945 (Feb. 3, 1992).
33. M. J. Molina and F. S. Rowland, *Nature* **249**, 810 (1974).
34. D. Slinn and S. Green, in Ref. 4, pp. 45–82.
35. W. H. Pearlson, *J. Fluorine Chem.* **32**, 29 (1986).
36. U.S. Pat. 4,972,002 (Nov. 20, 1990), O. Volkert (to BASF AG).
37. U.S. Pat. 5,089,152 (Feb. 18, 1992), R. M. Flynn, D. A. Johnson, and J. G. Owens (to Minnesota Mining & Mfg. Co.).

38. R. E. Tapscott, *1992 Halon Alternatives Technical Working Conference*, Albuquerque, N.M., May 12–14, 1992.
39. Eur. Pat. Appl. 390,206 (Oct. 3, 1990), F. Brunetta and G. Pantini (to Ausimont SRL).
40. *Fomblin MET Technical Bulletin*, Montefluos, Milan, Italy, Oct. 1987.
41. *Science* **206**, 205 (1979).
42. B. Mattiasson and P. Adlercreut, *Trends Biotechnol.* **5**, 250 (1987).
43. *Chem. Eng. News*, 12 (Nov. 26, 1979).

RICHARD M. FLYNN
3M Company

PERFLUOROEPOXIDES

Perfluoroepoxides were first prepared in the late 1950s by Du Pont Co. Subsequent work on these compounds has taken place throughout the world and is the subject of a number of reviews (1–5). The main use of these epoxides is as intermediates in the preparation of other fluorinated monomers. Although the polymerization of the epoxides has been described (6–12), the resulting homopolymers and their derivatives are not significant commercial products. Almost all the work on perfluoroepoxides has been with three compounds: tetrafluoroethylene oxide (TFEO), hexafluoropropylene oxide (HFPO), and perfluoroisobutylene oxide (PIBO). Most of this work has dealt with HFPO, the most versatile and by far the most valuable of this class of materials (4).

Physical Properties

In general, the perfluoroepoxides have boiling points that are quite similar to those of the corresponding fluoroalkenes. They can be distinguished easily from the olefins by ir spectroscopy, specifically by the lack of olefinic absorption and the presence of a characteristic band between 1440 and 1550 cm^{-1}. The nmr spectra of most of the epoxides have been recorded. Little physical property data concerning these compounds have been published (Table 1). The structure of HFPO by electron diffraction (13) as well as its solubility and heats of solution in some organic solvents have been measured (14,15).

Chemical Properties

There are three general reactions of perfluoroepoxides: pyrolyses (thermal reactions), electrophilic reactions, and by far the most important, reactions with nucleophiles and bases.

Thermal Reactions. Those perfluoroepoxides that contain a CF_2 group in the epoxide ring undergo a smooth decomposition at relatively mild, neutral conditions (140–220°C) to give a perfluorocarbonyl compound and difluorocarbene (16,17) (eq. 1).

Table 1. Physical Properties of Perfluorocarbon Epoxides

Material	CAS Registry Number	Molecular structure	Bp,[a] °C	Infrared absorption, μm	Reference
tetrafluoroethylene oxide	[694-117-7]	$\underset{\underset{O}{\diagdown\diagup}}{CF_2 - CF_2}$	−63.5[b]	6.21	67,68
hexafluoropropylene oxide	[428-59-1]	$\underset{\underset{O}{\diagdown\diagup}}{CF_3CF - CF_2}$	−27.4	6.43	69
trifluoroglycidyl fluoride	[24419-82-7]	$\underset{\underset{O}{\diagdown\diagup}}{CF_2 - CFCOF}$	16[c] (extrap.)	6.61 5.35 (COF)	70
perfluoro-1,2-epoxybutane	[3709-90-8]	$\underset{\underset{O}{\diagdown\diagup}}{CF_2 - CFCF_2CF_3}$		6.48	21
perfluoro-2,3-epoxybutane	[773-29-5]	$\underset{\underset{O}{\diagdown\diagup}}{CF_3CF - CFCF_3}$	0–1	6.63	32
perfluoroisobutylene oxide	[707-13-1]	$\underset{\underset{O}{\diagdown\diagup}}{(CF_3)_2C - CF_2}$	3[d]	6.66	69
perfluoro-1,2-epoxycyclobutane	[13324-28-2]	(cyclobutane epoxide structure)		6.3	16
perfluoro-4,5-epoxy-1-pentene	[15453-08-4]	$\underset{\underset{O}{\diagdown\diagup}}{CF_2 - CFCF_2CF=CF_2}$	37	6.45 5.65 (C=C)	72
perfluoro-1-oxaspiro[2,3]hexane	[53389-66-5]	(spiro epoxide structure)	18–21	6.55	73

Table 1. (Continued)

Material	CAS Registry Number	Molecular structure	Bp,[a] °C	Infrared absorption, µm	Reference
perfluoro-1,2-epoxycyclopentane	[710-70-3]	(cyclopentane epoxide structure)	26.5	6.55	61
perfluoro-4,5-epoxyvaleryl fluoride	[140173-04-2]	CF$_2$—CFCF$_2$CF$_2$COF, O	37–38	6.49	71
perfluoro-1,2-epoxyhexane	[72804-48-9]	CF$_2$—CFCF$_2$CF$_2$CF$_2$CF$_3$, O	55–56	6.45	32
perfluoro-1,2-epoxy-6-oxaheptane[e]	[71877-16-2]	CF$_2$—CFCF$_2$CF$_2$CF$_2$OCF$_3$, O	61–62	6.49	74
perfluoro-2-methyl-2,3-epoxypentane	[788-67-0]	(CF$_3$)$_2$C—CFCF$_2$CF$_3$, O	57	6.85	69
perfluoro-2,3-epoxy-4-methylpentane	[788-50-1]	CF$_3$CF—CFCF(CF$_3$)$_2$, O	53	6.61	69
2,3-bis(trifluoromethyl)perfluoro-2,3-epoxybutane	[1708-78-7]	(CF$_3$)$_2$C—C(CF$_3$)$_2$, O	53–54		75
perfluoro-5,6-epoxy-1-hexene	[15453-10-8]	CF$_2$—CFCF$_2$CF$_2$CF=CF$_2$, O	57–58	6.45 5.59 (C=C)	71
perfluoro-1,2;5,6-diepoxyhexane	[140173-03-1]	CF$_2$—CFCF$_2$CF$_2$CF—CF$_2$	57–58	6.49	71

Compound	CAS	Structure	bp (°C)	n_D^{20}	Ref
perfluoro-1,2-epoxy-5-methyl-4-oxahexane	[84424-45-3]	CF$_2$—CFCF$_2$OCF(CF$_3$)$_2$ \\ \\O	58.5–59	6.47	78
perfluoro-1,2-epoxycyclohexane	[5927-67-3]	(perfluorocyclohexane epoxide)	54	6.71	79
perfluoro-1-methyl-1,2:4,5-diepoxycyclohexane	[130482-35-8]	(perfluoromethyl diepoxycyclohexane)	77	6.69 \\ 6.92	80
perfluoro-7,8-epoxy-1-octene	[72264-78-9]	CF$_2$=CF(CF$_2$)$_4$CF—CF$_2$ \\ \\O	105		76
perfluoro-1,2:7,8-diepoxyoctane^f	[13714-88-0]	CF$_2$—CF(CF$_2$)$_4$CF—CF$_2$ \\ \\O \\ \\ O	104	6.45	76, 77
perfluorophenylglycidyl ether	[84329-68-0]	C$_6$F$_5$OCF$_2$CF—CF$_2$ \\ \\O	61–64 (4 kPa)^g	6.47	78
perfluoro-3,4-epoxy-2,3,5-trimethylhexane	[2355-27-3]	(CF$_3$)$_2$CF—C—CFCF(CF$_3$)$_2$ \\ with CF$_3$ and O	36–39 (0.13 kPa)^g	7.21	69
perfluoro-1,2:9,10-diepoxydecane^h	[13714-90-4]	CF$_2$—CF(CF$_2$)$_6$CF—CF$_2$ \\ \\O	88 (5.3 kPa)^g		76

^{a}At 101.3 kPa = 1 atm unless otherwise noted in parentheses. ^{b}Mp, –118°C. ^{c}Heat of vaporization = 28.9 kJ/mol (6.9 kcal/mol). ^{d}Mp, 122°C. $^e n_D^{20}$ 1.2560; d_4^{20} 1.6441. $^f n_D^{20}$ 1.2900; d_4^{20} 1.7220. ^{g}To convert kPa to mm Hg, multiply by 7.5. $^h n_D^{20}$ 1.3030; d_4^{20} 1.8260.

$$\underset{R_f'}{\overset{R_f}{>}}C\underset{O}{\overset{}{-}}CF_2 \rightarrow R_f\overset{O}{\overset{\|}{C}}R_f' + :CF_2 \quad (1)$$

where R_f = perfluoroalkyl; R_f' = perfluoroalkyl or fluorine

The difluorocarbene produced in this way may react with a variety of compounds (18). Epoxides of internal olefins which do not contain a CF_2 group have much greater stability (19).

Electrophilic Reactions. Perfluoroepoxides are quite resistant to electrophilic attack. However, they react readily with Lewis acids, for example SbF_5, to give ring-opened carbonyl compounds (20–22) (eq. 2).

$$R_f-CF-CF_2 \xrightarrow{SbF_5} R_f\overset{O}{\overset{\|}{C}}CF_3 \quad (2)$$

The structure of the ketones produced from unsymmetrical internal perfluoroepoxides has been reported (5). The epoxide ring may also be opened by strong protic acids such as fluorosulfonic acid or hydrogen fluoride at elevated temperatures (23–25). The ring opening of HFPO by sulfur trioxide at 150°C has been interpreted as an example of an electrophilic reaction (26) (eq. 3).

$$2\ CF_3CF-CF_2 + 2\ SO_3 \xrightarrow{150°C} CF_3CF-CF_2 + CF_3\overset{O}{\overset{\|}{C}}CF_2OSO_2F \quad (3)$$

Nucleophilic Reactions. The strong electronegativity of fluorine results in the facile reaction of perfluoroepoxides with nucleophiles. These reactions comprise the majority of the reported reactions of this class of compounds. Nucleophilic attack on the epoxide ring takes place at the more highly substituted carbon atom to give ring-opened products. Fluorinated alkoxides are intermediates in these reactions and are in equilibrium with fluoride ion and a perfluorocarbonyl compound. The process is illustrated by the reaction of methanol and HFPO to form methyl 2,3,3,3-tetrafluoro-2-methoxypropanoate (eq. 4).

$$CH_3OH + CF_3CF-CF_2 \rightarrow \begin{bmatrix} CF_3CFCF_2O^- \\ | \\ \overset{+}{H}OCH_3 \end{bmatrix} \rightarrow \begin{bmatrix} CF_3CFCOF \\ | \\ OCH_3 \end{bmatrix} + HF \xrightarrow{CH_3OH} CF_3CFCOOCH_3 \\ | \\ OCH_3 \quad (4)$$

TFEO is by far the most reactive epoxide of the series. However, all the reported perfluoroepoxides undergo similar ring-opening reactions. The most important reactions of these epoxides are those with the fluoride ion or perfluoroalkoxides. The reaction of PIBO and the fluoride ion is an example (27). It also illustrates the general scheme of oligomerization of perfluoroepoxides (eq 5).

$$(CF_3)_2C\underset{O}{\underbrace{}}CF_2 \xrightarrow{F^-} (CF_3)_2CFCF_2O^- \leftrightarrows (CF_3)_2CFCOF + F^- \xrightarrow{PIBO} (CF_3)_2CFCF_2OCCF_2O^- \leftrightarrows \text{etc} \tag{5}$$

with CF_3 groups on the central C

The direction of nucleophilic ring opening of unsymmetrical perfluoroepoxides has been shown to be a function of the nature of the nucleophile and the solvent (23,28). Although many oligomeric products have been prepared by this procedure and variations of it, no truly high polymers have been obtained (9).

Preparation

A large number of methods have been used to prepare perfluoroepoxides (5). All of these methods must contend with the great chemical reactivity of the epoxide product, especially with subsequent ionic and thermal reactions which result in the loss of the desired epoxide.

The reaction of perfluoroalkenes with alkaline hydrogen peroxide is a good general method for the preparation of the corresponding epoxides with the exception of the most reactive of the series, TFEO (eq. 6).

$$R_fCF{=}CF_2 + H_2O_2 \xrightarrow[\text{solvent}]{OH^-} R_fCF\underset{O}{\underbrace{}}CF_2 + H_2O \tag{6}$$

The alkene is allowed to react at low temperatures with a mixture of aqueous hydrogen peroxide, base, and a co-solvent to give a low conversion of the alkene (29). These conditions permit reaction of the water-insoluble alkene and minimize the subsequent ionic reactions of the epoxide product. Phase-transfer techniques have been employed (30). A variation of this scheme using a peroxycarbimic acid has been reported (31).

Reaction of perfluoroalkenes and hypochlorites has been shown to be a general synthesis of perfluoroepoxides (32) (eq. 7). This appears to be the method of choice for the preparation of epoxides from internal fluoroalkenes (38). Excellent yields of HFPO from hexafluoropropylene and sodium hypochlorite using phase-transfer conditions are claimed (34).

$$R_fCF{=}CFR_f + {}^-OCl \rightarrow R_fCF\underset{O}{\underbrace{}}CFR_f + Cl^- \tag{7}$$

The direct oxidation of fluoroalkenes is also an excellent general synthesis procedure for the preparation of perfluoroepoxides (eq. 8). This method exploits the low reactivity of the epoxide products to both organic and inorganic free radicals.

$$R_fCF{=}CF_2 + \tfrac{1}{2} O_2 \rightarrow R_fCF\underset{O}{-\!\!\diagdown\!\!\diagup\!\!-}CF_2 \qquad (8)$$

The oxidation may be carried out with an inert solvent thermally (35), with a sensitizer such as bromine (36), with uv radiation (37), or over a suitable catalyst (38). Principal by-products of all these oxidation processes are the acyl fluoride products derived from oxidative cleavage of the perfluoroalkene (eq. 9).

$$R_fCF{=}CF_2 + O_2 \rightarrow R_fCOF + COF_2 \qquad (9)$$

Perfluoroepoxides have also been prepared by anodic oxidation of fluoroalkenes (39), the low temperature oxidation of fluoroalkenes with potassium permanganate (40), by addition of difluorocarbene to perfluoroacetyl fluoride (41) or hexafluoroacetone (42), epoxidation of fluoroalkenes with oxygen difluoride (43) or peracids (44), the photolysis of substituted 1,3-dioxolan-4-ones (45), and the thermal rearrangement of perfluorodioxoles (46).

Tetrafluoroethylene Oxide

TFEO has only been prepared by a process employing oxygen or ozone because of its extreme reactivity with ionic reagents. This reactivity may best be illustrated by its low temperature reaction with the weak nucleophile, dimethyl ether, to give either of two products (47) (eq. 10).

$$CH_3OCF_2CF_2OCH_3 \xleftarrow{-20°C} CH_3OCH_3 + F_2C\underset{O}{-\!\!\diagdown\!\!\diagup\!\!-}CF_2 \xrightarrow{+25°C} CH_3OCF_2COF + CH_3F \qquad (10)$$

Reaction of TFEO with acid fluorides and the fluoride ion yields oligomers with the structure $R_fCF_2O(CF_2CF_2O)_nCF_2COF$ (47,48). The epoxide yields a waxy solid polymer when exposed to high energy radiation (47,49) or when treated with amines at low temperature (47,50). The extreme chemical reactivity and facile rearrangement to perfluoroacetyl fluoride have been deterrents to the large-scale development of TFEO. The structure of TFEO has been measured using microwave spectroscopy (51).

Hexafluoropropylene Oxide

HFPO is the most important of the perfluoroepoxides and has been synthesized by almost all of the methods noted. Many attempts have been made to polymerize HFPO (6,8). The most successful has been the reaction of HFPO with fluoride ion at low temperature to give a series of oligomeric acid fluorides which have been end capped to yield stable fluids (eq. 11, where X = H,F).

$$(n+2)CF_2\underset{O}{-\!\!\diagdown\!\!\diagup\!\!-}CFCF_3 \xrightarrow{F^-} CF_3CF_2CF_2O(\underset{\underset{CF_3}{|}}{C}FCF_2O)_n\underset{\underset{CF_3}{|}}{C}FCOF \rightarrow CF_3CF_2CF_2O(CFCF_2O)_n\underset{\underset{CF_3}{|}}{C}FXCF_3 \qquad (11)$$

FLUORINE COMPOUNDS, ORGANIC (PERFLUOROEPOXIDES)

Materials of this type have been sold by Du Pont Co. under the Freon E and Krytox trademarks. Perfluorinated materials structurally similar to those in equation 11 have been prepared by Ausimont by the low temperature irradiation of either hexafluoropropylene or tetrafluoroethylene with oxygen followed by heating and/or irradiation and have been sold as Fomblin liquids (52). An isomeric polyether, Demnum, prepared by the oligomerization of 2,2,3,3-tetrafluorooxetane followed by fluorination has been commercialized by Daikin (eq. 12).

$$\begin{array}{c} CH_2\text{---}CF_2 \\ | \quad\quad | \\ O\text{---}CF_2 \end{array} \rightarrow FCH_2CF_2CF_2O(CH_2CF_2CF_2O)_nOCH_2CF_2COF \rightarrow CF_3CF_2CF_2O(CF_2CF_2CF_2O)_nCF_2CF_3 \tag{12}$$

Higher molecular weight HFPO-based materials have been prepared by reaction with both ends of a perfluorodiacyl fluoride followed by coupling through triazine rings (9). Lower molecular weight HFPO oligomers have been coupled to give inert perfluorinated ethers which are sold as Hostinert liquids by Hoechst-Celanese (eq. 13).

$$2\ CF_3CF_2CF_2O(CFCF_2O)_nCFCOF \rightarrow C_3F_7O(CFCF_2O)_nCF\text{---}CF(OCF_2CF)_nOC_3F_7 \tag{13}$$
$$\quad\quad\quad\quad\quad\quad |\quad\quad\quad | \quad\quad\quad\quad\quad\quad\quad\quad |\quad\quad\quad | \quad | \quad\quad\quad |$$
$$\quad\quad\quad\quad\quad\quad CF_3 \quad\quad CF_3 \quad\quad\quad\quad\quad\quad\quad\quad\quad CF_3 \quad\quad CF_3\ CF_3 \quad\quad CF_3$$

HFPO reacts with a large number of acyl fluorides in a general reaction to give 2-alkoxytetrafluoropropionyl fluorides which in turn may be converted to trifluorovinyl ethers (eq. 14).

$$R_fCOF + F_2C\text{---}CFCF_3 \xrightarrow{F^-} R_fCF_2OCCFCOF \rightarrow R_fCF_2OCF\text{=}CF_2 \tag{14}$$
$$\quad\quad\quad\quad\quad\quad \backslash\ /\quad\quad\quad\quad\quad\quad\quad |$$
$$\quad\quad\quad\quad\quad\quad\ O\quad\quad\quad\quad\quad\quad\quad\quad\ CF_3$$

These ethers readily copolymerize with tetrafluoroethylene and other fluoroalkenes to commercially significant plastics, elastomers, and ion-exchange resins such as Teflon PFA, Kalrez, and Nafion (see FLUORINE COMPOUNDS, ORGANIC–TETRAFLUOROETHYLENE–PERFLUOROVINYL ETHER COPOLYMERS; ELASTOMERS, SYNTHETIC–FLUOROCARBON ELASTOMERS; IONIC POLYMERS).

Publications have described the use of HFPO to prepare acyl fluorides (53), fluoroketones (54), fluorinated heterocycles (55), as well as serving as a source of difluorocarbene for the synthesis of numerous cyclic and acyclic compounds (56). The isomerization of HFPO to hexafluoroacetone by hydrogen fluoride has been used as part of a one-pot synthesis of bisphenol AF (57). HFPO has been used as the starting material for the preparation of optically active perfluorinated acids (58). The nmr spectrum of HFPO is given in Reference 59. The molecular structure of HFPO has been determined by gas-phase electron diffraction (13).

Perfluoroisobutylene Oxide

PIBO has been prepared primarily by the addition of difluorocarbene to hexafluoroacetone or by the reaction of alkaline hydrogen peroxide with perfluoroisobutylene. The small amount of published work on PIBO deals with its oligomerization (60), isomerization to perfluoroisobutyryl fluoride (61), conversion to perfluoro-t-butyl alcohol (62), and reaction with nucleophiles (63). PIBO has been reported to be as toxic as perfluoroisobutylene (64). The nmr spectrum of PIBO is reported in Reference 65.

Other Epoxides

Large numbers of epoxides have been reported that contain only fluorine and carbon bound to the oxirane ring but which contain other halogens, oxygen, hydrogen, and other functional groups in one of the carbon side chains. Although these are not true perfluoroepoxides their syntheses and reactions of their epoxide rings are virtually identical to those of the perfluoro analogues. One example is the reaction of 3-chloropentafluoropropylene oxide with nucleophiles such as fluoride ion (66) (eq. 15).

$$CF_2\underset{O}{-}CFCF_2Cl \xrightarrow{F^-} CClF_2CF_2CF_2O(CFCF_2O)_n CFCOF \quad | \quad | \\ CF_2Cl \quad CF_2Cl \tag{15}$$

BIBLIOGRAPHY

"Perfluoroepoxides" under "Fluorine Compounds, Organic" in *ECT* 3rd ed., Vol. 10, pp. 956–962, by P. R. Resnick, E. I. du Pont de Nemours & Co., Inc.

1. D. Sianesi and co-workers, *Gazz. Chem. Ital.* **98**, 265, 277, 290 (1968).
2. P. Tarrant and co-workers, *Fluorine Chem. Rev.* **5**, 77 (1971).
3. N. Ishikawa, *Yuki Gosei Kagaku Kyokai Shi* **35**, 131 (1977).
4. H. Millauer, W. Schwertfeger, and G. Siegemund, *Angew. Chem. Int. Ed. Engl.* **24**, 161 (1985).
5. L. F. Sokolov, P. I. Valov, and S. V. Sokolov, *Usp. Khim.* **53**, 1222 (1984).
6. H. S. Eleuterio, *J. Macromol. Sci. Chem.* **A6**, 1027 (1972).
7. J. T. Hill and J. P. Erdman, *Am. Chem. Soc. Polym. Prepr.* **18**, 100 (1977).
8. J. T. Hill, *J. Macromol. Sci.* **A8**, 499 (1974).
9. Jpn. Kokai JP 02-202919 (Aug. 13, 1990), M. Ikeda and A. Aoshima (to Asahi Chemical).
10. Y. Duan, D. Ni, and Y. He, *Kao Fen Tzu T'ung Hsun*, 139 (1981).
11. P. P. Shpakov and co-workers, *Zh. Prikl. Khim.* **54**, 2726 (1981).
12. Eur. Pat. 154,297A (Sept. 11, 1985), T. R. Darling (to E. I. du Pont de Nemours & Co., Inc.).
13. B. Beagley, R. G. Pritchard, and R. E. Banks, *J. Fl. Chem.* **18**, 159 (1981).
14. V. A. Gubanov and co-workers, *Zh. Fiz. Khim.* **48**, 2084 (1974)
15. R. G. Makitra, T. I. Politanskaya, and F. B. Moin, *Zh. Prik. Khim.* **52**, 2623 (1979); **56**, 2724 (1983).
16. D. P. Carlson and A. S. Milian, *4th International Fluorine Symposium*, Estes Park, Colo., July 1967.
17. R. C. Kennedy and J. B. Levy, *J. Fluorine Chem.* **7**, 101 (1976).
18. P. B. Sargeant, *J. Am. Chem. Soc.* **91**, 3061 (1969); V. V. Karpov, V. E. Platonov, and G. G. Yakobson, *Izv. Akad. Nauk SSSR, Ser. Khim.*, 981 (1975); R. N. Beauchamp, C. W. Gillies, and N. C. Craig, *J. Am. Chem. Soc.* **109**, 1696 (1987).

19. P. L. Coe, A. W. Mott, and J. C. Tatlow, *J. Fluorine Chem.* **30**, 297 (1985).
20. Jpn. Kokai 78 25,512 (Mar. 9, 1978), Y. Osaka and H. Takashi (to Daiken Kogyo).
21. U.S. Pat. 3,213,134 (Oct. 19, 1965), D. E. Morin (to 3M).
22. U.S. Pat. 4,302,608 (Nov. 24, 1981), E. N. Squire (to E. I. du Pont de Nemours & Co., Inc.).
23. A. Ya. Zapevalov and co-workers, *Zh. Org. Khim.* **22**, 93 (1986).
24. T. I. Filyakova and co-workers, *Izv. Akad. Nauk SSSR, Ser. Khim.*, 1878 (1979).
25. U.S. Pat. 4,400,546 (Aug. 23, 1983), P. -P. Rammelt and G. Siegemund (to Hoechst).
26. I. L. Knunyants, V. V. Shokina, and E. I. Mysov, *Isv. Akad. Nauk SSSR Ser. Khim.*, 2725 (1973).
27. J. T. Hill, *J. Fl. Chem.* **9**, 97 (1972).
28. A. Ya. Zapevalov and co-workers, *Zh. Org. Khim.* **25**, 492 (1989).
29. U.S. Pat. 3,358,003 (Dec. 12, 1967), H. S. Eleuterio and R. W. Meschke (to E. I. du Pont de Nemours & Co., Inc.).
30. U.S. Pat. 4,954,643 (Sept. 4, 1990), G. Bornengo and co-workers.
31. Ger. Offen. 2,557,655 (June 30, 1977), R. A. Sulzbach and F. Heller (to Hoescht).
32. I. P. Kolenko and co-workers, *Izv. Akad. Nauk SSSR, Ser. Khim.*, 2509 (1979).
33. T. I. Filyakova and co-workers, *Zh. Org. Khim.* **24**, 371 (1988).
34. Eur. Pat. 64,293 (Dec. 10, 1986), M. Ikeda, M. Miura, and A. Aoshima (to Asahi Kasei Kogyo K. K.).
35. S. V. Kartsov and co-workers, *Izv. Akad. Nauk SSSR, Ser. Khim.*, 2230 (1975); U.S. Pat. 3,536,733 (Oct. 27, 1970), D. P. Carlson (to E. I. du Pont de Nemours & Co., Inc.).
36. Brit. Pat. 931,857 (July 17, 1963), (to E. I. du Pont de Nemours & Co., Inc.).
37. V. Caglioti, M. Lenzi, and A. Mele, *Nature* **201**, 610 (1964); D. Sianesi, A. Pasetti, and C. Corti, *Makromol. Chem.* **86**, 308 (1965).
38. U.S. Pat. 3,775,438 (Nov. 27, 1973, R. J. Cavanaugh (to E. I. du Pont de Nemours & Co., Inc.); U.S. Pat. 3,775,439 (Nov. 27, 1973), G. M. Atkins, Jr. (to E. I. du Pont de Nemours & Co., Inc.); U.S. Pat. 3,775,440 (Nov. 27, 1973), R. J. Cavanaugh and G. M. Atkins, Jr. (to E. I. du Pont de Nemours & Co., Inc.).
39. U.S. Pat. 4,014,762 (Mar. 29, 1977), H. Millauer (to Hoescht).
40. I. L. Knunyants and co-workers, *Izv. Akad. Nauk SSSR Ser. Khim.*, 2780 (1967).
41. W. Mahler and P. R. Resnick, *J. Fluorine Chem.* **3**, 451 (1973/74).
42. U.S. Pat. 3,338,978 (Aug. 29, 1967), E. P. Moore (to E. I. du Pont de Nemours & Co., Inc.).
43. U.S. Pat. 3,622,601 (Nov. 23, 1971), J. W. Dale (to Monsanto); U.S. Pat. 3,639,429 (Feb. 1, 1972), V. Weinmayr (to E. I. du Pont de Nemours & Co., Inc.).
44. Jpn. Kokai 77 108,914 (Sept. 12, 1977), Y. Oda, K. Uchida, and S. Morikawa (to Asahi Glass).
45. T. S. Croft, *J. Fluorine Chem.* **7**, 438 (1976).
46. M. -H. Hung and P. R. Resnick, *J. Am. Chem. Soc.* **112**, 9671 (1990).
47. J. L. Warnell in Ref. 16.
48. U.S. Pat. 3,250,806 (May 10, 1966), J. L. Warnell (to E. I. du Pont de Nemours & Co., Inc.).
49. P. Barnaba and co-workers, *Chim. Inc. (Milan)* **47**, 1060 (1965).
50. Can. Pat. 778,490 (Feb. 13, 1968), J. L. Warnell (to E. I. du Pont de Nemours & Co., Inc.).
51. J. W. Agopovich and co-workers, *J. Am. Chem. Soc.* **106**, 2251 (1984).
52. D. Sianesi, *Am. Chem. Soc. Polym. Prepr.* **12**, 411 (1971); D. Sianesi and co-workers, *Chim. Ind. (Milan)* **55**, 208 (1973).
53. N. Ishikawa and S. Sasaki, *Chem. Lett.* **483**, 1407 (1976).
54. T. Martini, *Tetrahedron Lett.*, 1857, 1861 (1976).
55. N. Ishikawa and S. Sasaki, *Bull. Chem. Soc. Jpn.* **50**, 2164 (1977).

56. V. M. Karpov, V. E. Platonov, and G. G. Yacobsen, *Izv. Akad. Nauk SSSR Ser. Khim.*, 2295 (1976).
57. Eur. Pat. Appl. EP54227A (June 23, 1982), P. P. Rammelt and G. Siegemund (to Hoechst).
58. N. Ishikawa, *J. Fl. Chem.* **25**, 17 (1984); H. Kawa and N. Ishikawa, *Chem. Lett.*, 843 (1980).
59. J. K. Ruff and R. F. Merritt, *J. Org. Chem.* **30**, 3968 (1965); K. W. Jolley, L. H. Sutcliffe, and K. L. Williamson, *Spectrochemica Acta* **30A**, 1455 (1974).
60. J. T. Hill, *8th International Fluorine Symposium*, Kyoto, Japan, Aug. 1977.
61. U.S. Pat. 3,321,515 (May 23, 1967), E. P. Moore and A. S. Milian (to E. I. du Pont de Nemours & Co., Inc.).
62. U.S. Pat. 3,385,904 (May 28, 1968), F. J. Pavlik (to Minnesota Mining and Manufacturing Co.).
63. I. L. Knunyants and co-workers, *Izv. Akad. Nauk SSSR Ser. Khim.*, 1133 (1972).
64. I. L. Knunyants, V. V. Shokina, and I. V. Salakhov, *Khim. Geterotsikl. Soedin.* **2**, 873 (1966).
65. F. J. Pavlik and P. E. Toren, *J. Org. Chem.* **35**, 2054 (1970).
66. Eur. Pat. Appl. 72579 (Feb. 23, 1983), B. R. Ezzell, W. P. Carl, and W. A. Mod (to Dow Chemical Co.).
67. Brit. Pat. 931, 857 (July 17, 1963), (to E. I. du Pont de Nemours & Co., Inc.).
68. V. Caglioti and co-workers, *J. Chem. Soc.*, 5430 (1964).
69. Brit. Pat. 904,877 (Sept. 5, 1962), (E. I. du Pont de Nemours & Co., Inc.).
70. W. Stuckey, J. Heicklen, and V. Knight, *Can. J. Chem.* **47**, 2329 (1969).
71. T. I. Filyakova, R. E. Ilatovskii, and A. Ya. Zapevalov, *Zh. Org. Khim.* **27**, 2055 (1991).
72. U.S. Pat. 3,366,610 (Jan. 30, 1968), B. C. Anderson (to E. I. du Pont de Nemours & Co., Inc.).
73. R. J. DePasquale, K. B. Baucom, and J. R. Patton, *Tetrahedron Lett.*, 1111 (1974).
74. V. V. Berenblit and co-workers, *Zh. Org. Khim.* **15**, 1417 (1979).
75. E. M. Rokhlin and co-workers, *Dokl. Akad. Nauk SSSR* **161**, 1356 (1965).
76. I. L. Knunyants and co-workers, *Zh. Obsch. Khim.* **36**, 1981 (1966).
77. USSR Pat. 545,645 (Mar. 25, 1977) D. S. Rondarev and co-workers.
78. U.S. Pat. 4,360,645 (Nov. 23, 1982), C. G. Krespan and A. P. King (to E. I. du Pont de Nemours & Co., Inc.).
79. P. L. Coe, A. W. Mott, and J. C. Tatlow, *J. Fl. Chem.* **20**, 243 (1982).
80. P. L. Coe, A. W. Mott, and J. C. Tatlow, *J. Fl. Chem.* **49**, 21 (1990).

PAUL R. RESNICK
E. I. du Pont de Nemours & Co., Inc.

FLUORINATED ACETIC ACIDS

Fluoroacetic acid [144-49-0], FCH_2COOH, is noted for its high toxicity to animals, including humans. It is sold in the form of its sodium salt as a rodenticide and general mammalian pest control agent. The acid has mp, 33°C; bp, 165°C; heat of combustion, −715.8 kJ/mol (−171.08 kcal/mol) (1); enthalpy of vaporization, 83.89 kJ/mol (20.05 kcal/mol) (2). Some thermodynamic and transport properties of its aqueous solutions have been published (3), as has the molecular structure of the acid as determined by microwave spectroscopy (4). Although first prepared in 1896 (5), its unusual toxicity was not published until 50 years later (6). The acid is the toxic constituent of a South African plant *Dichapetalum cymosum*, better known as gifblaar (7). At least 24 other poisonous plant species are known to contain it (8).

Chemically, fluoroacetic acid behaves like a typical carboxylic acid, although its acidity is higher ($K_a = 2.2 \times 10^{-3}$) than the average (9). It can be prepared from the commercially available sodium salt by distillation from sulfuric acid (10).

Sodium Fluoroacetate. Sodium fluoroacetate [62-74-8], FCH_2COONa, known as Compound 1080, is a hygroscopic white solid, mp, 200–202°C, which decomposes when heated above the melting point. Its solubility at 25°C in g/100 g solvent is water, 111; methanol, 5; ethanol, 1.4; acetone, 0.04; and carbon tetrachloride, 0.004. Because its carbon–fluorine bond is unreactive under most conditions, this salt can be converted by standard procedures to typical carboxylic acid derivatives such as fluoroacetyl esters (11,12), fluoroacetyl chloride [359-06-8] (13), fluoroacetamide (14), or fluoroacetonitrile [503-20-8] (14).

Sodium fluoroacetate is usually made by displacing the halogen from an ester of bromo- or chloroacetic acid with potassium fluoride or, in one instance, antimony fluoride, followed by hydrolysis with aqueous sodium hydroxide (15–17). A commercial process for its manufacture from ethyl chloroacetate and potassium fluoride has been described (18). The ester, purified by distillation to remove traces of acid and water, is treated with oven-dried, finely powdered potassium fluoride in a well-stirred autoclave at 200°C for 11 hours. The resulting ethyl fluoroacetate [459-72-3] is then distilled into an agitated tank containing sodium hydroxide dissolved in methanol. The solid product is isolated by centrifugation, followed by vacuum drying. Through this process, all liquids are handled in a closed system of pipes and vessels, carefully inspected for leaks before each run. This is important since the intermediate fluoroacetate is highly toxic, and the starting chloroacetate is a lacrimator.

Toxicity. Sodium fluoroacetate is one of the most effective all-purpose rodenticides known (18). It is highly toxic to all species of rats tested and can be used either in water solution or in bait preparations. Its absence of objectionable taste and odor and its delayed effects lead to its excellent acceptance by rodents. It is nonvolatile, chemically stable, and not toxic or irritating to the unbroken skin of workers. Rats do not appear to develop any significant tolerance to this compound from nonlethal doses. However, it is extremely dangerous to humans, to common household pets, and to farm animals, and should only be used by experienced personnel. The rodent carcasses should be collected and destroyed since they remain poisonous for a long period of time to any animal that eats them.

The unusually high toxicity of fluoroacetic acid and of other monofluorinated organic compounds that can be metabolized to fluoroacetate has stimulated much research into the mechanism of this toxicity (8,19–23). Fluoroacetate mimics acetate by being incorporated into the tricarboxylic acid cycle of cellular respiration where it becomes converted into fluorocitric acid. This acid inhibits the enzyme, aconitate hydratase, which normally catalyzes the dehydration of citric acid. As a result, citric acid accumulates in the organism and the energy-producing cycle is interrupted. Because of the time it takes for the fluorocitrate to form and accumulate, there is usually a latent time of at least an hour before the appearance of symptoms of fluoroacetate poisoning, eg, ventricular fibrillation or convulsions. This is advantageous in its use as a pesticide. One characteristic of fluoroacetate toxicity is the wide range in lethal doses for different species ranging from (LD_{50}, mg/kg) 0.06 in dogs, 0.2 in cats, 0.4 in sheep or rabbits, 2–10 in humans, 5 in rats, 7 in mice, to about 400 in toads (20,24). The only suggested antidotes for the poisoning are 1,2,3-propanetriol monoacetate (20,23), acetamide (20), and other acetate donors, but these only have an effect if administered before significant amounts of fluoroacetate have been converted to fluorocitrate. To determine if fluoroacetate poisoning has occurred, it is often desirable to detect the presence of small amounts of the poison in animal tissue. Although difficult, this can be done by spectrochemical methods (25), processes involving ion-selective fluoride electrodes (21,26), or gas chromatography often combined with mass spectrometry (27). A microbial detection of fluoroacetate utilizing DNA technology and bioluminescence has been reported (28).

Fluoroacetamide. Fluoroacetamide [640-19-7], FCH_2CONH_2, is a white water-soluble solid having mp 108°C (14). It has been used as a rodenticide and has been reported to have a better acceptability to rats than sodium fluoroacetate (29). However, like the latter compound, its misuse has caused deaths to farm animals and pets (20).

Tull Chemical Co. (Oxford, Alabama) is the only producer of sodium fluoroacetate. It is sometimes colored with the black dye nigrosine. It is usually packed in 8 oz (227 g) or 5 kg cans and is almost exclusively exported. There is very limited use in the United States.

Difluoroacetic Acid

Difluoroacetic acid [381-73-7], $F_2CHCOOH$, is a colorless liquid with a sharp odor; mp, 35°C; bp, 134°C; d_4^{10}, 1.539 g/mL; n_D^{20}, 1.3428 (30,31); flash point, 78°C (95% aqueous solution); enthalpy of vaporization, 67.82 kJ/mol (16.21 kcal/mol); and enthalpy of solution in water of the undissociated acid, -7.03 kJ/mol (-1.68 kcal/mol) (2). It is a moderately strong acid; determinations of its acid dissociation constant are 4.6×10^{-2} (32) and 3.5×10^{-2} (33). Its molecular structure in the gas phase has been determined by electron diffraction studies (34). Details of the acid's ir (35) and nmr (36) spectra also have been reported.

Difluoroacetic acid undergoes reactions typical of a carboxylic acid such as forming an ester when heated with an alcohol and sulfuric acid. Typical esters are methyl difluoroacetate [433-53-4], bp, 85.2°C, and ethyl difluoroacetate [454-31-9], bp, 99.2°C. It can also be photochemically chlorinated to chlorodifluoroacetic acid [76-04-0] or brominated in the presence of iron to bromodifluoroacetic acid [667-27-6] (37,38).

The acid can be synthesized in several different ways. The reaction of tetrafluoroethylene with ammonia to give 2,4,6-tris(difluoromethyl)-s-triazine, followed by its alkaline hydrolysis, has been reported to give the acid in 80% overall yield (31). The addition of diethylamine to tetrafluoroethylene gives, after partial hydrolysis, a 49% yield of amide N,N-diethyldifluoroacetamide [56425-08-2], $F_2CHCON(C_2H_5)_2$, which can be hydrolyzed in excellent yield to the acid (39). The same amide can be prepared in 60% yield by the addition of diethylamine to chlorotrifluoroethylene followed by hydrolysis and fluorination with KF in diethylene glycol. Another method that gives the acid in 86% yield is the permanganate oxidation of CHF_2CH=CCl_2 (40).

Difluoroacetic acid is much less toxic than fluoroacetic acid (LD_{50} = 180 mg/kg mouse iv) (41). It is available in research quantities for about $5/g (1992).

Trifluoroacetic Acid

Physical Properties. Trifluoroacetic acid [76-05-1], CF_3COOH, is a colorless liquid with a sharp odor resembling that of acetic acid. Its physical properties are shown in Table 1. It is a strong carboxylic acid with an acid dissociation constant at 25°C of 0.588 (9) or 0.32 (32). It is miscible with water, fluorocarbons, and most common organic solvents including methanol, benzene, carbon tetrachloride, acetone, ether, and hexane. Compounds with limited solubility in the acid include alkanes with more than six carbon atoms and carbon disulfide. It is a good solvent for proteins (50) and polyesters. The viscosities, densities, and conductivities of solutions of the acid in acetic acid, water, and several other liquids have been studied (46).

Chemical Properties. Trifluoroacetic acid undergoes reactions typical of a carboxylic acid. The trifluoromethyl group is inert to most common reducing agents, including lithium aluminum hydride, which give trifluoroacetaldehyde [75-90-1] and 2,2,2-trifluoroethanol [75-89-8] (51,52). Common oxidizing agents do not attack the acid at room temperature except for potassium permanganate, which slowly oxidizes the anhydrous acid to carbon dioxide and other products (53). The acid is also slowly attacked by boiling 25% aqueous sodium hydroxide to yield oxalate and fluoride ions (44). Although the acid is stable to temperatures above 250°C, its sodium salt decomposes above 205°C to give sodium fluoride, trifluoroacetyl fluoride [354-34-7], carbon monoxide, carbon dioxide, and other products (44). In ethylene glycol solution at 180°C the sodium salt can be made to decompose quantitatively to trifluoromethane [75-46-7] and carbon dioxide if a boric acid buffer is present (54). Except for a few instances like these, the reactions of trifluoroacetic acid closely parallel those of other carboxylic acids, but there are

Table 1. Physical Properties of Trifluoroacetic Acid

Property	Value	References
freezing point, °C	−15.36	42
boiling point, °C	71.8	43
water azeotrope (20.6% H_2O) bp, °C	105.5	44
density at 25°C, g/mL	1.4844	45
heat of vaporization, kJ/mol[a]	33.26	43
viscosity at 25°C, mPa·s(= cP)	0.813	46
dielectric constant at 25°, ε	42.1	47
conductivity at 25°C, 1/Ω·cm	2600	46
surface tension at 25°C mN/m(= dyn/cm)	13.44	48
heat of formation, liquid, kJ/mol[a]	−1060 ± 2	49

[a]To convert kJ to kcal, divide by 4.184.

important differences: eg, its amides and esters are more easily hydrolyzed than is typical for carboxylic acids. This has led to the use of the acid and its anhydride [407-25-0] (55) in making derivatives of carbohydrates (56), amino acids (57), and peptides (57) from which the trifluoroacetyl protective group can be removed with relative ease. Peroxytrifluoroacetic acid [359-48-8], formed from the reaction of trifluoroacetic anhydride and hydrogen peroxide, is a stronger oxidizing agent than other peroxycarboxylic acids and gives better yields of epoxides from alkenes (58), esters from ketones (59), and nitrobenzenes from anilines (60). Trifluoroacetic acid and its anhydride are also useful as catalysts for reactions involving other carboxylic acids such as esterifications of alcohols or acylations of aromatic or other unsaturated compounds (45). The acid has been reported to be superior to sulfuric acid as a catalyst for the Beckmann rearrangement of oximes to amides (61). Owing to its low nucleophilicity, the acid has been used as a solvent for basic research into solvolysis mechanisms (62).

Preparation. Because of its stability to further oxidation, trifluoroacetic acid can be prepared by the oxidation of compounds containing a trifluoromethyl group bonded to carbon. Although first prepared in 1922 by the oxidation of trifluoromethylcyclohexane or 3-aminobenzotrifluoride, later better results were obtained from the alkaline permanganate oxidation of olefins such as 1,1,2-trichloro-3,3,3-trifluoropropene (40), or more economically, 2,3-dichlorohexafluoro-2-butene which gives an 87% yield of the acid (63). The acid has been prepared by photochemical oxidation of ethanes such as 2-chloro-1,1,1-trifluoroethane or 2,2-dichloro-1,1,1-trifluoroethane with oxygen to give high yields of trifluoroacetyl chloride [354-32-5] which easily hydrolyzes to the acid (64,65). Another process involves the trimerization of trichloroacetonitrile to a triazine which can be fluorinated with a mixture of SbF_3 and $SbCl_5$ and then hydrolyzed to the acid (66). The reaction of trichloroacetyl chloride with hydrogen fluoride at 320°C over a chromium and nickel oxide catalyst has been reported to give trifluoroacetic acid in 92% yield (67).

Trifluoroacetic acid was produced commercially by 3M Co. by the electrolysis of mixtures containing acetyl fluoride, hydrogen fluoride, and sodium fluoride to give trifluoroacetyl fluoride, which upon hydrolysis gave the acid (68). Although a 71% yield is claimed, isolation of the low boiling acid fluoride product from by-product hydrogen is costly. Improvements in this process have been patented (69,70) as well as processes involving the electrochemical fluorination of 2-chloroethanol (71) or chloroacetyl fluoride (72).

Health and Safety Factors. Unlike fluoroacetic acid, trifluoroacetic acid presents no unusual toxicity problems. However, owing to its strong acidity, its vapors can be irritating to tissue, and the liquid acid can cause deep burns if allowed to contact the skin. The acid can be safely stored in containers made of glass or common corrosion-resistant alloys and metals such as stainless steel or aluminum.

Economic Aspects. Halocarbon Products Corp. is the largest producer of trifluoroacetic acid. The commercial grade is of very high purity with the main impurity being ca 0.2% water. A grade, which has a low residue specification, intended for use in protein synthesis (Biograde) is available. Other producers include Rhône-Poulenc and Solvay. The 1992 price was ca $15/kg.

BIBLIOGRAPHY

"Monofluoroacetic Acid" under "Fluorine Compounds, Organic," in *ECT* 1st ed., Vol. 6, pp. 764–766, by E. E. Hardy and J. H. Saunders, Monsanto Chemical Co.; "Difluoroacetic Acid" under "Fluorine Compounds, Organic," in *ECT* 1st ed., Vol. 6, pp. 766–767, by M. G. Gergel and M. Revelise, Columbia Organic Chemicals Co.; "Trifluoroacetic Acid" under "Fluorine Compounds, Organic," in *ECT* 1st ed., Vol. 6, pp. 767–768, by M. G. Gergel and M. Revelise, Columbia Organic Chemicals Co.; "Monofluoroacetic Acid" under "Fluorine Compounds, Organic," in *ECT* 2nd ed., Vol. 9, pp. 767–770, by E. Hardy, Monsanto Research Corp., J. H. Saunders, Mobay Chemical Co., and J. B. Hynes, Hynes Chemical Research Corp.; "Difluoroacetic Acid" under "Fluorine Compounds, Organic," in *ECT* 2nd ed., Vol. 9, pp. 770–771, by J. B. Hynes, Hynes Chemical Research Corp.; "Trifluoroacetic Acid" under "Fluorine Compounds, Organic," in *ECT* 2nd ed., Vol. 9, pp. 771–772, by C. Woolf, Allied Chemical Corp.; "Fluorinated Acetic Acids" under "Fluorine Compounds, Organic," in *ECT* 3rd ed., Vol. 10, pp. 891–896, by G. Astrologes, Halocarbon Products Corp.

1. F. Swarts, *Bull. Acad. R. Belg.* **35**, 849 (1898).
2. P. Haberfield and A. K. Rakshit, *J. Am. Chem. Soc.* **98**, 4393 (1976).
3. M. V. Kaulgud and G. H. Pandya, *Indian J. Chem. Sect. A.* **14A**(2), 91 (1976).
4. B. P. Van Eijck, P. Brandts, and J. P. M. Maas, *J. Mol. Struct.* **44**, 1 (1978)
5. F. Swarts, *Bull. Acad. R. Belg.* **31**, 675 (1896).
6. F. L. M. Pattison, *Toxic Aliphatic Fluorine Compounds*, Elsevier Publishing Co., New York, 1959, p. 16.
7. M. J. J. Meyer and N. Grobbelaar, *J. Plant Physiol.* **138**, 122 (1991).
8. G. W. Miller, M. H. Yu, and M. Psenak, *Fluoride* **6**(3), 203 (1973).
9. A. L. Henne and C. J. Fox., *J. Am. Chem. Soc.* **73**, 2323 (1953).
10. F. L. M. Pattison, J. B. Stothers, and R. G. Woolford, *J. Am. Chem. Soc.* **78**, 2255 (1956).
11. F. L. M. Pattison, S. B. D. Hunt, and J. B. Stothers, *J. Org. Chem.* **21**, 883 (1956)
12. C. C. Price and W. G. Jackson, *J. Am. Chem. Soc.* **69**, 1065 (1947).
13. F. L. M. Pattison and co-workers, *Can. J. Technol.* **34**, 21 (1956).
14. F. J. Buckle, R. Heap, and B. C. Saunders, *J. Chem. Soc.*, 912 (1949).
15. B. C. Saunders and G. J. Stacey, *J. Chem. Soc.*, 1773 (1948).
16. E. D. Bergmann and I. Blank, *J. Chem. Soc.*, 3786 (1953).
17. Ref. 6, pp. 21–22.
18. Ref. 6, p. 167.
19. D. D. Clarke, *Neurochem. Res.* **16**, 1055 (1991).
20. R. A. Peters, *Fluoride* **6**(3), 189 (1973).
21. M. N. Egyed, *Fluoride* **6**(3), 215 (1973).
22. P. Buffa, V. Guarriero-Bobyleva, and R. Costa-Tiozzo, *Fluoride* **6**, 224 (1973).
23. Ref. 6, pp. 27–56, 208–210.
24. Ref. 6, pp. 3–4.
25. I. Schoenfeld and M. Lidji, *J. Forsensic Sci.* **13**, 267 (1968).
26. J. A. Peters and K. J. Baxter, *Bull. Environ. Contam. Toxicol.* **11**(2), 177 (1974).
27. H. M. Stahr, W. B. Buck, and P. F. Ross, *J. Assn. Off. Anal. Chem.* **57**, 405 (1974).
28. S. Lee and co-workers, *Anal. Chim. Acta.* **244**, 201 (1991).
29. T. Kusano, *J. Fac. Agric. Tottori Univ.* **10**, 15 (1975).
30. F. Swarts, *Bull. Soc. Chim. Fr.*, 597 (1903).
31. A. L. Henne and R. L. Pelley, *J. Am. Chem. Soc.* **74**, 1426 (1952).
32. J. L. Kurz and J. M. Farrar, *J. Am. Chem. Soc.* **91**, 6057 (1969).
33. M. M. Kreevoy and co-workers, *J. Am. Chem. Soc.* **89**, 1201 (1967).
34. J. M. Bijen and J. L. Derissen, *J. Mol. Struct.* **27**, 233 (1975).
35. J. R. Barcelo and C. Otero, *Spectrochim. Acta* **18**, 1231 (1962).

36. V. Barboiu, *Rev. Roum. Chim.* **19**, 363 (1974).
37. F. Swarts, *Chem. Zentr.* **II**, 709 (1903).
38. F. Swarts, *Chem. Zentr.* **I**, 1237 (1906).
39. N. N. Yarovenko and co-workers, *Obschei Khim.* **27**, 2246 (1957).
40. A. L. Henne, T. Alderson, and M. S. Newman, *J. Am. Chem. Soc.* **67**, 918 (1945).
41. Ref. 6, pp. 62–63.
42. H. H. Cady and G. E. Cady, *J. Am. Chem. Soc.* **76**, 915 (1954).
43. M. D. Taylor and M. B. Templeman, *J. Am. Chem. Soc.* **78**, 2950 (1956).
44. F. Swarts, *Bull. Acad. R. Belg. Classe Sci.* **8**, 343 (1922).
45. *Trifluoracetic Acid Brochure*, Halocarbon Products Corp., Hackensack, N.J., 1967.
46. Y. Y. Fialkov and V. S. Zhikarev, *Zh. Obshch. Khim.* **33**, 3466, 3471, 3790 (1963).
47. J. H. Simons and K. E. Lorentzen, *J. Am. Chem. Soc.* **72**, 1426 (1950).
48. J. J. Jasper and H. L. Wedlick, *J. Chem. Eng. Data* **9**, 446 (1964).
49. V. P. Kolesov, G. M. Slavutskaya, and T. S. Papino, *Zh. Fiz. Khim.* **46**, 815 (1972).
50. J. J. Katz, *Nature* **174**, 509 (1954).
51. M. Braid, H. Iserson, and F. E. Lawlor, *J. Am. Chem. Soc.* **76**, 4027 (1954).
52. O. R. Pierce and T. G. Kane, *J. Am. Chem. Soc.* **76**, 300 (1954).
53. G. S. Fujioka and G. H. Cady, *J. Am. Chem. Soc.* **79**, 2451 (1957).
54. I. Auerbach, F. H. Verhoek, and A. L. Henne, *J. Am. Chem. Soc.* **72**, 299 (1950).
55. J. M. Tedder, *Chem. Rev.* **55**, 787 (1955).
56. E. J. Bourne and co-workers, *J. Chem. Soc.*, 2976 (1949).
57. F. Weygand, *Bull. Soc. Chim. Biol.* **43**, 1269 (1961).
58. W. D. Emmons and A. S. Pagano, *J. Am. Chem. Soc.* **77**, 89 (1955).
59. W. D. Emmons and G. B. Lucas, *J. Am. Chem. Soc.* **77**, 2287 (1955).
60. W. D. Emmons, *J. Am. Chem. Soc.* **76**, 3470 (1954).
61. U.S. Pat. 2,721,199 (Oct. 18, 1955), M. L. Huber (to E. I. du Pont de Nemours & Co., Inc.).
62. P. E. Peterson and co-workers, *J. Am. Chem. Soc.* **87**, 5169 1965).
63. A. L. Henne and P. Trott, *J. Am. Chem. Soc.* **69**, 1820 (1947).
64. R. N. Haszeldine and F. Nyman, *J. Am. Chem. Soc.*, 387 (1959).
65. U.S. Pat. 3,883,407 (May 13, 1975), A. L. Dittman (to Halocarbon Products Corp.).
66. T. R. Norton, *J. Am. Chem. Soc.* **72**, 3527 (1950).
67. Ger. Offen. 2,221,849 (Nov. 16, 1972), Ramanadin (to Rhone-Progil).
68. U.S. Pat. 2,717,871 (Sept. 13, 1965), H. M. Scholberg and H. G. Bryce (to Minnesota Mining and Manufacturing Co.).
69. U.S. Pat. 4,022,824 (May 10, 1977), W. V. Childs (to Phillips Petroleum Co.).
70. Jpn. Kokai 75 30,827 (Mar. 27, 1975), T. Suzuki and S. Yahara (to Mitsubishi Gas Chemical Co., Inc.).
71. USSR Pat. 329,165 (Feb. 9, 1972), N. M. Arakelyan and S. E. Isabekyan.
72. Czech. Pat. 119,682 (Sept. 15, 1966), D. Frantisek.

ARTHUR J. ELLIOTT
Halocarbon Products Corporation

FLUORINATED HIGHER CARBOXYLIC ACIDS

Perfluorinated carboxylic acids are corrosive liquids or solids. The acids are completely ionized in water. The acids are of commercial significance because of their unusual acid strength, chemical stability, high surface activity, and salt solubility characteristics. The perfluoroalkyl acids with six carbons or less are liquids; the higher analogues are solids (Table 1).

The higher members of the series decrease the surface tension of aqueous solutions well below the point possible with any type of hydrocarbon surfactant, although in practice because of their strong acid character and solubility characteristics, more commonly salts and other derivatives are employed. A 0.1% solution of $C_9F_{19}COOH$ has a surface tension of only 19 mN/m (dyn/cm) at 30°C (6).

Table 1. Properties of Perfluoroalkylcarboxylic Acids, $C_nF_{2n+1}COOH$[a]

Acid	CAS Registry Number	Bp, °C	Mp, °C	Density at 20°C, g/mL	Reference
perfluoropropanoic	[422-64-0]	96		1.561	1
perfluorobutanoic	[375-22-4]	120	−17.5	1.641	1,2
perfluoropentanoic	[2706-90-3]	139		1.713	1,3
perfluorohexanoic	[307-24-4]	157		1.762	1,3
perfluorocyclohexane carboxylic[b]	[374-88-9]	168		1.789	1,2
perfluoroheptanoic	[375-85-9]	175		1.792	1,2
perfluorooctanoic	[335-67-1]	189	52–54	1.792	1,2,4
perfluorodecanoic	[335-76-2]	218			1,2
perfluorotetradecanoic		270			1,5

[a]Except where noted.
[b]Cyclo-$C_6F_{11}COOH$.

Preparation

There are five methods for the preparation of long-chain perfluorinated carboxylic acids and derivatives: electrochemical fluorination, direct fluorination, telomerization of tetrafluoroethylene, oligomerization of hexafluoropropylene oxide, and photooxidation of tetrafluoroethylene and hexafluoropropylene.

Many of the perfluoroalkyl carboxylic acids were first prepared by the electrochemical fluorination (ECF) of the corresponding carboxylic acids (7). In ECF acid chlorides are converted to the corresponding perfluoroacid fluorides as shown in equation 1 for octanoyl chloride.

$$C_7H_{15}\overset{O}{\overset{\|}{C}}-Cl + 16\,HF \xrightarrow{ECF} C_7F_{15}\overset{O}{\overset{\|}{C}}-F + C_7F_{16} + \underset{C_4F_9}{\bigcirc_F} + \underset{C_3F_7}{\bigcirc_F} + HCl + H_2 \quad (1)$$

The principal by-products are cyclic perfluoroethers; the fluorine in the center of the ring denotes a perfluorinated ring structure. Octanoyl chloride gives only 10–15% of the perfluorooctanoyl fluoride (8), although the yield of fluorination can be improved by running under different cell conditions (9). As the chain length increases, formation of cyclic ethers and cleavage products becomes more prominent. In addition to cleavage and cyclization products, ECF results in isomerization of the carbon backbone in the product. The relative weight % of C_7F_{15} isomers in $C_7F_{15}COOH$ commercially produced by ECF is 78% linear, 9% internal branched, and 13% terminal (isopropyl) branched (4). Hydrolysis of the acid fluoride followed by removal of the fluoride ion and distillation yields the fluorinated acid.

Perfluorinated acid fluorides containing heteroatoms are also accessible by ECF. Long-chain perfluorinated acid fluorides produced by ECF containing nitrogen (10–12), oxygen (13), and sulfur (14,15) have been reported. The fluorinated mixed sulfonic acid–carboxylic acid precursors are also known. ECF of hydrocarbon sultones has led to formation of $FSO_2(CF_2)_nCOF$, where $n = 2,3$ (16).

Direct fluorination involves the treatment of an appropriate hydrocarbon precursor dissolved in an inert liquid with fluorine gas to yield a perfluorinated precursor to a long-chain carboxylic acid. Equations 2 and 3 illustrate the process for perfluorooctadecanoic acid (17).

$$C_{18}H_{37}O-\overset{O}{\underset{\|}{C}}-CF_3 + 37\ F_2 \rightarrow C_{18}F_{37}O-\overset{O}{\underset{\|}{C}}-CF_3 + 37\ HF \qquad (2)$$

$$C_{18}F_{37}O-\overset{O}{\underset{\|}{C}}-CF_3 + 2\ H_2O \rightarrow C_{17}F_{35}COOH + 2\ HF + HO-\overset{O}{\underset{\|}{C}}-CF_3 \qquad (3)$$

Unlike ECF, direct fluorination does not alter the carbon backbone; preparation of isomerically pure acids is possible (18). Both direct fluorination and ECF permit a great variety of structures to be made, but each method is better at certain types of structures than the other. Ether acids are produced in good yields, by direct fluorination (17), while ECF of ether-containing acids is fair to poor depending on the substrate. Despite much industrial interest, the costs and hazards of handling fluorine gas have prevented commercial application of this process.

Fluorinated carboxylic acids are also prepared by telomerization of tetrafluoroethylene, followed by oxidation (19–21).

$$C_2F_5I + n\ CF_2{=}CF_2 \rightarrow C_2F_5(CF_2CF_2)_nI \xrightarrow{SO_3} C_2F_5(CF_2CF_2)_{n-1}CF_2COOH \qquad (4)$$

This process yields a purely straight-chain acid of even carbon number. Typically, the value of n varies from two to six, and distillation yields the pure components. Du Pont pioneered the development of this technology. Allied has used hexafluoroacetone to produce telomer iodides containing the perfluoroisopropoxy end group, eg, $(CF_3)_2CFO(CF_2CF_2)_nI$ (22,23). Dichromate oxidation (24) or ozonolysis (25) of $C_2F_5(CF_2CF_2)_nCH{=}CH_2$, derived from reaction of the telomer iodide shown in equation 4, gives the acid $C_2F_5(CF_2CF_2)_nCOOH$.

Fluoride ion-catalyzed oligomerization of fluorinated epoxides leads to long-chain ether-containing acids. Equation 5 exemplifies this reaction for hexafluoropropylene oxide (HFPO). Tetrafluoroethylene oxide can undergo similar ring-opening reactions, but is seldom used because of its chemical instability. HFPO, however, is stable at room temperature in an anhydrous atmosphere, and in the absence of acid and base (26). Thermal decomposition occurs only at temperatures of 150°C or higher (27).

$$F^- + (n+2)\ \underset{F}{\overset{F}{\diagup}}\!\!\!\overset{O}{\triangle}\!\!\!\underset{CF_3}{\diagdown^F} \longrightarrow C_3F_7O(CFCF_2O)_n\!\!-\!\!\underset{CF_3}{\overset{\overset{O}{\|}}{CFC}}\!\!-\!\!F + F^- \quad (5)$$
$$\phantom{F^- + (n+2)\ \underset{F}{\overset{F}{\diagup}}\!\!\!\overset{O}{\triangle}\!\!\!\underset{CF_3}{\diagdown^F} \longrightarrow C_3F_7O(CFCF_2O)}\,\underset{CF_3}{|}$$

Hydrolysis of the acid fluoride, removal of fluoride ion, and distillation yield the perfluorinated acid. The value of n typically varies from one to six, depending on reaction conditions. Higher values of n are possible by employing more rigorous conditions. These acids are marketed by Du Pont under the trade name of Krytox acids. This process yields perfluoroether acids containing regular repeat units of perfluoroisopropoxy group and terminated by an alpha-branched carboxylate. The C_3F_7O end of the molecule can be varied by fluoride ion condensation of a perfluorinated acid fluoride with HFPO (26).

Photooxidation of tetrafluoroethylene (TFE) and hexafluoropropylene (HFP) yield peroxides that can be decomposed to esters and ultimately long-chain ether-containing carboxylic acids. Equation 6 shows a simplified version of what occurs during photooxidation and workup (TFE R = F, HFP R = CF_3) (28,29).

$$CFR\!=\!CF_2 + O_2 \xrightarrow{uv\ light} CF_3O(CF_2O)_n(CFCF_2O)_m[O(CF_2O)_n(CFCF_2O)_m]_zO(CFCF_2O)_m(CF_2O)_nCOF$$
$$\phantom{CFR=CF_2 + O_2 \xrightarrow{uv\ light} CF_3O(CF_2O)_n}\underset{R}{|}\underset{R}{|}\underset{R}{|}$$

$$\longrightarrow CF_3O(CF_2O)_n(CFCF_2O)_{m-1}CF_2CO_2CH_3 + CH_3OCOCF_2(CF_2O)_n(CFCF_2O)_{m-2}CF_2CO_2CH_3 \quad (6)$$
$$\underset{R}{|}\phantom{(CFCF_2O)_{m-1}CF_2CO_2CH_3 + CH_3OCOCF_2(CF_2O)_n}\underset{R}{|}$$

The acid is obtained by saponification of the ester. Photooxidation of TFE in practice yields mostly difunctional ether acids. Photooxidation of HFP yields mostly the monofunctional ether acid, but yields significant quantities of inert materials that are difficult to separate out. The segments ($CFRCF_2O$) and (CF_2O) are randomly arranged in the chain. The random arrangement of these units in the chain is thought to be responsible for the exceptional low temperature properties of polymers and other materials derived from this route (30). Molecular weight of the resulting materials is determined by process control parameters and subsequent distillation.

Derivatives

In general, the reactions of the perfluoro acids are similar to those of the hydrocarbon acids. Salts are formed with the ease expected of strong acids. The metal salts are all water soluble and much more soluble in organic solvents than the salts of the corresponding hydrocarbon acids. Esterification takes place readily with primary and secondary alcohols. Acid anhydrides can be prepared by distillation of the acids from phosphorus pentoxide. The amides are readily prepared by the ammonolysis of the acid halides, anhydrides, or esters and can be dehydrated to the corresponding nitriles (31).

The ammonium salts, $C_nF_{2n+1}COONH_4$, where n equals 7 and larger, are particularly useful as emulsifiers in the polymerization of fluorinated olefin monomers such as tetrafluoroethylene or vinylidene fluoride. Their surface activity, low rate of chain transfer, and the highly fluorochemical nature of micelles formed in aqueous media are unique features that give them broad utility in the emulsification and polymerization of fluorochemical monomers (32).

Amines of the formula $C_nF_{2n+1}CH_2NH_2$ can be prepared by the lithium aluminum hydride reduction of the corresponding amide, hydrogenolysis of the nitrile, or diborane reduction of the amide (33). The analogous alcohols, $C_nF_{2n+1}CH_2OH$, can be prepared by high pressure hydrogenation of an appropriate ester (ie, $C_nF_{2n+1}COOCH_3$) over a barium oxide stabilized chromite catalyst, lithium aluminum hydride reduction of the acid or esters, or sodium borohydride reduction of the esters (34,35) or acid fluorides (36,37). Acrylate esters of the dihydroalcohols copolymerize with hydrocarbon acrylates and other monomers and have been found to impart resistance to a variety of organic liquids at elevated temperatures (38).

$$C_nF_{2n+1}CH_2O\overset{O}{\overset{\|}{C}}C(R)=CH_2$$

The surface energy of the homopolymers of these acrylates, 10.6 mN/m(dyn/cm), is among the lowest ever recorded (39), lower even than Teflon, 18.5 mN/m(dyn/cm) (40).

Perfluorodicarboxylic Acids

The lowest members of the series of perfluoroalkanedicarboxylic acids have been prepared and are stable compounds. They have been synthesized by oxidation of the appropriate chlorofluoroolefin as well as by electrochemical fluorination and direct fluorination. Perfluoromalonic acid is an oxidation product of $CH_2=CHCF_2CH=CH_2$ (21). Perfluorosuccinic acid has been produced by oxidation of the appropriate olefin (see eq. 7) (5) or by electrochemical fluorination of succinyl chloride or butyrolactone (41) and subsequent hydrolysis.

$$\boxed{F}\genfrac{}{}{0pt}{}{-Cl}{-Cl} + \text{basic KMnO}_4 \rightarrow \genfrac{}{}{0pt}{}{CF_2COOH}{CF_2COOH} \tag{7}$$

Table 2 lists some typical properties of perfluoroalkanedicarboxylic acids and their esters along with references to their synthesis.

Table 2. Properties of Perfluroalkanedicarboxylic Acids, HOOC(CF$_2$)$_n$COOH

n	Acid	CAS Registry Number	Mp, °C	Bp, °C$_{kPa}$^{a}	Bp of ester,^{b} °C$_{kPa}$^{a}	Reference
1	perfluoromalonic	[1514-85-5]	117		58–58$_{1.2}$	35,42
2	perfluorosuccinic	[377-35-8]	115–116	150$_2$	173	35
3	perfluoroglutaric	[376-73-8]	78–88	134$_{0.4}$	100$_{4.5}$	5
4	perfluoroadipic	[336-08-3]	134		108–110$_4$	35,43,44
6	perfluorosuberic	[678-45-5]	154–158		156–159$_{3.6}$	45
8	perfluorosebacic	[307-78-8]			102–113$_{0.005}$	44
12	perfluorotetradecanedioic		191			35

^{a}To convert kPa to mm Hg, multiply by 7.5.
^{b}Methyl ester unless otherwise noted.
^{c}Ethyl ester.

Fluorinated ether-containing dicarboxylic acids have been prepared by direct fluorination of the corresponding hydrocarbon (17), photooxidation of tetrafluoroethylene, or by fluoride ion-catalyzed reaction of a diacid fluoride such as oxalyl or tetrafluorosuccinyl fluorides with hexafluoropropylene oxide (46,47). Equation 8 shows the reaction of oxalyl fluoride with HFPO. A difunctional ether-containing acid fluoride derived from HFPO contains regular repeat units of perfluoroisopropoxy group and is terminated by two alpha-branched carboxylates.

$$F\text{—}\overset{\overset{O}{\|}}{C}\text{—}\overset{\overset{O}{\|}}{C}\text{—}F + F^- + (n+2)\ \underset{F}{\overset{F}{\diagdown}}\!\!\!\underset{CF_3}{\overset{O}{\diagup}}\ \longrightarrow$$

$$\underset{CF_3}{\overset{\overset{O}{\|}}{FC}}\text{—}CFOCF_2CF_2O\underset{CF_3}{(CFCF_2O)_n}\text{—}\underset{CF_3}{\overset{\overset{O}{\|}}{CFC}}\text{—}F + F^- \quad (8)$$

Fluorinated diacids offer a convenient method for introducing a perfluoro moiety into organic molecules. They are of potential interest in the preparation of polyamides and other fluorinated polymers. A detailed description of the perfluorocarboxylic acids and their derivatives has been published (1), and a review article on polyfluorinated linear bifunctional compounds has appeared (35).

Derivatives similar to those mentioned for the monofunctional fluorinated carboxylic acids have been prepared: tetrahydrodiols, tetrahydrodiamines, diamides, and diesters.

Toxicology and Safety. Because of their strong acidity, the perfluorinated carboxylic acids themselves are corrosive to the skin and eyes. Protective clothing, ie, gloves and face shield/safety glasses, should be employed when handling them (48,49). Although perfluorooctanoic acid has been found to be corrosive to the eyes and severely irritating to the skin on contact, it is only slightly toxic on oral contact. The acute LD$_{50}$ (rat) of perfluorooctanoic acid was found to be slightly less than 1000 mg per kg of body weight. However, in feeding studies the test animals suffered erosion of the gastric mucous membrane because of corrosivity of the acid

(50). Perfluorodecanoic acid has been assessed for its genotoxic activity. It was found to test negative in the Ames test, Chinese hamster ovary gene mutation assay, sister chromatid exchange assay, chromosomal aberration assay, and *in vivo/in vitro* unscheduled DNA synthesis. Chromosomal aberrations were observed only when the S-9 fraction was incubated with perfluorodecanoic acid in the S-phase DNA synthesis assay (51).

The salts of the perfluorinated acids are not corrosive, so one is in a better position to discuss toxicity not related to corrosivity. The toxicity of the salts varies depending on the exact structure. The ammonium salt of perfluorooctanoic acid is nonirritating to the skin and moderately irritating to the eyes. Its oral toxicity is rated at moderate; the LD_{50} is 540 mg per kg of body weight (52). There has been some concern in the past that ammonium perfluorooctanoate was teratogenic. More recent results indicate that it is neither embryotoxic nor teratogenic (52,53). It was not found to be mutagenic in either the Ames assay or one employing *Saccharomyces cerevisiae* D4 yeast (52). It also did not cause cell transformation in a mammalian cell transformation assay (53). Although ammonium perfluorooctanoate was fed to albino rats for two years, no compound-induced carcinogenicity was found in the study. There were statistically significant compound-related benign testicular tumors (52,53). Prolonged or repeated exposure can cause liver damage which results in jaundice or tenderness of the upper abdomen (53). The dust from the ammonium salts of the perfluorinated acids is irritating to breathe and should only be handled in a well-ventilated area or preferably a hood.

BIBLIOGRAPHY

"Heptafluorobutyric Acid" under "Fluorine Compounds, Organic" in *ECT* 1st ed., Vol. 6, p. 769, by M. G. Gergel and M. Revelise, Columbia Organic Chemicals Co., Inc.; "Other Perfluorocarboxylic Acids" under "Fluorine Compounds, Organic" in *ECT* 1st ed., Vol. 6, pp. 769–771, by L. J. Hals and W. H. Pearlson, Minnesota Mining & Manufacturing Co.; "Heptafluorobutyric Acid" under "Fluorine Compounds, Organic" in *ECT* 2nd ed., Vol. 9, p. 773, by M. G. Gergel, Columbia Organic Chemicals Co., Inc.; "Other Perfluorocarboxylic Acids" under "Fluorine Compounds, Organic" in *ECT* 2nd ed., Vol. 9, pp. 773–775, by L. J. Hals and W. H. Pearlson, Minnesota Mining & Manufacturing Co.; "Fluorinated Higher Carboxylic Acids" under "Fluorine Compounds, Organic" in *ECT* 3rd ed., Vol. 10, pp. 897–900, by R. A. Guenthner, 3M Co.

1. A. M. Lovelace, W. Postelnek, and D. A. Rausch, *Aliphatic Fluorine Compounds*. ACS monograph 138 Reinhold Publishing Co., New York, 1958.
2. U.S. Pat. 2,567,011 (Sept. 4, 1951), A. R. Diesslin, E. A. Kauck, and J. H. Simons (to 3M Co.).
3. E. A. Kauck and A. R. Diesslin, *Ind. Eng. Chem.* **43**, 2332 (1952).
4. *Fluorad Fluorochemical Acid FC-26*, Minnesota Mining & Manufacturing Co., St. Paul, Minn., 1986.
5. A. L. Henne and W. J. Zimmerscheid, *J. Am. Chem.* **69**, 281 (1947).
6. H. M. Scholberg, "Surface Chemistry of Fluorocarbons and Their Derivatives," in *Abstracts of Papers, 116th Meeting, Am. Chem. Soc.* Atlantic City, N.J., Sept. 1949, p. 36K.
7. J. H. Simons and co-workers, *J. Electrochem. Soc.* **95**, 47–67 (1949).

8. H. C. Fielding, in R. E. Banks, ed., *Organofluorine Chemicals and their Application*, Ellis Howard, 1979, p. 216.
9. U.S. Pat. 3,919,057 (Nov. 11, 1975), E. Plattner, C. Comninellis, and P. Javet (to CIBA-GEIGY AG).
10. U.S. Pat. 3,471,484 (Oct. 7, 1969), R. A. Guenthner (to 3M Co.).
11. T. Abe and co-workers, *J. Fluorine Chem.* **48**, 257–279 (1990).
12. A. Dimitrov and S. T. Rudiger, *J. Fluorine Chem.* **50**, 197–205 (1990).
13. U.S. Pat. 2,826,564 (Mar. 11, 1958), F. A. Bovey, and J. F. Abere (to 3M Co.).
14. T. J. Brice, R. I. Coon, and W. A. Severson, "Properties of Some Fluorocarbon Derivatives of Sulfur Hexafluoride," paper presented at *American Chemical Society*, Minneapolis, Minn., 1955.
15. Eur. Pat. Appl. 444,822 (Feb. 28, 1991), J. C. Hansen and P. M. Savu (to 3M Co.).
16. Eur. Pat. Appl. 058,466 (Dec. 18, 1985), F. E. Behr and R. J. Koshar (to 3M Co.).
17. WO Pat. 90/06,296 (June 14, 1990), M. G. Costello and G. G. I. Moore (to 3M Co.).
18. R. M. Flynn, T. A. Kestner, and G. G. I. Moore, "Stereochemistry of the Direct Fluorination Process," *Tenth Winter Fluorine Conference*, St. Petersburg, Fla., 1991.
19. U.S. Pat. 3,132,185 (May 5, 1964), R. E. Parsons (to E. I. du Pont de Nemours and Co., Inc.); U.S. Pat. 3,226,449 (Dec. 28, 1965), W. A. Blanchard and J. C. Rhode (to E. I. du Pont de Nemours and Co., Inc.); U.S. Pat. 3,234,294 (Feb. 8, 1966), R. E. Parsons (to E. I. du Pont de Nemours and Co., Inc.); U.S. Pat. 4,425,199 (Jan. 10, 1984), M. Hamada, J. Ohmura, and F. Muranaka (to Asahi Kasei Kogyo Kabushi Kaisha).
20. M. Hauptschein, *J. Am. Chem. Soc.* **83**, 2500 (1961).
21. Brit. Pat. 1,004,575 (Sept. 15, 1962), M. Hauptschein and Parris (to Pennsalt Chem Corp.).
22. Brit. Pat. 1,165,912 (Oct. 1, 1969), Evans and Litt (to Allied Chem Corp.); U.S. Pat. 3,558,721 (Jan. 26, 1971), C. C. Y. Yao (to Allied Chem Corp.).
23. F. W. Evans and co-workers, *J. Org. Chem.* **33**, 1839 (1968).
24. U.S. Pat. 3,525,758 (Aug. 25, 1970), A. Katsushima (to Daikin Kogyo Kabushi Kaisha).
25. U.S. Pat. 4,138,417 (Feb. 6, 1979), H. Ukiihashi and co-workers, (to Asahi Glass Co.).
26. H. Millauer, W. Schwertfeger, and G. Siegemund, *Ang. Chem. Int. Ed. Eng.* **24**, 161–179 (1985).
27. H. S. Eleuterio, *J. Macmol. Sci.-Chem.* **A6**, 1027 (1972).
28. U.S. Pat. 3,442,942 (May 6, 1966), D. Sianesi and co-workers (to Montecatini Edison sPa).
29. U.S. Pat. 3,847,978 (Nov. 12, 1974), D. Sianesi and G. Caporiccio (to Montecatini Edison sPa).
30. U.S. Pat. 3,810,874 (May 14, 1974), R. A. Mitsch and J. L. Zollinger (to 3M Co.).
31. D. R. Husted and A. H. Diesslin, *J. Am. Chem. Soc.* **75**, 1605 (1953).
32. *Fluorad Fluorochemical Surfactant FC-143*, Minnesota Mining & Manufacturing Co., St. Paul, Minn., 1987.
33. T. Takakura and N. Sugiyama, *Asahi Garasu Kenkyu Hokosu* **37**, 257–262 (1987).
34. U.S. Pat. 4,156,791 (May 29, 1979), W. V. Childs (to Phillips Petroleum).
35. I. L. Knunyants, L. Chih-yuan, and V. V. Shokina, *Advances in Chem. (Uspekhi Khimi)* **32**, original 1502, Eng. trans. 461–476 (1963); translation RSIC-165, Redstone Information Center.
36. U.S. Pat. 3,293,306 (1966), R. E. Bleu and J. H. Fasenacht (to E. I. du Pont de Nemours and Co., Inc.).
37. U.S. Pat. 3,574,770 (Apr. 13, 1971), E. C. Stump, Jr., and S. E. Rochow (to NASA).
38. F. A. Bovey and co-workers, *J. Polym. Sci.* **XV**, 520–536 (1955).
39. M. K. Bernett and W. A. Zisman, *J. Chem. Physics* **66**, 1207 (1962).
40. H. W. Fox and W. A. Zisman, *J. Colloid Sci.* **7**, 109 (1952).

41. Jpn. Kokai Tokkyo Koho, JP 59,177,384 [87,177,384] (Oct. 8, 1984), M Hamada and F. Muranaka (to Asahi Chem. Industry Co.).
42. A. L. Henne and H. G. DeWitt, *J. Am. Chem. Soc.* **70**, 1548 (1948).
43. E. T. McBee, P. A. Wiseman, and G. B. Bachman, *Ind. Eng. Chem.* **39**, 415 (1947).
44. U.S. Pat. 5,093,432 (Mar. 3, 1992), T. R. Bierschenk and co-workers (to Exfluor Research Co.).
45. U.S. Pat. 2,606,206 (Aug. 5, 1952), R. Guenthner (to 3M Co.).
46. J. T. Hill, *J. Macromol. Sci-Chem* **A8**, 499–520 (1974).
47. U.S. Pat. 4,647,413 (Mar. 3, 1987), P. M. Savu (to 3M Co.).
48. *Material Safety Data Sheet Fluorad Brand Fluorochemical Acid FC-23*, Minnesota Mining & Manufacturing Co., St. Paul, Minn., July 22, 1992.
49. *Material Safety Data Sheet Fluorad Brand Fluorochemical Acid FC-26*, Minnesota Mining & Manufacturing Co., St. Paul, Minn., July 6, 1992.
50. *Product Toxicity Summary Sheet Fluorad Brand Fluorochemical Acid FC-26*, Minnesota Mining & Manufacturing Co., St. Paul, Minn., Dec. 1989.
51. C. S. Godin and co-workers, *Fundam. Appl. Toxicol.* **18**, 45431–0009 (1992).
52. *Product Toxicity Summary Sheet Fluorad Brand Fluorochemical Acid FC-143*, Minnesota Mining & Manufacturing Co., St. Paul, Minn., Apr. 26, 1990.
53. *Material Safety Data Sheet Fluorad Brand Fluorochemical Acid FC-143*, Minnesota Mining & Manufacturing Co., St. Paul, Minn., June 8, 1992.

PATRICIA M. SAVU
3M Company

PERFLUOROALKANESULFONIC ACIDS

Perfluoroalkanesulfonic acids and their derivatives are of commercial significance because of their unusual acid strength, chemical stability, and the surface activity of the higher members of the series (eight carbons and larger).

Preparation

The perfluoroalkane sulfonic acids were first reported in 1954. Trifluoromethanesulfonic acid was obtained by the oxidation of bis(trifluoromethyl thio) mercury with aqueous hydrogen peroxide (1). The preparation of a series of perfluoroalkanesulfonic acids derived from electrochemical fluorination (ECF) of alkane sulfonyl halides was also disclosed in the same year (2). The synthetic operations employed when the perfluoroalkanesulfonic acid is derived from electrochemical fluorination, which is the best method of preparation, are shown in equations 1–3.

$$R_h SO_2 F + HF \rightarrow R_f SO_2 F + H_2 \qquad (1)$$

where R_h is an alkyl group and R_f is a perfluoroalkyl group

$$R_f SO_2 F + KOH \rightarrow R_f SO_3 K + HF \qquad (2)$$
$$R_f SO_3 K + H_2 SO_4 \rightarrow R_f SO_3 H + KHSO_4 \qquad (3)$$

Perfluorosulfonyl fluorides can also be prepared by the electrochemical fluorination of saturated or unsaturated cyclic sulfones (3–5). Perfluorobutanesulfonyl fluoride can be prepared in 40–48% yield from sulfolane (eq. 4) (6).

$$\text{sulfolane} \xrightarrow{\text{ECF}} CF_3CF_2CF_2CF_2SO_2F \qquad (4)$$

Yields of sulfonyl fluorides prepared by ECF vary depending on the particular structure. Chain degradation becomes more important as the chain length increases (6). Yields can vary from 96% for perfluoromethanesulfonyl fluoride (7) to 43–50% for perfluorooctanesulfonyl fluoride (8).

Trifluoromethanesulfonic acid can be prepared via trifluoromethanesulfenyl chloride as shown in equations 5–7 (9).

$$CF_3S\text{–}SCF_3 + Cl_2 \rightarrow 2\ CF_3SCl \qquad (5)$$

$$CF_3SCl + 2\ HOCl \rightarrow CF_3SO_2Cl + 2\ HCl \qquad (6)$$

$$CF_3SO_2Cl + NaOH \rightarrow CF_3SO_3Na + HCl \qquad (7)$$

Other preparations of trifluoromethanesulfonic acid include oxidation of methyl-trifluoromethyl sulfide under a variety of conditions (10,11). Perfluorosulfonyl fluorides have also been prepared by reaction of fluoroolefins with sulfuryl fluoride (12,13). Chinese chemists have published numerous papers on the conversion of telomer-based alkyl iodides to sulfonyl fluorides (14,15) (eqs. 8 and 9):

$$2\ R_fI + Na_2S_2O_4 + NaHCO_3 \rightarrow 2\ R_fSO_2Na \qquad (8)$$

$$R_fSO_2Na + Cl_2 \rightarrow R_fSO_2Cl + NaCl \qquad (9)$$

Perfluorosulfonyl fluorides have also been prepared by direct fluorination, although in general yields are lower than preparation by ECF. Perfluoromethanesulfonyl fluoride has been produced in 15% yield from direct fluorination of dimethyl sulfone (16). Perfluoro-2-propanesulfonyl fluoride was prepared in 29% yield from propanesulfonyl fluoride (17). Direct fluorination of tetramethylene sulfone leads to the intact perfluorinated sulfone in 28% yield and the ring-opened product (perfluorobutanesulfonyl fluoride) in 10% yield (eq. 10) (17).

$$\text{sulfolane} \xrightarrow{F_2/He} \text{perfluorosulfolane} + CF_3CF_2CF_2CF_2SO_2F \qquad (10)$$

Currently, the commercially important methods of preparations of perfluorinated sulfonic acid derivatives are electrochemical fluorination and sulfur trioxide addition to tetrafluoroethylene with subsequent ring opening.

The boiling points of a series of perfluoroalkanesulfonic acids are listed in Table 1 (2).

FLUORINE COMPOUNDS, ORGANIC (ALKANESULFONIC)

Table 1. Boiling Points of Perfluoroalkanesulfonic Acids

Compound	CAS Registry Number	Bp, °C/kPa[a]	Bp, °C[b]
CF_3SO_3H	[1493-13-6]	60/0.4	166
$C_2F_5SO_3H$	[354-88-1]	81/2.9	175[c]
$C_4F_9SO_3H$	[59933-66-3]	76–84/0.017	200[c]
$C_5F_{11}SO_3H$	[3872-25-1]	110/0.67[d]	212[c,e]
$C_6F_{13}SO_3H$	[355-46-4]	95/0.47	225[c]
$C_8F_{17}SO_3H$	[1763-23-1]	133/0.8	249
4-CF_3(cyclo-C_6F_{10})SO_3H	[374-62-9][f]	120/0.4	241
4-C_2F_5(cyclo-C_6F_{10})SO_3H	[335-24-0][f]		254

[a]To convert kPa to mm Hg, multiply by 7.5.
[b]At 101.3 kPa = 1 atm.
[c]Estimated.
[d]The hydrate, $C_5F_{11}SO_3H \cdot H_2O$.
[e]$C_5F_{11}SO_3H$ anhydrous.
[f]Potassium salt.

Trifluoromethanesulfonic Acid

The first member of the series, CF_3SO_3H, has been extensively studied. Trifluoromethanesulfonic acid [1493-13-6] is a stable, hydroscopic liquid which fumes in air. Addition of an equimolar amount of water to the acid results in a stable, distillable monohydrate, mp 34°C, bp 96°C at 0.13 kPa (1 mm Hg) (18). Measurement of conductivity of strong acids in acetic acid has shown the acid to be one of the strongest protic acids known, similar to fluorosulfonic and perchloric acid (19).

Trifluoromethanesulfonic acid is miscible in all proportions with water and is soluble in many polar organic solvents such as dimethylformamide, dimethylsulfoxide, and acetonitrile. In addition, it is soluble in alcohols, ketones, ethers, and esters, but these generally are not suitably inert solvents. The acid reacts with ethyl ether to give a colorless, liquid oxonium complex, which on further heating gives the ethyl ester and ethylene. Reaction with ethanol gives the ester, but in addition dehydration and ether formation occurs.

Alkyl esters of trifluoromethanesulfonic acid, commonly called triflates, have been prepared from the silver salt and an alkyl iodide, or by reaction of the anhydride with an alcohol (18,20,21). Triflates of the 1,1-dihydroperfluoroalkanols, $CF_3SO_2OCH_2R_f$, can be prepared by the reaction of perfluoromethanesulfonyl fluoride with the dihydroalcohol in the presence of triethylamine (22,23). Triflates are important intermediates in synthetic chemistry. They are among the best leaving groups known, so they are commonly employed in anionic displacement reactions.

The metallic salts of trifluoromethanesulfonic acid can be prepared by reaction of the acid with the corresponding hydroxide or carbonate or by reaction of sulfonyl fluoride with the corresponding hydroxide. The salts are hydroscopic but can be dehydrated at 100°C under vacuum. The sodium salt has a melting point of 248°C and decomposes at 425°C. The lithium salt of trifluoromethanesulfonic acid [33454-82-9], CF_3SO_3Li, commonly called lithium triflate, is used as a battery electrolyte in primary lithium batteries because solutions of it exhibit high electrical conductivity, and because of the compound's low toxicity and excellent chemical stability. It melts at 423°C and decomposes at 430°C. It is quite soluble in polar organic solvents and water. Table 2 shows the electrical conductivities of lithium triflate in comparison with other lithium electrolytes which are much more toxic (24).

Due to the strong ionic nature of lithium trifluoromethanesulfonate, it can increase the conductivity of coating formulations, and thereby enhance the dissipation of static electricity in nonconducting substrates (see ANTISTATIC AGENTS) (25).

Trifluoromethanesulfonic acid anhydride, bp 84°C, is prepared by refluxing the acid over an excess of phosphorous pentoxide (18,26). The anhydride reacts instantaneously with ammonia or amines to form trifluoromethanesulfonamides. The anhydride reacts with most polar organic solvents. It polymerrmizes THF to give a living polyether having cationic activity at each chain end (27).

Several excellent review articles (28–31) cover the chemistry of the acid and its derivatives in great detail. Trifluoromethanesulfonic acid is available from the 3M Co. as Fluorochemical Acid FC-24; the lithium salt is available as Fluorochemical Specialties FC-122, FC-123, and FC-124 (32).

Table 2. Comparative Electrical Conductivity[a] of Lithium Salts

Concentration, M	CF_3SO_3Li	$LiClO_4$	$LiAsF_6$
0.5	24.4	29.9	26.3
0.1	5.81	7.4	6.94
0.05	3.12	3.83	3.57
0.01	0.70	0.83	0.79

[a] In water at 25°C, $ohm^{-1} cm^{-1} \times 10^{-3}$.

Higher Perfluoroalkanesulfonic Acids

The longer perfluoroalkanesulfonic acids are hydroscopic oily liquids. Distillation of the acid from a mixture of its salt and sulfuric acid gives a hydrated mixture with melting points above 100°C. These acids show the same general solubilities as trifluoromethanesulfonic acid, but are insoluble in benzene, heptane, carbon tetrachloride, and perfluorinated liquids. All of the higher perfluoroalkanesulfonic acids have been prepared by electrochemical fluorination (20).

The longer-chain acids and their salts, particularly $C_8F_{17}SO_3H$ and higher, are surface-active agents in aqueous media. They reduce the surface tension of water to levels not possible with hydrocarbon surfactants. The surfactant $C_8F_{17}SO_2N(C_2H_5)CH_2COOK$ [2991-51-7] lowers the surface tension of water to 17 mN/m(= dyn/cm) at 0.2 weight percent (33) and exhibits outstanding thermal and chemical stabilities. The potassium salt of perfluorooctanesulfonic acid [2795-39-3], $C_8F_{17}SO_3K$, or perfluoroethylcyclohexanesulfonic acid [335-24-0], C_2F_5-cyclo-(C_6F_{10})-SO_3K, can form a stable foam in hostile media such as chromium trioxide and sulfuric acid where conventional hydrocarbon and silicone surfactants would be destroyed (34). The ability of these materials to foam concentrated sulfuric acid is utilized to prevent sulfuric acid from aerosoling into the air in industrial situations where chrome plating is done. Instead of forming an aerosol, the sulfuric acid forms a foam blanket on top of the plating bath. The foam derived from $C_8F_{17}SO_3K$ is generally more stable and dense than that derived from C_2F_5-cyclo-(C_6F_{10})-SO_3K. As a result these fluorochemical surfactants are often used in combination to produce desired wetting and foaming activity.

Generally, derivatives of the longer-chain perfluoroalkanesulfonic acids have a number of unique surface-active properties and have formed a basis for a number of commercial products. Derivatives of N-alkyl perfluorooctanesulfonamidoethanol, $C_8F_{17}SO_2N(R)CH_2CH_2OH$, and polymers of N-alkyl perfluorooctanesulfonamidoethyl methacrylate, $C_8F_{17}SO_2N(R)CH_2CH_2OCOC(CH_3)=CH_2$, impart soil, oil, and water repellency to treated fabrics and paper; this forms the basis for 3M's Scotchguard and Scotchban products (35). Polymers of N-alkyl perfluorooctanesulfonamidoethyl acrylates, $C_8F_{17}SO_2N(R)CH_2CH_2OCOC(R')=CH_2$, with certain hydrocarbon acrylates and methacrylates have also been found to be surface-active agents in organic solvents and water. These polymers have applications in the areas of secondary crude oil recovery and wetting, leveling, and flow control agents (36,37).

Higher perfluoroalkanesulfonates are slightly more reactive than triflates toward nucleophilic displacements. The rate constants for acetolysis of methyl nonafluorobutanesulfonate [6401-03-2], methyl trifluoromethanesulfonate [333-27-7], and methyl toluenesulfonate [80-48-8] are 1.49×10^{-4}, 7.13×10^{-5}, and 3.1×10^{-9} s^{-1}, respectively. This means that the relative reactivities for nonafluorobutanesulfonate, trifluoromethanesulfonate, and toluenesulfonate are 48,000/22,900/1 (38).

Difunctional Perfluoroalkanesulfonic Acids

Alpha, omega-perfluoroalkanedisulfonic acids were first prepared by aqueous alkali permanganate oxidation of the bis-sulfone, $RSO_2(CF_2CF_2)_nSO_2R$ (39). Disulfonyl fluorides of the formula $FSO_2(CF_2)_nSO_2F$ have also been prepared by electrochemical fluorination where $n = 1$ to 5. These disulfonyl fluorides have been converted to the cyclic anhydrides by basic hydrolysis of the disulfonyl fluoride, acidification, and dehydration with phosphorous pentoxide (40). The alpha, omega-perfluoroalkanedisulfonic acids can also be prepared by the action of sodium dithionate on the diiodides to form the disulfinate salt, followed by chlorination and hydrolysis to give the disulfonic acids (14,15).

Carbonyl sulfonyl fluorides of the formula $FCO(CF_2)_nSO_2F$ have been prepared by electrochemical fluorination of hydrocarbon sultones (41,42). More commonly in a technology pioneered by Du Pont, perfluoroalkanecarbonyl sulfonyl fluorides are prepared by addition of SO_3 to tetrafluoroethylene followed by isomerization with a tertiary amine such as triethylamine (43).

$$CF_3{=}CF_2 + SO_3 \rightarrow \underset{O}{\overset{O}{\underset{\|}{S}}}\!\!\!\diagdown\!\!\!\overset{F_2C-CF_2}{\underset{O}{|\quad\quad|}}\quad \xrightarrow{(CH_3CH_2)_3N} \quad F-\underset{\underset{O}{\|}}{\overset{\overset{O}{\|}}{S}}-CF_2\overset{O}{\overset{\|}{C}}-F \quad (11)$$

Fluorosulfonyldifluoroacetyl fluoride [677-67-8] is an important industrial intermediate used in the production of Du Pont's Nafion ion-exchange membrane. Nafion is an ion-exchange membrane used under the extreme conditions in electrolytic cells, especially for the electrolysis of sodium chloride to produce chlorine and caustic soda. Other commercial fluorinated ion-exchange resins utilize perfluorinated carboxylate groups instead of sulfonate groups in the monomer in order to make the polymer conductive. Shown in equation 12 are the synthetic operations used to convert fluorosulfonyldifluoroacetyl fluoride to its polymerizable form.

$$\underset{\underset{O}{\overset{O}{\|}}}{\overset{\overset{O}{\|}}{F-S}}-CF_2\overset{\overset{O}{\|}}{C}-F + 2\ \underset{F}{\overset{F}{\diagdown}}\!\!\overset{O}{\triangle}\!\!\underset{CF_3}{\diagup} \xrightarrow{F^-} \underset{\underset{O}{\overset{O}{\|}}}{\overset{\overset{O}{\|}}{F-S}}-CF_2CF_2OCFCF_2O-\underset{CF_3}{\overset{\overset{O}{\|}}{C}FC}-F \xrightarrow{Na_2CO_3}$$

$$\underset{\underset{O}{\overset{O}{\|}}}{\overset{\overset{O}{\|}}{F-S}}-CF_2CF_2OCFCF_2O-\underset{CF_3}{CF}=CF_2 + 2\ NaF + 2\ CO_2 \quad (12)$$

The vinyl ether in the latter part of the equation is copolymerized with tetrafluoroethylene, and then the sulfonyl fluoride group is hydrolyzed under basic conditions in order to produce the ion-exchange membrane (44–46).

BIBLIOGRAPHY

"Perfluoroalkane Sulfonic Acids" under "Fluorine Compounds, Organic" in *ECT* 3rd ed., Vol. 10, pp. 952–955, by R. A. Guenthner, 3M Co.

1. R. N. Hazeline and J. M. Kidd, *J. Chem. Soc.*, 4228 (1954).
2. P. W. Trott and co-workers, *126th National Meeting of the American Chemical Society*, New York, 1954, abstract p. 42-M.
3. U.S. Pat. 3,623,963 (1971), P. Voss (to Bayer AG); U.S. Pat. 3,951,762 (1972), P. Voss (to Bayer AG).
4. Brit. Pat. 1,099,240 (1968) (to Dow Corning).
5. I. N. Rozhkow, A. V. Bukhtiarov, and I. L. Knunyants, *Ikv. Akad. Nauk SSR, Ser. Khim.* **4**, 945 (1969).
6. T. Abe and S. Nagase, in R. E. Banks, ed., *Preparation, Properties, and Industrial Applications of Organofluorine Compounds*, Ellis Howard, 1982, p. 37.
7. T. Gramstad and R. N. Hazeldine, *J. Chem. Soc.*, 173 (1956); *Ibid.*, 2640 (1957).
8. Ger. Offen. 2,201,649 (1973), P. Voss, H. Niederprum, and M. Wechsberg; Ger. Offen. 2,234,837 (1974), P. Heinze and M. Schwarzmann; Ch. Comninellis, Ph. Javet, and E. Platterner, *J. Appl. Electrochem.* **4**, 287 (1974).
9. R. N. Hazeltine and J. M. Kidd, *J. Chem. Soc.*, 2901 (1955).
10. R. N. Hazeltine and co-workers, *Chem. Commum.*, 249 (1972).
11. R. B. Ward, *J. Org. Chem.* **30**, 3009 (1965).
12. U.S. Pat. 3,542,864 (Nov. 24, 1970), R. J. Koshar (to 3M Co.).
13. Brit. Pat. 1,189,561 (1970) (to Du Pont).
14. H. Weiyuan and C. Qingyun, *Chemistry* **2**, 31–76 (1987).
15. W. Y. Huang, *J. Fluorine Chem.* **32**, 179–195 (1986).

16. R. J. Lagow and co-workers, *J. C. S. Perkin I* **11**, 2675–2678 (1979).
17. H. Huang, H. Roesky, and R. J. Lagow, *Inorg. Chem.* **30**, 789–794 (1991).
18. T. Gramstad and R. N. Hazeldine, *J. Chem. Soc.*, 4069 (1957).
19. T. Gramstad, *Tidsskr, Kremi, Bergres, Metall.* **19**, 62 (1959).
20. U.S. Pat. 2,732,398 (Jan. 12, 1956), T. J. Brice and P. W. Trott (to 3M Co.).
21. P. G. Gassman and C. K. Harrington, *J. Org. Chem.* **49**, 2258–2273 (1984).
22. K. A. Epstein and co-workers, "Fluorinated Ferroelectric Liquid Crystals: Overview and Synthesis," *Eleventh Winter Fluorine Conference*, St. Petersburg, Fla., 1993.
23. U.S. Pat. 3,419,595 (Dec. 31, 1968), R. L. Hansen (to 3M Co.).
24. *Fluorad Fluorochemical Specialties FC-122*, and *Fluorad Fluorochemical Specialties FC-124*, Minnesota Mining & Manufacturing Co., St. Paul, Minn., 1986.
25. *Fluorad Fluorochemical Specialties FC-123*, Minnesota Mining & Manufacturing Co., St. Paul, Minn., 1992.
26. J. Burden and co-workers, *J. Chem. Soc.*, 2574 (1957).
27. S. Smith and A. J. Hubin, *J. Macromol. Sci.-Chem.* **A7**, 1399–1413 (1973).
28. A. Senning, *Chem. Rev.* **65**, 385 (1965).
29. R. D. Howells and J. D. McCown, *Chem. Rev.* **77**, 69 (1977).
30. P. J. Stang and M. R. White, *Aldrichimica Acta* **16**, 15 (1983).
31. P. J. Stang, M. Hanack, and L. R. Subramanian, *Synthesis*, 85 (1982).
32. *Fluorad Fluorochemical Acid FC-24*, Minnesota Mining & Manufacturing Co., St. Paul, Minn., 1978.
33. *Fluorad Fluorochemical Specialties FC-129*, Minnesota Mining & Manufacturing Co., St. Paul, Minn., 1991.
34. *Fluorad Fluorochemical Specialties FC-95*, Minnesota Mining & Manufacturing Co., St. Paul, Minn., 1987.
35. H. C. Fielding, in Ref. 6, pp. 226–232.
36. *Fluorad Fluorochemical Specialties FC-430/431*, Minnesota Mining & Manufacturing Co., St. Paul, Minn., 1990.
37. *Fluorad Fluorochemical Specialties FC-740*, Minnesota Mining & Manufacturing Co., St. Paul, Minn., 1987.
38. R. L. Hansen, *J. Org. Chem.* **30**, 4322–4324 (1965).
39. U.S. Pat. 3,346,606 (Oct. 10, 1967), R. B. Ward (to Du Pont).
40. U.S. Pat. 4,329,478 (May 11, 1982), F. E. Behr (to 3M Co.).
41. U.S. Pat. 4,332,954 (June 1, 1982), R. J. Koshar (to 3M Co.).
42. Eur. Pat. Appl. 062,430 (1982), M. Hamada, J. Ohmura, and F. Muranaka.
43. U.S. Pat. 2,852,554 (Sept. 16, 1958), D. C. England and H. Oak (to Du Pont).
44. Y. Yen, C. Nieh, and L. C. Hsu, *Fluorinated Polymers, Process Economics Report #166*, SRI International, Menlo Park, Calif., 1983, p. 165.
45. A. Eisenberg and H. L. Yeager, eds., *Perfluorinated Ionomer Membranes, ACS Series 180*, American Chemical Society, Washington, D.C., 1982.
46. L. A. Ultracki and R. A. Weiss, eds., *Multiphase Polymers: Blends and Ionomers, ACS Symposium Series 395*, American Chemical Society, Washington, D.C., 1989.

PATRICIA SAVU
3M Center

FLUORINATED AROMATIC COMPOUNDS

Preparative methods, **325**
Ring-fluorinated benzenes, **330**
Fluoronaphthalenes and other fused-ring fluoroaromatics, **347**
Side-chain fluorinated aromatics, **347**
Arylfluoroalkyl ethers, **356**
Fluorinated nitrogen heterocyclics, **357**

Aromatic fluorine compounds have been known for nearly a century, but numerous applications have surfaced only in recent years. The special properties conferred by fluorine justify the higher costs required to produce fluoroaromatics. The unusual physiochemical and biological properties that fluorine imparts to aromatics result from the small size of fluorine (it is bioisosteric with both the hydrogen atom and the hydroxyl group) and from its striking electronic properties, including high electronegativity and the ability to alter polarity of adjacent groups, as well as to donate electrons by resonance. Other significant properties are the enhanced stability of the C–F bond in the absence of activating groups, high lipid solubility, hydrogen-bonding potential (acceptor role), and enzyme inhibition. The carbon–fluorine link in fluoroaromatics has been of considerable value as a label for metabolic, mechanistic, and structural studies.

Depending on which substituents are present, fluoroaromatic intermediates can be converted into fluorinated or fluorine-free products. Fluorine substitution can affect the biological spectrum of the parent aromatic or heterocyclic compound by enhancement of desired properties or by suppression of undesired properties. Fluorine-containing aromatics have been incorporated into drugs (hypnotics, tranquilizers, antiinflammatory agents, analgesics, antibacterials, etc) and into crop protection chemicals (herbicides, insecticides, fungicides). Liquid crystals, positron emission tomography, and imaging systems are newer use areas for fluoroaromatics and fluoroheterocyclics.

For fluorine-free products, the lability of fluorine in fluoronitrobenzenes and other activated molecules permits it to serve as a handle in hair-dye manufacturing operations, high performance polymers such as polyetheretherketone (PEEK), production of drugs such as diuretics, and fiber-reactive dyes. Labile fluorine has also been used in analytical applications and biological diagnostic reagents.

Preparative Methods

Ring-Fluorinated Aromatics and Heterocyclics. In contrast with other molecular halogens (Cl_2, Br_2), early attempts at direct aromatic substitution with fluorine (F_2) gave violent reactions involving ring scission, addition, coupling, and polymerization. Consequently, indirect fluorination techniques based on diazotization of anilines or exchange fluorination of activated haloaromatics were developed. Recent advances in synthetic methods include discoveries of new fluorinating agents and modifications of known methods. Some of these efforts were stimulated by objectives to effect selective fluorination of natural or biologically active compounds. The need to prepare ^{18}F-labeled pharmaceuticals for use in positron emission tomography also accelerated the need for improved aromatic fluorination techniques (1).

Substitutive Aromatic Fluorination. The search for improved substitutive aromatic fluorination tools based on tamed fluorine continues (2). These reagents include elemental fluorine (F_2) (3–7), chlorine trifluoride (8) and pentafluoride (9), xenon fluorides (eg, XeF_2) (10), silver difluoride (11), cesium fluoroxysulfate (12), trifluoromethyl hypofluorite (CF_3OF) (13), bis(fluoroxy)difluoromethane ($CF_2(OF)_2$) (13), and acetyl hypofluorite (CH_3CO_2F) (14,15). Substitutive aromatic fluorination with elemental fluorine is commercially practiced for the manufacture of the antineoplastic, 5-fluorouracil (5-FU) from uracil. Nitrogen–fluorine reagents can also effect substitutive aromatic fluorination, for example, N-fluorobis[(trifluoromethyl)sulfonyl]imide (16); N-fluorobenzenesulfonimides (NFSi) (17); N-fluoropyridinium trifluoromethanesulfonates (18); and 1-alkyl-4-fluoro-1,4-diazoniabicyclo[2·2·2]octane salts, marketed as SELECTOFLUOR reagents (19,20).

$(C_6H_5SO_2)_2NF$

Diazotization Routes. Conventional Sandmeyer reaction conditions are not suitable to make fluoroaromatics. Phenols primarily result from high solvation of fluoride ion in aqueous media.

Fluoroaromatics are produced on an industrial scale by diazotization of substituted anilines with sodium nitrite or other nitrosating agents in anhydrous hydrogen fluoride, followed by *in situ* decomposition (fluorodediazoniation) of the aryldiazonium fluoride (21). The decomposition temperature depends on the stability of the diazonium fluoride (22,23). A significant development was the addition of pyridine (24), tertiary amines (25), and ammonium fluoride (or bifluoride) (26,27) to permit higher decomposition temperatures (>50°C) under atmospheric pressure with minimum hydrogen fluoride loss.

The Balz-Schiemann reaction is a useful laboratory and industrial method for the preparation of fluoroaromatics. The water-insoluble diazonium fluoroborate is filtered, dried, and thermally decomposed to give the aryl fluoride, nitrogen, and boron trifluoride (28–30).

Extreme caution must be exercised in the handling of nitroaryldiazonium fluoroborates because of unruly decomposition (29,30). Water-insoluble aryl diazonium hexafluorophosphates, ArN_2PF_6, frequently give higher yields of the fluoroaromatic (31). Substitution of aqueous sodium nitrite by nitrite esters–boron trifluoride in organic solvents gives high yields of aryl diazonium fluoroborate (32,33). A variant of the Balz-Schiemann reaction features diazotization by nitrosonium tetrafluoroborate, $NO^+BF_4^-$, in organic solvents followed by *in situ* decomposition to give high yields of aryl fluoride (34). A single fluorine atom can be introduced sequentially by the Balz-Schiemann reaction (via successive nitration, reduction, and diazotization) for a total of up to four fluorine atoms, eg, 1,2,4,5-tetrafluorobenzene. The Balz-Schiemann process is used to manufacture fluoroaromatics, eg, *o*- and *p*-difluorobenzene, not readily accessible by standard aniline–hydrogen fluoride diazotization or exchange-fluorination (Halex) routes. Estimates of producer capacities utilizing Balz-Schiemann technology range from 50–100 t/yr (35) to hundreds of t/yr (36). A continuous feed aryl diazonium fluoroborate decomposition step has been patented (37) and commercial details described (38).

The discovery of the Balz-Schiemann reaction in 1927 replaced the earlier Wallach procedure (1886) based on fluorodediazoniation of arenediazonium piperidides (aryltriazenes) in aqueous hydrogen fluoride (39,40). The Wallach aryltriazene fluorodediazoniation technique has found new utility in agrochemicals (41), pharmaceuticals (42), and positron emission tomography (43). This is illustrated in the synthesis of 2,4-dichloro-5-fluorotoluene [*86522-86-3*], an intermediate to the fluoroquinolone antibacterial ciprofloxacin, by heating *N*-(2,4-dichloro-5-methylphenyl)-*N'*,*N'*-dimethyltriazene in anhydrous fluoride (42).

Exchange Fluorination. Fluorobenzene cannot be made from chlorobenzene and potassium fluoride because of absence of substrate activation. The halogen exchange (Halex) reaction of activated haloaromatics and haloheterocyclics with potassium fluoride is a primary industrial fluoroaromatics synthesis tool (44,45). Early work featured preparation of *o*- and *p*-fluoroaromatics activated by nitro or cyano groups by exchange fluorination in dipolar aprotic solvents (46). Features of this technique include good fluorine utilization (1:1 stoichiometry), facile product separation, and potential recycling of potassium chloride (as KF) by treatment with hydrogen fluoride or fluorine. Aprotic solvents permit less solvation of fluoride ion (as compared with protic solvents), a kinetically significant amount of fluoride ion in solution, and greater insolubility of potassium chloride which, in turn, provides a further reaction driving force.

$$p\text{-ClC}_6\text{H}_4\text{NO}_2 + \text{KF} \xrightarrow{\text{solvent}} p\text{-FC}_6\text{H}_4\text{NO}_2$$

The degree of fluorination can be limited by the thermal stability of the solvent or by its reaction with basic potassium fluoride through proton abstraction. Such solvent-derived by-products can subsequently react with the starting material and/or main product.

Of the alkali metal fluorides, potassium fluoride offers the best compromise between cost and effectiveness. Although cesium fluoride generally gives higher yields, its higher cost may be a potential drawback as an industrial fluorination tool except for those substrates resistant to potassium fluoride. In contrast, inexpensive sodium fluoride consistently gives lower yields than potassium fluoride. Tetra-*n*-butylammonium fluoride (TBAF) (47), tetra-*n*-butylphosphonium hydrogen difluoride, and dihydrogen trifluoride (48) have been successfully employed in Halex reactions. Mode of potassium fluoride preparation (spray-, calcine- or freeze-dried) can affect exchange fluorination activity (49,50). Enhanced fluorination rates are associated with decreasing particle size and increasing surface area.

Halex rates can also be increased by phase-transfer catalysts (PTC) with widely varying structures: quaternary ammonium salts (51–53); 18-crown-6-ether (54); pyridinium salts (55); quaternary phosphonium salts (56); and poly(ethylene glycol)s (57). Catalytic quantities of cesium fluoride also enhance Halex reactions (58).

The inertness of chlorine in the meta position in Halex reactions is of commercial value. For example, 3,4-dichloronitrobenzene [99-54-7] forms 3-chloro-4-fluoronitrobenzene [350-30-1], which is then reduced to 3-chloro-4-fluoroaniline [367-21-5] for incorporation in the herbicide flamprop–isopropyl or the fluoroquinolone antibacterials, norfloxacin and pefloxacin.

Activating groups other than nitro or cyano have extended the versatility of exchange-fluorination reactions: —CHO (59,60); —COCl (60); —CO$_2$R (61); —(CONRCO)— (62); —SO$_2$Cl (63); and —CF$_3$ (56,64).

Explosions have been reported during preparation of fluoronitroaromatics by the Halex reaction on a laboratory or industrial scale: o-fluoronitrobenzene (65); 2,4-dinitrofluorobenzene (66); 2,4-difluoronitrobenzene (67); and 1,5-difluoro-2,4-dinitrobenzene (68).

Fluorodenitration of nitroaromatics represents an exchange fluorination technique with commercial potential. For example, m-fluoronitrobenzene [402-67-5] from m-dinitrobenzene [99-65-0] and KF in the presence of various promoters can be realized (69–72). This is not feasible under Halex conditions with m-chloronitrobenzene [121-73-3]

Saturation–Rearomatization. The first commercial route to perfluorinated aromatics such as hexafluorobenzene, octafluorotoluene [434-64-0] and fused-ring polycyclics was based on a multistage saturation–rearomatization process (73). In the first stage, benzene is fluorinated by a high valency oxidative metal fluoride (cobalt trifluoride) to give a mixture of polyfluorocyclohexanes. The latter is subjected to a combination of dehydrofluorination (with alkali) and/or defluorination (with heated iron, iron oxide, or nickel packing) to give hexa-, penta-, and tetrafluorobenzenes. Modifications of the first stage include the use of complex metal fluorides, eg, potassium tetrafluorocobaltate(III), that have been found to be milder and more selective fluorinating agents than cobalt trifluoride (74). A related process features successive treatment of hexachlorobenzene with chlorine trifluoride (75) or fluorine (76), followed by dehalogenation with iron powder at 300°C. More emphasis is now given to Halex processes for perfluoroaromatics manufacture rather than saturation–rearomatization routes.

Fluoroaliphatic Thermolytic Routes. The reaction of difluorocarbene (generated from CHClF$_2$ at 600°C) with cyclopentadiene to give fluorobenzene (70% yield) has been scaled up in a pilot-plant/semiworks facility (capacity = several dozen t/yr) (77,78). The same process can now be effected under liquid-phase conditions in the presence of phase-transfer catalysts (79,80).

Miscellaneous Methods. Exhaustive evaluation of the decarbonylation of benzoyl fluorides, ArCOF, by Wilkinson's catalyst [14694-95-2], Rh[(C$_6$H$_5$)$_3$P]$_3$Cl, to give aryl fluorides has established (81) that previous claims (82) cannot be reproduced.

One approach to aryl fluorides (83) based on phenolic derivatives features more moderate thermal decarboxylation of phenyl fluoroformates employing alumina-impregnated platinum group catalysts (84). Treatment of phenyl chloroformates with hydrogen fluoride using Lewis acid catalysts to give aryl fluorides may have potential industrial importance (85,86).

Fluorodesulfonylation represents a complementary extrusion technique to aryl fluorides (87) which has attracted interest (88,89). For example, 2-fluorobenzonitrile [394-47-8] was obtained in 84% yield from 2-cyanobenzenesulfonyl fluoride and potassium fluoride in sulfolane (88).

The electrochemical route to fluoroaromatics (90) based on controlled potential electrolysis in the absence of hydrogen fluoride (platinum anode, +2.4 V; acetonitrile solvent; tetraalkylammonium fluoride electrolyte) has not been commercialized. However, considerable industrial interest in the electrochemical approach still exists (91–93).

The single-step p-fluoroaniline [31-40-4] process based on fluorodeoxygenation of nitrobenzene (via in situ generation of N-phenylhydroxylamine) in anhydrous hydrogen fluoride (94–96) has not been commercialized primarily due to concurrent formation of aniline, as well as limited catalyst life. The potential attractiveness of this approach is evidenced by numerous patents (97–101). Concurrent interest has been shown in the two-step process based on N-phenylhydroxylamine (HF-Bamberger reaction) (102–104).

Side-Chain Fluorinated Aromatics and Heterocyclics. Benzotrifluorides generally are prepared from trichloromethylaromatics with metal fluorides or hydrogen fluoride. Industrial processes feature reaction with hydrogen fluoride under high pressure, atmospheric pressure, or vapor-phase conditions. A potential simplification is the single-step conversion of toluene to benzotrifluoride employing chlorine–hydrogen fluoride (CCl_4 diluent, 460°C) (105).

Sulfur Tetrafluoride and Aromatic Carboxylic Acids. Benzotrifluorides also are prepared from aromatic carboxylic acids and their derivatives with sulfur tetrafluoride (SF_4) (106,107). Hydrogen fluoride is frequently used as a catalyst. Two equivalents of sulfur tetrafluoride are required:

$$ArCOOH + SF_4 \xrightarrow{HF} ArCOF + HF + SOF_2$$

$$ArCOF + SF_4 \longrightarrow ArCF_3 + SOF_2$$

The high cost of SF_4 and the incomplete use of fluorine justify its use only for inaccessible benzotrifluorides. The related liquid S–F reagent, (diethylamino)sulfur trifluoride (DAST), $(C_2H_5)_2NSF_3$, also effects similar transformations with aromatic carboxylic acids (108).

Perfluoroalkylation. A significant technical advance features perfluoroalkylation of aromatics (devoid of electron-withdrawing groups) with carbon tetrachloride–hydrogen fluoride to give high selectivity of benzotrifluorides (109,110). Hydrogen fluoride performs a threefold role: solvent, Friedel-Crafts alkylation catalyst, and fluorinating agent.

$$\underset{HF}{\overset{CCl_4}{\longrightarrow}} \text{R-C}_6H_5 \longrightarrow [\text{R-C}_6H_4\text{-CCl}_3] \longrightarrow \text{R-C}_6H_4\text{-CF}_3$$

Aromatic perfluoroalkylation can be effected by fluorinated aliphatics via different techniques. One category features copper-assisted coupling of aryl halides with perfluoroalkyl iodides (eg, CF_3I) (111,112) or difluoromethane derivatives such as CF_2Br_2 (Burton's reagent) (113,114), as well as electrochemical trifluoromethylation using CF_3Br with a sacrificial copper anode (115). Extrusion of spacer groups attached to the fluoroalkyl moiety, eg, CF_3COONa and higher perfluorocarboxylated salts (116,117), CF_3SO_2Na (118), and esters such as $CF_2ClCOOCH_3$ (119) or $FSO_2CF_2COOCH_3$ (120), represents a novel trifluoromethylation concept.

$$\text{ArI} + CF_3COONa \xrightarrow{\text{CuI/NMP}} \text{Ar—CF}_3$$

Aromatic perfluoroalkylation can also be performed in the absence of copper employing $(CF_3COO)_2$ (121) or $R_fI(C_6H_5)OSO_2CF_3$ (FITS reagents) (122). Aluminum chloride-catalyzed alkylation of fluorobenzene with hexafluoroacetone, CF_3COCF_3, gave 66% yield of p-fluoro-α,α-bis(trifluoromethyl)benzyl alcohol [2402-74-6] (123).

Oxidative Fluorination of Aromatic Hydrocarbons. The economically attractive oxidative fluorination of side chains in aromatic hydrocarbons with lead dioxide or nickel dioxide in liquid HF stops at the benzal fluoride stage (67% yield) (124).

$$p\text{-CH}_3C_6H_4NO_2 \xrightarrow{\text{PbO}_2/\text{HF}} p\text{-CHF}_2C_6H_4NO_2$$

Cyclization. Construction of benzotrifluorides from aliphatic feedstocks represents a new technique with economic potential. For example, 1,1,1-trichloro-2,2,2-trifluoroethane [354-58-5] and dimethyl itaconate [617-52-7] form 4-methoxy-6-trifluoromethyl-2H-pyran-2-one [101640-70-4], which is converted to methyl 3-(trifluoromethyl)benzoate [2557-13-3] with acetylene or norbornadiene (125).

Ring-Fluorinated Benzenes

FLUOROBENZENE

Properties. Fluorobenzene [462-06-6] (monofluorobenzene), C_6H_5F, has a molecular weight of 96.1, and is a colorless mobile liquid with a pleasant aromatic odor (Table 1). Its thermal stability is of a high order; fluorobenzene undergoes no detectable decomposition when kept at 350°C for 24 h at pressures of up to

Table 1. Physical Properties of Fluorobenzene

Property	Value
melting point, °C	−42.22
boiling point, °C	84.73
density, 25°C, g/mL	1.0183
coefficient of expansion	0.00116
refractive index, n_D^{25}	1.4629
viscosity, mPa·s(=cP)	
9.3°C	0.653
19.9°C	0.585
80.9°C	0.325
surface tension, mN/m(=dyn/cm)	
9.3°C	28.49
20.0°C	27.71
34.5°C	25.15
latent heat of fusion, J/mol^a	11,305.2
latent heat of vaporization, 25°C, J/mol^a	34,576.6
specific heat, 25°C, J/mol^a	146.3
critical temperature, °C	286.94
critical pressure, kPa^b	4550.9
critical density, g/mL	0.269
dielectric constant, 30°C	5.42
dipole moment, C·m^c	4.90×10^{-30}
heat of combustion, J/g^a	−32,273.3
heat of formation, kJ/mol^a	
vapor	−110.5
liquid	−145.2
solubility in water, 30°C, g/100 g	0.154
solubility of water in fluorobenzene, 25°C, g/100 g	0.031
boiling point of binary azeotrope, °C	
with 31 wt % tert-butyl alcohol	76.0
with 32 wt % methanol	59.7
with 30 wt % isopropyl alcohol	74.5
flash point (Tag open cup), °C^d	−13
vapor pressure, in °C and kPa^b	Antoine equation^e

^{a}To convert J to cal, divide by 4.184.
^{b}To convert kPa to mm Hg, multiply by 7.5; log kPa = log mm Hg −0.895.
^{c}To convert C·m to debye (D), divide by 3.336×10^{-30}.
^{d}Ref. 126.
^{e}$\mathrm{Log}_{10}P = 6.07687 - \dfrac{1248.083}{(t + 221.827)}$.

40.5 MPa (400 atm). Toxicity: oral (rat), $LD_{50} > 4$ g/kg; inhalation (mouse), LD_{50} 45 g/m^3 (2 h) (127).

Reactions. *Electrophilic Substitution.* Fluorobenzene electrophilic substitution reactions are more para directing than are the same chlorobenzene reactions (128). Nitration of fluorobenzene with concentrated nitric and sulfuric acid gives a 92:8 mixture of *p*- and *o*-fluoronitrobenzene [1493-27-2] which can be separated by distillation. The other commercial route to *o*- and *p*-fluoronitrobenzene

[350-46-9] is based on exchange fluorination (KF) in a polar solvent; phase-transfer catalysts are frequently employed.

The Friedel-Crafts ketone synthesis is of commercial importance in upgrading fluorobenzene for drug, polymer, and electronic applications (Table 2).

Nucleophilic Displacement Reactions. The presence of activating groups, eg, *o, p* nitro groups, makes aromatic fluorine reactive in nucleophilic displacement reactions. This has been demonstrated by determination of the relative fluorine–chlorine displacement ratios from the reaction of halonitrobenzenes with sodium methoxide in methanol (137); F is displaced 200–300 times more readily than Cl.

Numerous applications have been developed based on the lability of fluoronitroaromatics; 4-fluoro-3-nitroaniline [364-76-1] and 4-fluoro-3-nitro-*N,N*-bis(hydroxyethyl)aniline [29705-38-2], commercial hair dye intermediates (138–140); 2,4-dinitrofluorobenzene [70-34-8] (Sanger's reagent), for amino acid characterization (141); 4-fluoro-3-nitrophenyltrimethylammonium iodide [39508-27-5], a protein solubilizing reagent (142); and 4-fluoro-3-nitrophenylazide [28166-06-5], an antibody tagging reagent (143) also used for industrial immobilization of enzymes (144). Other examples of biochemical applications (amino acid or peptide characterization, protein cross-linking reagent) include 2,4-dinitro-5-fluoroaniline [361-81-7] (Bergmann's reagent) (145); 4-fluoro-3-nitrobenzoates (146,147); 4-fluoro-3-nitrobenzenesulfonic acid [349-05-3] (148); 4-fluoro-3-nitrophenyl sulfone [51451-34-4] (149); 1,5-difluoro-2,4-dinitrobenzene [327-92-4] (150); 3,5-dinitro-2-fluoroaniline [18646-02-1] (151); 4-fluoro-7-nitrobenzofurazan [29270-56-2] (NBD-F) (152); and 1-fluoro-2,4-dinitrophenyl-5-L-alanine amide

Table 2. Friedel-Crafts Ketone Synthesis with Fluorobenzene[a]

Acylating agent	Reference	Product	CAS Registry Number	End use
4-chlorobutyryl chloride	129	4-chloro-4'-fluorobutyrophenone	[3874-54-2]	haloperidol[b] (tranquilizer)
acetyl chloride or acetic anhydride	130	4-fluoroacetophenone	[403-42-9]	flazalone[b] (anti-inflammatory)
4-fluorobenzoyl chloride	131	4,4'-difluorobenzophenone	[345-92-6]	polyetheretherketone (PEEK), a high performance thermoplastic
isophthaloyl chloride	132, 133	1,3-bis(4-fluorobenzoyl)benzene	[108464-88-6]	poly(arylene ethers) (PAE); polyimides
oxalyl chloride	134	4-fluorobenzoyl chloride	[403-43-0]	liquid crystal intermediate
chloroacetyl chloride	135	2-chloro-4'-fluoroacetophenone	[456-04-2]	flutriafol[c] (fungicide)
2,3-naphthalenedicarboxylic anhydride	136	3-(4-fluorobenzoyl)-2-naphthalenecarboxylic acid	[91786-16-2]	organo-selenium metallic conductors

[a]AlCl$_3$ catalyst.
[b]See Table 4.
[c]See Table 3.

[*95713-52-3*] (Marfey's reagent) (153). Labile fluorine in 4-fluoronitrobenzene can be used to form piperidinylimino-linked polar chromophores for nonlinear optical (NLO) materials (154).

Examples of commercial reactive fluoroaromatics are not restricted to fluoronitrobenzenes. The fluorine-free diuretic, furosemide [*54-31-9*], is prepared in 85% yield from 2-fluoro-4-chloro-5-sulfamoylbenzoic acid and furfurylamine at 95°C for 2 h (155).

Cyclothiazide [*2259-96-3*] is another example of a fluorine-free pharmaceutical (diuretic, antihypertensive) based on *m*-chlorofluorobenzene [*625-98-9*] where fluorine activation is subsequently provided by two sulfonamide groups (156).

Another commercial application of nucleophilic reactions of nitro-free fluoroaromatics is the manufacture of polyetheretherketone (PEEK) high performance polymers from 4,4'-difluorobenzophenone [*345-92-6*], and hydroquinone [*121-31-9*] (131) (see POLYETHERS, AROMATIC).

Polyether sulfones (PES) prepared from 4,4'-difluorodiphenyl sulfone and bisphenol A (potassium salt, DMSO) react faster than the corresponding reaction with 4,4'-dichlorodiphenyl sulfone (157) (see POLYMERS CONTAINING SULFUR, POLYSULFONES). Poly(ether sulfone)s prepared from sodium 4-fluorobenzenethiolate, α,ω-diiodoperfluoroalkanes, and bisphenol A exhibit good permeability and selectivity for O_2–N_2 gas separations (158,159). Fluorine-free membranes based on 2,6-difluorobenzonitrile and bisphenol A can also be used to separate gas mixtures (160,161).

Less activated substrates such as fluorohalobenzenes also undergo nucleophilic displacement and thereby permit entry to other useful compounds. Bromine is preferentially displaced in *p*-bromofluorobenzene [*460-00-4*] by hydroxyl ion under the following conditions: calcium hydroxide, water, cuprous oxide catalyst, 250°C, 3.46 MPa (500 psi), to give *p*-fluorophenol [*371-41-5*] in 79% yield (162,163). This product is a key precursor to sorbinil, an enzyme inhibitor (aldose reductase).

Fluoroaryl Organometallics. Fluorobenzene does not form a Grignard reagent with magnesium (164). 4-Bromofluorobenzene [*460-00-4*] can be selectively converted to 4-fluorophenylmagnesium bromide [*352-13-6*] for subsequent incorporation into the silicon-containing fungicide, flusilazole [*85509-19-9*] (165).

$$\left(F-\underset{2}{\underbrace{\bigcirc}}\right)_2 Si(CH_3)-CH_2N\underset{N=\!\!=\!\!N}{\overset{N=\!\!=\!\!N}{\diagdown}}$$

This represents the first large-scale application of a fluoroaryl organometallic. Other silicon-containing aryl fluorides such as pentafluorophenyldimethyl silanes, $C_6F_5Si(CH_3)_2X$ (X = Cl; NH_2; $N(C_2H_5)_2$), are offered commercially as Flophemsyl reagents for derivatization of sterols in chromatographic analysis (166).

Phenyllithium cannot be formed from fluorobenzene. Instead, the electronegativity of fluorine makes the ortho hydrogen sufficiently acidic to permit reaction with n-butyllithium in tetrahydrofuran at −50°C to give 2-fluorophenyllithium [348-53-8]. An isomer, 4-fluorophenyllithium [1493-23-8], was reported to be explosive in the solid state (167).

The chelate, cobalt bis(3-fluorosalicylaldehyde)ethyleneimine [6220-65-5] (fluomine) had been under active evaluation for an oxygen-regenerative system in aircraft (168). Boron-containing fluoroaromatics are commercially offered as laboratory reagents: tetrakis(4-fluorophenyl)boron sodium·2H_2O, a titration agent for nonionic surfactants (169); and 4-fluorobenzeneboronic acid [1765-93-1], a glc reagent for derivatization of diols.

Biotransformation Reactions. Enzymatic oxygenation of aryl fluorides without ring opening provides a new production tool to fluoroaromatic fine chemicals. Microbial oxidation of fluorobenzene forms 3-fluoro-*cis*-1,2-dihydrocatechol, followed by chemical rearomatization to give 3-fluorocatechol [363-52-0] (170–172). This technique represents a significant improvement over the standard four-step chemical route based on 3-fluoroanisole [456-49-5]. Dehydration (acid pH) of the fluorodihydrocatechol also provides a new route to 2-fluorophenol [367-12-4]. Biological oxidation of fluoroaromatics has been demonstrated at the tonnage scale in up to 20-m^3 reactors (171).

Manufacture. Fluorobenzene is produced by diazotization of aniline in anhydrous hydrogen fluoride at 0°C, followed by *in situ* decomposition of benzenediazonium fluoride at 20°C (21). According to German experience during World War II, the yield for 750-kg batches was 75–77%. Aryldiazonium fluoride–hydrogen fluoride solutions can also be decomposed by continuous feed through a heated reaction zone (173,174) or under super atmospheric pressure conditions (175).

The spent hydrogen fluoride layer, which contains water and sodium bifluoride, from this process is treated with sulfur trioxide or 65% oleum, and hydrogen fluoride is distilled for recycle to the next batch (176,177).

Nitrosyl chloride (178), nitrosyl chloride–hydrogen fluoride (NOF·3HF, NOF·6HF) (179), nitrous acid–hydrogen fluoride solutions (180,181), or nitrogen trioxide (prepared *in situ* from nitric oxide and oxygen) (27) can be used in place of sodium nitrite in the diazotization step.

Firms producing fluorobenzene and other ring-fluorinated aromatics by the diazotization of anilines in hydrogen fluoride include Rhône-Poulenc, ICI, Du Pont, Mallinckrodt, MitEni, and Riedel de Haën/Hoechst (182). With announcements of plant expansions and entry of new manufacturers, surplus capacity in basic fluoroaromatics exists. Prices (1991) are quoted at $15–18/kg delivered in the United States for basic intermediates such as fluorobenzene and the fluorotoluenes (182). Emphasis has now been placed on higher value downstream derivatives development programs.

Applications. *Crop Protection Chemicals.* The fluorinated analogue of DDT (GIX, DFDT), 1,1-bis(4-fluorophenyl)-2,2,2-trichloroethane [475-26-3], was produced from chloral and fluorobenzene as an insecticide in Germany during World War II. Other agricultural applications did not subsequently materialize since lower manufacturing costs of chlorinated aromatic crop-protection chemicals represented an advantage over ring-fluorinated analogues. However, chloroaromatics pose ecological problems such as pesticide persistency, toxicity, etc. Because fluoroaromatics offer agronomic advantages, eg, dosage, selectivity, and crop safety, significant commercialization of these compounds as crop protection chemicals (herbicides, fungicides, and insecticides) has occurred. Table 3 lists representative examples.

Drugs. Ring-fluorinated aromatics have found broad pharmaceutical applications, eg, in tranquilizers, hypnotics, sedatives, antibacterial agents (qv), etc. Representative monofluorinated drugs are listed in Table 4. Arprinocid [5579-18-15] is a fluoroaromatic-based veterinary drug that has found wide acceptance as a coccidiostat for chicken feed.

Other Medical Applications. Positron emission tomography, a noninvasive technique for monitoring biochemical functions in humans, represents a significant advance in medical diagnosis. Synthetic methods have been developed for incorporation of the ^{18}F isotope ($t_{1/2}$, 109.27 ± 0.06 min) into numerous biologically active radiopharmaceuticals (1). One example is the stereospecific and regiospecific synthesis of 6-[^{18}F]fluoroDOPA (6-fluoro-3,4-dihydroxyphenylalanine) by fluorodemetallation of a trimethylsilyl precursor using $^{18}F_2$ for Parkinson's disease research.

Labeling aromatics with fluorine using ^{19}F-nmr as a probe for product identification has been a useful analytical tool (183) which has been extended to med-

Table 3. Monofluoroaromatic Crop Protection Chemicals

Common name	CAS Registry Number	Structure	Application
flamprop-isopropyl	[52756-22-6]		post-emergent herbicide
fluoronitrofen	[13738-63-1]		post-emergent herbicide
fluoroimide	[41205-21-4]		fungicide
cyfluthrin	[68359-37-5]		insecticide
flutriafol	[76674-21-0]		fungicide

ical diagnosis. Magnetic resonance imaging (mri) is a noninvasive technique complementary to x-ray contrast agents, ultrasound devices, and computerized tomography, without the need of radioisotopes (184). For example, the mri technique has been applied to the interaction of fluoroquine, a fluorine analogue of the antimalarial drug, chloroquine, with DNA and t-RNA (185).

Liquid Crystals. Based on worldwide patent activity, numerous compounds containing fluoroaromatic moieties have been synthesized for incorporation into liquid crystals. For example, fluoroaromatics are incorporated in ZLI-4792 and

Table 4. Monofluorinated Aromatic Drugs

Common name	CAS Registry Number	Structure	Application
haloperidol	[52-86-8]		tranquilizer
flurazepam hydrochloride	[1172-39-5]		hypnotic, sedative
floxacillin	[5250-39-5]		antibacterial
flazalone	[21221-18-1]		anti-inflammatory
fluspirilene	[1841-19-6]		tranquilizer
lidoflazine	[3416-26-0]		vasodilator (coronary)
sorbinil	[68367-52-2]		enzyme inhibitor (aldose reductase)

ZLI-4801-000/-100 for active matrix displays (AMD) containing super fluorinated materials (SFM) (186,187). Representative structures are as follows.

Photoconductive Imaging. Considerable attention has been placed on the xerographic properties of fluorosquaraines based on N,N-dimethyl-3-fluoroaniline and other 3-fluoroaniline derivatives for imaging applications (188–191). A typical structure of a fluorosquaraine is as follows:

Dyes. In contrast to benzotrifluorides and fluoropyrimidines, limited commercialization has developed for dyes containing a fluoroaromatic group. Fluorophenylhydrazines have been converted to (fluorophenyl)pyrazolones, which are disperse dyes for cellulose acetate and nylon (192).

DIFLUOROBENZENES

Interest in the commercialization of difluoroaromatics in crop protection chemicals and drugs (Table 5) continues to be strong. Numerous liquid crystals containing the 1,2-difluorobenzene moiety have been synthesized. Table 6 lists physical properties of commercially significant intermediates such as *o*-, *m*-, and *p*-difluorobenzene, 2,4-difluoroaniline and 2,6-difluorobenzonitrile. The LD_{50} values for the three isomeric difluorobenzenes are identical: 55 g/m^3 for 2 h (inhalation, mouse) (127).

1,2-Difluorobenzene. Tetrazotization-fluorination of *o*-phenylenediamine [95-54-5] in hydrogen fluoride or by the Balz-Schiemann reaction is not a practical route to 1,2-difluorobenzene but this product can be prepared from 2-fluoroaniline [348-54-9] by the Balz-Schiemann reaction (193); heating the diazonium fluoroborate in organic solvents increases the yield to 78% (194). Electrophilic substitution reactions are site-specific: nitration gives 3,4-difluoronitrobenzene [369-34-6], and bromination forms 3,4-difluorobromobenzene [348-61-8], a precursor to dicyclohexylethylene liquid crystals (195). Vicinal metallation (*n*-butyllithium, −78°C) of 1,2-difluorobenzene is also employed to prepare *trans*-4-alkyl cyclohexyl-substituted 2,3-difluorobiphenyls for liquid crystal applications (196).

1,3-Difluorobenzene. This isomer has been prepared in 78% yield by tetrazotization-fluorination of *m*-phenylenediamine [108-45-2] in pyridine–hydrogen fluoride at 100°C (23,197). Balz-Schiemann yields for the corresponding reaction vary from 31 to 49% (198,199). Diazotization of *m*-fluoroaniline [372-19-0] in the presence of ammonium bifluoride, tertiary amines, or dimethyl sulfoxide gave 46–

Table 5. Difluoroaromatic Applications

Common name	CAS Registry Number	Structure
Crop protection chemicals[a]		
diflubenzuron	[35367-38-5]	2,6-difluorophenyl-C(O)NHC(O)NH-(4-chlorophenyl)
diflufenican[b]	[83164-33-4]	pyridine-3-C(O)NH-(2,4-difluorophenyl); pyridine-2-O-(3-CF₃-phenyl)
flufenoxuron	[101463-69-8]	2,6-difluorophenyl-C(O)NHC(O)NH-(3-fluoro-4-(2-chloro-4-CF₃-phenoxy)phenyl)
teflubenzuron	[83121-18-0]	2,6-difluorophenyl-C(O)NHC(O)NH-(2,3,5,6-tetrasubstituted phenyl: F, Cl, F, Cl)
Drugs		
diflunisal[c]	[22494-42-4]	4-fluoro-biphenyl with 3-F, 3'-COOH, 4'-OH
fluconazole[d]	[86386-73-4]	bis(1,2,4-triazol-1-ylmethyl) with (2,4-difluorophenyl) and OH on central C

[a] Insecticide unless otherwise noted.
[b] Herbicide.
[c] Analgesic; antiinflammatory.
[d] Antifungal.

Table 6. Properties of Fluorinated Aromatic Compounds[a]

Component	CAS Registry Number	Mol wt	Mp, °C	Bp, °C[b]	Refractive index, n_D^t	Specific gravity, d_4^t	Surface tension, 20°C mN/m (=dyn/cm)	Flash point,[c] °C
$C_6H_4F_2$								
1,2-difluorobenzene	[367-11-3]	114.09	−34	91–92	1.4452^{20}	1.1496^{25}		7.2^d
1,3-difluorobenzene	[372-18-9]	114.09	−59.3	82–83	1.4410^{20}	1.1572^{20}	25.93	-11.1^d
1,4-difluorobenzene	[540-36-3]	114.09	−13	88–89	1.4421^{20}	1.1716^{20}	27.05	-11.7^d
$C_6H_3F_3$			*Trifluorobenzenes*					
1,2,3-trifluorobenzene	[1489-53-8]	132.08		94–95	1.4230^{20}	1.280^{20}		−3
1,2,4-trifluorobenzene	[367-23-7]	132.08		88	1.4230^{20}	1.264^{20}	26.2	-5^d
1,3,5-trifluorobenzene	[372-38-3]	132.08	−5.5	75.5	1.4140^{20}	1.277^{20}	27.16	−7
$C_6H_2F_4$			*Tetrafluorobenzenes*					
1,2,3,4-tetrafluorobenzene	[551-62-2]	150.08	−42	95	1.4069^{20}	1.422^{25}	23.99	20
1,2,3,5-tetrafluorobenzene	[2367-82-0]	150.08	−48	83	1.4011^{25}	1.393^{20}		4
1,2,4,5-tetrafluorobenzene	[327-54-8]	150.08	4	90	1.4045^{20}	1.424^{25}	24.9	16
C_6HF_5			*Pentafluorobenzene*					
	[363-72-4]	168.07	−48	85	1.3881^{25}	1.531^{20}		13

Fluorotoluenes							
2-fluorotoluene[e]	[95-52-3]	110.13		113–114	1.4704^{25}	1.001^{20}	12
3-fluorotoluene	[352-70-5]	110.13		115	1.4691^{20}	0.991^{20}	9
4-fluorotoluene	[352-32-9]	110.13		116	1.4690^{20}	1.000^{20}	17
2-chloro-6-fluorotoluene	[443-83-4]	144.58		155	1.5026^{20}	1.129^{20}	48
$C_6H_4F(NH_2)$			*Fluoroanilines*				
2-fluoroaniline[f]	[348-54-9]	111.12	−29	175	1.5406^{25}	1.152^{25}	60
3-fluoroaniline[g,h]	[372-19-0]	111.12		186	1.5445^{25}	1.152^{25}	77
4-fluoroaniline[e,f,i]	[371-40-4]	111.12	−1.9	187	1.5375^{25}	1.158^{25}	73
$C_6H_3F_2(NH_2)$	[367-25-9]	129.11	−7.5	169.5	1.5043^{25}	1.268^{25}	62
2,4-difluoroaniline							
$C_6H_3(Cl)(F)(NH_2)$	[367-21-5]	145.57	44–47	227–228		1.42^{20} (solid)	110
3-chloro-4-fluoroaniline						1.3^{60} (liquid)	
			Fluorobenzonitriles				
2,6-difluorobenzonitrile	[1897-52-5]	139.11	30–32	99[j]	1.4875^{25}	1.236^{40}	80

[a] Colorless unless otherwise noted.
[b] At 101.1 kPa = 1 atm unless otherwise noted.
[c] Closed cup (ASTM Procedure D3278) unless otherwise noted.
[d] Open cup.
[e] LD_{50} = 100 mg/kg (oral, wild bird) (127).
[f] Pale yellow.
[g] Amber.
[h] LD_{50} = 56 mg/kg (oral, wild bird) (127).
[i] LD_{50} = 50 mg/kg (oral, rat) (127).
[j] At 2.67 kPa = 20 mm Hg.

73% yields of 1,3-difluorobenzene (25,26). The latter can also be made by reductive-dediazoniation of 2,4-difluoroaniline [367-25-9] in 77% yield from sodium nitrite, hydrochloric acid, and hypophosphorus acid (200). A 95% yield was realized by treatment of 2,4-difluorobenzenediazonium fluoroborate with copper powder in the presence of 18-crown-6 ether in dichloromethane (201).

Nitration of 1,3-difluorobenzene at 0°C forms 2,4-difluoronitrobenzene [446-35-5] in 92% yield. The latter can also be prepared from 2,4-dichloronitrobenzene and potassium fluoride in polar solvents (46,202); phase-transfer catalysts, eg, quaternary ammonium salts, serve to both lower reaction temperature and enhance fluorination rates (203). Reduction gives 2,4-difluoroaniline, a precursor to the analgesic/antiinflammatory diflunisal, and the herbicide diflufenican.

1,4-Difluorobenzene. This compound has been prepared in 65% yield by tetrazotization-fluorination of p-phenylenediamine [106-50-3] in pyridine–hydrogen fluoride at 120°C (23,197); 27–40% yields are obtained by the Balz-Schiemann reaction with p-phenylenediamine or p-fluoroaniline [371-40-4] (198).

TRIFLUOROBENZENES

Table 6 lists physical properties of representative trifluorobenzenes.

1,2,3-Trifluorobenzene. This compound is formed in low yield (13–24%) from 1,2,3-trichlorobenzene or 2,3-difluorochlorobenzene and KF/CsF in dimethyl sulfone (204). Likewise, low yields are realized when the Balz-Schiemann reaction is applied to 2,3-difluoroaniline or 2,6-difluoroaniline (205). Pyrolysis (520°C, iron gauze) of 1H, 2H, 3H-pentafluorocyclohexa-1,3-diene forms 1,2,3-trifluorobenzene (206). Derivatives such as 2,3,4-trifluoronitrobenzene [393-79-3] and 2,3,4-trifluoroaniline [3862-73-5] have been used to prepare fluoroquinolone antibacterials such as ofloxacin (207) and lomefloxacin (208), respectively.

1,2,4-Trifluorobenzene. This isomer can be prepared in good yield from 2,4-difluoroaniline by the standard Balz-Schiemann route (209) or modifications using nitrite esters–boron trifluoride (210). Its ionization potential is 9.37 V. Electrophilic substitution reactions of 1,2,4-trifluorobenzene provide useful routes to 2,4,5-trifluorobenzoic acid [446-17-3], a key precursor to fluoroquinolone antibacterials: bromination forms 1-bromo-2,4,5-trifluorobenzene [327-52-6] (211), followed by exchange cyanation–hydrolysis (212); acetylation gives 2,4,5-trifluoroacetophenone [129322-83-4], followed by oxidation with commercial bleach (213). New routes to 2,4,5-trifluorobenzoic acid are also based on exchange-fluorination of chloroaromatic feedstocks such as 3,4,6-trichlorophthalic acid (62) and tetrachloroisophthalonitrile (214). Other 1,2,4-trifluorobenzene derivatives such as 3-chloro-2,4,5-trifluorobenzoic acid have also been converted to fluoroquinolone antibacterials (215).

1,3,5-Trifluorobenzene. This isomer, s-trifluorobenzene, has been prepared in 63% yield by the Balz-Schiemann reaction with 3,5-difluoroaniline [372-39-4] (216). By modification of exchange fluorination conditions, tetrachloroisophthalonitrile [1897-45-6] was converted to 1,3,5-trifluorobenzene by a four-step process (217).

TETRAFLUOROBENZENES

Interest in tetrafluoroaromatics includes crop protection and as intermediates to fluoroquinolone antibacterials. Physical properties of tetrafluorobenzenes are listed in Table 6. A useful compilation of recipes for 35 tetrafluorinated aromatics has been published (199).

1,2,3,4-Tetrafluorobenzene. This compound has been prepared by fluorination of benzene with cobalt trifluoride and subsequent combination of the dehydrofluorination and defluorination steps. Its ionization potential is 9.01 V. Nitration gives 2,3,4,5-tetrafluoronitrobenzene [5580-79-0] in 75% yield, an intermediate to fluoroquinolone antibacterials (218).

Halex technology has also been employed to prepare 1,2,3,4-tetrafluorobenzene derivatives, eg, tetrachlorophthalic anhydride [117-08-8] was converted to 2,3,4,5-tetrafluorobenzoic acid [1201-31-6] for use in fluoroquinolone antibacterials (219,220).

1,2,3,5-Tetrafluorobenzene. This isomer has been prepared from 2,3,5-trifluoroaniline [363-80-4] in 43% yield by the Balz-Schiemann reaction.

1,2,4,5-Tetrafluorobenzene. This compound has been prepared from 2,4,5-trifluoroaniline [57491-45-9] by the Balz-Schiemann reaction in 38–46% yield or from pentafluorophenylhydrazine [828-39-9] with aqueous sodium hydroxide in 90–95% yield (221). Its ionization potential is 9.39 V.

Derivatives of 1,2,4,5-tetrafluorobenzene such as 2,3,5,6-tetrafluorobenzoic acid have been converted into fluoroquinolone antibacterials (222–224). The synthetic pyrethroid, tefluthrin [795-38-2], is prepared from 2,3,5,6-tetrafluoro-4-methylbenzyl alcohol.

Isomeric dichlorotetrafluorobenzenes have been studied for Rankine-cycle external combustion engines (225).

PENTAFLUOROBENZENE AND PENTAFLUOROPHENYL COMPOUNDS

Pentafluorobenzene. Pentafluorobenzene has been prepared by several routes: multistage saturation–rearomatization process based on fluorination of benzene with cobalt trifluoride; reductive dechlorination of chloropentafluorobenzene with 10% palladium-on-carbon in 82% yield (226,227); and oxidation of pentafluorophenylhydrazine in aqueous copper sulfate at 80°C in 77% yield (228). Its ionization potential is 9.37 V. One measure of toxicity is LD_{50} = 710 mg/kg (oral, mouse) (127).

Nucleophiles react with pentafluorobenzene to give para-substituted (relative to the hydrogen atom) tetrafluorophenyl products, p-XC_6F_4H (X = H, NH_2, $NHNH_2$, SH, OCH_3, SC_6H_5, OH). Nitration of pentafluorobenzene with concentrated nitric acid and boron trifluoride in sulfolane gave pentafluoronitrobenzene [880-78-4] in 82% yield (229).

Pentafluoroaniline. Pentafluoroaniline [771-60-8] has been prepared from amination of hexafluorobenzene with sodium amide in liquid ammonia or with ammonium hydroxide in ethanol (or water) at 167–180°C for 12–18 h. It is weakly basic (pK_a = 0.28) and dissolves only in concentrated acids. Liquid crystals have been prepared from Schiff bases derived from pentafluoroaniline (230).

Pentafluorophenol. This compound has been prepared from the reaction of hexafluorobenzene with potassium hydroxide in t-butyl alcohol. Pentafluorophenyl esters prepared from pentafluorophenol [771-61-9] illustrate the key features of a rapid stepwise peptide synthesis technique (231). Commercial high performance elastomers based on copolymerization of tetrafluoroethylene, perfluoro(methyl vinyl ether), and a third monomer incorporating a pentafluorophenoxy group as a cure site, give vulcanizates with good chemical and fluid resistance and high temperature oxidative resistance (232,233).

Pentafluorotoluene. Pentafluorotoluene [771-56-2] has been prepared from the reaction of methyllithium with hexafluorobenzene or from pentafluorophenylmagnesium bromide with dimethyl sulfate. Derivatives such as 2,3,4,5,6-pentafluorobenzyl bromide [1765-40-8] are used to derivatize organic acids as esters for determination by electron-capture gas chromatography (234). The synthetic pyrethroid, fenfluthrin [75867-00-4], is an insecticide containing a pentafluorobenzyl group.

Bromopentafluorobenzene. Aluminum bromide-catalyzed bromination of pentafluorobenzene in 20% oleum gives bromopentafluorobenzene [1765-40-8]. It is readily converted to pentafluorophenylmagnesium bromide [879-05-0]; the latter undergoes conventional Grignard reactions (qv). Pentafluorophenyllithium [1076-44-4] can be synthesized from bromopentafluorobenzene and n-butyllithium or lithium amalgam in ether at 0°C. The preferred route is metallation of pentafluorobenzene with n-butyllithium at $-65°C$ (235). A serious explosion has been reported during hydrolysis (D_2O) of pentafluorophenyllithium (236).

Pentafluorophenylmagnesium bromide or lithium can be converted to other pentafluorophenyl organometallics by reaction with the corresponding metal chloride (237). Bis(pentafluorophenyl)phenylphosphine [5074-71-5] (Ultramark 443), $(C_6F_5)_2C_6H_5P$, is offered commercially as a marker for mass spectral standardization (238).

Pentafluorobenzoic Acid. Standard routes to pentafluorobenzoic acid [602-94-8] include chloropentafluorobenzene (n-butyllithium or magnesium, carbonation, hydrolysis); pentafluorobenzene (phosgene, hydrolysis); octafluorotoluene (hydrolysis). Of potential economic significance is a new route based on benzonitrile: chlorination to pentachlorobenzonitrile, exchange fluorination, and hydrolysis (239). Pentafluorobenzoyl chloride [2251-50-8] has been used to derivatize anticonvulsants such as ethosuximide, carbamazepine, and primidone in electron-capture gas chromatography (240,241).

Pentafluorobenzaldehyde. Pentafluorobenzaldehyde [653-37-2] can be prepared by reaction of N-methylformanilide with pentafluorophenylmagnesium bromide or pentafluorophenyllithium. One process is based on the catalytic hydrogenation of pentafluorobenzonitrile (239). Pentafluorobenzaldehyde is used as a reagent for gas chromatographic assay of biological amines such as catecholamines by conversion to the pentafluorobenzylimine–trimethylsilyl derivatives (242). Catalytic hydrogenation gives pentafluorobenzyl alcohol [440-60-8]. Derivatives of the latter are employed in gas chromatography (electron capture): O-(2,3,4,5,6-pentafluorobenzyl)hydroxylamine hydrochloride [57981-02-9] (Florox reagent) $C_6F_5CH_2ONH_2 \cdot HCl$, for assay of ketosteroids (243,244); 2,3,4,5,6-pentafluorobenzyl chloroformate [53526-74-2], for assay of physiologically active tertiary amines (245).

Numerous examples for the incorporation of the pentafluorophenyl group in chromatographic derivatization of biologically active compounds have been compiled in a monograph (246). A review on the effects of the pentafluorophenyl group on the reactivity of organic compounds has been published (247).

HEXAFLUOROBENZENE

The development of commercial routes to hexafluorobenzene [392-56-3] included an intensive study of its derivatives. Particularly noteworthy was the development of high temperature lubricants, heat-transfer fluids, and radiation-resistant polymers (248).

Hexafluorobenzene. Hexafluorobenzene [392-56-3] C_6F_6, is a colorless liquid with a sweet odor. Hexafluorobenzene (perfluorobenzene) has a good thermal stability; slight decomposition occurs at 500°C in Nimonic 75 (alloy containing 85% nickel and 20% chromium) after three weeks. Toxicity: inhalation (mouse), LD_{50}-95 g/m³ (2 h) (127). Physical properties of hexafluorobenzene are given in Table 7.

Manufacture. One commercial process features a three-stage saturation–rearomatization technique using benzene and fluorine gas as raw materials (73). Principal problems with this method are the complex nature of the process, its dependence on fluorine gas which is costly to produce, and the poor overall utili-

Table 7. Physical Properties of Hexafluorobenzene

Property	Value
mol wt	186.06
melting point, °C	5.10
boiling point, °C	80.261
density, 25°C, g/mL	1.60682
refractive index, n_D^{25}	1.3761
latent heat of fusion, kJ/mol[a]	11.59
latent heat of vaporization	
at 25°C, kJ/mol[a]	35.69 ± 0.084
at bp, kJ/mol[a]	32.69
specific heat, 23°C, kJ/mol·°C)[a]	0.221
critical temperature, °C	243.57 ± 0.03
critical pressure, MPa[b]	3.304 ± 0.005
coefficient of cubical expansion at 25°C	0.001412
heat of combustion, kJ/mol[a]	−2444.0 ± 1.2
heat of formation at 25°C, kJ/mol[a]	
liquid	−958.30
gas	−922.15
vapor pressure in °C and kPa[c]	Antoine equation[d]

[a] To convert J to cal, divide by 4.184.
[b] To convert MPa to atm, divide by 0.101.
[c] To convert kPa to mm Hg, multiply by 7.5; log kPa = log mm Hg − 0.875.
[d] $\mathrm{Log}_{10}P = 6.1422 - \dfrac{1219.410}{(t + 214.525)}$.

zation of fluorine, because nearly one-half of the input fluorine is removed during the process.

An alternative hexafluorobenzene process features exchange fluorination (KF) of hexachlorobenzene in the presence of polar solvents (226,249) or under solvent-free conditions (450–540°C, autoclave) (250). Intermediates such as chloropentafluorobenzene can be further fluorinated to hexafluorobenzene (42–51% yield) by cesium fluoride in sulfolane (226,249).

Pyrolytic routes to hexafluorobenzene have also attracted attention but have not been commercialized. Pyrolysis of tribromofluoromethane [353-54-8], CBr_3F, at 630–640°C in a platinum tube gives hexafluorobenzene in 55% yield (251–253). The principal disadvantage of this process is the low weight yield of product; 90% of the costly CBr_3F that is charged is lost as bromine. Of economic potential is the related copyrolysis of dichlorofluoromethane [754-34-0] and chlorofluoromethane [593-70-4] (254,255).

Reactions. Hexafluorobenzene is susceptible to attack by nucleophilic agents to give pentafluorophenyl compounds of the general formula C_6F_5X, where X is OCH_3, NH_2, OH, SH, $NHNH_2$, $NHCH_3$, $N(CH_3)_2$, H, C_6H_5, CH_3, $CH_3CH=CH$, n-C_4H_9, C_6H_5S, etc (256).

FLUOROBIPHENYLS

Fluorobiphenyls are incorporated into the analgesic and antiinflammatory drugs diflunisal [22494-42-4] and flurbiprofen [5104-49-4]. The first is a difluoro compound and the other monofluoro.

Fluorinated biphenyls have been incorporated into numerous liquid crystal structures as attested by patents and publications from the following organizations: E. Merck GmbH (257,258); Sharp (258); Hoffmann-LaRoche (259); Kanto Chemical (260); U.K. Defence Secretariat (261); Dainippon (262); Chisso (263); Sanyo Chemical (264); and the University of Hull (265). Seiko Epson has also patented fluorinated terphenyls for liquid crystal applications (266).

Fluorinated biphenyls can be synthesized by diazotization–fluorination, Gomberg-Bachmann arylation, or Ullmann coupling reactions. Mono- and difluorophenyls can be prepared by the Balz-Schiemann reaction (or modification in HF), eg, 4,4'-difluorobiphenyl was formed in 80% yield from 4,4'-diaminobiphenyl by the Balz-Schiemann reaction (267). 2,4-Difluorobiphenyl [2285-28-1], a key precursor to diflunisal, is formed by successive diazotization of 2,4-difluoroaniline and coupling with benzene (268). Similar diazotization-coupling of 4-bromo-2-fluoroaniline [367-24-8] with benzene gives 4-bromo-2-fluorobiphenyl [41604-19-7], a key intermediate to flurbiprofen (269).

Decafluorobiphenyl [434-90-2], $C_6F_5C_6F_5$ (mol wt, 334.1; mp, 68°C; bp, 206°C), can be prepared by Ullmann coupling of bromo- [344-04-7], chloro- [344-

07-0], or iodopentafluorobenzene [*827-15-6*] with copper. This product shows good thermal stability; decafluorobiphenyl was recovered unchanged after 1 h below 575°C (270). Decafluorobiphenyl-based derivatives exhibit greater oxidative stability than similar hydrocarbon compounds (271). Thermally stable poly-(fluorinated aryl ether) oligomers prepared from decafluorobiphenyl and bisphenols show low dielectric constant and moisture absorption which are attractive for electronic applications (272).

Fluoronaphthalenes and Other Fused-Ring Fluoroaromatics

Few applications for fluoronaphthalenes and related polycyclic structures have materialized. The fused-ring bicyclic, sulindac [*38194-50-2*], a monofluorinated indene-3-acetic acid, is used as an antiinflammatory agent.

1-Fluoronaphthalene [*321-38-0*] is prepared from 1-naphthylamine by the Balz-Schiemann reaction in 52% yield or by diazotization in anhydrous hydrogen fluoride in 82% yield. Electrophilic substitution occurs at the 4-position, eg, nitration with fuming nitric acid in acetic acid gave 88% yield of 1-fluoro-4-nitronaphthalene [*341-92-4*].

2-Fluoronaphthalene [*323-09-1*] is prepared in 54–67% yield from 2-naphthylamine by the Balz-Schiemann reaction or in 51% yield by pyrolysis of indene and chlorofluoromethane at 600°C (77).

1,4-Difluoronaphthalene [*315-52-6*] is prepared from 4-fluoro-1-naphthylamine by the Balz-Schiemann reaction. 1,4-Difluoronaphthalene is used in chemical carcinogenesis studies as a synthon for highly condensed difluoro–polycyclic aromatic hydrocarbons (273).

Octafluoronaphthalene [*313-72-4*] is prepared in 53% yield by defluorination of perfluorodecahydronaphthalene [*306-94-5*] over iron or nickel at 500°C. Exchange fluorination of octachloronaphthalene with KF in sulfolane (235°C) gave 60% yield of octafluoronaphthalene. This product exhibits good stability to ionizing radiation (274).

Fused-ring polycyclic fluoroaromatics can be made from the corresponding amino fused-ring polycyclic or from preformed fluoroaromatics, eg, 4-fluorophenylacetonitrile [*459-22-3*] (275). Direct fluorination techniques have been successfully applied to polycyclic ring systems such as naphthalene, anthracene, benzanthracenes, phenanthrene, pyrene, fluorene, and quinolines with a variety of fluorinating agents: xenon fluorides (10), acetyl hypofluorite (276), cesium fluoroxysulfate (277), and electrochemical fluorination (278,279).

Side-Chain Fluorinated Aromatics

Trifluoromethyl aromatics are used widely in the production of drugs, cropprotection chemicals, germicides, dyes, etc.

General Properties. The trifluoromethyl group is stable under different reaction conditions, eg, the multistep classical transformation of benzotrifluoride to trifluoroacetic acid features successive nitration, reduction, and oxidation.

Thermal Stability. Benzotrifluoride is stable at 350°C in the presence of iron or copper. Working fluids for external combustion engines (Rankine cycle) must exhibit thermal stability with engine materials at high temperatures. Some of the promising working fluids include 1,3-bis(trifluoromethyl)benzene [402-31-3] (280) or a mixture of perfluorotoluene [434-64-0], $CF_3C_6F_5$, and hexafluorobenzene [392-56-3] (281). The stability of the isomeric CF_3-substituted anilines has been established by differential thermal analysis (dta) (282): m-$CF_3C_6H_4NH_2$ [98-15-7], 223°C > o-$CF_3C_6H_4NH_2$ [88-17-5], 187°C > p-$CF_3C_6H_4NH_2$ [455-14-1], 155°C.

Hydrolytic Stability. The trifluoromethyl group is sensitive to hydrolysis in acidic media. Benzotrifluoride is hydrolyzed to benzoic acid by heating with hydrobromic, hydrofluoric, or > 80% sulfuric acid. Reaction conditions and structural features for this reaction have been summarized (283). Benzotrifluorides are generally stable to base. Benzotrifluoride was recovered unchanged after heating (120–130°C) with sodium hydroxide. Although m-hydroxybenzotrifluoride [98-17-9] is stable to refluxing 50% sodium hydroxide, cold dilute alkali polymerizes the para isomer [402-45-9] (284). Similar polymers are formed from 2,3,5,6-tetrafluoro-4-trifluoromethylphenol [2787-79-3] (285). Photohydrolysis of hydroxy- and aminobenzotrifluorides in dilute acid and alkali, respectively, gives the corresponding hydroxy- and aminobenzoic acid in high yield (286). Benzotrifluoride is inert under these photohydrolytic conditions.

Oxidative Stability. Benzotrifluoride resists ring oxidation. In contrast, chromic acid readily oxidizes 3-aminobenzotrifluoride to trifluoroacetic acid in 95% yield (287).

Stability to Reducing Agents. The trifluoromethyl group is inert to numerous reducing agents. Catalytic hydrogenation (platinum black) of benzotrifluoride gives trifluoromethylcyclohexane [401-75-2] (288). Benzotrifluoride was not reduced by lithium aluminum hydride (289). However, o-trifluoromethylbenzoic acid [433-97-6] and m-trifluoromethylbenzoic acid [454-92-2] are catalytically reduced (Raney nickel or cobalt alloys) to o- and m-toluic acid, respectively (290).

Instability of Trifluoromethylphenyl Organometallics. Care must be exercised in handling trifluoromethylphenyl organometallics. o-Trifluoromethylphenyllithium [49571-35-5] has exploded during reflux in diethyl ether under nitrogen (291). Both m- [368-49-0] and p-trifluoromethylphenyllithium [2786-01-8] are explosive in the solid state (292). Explosions have been reported in the preparation of o- [395-47-1], m- [402-26-6], and p-trifluoromethylphenylmagnesium bromide [402-51-7] (292,293). A violent explosion accompanied by loss of life and destruction of a chemical plant during preparation of p-trifluoromethylphenylmagnesium chloride [2923-41-3] has been reported (294). A compilation of reactive chemical hazards of trifluoromethylphenyl organometallics was published in 1990 (68).

Reactions. Benzotrifluoride undergoes electrophilic substitution reactions, eg, halogenation, nitration, typical of an aromatic containing a strong electron-withdrawing group. The trifluoromethyl group (sometimes referred to as a pseudohalogen) is meta directing.

Halogenation. Liquid-phase monochlorination of benzotrifluoride gives pronounced meta orientation (295); in contrast, vapor-phase halogenation favors para substitution (296). Sealed tube, photochemical, or dark chlorination (radical initiator) forms hexachloro(trifluoromethyl)cyclohexane; thermal dehydrochlorination (550°C) gives 2,4,6-trichlorobenzotrifluoride [567-59-9] (297). Liquid-phase bromination (Br$_2$) provides 3-bromobenzotrifluoride [401-78-5] in 60% yield. Catalyst performance decreases in the order FeCl$_3$ > Fe > SbCl$_5$ > I$_2$ (298). Silica gel (299) and calcium chloride (300) serve as hydrogen fluoride scavengers to suppress corrosion of glass reactors during halogenation. Bromination of benzotrifluoride can also be accomplished with bromine chloride in the presence of a halogen carrier, eg, antimony pentachloride; this technique permits complete utilization of bromine (301).

Hydrogen peroxide–hydrochloric acid reagent converts 2-aminobenzotrifluoride to 2-amino-5-chlorobenzotrifluoride [121-50-6], a dye intermediate (CI Azoic Diazo Component 17), without contamination by the 3-chloro isomer such as is observed with molecular chlorine (Cl$_2$) (302).

Nitration. Nitration of benzotrifluorides is an important industrial reaction. Mononitration of benzotrifluoride gives pronounced meta-orientation: 91% meta [98-46-4]; 6% ortho [384-22-5]; and 3% para [402-54-0] (296). Further nitration to 3,5-dinitrobenzotrifluoride [401-99-0] can be effected under forcing conditions at 100°C.

Alkylation. Benzotrifluoride can also be alkylated, eg, chloromethyl methyl ether–chlorosulfonic acid forms 3-(trifluoromethyl)benzyl chloride [705-29-3] (303,304), which can also be made from *m*-xylene by a chlorination–fluorination sequence (305). Exchange cyanation of this product in the presence of phase-transfer catalysts gives 3-(trifluoromethylphenyl)acetonitrile [2338-76-3] (304,305), a key intermediate to the herbicides flurtamone [96525-23-4] (306) and fluridone [59756-60-4].

flurtamone

fluridone

Nucleophilic Displacement Reactions. The strong electron-withdrawing effect of a trifluoromethyl group activates ortho and para halogen toward nucleophilic attack. Such chlorine lability is utilized in the manufacture of crop control chemicals containing trifluoromethyl and nitro groups.

Reactions Involving the Trifluoromethyl Group. Aluminum chloride effects chlorinolysis of benzotrifluoride to give benzotrichloride (307). High yields of volatile acid fluorides are formed from benzotrifluoride and perfluorocarboxylic acids (308).

$$\text{HCF}_2\text{CF}_2\text{CO}_2\text{H} + \text{C}_6\text{H}_5\text{CF}_3 \xrightarrow{\text{Lewis acid}} \text{HCF}_2\text{CF}_2\overset{\text{O}}{\underset{\|}{\text{C}}}\text{F} + \text{C}_6\text{H}_5\overset{\text{O}}{\underset{\|}{\text{C}}}\text{F} + \text{HF}$$

4,4'-Difluorobenzophenone, the key precursor to PEEK high performance resins, can be prepared by sequential (*1*) Friedel-Crafts coupling of 3,4-dichlorobenzotrifluoride, (2) exchange fluorination using KF/CH$_3$SO$_2$CH$_3$, and (3) reductive dechlorination with HCOONa/Pd-C (309).

$$\underset{\text{Cl}}{\underset{|}{\text{CF}_3\text{-C}_6\text{H}_3\text{-Cl}}} + \text{C}_6\text{H}_4\text{Cl}_2 \xrightarrow{\text{steps }(1)-(3)} \text{F-C}_6\text{H}_4\text{-C(O)-C}_6\text{H}_4\text{-F}$$

Electroreductive coupling of benzotrifluorides with sacrificial aluminum or magnesium anodes in the presence of acetone, carbon dioxide, or N,N-dimethylformamide provides a novel route to ArCF$_2$-derivatives (310).

$$\text{C}_6\text{H}_5\text{CF}_3 + \text{CH}_3\text{CCH}_3 \xrightarrow{\text{Al anode}} \text{C}_6\text{H}_5\text{CF}_2\text{C(CH}_3)_2\text{OH}$$
$$\underset{\text{O}}{\|} \qquad\qquad (80\%)$$

Biotransformation. Enzymatic oxidation of benzotrifluoride forms 3-trifluoromethyl-*cis*-1,2-dihydrocatechol; dehydration (acid pH) provides a novel route to 3-hydroxybenzotrifluoride [98-17-9] (171).

Benzotrifluoride. Benzotrifluoride [98-08-8] (α,α,α-trifluorotoluene), C$_6$H$_5$CF$_3$ (mol wt, 146.11), is a colorless liquid (Table 8). Toxicity: oral (rat), LD$_{50}$, 1500 mg/kg; oral (mouse), LD$_{LO}$, 10,000 mg/kg; subcutaneous (frog), LD$_{LO}$, 870 mg/kg; intraperitoneal (mouse), LD$_{LO}$, 100 mg/kg (127).

Benzotrifluoride was first synthesized in 1898 via the reaction of benzotrichloride and antimony trifluoride (313). Benzotrifluoride can be produced by the high pressure reaction of benzotrichloride with anhydrous hydrogen fluoride (AHF). Typical conditions include a 4:1 AHF–benzotrichloride mole ratio at 80–110°C and 1.52–1.55 MPa (220–225 psi) for 2–3 h to give 70–75% yields of benzotrifluoride (314,315). The pressure fluorination can be performed continuously in a series of autoclaves (316) or through a nickel reaction tube at 90–130°C at 3–5 MPa (435–725 psi) (317). Batch liquid-phase catalyzed processes at atmospheric pressure (318,319) and continuous processes have been developed (320). High temperature vapor-phase fluorination processes have also been described (321).

Benzotrifluoride Derivatives. Laboratory recipes for 45 benzotrifluorides have been published (322). Physical properties and toxicity of commercially significant benzotrifluoride derivatives are listed in Table 9; the amino compounds are colorless to yellow and other derivatives are colorless.

2-Chlorobenzotrifluoride. This compound is produced from 2-chlorobenzotrichloride and anhydrous hydrogen fluoride under atmospheric or high pressure conditions. Nitration forms 2-chloro-5-nitrobenzotrifluoride [777-37-7], a dye and germicide precursor.

4-Chlorobenzotrifluoride. This isomer is produced from 4-chlorobenzotrichloride and anhydrous hydrogen fluoride. Nitration provides either 4-chloro-3-

Table 8. Physical Properties of Benzotrifluoride[a]

Property	Value
mol wt	146.11
color	colorless
melting point, °C	−29.02
boiling point, °C	102.05
density, 25°C, g/mL	1.1814
coefficient of expansion (30–40°C)	0.00121
refractive index, n_D^{25}	1.4114
viscosity, mPa·s(=cP)	
38°C	0.488
99°C	0.282
surface tension, mN/m(=dyn/cm)	
20°C	23.39
latent heat of fusion, J/mol[b]	13,782.1
latent heat of vaporization, 102.05°C	
J/mol[b]	32.635.2
specific conductivity at 25°C, S/cm	1×10^{-7}
dielectric constant, 30°C	9.18
dipole moment, C·m[c]	8.54×10^{-30}
heat of combustion, J/g[b]	23,064.3
heat of formation, kJ/mol[b]	
vapor	−580.7
liquid	−618.4
boiling point of binary azeotrope, °C with 96.7 mol % bromine	58.1
solubility in water, g/100 g at room temperature	0.045
flash point (Cleveland open cup), °C[d]	15.6
fire point (Cleveland open cup), °C[d]	15.6
vapor pressure, in °C and kPa[e]	Antoine equation[f]

[a]Other properties of this compound and its derivatives have been reviewed (311).
[b]To convert J to cal, divide by 4.184.
[c]To convert C·m to debye (D), divide by 3.336×10^{-30}.
[d]Ref. 312.
[e]To convert kPa to mm Hg, multiply by 7.5; log kPa = log mm Hg − 0.875.
[f]$\text{Log}_{10} P = 6.0939 - \dfrac{1305.509}{(t + 217.280)}$.

nitrobenzotrifluoride [*121-17-5*] (one-step) or 4-chloro-3,5-dinitrobenzotrifluoride [*393-75-9*] (two-step) for use in crop protection applications. Dinitration can also be accomplished in one step (85% yield) with 90% nitric acid/20% oleum (325). Single-step dehalogenation-reduction of 4-chloro-3,5-dinitrobenzotrifluoride provides a 96% yield of 3,5-diaminobenzotrifluoride [*368-53-6*] (326), an intermediate to specialty polymers.

2,4-Dichlorobenzotrifluoride. This dichloro compound is produced from 2,4-dichlorobenzotrichloride and hydrogen fluoride. One commercial application is the manufacture of the pre-emergent herbicide, dinitramine [*29091-05-2*].

3,4-Dichlorobenzotrifluoride. This compound is produced by chlorination of 4-chlorobenzotrifluoride and exhibits sufficient activation to undergo nucleophilic

Table 9. Properties of Benzotrifluoride Derivatives

Component	CAS Registry Number	Mol wt	Mp, °C	Bp, °C$_{kPa}$[a]	Refractive index n_D^t	Specific gravity, d_4^t	Flash point,[b] °C	Toxicity, LD$_{50}$
$H_2NC_6H_4CF_3$				*Aminobenzotrifluoride*				
2-aminobenzotrifluoride	[88-17-5]	116.13	34	174–175$_{100.4}$	1.4800^{25}	1.290^{25}	55	440[c], 690[d,e]
3-aminobenzotrifluoride	[98-16-3]	116.13	5–6	187–188 86$_{2.67}$	1.4788^{20}	1.305^{25}	85	220[f]
$CF_3C_6H_4Cl$				*Monochlorobenzotrifluorides*				
2-chlorobenzotrifluoride	[88-16-4]	180.56		152.5	1.4550^{20}	1.367^{20}	98	>6.8[g]
4-chlorobenzotrifluoride	[98-15-7]	180.56	−36	139 29.5$_{1.33}$	1.4444^{25}	1.338^{25}	110	>2.7[h]
$CF_3C_6H_3Cl_2$				*Dichlorobenzotrifluorides*				
2,4-dichlorobenzotrifluoride	[320-60-5]	215.00	−26	177.5	1.4793^{25}	1.501$^{15.5}$	72	2900[i]
3,4-dichlorobenzotrifluoride	[328-84-7]	215.00	−12.4	173.5	1.4736^{25}	1.478^{25}	65	>2[j]

[a] To convert kPa to mm Hg, multiply by 7.5.
[b] Closed cup (ASTM procedure D3278).
[c] Inhalation (rat), LC$_{50}$ mg/m^3 for 4 h (127).
[d] Inhalation (mouse), g/m^3 for 2 h (127).
[e] LC$_{50}$.
[f] Oral (mouse), mg/kg (127).
[g] Acute oral (rat), g/kg (323).
[h] Acute dermal (rabbit), g/kg (323).
[i] Acute oral (rat), mg/kg (324).
[j] Acute dermal (rabbit), g/kg (324).

displacement with phenols to form diaryl ether herbicides, eg, acifluorofen sodium [62476-59-9].

$$O_2N-\underset{}{\bigcirc}(COONa)-O-\underset{}{\bigcirc}(Cl)-CF_3$$

3-Aminobenzotrifluoride. The standard manufacturing route to 3-aminobenzotrifluoride involves nitration of benzotrifluoride to 3-nitrobenzotrifluoride [98-46-4], followed by hydrogenation. A comprehensive study on materials of construction to minimize corrosion during catalytic hydrogenation led to the recommendation of Cr–Ni–Mo steel (with ≥ 3% Mo) (327). Gas chromatographic details for monitoring this two-step process have been described (328), as well as analytical methods for assay of the 2- and 4-isomer impurities in 3-aminobenzotrifluoride (329). Environmental aspects of the manufacture of this product in Germany have been published (330). A novel process based on the one-step *in situ* fluorination-reduction of 3-nitrobenzotrichloride with ammonium bifluoride–hydrogen fluoride has been described (331,332).

The amine group of 3-aminobenzotrifluoride can be replaced by Cl, Br, I, F, CN, or OH groups by standard diazotization reactions. Phosgenation gives 3-trifluoromethylphenylisocyanate [329-01-1], which can then be converted to the selective herbicide fluometuron [2164-17-2], a substituted urea.

Application. *Crop Protection Chemicals.* Benzotrifluoride derivatives have gained wide acceptance as herbicides, insecticides, and fungicides (Table 10).

Drugs. Trifluoromethyl-based pharmaceuticals had been limited to phenothiazine tranquilizers and benzothiadiazine 1,1-dioxide diuretics (qv). However, new drugs have been developed (Table 11). One of the key properties of the CF_3 group is its high lipophilicity; it increases the lipid solubility of the pharmaceutical and thus accelerates absorption and transport within the host organism.

Germicides. Benzotrifluoride derivatives have also found wide use as antimicrobial agents in soaps, eg, the brominated and chlorinated materials, fluorosalan [4776-06-1] and cloflucarban [369-77-7].

Other Biological Applications. 4-Nitro-3-(trifluoromethyl)phenol [88-30-2] (TFM) is still employed by the Canadian Bureau of Fisheries and the U.S. Fish and Wildlife Service as a lampricide for the control of parasitic sea lamprey in the Great Lakes (see AQUACULTURE).

Dyes. Several reviews on fluorine-containing dyes have been published (333–335). The relative accessibility of benzotrifluorides has reflected the wide incorporation of the trifluoromethylphenyl group into azo, anthraquinone, and triphenylmethane dyes (qv). The trifluoromethyl group is claimed to improve the

Table 10. Benzotrifluoride Crop-Protection Chemicals

Common name	CAS Registry Number	Structure	Application
trifluralin	[1582-09-8]		pre-emergent herbicide
lactofen	[77501-63-4]		herbicide
norflurazon	[27314-13-2]		selective herbicide
fluvalinate	[69409-94-5]		insecticide
hydramethylnon	[67485-29-4]		insecticide
fentrifanil	[62441-54-7]		acaricide
triflumizole	[68694-11-1]		fungicide

Table 11. Benzotrifluoride Drugs

Common name	CAS Registry Number	Structure	Application
triflupromazine	[146-54-3]		tranquilizer
flumethiazide	[148-56-1]		diuretic
fenfluramine	[458-24-2]		anorexigen
flufenamic acid	[53-78-9]		analgesic
trifluperidol	[749-13-3]		tranquilizer
flumetramide	[7125-73-7]		muscle relaxant
fluoxetine	[54910-89-3]		antidepressant

brightness, tinctorial values, and lightfastness of dyes. The electron-withdrawing effects of this group also tend to modify the absorption of light by a dye in the visible and uv region. In azo dyes (qv), aminobenzotrifluorides (Fast Base) are diazotized for subsequent coupling: 2-aminobenzotrifluoride CI Pigment Yellow 154 [88-17-5]; 3,5-bis(trifluoromethyl)aniline [328-74-5], CI Azoic Diazo Component 16; 2-amino-5-chlorobenzotrifluoride [445-03-4], CI Azoic Diazo Component 17; 3-amino-4-ethylsulfonylbenzotrifluoride [382-85-4], CI Azoic Diazo Component 19; and 3-amino-4-chlorobenzotrifluoride, CI Azoic Diazo Component 49

[*121-50-6*]. 3-Trifluoromethylbenzoyl halides are used to make anthraquinone vat dyes, eg, Indanthrene blue CLB [*6942-78-0*] (see DYES, ANTHRAQUINONE).

Miscellaneous Applications. Benzotrifluoride derivatives have been incorporated into polymers for different applications. 2,4-Dichlorobenzotrifluoride or 2,3,5,6-tetrafluorobenzotrifluoride [*651-80-9*] have been condensed with bisphenol A [*80-05-7*] to give benzotrifluoride aryl ether semipermeable gas membranes (336,337). 3,5-Diaminobenzotrifluoride [*368-53-6*] and aromatic dianhydrides form polyimide resins for high temperature composites (qv) and adhesives (qv), as well as in the electronics industry (338,339).

Photoresist applications in the microelectronics industry have also been disclosed (340). Thermally stable benzyl sulfonate esters based on 2-methyl-3-nitrobenzotrifluoride [*6656-49-1*] can serve as nonionic photoacid generators to promote a cascade of reactions during irradiation of the resist.

Liquid crystal applications include esters based on *m*- or *p*-hydroxybenzotrifluoride (341,342), hydroxytrifluoromethylbiphenyls (343) or hydroxytrifluoromethylphenyl Schiff bases (344).

Inorganic analytical applications for benzotrifluoride derivatives include sodium tetrakis[3,5-bis(trifluoromethyl)phenyl]borate (Kobayashi's reagent) (345), and 4-(2,6-dinitro-4-trifluoromethylphenyl)aminobenzo-15-crown-5 (modified Takagi reagent (346).

Benzotrifluoride has been recommended as a fuel additive for internal combustion engines (347).

Arylfluoroalkyl Ethers

α,α,α-Trifluoromethoxybenzene [*456-55-3*], $C_6H_5OCF_3$, and other arylfluoroalkyl ethers and thioethers ($HCF_2O—$, $HCF_2CF_2O—$, $CF_3CH_2O—$, and $CF_3S—$), are assuming greater importance as crop-protection chemicals and pharmaceuticals.

Properties. The trifluoromethoxy ($CF_3O—$) group in $ArOCF_3$ exhibits unusual stability to strong acids and bases (including organometallic reagents), as well as to strong oxidizing and reducing conditions (348). The thermal stability is exceptional; extensive degradation in the gas phase was not observed (mass spectroscopy) in a sealed nickel tube until 600°C (348). Nuclear chlorinated trifluoromethoxy and bis(trifluoromethoxy)benzenes have exhibited moderate thermal stability for use in transformers and Rankine cycle engines (349).

Reactions. The $CF_3O—$ group exerts predominant para orientation in electrophilic substitution reactions such as nitration, halogenation, acylation, and alkylation (350).

Trifluoromethoxybenzenes ($ArOCF_3$). Trifluoromethoxybenzene (α,α,α-trifluoroanisole, phenyl trifluoromethyl ether [*456-55-3*]), $C_6H_5OCF_3$ (mol wt 162.11), is a colorless liquid, bp 102°C, mp -50°C, n_D^{20} 1.4060, d_4^{25} 1.226, flash point (closed cup), 12°C.

Depending on the ring substituent, trifluoromethoxybenzenes can be made by the sequential chlorination–fluorination of anisole(s) (351–354). A one-step process with commercial potential is the BF_3 (or SbF_3)-catalyzed reaction of phenol with carbon tetrachloride/hydrogen fluoride (355). Aryl trifluoromethyl ethers, which may not be accessible by the above routes, may be made by fluorination of aryl fluoroformates or aryl chlorothioformates with sulfur tetrafluoride (348) or molybdenum hexafluoride (356).

$$\underset{\underset{O}{\parallel}}{(ArOCF)} \quad \underset{\underset{S}{\parallel}}{(ArOCCl)}$$

Trifluoromethylthioaromatics (ArSCF$_3$). Trifluoromethylthioaromatics (aryl trifluoromethyl sulfides) can be made by sequential chlorination–fluorination (SbF$_3$ or HF) of the corresponding thioanisole (351,357). In the case of 4-trifluoromethylmercaptophenol [825-83-2], 4-CF$_3$C$_6$H$_4$SH, used for the production of the coccidiostat toltrazuril [69004-03-1] (Table 12), protection of the phenolic group as the carbonate ester is required prior to chlorination (358). Coupling of aryl halides with trifluoromethylthiocopper, CF$_3$SCu, provides an alternative entry to trifluoromethylthioaromatics (359,360).

Difluoromethoxyaromatics (ArOCHF$_2$) and Sulfur Analogues (ArSCHF$_2$). Difluoromethyltion of phenol (or thiophenols) with chlorodifluoromethane, CHClF$_2$, and aqueous caustic in dioxane gives good yields of aryldifluoromethyl ethers (361). A modification features the use of phase-transfer catalysts such as tris(3,6-dioxaheptyl)amine (TDA-1) (362).

Tetrafluoroethoxyaromatics (ArOCF$_2$CF$_2$H). Tetrafluoroethoxyaromatics are produced by base-catalyzed addition of tetrafluoroethylene to phenols (348,363).

Aryltrifluoroethyl Ethers (ArOCH$_2$CF$_3$). 2,2,2-Trifluoroethoxybenzenes are obtained from the reaction of activated haloaromatics with sodium 2,2,2-trifluoroethoxide in polar solvents (364); phase-transfer catalysts are also employed (365). Nitro groups can also be displaced by the fluoroalkylation technique, eg, 4-nitrobenzonitrile was converted to 4-(2,2,2-trifluoroethoxy)benzonitrile (366). Trifluoroethoxybenzene pharmaceutical intermediates can be prepared by the base-catalyzed reaction of 2,2,2-trifluoroethyl trifluoromethanesulfonate (a trifluorethyl-transfer agent), CF$_3$SO$_3$CH$_2$CF$_3$, with phenols (367).

Applications. Table 12 lists crop-protection chemicals and pharmaceuticals containing the aryl fluoroalkyl ether group. Trifluoromethoxybenzene (ArOCF$_3$) derivatives (186,187) and related arylfluoroalkyl ethers, ArO(CH$_2$)$_n$R$_f$ (n = 1–6) (368,369) are of use in liquid crystal applications.

Fluorinated Nitrogen Heterocyclics

Ring- or side-chain fluorinated nitrogen heterocyclics have been incorporated into crop-protection chemicals, drugs, and reactive dyestuffs. Key intermediates include fluorinated pyridines, quinolines, pyrimidines, and triazines. Physical properties of some fluorinated nitrogen heterocyclics are listed in Table 13.

RING-FLUORINATED PYRIDINES

Exchange fluorination of chloropyridines is the principal tool for production of ring-fluorinated pyridines. Diazotization of aminopyridines in pyridine–hydrogen fluoride (Olah's reagent) (370) or ammonium fluoride–hydrogen fluoride (371) has also been used. An emerging synthesis tool is the use of fluorinated aliphatic

Table 12. Aryl Fluoroalkyl Ether Applications

Common name	CAS Registry Number	Structure
Insecticides		
triflumuron	[64628-44-0]	CF_3O–C$_6H_4$–NHC(O)NHC(O)–C$_6H_4$–Cl
toltrazuril[a]	[69004-03-1]	CF_3S–C$_6H_4$–O–C$_6H_4$–N(triazinetrione)
flucythrinate	[70124-77-5]	HCF_2O–C$_6H_4$–CH(CH(CH$_3$)$_2$)–COOCH(CN)–C$_6H_4$–O–C$_6H_5$
hexaflumuron	[86479-06-3]	HCF_2CF_2O–C$_6H_2$Cl$_2$–NHC(O)NHC(O)–C$_6H_3$F$_2$
Herbicides		
primsulfuronmethyl	[86209-51-0]	C$_6H_4$(COOCH$_3$)–SO$_2$NHC(O)NH–pyrimidine(OCHF$_2$)$_2$
tetrafluron	[27954-37-6]	HCF_2CF_2O–C$_6H_4$–N(CH$_3$)C(O)NHC(CH$_3$)$_2$
Other		
flurprimidol[b]	[56425-91-3]	CF_3O–C$_6H_4$–C(OH)(CH(CH$_3$)$_2$)(pyrimidinyl)
flecainide acetate[c]	[54143-56-5]	CF_3CH_2O–C$_6H_3$(OCH$_2$CF$_3$)–CONH–CH$_2$–piperidinyl · CH$_3$COOH

[a] Coccidiostat.
[b] Plant growth regulator.
[c] Cardiac depressant.

Table 13. Properties of Miscellaneous Fluorinated Heterocyclic Compounds

Component	CAS Registry Number	Mol wt	Mp, °C	Bp, °C/kPa[a]	Refractive index, n_D^t	Specific gravity, d_4^t	Flash point, °C
C_5H_4FN				*Fluoropyridines*			
2-fluoropyridine	[372-48-5]	97.09		$126_{100.4}$	1.4678^{20}	1.1281^{20}	24^b
3-fluoropyridine	[372-47-4]	97.09		$105–107_{100.3}$	1.4700^{20}	1.125^{25}	13^b
4-fluoropyridine	[694-52-0]	97.09	100^c	$108_{100.0}$	1.4730^{20}		
$C_5H_3F_2N$							
2,4-difluoropyridine	[3491-90-7]	115.08	134–135^c	104–105			
2,6-difluoropyridine	[1513-65-1]	115.08		$124.5_{99.1}$	1.4349^{25}	1.265^{25}	32^b
$C_5H_2F_3N$							
2,4,6-trifluoropyridine	[3512-17-2]	133.07		94–95			
C_5HF_4N							
2,3,5,6-tetrafluoropyridine	[2875-18-5]	151.06		102			
$C_5H_5N^d$			*Perfluoroalkylpyridines*				
pentafluoropyridine	[70-16-3]	169.05	−41.5	83.3	1.3856^{20}		
2-trifluoromethylpyridine	[368-48-9]	147.10		$143_{99.4}$	1.4144^{25}		
3-trifluoromethylpyridine	[3796-23-4]	147.10		113–115	1.4150^{25}		
4-trifluoromethylpyridine	[3796-24-5]	147.10		108–110	1.4144^{25}		
2-chloro-5-trifluoromethylpyridine	[52334-81-3]	181.55	32–34	152_{100}		1.417^{20}	110^e
			Fluoropyrimidines				
2,4,6-trifluoropyrimidine	[696-82-2]	134.06		$98;60_{24}$	1.4015^{25}		
2,4,5,6-tetrafluoropyrimidine	[767-79-3]	152.06		89	1.3875^{25}		
5-chloro-2,4,6-trifluoropyrimidine	[697-83-6]	168.51		114.5	1.4390^{20}		
			Fluorotriazines				
2,4,6-trifluoro-1,3,5-triazine	[675-14-9]	135.05	−38	$72.4_{101.67}$	1.3844^{24}	1.60^{25}	
2,4,6-*tris*-(trifluoromethyl)-1,3,5-triazine	[368-66-1]	285.07	−24.8	95–96	1.3161^{25}	1.593^{25}	

^{a}To convert kPa to mm Hg, multiply by 7.5. ^{b}Tag closed cup. ^{c}Mp of HCl salt.
^{d}Trouton's constant, 24.9; latent heat of vaporization, 36, 338 J/mol. ^{e}Closed cup (ASTM procedure D3278).

building blocks to make fluoropyridines. Early studies on the substitutive fluorination (F_2) of pyridine gave 2-fluoropyridine in low yields and posed severe reaction hazards (372,373). Modifications featuring low temperature fluorination of substituted pyridines (alkyl, halogen, ester, or ketone functions) in 1,1,2-trichloro-1,2,2-trifluoroethane, $CF_2ClCFCl_2$, solvent give good yields of the corresponding 2-fluoropyridine (374,375). Tamed fluorine reagents such as xenon difluoride (376) and cesium fluoroxysulfate (377) can also fluorinate pyridine. A promising substitutive fluorination technique is the base-catalyzed decomposition of N-fluoropyridinium salts, BF_4^-, PF_6^-, or SbF_6^-, to give high yields of the substituted 2-fluoropyridine (378). The salt is made in 80% yield; the decomposition yields 72–91% of product.

Monofluoropyridines. *2-Fluoropyridine.* Diazotization of 2-aminopyridine in anhydrous hydrogen fluoride forms 2-fluoropyridine in high yield (178,370). Modifications include fluorodediazonization of substituted 2-aminopyridines in ammonium fluoride–hydrogen fluoride (371) or pyridine–hydrogen fluoride (370) media. Exchange-fluorination of 2-chloropyridine with potassium fluoride in polar solvents is sluggish (210°C for 21 d; 50–58% yield) (379). The solvent-free exchange-fluorination employing potassium bifluoride (KHF_2) and 2-chloropyridine (315°C/4 h) gave 2-fluoropyridine in 74% yield (380). A new development features exchange fluorination of 2-chloropyridine with hydrogen fluoride–γ-collidine at 150–200°C to give 94% yield of product (381).

Fluorine at the 2 position ($pK_a = -0.44$) significantly reduces pyridine ($pK_a = 5.17$) basicity more than at the 3 position ($pK_a = 2.97$) (382). 2-Fluoropyridine is readily hydrolyzed to 2-pyridone in 60% yield by reflux in 6 N hydrochloric acid (383). It is quite reactive with nucleophiles. For example, the halogen mobility ratio from the comparative methoxydehalogenation of 2-fluoropyridine and 2-chloropyridine was 85.5/1 at 99.5°C (384). This lability of fluorine has been utilized to prepare fluorine-free O-2-pyridyl oximes of 3-oxo steroids from 2-fluoropyridine for possible use as antifertility agents (385).

2-Fluoropyridine is a useful reagent for synthetic applications. It reacts with methyl p-toluenesulfonate to give 1-methyl-2-fluoropyridinium p-toluenesulfonate [58086-67-2] (Mukaiyama's reagent). Such onium salts are used in the oxidation of alcohols, cross-coupling of Grignard reagents, Beckmann rearrangements of ketoximes, nonenzymatic biogenetic-like synthesis of terpenes (386), and preparation of new synthetic penicillins (387).

Metallation of 2-fluoropyridine with lithium diisopropylamide (LDA) gives 2-fluoro-3-lithiopyridine, thereby providing entry to 3-substituted pyridines (388). This technique has been used to make fluorine analogues of the antitumor ellipticines (389).

$$\underset{\text{N}}{\bigcirc}\!\!-\text{F} \xrightarrow[-70°C]{LDA} \left[\underset{\text{N}}{\bigcirc}\!\!\overset{\text{Li}}{-}\text{F}\right] \xrightarrow[2.\ H_2O]{1.\ R_1COR_2} \underset{\text{N}}{\bigcirc}\!\!\overset{-CH(OH)R_1R_2}{-\text{F}}$$

3-Fluoropyridine. Diazotization of 3-aminopyridine(s) in hydrogen fluoride (390), pyridine–hydrogen fluoride (370), or ammonium fluoride–hydrogen fluoride (371) can be effected in good yield. 3-Fluoropyridine can also be made by the Balz-Schiemann technique in 50% yield. Earlier warnings concerning the instability of 3-pyridyldiazonium fluoroborate (391) were confirmed by later reports on detonations involving this salt (392). Related compounds such as 2-chloro-3-pyridyldiazonium fluoroborate also decomposed with explosive violence (393).

3-Fluoropyridine derivatives can be constructed from fluoroaliphatic feedstocks. 5-Fluoro-2,6-dihydroxynicotinamide [655-13-0], a precursor to the antibacterial, enoxacin [74011-58-8], was prepared in 63% yield from ethyl fluoroacetate [459-72-3], ethyl formate [109-94-4], and malonamide [108-13-4] (394).

$$FCH_2COOC_2H_5 \xrightarrow[NaOC_2H_5]{HCOOC_2H_5} \left[\begin{array}{c} F\diagdown\!\!\!\!\!\!{}_C\!\diagup COOC_2H_5 \\ \parallel \\ H\diagup{}^C\diagdown ONa \end{array}\right] \xrightarrow{CH_2(CONH_2)_2} \underset{HO\diagdown N\diagup OH}{F-\bigcirc-CONH_2}$$

A complementary cyclization technique based on dichlorofluoroacetonitrile [83620-05-7], $FCCl_2CN$, was employed to form 2-chloro-3-fluoro-5-methylpyridine [34552-15-3] (395).

Derivatives such as 3-fluoro-4-nitropyridine [13505-01-6] (396) or the 1-oxide [769-54-0] (397) have been used to characterize amino acids and peptides. 5-Fluoro-3-pyridinemethanol [22620-32-2] has been patented as an antilipolytic agent (398). A promising antidepressant, 1-(3-fluoro-2-pyridyl)piperazine hydrochloride [85386-84-1] is based on 2-chloro-3-fluoropyridine [17282-04-1] (399).

4-Fluoropyridine. This isomer can be prepared in 54–81% yield by diazotization of 4-aminopyridine in anhydrous hydrogen fluoride (370,371,400). Free 4-fluoropyridine readily undergoes self-quaternization to give pyridyl pyridinium salts (401); stabilization can be effected as the hydrochloride salt (371,400). Numerous 4-fluoropyridinium salts, eg, 4-fluoro-1-methylpyridinium iodide, have been converted to novel penicillins (387,402).

Difluoropyridines. 2,4-Difluoropyridine can be prepared (26% yield) from 2,4-dichloropyridine and potassium fluoride in sulfolane and ethylene glycol initiator (403). The 4-fluorine is preferentially replaced by oxygen nucleophiles to give 2-fluoro-4-hydroxypyridine derivatives for herbicidal applications (404).

Fluorination of 2,6-diaminopyridine in anhydrous hydrogen fluoride gave a 62% yield of 2,6-difluoropyridine (26,371,405). 2,6-Difluoropyridine is also prepared in 52% yield (200°C, 100 h) from 2,6-dichloropyridine and potassium fluoride in dimethyl sulfone or sulfolane (406). The reaction can be performed in dimethyl sulfoxide with shorter reaction times (9 h, 186°C) (407); addition of tetramethylammonium chloride (TMAC) catalyst lowers reaction temperature to 150°C, thereby minimizing solvent degradation (408). Solvent-free exchange-fluorination (KF) at 400°C (16 h) gave 80% yield of 2,6-difluoropyridine which attests to its high thermal stability (409).

Displacement reactions with oxygen nucleophiles are of potential commercial interest. Alkaline hydrolysis provides 2-fluoro-6-hydroxypyridine [55758-32-

2], a precursor to 6-fluoropyridyl phosphorus ester insecticides (410–412). Other oxygen nucleophiles such as bisphenol A and hydroquinone have been used to form aryl–pyridine copolymers (413).

$$F-\text{[pyridine]}-F + HO-\text{[C}_6\text{H}_4\text{]}-C(CH_3)_2-\text{[C}_6\text{H}_4\text{]}-OH \longrightarrow \left(O-\text{[pyridine]}-O-\text{[C}_6\text{H}_4\text{]}-C(CH_3)_2-\text{[C}_6\text{H}_4\text{]} \right)_x$$

Nitration with mixed nitric and sulfuric acids provides 79% yield of 3-nitro-2,6-difluoropyridine [5860-02-1], bp 218–220°C (414).

3-Bromo or chloro-2,6-difluoropyridines can be prepared in 50% yield by diazotization of the corresponding 3-halo-2,6-diaminopyridine in ammonium fluoride–hydrogen fluoride solvent (371). 5-Chloro-2,3-difluoropyridine [89402-43-7], a precursor to the herbicide pyroxofop [105512-06-9], was synthesized by a multistep sequence based on allyl chlorodifluoroacetate [118337-48-7], $ClCF_2CO_2$-$CH_2CH{=}CH_2$ (415).

Tri-, Tetra-, and Pentafluoropyridines. 2,4,6-Trifluoropyridine can be prepared in 75% yield by catalytic hydrogenolysis (palladium-on-carbon, 280°C) of 3,5-dichloro-2,4,6-trifluoropyridine [1737-93-5] (416). The latter is synthesized by exchange fluorination of pentachloropyridine with potassium fluoride in polar solvents such as N-methylpyrrolidinone (417,418). 3,5-Dichloro-2,4,6-trifluoropyridine is used to prepare the herbicides haloxydine [2693-61-0] and fluroxypyr-(1-methylheptyl) [81406-37-3].

2,3,5,6-Tetrafluoropyridine can be prepared in 75% yield from the hydrogenation of pentafluoropyridine under free-radical (catalytic) or nucleophilic (lithium aluminum hydride) conditions (416,419). No practical uses for 2,3,5,6-tetrafluoropyridine are known.

Pentafluoropyridine was first synthesized in 1960 in 27% yield by the defluorination of undecafluoropiperidine (prepared in low yield by the electrochemical fluorination of pyridine) over a nickel or iron surface at 560–610°C (420,421). The preferred route is the solvent-free exchange-fluorination (KF) of pentachloropyridine at 480–500°C to give 69–83% yields of pentafluoropyridine (418,422). Pentafluoropyridine is a weak base, does not form a hydrochloride salt, and is more volatile (bp 83.3°C) than pyridine (bp 115°C). Pentafluoropyridine readily undergoes reaction with nucleophilic agents to give 4-substituted-2,3,5,6-tetrafluoropyridines. More than 30 examples of these 4-substitution reactions have been compiled (423). Derivatives of 4-hydroxytetrafluoropyridine [2693-66-5] and related compounds exhibit herbicidal properties (424,425).

The nucleophilic equivalent of the Friedel-Crafts reaction of pentafluoropyridine with hexafluoropropene–potassium fluoride in sulfolane gave perfluoro-(4-isopropyl)pyridine in 94% yield (426).

Pentafluoropyridine–hexafluorobenzene working fluids show the requisite stability at 382°C for automotive Rankine-cycle power units (427). Hydroxyl and related functions in steroids can be selectively protected as tetrafluoro-4-pyridyl ethers by pentafluoropyridine (428).

Applications. Until recently, haloxydine, a herbicide, was one of the few early examples of crop-protection chemicals containing ring-fluorinated pyridines. Fluroxypyr-(1-methylheptyl) and pyroxofop are new herbicides that are being commercialized (Table 14).

Several 3-fluoropyridine derivatives are employed to produce enoxacin, tosufloxacin, and other naphthyridine antibacterials (Table 14). Examples of such intermediates include 2,6-dichloro-5-fluoronicotinonitrile (429), ethyl 2,6-dichloro-5-fluoronicotinate (430), 2-chloro-3-fluoropyridine (393), 6-acetyl-2-(4-acetyl-1-piperazinyl)-3-fluoropyridine (431), and 5-fluoro-2,6-dihydroxynicotinamide (394).

Table 14. Applications of Ring-Fluorinated Pyridines

Common name	CAS Registry Number	Structure
Herbicides		
haloxydine	[2693-61-0]	
fluroxypyr-(1-methylheptyl)	[81406-37-3]	
pyroxofop	[105512-06-9]	
Antibacterials		
enoxacin	[74011-58-8]	
tosufloxacin	[108138-46-1]	

PERFLUOROALKYLPYRIDINES

New developments in trifluoromethylpyridine technology are associated with the commercialization of numerous crop-protection chemicals as herbicides, fungicides, and insecticides (Table 15). Physical properties for representative trifluoromethylpyridines are listed in Table 13.

The standard synthesis method features side-chain chlorination of a methylpyridine (picoline), followed by exchange-fluorination with hydrogen fluoride or antimony fluorides (432,433). The fluorination of pyridinecarboxylic acids by sul-

Table 15. Trifluoromethylpyridine-Based Crop-Protection Chemicals[a]

Common name	CAS Registry Number	Structure
dithiopyr	[97886-45-8]	
fluazilop–butyl	[69806-50-4]	
haloxyfop–methyl	[69806-40-2]	
flazasulfuron	[104040-78-0]	
chlorfluazuron[b]	[71422-67-8]	
fluazinam[c]	[79622-59-6]	

[a]Herbicide unless otherwise noted.
[b]Insecticide.
[c]Fungicide.

fur tetrafluoride (434) or molybdenum hexafluoride (435) is of limited value for high volume production operations due to high cost of fluorinating agent.

A significant development in trifluoromethylpyridine synthesis strategy is the use of fluorinated aliphatic feedstocks for the ring-construction sequence. Examples include the manufacture of the herbicide dithiopyr, utilizing ethyl 4,4,4-trifluoroacetoacetate [372-31-6], $CF_3COCH_2COOC_2H_5$ (436,437). 2,3-Dichloro-5-trifluoromethylpyridine [69045-84-7], a precursor to several crop-protection chemicals (see Table 15), can be prepared by conversion of 1,1,1-trichloro-2,2,2-trifluoroethane [354-58-5], CF_3CCl_3, to 2,2-dichloro-3,3,3-trifluoropropionaldehyde [82107-24-2], CF_3CCl_2CHO, followed by cyclization with acrylonitrile [107-13-1] (415).

2-Trifluoromethylpyridine can be prepared in 54% yield from picolinic acid and sulfur tetrafluoride–hydrogen fluoride (434). 2-Trifluoromethylpyridine is a weak base; no hydrochloride salt is formed. However, 2-trifluoromethylpyridine 1-oxide [22253-71-0] (bp 132–133°C/2.7 kPa (20 mm Hg)) is prepared in 81% yield using 30% hydrogen peroxide–acetic acid (438).

3-Trifluoromethylpyridine can be prepared in 25–65% yield from nicotinic acid and sulfur tetrafluoride (434,439). An alternative method is the passage of chlorine into a mixture of β-picoline and hydrogen fluoride in an autoclave (190°C, 3 MPa) (440). 4-Trifluoromethylpyridine is prepared in 57% yield from isonicotinic acid and sulfur tetrafluoride.

2-Chloro-5-trifluoromethylpyridine, an intermediate to the herbicide fluazilop–butyl, can be made from β-picoline by two processes. β-Picoline is chlorinated to 2-chloro-5-trichloromethylpyridine [69405-78-9], followed by fluorination with hydrogen fluoride under pressure (200°C, 10 h) (441) or vapor-phase (350°C, CCl_4 diluent) conditions (442). An alternative process features the single-step vapor-phase reaction of β-picoline with chlorine–hydrogen fluoride (400°C, N_2 or CCl_4 diluent) (443).

FLUOROQUINOLINES

The standard routes to monofluoroquinolines have been the Balz-Schiemann reaction from the corresponding aminoquinoline or the Skraup reaction from glycerol and a fluoroaniline. Exchange-fluorination also has been used. 2-Chloroquinoline and potassium fluoride in dimethyl sulfone gave 60% yield of 2-fluoroquinoline [580-21-2], C_9H_5FN; bp 133°C at 4 kPa (30 mm Hg); 75°C at 0.3 kPa (2 mm Hg); n_D^{25} 1.5827 (406). Likewise, heptachloroquinoline and potassium fluoride at 470°C for 17 h gave a 71% yield of heptafluoroquinoline [13180-38-6], C_9F_7N; mp 95–95.5°C, bp 205°C (444).

The main preparative techniques used to make all seven trifluoromethylquinoline isomers include copper-assisted coupling of the haloquinoline with trifluoromethyl iodide (112); quinolinecarboxylic acid with sulfur tetrafluoride–hydrogen fluoride (434,445); and aminobenzotrifluoride and glycerol (Skraup reaction) (446,447).

Commercial trifluoromethylquinoline-based products are mefloquine [53230-10-7], an antimalarial, and floctafenine [23779-99-9], an analgesic. The cyclization step to construct the 2,8-bis(trifluoromethyl)quinoline nucleus in mefloquine employs 2-aminobenzotrifluoride [88-17-5] and ethyl 4,4,4-trifluoroacetoacetate [372-31-6] (448).

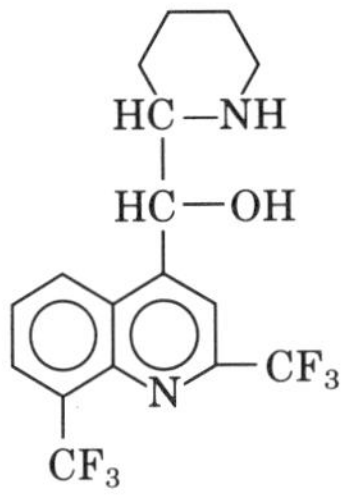

mefloquine

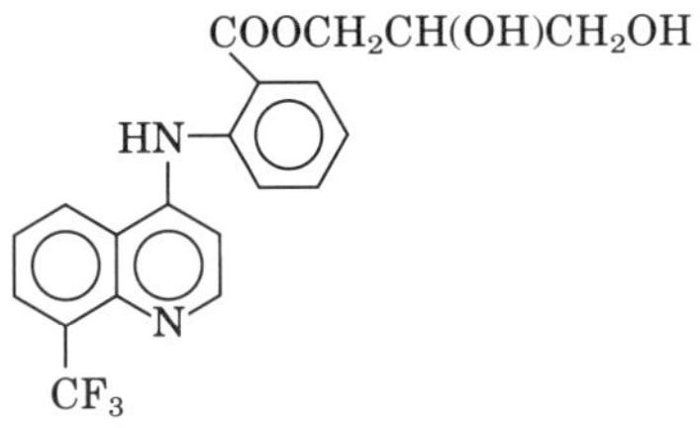

floctafenine

FLUOROQUINOLONES

A primary development has been the rapid commercialization of fluoroquinolone antibacterials (427). The single-fluorinated quinolones (second generation) constitute 10% of the worldwide prescriptions for antibiotics: norfloxacin [70458-96-7], enoxacin [74011-58-8], perfloxacin [70458-92-3], ciprofloxacin [85721-33-1], and ofloxacin [82419-36-1] (449) (see ANTIBACTERIAL AGENTS, QUINOLONES). Annual sales (1992) were estimated at $800 million. As new agents are introduced, fluoroquinolones are expected to maintain an average growth rate of 30% during the 1991–1997 time period (450). New synthesis strategy includes multiple-fluorinated quinolones (third generation) such as lomefloxacin [98079-51-7], fleroxacin [79660-72-3], temafloxacin hydrochloride [105784-61-0], and tosufloxacin [108138-46-1].

The discovery of new broad spectrum antibiotics has been accompanied by the development of processes for fluorinated feedstocks: ring-fluorinated aromatics for those quinolones containing a fluorobenzopyridone group, and fluorinated pyridine precursors for those antibiotics containing a naphthyridine nucleus (enoxacin, tosufloxacin) (see Table 14).

FLUOROPYRIMIDINES

Fluoropyrimidines find diverse use in cancer chemotherapy and other drug applications, as well as in fiber-reactive dyes. Table 13 lists physical properties of representative fluoropyrimidines.

5-Fluoropyrimidine derivatives are of tremendous importance in cancer chemotherapy, eg, 5-fluorouracil [51-21-8] (5-FU). The original 5-fluorouracil process featured a multistep low yield route based on ethyl fluoroacetate (451). Direct fluorination (fluorine) of uracil [66-22-8] gives high yields of 5-FU (452–455). This process has now been commercialized.

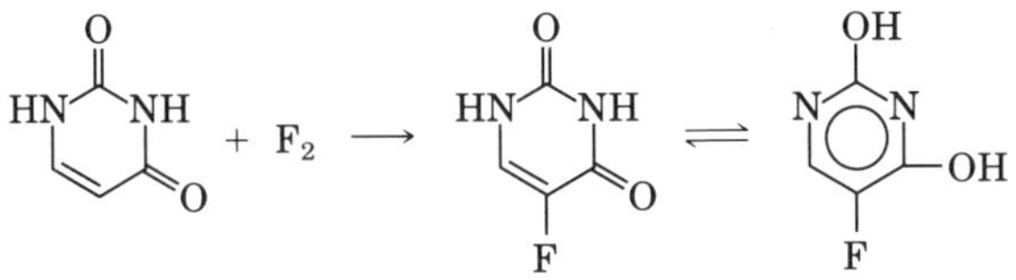

Other monofluoropyrimidines of biological interest are 5-fluorocytosine [2022-85-7], an antifungal agent; 2'-deoxy-5-fluorouridine [50-91-9] (5-FUDR), an antiviral and antineoplastic agent; 5-fluoroorotic acid [703-95-7] (5-FOA), used in yeast molecular genetics (456); and tegafur [17902-23-7] (Ftorafur), an antineoplastic agent which releases 5-FU *in vivo* (see CHEMOTHERAPEUTICS, ANTICANCER).

5-fluorocytosine   5-FOA   tegafur

Exchange fluorination is the main synthetic tool to prepare polyfluoropyrimidines. It was established that choice of fluorinating agent permits selectivity during exchange fluorination of nuclear- and side-chain chlorinated pyrimidines: NaF and KF fluorinate only in the heterocyclic nucleus; HF in the nucleus and in the chlorinated methyl group; and SbF_3 only in the chlorinated methyl group (457).

2,4,6-Trifluoropyrimidine can be prepared in 85% yield from 2,4,6-trichloropyrimidine [3764-01-0] and potassium fluoride in sulfolane or solvent-free conditions (458,459). Derivatives such as 1,1,1-trichloro-3-[5-(2,4,6-trifluoropyrimidyl)]-3,4-epoxybutane [121058-68-2] have been prepared as potential herbicides (460).

2,4,5,6-Tetrafluoropyrimidine has been prepared by direct fluorination of 2,4,6-trifluoropyrimidine with silver difluoride in perfluorobutylamine solvent (461,462). A more direct route (85% yield) is the reaction of tetrachloropyrimidine and potassium fluoride in an autoclave at 480°C for 42 h (463).

Tetrafluoropyrimidine was converted to the antineoplastic 5-fluorouracil (5-FU) by a novel process based on the sequence: partial exchange chlorination (61% yield), selective hydrogenolysis in triethylamine (71% yield) and hydrolysis (85–93% yield) (464).

5-Chloro-2,4,6-trifluoropyrimidine [697-83-6] has gained commercial importance for the production of fiber-reactive dyes (465,466). It can be manufactured by partial fluorination of 2,3,5,6-tetrachloropyrimidine [1780-40-1] with anhydrous hydrogen fluoride (autoclave or vapor phase) (467) or sodium fluoride (autoclave, 300°C) (468). 5-Chloro-2,4,6-trifluoropyrimidine is condensed with amine chromophores to provide the 5-chloro-2,4-difluoropyrimidyl group; the fluorine atom of the latter then reacts with a nucleophilic site in the fabric. Commercial reactive dyes for cottons and cellulosics include Levafix EA and PA Dyestuffs and Drimarene K and R Dyestuffs. For wool, the following 5-chloro-2,4-difluoropyrimidyl reactive dyes are offered: Verofix Dyestuffs and Drimalene Dyestuffs (see DYES, REACTIVE).

FLUOROTRIAZINES

Ring-fluorinated triazines are used in fiber-reactive dyes. Perfluoroalkyl triazines are offered commercially as mass spectral markers and have been intensively evaluated for elastomer and hydraulic fluid applications. Physical properties of representative fluorotriazines are listed in Table 13. Toxicity data are available. For cyanuric fluoride, LD_{50} = 3.1 ppm for 4 h (inhalation, rat) and 160 mg/kg (skin, rabbit) (127).

2,4,6-Trifluoro-1,3,5-Triazine. Cyanuric fluoride [675-14-9] can be produced from 2,4,6-trichloro-*s*-1,3,5-triazine [108-77-0] (cyanuric chloride) with hydrogen fluoride under autoclave (469,470) or vapor-phase (471) conditions. Sodium fluoride (in sulfolane solvent) can also be used to manufacture cyanuric fluoride (472,473).

Cyanuric fluoride is readily hydrolyzed to 2,4,6-trihydroxy-1,3,5-triazine [108-80-5] (cyanuric acid). Cyanuric fluoride reacts faster with nucleophilic agents such as ammonia and amines than cyanuric chloride.

Fiber-reactive dyes containing the fluorotriazinyl group are based on the condensation of chromophores containing amino groups with 6-substituted-2,4-difluorotriazines. The latter can be prepared from cyanuric fluoride or from the reaction of alkali metal fluorides with 6-substituted-2,4-dichlorotriazines. Comparative advantages of monofluorotriazinyl dyes over commercial monochlorotriazinyl analogues have been reviewed (466).

Cyanuric fluoride has been employed as a specific reagent for tyrosine residues in enzymes (474). Cyanuric fluoride can also serve as a fluorinating agent in fluorodehydroxylation reactions, eg, the conversion of 2-hydroxypyridine [142-08-5] to 2-fluoropyridine (475). This technique was subsequently extended to the preparation of acid fluorides from the corresponding carboxylic acid (476). It has found application in peptide synthesis from amino acids through the corresponding acid fluoride (477).

2,4,6-tris-(Trifluoromethyl)-1,3,5-Triazine. This compound can be prepared by trimerization of trifluoroacetonitrile (478) or fluorination of 2,4,6-*tris*-(trichloromethyl)-1,3,5-triazine with hydrogen fluoride–antimony pentachloride or antimony trifluoride–antimony pentafluoride (479). LC_{50} = 1400 ppm for 4 h (inhalation, rat) (127).

tris-(Trifluoromethyl)-*s*-triazine [368-66-1], as well as the *tris*-perfluoroethyl [858-46-8], propyl [915-22-9], heptyl [21674-38-4], and nonyl [57104-59-4] *s*-triazines are commercially offered as mass spectrometry internal reference standards for a wide mass range, 285–1485. The perfluoroalkylene (perfluoroalkyl)-*s*-triazines and perfluoroalkylene(perfluoroalkyloxy)-*s*-triazines were found to be suitable nonflammable hydraulic fluids in the −25 to +300°C temperature range. Numerous laboratories have investigated the synthesis and properties of perfluoroalkylene elastomers containing the *s*-triazine functionality (480,481).

MISCELLANEOUS FLUORINATED NITROGEN HETEROCYCLICS

Two reviews (1981, 1990) include nitrogen heterocyclics not covered in the present survey (482,483). The 1990 review dealing with four-, five-, and six-membered ring heterocyclic compounds emphasizes biological properties (482).

BIBLIOGRAPHY

"Fluorinated Aromatic Compounds" under "Fluorine Compounds, Organic," in *ECT* 2nd ed., Vol. 9, pp. 775–802, by A. K. Barbour, M. W. Buxton, and G. Fuller, Imperial Smelting Corp., Ltd.; in *ECT* 3rd ed., Vol. 10, pp. 901–936, by M. M. Boudakian, Olin Chemicals; "Perfluoroalkylene Triazines" under "Fluorine Compounds, Organic," *ECT* 3rd ed., Vol. 10, pp. 948–951, by W. R. Griffen, Air Force Materials Laboratory.

1. M. R. Kilbourn, *Fluorine-18 Labeling of Radiopharmaceuticals*, National Academy Press, Washington, D.C., 1990.
2. G. Furin, in L. German and S. Zemskov, eds., *New Fluorinating Agents in Organic Synthesis*, Springer-Verlag, Berlin, 1989, Chapt. 9, p. 35.
3. V. Grakauskas, *J. Org. Chem.* **35**, 723 (1970); *Ibid.*, **34**, 2835 (1969).
4. N. B. Kaz'mina and co-workers, *Dokl. Akad. Nauk SSSR* **194**, 1329 (1970).
5. F. Cacace and co-workers, *J. Am. Chem. Soc.* **102**, 3511 (1980).
6. S. T. Purrington and D. L. Woolard, *J. Org. Chem.* **56**, 142 (1991).
7. S. Misaki, *J. Fluorine Chem.* **21**, 191 (1982); *Ibid.*, **17**, 159 (1981).
8. J. Ellis and W. K. R. Musgrave, *J. Chem. Soc.*, 3608 (1950).
9. M. M. Boudakian and G. A. Hyde, *J. Fluorine Chem.* **25**, 435 (1984).
10. R. Filler, *Isr. J. Chem.* **17**, 71 (1978).
11. A. Zweig and co-workers, *J. Org. Chem.* **45**, 3597 (1980).
12. S. Stavber and M. Zupan, *J. Org. Chem.* **50**, 3609 (1985).
13. M. J. Fifolt and co-workers, *J. Org. Chem.* **50**, 4576 (1985).
14. S. Rozen and co-workers, *J. Org. Chem.* **49**, 806 (1984); *Ibid.*, **46**, 4629 (1981).
15. D. Hebel and co-workers, *Tetrahedron Lett.* **31**, 619 (1990).
16. S. Singh and co-workers, *J. Am. Chem. Soc.* **109**, 7194 (1987).
17. E. Differding and H. Ofner, *SYNLETT*, 187 (Mar. 1991); *Chem. Eng. News*, 33 (Sept.7, 1992).
18. T. Umemoto and co-workers, *J. Am. Chem. Soc.* **112**, 8563 (1990).
19. A. G. Gilicinski and co-workers, *J. Fluorine Chem.* **59**, 157 (1992).
20. R. E. Banks and co-workers, *J. Chem. Soc., Chem. Commun.*, (8), 595 (1992).
21. Ger. Pat. 600,706 (July 30, 1934), P. Osswald and O. Scherer (to I. G. Farbenind.); F. O. Robitschek and B. H. Wilcoxon, *FIAT Final Report*, No. 998, Mar. 31, 1947.
22. R. L. Ferm and C. A. VanderWerf, *J. Am. Chem. Soc.* **72**, 4809 (1950).
23. T. Fukuhara and co-workers, *Synth. Commun.* **17**, 685 (1987).
24. G. A. Olah and J. Welch, *J. Am. Chem. Soc.* **97**, 208 (1975).
25. U.S. Pat. 4,096,196 (June 20, 1978), M. M. Boudakian (to Olin).
26. U.S. Pat. 4,075,252 (Feb. 21, 1978), M. M. Boudakian (to Olin).
27. U.S. Pat. 4,912,268 (Mar. 27, 1990), M. H. Krackov and C. H. Ralston (to Du Pont).
28. D. T. Flood, *Org. Synth., Coll. Vol.* **II**, 295 (1943).
29. A. Roe, in R. Adams, ed., *Organic Reactions*, Vol. 5, John Wiley & Sons, Inc., New York, 1949, Chapt. 4, p. 193.
30. H. Suschitzky, in M. Stacey, J. C. Taylor, and A. G. Sharpe, eds., *Advances in Fluorine Chemistry*, Vol. 4, Butterworths, Washington, D.C., 1964, p. 1.
31. K. G. Rutherford and co-workers, *J. Org. Chem.* **26**, 5149 (1961).
32. M. P. Doyle and W. J. Bryker, *J. Org. Chem.* **44**, 1572 (1979).

33. U.S. Pat. 4,476,320 (Oct. 9, 1984), H. Diehl, H. Pelster, and H. Habetz (to Bayer).
34. D. J. Milner, *Synth. Commun.* **22**, 73 (1992).
35. L. Treschanke, Riedel-de Haën, personal communication, Feb. 5, 1992.
36. J. P. Regan, *Specialty Chemicals* **21**, 56 (Jan. 1986).
37. U.S. Pat. 2,705,730 (Apr. 5, 1955), J. D. Head (to Dow).
38. *Chem. Eng. News*, 65 (Apr. 24, 1961).
39. O. Wallach, *Ann. Chem.* **235**, 255 (1886).
40. O. Wallach and F. Heusler, *Ann. Chem.* **243**, 219 (1888).
41. U.S. Pat. 4,194,054 (Mar. 18, 1980), H. Förster and co-workers (to Bayer).
42. Ger. Pat. 3,142,856 (May 11, 1983), E. Klauke and K. Grohe (to Bayer).
43. T. J. Tewson and M. Welch, *J. Chem. Soc., Chem. Commun.*, 1149 (1979).
44. W. Prescott, *Chem. Ind. London*, 56 (1978); L. Dolby-Glover, *Chem. Ind. London*, 518 (1986).
45. Y. Kimura, *J. Synth. Org. Chem., Japan* **47**, 258 (1989).
46. G. C. Finger and co-workers, *J. Am. Chem. Soc.* **78**, 6034 (1956); *Chem. Ind. London*, 1328 (1962).
47. A. J. Beaumont and J. H. Clark, *J. Fluorine Chem.* **32**, 295 (1991).
48. Y. Uchibori and co-workers, *SYNLETT*, 345 (Apr. 1992).
49. N. Ishikawa and co-workers, *Chem. Lett.*, 761 (1981).
50. Y. Kimura and H. Suzuki, *Tetrahedron Lett.* **30**, 1271 (1989).
51. U.S. Pat. 4,069,262 (Jan. 17, 1978), R. A. Kunz (to Du Pont).
52. U.S. Pat. 4,642,399 (Feb. 10, 1987), C. R. White (to Mallinckrodt).
53. U.S. Pat. 4,287,374 (Sept. 1, 1981), R. A. North (to Boots).
54. C. Liotta and H. P. Harris, *J. Am. Chem. Soc.* **96**, 2250 (1974).
55. U.S. Pat. 4,642,398 (Feb. 10, 1987), G. L. Cantrell (to Mallinckrodt).
56. S. Kumai and co-workers, *Reports Res. Lab. Asahi Glass Co.* **39**, 317 (1989).
57. J. Deutsch and H.-J. Niclas, *Synth. Commun.* **21**, 205 (1991).
58. U.S. Pat. 4,429,365 (Oct. 21, 1980), H. -G. Oeser and co-workers (to BASF).
59. Y. Yoshida and Y. Kimura, *J. Fluorine Chem.* **44**, 291 (1989).
60. R. E. Banks and co-workers, *J. Fluorine Chem.* **46**, 529 (1990).
61. Y. Yoshida and co-workers, *J. Fluorine Chem.* **53**, 301 (1991).
62. N. C. O'Reilly and co-workers, *SYNLETT*, 609 (Oct. 1990).
63. Y. Yoshida and co-workers, *J. Fluorine Chem.* **53**, 335 (1991).
64. U.S. Pat. 4,937,396 (June 26, 1990), R. G. Pews and co-workers (to Dow).
65. U.S. Pat. 3,240,824 (Mar. 15, 1966), M. M. Boudakian and E. R. Shipkowski (to Olin).
66. *Chem. Eng. News* **29**, 2666 (1951).
67. D. G. Mooney, *Inst. Chem. Eng. Symp. Series* **124**, 381 (1991); A. T. Cates, *J. Hazard. Mater.* **32**, 1 (1992).
68. L. Bretherick, *Handbook of Reactive Chemical Hazards*, 4th ed., Butterworths, London, 1990.
69. G. Bartoli and co-workers, *J. Chem. Soc. Perkin Trans.* **1**, 2671 (1972).
70. F. Effenberger and W. Streicher, *Chem. Ber.* **124**, 157 (1991).
71. S. Kumai and co-workers, *Reports Res. Lab. Asahi Glass Co.* **35**, 153 (1985).
72. U.S. Pat. 4,568,781 (Feb. 4, 1986), F. Effenberger and W. Streicher (to BASF); H. Suzuki and co-workers, *Bull. Chem. Soc. Jpn.* **63**, 2010 (1990).
73. M. Stacey and J. C. Tatlow, in M. Stacey, J. C. Tatlow, and A. G. Sharpe, eds., *Advances in Fluorine Chemistry*, Vol. 1, Butterworths, London, 1960, p. 166.
74. P. A. Coe and co-workers, *J. Chem. Soc. C*, 1060 (1969).
75. R. D. Chambers and co-workers, *Tetrahedron* **19**, 891 (1963).
76. G. M. Brooks and co-workers, *J. Chem. Soc.*, 729 (1964).
77. U.S. Pat. 3,499,942 (Mar. 10, 1970), O. M. Nevedov and A. Ivashenko.

78. N. Ishikawa, *Senryo to Yakuhin* **26**(6), 106 (1981); *J. Synth. Org. Chem., Jpn.* **40**(2), 158 (1982).
79. N. V. Volchkov and co-workers, *Bull. Acad. Sci. USSR* **38**, 1782 (1989). (Engl. trans.).
80. U.S. Pat. 4,390,740 (June 28, 1983), I. Tabushi and co-workers (to Daikin Kogyo).
81. R. E. Ehrenkaufer and co-workers, *J. Org. Chem.* **47**, 2489 (1982).
82. G. A. Olah and P. Kreienbühl, *J. Org. Chem.* **32**, 1614 (1967).
83. K. O. Christe and A. E. Pavlath, *J. Org. Chem.* **31**, 559 (1966).
84. D. P. Ashton and co-workers, *J. Fluorine Chem.* **27**, 263 (1985); U.S. Pat. 4,745,235 (May 17, 1988), D. P. Ashton and co-workers (to ICI).
85. Fr. Demande Pat. 2,647,106 (Nov. 23, 1990), L. Gilbert and co-workers (to Rhône-Poulenc).
86. Eur. Pat. Appl. 427,603 (May 15, 1991), H. Garcia and co-workers (to Rhône-Poulenc).
87. G. G. Yakobson and co-workers, *Zh. Vses. Khim. Obshchestva im. D. I. Mendeleeva* **10**, 466 (1965).
88. M. Van Der Puy, *J. Org. Chem.* **53**, 4398 (1988); U.S. Pat. 5,081,275 (Jan. 14, 1992) (to AlliedSignal).
89. Y. Yazawa and co-workers, *Chem. Lett.*, 2213 (1989).
90. I. N. Rozhkov and co-workers, *Isv. Akad. Nauk SSSR Ser. Khim.*, 1130 (1972); *Russian Chem. Rev.* **45**, 615 (1976) (Engl. trans.).
91. Ger. Offen. 2,516,355 (Oct. 28, 1976), F. Beck (to BASF).
92. Jpn. Kokai Tokkyo Koho JP 63,111,192 (May 16, 1988), A. Shimizu and K. Yamatka (to Asahi Chemical Ind.); Jpn. Kokai Tokkyo Koho 62,127,488 (June 9, 1987), A. Shimizu and K. Yamatka (to Asahi Chemical Ind.).
93. J. H. H. Meurs and co-workers, *Angew. Chem. Int. Ed. Engl.* **28**, 927 (1989).
94. U.S. Pat. 2,884,458 (Apr. 28, 1959), D. A. Fidler (to Olin).
95. D. A. Fidler, J. S. Logan, and M. M. Boudakian, *J. Org. Chem.* **26**, 4014 (1961).
96. P. H. Scott and co-workers, *Tetrahedron Lett.*, 1153 (1970); U.S. Pat. 3,558,707 (Jan. 26, 1971), J. W. Churchill and E. H. Kober (to Olin).
97. Brit. Pat. Appl. 2,241,952 (Sept. 18, 1991), D. Levin and J. S. Moillet (to ICI).
98. Jpn. Kokai Tokkyo Koho 01,238,560 (Sept. 22, 1989), J. Negishi and T. Kawai (to Central Glass).
99. Eur. Pat. 248,746 (Apr. 25, 1990), M. Desbois (to Rhône-Poulenc).
100. U.S. Pat. 3,910,985 (Oct. 7, 1975), P. P. Montijn (to Shell Oil).
101. Span. ES 493,977 (Aug. 1, 1981), A. L. Porta (to Dr. Andreu S. A.).
102. A. I. Titov and A. N. Baryshnikova, *J. Gen. Chem. (USSR)* **23**, 346 (1953) (Engl. trans.).
103. U.S. Pat. 4,391,991 (July 5, 1983), P. F. Mundhenke and M. J. Fifolt (to Occidental Chemical).
104. PCT Int. Appl. WO 91 17,138 (Nov. 14, 1991), T. F. Braish (to Pfizer).
105. U.S. Pat. 4,367,350 (Jan. 4, 1983), U. Hiramatsu and co-workers (to Daikin Kogyo); U.S. Pat. 4,400,563 (Apr. 23, 1983), Y. Ohsaka and co-workers (to Daikin Kogyo).
106. G. A. Boswell and co-workers, *Organic Reactions* **21**, 1 (1974).
107. C. -L. Wang, *Organic Reactions* **34**, 319 (1985).
108. U.S. Pat. 3,976,691 (Aug. 24, 1976), W. J. Middleton (to Du Pont).
109. U.S. Pat. 4,533,777 (Aug. 6, 1985), A. Marhold and E. Klauke (to Bayer).
110. A. Marhold and E. Klauke, *J. Fluorine Chem.* **18**, 281 (1981).
111. V. C. R. McLaughlin and J. Thrower, *Tetrahedron* **25**, 5921 (1969).
112. Y. Kobayashi and I. Kumadaki, *Tetrahedron Lett.*, 4095 (1969); *Chem. Pharm. Bull.* **18**, 2334 (1970).
113. D. M. Wiemers and D. J. Burton, *J. Am. Chem. Soc.* **108**, 832 (1986).
114. J. H. Clark and co-workers, *J. Fluorine Chem.* **50**, 411 (1950).
115. J. M. Paratian and co-workers, *J. Chem. Soc., Chem. Commun.*, 53 (1992).

116. K. Matsui and co-workers, *Chem. Lett.*, 1719 (1981).
117. J. N. Freskos, *Synth. Commun.* **18**, 965 (1988).
118. B. R. Langlois and co-workers, *Tetrahedron Lett.* **32**, 7525 (1991).
119. J. G. MacNeil and D. J. Burton, *J. Fluorine Chem.* **55**, 225 (1991).
120. Q.-Y. Chen and co-workers, *J. Fluorine Chem.* **55**, 291 (1991).
121. H. Sawada and co-workers, *J. Fluorine Chem.* **46**, 423 (1990).
122. T. Umemoto and co-workers, *Chem. Lett.*, 1663 (1981).
123. W. A. Sheppard, *J. Am. Chem. Soc.* **87**, 2410 (1965).
124. A. F. Feiring, *J. Fluorine Chem.* **10**, 375 (1977); *J. Org. Chem.* **44**, 1252 (1979).
125. P. Martin, *Tetrahedron* **41**, 4057 (1985).
126. *Handbook of Aromatic Fluorine Compounds*, Olin Corp., Stamford, Conn., 1976.
127. D. V. Sweet, ed., *Registry of Toxic Effects of Chemical Substances-1985/1986 Edition*, Vols. 1–5, DHHS (NIOSH) publication no. 87-114, Washington, D.C., 1987.
128. G. A. Olah and co-workers, *J. Chem. Soc.*, 1823 (1957).
129. P. A. Janssen and co-workers, *J. Med. Pharm. Chem.* **1**, 281 (1959).
130. H. C. Brown and G. Marino, *J. Am. Chem. Soc.* **84**, 1658 (1962).
131. P. A. Staniland, in G. Allen and J. C. Bevington, eds., *Comprehensive Polymer Science*, Vol. 5, Pergamon Press, New York, 1989, Chapt. 29, p.490.
132. P. M. Hergenrother and co-workers, *Polymer* **29**, 358 (1988).
133. P. M. Hergenrother and co-workers, *J. Polym. Sci. Part A: Polym. Chem.* **29**, 1483 (1991).
134. M. E. Neubert and D. L. Fishel, *Mol. Cryst. Liq. Cryst.* **53**, 101 (1979).
135. P. A. Worthington, in D. R. Baker and co-workers, eds., *Synthesis and Chemistry of Agrochemicals*, ACS Symposium Series no. 355. American Chemical Society, Washington, D.C. 1987, Chap. 27, p.302.
136. U.S. Pat. 4,522,754 (June 11, 1985), B. Hilti and co-workers (to Ciba-Geigy).
137. R. E. Parker, in M. Stacey, J. C. Tatlow, and A. G. Sharpe, eds., *Advances in Fluorine Chemistry*, Vol. 3, Butterworths, Washington, D.C., 1963, p.63.
138. M. Bil, *J. Appl. Chem. Biotechnol.* **22**, 853 (1972); *Chem. Ind. London*, 656 (1971).
139. U.S. Pat. 3,959,377 (May 25, 1976) and U.S. Pat. 3,632,582 (Jan. 4, 1972), M. Bil (to Clairol).
140. Ger. Offen. 3,534,369 (Apr. 3, 1986), J. F. Grollier and co-workers (to Oreal).
141. F. Sanger, *Biochem. J.* **39**, 507 (1945).
142. S. E. Drewes and co-workers, *J. Chem. Soc., Perkin Trans.* **1**, 1283 (1975).
143. G. W. J. Fleet and R. Porter, *Nature*, 511 (1970).
144. K. J. Skinner, *Chem. Eng. News*, 23 (Aug. 18, 1975).
145. E. D. Bergmann and M. Bentov, *J. Org. Chem.* **26**, 1480 (1961).
146. R. W. Holley and A. D. Holley, *J. Am. Chem. Soc.* **74**, 1110 (1952).
147. F. F. Micheel and co-workers, *Ann. Chem.* **581**, 238 (1953).
148. H. Zahn and K. H. Lebkücher, *Biochem. Z.* **334**, 133 (1961).
149. H. Zahn and H. Zuber, *Ber.* **86**, 172 (1953).
150. H. Zahn and J. Meinhoffer, *Makromol. Chem.* **26**, 126 (1958).
151. K. L. Kirk and L. A. Cohen, *J. Org. Chem.* **34**, 395 (1969).
152. Y. Watanabe and Y. Imai, *J. Chromatogr.* **239**, 723 (1982).
153. P. Marfey, *Carlsberg Res. Commun.* **49**, 591 (1984).
154. A. E. Katz and co-workers, *Polym. Prepr. (Am. Chem. Soc., Div. Polym. Chem.)* **33**(3), 144 (1991).
155. K. Sturm and co-workers, *Chem. Ber.* **99**, 328 (1966).
156. U.S. Pats. 3,668,248 (June 6, 1972) and 3,419,552 (Dec. 31, 1968), C. W. Whitehead and J. J. Traverso (to Eli Lilly).
157. R. N. Johnson and co-workers, *J. Polym. Sci. Part A-1* **5**, 2375 (1967).
158. A. E. Feiring and co-workers, *J. Polym. Sci. Part A: Polym. Chem.* **28**, 2809 (1990).

159. U.S. Pat. 5,084,548 (Jan. 28, 1992), A. E. Feiring and S. D. Archer (to Du Pont).
160. U.S. Pat. 5,034,034 (July 23, 1991), E. S. Sanders and T. L. Parker (to Dow).
161. H. R. Kricheldorf and M. Bergahn, *Makromol. Chem., Rapid Commun.* **12**, 529 (1991).
162. M. M. Boudakian and co-workers, *J. Org. Chem.* **26**, 4641 (1961).
163. G. G. Yakobson and co-workers, *Dokl. Akad. Nauk SSSR* **141**, 1395 (1961).
164. S. H. Yu and E. C. Ashby, *J. Org. Chem.* **36**, 2123 (1971).
165. W. K. Moberg and co-workers, in Ref. 135, Chapt. 26, p. 288.
166. E. D. Morgan and C. F. Poole, *J. Chromatogr.* **104**, 351 (1975).
167. L. Bretherick, *Chem. Ind. London*, 1017 (1971).
168. A. J. Adduci, *Chemtech*, 575 (Sept. 1976).
169. M. Tsubouchi and co-workers, *Anal. Chem.* **57**, 783 (1985).
170. D. W. Ribbons and co-workers, *J. Fluorine Chem.* **37**, 299 (1987).
171. S. C. Taylor and co-workers, *Performance Chemicals*, 18 (Nov. 1986); *Specialty Chemicals* **8**(3), 236 (June 1988).
172. Eur. Pat. Appl. 253,438 (Jan. 20, 1988), J. A. Schofield and co-workers (to Shell Int. Research).
173. Ger. Offen. 3,520,316 (Dec. 11, 1986), E. Begemann and H. Schmand (to Riedel-de Haën).
174. U.S. Pat. 4,822,927 (Apr. 18, 1989), N. J. Stepaniuk and B. J. Lamb (to Mallinckrodt).
175. U.S. Pat. 4,812,572 (Mar. 14, 1989), M. S. Howarth and D. M. Tomkinson (to ICI).
176. U.S. Pat. 2,939,766 (June 7, 1960), J. W. Churchill (to Olin).
177. U.S. Pat. 5,032,371 (July 16, 1991), H. J. Buehler (to Mallinckrodt).
178. U.S. Pat. 2,563,796 (Aug. 7, 1951), W. H. Shenk, Jr. and G. R. Mellon (to Harshaw Chemical).
179. U.S. Pat. 3,160,623 (Dec. 8, 1964), L. G. Anello and C. Woolf (to Allied Chemical).
180. F. Seel, *Angew. Chem. Int. Ed. Engl.* **4**, 635 (1965).
181. Jpn. Kokai 74 81,330 (Aug. 6, 1974), S. Misaki and M. Okamoto (to Daikin Kogyo).
182. D. Hunter, *Chem. Week* **148**(27), 36 (July 24, 1991); *Ibid*, **146**(9), 41 (Mar. 1, 1989).
183. F. J. Weigert and W. A. Sheppard, *J. Org. Chem.* **25**, 4006 (1976).
184. R. Filler and S. M. Naqvi, in R. Filler and Y. Kobayashi, eds., *Biomedicinal Aspects of Fluorine Chemistry*, Kodansha Biomedical Press, Tokyo, 1982, Chapt. 1, p. 28.
185. P. H. Bolton and co-workers, *Biopolymers* **20**, 435 (1981).
186. H. J. Plach, *New LC Mixtures for STN and Active Matrix Displays Using New Terminally Fluorinated Compounds*, technical literature, E. Merck & Co., Darmstadt, Oct. 1990.
187. R. Neef, *The Merck Group Liquid Crystal Newsletter*, (7) (Feb. 1991), (8) (Sept. 1991), Darmstadt.
188. U.S. Pat. 5,077,160 (Dec. 31, 1991), K. -Y. Law and F. C. Bailey (to Xerox).
189. K.-Y. Law and F. C. Bailey, *J. Imaging Sci.* **31**, 172 (1987).
190. K.-Y. Law and co-workers, *Dyes and Pigments* **9**, 187 (1988).
191. P. M. Kazmeier and co-workers, *J. Imaging Sci.* **32**, 1 (1988).
192. H. Hopff, *Chimia* **15**, 193 (1961).
193. J. T. Minor and C. A. VanderWerf, *J. Org. Chem.* **17**, 1425 (1952).
194. Jpn. Kokai Tokkyo Koho JP 03,232,828 (Oct. 16, 1991), K. Momota and T. Yonezawa (to Morita Kagaku).
195. U.S. Pat. 5,055,220 (Oct. 8, 1991), M. Uchida and co-workers (to Chisso).
196. M. Hird and co-workers, *Liquid Crystals* **11**, 531 (1992).
197. Jpn. Kokai Tokkyo Koho JP 61 63,627 (Apr. 1, 1986), N. Yoneda and co-workers (to Mitsubishi Chemical).
198. G. Schiemann and R. Pillarsky, *Chem. Ber.* **62**, 3035 (1929).
199. G. Furin and co-workers, in I. L. Knunyants and G. G. Yakobson, eds., *Syntheses of Fluoroorganic Compounds*, Springer-Verlag, Berlin, 1985, Chapt. 2, p. 109.

200. Jpn. Kokai Tokkyo Koho JP 03 34,944 (Feb. 14, 1991), T. Mogi (to Tokemu Product K.K.).
201. G. D. Hartman and S. E. Biffar, *J. Org. Chem.* **42**, 1468 (1977).
202. Brit. Pat. 1,514,082 (June 14, 1978), G. Fuller (to ISC).
203. Brit. Pat. Appl. 2,058,067 (Apr. 8, 1988), D. Wotton (to ISC).
204. R. H. Shiley, D. R. Dickerson and G. C. Finger, *J. Fluorine Chem.* **2**, 19 (1972/73).
205. A. M. Roe, R. A. Burton, and D. R. Reavill, *J. Chem. Soc., Chem. Commun.*, 582 (1965).
206. W. J. Feast and R. Stephens, *J. Chem. Soc.*, 3502 (1965).
207. I. Hayakawa and co-workers, *Chem. Pharm. Bull.* **32**, 4907 (1984).
208. U.S. Pat. 4,528,287 (July 9, 1985), Y. Itoh and co-workers (to Hokuriku Pharmaceutical).
209. G. C. Finger and R. E. Oesterling, *J. Am. Chem. Soc.* **78**, 2593 (1956).
210. T. F. Braish and D. E. Fox, *Org. Prep. Proced. Int.* **23**, 655 (1991).
211. G. C. Finger and co-workers, *J. Am. Chem. Soc.* **73**, 145 (1951).
212. J. P. Sanchez and co-workers, *J. Med. Chem.* **31**, 983 (1988); *Ibid.* **35**, 361 (1992).
213. D. T. W. Chu and co-workers, *Can. J. Chem.* **70**, 1323 (1992).
214. U.S. Pat. 5,021,605 (June 4, 1991), H. Kobayashi and M. Shimizu (to SDS Biotech).
215. U.S. Pat. 4,885,386 (Dec. 5, 1989), J. N. Wemple and co-workers (to Warner-Lambert).
216. G. C. Finger and co-workers, *J. Am. Chem. Soc.* **73**, 153 (1951).
217. Eur. Pat. Appl. 460,639 (Dec. 11, 1991), Y. Kobayashi and co-workers (to SDS Biotech).
218. T. Ishizaki and co-workers, *Nippon Kagaku Kaishi*, 2054 (1985).
219. U.S. Pat. 4,782,180 (Nov. 1, 1988), J. N. Wemple and co-workers (to Warner-Lambert).
220. U.S. Pat. 5,047,553 (Sept. 10, 1991), D. M. Novak and H. C. Lin (to Occidental Chemical).
221. D. G. Holland and co-workers, *J. Org. Chem.* **29**, 3042 (1964).
222. U.S. Pat. 4,952,695 (Aug. 28, 1990), K. Grohe and co-workers (to Bayer).
223. D. T. W. Chu and co-workers, *J. Med. Chem.* **30**, 504 (1987).
224. H. Egawa and co-workers, *Chem. Pharm. Bull.* **34**, 4098 (1986).
225. S. K. Ray, *Adv. Energy Convers.* **6**, 89 (1966).
226. G. Fuller, *J. Chem. Soc.*, 6264 (1965).
227. Brit. Pat. 996,498 (June 30, 1965), G. Fuller (to ISC).
228. J. M. Birchall and co-workers, *J. Chem. Soc.*, 4966 (1962).
229. P. L. Coe and co-workers, *J. Chem. Soc. C*, 2323 (1966).
230. M. M. Murza and co-workers, *Zh. Org. Khim.* **13**, 1046 (1977).
231. I. Kisfaludy and co-workers, *Tetrahedron Lett.*, 1785 (1974).
232. G. H. Kalb and co-workers, *Appl. Polym. Symposium*, (22), 127 (1973).
233. G. H. Kalb and co-workers, in N. A. J. Platzer ed., *Polymerization Reactions and New Polymers*, Advances in Chemistry Series no. 129, American Chemical Society, Washington, D.C., 1973, Chapt. 2, p. 13.
234. F. K. Kawahara, *Anal. Chem.* **40**, 2073 (1968).
235. R. J. Harper and co-workers, *J. Org. Chem.* **29**, 3042 (1964).
236. E. Kinsella and A. G. Massey, *Chem. Ind. London*, 1017 (1971).
237. S. C. Cohen and A. G. Massey, in J. C. Tatlow, R. D. Peacock, and H. H. Hyman, eds., *Advances in Fluorine Chemistry*, Vol. 6, CRC Press, Cleveland, Ohio, 1970, p. 83.
238. J. W. Eichelberger and co-workers, *Anal. Chem.* **47**, 995 (1975).
239. T. Nakamura and O. Kaieda, *J. Synth. Org. Chem., Jpn.* **47**, 20 (1989).
240. J. Wallace and co-workers, *Clin. Chem.* **25**, 252 (1979); *Ibid* **24**, 895 (1978).
241. J. Wallace and co-workers, *Anal. Chem.* **49**, 903 (1977).
242. J. C. Lhuguenot and B. F. Baume, *J. Chromatogr. Sci.* **12**, 411 (1974).
243. K. T. Koshy and co-workers, *J. Chromatogr. Sci.* **13**, 97 (1975).

244. T. Nambara and co-workers, *J. Chromatogr.* **114**, 81 (1975).
245. U.S. Pat. 3,946,063 (May 23, 1976), J. Vessman and co-workers (to Pierce Chemical).
246. K. Blau and G. S. King, eds., *Handbook of Derivatives for Chromatography*, Heyden, London, 1977.
247. R. Filler, *Fluorine Chem. Rev.* **8**, 1 (1977).
248. L. A. Wall, ed., *Fluoropolymers*, Vol. 25, Wiley-Interscience, New York, 1972.
249. G. W. Holbrook, L. A. Loree, and O. R. Pierce, *J. Org. Chem.* **31**, 1259 (1966).
250. N. N. Vorozhtsov, Jr. and co-workers, *Isv. Akad. Nauk SSSR, Ser. Khim.* (8), 1524 (1963).
251. Y. Désirant, *Bull. Acad. Sci. Roy. Belg.* **41**, 759 (1955).
252. J. M. Birchall and R. N. Haszeldine, *J. Chem. Soc.*, 13 (1959).
253. L. A. Wall and co-workers, *J. Research NBS* **65A**, 239 (1961).
254. R. A. Falk, *Sperry Eng. Rev.* **16**(3), 24 (Fall 1963).
255. U.S. Pat. 3,158,657 (Nov. 24, 1964), F. R. Callihan and C. L. Quateta (to Sperry Rand).
256. L. S. Kobrina, *Fluorine Chem. Rev.* **7**, 1 (1974).
257. U.S. Pat. 5,087,764 (Feb. 11, 1992), V. Reiffenrath and J. Krause (to E. Merck GmbH).
258. U.S. Pat. 5,064,567 (Nov. 12, 1991), F. Funada and co-workers (to Sharp and E. Merck GmbH).
259. A. Villiger and F. Leenhouts, *Mol. Cryst. Liq. Cryst.* **209**, 297 (1991).
260. Jpn. Kokai Tokkyo Koho JP 03,197,438 (Aug. 28, 1991), N. Satake and co-workers (to Kanto Chemical).
261. Brit. Pat. Appl. 2,232,416 (Dec. 12, 1990), D. Coates and co-workers (to U.K. Secretary of State for Defence).
262. Jpn. Kokai Tokkyo Koho JP 03,176,445 (July 31, 1991), T. Kuriyama and co-workers (to Dainippon).
263. U.S. Pat. 5,032,314 (July 16, 1991), M. Ushioda and co-workers (to Chisso).
264. Jpn. Kokai Tokkyo Koho JP 04 29,951 (Jan. 31, 1992), T. Watanabe and M. Sato (to Sanyo Chemical).
265. M. Hird and co-workers, *Liquid Crystals* **11**, 531 (1992).
266. U.S. Pat. 5,061,400 (Oct. 29, 1991), T. Obikawa (to Seiko Epson).
267. G. Schiemann and E. Bolstead, *Chem. Ber.* **61B**, 1403 (1928).
268. J. Hannah and co-workers, *J. Med. Chem.* **21**, 1093 (1978).
269. U.S. Pat. 4,443,631 (Apr. 13, 1984), A. G. Padilla (to Upjohn).
270. W. J. Pummer and L. A. Wall, *J. Chem. Eng. Data* **6**, 76 (1961).
271. M. W. Buxton and co-workers, *J. Fluorine Chem.* **2**, 387 (1972/73).
272. F. Mercer and co-workers, *J. Polym. Sci., Part A: Polym. Chem.* **30**, 1767 (1992); U.S. Pat. 5,114,780 (May 19, 1992), F. Mercer and co-workers (to Raychem).
273. M. S. Newman, in P. O. P. Ts'o and J. D. DiPaolo, eds., *Chemical Carcinogenesis*, Marcel Dekker, Inc., New York, 1974, p. 177.
274. D. R. MacKenzie and co-workers, *J. Phys. Chem.* **69**, 2526 (1965).
275. E. Boger and co-workers, *J. Fluorine Chem.* **8**, 513 (1976).
276. O. Lerman and co-workers, *J. Org. Chem.* **49**, 806 (1984).
277. S. Stavber and M. Zupan, *J. Fluorine Chem.* **17**, 597 (1981).
278. I. N. Rozhkov and co-workers, *Dokl. Akad. Nauk SSSR, Chem. Sect.* **193**, 618 (1970) (Engl. trans.).
279. R. F. O'Malley and co-workers, *J. Org. Chem.* **46**, 2816 (1981); *J. Electrochem. Soc.* **130**, 2170 (1983).
280. U.S. Pat. 3,707,843 (Jan. 2, 1973), R. C. Connor and L. L. Ferstandig (to Halocarbon Products).
281. U.S. Pat. 3,753,345 (Aug. 21, 1973), F. H. Cassidy and R. Garcia (to Aerojet-General).
282. E. Kühle and E. Klauke, *Angew. Chem. Int. Ed. Engl.* **16**, 735 (1977).
283. R. Filler and H. Novar, *Chem. Ind. London*, 1273 (1960).

284. R. G. Jones, *J. Am. Chem. Soc.* **69**, 2346 (1947).
285. V. C. R. McLoughlin and J. Thrower, *Chem. Ind. London*, 1557 (1964).
286. R. Grinter and co-workers, *Tetrahedron Lett.*, 3845 (1968).
287. R. Wächter, *Angew. Chem.* **67**, 305 (1955).
288. F. Swarts, *Bull. Sci. Acad. Roy. Belg.*, 399 (1920).
289. Y. Kobayashi and I. Kumadaki, *Acc. Chem. Res.* **11**, 197 (1978).
290. N. P. Buu-Hoi and co-workers, *Compt. Rend.* **257** (21), 3182 (1963).
291. *Product List No. 4*, Marshallton Research Labs, West Chester, Pa., Jan. 1973, p. 27.
292. L. Bretherick, *Chem. Ind. London*, 1017 (1971).
293. I. C. Appleby, *Chem. Ind. London*, 120 (1971).
294. E. C. Ashby and D. M. Al-Fekri, *J. Organometallic Chem.* **390**, 275 (1990).
295. U.S. Pat. 3,234,292 (Feb. 8, 1966), S. Robota and E. A. Belmore (to Hooker Chemical).
296. R. J. Albers and E. C. Kooyman, *Rec. Trav. Chim.* **83**, 930 (1964).
297. A. A. Ushakov and co-workers, *Zh. Org. Khim.* **12**, 158 (1963).
298. K. Inukai and T. Ueda, *Kogyo Kagaku Zasshi* **64**, 2156 (1961).
299. Jpn. Kokai 75 76,029 (June 21, 1975), S. Misaki and co-workers (to Daikin Kogyo).
300. U.S. Pat. 4,401,623 (Aug. 30, 1983), T. Giacobbe and G. Tsien (to Rhône-Poulenc).
301. E. T. McBee and co-workers, *J. Am. Chem. Soc.* **72**, 1651 (1950).
302. U.S. Pat. 4,008,278 (Feb. 15, 1977), M. M. Boudakian (to Olin).
303. Ger. Pat. 1,568,938 (Dec. 12, 1974), H. Treiber (to Knoll).
304. U.S. Pat. 4,144,265 (Mar. 13, 1979), W. Dowd and T. H. Fisher (to Dow).
305. U.S. Pat. 4,966,988 (Oct. 30, 1990), W. I. Schinski and P. Denisevich (to Chevron).
306. C. E. Ward and co-workers, in Ref. 135, Chapt. 6, p.65.
307. A. L. Henne and M. S. Newman, *J. Am. Chem. Soc.* **60**, 1697 (1938).
308. W. Schwertfeger and G. Siegemund, *J. Fluorine Chem.* **36**, 237 (1987).
309. U.S. Pat. 4,978,798 (Dec. 18, 1990), J. S. Stults (to Occidental Chemical).
310. C. Saboureau and co-workers, *J. Chem. Soc., Chem. Commun.*, 1138 (1989).
311. R. Filler, in Ref. 237, p. 1.
312. *Benzotrifluoride*, data sheet no. 778-D, Hooker Chemical Corp., Niagara Falls, N.Y., 1969.
313. F. Swarts, *Bull. Acad. Roy. Belg. Cl. Sci.* **35**, 375 (1898).
314. A. K. Barbour, L. J. Belf, and M. W. Buxton, in M. Stacey, J. C. Tatlow, and A. G. Sharpe, eds., *Advances in Fluorine Chemistry*, Vol. 3, Butterworths, London, 1963, p. 181.
315. J. H. Brown and co-workers, *J. Chem. Soc. Suppl. Issue* **1**, S95 (1949).
316. U.S. Pat. 3,966,832 (June 26, 1978), R. Lademann and co-workers (to Hoechst).
317. Ger. Pat. 1,618,390 (Sept. 28, 1972), O. Scherer and co-workers (to Hoechst).
318. U.S. Pat. 4,183,873 (Jan. 15, 1980), Y. Baxamusa and S. Robota (to Hooker Chemical).
319. U.S. Pat. 4,130,594 (Dec. 19, 1978) and U.S. Pat. 4,129,602 (Dec. 12, 1978), L. Sendlak (to Hooker Chemical).
320. U.S. Pat. 4,462,937 (July 31, 1984), A. Ramanadin and L. Seigneurin (to Rhône-Poulenc).
321. U.S. Pat 3,859,372 (Jan. 7, 1975), S. Robota (to Hooker Chemical).
322. Y. A. Fialkov and L. M. Yagupolski, in Ref. 199, Chapt. 3, p.233.
323. *para-Chlorobenzotrifluoride (OXSOL 100)*, data sheet no. 784 588, MSDS No. M29358, Occidental Chemical Co., Niagara Falls, N.Y., Dec. 26, 1991.
324. *3,4-Dichlorobenzotrifluoride (OXSOL 1000)*, data sheet no. 343, 588, MSDS No. M29478, Occidental Chemical Co., Niagara Falls, N.Y., Jan. 16, 1992.
325. U.S. Pat. 4,110,405 (Aug. 29, 1978), M. Bornengo (to Montedison).
326. Eur. Pat. Appl. 490,115 (June 17, 1992) and Eur. Pat. Appl. 484,767 (May 13, 1992), R. Krishnamurti and co-workers (to Occidental Chemical).
327. A. Pozdeeva, *Zh. Prikl. Khim. (Leningrad)* **52**, 1119 (1979).

328. G. A. Mandrov and co-workers, *Khim. Prom-st., Ser.: Metody Anal. Kontrolya Kach. Prod. Khim. Prom-sti.* (8), 10 (1981).
329. B. W. Lawrence and co-workers, *Manuf. Chem. Aerosol News* **41**, 37 (Jan. 1970).
330. *3-(Trifluoromethyl)aniline, BUA Substance Report 44 (Sept. 1989)*, Advisory Board for Environmental Relevant Waste of the German Chemical Society, VCH Publishers, Weinheim, Germany, 1990, 58 pp.
331. U.S. Pat. 4,582,935 (Apr. 15, 1986), M. M. Boudakian (to Olin).
332. M. M. Boudakian, *J. Fluorine Chem.* **36**, 283 (1987).
333. L. M. Yagupolski and co-workers, *Russian Chem. Rev.* **52**, 993 (1983) (Engl. trans.).
334. N. Ishikawa, *Senryo to Yakuhin* **28**, 52 (1983).
335. G. Wulfram, in R. E. Banks, ed., *Organofluorine Chemicals and Their Industrial Applications*, E. Horwood, Chichester, U.K., 1979, Chapt. 10, p.208.
336. U.S. Pat. 5,030,252 (July 9, 1991), E. Sanders and T. L. Parker (to Dow).
337. U.S. Pat. 5,082,921 (Jan. 21, 1992), T. L. Parker (to Dow).
338. U.S. Pat. 4,876,329 (Oct. 24, 1989), W. L. Chiang and co-workers (to Amoco).
339. U.S. Pat. 5,021,540 (June 4, 1991), A. Leone-Bay and co-workers (to American Cyanamid).
340. F. M. Houlihan and co-workers, *Chem. Mater.* **3**, 462 (1991).
341. S. Misaki and co-workers, *Mol. Cryst. Liq. Cryst.* **66**, 443 (1981).
342. Jpn. Kokai Tokkyo Koho JP 59,16750 (Sept. 21, 1984) (to Chisso).
343. Jpn. Kokai Tokkyo Koho JP 03,200,737 (Sept. 2, 1991), S. Takehara and co-workers (to Dainippon).
344. A. C. Griffin and N. W. Buckley, *Mol. Cryst. Liq. Cryst.* **41**, 141 (1978).
345. H. Kobayashi and co-workers, *Chem. Lett.*, 1185 (1982).
346. G. E. Pacey and co-workers, *Analyst* **106**, 636 (1981).
347. U.S. Pat. 3,947,257 (Mar. 30, 1976), R. S. Johannsen and W. M. Moyer (to Raychem).
348. W. A. Sheppard, *J. Org. Chem.* **29**, 1 (1964).
349. F. E. Herkes, *J. Fluorine Chem.* **9**, 113 (1977).
350. G. A. Olah and co-workers, *J. Am. Chem. Soc.* **109**, 3708 (1987).
351. B. Langlois and M. Desbois, *Ann. Chim. Fr.*, 729 (1984).
352. Brit. Pat. 765,527 (Jan. 9, 1957) (to Hoechst).
353. U.S. Pat. 4,620,040 (Oct. 28, 1986), D. J. Alsop (to Occidental Chemical).
354. L. M. Yagupolski, *Dokl. Akad. Nauk (SSSR)* **105**, 100 (1955).
355. A. E. Feiring, *J. Org. Chem.* **44**, 2907 (1979).
356. F. Mathey and J. Bensoam, *Tetrahedron Lett.*, 2253 (1973).
357. L. M. Yagupolski and V. V. Orda, *Zh. Obsch. Khim.* **34**, 1979 (1964).
358. R. P. J. Braden and E. Klauke, *Pesticide Sci.* **17**, 418 (1986).
359. L. M. Yagupolski and co-workers, *Synthesis*, 721 (1975).
360. D. C. Remy and co-workers, *J. Org. Chem.* **41**, 1644 (1976).
361. T. G. Miller and J. W. Thanasi, *J. Org. Chem.* **25**, 2009 (1960).
362. B. R. Langlois, *J. Fluorine Chem.* **41**, 247 (1988).
363. D. C. England and co-workers, *J. Am. Chem. Soc.* **82**, 5116 (1960).
364. J. P. Idoux and co-workers, *J. Org. Chem.* **48**, 3771 (1983).
365. J. T. Gupton and co-workers, *Synth. Commun.* **15**, 431 (1985).
366. J. P. Idoux and co-workers, *J. Org. Chem.* **50**, 1876 (1985).
367. E. H. Banitt and co-workers, *J. Med. Chem.* **18**, 1130 (1975); *Ibid.* **20**, 821 (1977).
368. U.S. Pat. 5,082,587 (Jan. 21, 1992), E. P. Janulis (to 3M Co.).
369. World Pat. 91/00897 (Jan. 24, 1991). A. Wächtler and co-workers (to E. Merck GmbH).
370. N. Yoneda and co-workers, *J. Fluorine Chem.* **38**, 435 (1988).
371. M. M. Boudakian, *J. Fluorine Chem.* **18**, 497 (1981).
372. U.S. Pat. 2,447,717 (Aug. 24, 1948), J. H. Simons (to Minnesota Mining and Manufacturing).

373. H. Meinert, *Z. Chem.* **5**, 64 (1965).
374. M. Van der Puy, *Tetrahedron Lett.* **28**, 255 (1987).
375. U.S. Pat. 4,786,733 (Nov. 22, 1988), M. Van der Puy and R. E. Eibeck (to Allied-Signal).
376. S. P. Anand and R. Filler, *J. Fluorine Chem.* **7**, 179 (1976).
377. S. Stavber and M. Zupan, *Tetrahedron Lett.* **31**, 775 (1990).
378. T. Umemoto and G. Tomizawa, *J. Org. Chem.* **54**, 1726 (1989).
379. G. C. Finger and co-workers, *J. Org. Chem.* **28**, 1666 (1963).
380. M. M. Boudakian, *J. Heterocyclic Chem.* **4**, 381 (1967); U.S. Pat. 3,296,269 (Jan. 3, 1967), M. M. Boudakian (to Olin).
381. Jpn. Kokai Tokkyo Koho JP 04,124,176 (Apr. 24, 1992), T. Fukuhara and co-workers (to Tohkem Products).
382. H. C. Brown and D. McDaniel, *J. Am. Chem. Soc.* **77**, 3752 (1955).
383. L. Bradlow and C. VanderWerf, *J. Org. Chem.* **14**, 509 (1949).
384. G. Bressan and co-workers, *J. Chem. Soc. B.*, 225 (1971).
385. U.S. Pats. 3,873,701 (Mar. 25, 1975), and 3,816,406 (June 11, 1974), A. F. Hirsch (to Ortho Pharmaceuticals).
386. T. Mukaiyama, *Angew. Chem. Int. Ed. Engl.* **18**, 707 (1979).
387. J. Hannah and co-workers, *J. Med. Chem.* **25**, 457 (1982).
388. T. Güngör, *J. Organometallic Chem.* **215**, 139 (1981).
389. F. Marsais and co-workers, *J. Org. Chem.* **57**, 565 (1992).
390. R. D. Beaty and W. K. R. Musgrave, *J. Chem. Soc.*, 875 (1952).
391. A. Roe and C. Hawkins, *J. Am. Chem. Soc.* **69**, 2443 (1947).
392. *Chem. Eng. News*, 44 (Oct. 16, 1967); *Ibid.*, 8 (Dec. 8, 1967).
393. Span. Pat. ES 548,693 (May 16, 1986), J. C. Verde.
394. T. Miyamoto and co-workers, *Chem. Pharm. Bull.* **35**, 2280 (1987).
395. R. G. Pews and Z. Lysenko, *J. Org. Chem.* **50**, 5115 (1985).
396. T. Talik and Z. Talik, *Rocz. Chem.* **40**, 1187 (1966).
397. *Ibid.*, **38**, 785 (1964).
398. U.S. Pat. 3,637,714 (Jan. 25, 1972), L. A. F. Carlsson and co-workers (to Aktiebolaget Astra).
399. W. S. Saari and co-workers, *J. Med. Chem.* **26**, 1696 (1983).
400. U.S. Pat. 3,703,521 (Nov. 21, 1972), M. M. Boudakian (to Olin).
401. P. B. Desai, *J. Chem. Soc. Perkin Trans.* 1, 1866 (1973).
402. U.S. Pats. 4,283,397 (Aug. 11, 1981), 4,255,424 (Mar. 10, 1981), 4,241,062 (Dec. 23, 1980), 4,263,306 (Apr. 21, 1981), and 4,282,219 (Aug. 4, 1981), J. Hannah (to Merck).
403. Ger. Offen. 2,128,540 (Dec. 16, 1971), A. Nicolson (to ICI).
404. Ger. Offen. 2,425,239 (May 24, 1973), D. W. R. Headford and co-workers (to ICI).
405. U.S. Pat. 3,798,228 (Mar. 19, 1974), M. M. Boudakian (to Olin).
406. J. Hamer and co-workers, *Rec. Trav. Chim.* **81**, 1059 (1962).
407. U.S. Pat. 4,071,521 (Jan. 31, 1978), T. G. Muench (to Dow).
408. U.S. Pat. 4,031,100 (June 21, 1977), T. L. Giacobbe (to Dow).
409. M. M. Boudakian, *J. Heterocyclic Chem.* **5**, 683 (1968).
410. U.S. Pat. 4,320,122 (Mar. 16, 1982), H. Theobold and co-workers (to BASF).
411. U.S. Pat. 3,810,902 (May 14, 1974), R. H. Rigterink (to Dow).
412. U.S. Pat. 4,115,557 (Sept. 19, 1978), C. E. Pawloski (to Dow).
413. H. R. Kricheldorf and co-workers, *Makromol. Chem.* **189**, 2255 (1988); *Ibid.* **191**, 2027 (1990).
414. F. Mutterer and C. D. Weis, *Helv. Chim. Acta* **59**, 229 (1976).
415. E. Differding and co-workers, *Bull. Soc. Chim. Belg.* **99**, 647 (1990).
416. R. D. Chambers and co-workers, *J. Chem. Soc.*, 5045 (1965).
417. U.S. Pat. 4,746,744 (May 24, 1988), C. A. Wilson and A. Fung (to Dow).

418. R. E. Banks and co-workers, *J. Chem. Soc.*, 594 (1965).
419. *Ibid.*, 575 (1965).
420. *Ibid.*, 1740 (1961).
421. J. A. Burdon and co-workers, *Nature* **196**, 231 (1960).
422. R. D. Chambers and co-workers, *J. Chem. Soc.*, 3573 (1964).
423. G. G. Yakobson and co-workers, *Fluorine Chem. Rev.* **7**, 115 (1974).
424. U.S. Pat. 3,850,943 (Nov. 26, 1974), R. D. Bowden and co-workers (to ICI).
425. U.S. Pat. 4,063,926 (Dec. 20, 1977), C. D. S. Tomlin and co-workers (to ICI).
426. R. D. Chambers and co-workers, *J. Chem. Soc. C*, 2221 (1968); *J. Chem. Soc., Chem. Commun.*, 384 (1966).
427. L. A. Mitscher and co-workers, *Chemtech*, 50 (Jan. 1991); *Ibid.*, 249 (Apr. 1991).
428. M. Jarman and R. McCague, *J. Chem. Soc., Chem. Commun.*, 125 (1984).
429. T. Miyamoto and co-workers, *J. Heterocyclic Chem.* **24**, 1333 (1987).
430. D. T. W. Chu and co-workers, *J. Med. Chem.* **29**, 2363 (1986).
431. J. Matsumoto and co-workers, *J. Heterocyclic Chem.* **21**, 673 (1984).
432. E. T. McBee and co-workers, *Ind. Eng. Chem.* **39**, 389 (1947).
433. U.S. Pat. 2,516,402 (July 25, 1950), E. T. McBee and E. M. Hodnett (to Purdue Research Foundation).
434. M. S. Raasch, *J. Org. Chem.* **27**, 1406 (1962).
435. I. D. Shustov and co-workers, *Zh. Obshch. Khim.* **53**, 103 (1983).
436. L. F. Lee and co-workers, *J. Org. Chem.* **55**, 2872 (1990).
437. L. F. Lee and co-workers, in D. R. Baker, J. G. Fenyes, and W. K. Moberg, eds., *Synthesis and Chemistry of Agrochemicals, II*. ACS Symposium Series No. 443, American Chemical Society, Washington, D.C., 1991, Chapt. 16, p.195.
438. Y. Kobayashi and I. Kumadaki, *Chem. Pharm. Bull.* **17**, 510 (1969).
439. Y. Kobayashi and E. Chinen, *Chem. Pharm. Bull.* **17**, 510 (1967).
440. U.S. Pat. 4,259,496 (Mar. 31, 1981), G. Whittaker (to ICI).
441. U.S. Pat. 4,324,627 (Apr. 13, 1982), D. Cartwright (to ICI).
442. U.S. Pat. 4,266,064 (May 5, 1981), R. Nishiyama and co-workers, (to Ishihara).
443. U.S. Pats. 4,417,055 (Nov. 22, 1983), and 4,288,599 (Sept. 8, 1981), R. Nishiyama and co-workers (to Ishihara).
444. R. D. Chambers and co-workers, *J. Chem. Soc. C*, 2328 (1966).
445. Y. Kobayashi and co-workers, *Chem. Pharm. Bull.* **17**, 2335 (1969).
446. H. Gilman and D. Blume, *J. Am. Chem. Soc.* **65**, 2467 (1943).
447. E. Pouterman and A. Girardet, *Helv. Chim. Acta* **30**, 107 (1947).
448. C. J. Ohnmacht and co-workers, *J. Med. Chem.* **14**, 926 (1971).
449. H. Vergin and R. Metz, *Drugs of Today* **27**(3), 177 (1991).
450. *Chem. Mktg. Rep.*, 5 (Dec. 15, 1992).
451. R. Duschinsky and co-workers, *J. Am. Chem. Soc.* **79**, 4559 (1957).
452. U.S. Pat. 3,682,917 (Aug. 8, 1972), I. L. Knunyants and co-workers.
453. U.S. Pat. 3,846,429 (Nov. 5, 1974), S. A. Giller and co-workers.
454. U.S. Pats. 3,954,758 (May 4, 1976), and 4,113,949 (Sept. 12, 1978), P. D. Schuman and co-workers (to PCR).
455. U.S. Pat. 4,082,752 (Apr. 4, 1978), T. Takahara and S. Misaki (to Daikin Kogyo).
456. F. Winston, *PCR Research Chemicals Catalog, 1990–91*. Gainesville, Fla., p.174.
457. E. Klauke and co-workers, *J. Fluorine Chem.* **21**, 495 (1982).
458. U.S. Pat. 3,314,955 (Apr. 18, 1967), M. M. Boudakian and C. W. Kaufman (to Olin).
459. U.S. Pat. 3,280,124 (Oct. 18, 1966), M. M. Boudakian and co-workers (to Olin).
460. R. G. Pews and W. E. Puckett, *J. Fluorine Chem.* **42**, 179 (1989).
461. H. Schroeder, *J. Am. Chem. Soc.* **82**, 4115 (1960).
462. H. Schroeder and co-workers, *J. Org. Chem.* **27**, 2580 (1962).
463. R. E. Banks and co-workers, *J. Chem. Soc. C*, 1822 (1967).

464. B. Baasner and E. Klauke, *J. Fluorine Chem.* **45**, 417 (1989).
465. D. R. Hildebrand, *Chemtech*, 224 (Apr. 1978).
466. W. Harms, in Ref. 335, Chapt. 9, p. 188.
467. U.S. Pat. 3,694,444 (Sept. 26, 1972), E. Klauke and H.-S. Bien (to Bayer); Brit. Pat. 1,273,914 (June 12, 1970), H.-U. Alles and co-workers (to Bayer).
468. Can. Pat. 844,625 (June 16, 1970), E. Klauke and H.-S. Bien (to Bayer).
469. Brit. Pat. 873,251 (July 19, 1961), A. Dorlars (to Bayer).
470. Ger. Offen. 2,729,762 (Jan. 18, 1979), F. Kysela and co-workers (to Bayer).
471. U.S. Pat. 4,332,939 (June 1, 1982), G. Seifert and S. Stäubli (to Ciba-Geigy).
472. C. W. Tullock and D. D. Coffman, *J. Org. Chem.* **25**, 2016 (1960).
473. U.S. Pat. 4,329,458 (May 11, 1982), E. Klauke and co-workers (to Bayer).
474. K. Kuriha and co-workers, *Biochem. Biophys. Acta* **384**, 127 (1975).
475. U.S. Pat. 2,975,179 (Mar. 14, 1961), A. Dorlars (to Bayer).
476. G. A. Olah and co-workers, *Synthesis*, 487 (1973).
477. L. A. Carpino and co-workers, *J. Am. Chem. Soc.* **112**, 9651 (1990); *J. Org. Chem.* **56**, 2611 (1991).
478. W. L. Reilly and H. C. Brown, *J. Org. Chem.* **22**, 698 (1967).
479. E. T. McBee and co-workers, *Ind. Eng. Chem.* **39**, 391 (1947).
480. S. Smith, in R. E. Banks, ed., *Preparation, Properties and Industrial Applications of Organofluorine Compounds*, E. Horwood, Chichester, U.K., 1982, Chapt. 8, p. 234.
481. J. A. Young, in Ref. 248, Chapt. 9.
482. K. Tanaka, *J. Synth. Org. Chem., Japan* **48**, 16 (1990).
483. R. D. Chambers and C. R. Sargent, in A. R. Katritzky and A. J. Boulton, eds., *Advances in Heterocyclic Chemistry*, Academic Press, Inc., New York, 1981, Vol. 28, p. 1.

General References

A. E. Pavlath and A. L. Leffler, *Aromatic Fluorine Compounds*, ACS Monograph No. 155, Reinhold, N.Y., 1962.

W. A. Sheppard and C. M. Sharts, *Organic Fluorine Chemistry*, W. A. Benjamin, N.Y., 1969.

M. Hudlicky, *Chemistry of Organic Fluorine Compounds*, 2nd ed., Ellis Horwood, Chichester, U.K., 1976.

G. Schiemann and B. Cornils, *Chemie und Technologie Cyclischer Fluorverbindungen*, F. Enke Verlag, Stuttgart, 1969.

Handbook of Aromatic Fluorine Compounds, Olin Corporation, Stamford, Conn., 1976.

R. D. Chambers, *Fluorine in Organic Chemistry*, John Wiley & Sons, Inc., New York, 1973.

E. Forche, "Houben-Weyl, Methoden der Organischen Chemie," in E. Mueller, ed., *Halogen-Verbindungen*, Georg Thieme Verlag, Stuttgart, 1962, pp. 1–502.

R. Filler, in J. C. Tatlow, R. D. Peacock, and H. H. Hyman, eds., *Advances in Fluorine Chemistry*, Vol. 6, CRC Press, Cleveland, Ohio, 1970, p.1.

G. Fuller, *Manuf. Chem. Aerosol News* **35**, 43 (May 1964); *Ibid.* **35**, 45 (June 1964).

A. K. Barbour, L. J. Belf, and M. W. Buxton, in M. Stacey, J. C. Tatlow, and A. G. Sharpe, eds., *Advances in Fluorine Chemistry*, Vol. 3, Butterworths, Washington, D.C., 1963, p.181.

R. H. Shiley, D. R. Dickerson, and G. C. Finger, *Aromatic Fluorine Chemistry at the Illinois State Geological Survey*, circular 501, Urbana, Ill., 1978.

M. M. Boudakian, "Halopyridines," in R. A. Abramovitch, ed., *The Chemistry of Heterocyclic Compounds*, Vol. 14, supplement part 2, *Pyridine and Its Derivatives*, Wiley-Interscience, New York, 1974, Chapt. 6, p. 407.

A. Roe, in R. Adams ed., *Organic Reactions*, Vol. 5, John Wiley & Sons, Inc., New York, 1949, Chapt. 4, p. 193.

H. Suschitzky, in M. Stacey, J. C. Tatlow, and A. G. Sharpe, eds., *Advances in Fluorine Chemistry*, Vol. 4, Butterworths, Washington, D.C., 1964, p. 1.

R. Filler and Y. Kobayashi, eds., *Biomedical Aspects of Fluorine Chemistry*, Kodansha, Tokyo, 1982.

R. E. Banks, D. W. A. Sharp, and J. C. Tatlow, eds., *Fluorine: The First Hundred Years*, Elsevier, Lausanne, 1986.

L. German and S. Zemzkov, eds., *New Fluorinating Agents in Organic Synthesis*, Springer-Verlag, Berlin, 1989.

I. L. Knunyants and G. G. Yakobson, eds., *Syntheses of Organic Fluorine Compounds*, Springer-Verlag, Berlin, 1985.

S. B. Walker, ed., *Fluorine in Agrochemicals*, Fluorochem, Ltd., Old Glossop, U.K., 1990.

R. E. Banks, ed., *Organofluorine Chemicals and Their Industrial Applications*, E. Horwood, Chichester, U.K., 1979.

R. E. Banks, ed., *Preparation, Properties and Industrial Applications of Organofluorine Compounds*, Ellis Horwood, Chichester, U.K., 1982.

MAX M. BOUDAKIAN
Chemical Consultant

POLYTETRAFLUOROETHYLENE

Polytetrafluoroethylene [*9002-84-0*] (PTFE), more commonly known as Teflon (Du Pont), a perfluorinated straight-chain high polymer, has a most unique position in the plastics industry due to its chemical inertness, heat resistance, excellent electrical insulation properties, and low coefficient of friction over a wide temperature range. Polymerization of tetrafluoroethylene (TFE) monomer gives this perfluorinated straight-chain high polymer with the formula $-(CF_2-CF_2)_n-$. The white to translucent solid polymer has an extremely high molecular weight, in the 10^6-10^7 range, and consequently has a viscosity in the range of 1 to 10 GPa·s ($10^{10}-10^{11}$ P) at 380°C. It is a highly crystalline polymer and has a crystalline melting point. Its high thermal stability results from the strong carbon–fluorine bond and characterizes PTFE as a useful high temperature polymer.

The discovery of PTFE (1) in 1938 opened the commercial field of perfluoropolymers. Initial production of PTFE was directed toward the World War II effort, and commercial production was delayed by Du Pont until 1947. Commercial PTFE is manufactured by two different polymerization techniques that result in two different types of chemically identical polymer. Suspension polymerization produces a granular resin, and emulsion polymerization produces the coagulated dispersion that is often referred to as a fine powder or PTFE dispersion.

Because of its chemical inertness and high molecular weight, PTFE melt does not flow and cannot be fabricated by conventional techniques. The suspension-polymerized PTFE polymer (referred to as granular PTFE) is usually fabricated by modified powder metallurgy techniques. Emulsion-polymerized PTFE behaves entirely differently from granular PTFE. Coagulated dispersions are processed by a cold extrusion process (like processing lead). Stabilized PTFE dispersions, made by emulsion polymerization, are usually processed according to latex processing techniques.

Manufacturers of PTFE include Daikin Kogyo (Polyflon), Du Pont (Teflon), Hoechst (Hostaflon), ICI (Fluon), Ausimont (Algoflon and Halon), and the CIS (Fluoroplast). India and The People's Republic of China also manufacture some PTFE products.

Monomer

Preparation. The manufacture of tetrafluoroethylene [116-14-3] (TFE) involves the following steps (2–9). The pyrolysis is often conducted at a PTFE manufacturing site because of the difficulty of handling TFE.

$$CaF_2 + H_2SO_4 \longrightarrow CaSO_4 + 2\ HF$$

$$CH_4 + 3\ Cl_2 \longrightarrow CHCl_3 + 3\ HCl$$

$$CHCl_3 + 2\ HF \xrightarrow{SbF_3} CHClF_2 + 2\ HCl$$

$$2\ CHClF_2 \xrightarrow{\Delta} CF_2{=}CF_2 + 2\ HCl$$

Pyrolysis of chlorodifluoromethane is a noncatalytic gas-phase reaction carried out in a flow reactor at atmospheric or subatmospheric pressure; yields can be as high as 95% at 590–900°C. The economics of monomer production is highly dependent on the yields of this process. A significant amount of hydrogen chloride waste product is generated during the formation of the carbon–fluorine bonds.

A large number of by-products are formed in this process, mostly in trace amounts; more significant quantities are obtained of hexafluoropropylene, perfluorocyclobutane, 1-chloro-1,1,2,2-tetrafluoroethane, and 2-chloro-1,1,1,2,3,3-hexafluoropropane. Small amounts of highly toxic perfluoroisobutylene, $CF_2{=}C(CF_3)_2$, are formed by the pyrolysis of chlorodifluoromethane.

In this pyrolysis, subatmospheric partial pressures are achieved by employing a diluent such as steam. Because of the corrosive nature of the acids (HF and HCl) formed, the reactor design should include a platinum-lined tubular reactor made of nickel to allow atmospheric pressure reactions to be run in the presence of a diluent. Because the pyrolysate contains numerous by-products that adversely affect polymerization, the TFE must be purified. Refinement of TFE is an extremely complex process, which contributes to the high cost of the monomer. Inhibitors are added to the purified monomer to avoid polymerization during storage; terpenes such as d-limonene and terpene B are effective (10).

Tetrafluoroethylene was first synthesized in 1933 from tetrafluoromethane, CF_4, in an electric arc furnace (11). Since then, a number of routes have been developed (12–18). Depolymerization of PTFE by heating at ca 600°C is probably the preferred method for obtaining small amounts of 97% pure monomer on a laboratory scale (19,20). Depolymerization products contain highly toxic perfluoroisobutylene and should be handled with care.

Properties. Tetrafluoroethylene (mol wt 100.02) is a colorless, tasteless, odorless, nontoxic gas (Table 1). It is stored as a liquid; vapor pressure at

Table 1. Physical Properties of Tetrafluoroethylene[a]

Property	Value
boiling point at 101.3 kPa,[b] °C	−76.3
freezing point, °C	−142.5
liquid density at t °C, g/mL	
$-100 < t < -40$	$1.202 - 0.0041\,t$
$-40 < t < 8$	$1.1507 - 0.0069\,t - 0.000037\,t^2$
$8 < t < 30$	$1.1325 - 0.0029\,t - 0.00025\,t^2$
vapor pressure at T K, kPa[c]	
$196.85 < T < 273.15$	$\log_{10} P_{kPa} = 6.4593 - 875.14/T$
$273.15 < T < 306.45$	$\log_{10} P_{kPa} = 6.4289 - 866.84/T$
critical temperature, °C	33.3
critical pressure, MPa[d]	39.2
critical density, g/mL	0.58
dielectric constant at 28°C	
at 101.3 kPa[b]	1.0017
at 858 kPa[b]	1.015
thermal conductivity at 30°C, mW/(m·K)	15.5
heat of formation for ideal gas at 25°C, ΔH, kJ/mol[e,f]	−635.5
heat of polymerization at 25°C to solid polymer ΔH, kJ/mol[e,g]	−172.0
flammability limits in air at 101.3 kPa,[c] vol %	14–43

[a] From Ref. 21, unless otherwise stated.
[b] To convert kPa to atm, multiply by 0.01.
[c] To convert kPa to psi, multiply by 0.145.
[d] To convert MPa to atm, divide by 0.101.
[e] To convert J to cal, divide by 4.184.
[f] Ref. 22.
[g] Ref. 23.

−20°C = 1 MPa (9.9 atm). It is usually polymerized above its critical temperature and below its critical pressure. The polymerization reaction is highly exothermic.

Tetrafluoroethylene undergoes addition reactions typical of an olefin. It burns in air to form carbon tetrafluoride, carbonyl fluoride, and carbon dioxide (24). Under controlled conditions, oxygenation produces an epoxide (25) or an explosive polymeric peroxide (24). Trifluorovinyl ethers, RO—CF=CF$_2$, are obtained by reaction with sodium salts of alcohols (26). An ozone–TFE reaction is accompanied by chemiluminescence (27). Dimerization at 600°C gives perfluorocyclobutane, C$_4$F$_8$; further heating gives hexafluoropropylene, CF$_2$=CFCF$_3$, and eventually perfluoroisobutylene, CF$_2$=C(CF$_3$)$_2$ (28). Purity is determined by both gas–liquid and gas–solid chromatography; the ir spectrum is complex and therefore of no value.

Uses. Besides polymerizing TFE to various types of high PTFE homopolymer, TFE is copolymerized with hexafluoropropylene (29), ethylene (30), perfluorinated ether (31), isobutylene (32), propylene (33), and in some cases it is used as a termonomer (34). It is used to prepare low molecular weight polyfluorocarbons (35) and carbonyl fluoride (36), as well as to form PTFE *in situ* on metal

FLUORINE COMPOUNDS, ORGANIC (POLYMERS)

surfaces (37). Hexafluoropropylene [116-15-4] (38,39), perfluorinated ethers, and other oligomers are prepared from TFE.

In the absence of air, TFE disproportionates violently to give carbon and carbon tetrafluoride; the same amount of energy is generated as in black powder explosions. This type of decomposition is initiated thermally and equipment hot spots must be avoided. The flammability limits of TFE are 14–43%; it burns when mixed with air and forms explosive mixtures with air and oxygen. It can be stored in steel cylinders under controlled conditions inhibited with a suitable stabilizer. The oxygen content of the vapor phase should not exceed 10 ppm. Although TFE is nontoxic, it may be contaminated by highly toxic fluorocarbon compounds.

Manufacture of PTFE

Engineering problems involved in the production of TFE seem simple compared with those associated with polymerization and processing of PTFE resins. The monomer must be polymerized to an extremely high molecular weight in order to achieve the desired properties. The low molecular weight polymer does not have the strength needed in end use applications.

Polytetrafluoroethylene is manufactured and sold in three forms: granular, fine powder, and aqueous dispersion; each requires a different fabrication technique. Granular resins are manufactured in a wide variety of grades to obtain a different balance between powder flows and end use properties (Fig. 1). Fine pow-

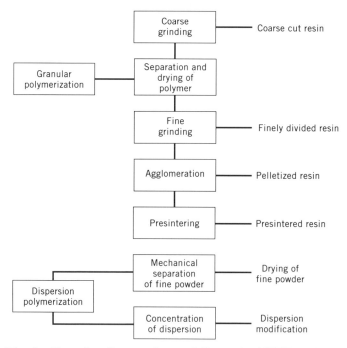

Fig. 1. Granular, fine powder, and dispersion PTFE products.

ders that are made by coagulating aqueous dispersions also are available in various grades. Differences in fine powder grades correspond to their usefulness in specific applications and to the ease of fabrication. Aqueous dispersions are sold in latex form and are available in different grades. A variety of formulation techniques are used to tailor these dispersions for specific applications.

Polymerization. In aqueous medium, TFE is polymerized by two different procedures. When little or no dispersing agent is used and vigorous agitation is maintained, a precipitated resin is produced, commonly referred to as granular resin. In another procedure, called aqueous dispersion polymerization, a sufficient dispersing agent is employed and mild agitation produces small colloidal particles dispersed in the aqueous reaction medium; precipitation of the resin particles is avoided. The two products are distinctly different, even though both are high molecular weight PTFE polymers. The granular product can be molded in various forms, whereas the resin produced by the aqueous dispersion cannot be molded, but is fabricated by dispersion coating or conversion to powder for paste extrusion with a lubricant medium. Granular resin cannot be paste extruded or dispersion coated.

Granular Resins. Granular PTFE is made by polymerizing TFE alone or in the presence of trace amounts of comonomers (40,41). An initiator, a small amount of dispersing agent, and other additives (42) may be present; an alkaline buffer is occasionally used (43). In the early stages of polymerization, an unstable dispersion is formed, but lack of dispersing agent and vigorous agitation cause the polymer to partially coagulate; the remainder of the process is fairly complex. The polymerized product is stringy, irregular, and variable in shape. The dried granular polymer is ground to different average particle sizes, depending on the product requirements, eg, the flow and other properties. Coarser fabrication of particles leaves a higher void in the sintered article. A better balance between handleability and moldability (ability to mold and sinter in the absence of voids) is achieved by agglomerating the finely divided resin to ca 400–800 μm (44). For ram extrusion of granular resin into long tubes and rods, a partially presintered resin is preferred. Granular PTFE resin is nonflammable.

Fine Powder Resins. Fine powder resins are made by polymerizing TFE in an aqueous medium with an initiator and emulsifying agents (45). The polymerization mechanism is not a typical emulsion type, but is subject to some of the principles of emulsion polymerization. The process and ingredients have a significant effect on the product. It is extremely important that the dispersion remains sufficiently stable throughout polymerization, avoiding premature coagulation (46), but unstable enough to allow subsequent coagulation into a fine powder. Gentle stirring ensures dispersion stability. The amount of emulsifying agent in the polymerization process is usually less than its critical micelle concentration. The rate of polymerization and the particle shape are influenced by the amount of the emulsifying agent (47–50). The particle structure can be influenced by the polymerization process. Most of the particles are formed in the early stages of the polymerization process and the particles grow as the batch progresses; hence, the radial variation in molecular weight and polymer composition within the dispersion particle can be achieved by controlling the polymerization variables, including ingredients and operating conditions (51–57).

The thin dispersion rapidly thickens into a gelled matrix and coagulates into a water-repellent agglomeration that floats on the aqueous medium as the mechanical agitation is continued. The agglomeration is dried gently; shearing must be avoided.

Aqueous Dispersions. The dispersion is made by the polymerization process used to produce fine powders of different average particle sizes (58). The most common dispersion has an average particle size of about 0.2 μm, probably the optimum particle size for most applications. The raw dispersion is stabilized with a nonionic or anionic surfactant and concentrated to 60–65 wt % solids by electrodecantation, evaporation, or thermal concentration (59). The concentrated dispersion can be modified further with chemical additives. The fabrication characteristics of these dispersions depend on polymerization conditions and additives.

Filled Resins. Fillers such as glass fibers, graphite, asbestos, or powered metals are compounded into all three types of PTFE. Compounding is achieved by intimate mixing. Coagulation of the polymer with a filler produces a filled fine powder.

Properties

The properties described herein are related to the basic structure of polytetrafluoroethylene and are exhibited by both granular and fine powder products. The carbon–carbon bonds, which form the backbone of the PTFE chain, and the carbon–fluorine bonds are extremely strong and are the key contributors in imparting an outstanding combination of properties. The fluorine atoms form a protective sheath over the chain of carbon atoms. If the atoms attached to the carbon-chain backbone were smaller or larger than fluorine, the sheath would not form a regular uniform cover. This sheath shields the carbon chain from attack and confers chemical inertness and stability. It also reduces the surface energy resulting in low coefficient of friction and nonstick properties.

Polytetrafluoroethylene does not dissolve in any common solvent; therefore, its molecular weight cannot be measured by the usual methods. A number-average molecular weight has been estimated by determining the concentration of end groups derived from the initiator. Earlier estimates, based on an iron bisulfite system containing radioactive sulfur, ^{35}S, ranged from 142×10^3 to 534×10^3 for low molecular weight polymer. The same technique applied to polymers of industrial interest gave molecular weights of 389×10^3 to 8900×10^3 (60,61). In the absence of a normal molecular weight determination method, an estimated relative molecular weight is used for all practical purposes. It is obtained by measuring the specific gravity following a standardized fabricating and sintering procedure (ASTM D1457-83). Because the rate of crystallization decreases with increasing molecular weight, samples prepared from the high molecular weight polymer and cooled from the melt at a constant slow rate have lower standard specific gravities than those prepared from low molecular weight polymer cooled at the same rate (62). The correlation between number-average molecular weight (M_n) based on end group estimations, and standard specific gravity (SSG) is given by

$$\text{SSG} = 2.612 - 0.058 \log_{10} M_n$$

The SSG procedure assumes absence of voids (or constant void content). Voids depress the values of the measured specific gravity. The inaccuracies that result from voids can be corrected by applying ir techniques (63).

Melting and recrystallization behavior of virgin PTFE has been studied by dsc (64). A quantitative relationship was found between M_n and the heat of crystallization (ΔH_c) in the molecular weight range of 5.2×10^5 to 4.5×10^7, where H_c is heat of crystallization in J/g, which is independent of cooling rates of 4–32°C/min.

$$M_n = 2.1 \times 10^{10} \cdot \Delta H_c^{-5.16}$$

At ca 342°C, virgin PTFE changes from white crystalline material to almost transparent amorphous gel. Differential thermal analysis indicates that the first melting of virgin polymer is irreversible and that subsequent remeltings occur at 327°C, which is generally reported as the melting point. Most of the studies reported in the literature are based on previously sintered (ie, melted and recrystallized) polymer; very little work is reported on the virgin polymer. Melting is accompanied by a volume increase of ca 30%. Because the viscosity of the polymer at 380°C is 10 GPa·s (10^{11} P), the shape of the melt is stable. The melting point increases with increasing applied pressure at the rate of 1.52°C/MPa (0.154°C/atm) (65).

Virgin PTFE has a crystallinity in the range of 92–98%, which indicates an unbranched chain structure. The fluorine atoms are too large to allow a planar zigzag structure, which would permit chain flexibility; therefore the chains are rigid (66). Electron micrographs and diffraction patterns (67) of PTFE dispersion particles indicate that the rod-like particles present in virgin PTFE dispersions are fully extended chain crystals containing few defects. The spherical particles appear to be composed of similar rod-like entities that are wrapped around themselves in a more or less random fashion.

Between 50 and 300°C, PTFE obeys the relationship between stress τ and the apparent shear rate γ: $\tau = K\gamma^{1/4}$. Melting of PTFE begins near 300°C. Above this temperature, the shear stress at constant shear rate increases and the rheological exponent rises from 0.25 toward 0.5 at the final melting point (68).

Transitions. Transitions observed by various investigators (69–74), their interpretation, and the modes of identification are shown in Table 2. Besides the transition at the melting point, the transition at 19°C is of great consequence because it occurs at ambient temperature and significantly affects the product behavior. Above 19°C, the triclinic pattern changes to a hexagonal unit cell. Around 19°C, a slight untwisting of the molecule from a 180° twist per 13 CF_2 groups to a 180° twist per 15 CF_2 groups occurs. At the first-order transition at 30°C, the hexagonal unit cell disappears and the rod-like hexagonal packing of the chains in the lateral direction is retained (69). Below 19°C there is almost perfect three-dimensional order; between 19 and 30°C the chain segments are disordered; and above 30°C, the preferred crystallographic direction is lost and the molecular segments oscillate above their long axes with a random angular orientation in the lattice (70,71).

Table 2. Transitions in Polytetrafluoroethylene

Temperature, °C	Region affected	Technique	Reference
	1st order		
19	crystalline, angular displacement causing disorder	thermal methods, x-ray, nmr	70
30	crystalline, crystal disordering	thermal methods, x-ray, nmr	70
90 (80 to 110)	crystalline	stress relaxation, Young's modulus, dynamic methods	73
	2nd order		
−90 (−110 to −73)	amorphous, onset of rotational motion around C—C bond	thermal methods, dynamic methods	74
−30 (−40 to −15)	amorphous	stress relaxation, thermal expansion, dynamic methods	73
130 (120 to 140)	amorphous	stress relaxation, Young's modulus, dynamic methods	73

The dynamic mechanical properties of PTFE have been measured at frequencies from 0.033 to 90 Hz. Abrupt changes in the distribution of relaxation times are associated with the crystalline transitions at 19 and 30°C (75). The activation energies are 102.5 kJ/mol (24.5 kcal/mol) below 19°C, 510.4 kJ/mol (122 kcal/mol) between the transitions, and 31.4 kJ/mol (7.5 kcal/mol) above 30°C.

Polytetrafluoroethylene transitions occur at specific combinations of temperature and mechanical or electrical vibrations. Transitions, sometimes called dielectric relaxations, can cause wide fluctuations in the dissipation factor.

Mechanical Properties. Mechanical properties of PTFE depend on processing variables, eg, preforming pressure, sintering temperature and time, cooling rate, void content, and crystallinity. Properties, such as the coefficient of friction, flexibility at low temperatures, and stability at high temperatures, are relatively independent of fabrication. Molding and sintering conditions affect flex life, permeability, stiffness, resiliency, and impact strength. The physical properties of PTFE have been reviewed and compiled (72,76,77) (Table 3).

A marked change in volume of 1.0–1.8% is observed for PTFE in the transition zone from 18 to 25°C. An article that has been machined on either side of this zone changes dimensions when passing through the transition zone; hence, the final operating temperature of a precision part must be accurately determined. Articles fabricated of PTFE resins exhibit high strength, toughness, and self-lubrication at low temperatures. They are useful from 5 K and are highly flexible from 194 K. They tend to return to their original dimensions after a deformation. At sintering temperature, they rapidly recover their original shapes. For most

Table 3. Typical Mechanical Properties of Molded and Sintered PTFE Resins[a]

Property	Granular resin	Fine powder	ASTM method
tensile strength at 23°C, MPa[b]	7–28	17.5–24.5	D638-61T
elongation at 23°C, %	100–200	300–600	D628-61T
flexural strength at 23°C, MPa[b]	does not break		D790-61
flexural modulus at 23°C, MPa[b]	350–630	280–630	D747-61T
impact strength, J/m[c]			
21°C	106.7		D256-56
24°C	160		
77°C	>320		
hardness durometer, D	50–65	50–65	D1706-59T
compression stress, MPa[b]			
at 1% deformation at 23°C	4.2		D695-52T
at 1% offset at 23°C	7.0		D695-52T
coefficient of linear thermal expansion per °C, 23–60°C	12×10^{-5}		D696-44
thermal conductivity, 4.6-mm thickness, W/(m·K)	0.24		Cenco-Fitch
deformation under load, at 26°C, 24 h, %			D621-59
6.86 MPa[b]		2.4	
13.72 MPa[b]	15		
water absorption, %	<0.01	<0.01	D570–54T
flammability	nonflammable		D635-56T
static coefficient of friction with polished steel	0.05–0.08		

[a] Ref. 77.
[b] To convert MPa to psi, multiply by 145.
[c] To convert J/m to ft·lbf/in., divide by 53.38.

applications no special precautions are necessary because decomposition rates below the recommended maximum service temperature of 260°C are very low. Impact strength is excellent over a wide range of temperatures. Static friction decreases with an increase in load. Static coefficient of friction is lower than the dynamic coefficient and therefore reduces stick-slip problems.

The surface of PTFE articles is slippery and smooth. Liquids with surface tensions below 18 mN/m(= dyn/cm) are spread completely on the PTFE surface; hence, solutions of various perfluorocarbon acids in water wet the polymer (78). Treatment with alkali metals promotes the adhesion between PTFE and other substances (79) but increases the coefficient of friction (80).

Filled Resins. Filled compositions meet the requirements of an increased variety of mechanical, electrical, and chemical applications. Physical properties of filled granular compounds are shown in Table 4 (81).

Chemical Properties. Vacuum thermal degradation of PTFE results in monomer formation. The degradation is a first-order reaction (82). Mass spectroscopic analysis shows that degradation begins at ca 440°C, peaks at 540°C, and continues until 590°C (83).

FLUORINE COMPOUNDS, ORGANIC (POLYMERS)

Table 4. Properties of Filled PTFE Compounds[a]

Property	Unfilled	Glass fiber, wt %		Graphite, 15 wt %	Bronze, 60 wt %
		15	25		
specific gravity	2.18	2.21	2.24	2.16	3.74
tensile strength, MPa[b]	28	25	17.5	21	14
elongation, %	350	300	250	250	150
stress at 10% elongation, MPa[b]	11	8.5	8.5	11	14
thermal conductivity, mW/(m·K)	0.244	0.37	0.45	0.45	0.46
creep modulus, kN/m[c]	2	2.21	2.1	3.4	6.2
hardness, Shore durometer, D	51	54	57	61	70
Izod impact, J/m[d]	152	146	119		
PV,[e] (kPa·m)/s[f]	0.70	106	177	52	281
wear factor, 1/Pa[g]	5×10^{-14}	28×10^{-17}	26×10^{-17}	100×10^{-17}	12×10^{-17}
coefficient of friction					
static, 3.4 MPa[b] load	0.08	0.13	0.13	0.10	0.10
dynamic at					
PV = 172, (kPa·m)/s[f]		0.15–0.24	0.17	0.15	0.15
V = 900 m/s	0.01		−0.24	−0.18	−0.22

[a] Ref. 81.
[b] To convert MPa to psi, multiply by 145.
[c] To convert kN/m to lbf/in., divide by 0.175.
[d] To convert J/m to ft·lbf/in., divide by 53.38.
[e] PV = pressure × velocity. For 0.13-mm radial wear in 1000 h, unlubricated.
[f] To convert kPa to psi, multiply by 0.145.
[g] To convert 1/Pa to (in.^{3}·min)/(ft·lbf·h), divide by 2×10^{-7}.

Radiation Effects. Polytetrafluoroethylene is attacked by radiation. In the absence of oxygen, stable secondary radicals are produced. An increase in stiffness in material irradiated in vacuum indicates cross-linking (84). Degradation is due to random scission of the chain; the relative stability of the radicals in vacuum protects the materials from rapid deterioration. Reactions take place in air or oxygen and accelerated scission and rapid degradation occur.

Crystallinity has been studied by x-ray irradiation (85). An initial increase caused by chain scission in the amorphous phase was followed (above 3 kGy or 3×10^5 rad) by a gradual decrease associated with a disordering of the crystallites. The amorphous component showed a maximum of radiation-induced broadening in the nmr at 7 kGy (7×10^5 rad).

In air, PTFE has a damage threshold of 200–700 Gy ($2 \times 10^4 - 7 \times 10^4$ rad) and retains 50% of initial tensile strength after a dose of 10^4 Gy (1 Mrad), 40% of initial tensile strength after a dose of 10^5 Gy (10^7 rad), and ultimate elongation of 100% or more for doses up to 2–5 kGy (2×10^5–5×10^5 rad). During irradiation, resistivity decreases, whereas the dielectric constant and the dissipation factor increase. After irradiation, these properties tend to return to their preexposure values. Dielectric properties at high frequency are less sensitive to radiation than are properties at low frequency. Radiation has very little effect on dielectric strength (86).

Absorption, Permeation, and Interactions. Polytetrafluoroethylene is chemically inert to industrial chemicals and solvents even at elevated temperatures and pressures (87). This compatibility is due to the strong interatomic bonds, the almost perfect shielding of the carbon backbone by fluorine atoms, and the high molecular weight of the polymer. Under some severe conditions PTFE is not compatible with certain materials. It reacts with molten alkali metals, fluorine, strong fluorinating agents, and sodium hydroxide above 300°C. Shapes of small cross section burn vertically upward after ignition in 100% oxygen. Because gases may be evolved, the weight loss during sintering of a blend of PTFE and white asbestos is many times greater than loss from pure PTFE. Finely divided aluminum and magnesium thoroughly mixed with finely divided PTFE react vigorously after ignition or at high temperatures (87).

Absorption of a liquid is usually a matter of the liquid dissolving in the polymer; however, in the case of PTFE, no interaction occurs between the polymer and other substances. Submicroscopic voids between the polymer molecules provide space for the material absorbed; which is indicated by a slight weight increase and sometimes by discoloration. Common acids or bases are not absorbed up to 200°C. Aqueous solutions are scarcely absorbed at atmospheric pressure. Even the absorption of organic solvents is slight, partially resulting from the low wettability of PTFE. Since absorption of chemicals or solvents has no substantial effect on the chemical bond within the fluorocarbon molecule, absorption should not be confused with degradation; it is a reversible physical process. The polymer does not suffer loss of mechanical or bulk electrical properties unless subjected to severely fluctuating conditions (87).

Dynamic mechanical measurements were made on PTFE samples saturated with various halocarbons (88). The peaks in loss modulus associated with the amorphous relaxation near $-90°C$ and the crystalline relaxation near room temperature were not affected by these additives. An additional loss peak appeared near $-30°C$, and the modulus was reduced at all higher temperatures. The amorphous relaxation that appears as a peak in the loss compliance at 134°C is shifted to 45–70°C in the swollen samples.

The sorption behavior of perfluorocarbon polymers is typical for nonpolar partially crystalline polymers (89). The weight gain strongly depends on the solubility parameter. Little sorption of substances such as hydrocarbons and polar compounds occurs.

As an excellent barrier resin, PTFE is widely used in the chemical industry. However, it is a poor barrier for fluorocarbon oils because similarity in the chemical composition of a barrier and a permeant increases permeation. Most liquids and gases (other than fluorocarbons) do not permeate highly crystalline PTFE. Permeabilities at 30°C (in $mol/(m \cdot s \cdot Pa) \times 10^{15}$) are as follows: CO_2, 0.93; N_2, 0.18; He, 2.47; anhydrous HCl, < 0.01 (89).

Gases and vapors diffuse through PTFE more slowly than through most other polymers (Table 5). The higher the crystallinity, and the less space between polymer molecules, the slower the permeation. Voids greater than molecular size cause an increase in permeability. However, the permeability of the finished article can be controlled by molding the resin to low porosity and high density. The

Table 5. Permeability of PTFE Resin to Vapors

Permeant	Permeability constant,[a,b] mol/(m·s·Pa) × 10^{15}	
	23°C	30°C
benzene	1.81	2.93
carbon tetrachloride	0.13	
ethanol	1.88	
HCl, 20%	<0.71	
piperidine	0.96	
H_2SO_4, 98%	54.20	
water		20.70

[a] Ref. 87. Test method ASTM E96-35T (at vapor pressure; for 25.4 μm film thickness). Values are averages only and not for specification purposes.
[b] Original data converted to SI units using vapor pressure data from Ref. 90.

optimum specific gravity for low permeability and good flexural properties is 2.16–2.195. Permeability increases with temperature as a result of the increase in activity of the solvent molecules and because of the increase in vapor pressure of the liquids. Swelling of PTFE resins and film is very low.

Electrical Properties. Polytetrafluoroethylene is an excellent electrical insulator because of its mechanical strength and chemical and thermal stability as well as excellent electrical properties (Table 6). It does not absorb water and volume resistivity remains unchanged even after prolonged soaking. The dielectric constant remains constant at 2.1 for a temperature range of −40 to 250°C and a frequency range of 5 Hz to 10 GHz.

Articles fabricated according to standard practice should have dielectric constants in the range of 2.05 ± 0.5 when tested at RT. The dielectric constant varies with density and factors that affect density. Machined components can be fabricated to a predetermined dielectric constant by controlling the rod density during processing by adjusting the preforming pressure on the resin and cooling after

Table 6. Electrical Properties of Polytetrafluoroethylene[a]

Property	Granular	Fine powder	ASTM method
dielectric strength, short time, 2-mm thickness, V/mm	23,600	23,600	D149-55T
surface arc-resistance, s	>300	>300	D495-55T
volume resistivity, Ω·cm	>10^{18}	>10^{18}	D257-57T
surface resistivity at 100% rh, Ω/sq	>10^{16}		D257-57T
dielectric constant, at 60 to 2 × 10^9 Hz	2.1	2.1	D150-59T
dissipation factor, at 60 to 2 × 10^9 Hz	0.0003		D150-59T

[a] Ref. 77.

sintering. The dielectric constant and the density have a linear relationship. Predictable variations in the dielectric constant result from density changes that accompany thermal expansion occuring with increasing temperature. The dielectric constant did not change over two to three years of measurements.

The dissipation factor (the ratio of the energy dissipated to the energy stored per cycle) is affected by the frequency, temperature, crystallinity, and void content of the fabricated structure. At certain temperatures and frequencies, the crystalline and amorphous regions become resonant. Because of the molecular vibrations, applied electrical energy is lost by internal friction within the polymer which results in an increase in the dissipation factor. The dissipation factor peaks for these resins correspond to well-defined transitions, but the magnitude of the variation is minor as compared to other polymers. The low temperature transition at $-97°C$ causes the only meaningful dissipation factor peak. The dissipation factor has a maximum of 10^8-10^9 Hz at RT; at high crystallinity (93%) the peak at 10^8-10^9 Hz is absent.

As crystallinity increases, the internal molecular friction and the dissipation factor decrease. Voids reduce the dissipation factor in proportion to the percentage of microvoids present. Certain extruded shapes utilize air to reduce the effective dielectric constant and dissipation factor of a coaxial cable. The dielectric strength of these resins is high and is unaffected by thermal aging at 200°C. Frequency has a marked effect on the dielectric strength because corona discharge becomes more continuous as frequency increases. If the voltage stress is not high enough to cause corona ignition, a very long dielectric life is anticipated at any frequency. Corona discharges on the surface or in a void initiate dielectric breakdown (91). Surface arc resistance of these resins is high and not affected by heat aging. The resins do not track or form a carbonized conducting path when subjected to a surface arc in air. Polytetrafluoroethylene resins are capable of continuous service up to 260°C and can withstand much higher temperatures for limited periods of time. They do not melt or flow and retain some strength even in the gel state which begins at 327°C.

Fabrication

Granular Resins. These resins are sold in different forms; an optimum balance between handleability and product properties is desired. A free-flowing resin is used in small and automatic moldings. A finely divided resin is more difficult to handle but it distributes evenly in large moldings and has superior properties in sintered articles; it is used for large billet- and sheet-molding operations. A presintered resin with low crystallinity and superior handleability is highly suitable for ram extrusion.

Virgin PTFE melts at about 342°C; viscosity, even at 380°C, is 10 GPa·s (10^{11} P). This eliminates processing by normal thermoplastic techniques, and other fabrication techniques had to be developed: the dry powder is compressed into handleable form by heating above the melting point. This coalesces the particles into a strong homogeneous structure; cooling at a controlled rate achieves the desired degree of crystallinity.

Molding. Many PTFE manufacturers give detailed descriptions of molding equipment and procedures (92–98). Round piston molds for the production of solid or hollow cylinders are the most widely used. Because preforming usually takes place below 100°C, carbon steel is a suitable material of construction. The compression ratio (ie, the bulk volume of the powder to the specific volume of the unsintered molding) for granular resins is 3:1 to 6:1. For large-area, thin-walled moldings of unfilled polymer, a short-stroke press with a working capacity of 19.6–34.3 MPa (194–339 atm) mold pressure is sufficient; for tall moldings of filled compounds with a small cross-sectional area, a long-stroke press with a low thrust is required. The powder should be evenly distributed and leveled in the mold (92). To ensure adequate compression uniformly throughout the preform, maximum pressure should be maintained for a sufficient length of time, and then be released slowly.

Automatic molding permits high speed mass production; it is preferable to machining finished material. Automatic presses can be operated mechanically, pneumatically, or hydraulically. The mold is filled by means of a special metering system from a storage hopper containing a free-flowing resin. Loading buckets that shuttle back and forth over the single-cavity mold are also used. Because automatic molding requires short cycles, the powder is usually compressed at high speed with a high preform pressure. Small articles such as rings, bushings, washers, gaskets, and ball-valve seats can be molded by this technique.

Isostatic molding allows uniform compression from all directions. A flexible mold is filled with a free-flowing granular powder and evacuated, tightly sealed, and placed in an autoclave containing a liquid that can be raised to the pressure required for preforming. The moldings require subsequent finishing because close tolerance cannot be achieved.

Sintering. Electrical ovens with air circulation and service temperatures up to 400°C are satisfactory for sintering. In free sintering, the cheapest and most widely used process, a preformed mold is placed in an oven with a temperature variation of ± 2°C. In pressure sintering, the preform is not removed from the mold; instead the mold containing the preform is heated in an oven until the sintering temperature is reached. During sintering and cooling, the mold is again placed under pressure but lower than the preform pressure. Pressure-sintered products have internal stresses that can be relieved by subsequent annealing. In the pressure-cooling process, pressure is applied on the molded article after it has reached sintering temperature and is maintained throughout the cooling period. The final product has a lower void content than the free-sintered mold.

To improve homogeneity, the preformed article is heated to 370–390°C. The time required for heating and sintering depends on the mold dimensions; cooling, which affects the crystallinity and product properties, should be slow.

Free-sintered articles do not have the same dimensions as the mold cavity because they shrink at right angles to the direction of the preform pressure and grow in the direction of the applied pressure.

For processing after sintering, in the least expensive method for sintered PTFE tape or sheet, a large billet is skived on a lathe after it has been sintered and cooled. High precision articles are machined from ram-extruded rods.

Articles that are too complicated to be made by machining are made by coining. A sintered molding is heated to its melting point, transferred to a mold, and quickly deformed at low pressure, where it is held until it has cooled sufficiently to retain the improved shape. However, the coined molding, if reheated to a high temperature, returns to its original shape, and hence there is a limit on the maximum temperature to which coined moldings can be heated.

Ram Extrusion. Compression molding is not suitable for the manufacture of continuous long moldings such as pipes or rods. In ram extrusion, a small charge of PTFE powder is preformed by a reciprocating ram and sintered. Subsequent charges are fused into the first charge, and this process continues to form homogeneous long rods (92,99–101). The die tube, which is made of a corrosion-resistant material, is heated by resistance heating. Good temperature control is essential, and the melted and compacted powder must not pass any constrictions in its path. Thermal expansion and friction produce great resistance to movement, and as a result, a considerable force is required to push the polymer through the tube. A high quality surface finish on the inside of the tube reduces the pressure. If adequate bond strength between successive charges is not developed, the extrudate may break at the interface (poker chipping). Free-flowing powders and presintered resins are preferred for ram extrusion. Ram-extruded rods are used for automatic screw machining. Tubing is used as pipe liners or stock from which seals, gaskets, and bellows are machined.

Fine Powder Resins. Fine powder PTFE resins are extremely sensitive to shear. They must be handled gently to avoid shear, which prevents processing. However, fine powder is suitable for the manufacture of tubing and wire insulation for which compression molding is not suitable. A paste-extrusion process may be applied to the fabrication of tubes with diameters from fractions of a millimeter to about a meter, walls from thicknesses of 100–400 μm, thin rods with up to 50-mm diameters, and cable sheathing. Calendering unsintered extruded solid rods produces thread-sealant tape and gaskets.

The paste-extrusion process includes the incorporation of ca 16–25 wt % of the lubricant (usually a petroleum fraction); the mixture is rolled to obtain uniform lubricant distribution. This wetted powder is shaped into a preform at low pressure (2.0–7.8 MPa or 19–77 atm) which is pushed through a die mounted in the extruder at ambient temperature. The shear stress exerted on the powder during extrusion confers longitudinal strength to the polymer by fibrillation. The lubricant is evaporated and the extrudate is sintered at ca 380°C.

The exact amount of lubricant required for extrusion depends on the design of the extruder, the reduction ratio (ie, ratio of the cross-sectional preform area to the cross-sectional area in the die), and the quality of the lubricant. A low lubricant content results in a high extrusion pressure, whereas a high lubricant content causes a poor coalescence and generates defects in the extrudate.

Fine powder resins can be colored with pigments that can withstand the sintering temperature. The pigment should be thoroughly mixed with the powder by rolling the mixture before adding the lubricant. Detailed design parameters of the paste extruder are given in Reference 102–108.

The extrudate is dried and sintered by passing it through a multistage oven located immediately after the extruder. Pipes and rods may be heated up to 380°C. The throughput rate depends on the length of the sintering oven. Residence time varies from a few seconds for thin-walled insulations on a wire to a few minutes for large diameter tubing. For short residence times temperatures may be as high as ca 480°C. The extrusion pressure depends on the reduction ratio, the extrusion rate, the lubricant content, and the characteristics of the extruder.

To produce unsintered tape by paste extrusion, the fine powder is lubricated and preformed according to the procedure described above. The preform is extruded in the form of rods, which are calendered on hot rolls to the desired width and thickness (109,110).

Different resins have been developed for use in different reduction–ratio application ranges (111,112). The powders suitable for high reduction–ratio applications, such as wire coatings, are not necessarily suitable for the medium reduction–ratio applications, such as tubings, or the low reduction–ratio applications, such as thread-sealant tapes or pipe liners. Applications and processing techniques are being used, which utilize the unique combination of properties offered by PTFE in fine powder form (113–115).

Dispersion Resins. Polytetrafluoroethylene dispersions in aqueous medium contain 30–60 wt % polymer particles and some surfactant. The type of surfactant and the particle characteristics depend on the application. These dispersions are applied to various substrates by spraying, flow coating, dipping, coagulating, or electrodepositing.

Aqueous dispersion is sprayed on metal substrates to provide chemical resistance, nonstick, and low friction properties. The coated surface is dried and sintered. Impregnation of fibrous or porous materials with these dispersions combines the properties of the materials with those of PTFE. Some materials require only a single dipping, eg, asbestos. The material is usually dried after dipping. For high pressure sealing applications, sintering at 380–400°C increases strength and dimensional stability. For film castings, the dispersion is poured on a smooth surface; the formed film is dried and sintered and peeled from the supporting surface.

Aqueous dispersions are used for spinning PTFE fibers. The dispersion is mixed with a matrix-forming medium (116,117) and forced through a spinneret into a coagulating bath. The matrix material is removed by heating and the fibers are sintered and drawn molten to develop their full strength.

Effects of Fabrication on Physical Properties of Molded Parts. The physical properties are affected by molecular weight, void content, and crystallinity. Molecular weight can be reduced by degradation but not increased during processing. These factors can be controlled during molding by the choice of resin and fabricating conditions. Void distribution (or size and orientation) also affects properties; however, it is not easily measured.

Preforming primarily affects void content, sintering controls molecular weight, and cooling determines crystallinity. Voids caused by insufficient consolidation of particles during preforming may appear in the finished articles. Densities below 2.10 g/cm^3 indicate a high void content. Electrical and chemical applications require a minimum density of 2.12–2.14 g/cm^3. Particle size, shape, and porosity are also important in determining void content. Although void content is determined largely by particle characteristics and preforming conditions, sintering conditions can also have an effect. Temperatures too high or too low increase void content. Excessively high sintering temperature can decrease the molecular weight. The final crystallinity of a molding depends on the initial molecular weight of the polymer, the rate of cooling of the molding, and to a lesser extent on sintering conditions. The degree of crystallinity of moldings is affected by the cooling or annealing conditions.

Flexural modulus increases by a factor of five as crystallinity increases from 50 to 90% with a void content of 0.2%; however, recovery decreases with increasing crystallinity. Therefore, the balance between stiffness and recovery depends on the application requirements. Crystallinity is reduced by rapid cooling but increased by slow cooling. The stress–crack resistance of various PTFE insulations is correlated with the crystallinity and change in density due to thermal mechanical stress (118).

Applications

Consumption of PTFE increases continuously as new applications are being developed. Electrical applications consume half of the PTFE produced; mechanical and chemical applications share equally the other half. Various grades of PTFE and their applications are shown in Table 7.

Electrical Applications. The largest application of PTFE is for hookup and hookup-type wire used in electronic equipment in the military and aerospace industries. Coaxial cables, the second largest application, use tapes made from fine powder resins and some from granular resin. Interconnecting wire applications include airframes. Other electrical applications include computer wire, electrical tape, electrical components, and spaghetti tubing.

Mechanical Applications. Seals and piston rings, basic shapes, and antistick uses constitute two-thirds of the resin consumed in mechanical applications. Bearings, mechanical tapes, and coated glass fabrics also consume a large amount of PTFE resins. Seals and piston rings, bearings, and basic shapes are manufactured from granular resins, whereas the dispersion is used for glass–fabric coating and antistick applications. Most pressure-sensitive mechanical tapes are made from granular resins.

Chemical Applications. The chemical processing industry uses large amounts of granular and fine powder PTFE. Soft packing applications are manufactured from dispersions, and hard packings are molded or machined from stocks and shapes made from granular resin.

Overbraided hose liners are made from fine powder resins by paste extrusion, and thread-sealant tapes are produced from fine powder by calendering. Fabricated gaskets are made from granular resins and pipe liners are produced from fine powder resins. Fibers and filament forms are also available.

Highly porous fabric structures, eg, Gore-Tex, that can be used as membranes have been developed by exploiting the unique fibrillation capability of dispersion-polymerized PTFE (113).

Micropowders. The PTFE micropowders, also called waxes, are tetrafluoroethylene homopolymers with molecular weights significantly lower than that of normal PTFE. The molecular weight for micropowders varies from 2.5×10^4 to 25×10^4, whereas that of normal PTFE is of the order of 10×10^6. Micropowders are generally white in color and are friable. The average agglomerate particle size is between 5 to 10 μm and is composed of smaller, "as polymerized" primary particles which are approximately 0.2 μm in diameter. The dsc curves of

Table 7. Applications of Polytetrafluoroethylene Resins

Resin grade	Processing	Description	Main uses
Granular			
agglomerates	molding, preforming, sintering, ram extrusion	free-flowing powder	gaskets, packing seals, electronic components, bearings, sheet, rod, heavy-wall tubing; tape and molded shapes for nonadhesive applications
coarse	molding, preforming, sintering	granulated powder	tape, molded shapes, nonadhesive applications
finely divided	molding, preforming, sintering	powder for highest quality, void-free moldings	molded sheets, tape, wire wrapping, tubing, gaskets
presintered	ram extrusion	granular, free-flowing powder	rods and tubes
Fine powder			
high reduction ratio	paste extrusion	agglomerated powder	wire coating, thin-walled tubing
medium reduction ratio	paste extrusion	agglomerated powder	tubing, pipe, overbraided hose, spaghetti tubing
low reduction ratio	paste extrusion	agglomerated powder	thread-sealant tape, pipe liners, tubing, porous structures
Dispersion			
general-purpose	dip coating	aqueous dispersion	impregnation, coating, packing
coating	dip coating	aqueous dispersion	film coating
stabilized	coagulation	aqueous dispersion	bearings

lower molecular weight micropowder show a higher heat of crystallization and melting (second heating) than normal PTFE. This is due to the higher crystallinity of the micropowder.

The production of micropowders involves the scission of the high molecular weight PTFE chain by gamma or electron beam irradiation at a variety of dosage levels. An increase in dosage reduces the molecular weight. The irradiated low molecular weight material is ground to a particle size ranging from 1 to 25 μm in the final product.

Economic Aspects

Polytetrafluoroethylene homopolymers are more expensive than most other thermoplastics because of high monomer refining costs. For extremely high molecular weights, ingredients and manufacturing process must be free of impurities, which increases costs. In the United States, the 1992 list prices from primary producers were between 16.3 and 23.5 $/kg, depending on the resin type. For example, granular PTFE resins cost 16.3–18.0 $/kg supplied in 22.5-kg containers. The coagulated fine powders cost 19.10–22.10 $/kg packaged in 22.5-kg containers. Formulated dispersions are 20.00–23.5 $/kg in 19-L or 113-L containers. Although fine powder sales have increased in recent years, the sales of granular PTFE are the highest on a worldwide basis. Most of the resin is consumed in the United States (ca 9000 t in 1991), followed by Europe and Japan.

Testing and Standards

A description of PTFE resins and their classification are given in ASTM D1457-83. A comprehensive listing of industrial and military specifications covering mechanical, electrical, and chemical applications of PTFE can be found in Reference 119.

Health and Safety

Exposure to PTFE can arise from ingestion, skin contact, or inhalation. The polymer has no irritating effect to the skin, and test animals fed with the sintered polymer have not shown adverse reactions. Dust generated by grinding the resin also has no effect on test animals. Formation of toxic products is unlikely. Only the heated polymer is a source of a possible health hazard (120).

Because PTFE resins decompose slowly, they may be heated to a high temperature. The toxicity of the pyrolysis products warrants care where exposure of personnel is likely to occur (120). Above 230°C decomposition rates become measurable (0.0001% per hour). Small amounts of toxic perfluoroisobutylene have been isolated at 400°C and above; free fluorine has never been found. Above 690°C the decomposition products burn but do not support combustion if the heat is removed. Combustion products consist primarily of carbon dioxide, carbon tetrafluoride, and small quantities of toxic and corrosive hydrogen fluoride. The PTFE resins are nonflammable and do not propagate flame.

Prolonged exposure to thermal decomposition products causes so-called polymer fume fever, a temporary influenza-like condition. It may be contracted by smoking tobacco that has been contaminated with the polymer. It occurs several hours after exposure and passes within 36–48 hours; the temporary effects are not cumulative.

Large quantities of PTFE resins have been manufactured and processed above 370°C. In various applications they are heated above the recommended use temperatures. No cases of serious injury, prolonged illness, or death have been reported resulting from the handling of these resins. However, when high molecular weight PTFE is converted to micropowder by thermal degradation, highly toxic products result.

Micropowders are added to a wide variety of material used in industry, where they provide nonstick and sliding properties. They are incorporated into the product by blending and grinding. To disperse well, the powder must have good flow properties. Conditions that make the powder sticky should be avoided.

The PTFE micropowders are commonly used in plastics, inks, lubricants, and finishes such as lacquer. Lubricants containing micropowders are used for bearings, valve components, and other moving parts where sliding friction must be minimized or eliminated. Nonstick finishes that require good release properties, for example, in the food and packaging industry, commonly use PTFE micropowders.

In some applications the high heat stability of the micropowder can be utilized over a reasonably wide temperature range. A maximum service temperature is normally 260°C, provided the crystalline melting point is between 320 and 335°C. Exposure above 300°C leads to degradation and possible evolution of toxic decomposition products.

The particulate morphology of PTFE micropowder in printing inks provides desirable gloss to the printed product. Its inherent lubricity results in good wear and slip properties and surface smoothness. The chemical resistance of the micropowder is as high as that of high molecular weight PTFE. It is therefore used in applications requiring service in strong or corrosive chemical environments such as concentrated mineral acids and alkalies.

BIBLIOGRAPHY

"Tetrafluorethylene Resins" in *ECT* 1st ed., Vol. 11, pp. 687–691, by B. E. Ely, E.I. du Pont de Nemours & Co., Inc.,; "Polytetrafluoroethylene" under "Fluorine Compounds, Organic" in *ECT* 2nd ed., Vol. 9, pp. 805–831, by S. Sherratt, Imperial Chemical Industries, Ltd., Plastics Division; in *ECT* 3rd ed., Vol. 11, pp. 1–24, by S. V. Gangal, E.I. du Pont de Nemours & Co., Inc.

1. U. S. Pat. 2,230,654 (Feb. 4, 1941), R. J. Plunkett (to Kinetic Chemicals, Inc.).
2. J. D. Park and co-workers, *Ind. Eng. Chem.* **39**, 354 (1947).
3. J. M. Hamilton, in M. Stacey, J. C. Tatlow, and A. G. Sharpe, eds., *Advances in Fluorine Chemistry*, Vol. 3, Butterworth & Co., Ltd., Kent, U.K., 1963, p. 117.
4. J. W. Edwards and P. A. Small, *Nature* **202**, 1329 (1964); J. W. Edwards and P. A. Small, *Ind. Eng. Chem. Fundam.* **4**, 396 (1965).
5. F. Gozzo and C. R. Patrick, *Nature* **202**, 80 (1964).
6. Jpn. Pat. 60 15,353 (Oct. 14, 1960), M. Hisazumi and H. Shingu.
7. U.S. Pat. 2,994,723 (Aug. 1, 1961), O. Scherer and co-workers (to Farbewerke Hoechst).
8. Brit. Pat. 960,309 (June 10, 1964), J. W. Edwards, S. Sherratt, and P. A. Small (To ICI).
9. U.S. Pat. 3,459,818 (Aug. 5, 1969), H. Ukahashi and M. Hisasne (to Asahi Glass Co.).
10. U.S. Pat. 2,407,405 (Sept. 10, 1946), M. A. Dietrich and R. M. Joyce (to E.I. du Pont de Nemours & Co., Inc.).
11. O. Ruff and O. Bretschneider, *Z. Anorg. Allg. Chem.* **210**, 173 (1933).
12. E. G. Locke, W. R. Brode, and A. L. Henne, *J. Am. Chem. Soc.* **56**, 1726 (1934).

13. O. Ruff and W. Willenberg, *Chem. Ber.* **73**, 724 (1940).
14. L. T. Hals, T. S. Reid, and G. H. Smith, *J. Am. Chem. Soc.* **73**, 4054 (1951); U.S. Pat. 2,668,864 (Feb. 9, 1954), (to Minnesota Mining and Manufacturing Co.).
15. U.S. Pat. 3,009,966 (Nov. 21, 1961), M. Hauptschein and A. H. Fainberg (to Pennsalt Chemical Corp.).
16. U.S. Pat. 3,471,546 (Oct. 7, 1969), G. Bjornson (to Phillips Petroleum Co.).
17. U.S. Pat. 3,662,009 (May 9, 1972), W. M. Hutchinson (to Phillips Petroleum Co.).
18. U.S. Pat. 3,799,996 (Mar. 26, 1974), H. S. Bloch (to Universal Oil Products).
19. E. E. Lewis and M. A. Naylor, *J. Am. Chem. Soc.* **69**, 1968 (1947).
20. U.S. Pat. 3,832,411 (Aug. 27, 1974), B. C. Arkles and R. N. Bonnett (to Liquid Nitrogen Processing Co.).
21. M. M. Renfrew and E. E. Lewis, *Ind. Eng. Chem.* **38**, 870 (1946).
22. H. C. Duus, *Ind. Eng. Chem.* **47**, 1445 (1955).
23. W. M. D. Bryant, *J. Polym. Sci.* **56**, 277 (1962).
24. A. Pajaczkowski and J. W. Spoors, *Chem. Ind.* **16**, 659 (1964).
25. Brit. Pat. 931,587 (July 17, 1963), H. H. Gibbs and J. L. Warnell (to E.I. du Pont de Nemours & Co., Inc.).
26. U.S. Pat. 3,159,609 (Dec. 1, 1964), J. F. Harris, Jr., and D. I. McCane (E.I. du Pont de Nemours & Co., Inc.).
27. F. S. Toby and S. Toby, *J. Phys. Chem.* **80**, 2313 (1976).
28. B. Atkinson and V. A. Atkinson, *J. Chem. Soc. Part II*, 2086 (1957).
29. U.S. Pat. 2,946,763 (July 26, 1960), M. I. Bro and B. W. Sandt (to E.I. du Pont de Nemours & Co., Inc.).
30. U.S. Pat. 3,847,881 (Nov. 12, 1974), M. Mueller and S. Chandrasekaran (to Allied Chemicals Co.).
31. U.S. Pat. 3,528,954 (Sept. 15, 1970), D. P. Carlson (to E.I. du Pont de Nemours & Co., Inc.).
32. U.S. Pat. 3,475,391 (Oct. 28, 1969), J. N. Coker (to E.I. du Pont de Nemours & Co., Inc.).
33. U.S. Pat. 3,846,267 (Nov. 5, 1974), Y. Tabata and G. Kojima (to Japan Atomic Energy Research Institute).
34. U.S. Pat. 3,467,636 (Sept. 16, 1969), A. Nersasian (to E.I. du Pont de Nemours & Co., Inc.).
35. U.S. Pat. 3,403,191 (Sept. 24, 1968), D. P. Graham (to E.I. du Pont de Nemours & Co., Inc.).
36. U.S. Pat. 3,404,180 (Oct. 1, 1969), K. L. Cordes (to E.I. du Pont de Nemours & Co., Inc.).
37. U.S. Pat. 3,567,521 (Mar. 2, 1971), M. S. Toy and N. A. Tiner (to McDonnell Douglas).
38. U.S. Pat. 3,446,858 (May 27, 1969), H. Shingu and co-workers (to Daikin Kogyo Co.).
39. U.S. Pat. 3,873,630 (Mar. 25, 1975), N. E. West (to E.I. du Pont de Nemours & Co., Inc.).
40. U.S. Pat. 3,855,191 (Dec. 17, 1974), T. R. Doughty, C. A. Sperati, and H. Un (to E.I. du Pont de Memours & Co., Inc.).
41. U.S. Pat. 3,655,611 (Apr. 11, 1972), M. B. Mueller, P. O. Salatiello, and H. S. Kaufman (to Allied Chemicals Co.).
42. U.S. Pat 4,189,551 (Feb. 19, 1980), S. V. Gangal (to E.I. du Pont de Nemours & Co., Inc.).
43. U.S. Pat. 3,419,522 (Dec. 31, 1968), P. N. Plimmer (to E.I. du Pont de Nemours & Co., Inc.).
44. U.S. Pat. 3,766,133 (Oct. 16, 1973) R. Roberts and R. F. Anderson (to E.I. du Pont de Nemours & Co., Inc.).

45. U.S. Pat. 2,612,484 (Sept. 30, 1952), S. G. Bankoff (to E.I. du Pont de Nemours & Co., Inc.).
46. U.S. Pat 4,186,121 (Jan. 29, 1980), S. V. Gangal (to E.I. du Pont de Nemours & Co., Inc.).
47. U.S. Pat. 4,725,644 (1988), S. Malhotra (to E.I. du Pont de Nemours & Co., Inc.).
48. T. Folda and co-workers, *Nature* **333**, 55 (1988).
49. B. Luhmann and A. E. Feiring, *Polymer* **30**, 1723 (1989).
50. B. Chu, C. Wu, and W. Buck, *Macromolecules* **22**, 831 (1989).
51. U.S. Pat. 4,576,869 (Mar. 18, 1986), S. C. Malhotra (to E.I. du Pont de Nemours & Co., Inc.).
52. U.S. Pat. 4,363,900 (Dec. 14, 1982), T. Shimizu and S. Koizumi (to Daikin Kogyo Co.).
53. U.S. Pat. 4,766,188 (Aug. 23, 1988), T. E. Attwood and R. F. Bridges (to ICI).
54. U.S. Pat. 4,036,802 (July 19, 1977), R. V. Poirier (to E.I. du Pont de Nemours & Co., Inc.).
55. U.S. Pat. 4,129,618 (Dec. 12, 1978), J. M. Downer, W. G. Rodway, and L. S. J. Shipp (to ICI).
56. U.S. Pat. 4,840,998 (June 6, 1989), T. Shimizu and K. Hosokawa (to Daikin Kogyo Co.).
57. U.S. Pat. 4,879,362 (Nov. 7, 1979), R. A. Morgan (to E.I. du Pont de Nemours & Co., Inc.).
58. U.S. Pat. 4,342,675 (Aug. 3, 1982), S. V. Gangal (to E.I. du Pont de Nemours & Co., Inc.).
59. U.S. Pat. 2,478,229 (Aug. 9, 1949), K. L. Berry (to E.I. du Pont de Nemours & Co., Inc.).
60. K. L. Berry and J. H. Peterson, *J. Am. Chem. Soc.* **73**, 5195 (1951).
61. R. C. Doban and co-workers, paper presented at *130th Meeting of the American Chemical Society*, Atlantic City, N. J., Sept. 1956.
62. C. A. Sperati and H. W. Starkweather, *Fortschr. Hochpolym. Forsch.* **2**, 465 (1961).
63. R. E. Moynihan, *J. Am. Chem. Soc.* **81**, 1045 (1959).
64. T. Suwa, M. Takehisa, and S. Machi, *J. Appl. Polym. Sci.* **17**, 3253 (1973).
65. P. L. McGeer and H. C. Duus, *J. Chem. Phys.* **20**, 1813 (1952).
66. C. W. Bunn, *J. Polym. Sci.* **16**, 332 (1955).
67. H. D. Chanzy, P. Smith, and J. Revol, *J. of Polym. Sci. Polym. Lett. Ed.* **24**, 557 (1986).
68. H. W. Starkweather, Jr., *J. Polym. Sci. Polym. Phys. Ed.* **17**, 73–79 (1979).
69. R. H. H. Pierce and co-workers, in Ref. 61.
70. E. S. Clark and L. T. Muus, paper presented at *133rd Meeting of the American Chemical Society*, New York, Sept. 1957.
71. E. S. Clark, paper presented at *Symposium on Helices in Macromolecular Systems*, Polytechnic Institute of Brooklyn, Brooklyn, N.Y., May 16, 1959.
72. C. A. Sperati, in J. Brandrup and E. H. Immergut, eds., *Polymer Handbook*, 2nd ed., John Wiley & Sons, Inc., New York, 1975 pp. V-29–36.
73. Y. Araki, *J. Appl. Polym. Sci.* **9**, 3585 (1965).
74. N. G. McCrum, *J. Polym. Sci.* **34**, 355 (1959).
75. H. W. Starkweather, Jr., *Macromolecules* **19**, 2541 (1986).
76. J. T. Milek, *A Survey Materials Report on PTFE Plastics*, AD 607798, U.S. Dept. of Commerce, Washington, D.C., Sept. 1964.
77. *Mechanical Design Data, Teflon Fluorocarbon Resins*, bulletin, E.I. du Pont de Nemours & Co., Inc. Wilmington, Del. Sept. 1964.
78. M. K. Bernett and W. A. Zisman, *J. Phys. Chem.* **63**, 1911 (1959).
79. U.S. Pat. 2,871,144 (Jan. 27, 1959), R. C. Doban (to E.I. du Pont de Nemours & Co., Inc.).
80. A. J. G. Allan and R. Roberts, *J. Polym. Sci.* **39**, 1 (1959).

81. *J. Teflon (Du Pont)* **13**(2), 3 (1972).
82. J. C. Siegle and co-workers, *J. Polym. Sci.* **Part A2**, 391 (1964).
83. G. P. Shulman, *Polym. Lett.* **3**, 911 (1965).
84. L. A. Wall and R. E. Florin, *J. Appl. Polym. Sci.* **2**, 251 (1959).
85. W. M. Peffley, V. R. Honnold, and D. Binder, *J. Polym. Sci.* **4**, 977 (1966).
86. *J. Teflon (Du Pont)* **10**(1), (Jan.–Feb. 1969).
87. *J. Teflon (Du Pont)* **11**(1), (Jan.–Feb. 1970).
88. H. W. Starkweather, Jr., *Macromolecules* **17**, 1178 (1984).
89. *Ibid.*, **10**, 1161 (1977).
90. D. W. Green, ed., *Perry's Chemical Engineers' Handbook*, 6th ed., McGraw-Hill Book Co., Inc., New York, 1984.
91. J. C. Reed, E. J. McMahon, and J. R. Perkins, *Insulation (Libertyville, Ill.)* **10**, 35 (1964).
92. *Hostaflon*, TF product information booklet, American Hoechst Corp., Somerville, N.J., 1970.
93. *Soreflon*, products information booklet, Ugine Kuhlmann, France.
94. *Teflon TFE-Fluorocarbon Resins Molding Techniques*, 2nd ed., bulletin, E.I. du Pont de Nemours & Co., Inc., Wilmington, Del., 1966.
95. *The Moulding of Granular Polymers*, technical service note, ICI, Wilmington, Del., 1966.
96. J. A. Ross, *High Speed Molding of Teflon Tetrafluoroethylene Resins*, technical release, E.I. du Pont de Nemours & Co., Inc., Wilmington, Del., 1963.
97. *Automatic Molding with "Halon" TFE*, Allied Chemical Corp., Morristown, N.J., 1967.
98. *Isostatic Molding of Teflon TFE-Fluorocarbon Resins*, preliminary information bulletin, E.I. du Pont de Nemours & Co., Wilmington, Del., 1969.
99. R. J. Dahlen, *The Ram Extrusion of Teflon*, technical release, E.I. du Pont de Nemours & Co., Inc., Wilmington, Del., 1969.
100. *The Granular Extrusion of "Halon" TFE*, Allied Chemical Corp., Morristown, N.J., 1968.
101. *The Extrusion of Granular Polymers*, 2nd ed., technical service note, ICI, Wilmington, Del., 1966.
102. V. Adamec, *Nature* **200**, 1196 (1963).
103. Technical brochure, *Teflon 62 Hose and Tubing*, H-11959, E. I. du Pont de Nemours & Co., Inc., Wilmington, Del., Feb. 1991.
104. J. F. Lontz and co-workers, *Ind. Eng. Chem.* **44**, 1805 (1952).
105. *TFE-Fluorocarbon Resins Paste Extrusion of Wire Insulations*, bulletin, E.I. du Pont de Nemours & Co., Inc., Wilmington, Del., 1961.
106. *The Extrusion Coating of Wire*, 2nd ed., technical service note, Wilmington, Del., 1969.
107. *Teflon TFE-Fluorocarbon Resins Extrusion of Thin-Walled Tubing*, bulletin, E.I. du Pont de Nemours & Co., Inc., Wilmington, Del., 1957.
108. *The Extrusion of Thin Sections*, 3rd ed., technical service note, ICI, Wilmington, Del., 1969.
109. R. C. Ribbans, *Unsintered Tape Manufactured Calendering Round Rods*, technical release, E.I. du Pont de Nemours & Co., Inc., Wilmington, Del., 1966.
110. *The Manufacture of Unsintered Tape*, 2nd ed., technical data, ICI, Wilmington, Del., 1966.
111. U.S. Pat. 3,142,665 (July 28, 1964), A. J. Cardinal, W. L. Edens, and J. W. Van Dyk (to E.I. du Pont de Nemours & Co., Inc.).
112. U.S. Pat. 4,038,231 (July 26, 1977), J. M. Douner, W. G. Rodway, and L. S. J. Shipp (to ICI).
113. U.S. Pat. 3,962,153 (June 8, 1976), R. W. Gore (to W. L. Gore and Assoc.).
114. U.S. Pat. 3,993,584 (Nov. 23, 1976), J. E. Owen and J. W. Vogt (to Kewanee Oil Co.).

115. U.S. Pat. 3,704,171 (Nov. 28, 1972), H. P. Landi (to American Cyanamid Co.).
116. U.S. Pat. 3,051,545 (Aug. 28, 1962), W. Steuber (to E.I. du Pont de Nemours & Co., Inc.).
117. P. E. Frankenburg, in *Ullmann's Encyclopedia of Industrial Chemistry*, Vol. A-10, 5th ed., VCH Publishing, Inc., New York, 1987, pp. 649–650.
118. R. L. Baillie, J. J. Bednarczyk, and P. M. Mehta, paper presented at *35th International Wire and Cable Symposium*, Nov. 18–20, 1986.
119. *J. Teflon (Du Pont)* **8**, 6 (Nov. 1967).
120. *Teflon Occupational Health Bull.* **17**(2), (1962) (published by Information Service Division, Dept. of National Health and Welfare, Ottawa, Canada).

Subhash V. Gangal
E.I. du Pont de Nemours & Co., Inc.

PERFLUORINATED ETHYLENE–PROPYLENE COPOLYMERS

Perfluorinated ethylene–propylene (FEP) resin [*25067-11-2*] is a copolymer of tetrafluoroethylene [*116-14-3*] (TFE) and hexafluoropropylene [*116-15-4*] (HFP); thus its branched structure contains units of —CF_2—CF_2— and units of —CF_2—$CF(CF_3)$—. It retains most of the desirable characteristics of polytetrafluoroethylene (PTFE) but with a melt viscosity low enough for conventional melt processing. The introduction of hexafluoropropylene lowers the melting point of PTFE from 325°C to about 260°C.

The desire for a resin with polytetrafluoroethylene properties yet capable of being fabricated by conventional melt processing led to the discovery of this product (1). It allows melt extrusion of wire insulations of longer continuous lengths than the batchwise paste extrusion of PTFE as well as the injection molding of intricately shaped parts. The FEP polymer is melt-fabricable without severe sacrifice in mechanical properties because the perfluoromethyl side groups on the main polymer chain reduce crystallinity, which varies between 30 and 45%. This change in the crystallinity causes FEP and other copolymer particles to behave differently form PTFE particles; they do not fibrillate like PTFE particles and therefore do not agglomerate easily.

As a true thermoplastic, FEP copolymer can be melt-processed by extrusion and compression, injection, and blow molding. Films can be heat-bonded and sealed, vacuum-formed, and laminated to various substrates. Chemical inertness and corrosion resistance make FEP highly suitable for chemical services; its dielectric and insulating properties favor it for electrical and electronic service; and its low frictional properties, mechanical toughness, thermal stability, and nonstick quality make it highly suitable for bearings and seals, high temperature components, and nonstick surfaces.

Mechanical properties are retained up to 200°C, even in continuous service, which is better than with most plastics. At high temperatures, these copolymers react with fluorine, fluorinating agents, and molten alkali metals. They are commercially available under the Du Pont trademark Teflon FEP fluorocarbon resin. A similar product is manufactured by Daikin Kogyo of Japan and sold under the trademark Neoflon. The People's Republic of China also manufactures some FEP products.

Monomers

Preparation. The preparation, properties, and uses of tetrafluoroethylene have been described (see FLUORINE COMPOUNDS, ORGANIC–POLYTETRAFLUORO-ETHYLENE).

Hexafluoropropylene (HFP) was initially prepared by pyrolysis of PTFE (2,3) and by fluorination of 1,2,3-trichloropropane followed by dehalogenation (4). A number of other routes are described in the patent literature (5–10). Hexafluoropropylene can be prepared in high yield by thermally cracking TFE at reduced pressure at 700–800°C (11,12). Pyrolysis of PTFE at 860°C under vacuum gives a 58% yield of HFP (13). Fluorination of 3-chloropentafluoro-1-propene [79-47-0] at 200°C over activated carbon catalyst yields HFP (14). Decomposition of fluoroform [75-46-7] at 800–1000°C in a platinum-lined nickel tube is another route (15). The thermal decomposition of sodium heptafluorobutyrate [2218-84-4], $CF_3CF_2CF_2CO_2Na$ (16), and copyrolyses of fluoroform and chlorotrifluoroethylene [79-38-9] (17), and chlorodifluoromethane [75-45-6] and 1-chloro-1,2,2,2-tetrafluoroethane [2837-89-0] (18) give good yields of HFP.

Properties and Reactions. The properties of HFP are shown in Table 1. It does not homopolymerize easily and hence can be stored as a liquid. It undergoes many addition reactions typical of an olefin. Reactions include preparation of linear dimers and trimers and cyclic dimers (21,22); decomposition at 600°C with

Table 1. Properties of Hexafluoropropylene[a]

Property	Value
molecular weight	150.021
boiling point at 101 kPa,[b] °C	−29.4
freezing point, °C	−156.2
critical temperature, °C	85
critical pressure, kPa[b]	3254
critical density, g/cm^3	0.60
vapor pressure at K, kPa[b]	
243.75 < T < 358.15	$\log P(kPa) = 6.6938 - 1139.156/T$
liquid density, g/cm^3	
60°C	1.105
20°C	1.332
0°C	1.419
−20°C	1.498
heat of formation for ideal gas at 25°C, ΔH, kJ/mol[c,d]	−1078.6
flammability limits in air at 101 kPa[b]	nonflammable for all mixtures of air and hexafluoropropylene
heat of combustion, kJ/mol[c,d]	879
toxicity, LC_{50} (rat), 4 h, ppm[e]	3000

[a]Ref. 4.
[b]To convert kPa to mm Hg, multiply by 7.5.
[c]To convert kJ to kcal, divide by 4.184.
[d]Ref. 19.
[e]Ref. 20.

subsequent formation of octafluoro-2-butene and octafluoroisobutylene (23); oxidation with formation of an epoxide (24), an intermediate for a number of perfluoroalkyl perfluorovinyl ethers (25,26); and homopolymerization to low molecular weight liquids (27,28) and high molecular weight solids (29,30). Hexafluoropropylene reacts with hydrogen (31), alcohols (32), ammonia (33), and the halogens and their acids, except I_2 and HI (31,34–36). It is used as a comonomer to produce elastomers and other copolymers (37–41). The toxicological properties are discussed in Reference 42.

Copolymers

Hexafluoropropylene and tetrafluoroethylene are copolymerized, with trichloracetyl peroxide as the catalyst, at low temperature (43). Newer catalytic methods, including irradiation, achieve copolymerization at different temperatures (44,45). Aqueous and nonaqueous dispersion polymerizations appear to be the most convenient routes to commercial production (1,46–50). The polymerization conditions are similar to those of TFE homopolymer dispersion polymerization. The copolymer of HFP–TFE is a random copolymer; that is, HFP units add to the growing chains at random intervals. The optimal composition of the copolymer requires that the mechanical properties are retained in the usable range and that the melt viscosity is low enough for easy melt processing.

Hexafluoropropylene–tetrafluoroethylene copolymers are available in low melt viscosity, extrusion grade, intermediate viscosity, high melt viscosity, and as dispersions. The low melt viscosity (MV) resin can be injection molded by conventional thermoplastic molding techniques. It is more suitable for injection molding than other FEP resins (51).

The extrusion grade is suitable for tubing, wire coating, and cable jacketing. It is less suitable for injection molding than the low MV resin because of its relatively high melt viscosity. The intermediate MV (Teflon FEP-140) resin is used for insulation of wires larger than AWG 12 (American wire gauge) and applications involving smaller wire sizes, where high current loads or excessive thermal cycling may occur. It is also ideal for jacketing wire braid construction, such as coaxial cables, and for heater cable jackets.

The high MV resin is used as liners for process equipment. Its melt viscosity is significantly higher than that of other resins and therefore it is unsuitable for conventional injection molding. Stress-crack resistance and mechanical properties are superior to those of the other three products (52) (Table 2).

Both high and low color concentrates are available for pigmenting extruded coatings of FEP resins. The concentrates are prepared for melt dispersion in extrusion applications. The pigments are purified, thermally stable, and carefully selected to meet electrical, mechanical, and thermal end use specifications. Color concentrate pellets are easily dispersed among clear pellets by conventional tumbling. The ratio of concentrate to natural resin varies, depending on the wire size, insulation thickness, and color intensity desired.

An FEP copolymer dispersion is available as a 55-wt % aqueous dispersion containing 6% nonionic surfactant (on a solids basis) and a small amount of anionic dispersing agent. Its average particle size is ca 0.2 μm.

Table 2. Properties of Teflon FEP Fluorocarbon Resin[a]

Mechanical property	ASTM method	Teflon 110	Teflon 100	Teflon 140	Teflon 160
melt flow number, g/10 min	D2116		7.0	3.0	1.5
specific gravity	D792	2.13–2.17	2.13–2.17	2.13–2.17	2.13–2.17
tensile strength,[b] MPa[c]	D1708	20	23	30	31
elongation,[b] %	D1708	300	325	325	305
compressive strength, MPa[c]	D695		21	21	23
flexural strength,[b] MPa[c]	D790		18	18	18
impact strength,[b] J/m[d]	D256		no break	no break	no break
flexural modulus,[b] MPa[c]	D790	655	620	620	586
hardness durometer, Shore D	D2240	55	56	56	57
coefficient of friction, metal–film	D1894		0.27	0.27	0.235
deformation under load,[e] %	D621	1.8	0.5	0.5	0.5
water absorption, 24 h, %	D570	<0.01	0.004	0.004	0.004
linear coefficient of expansion per °C $\times 10^{-5}$	E381				
0–100°C			13.5	13.9	7.6
100–150°C			20.8	21.2	11.5
150–200°C			26.6	27.0	14.2

[a]Compression-molded specimens; property data on extruded wire specimens are similar.
[b]At 23°C
[c]To convert MPa to psi, multiply by 145.
[d]To convert J/m to ft·lbf/in., divide by 53.38.
[e]At 23°C, 6.9 MPa,[c] 23 h.

Properties. The crystallinity of FEP polymer is significantly lower than that of PTFE (70 vs 98%). The structure resembles that of PTFE, except for a random replacement of a fluorine atom by a perfluoromethyl group (CF_3). The crystallinity after processing depends on the rate of cooling the molten polymer. The presence of HFP in the polymer chain tends to distort the highly crystallized structure of the PTFE chain and results in a higher amorphous fraction.

In the free-radical polymerization of FEP copolymers, chain termination occurs by binary coupling of chain ends, thus contributing to high molecular weights. Linear viscoelastic properties of these polymers in the amorphous melts were measured by dynamic rheometry. The FEP samples had high molecular weights and were found to verify the relation of zero shear viscosity vs $(\text{mol wt})^3$ predicted by the reptation theory. At lower molecular weights, the empirical relation of viscosity vs $(\text{mol wt})^{3.4}$ holds (53).

Transitions and Relaxations. Only one first-order transition is observed, the melting point. Increasing the pressure raises mp. At low pressure, the rate of increase in the melting point is ca 1.74°C/MPa (0.012°C/psi); at high pressures this rate decreases to ca 0.725°C/MPa (0.005°C/psi). Melting increases the volume by 8%. In the presence of the HFP comonomer, crystal distortion occurs with an increase in intramolecular distance that, in turn, reduces the melting point (54).

The relaxation temperature appears to increase with increasing HFP content. Relaxation involves 5–13 of the chain carbon atoms. Besides α and γ relaxations, one other dielectric relaxation was observed below −150°C, which did not vary in temperature or in magnitude with comonomer content or copolymer density (55). The α relaxation (also called Glass I) is a high temperature transition (157°C) and γ relaxation (Glass II) (internal friction maxima) occurs between −5 and 29°C.

Thermal Stability. The polymer is thermally stable and can be processed at ca 270°C. Thermal degradation is a function of temperature and time, and the stability is therefore limited. The melt-flow rate (thermal degradation) increases significantly for short periods above 280°C, and degradation occurs at lower temperatures with longer hold times. The hourly weight loss is 0.0004% at 230°C, 0.001% at 260°C, 0.01% at 290°C, 0.02% at 320°C, 0.08% at 340°C, and 0.3% at 370°C. Degradation is not significant if the change in melt-flow rate during molding is < 10%. Physical strength decreases after prolonged exposure above 205°C, which accounts for the lower temperature rating of FEP resins (56).

Radiation Effects. The primary effect of radiation is the degradation of large molecules to small molecules. Molecular weight reduction can be minimized by excluding oxygen. If FEP is lightly irradiated at elevated temperatures in the absence of oxygen, cross-linking offsets molecular breakdown (55,57).

The degree to which radiation exposure affects FEP resins is determined by the energy absorbed, regardless of the type of radiation. Changes in mechanical properties depend on total dosage, but are independent of dose rate. The radiation tolerance of FEP in the presence or absence of oxygen is higher than that of PTFE by a factor of 10:1.

Mechanical Properties. Extensive lists of the physical properties of FEP copolymers are given in References 58–63. Mechanical properties are shown in Table 3. Most of the important properties of FEP are similar to those of PTFE; the main difference is the lower continuous service temperature of 204°C of FEP compared to that of 260°C of PTFE. The flexibility at low temperatures and the low coefficients of friction and stability at high temperatures are relatively independent of fabrication conditions. Unlike PTFE, FEP resins do not exhibit a marked change in volume at room temperature, because they do not have a first-order transition at 19°C. They are useful above −267°C and are highly flexible above −79°C (64).

Static friction decreases with an increase in load, and the static coefficient of friction is lower than the dynamic coefficient. The tendency to creep must be considered carefully in FEP products designed for service under continuous stresses. Creep can be minimized by suitable fillers. Fillers are also used to improve wear resistance and stiffness. Compositions such as 30% bronze-filled FEP, 20% graphite-filled FEP, and 10% glass-fiber-filled FEP offer high PV values (~400(kPa·m)/s) and are suitable for bearings.

Articles fabricated from FEP resins can be made bondable by surface treatment with a solution of sodium in liquid ammonia, or naphthalenyl sodium in tetrahydrofuran (64) to facilitate subsequent wetting. Exposing the surface to corona discharge (65) or amines at elevated temperatures in an oxidizing atmosphere (66) also makes the resins bondable. Some of the more recent work is described in References 67–69.

Table 3. Mechanical Properties of FEP[a]

Property	Value	ASTM method
specific gravity	2.14–2.17	D792-50
thermal conductivity, W/(m·K)		Cenco-Fitch
−129 to 182°C	2.4	
−253°C	1.4	
water absorption in 24 h, 3.175-mm thick sample		D570-547
% wt increase	<0.1	
dimensional change at 23°C	none	
coefficient of thermal expansion per °C		D696-44
> 23°C	9.3×10^{-5}	
< 23°C	5.7×10^{-5}	
specific heat, kJ/(kg·K)[b]		
20°C	1.09	
100°C	1.17	
260°C	1.30	
heat distortion, °C		D648-56
455 kPa[c]	70	
1820 kPa[c]	51	
tensile yield strength, av, MPa[d]		D638-527
−251°C	165	
−73°C	62	
23°C	12	
121°C	3.5	
tensile modulus, MPa[d]		
−251°C	57	
−73°C	24	
23°C	4	
100°C	1	
tensile elongation, %		D638-527
−251°C	4	
−73°C	200	
23°C	350	
flexural modulus, MPa[d]		D747-50
−251°C	5300	
−101°C	3200	
23°C	660	
55°C	340	
compressive strength, MPa[d]		D695
−251°C	251	
23°C	15	
100°C	3.4	
Izod impact strength, notched, J/m[e]		D256-56
23°C	no break	
hardness, Durometer		D2240-T
23°C	D59	
Taber abrasion, g/MHz, 100-g load		
CS-17 wheel	7.5	

[a] Measured on Teflon FEP T-100. [b] To convert kJ to kcal, divide by 4.184.
[c] To convert kPa to atm, multiply by 0.01. [d] To convert MPa to psi, multiply by 145.
[e] To convert J/m to ft·lbf/in., divide by 53.38.

Vibration-dampening properties at sonic and ultrasonic frequencies are excellent. However, the thickness of the resin must be sufficient to absorb the energy produced; this is usually determined experimentally.

Electrical Properties. Because of excellent electrical properties, FEP is a valuable and versatile electrical insulator. Within the recommended service temperature range, PTFE and FEP have identical properties as electrical insulators. Volume resistivity, which is $> 10^{17}$ Ω/cm, remains unchanged even after prolonged soaking in water; surface resistivity is $> 10^{15}$ Ω/sq.

At low frequencies, the dielectric constant of FEP remains the same (~2). However, at > 100 MHz the constant drops slightly with increasing frequency. As a true thermoplastic, FEP has a void content of zero and most of the fabricated material has a density of 2.14–2.17 g/cm^3. The National Bureau of Standards has selected Teflon FEP resins for dielectric reference specimens because of the stability of their dielectric constant. The dissipation factor has several peaks as a function of temperature and frequency (3×10^{-4} at 100 kHz; 7×10^{-4} at 1 MHz). The magnitude of the dissipation factor peak is greater for FEP than for PTFE because the molecular structure of the former is less symmetrical. The dissipation factor is hardly affected by irradiation annealing (70) and unaffected by humidity. The dielectric strength is high (80 GV/mm for 0.25 mm film at 23°C) and unaffected by thermal aging at 200°C. At high frequencies, the dielectric properties deteriorate in the presence of corona. If the voltage stress is not high enough to cause corona ignition, an infinitely long dielectric life is expected at any frequency. Corona discharges on the surface or in a void initiate dielectric breakdown (71). The FEP resins are recommended for continuous service up to 205°C. Although they begin to melt flow at 270°C, they retain some structural integrity up to 250°C (70).

Chemical Properties. The FEP resin is inert to most chemicals and solvents, even at elevated temperatures and pressures. However, it reacts with fluorine, molten alkali metal, and molten sodium hydroxide. Acids or bases are not absorbed at 200°C and exposures of one year. The absorption of organic solvents is less than 1% at elevated temperatures and long exposure times. Absorption of chemicals or solvents has no effect on the chemical integrity of the FEP molecule and is a reversible physical process.

Gases and vapors permeate FEP resin at a rate that is considerably lower than that of most plastics. Because FEP resins are melt processed, they are void-free and permeation occurs only by molecular diffusion. Variation in crystallinity and density is limited, except in unusual melt-processing conditions.

Because of its low permeability, FEP polymer is used extensively in the chemical industry. Its permeation characteristics are similar to those of PTFE (Table 4). An inverse relationship between permeability and film thickness applies to FEP.

Weathering. Articles fabricated from FEP are unaffected by weather, and their resistance to extreme heat, cold, and uv irradiation suits them for applications in radar and other electronic components. For example, after 15 years of solar exposure in Florida, the tensile strength (73) and light transmission (96%) of a 25-µm thick film was unchanged and the film remained crystal clear. Elongation increased slightly for the first 5 to 7 years of outdoor exposure, probably as a result of stress relaxation. Beyond 10 years, a small decrease was observed.

Table 4. Permeability of FEP Fluorocarbon Resins to Liquid Vapors and Gases

Permeant	Permeability constant,[a,b] mol/(m·s·Pa) × 10^{15}		
	23°C	35°C	50°C
Liquid vapors			
acetic acid		9.07	
acetone	0.37		3.23
benzene	0.75		
carbon tetrachloride	0.24	0.41	
decane	112.18		33.48
dipentene	23.50		10.67
ethyl acetate	0.27	2.06	4.09
ethanol	1.61	4.66	
H_2SO_4, 98%	21.70		
toluene	5.38		
water	8.14	20.32	18.26
Gases[c]			
oxygen	18.69		
helium	113.47		
nitrogen	6.10		
hydrogen	40.15		
methane	3.17		

[a]Ref. 60. Test method ASTM E96-35T (at vapor pressure; for 25.4-μm film thickness). Values are averages only and not for specification purposes.
[b]Original data converted to SI units using vapor pressure data from Ref. 72.
[c]At 20°C.

Optical Properties. Teflon FEP fluorocarbon film transmits more ultraviolet, visible light, and infrared radiation than ordinary window glass. The refractive index of FEP film is 1.341–1.347 (74).

Fabrication

Standard thermoplastic processing techniques can be used to fabricate FEP. Thermal degradation must be avoided, and a homogeneous structure and good surface quality must be maintained.

Injection Molding. Compared to most thermoplastic products, even the low MV resin has a significantly higher melt viscosity and therefore requires higher processing temperatures, slower injection rates, special mold design, and corrosion-resistant material of construction. When the flow velocity in melt processing exceeds a critical value, melt fracture occurs. The critical shear rate of FEP is much lower than that of other thermoplastics. Recommendations for materials of construction and the screw design, valves, smear heads, nozzle, operating conditions, and mold design are given in Reference 52.

Pigments (thermally stable at processing temperature) are dry blended with the resin before molding. At loadings of 0.1–1%, pigments have no appreciable effect on the dielectric strength, dielectric constant, or mechanical properties. The dissipation factor of pigmented resin varies with the pigment and its amount (75).

Extrusion. Conventional melt-extrusion equipment is used in processing FEP resins. Commercial pigments are mixed with the resin before extrusion into wire coating, tubing, rods, molding, beading channels, etc. Coating thicknesses of 0.076–2.54 mm have been extruded over such materials as silicone rubber, poly(vinyl chloride), glass braid, metal-shielded cables, twisted conductors, and parallel multiconductor cables.

For primary insulation or cable jackets, high production rates are achieved by extruding a tube of resin with a larger internal diameter than the base wire and a thicker wall than the final insulation. The tube is then drawn down to the desired size. An operating temperature of 315–400°C is preferred, depending on holdup time. The surface roughness caused by melt fracture determines the upper limit of production rates under specific extrusion conditions (76). Corrosion-resistant metals should be used for all parts of the extrusion equipment that come in contact with the molten polymer (77).

Tubing is made in a wide range of sizes and is used as slip-on electrical insulation, instrument tubing, and for hoses. Small tubing, called spaghetti tubing, can be produced by a free-extrusion technique, whereas hose-size tubing is produced by conventional forming-box techniques; FEP also is extruded into films.

Dispersion Processing. The commercial aqueous dispersion of FEP contains 55 wt % of hydrophobic, negatively charged FEP particles and ca 6 wt % (based on FEP) of a mixture of nonionic and anionic surface-active agents. The average particle size is ca 0.2 μm. The dispersion is processed by the same technique used for PTFE dispersion. For example, the fabric is coated with FEP dispersion, the water is evaporated from the coating, the wetting agent is removed, and the FEP layer is fused with the fabric.

Dispersion is used as a coating for glass fabric, chemical barriers, and wire-insulating tapes; as adhesive coatings for bonding seals and bearings of PTFE to metallic and nonmetallic components; and as antifriction or antistick coatings for metals. The fusion of FEP to provide a continuous film depends on a time–temperature relationship; 1 min at 400°C or 40 min at 290°C are sufficient to achieve good fusion (78).

Other Techniques. The FEP resin is bonded to metal surfaces by the application of heat and pressure; it can be heat sealed or hot-gas welded. Heating FEP at 260°C and allowing it to cool slowly results in stress relieving, or annealing. The FEP film is used to weld PTFE-coated surfaces.

Effects of Fabrication on Product Properties. Extrusion conditions have a significant effect on the quality of the product (77). Contamination can be the result of corrosion, traces of another resin, or improper handling. Corrosion-resistant Hastelloy C parts should be used in the extruder. Surface roughness is the result of melt fracture or mechanical deformation. Melt fracture can be eliminated by increasing the die opening, die temperature, and the melt temperature and reducing the extrusion rate. Bubbles and discoloration are caused by resin degradation, air entrapment, or condensed moisture. Excessive drawdown, resin degradation, or contamination can result in pinholes, tears, and cone breaks. The blisters are caused by degassing of primary coatings, and loose coatings are caused by rapid cooling and long cones.

Testing and Standards. Requirements for extrusion and molding grades are cited in ASTM specifications (79) and in Federal specification LP-389A of May 1964 (80). For fabricated shapes, FEP film and sheet are covered by Aeronautical Material Specifications (AMS) 3647 and LP-523 (81). Besides the specifications covered by the Fluorocarbons Division of the Society of the Plastics Industry, Inc. (82), other specifications are listed in Reference 83.

Economic Aspects

Because of the high cost of hexafluoropropylene, FEP is more expensive than PTFE. In the United States in 1992, FEP sold at prices up to $28.3 kg, depending on the type and quantity. Most grades are marketed in a colorless, translucent, extruded pellet form. The dispersion containing about 55% solids is priced at ca $33 kg. During the 1980s FEP sales increased rapidly because of usage in plenum cable, but since there are other polymers that can be used in this application the growth rate for FEP is expected to slow down.

Health and Safety

The safety precautions required in handling TFE–HFP copolymers are the same as those applied to handling PTFE. Large quantities have been processed safely by many different fabricators in a variety of operations. With proper ventilation, the polymer can be processed and used at elevated temperatures without hazard. The fumes from heated FEP or its thermal decomposition products are toxic in high concentrations, like the fumes or decomposition products of other polymers. Ventilation should be provided in areas where the resin is at processing temperature (270–400°C). At ambient temperatures, FEP resin is essentially inert. Inhalation of fumes given off by heated FEP resin may result in influenza-like symptoms. They may occur several hours after exposure and disappear within 35–48 hours, even in the absence of treatment; the effects are not cumulative (52). Such attacks usually follow exposure to vapors evolved from the polymer without adequate ventilation or from smoking tobacco or cigarettes contaminated with the polymer. Toxicology study of the particulates and fumes is reported in Reference 84.

Applications

The principal electrical applications include hook-up wire, interconnecting wire, coaxial cable, computer wire, thermocouple wire, plenum cable, and molded electrical parts. Principal chemical applications are lined pipes and fittings, overbraided hose, heat exchangers, and laboratory ware. Mechanical uses include antistick applications, such as conveyor belts and roll covers. A recent development of FEP film for solar collector windows takes advantage of light weight, excellent weatherability, and high solar transmission. Solar collectors made of FEP film are efficient, and installation is easy and inexpensive.

BIBLIOGRAPHY

"Fluorinated Ethylene–Propylene Copolymers," under "Fluorine Compounds, Organic," in *ECT* 3rd ed., Vol. 11, pp. 24–35, by S. V. Gangal, E. I. du Pont de Nemours & Co., Inc.

1. U.S. Pat. 2,946,763 (July 26, 1960), M. I. Bro and B. W. Sandt (to E. I. du Pont de Nemours & Co., Inc.).
2. U.S. Pat. 2,394,581 (Feb. 12, 1946), A. F. Benning, F. B. Dowing, and J. D. Park (to Kinetic Chemicals, Inc.).
3. E. G. Young and W. S. Murray, *J. Am. Chem. Soc.* **70**, 2814 (1949).
4. A. L. Henne and T. P. Waalkes, *J. Am. Chem. Soc.* **68**, 496 (1946).
5. U.S. Pat. 3,446,858 (May 27, 1969), H. Shinzu and co-workers (to Daikin Kyogo Co.).
6. U.S. Pat. 3,459,818 (Aug. 15, 1969), H. Ukihashi and M. Hisasue (to Asahi Glass Co.).
7. U.S. Pat. 3,873,630 (Mar. 25, 1975), N. E. West (to E. I. du Pont de Nemours & Co., Inc.).
8. U.S. Pat. 5,043,491 (Aug. 27, 1991), J. Webster and co-workers, (to E. I. du Pont de Nemours & Co., Inc.).
9. U.S. Pat. 5,057,634 (Oct. 15, 1991), J. Webster and co-workers (to E. I. du Pont de Nemours & Co., Inc.).
10. U.S. Pat. 5,068,472 (Nov. 26, 1991), J. Webster and co-workers (to E. I. du Pont de Nemours & Co., Inc.).
11. U.S. Pat. 3,758,138 (Aug. 7, 1956), D. A. Nelson (to E. I. du Pont de Nemours & Co., Inc.).
12. B. Atkinson and A. B. Trenwith, *J. Am. Chem. Soc., Pt. II*, 2082 (1953).
13. U.S. Pat. 2,759,983 (Aug. 21, 1956), J. S. Waddell (to E. I. du Pont de Nemours & Co., Inc.).
14. U.S. Pat. 3,047,640 (July 31, 1962), R. F. Sweeny and C. Woolf (to Allied Chemical Corp.).
15. U.S. Pat. 3,009,966 (Nov. 21, 1961), M. Hauptschein and A. Fainberg (to Pennsalt Chemicals Corp.).
16. L. T. Hals, T. S. Reid, and G. H. Smith, *J. Am. Chem. Soc.* **73**, 4054 (1951); U.S. Pat. 2,668,864 (Feb. 9, 1954), J. T. Hals, T. S. Reid, and G. H. Smith (to Minnesota Mining and Manufacturing Co.).
17. G. Pass, *J. Am. Chem. Soc., Pt. I*, 824 (Jan. 1965); Fr. Pat. 1,399,414 (May 14, 1965), (to Imperial Chemical Industries, Ltd.).
18. Ger. Pat. 1,236,497 (Mar. 16, 1967), W. Oese, H. Dude, and F. Reinke (to VEB Fluorwerke Dohma).
19. H. C. Duus, *Ind. Eng. Chem.* **47**, 1445 (1955).
20. J. W. Clayton, *Occup. Med.* **4**, 262 (1962).
21. U.S. Pat. 2,918,501 (Dec. 22, 1959), W. J. Brehm and co-workers (to E. I. du Pont de Nemours & Co., Inc.).
22. U.S. Pat. 3,316,312 (Apr. 25, 1967), D. I. McCane and I. M. Robinson (to E. I. du Pont de Nemours & Co., Inc.).
23. R. A. Matula, *J. Phys. Chem.* **72**, 3054 (1968).
24. U.S. Pat. 3,358,003 (Dec. 12, 1967), H. S. Eleuterio and R. W. Meschke (to E. I. du Pont de Nemours & Co., Inc.).
25. U.S. Pat. 3,180,895 (Apr. 27, 1965), J. F. Harris, Jr. and D. I. McCane (to E. I. du Pont de Nemours & Co., Inc.).

26. U.S. Pat. 3,291,843 (Dec. 13, 1966), C. G. Fritz and S. Selman (to E. I. du Pont de Nemours & Co., Inc.).
27. E. V. Volkova and A. E. Skobina, *Vysokomol. Soedin.* **6**(5), 964 (1964).
28. Fr. Pat. 1,524,571 (May 10, 1968), S. W. Osborn and E. Broderich (to Thiokol Chemical Corp.).
29. U.S. Pat. 2,983,764 (May 9, 1961), D. F. Knaack (to E. I. du Pont de Nemours & Co., Inc.).
30. U.S. Pat. 2,958,685 (Nov. 1, 1960), H. S. Eleuterio (to E. I. du Pont de Nemours & Co., Inc.).
31. I. L. Knunyants, E. I. Mysov, and M. P. Krasuskaya, *Izv. Akad. Nauk. SSSR Otd. Khim. Nauk.*, 906 (1958).
32. I. L. Knunyants, A. E. Shchekotikhin, and A. V. Fakin, *Izv. Akad. Nauk. SSSR Otd. Khim. Nauk.*, 282 (1953).
33. I. L. Knunyants, L. S. German, and B. L. Dyatkin, *Izu. Akad. Nauk. SSSR Otd. Khim. Nauk.*, 1353 (1956).
34. R. N. Haszeldine and B. R. Steele, *J. Chem. Soc.*, 1592 (1953).
35. W. T. Miller, Jr., E. Bergman, and A. H. Fainberg, *J. Am. Chem Soc.* **79**, 4159 (1957).
36. I. L. Knunyants, V. V. Shokina, and N. D. Kuleshova, *Izv. Akad. Nauk. SSSR Otd. Khim. Nauk.*, 1936 (1960).
37. U.S. Pat. 3,467,636 (Sept. 16, 1969), A. Nersasian (to E. I. du Pont de Nemours & Co., Inc.).
38. U.S. Pat. 3,536,683 (Oct. 27, 1970), F. V. Bailor and J. R. Cooper (to E. I. du Pont de Nemours & Co., Inc.).
39. U.S. Pat. 3,790,540 (Feb. 5, 1974), J. E. Dohany and A. C. Whiton (to Pennwalt Corp.).
40. U.S. Pat. 3,817,951 (June 18, 1974), D. N. Robinson (to Pennwalt Corp.).
41. U.S. Pat. 3,868,337 (Feb. 25, 1975), P. Gros (to Society Superflexit).
42. G. L. Kennedy, Jr., *Crit. Revs. Toxicol.* **21**(2), 149–170 (1990).
43. U.S. Pat. 2,598,283 (May 27, 1952), W. T. Miller (to U.S. Atomic Energy Commission).
44. R. A. Naberezhnykh and co-workers, *Dokl. Akad. Nauk. SSSR* **214**, 149 (1974).
45. A. S. Kabankin, S. A. Balabanova, and A. M. Markevich, *Vysokomol. Soedin. Ser. A* **12**, 267 (1970).
46. Br. Pat. 781,532 (Aug. 21, 1957), C. G. Krespan (to E. I. du Pont de Nemours & Co., Inc.).
47. U.S. Pat. 3,132,124 (May 5, 1964), M. J. Couture, D. L. Schindler, and R. B. Weiser (to E. I. du Pont de Nemours & Co., Inc.).
48. U.S. Pat. 4,380,618 (1983), A. Khan and R. Morgan (to E. I. du Pont de Nemours & Co., Inc.).
49. U.S. Pat. 4,384,092 (1983), J. Herison (to Ugine Kuhlmann).
50. U.S. Pat. 4,861,845 (1989), E. Slocum, A. Sobrero, and R. Wheland (to E. I. du Pont de Nemours & Co., Inc.).
51. R. S. Atland, *Modern Plast.* **62**, 200 (1985).
52. *Teflon-FEP Fluorocarbon Resin, Techniques for Injection Molding*, information bulletin 95d, E. I. du Pont de Nemours & Co., Inc., Wilmington, Del., 1969.
53. S. Wu, *Macromolecules* **18**, 2023–2030 (1985).
54. R. K. Eby, *J. Appl. Phys.* **34**, 2442 (1963).
55. R. K. Eby and F. C. Wilson, *J. Appl. Phys.* **33**, 2951 (1962).
56. *Safe Handling Guide, Teflon Fluorocarbon Resins, du Pont Materials for Wire and Cable*, bulletin E-85433, E. I. du Pont de Nemours & Co., Inc., Wilmington, Del., 1986.
57. R. Y. M. Huang and P. J. F. Kanitz, *Polym. Prepr. Am. Chem. Soc. Div. Polym. Chem.* **10**(2), 1087 (1969).
58. R. J. Diamond, *Plastics* **27**, 109 (1962).

59. J. Frados, ed., *Modern Plastics Encyclopedia*, Vol. 46, No. 10A, McGraw-Hill Book Co., Inc., New York, 1969, p. 974.
60. *J. Teflon* **11**(1), 8 (1970).
61. J. A. Brydson, *Plastics Materials*, Iliffe Books, Ltd., London, 1966, Chapt. 10, pp. 203–218.
62. *Teflon 100 FEP–Fluorocarbon Resin—Melt Processible Resin*, information bulletin X-90a, E. I. du Pont de Nemours & Co., Inc., Wilmington, Del., 1960.
63. *Teflon Fluorocarbon Resins, Mechanical Design Data*, 2nd ed., E. I. du Pont de Nemours & Co., Inc., Wilmington, Del., 1965.
64. A. A. Benderly, *J. Appl. Polym. Sci.* **6**, 221 (1962).
65. Brit. Pat. 890,466 (Feb. 28, 1962), D. L. Ryan (to E. I. du Pont de Nemours & Co., Inc.).
66. U. S. Pat. 3,063,882 (Nov. 13, 1962), J. R. Chesire (to E. I. du Pont de Nemours & Co., Inc.).
67. R. R. Rye and G. W. Arnold, *Langmuir* **5**, 1331 (1989).
68. D. T. Clark and D. R. Hutton, *J. Polym. Sci., Polym. Chem. Ed.* **25**, 2643 (1987).
69. R. C. Bening and J. J. McCarthy, *Polym. Prep.* **29**, 336 (1988).
70. *Electrical/Electronic Design Data for Teflon*, E. I. du Pont de Nemours & Co., Inc., Wilmington, Del.
71. J. C. Reed, E. J. McMahon, and J. R. Perkins, *Insulation (Libertyville, Ill.)* **10**, 35 (1964).
72. D. W. Green, ed., *Perry's Chemical Engineers' Handbook*, 6th ed., McGraw-Hill Book Co., New York, 1984.
73. *Teflon Solar Film for Solar Collectors*, E. I. du Pont de Nemours & Co., Inc., Wilmington, Del.
74. *Teflon FEP–Fluorocarbon Film*, bulletin T-5A, Optical, E. I. du Pont de Nemours & Co., Inc., Wilmington, Del.
75. L. H. Gillespe, D. O. Saxton, and F. M. Chapman, *New Design Data for Teflon*, E. I. du Pont de Nemours & Co., Inc., Wilmington, Del., 1960.
76. *J. Teflon* **18**(1), 8(1977).
77. *Teflon FEP–Fluorocarbon Resin—Techniques for Processing by Melt Extrusion*, 2nd ed., information bulletin X-82, E. I. du Pont de Nemours & Co., Inc., Wilmington, Del., 1960.
78. *Properties and Processing Techniques for Teflon 120 FEP–Fluorocarbon Resin Dispersion*, preliminary information bulletin no. 20, E. I. du Pont de Nemours & Co., Inc., Wilmington, Del., 1961.
79. *ASTM Annual Book of ASTM Standards*, Vol. 08.01, American Society for Testing and Materials, Philadelphia, Pa., 1993.
80. *Federal Supply Service Bureau Specification L-P-389A*, Section SW, 470E L'Enfant Plaza, Washington, D.C.
81. Technical data, SAE International, Inc., Warrendale, Pa., 1993.
82. *ASTM Annual Book of ASTM Standards*, Vol. 08.03, American Society for Testing and Materials, Philadelphia, Pa., 1993.
83. Technical data, Society of the Plastics Industry, Inc., Fluorocarbon Division, Washington, D.C., 1993; *J. Teflon* **15**(1), 10 (1974).
84. K. P. Lees and W. C. Seidal, *Inhalation Toxicol.* **3**(3), 237 (1991).

General References

S. V. Gangal, "Tetrafluoroethylene Polymers, Tetrafluoroethylene–Hexafluoropropylene Copolymers," in J. I. Kroschwitz, ed., *Encyclopedia of Polymer Science and Engineering*, 2nd ed., Vol. 16, John Wiley & Sons, Inc., New York, 1989, pp. 601–613.

SUBHASH V. GANGAL
E. I. du Pont de Nemours & Co., Inc.

TETRAFLUOROETHYLENE–ETHYLENE COPOLYMERS

Copolymers of ethylene [74-85-1] and tetrafluoroethylene [116-14-3] (ETFE) have been a laboratory curiosity for more than 40 years. These polymers were studied in connection with a search for a melt-fabricable PTFE resin (1–5); interest in them fell with the discovery of TFE–HFP (FEP) copolymers (6). In the 1960s, however, it became evident that a melt-fabricable fluorocarbon resin was needed with higher strength and stiffness than those of PTFE resins. Earlier studies indicated that TFE–ethylene copolymers [11939-51-6] might have the right combination of properties. Subsequent research efforts (7) led to the introduction of modified ethylene–tetrafluoroethylene polymer [25038-71-5] (Tefzel) by E. I. du Pont de Nemours & Co., Inc. in 1970.

Modified ethylene–tetrafluoroethylene copolymers are the products of real commercial value because they have good tensile strength, moderate stiffness, high flex life, and outstanding impact strength, abrasion resistance, and cut-through resistance. Electrical properties include low dielectric constant, high dielectric strength, excellent resistivity, and low dissipation factor. Thermal and cryogenic performance and chemical resistance are good. These properties, combined with elasticity, make this material an ideal candidate for heat-shrinkable film and tubing. This family of copolymers can be processed by conventional methods such as melt extrusion, injection molding, transfer molding, and rotational molding. The properties of the copolymers vary with composition; polymers containing 40–90% tetrafluoroethylene (by weight) soften between 200 and 300°C, depending on composition (1). The tetrafluoroethylene segments of the molecules account for >75% of the weight of an approximately 1:1 mole ratio copolymer. The two monomers combine readily into a nearly 1:1 alternating structure. Such polymers exhibit a unique combination of mechanical, chemical, and electrical properties as well as excellent weatherability. However, thermal stress-crack resistance is poor. The copolymer can be modified with a termonomer that undergoes free-radical polymerization and does not cause undesirable chain transfer or termination during polymerization. The modified copolymer exhibits almost the identical physical, chemical, and electrical properties characteristic of the 1:1 alternating copolymer, but retains high ultimate elongation up to 200°C.

Ethylene and tetrafluoroethylene are copolymerized in aqueous, nonaqueous, or mixed medium with free-radical initiators. The polymer is isolated and converted into extruded cubes, powders, and beads, or a dispersion. This family of products is manufactured by Du Pont, Hoechst, Daikin, Asahi Glass, and Ausimont and sold under the trade names of Tefzel, Hostaflon ET, Neoflon EP, Aflon COP, and Halon ET, respectively.

Monomers

Tetrafluoroethylene of purity suitable for granular or dispersion polymerizations is acceptable for copolymerization with ethylene. Polymerization-grade ethylene is suitable for copolymerization with tetrafluoroethylene. Modifying termonomers, eg, perfluorobutylethylene and perfluoropropylene, are incorporated by free-radical polymerization.

Manufacture

Tetrafluoroethylene–ethylene copolymers have tensile strengths two to three times as high as the tensile strength of polytetrafluoroethylene or of the ethylene homopolymer (1). Because these copolymers are highly crystalline and fragile at high temperature, they are modified with a third monomer, usually a vinyl monomer free of telegenic activity. The termonomer provides the copolymer with side chains of at least two carbon atoms, such as perfluoroalkylvinyl or vinylidene compounds, perfluoroalkyl ethylenes, and perfluoroalkoxy vinyl compounds. For high tensile properties and cut-through resistance, a molar ratio of ethylene and tetrafluoroethylene between 60:40 and 40:60 is required (8,9).

Copolymerization is effected by suspension or emulsion techniques under such conditions that tetrafluoroethylene, but not ethylene, may homopolymerize. Bulk polymerization is not commercially feasible, because of heat-transfer limitations and explosion hazard of the comonomer mixture. Polymerizations typically take place below 100°C and 5 MPa (50 atm). Initiators include peroxides, redox systems (10), free-radical sources (11), and ionizing radiation (12).

Purely aqueous polymerization systems give copolymers that are not wetted by the reaction medium. The products agglomerate and plug valves, nozzles, and tubing, and adhere to stirrer blades, thermocouples, or reactor walls. These problems do not occur in organic media or mixtures of these with water.

Aqueous emulsion polymerization is carried out using a fluorinated emulsifier, a chain-transfer agent to control molecular weight, and dispersion stabilizers such as manganic acid salts and ammonium oxalate (13,14).

Reactivity ratios of ethylene and tetrafluoroethylene are

Temperature, °C	r_{TFE}	r_E
-35	0.014 ± 0.008	0.010 ± 0.02
65	0.045 ± 0.010	0.14 ± 0.03

These values indicate strong alternation tendencies that decrease with increasing temperature. Computations show that 1:1 ETFE copolymers obtained at -30 and 65°C should have about 97 and 93%, respectively, of alternating sequences (15).

Properties

The equimolar copolymer of ethylene and tetrafluoroethylene is isomeric with poly(vinylidene fluoride) but has a higher melting point (16,17) and a lower dielectric loss (18,19) (see FLUORINE COMPOUNDS, ORGANIC–POLY(VINYLIDENE FLUORIDE)). A copolymer with the degree of alternation of about 0.88 was used to study the structure (20). Its unit cell was determined by x-ray diffraction. Despite irregularities in the chain structure and low crystallinity, a unit cell and structure was derived that gave a calculated crystalline density of 1.9 g/cm^3. The unit cell is believed to be orthorhombic or monoclinic ($a = 0.96$ nm, $b = 0.925$ nm, $c = 0.50$ nm; $\gamma = 96°$).

ethylene–tetrafluoroethylene unit

poly(vinylidene fluoride) segment

The molecular conformation is that of extended zigzag. Molecular packing appears to be orthorhombic, each molecule having four nearest neighbors with the CH_2 groups of one chain adjacent to the CF_2 groups of the next. The x-ray spectrum of a 1:1 copolymer has two main peaks at $Z_0 = 19.63°$ and $Z_0 = 21.00°$, corresponding to Bragg distances of 0.45 and 0.42 nm, respectively. Compression-molded samples are 50–60% crystalline; however, crystallinity is greatly affected by composition, quench rate, and temperature.

Alternation is usually above 90%. Nearly perfect alternation of isomeric units in a ca 1:1 monomer ratio has been confirmed by infrared spectroscopy. Bands at 733 and 721 cm^{-1} have an intensity proportional to the concentration of $(CH_2)_n$ groups ($n = 4$ and <6, respectively) present in a copolymer containing 46 mol % tetrafluoroethylene; intensity decreases with increasing concentration of fluorinated monomer.

The molecular weight and its distribution have been determined by laser light scattering, employing a new apparatus for ETFE dissolution and solution clarification at high temperature; diisobutyl adipate is the solvent at 240°C. The molecular weight of molten ETFE is determined by high temperature rheometry (21).

This polymer can be dissolved in certain high boiling esters at temperatures above 230°C (22), permitting a weight-average molecular weight determination by light scattering. Solution viscosity data suggest that the polymer exists as a slightly expanded coil under similar conditions (23).

Transitions. Samples containing 50 mol % tetrafluoroethylene with ca 92% alternation were quenched in ice water or cooled slowly from the melt to minimize or maximize crystallinity, respectively (19). Internal motions were studied by dynamic mechanical and dielectric measurements, and by nuclear magnetic resonance. The dynamic mechanical behavior showed that the α relaxation occurs at 110°C in the quenched sample; in the slowly cooled sample it is shifted to 135°C. The β relaxation appears near −25°C. The γ relaxation at −120°C in the quenched sample is reduced in peak height in the slowly cooled sample and shifted to a slightly higher temperature. The α and γ relaxations reflect motions in the amorphous regions, whereas the β relaxation occurs in the crystalline regions. The γ relaxation at −120°C in dynamic mechanical measurements at 1 Hz appears at −35°C in dielectric measurements at 10^5 Hz. The temperature of the α relaxation varies from 145°C at 100 Hz to 170°C at 10^5 Hz. In the mechanical measurement, it is 110°C. There is no evidence for relaxation in the dielectric data.

The activation energy is 318.1 kJ/mol (76 kcal/mol) for the α relaxation and 44.3 kJ/mol (10.6 kcal/mol) for the γ relaxation. These relaxations are attributed to the motion of long and short segments in the amorphous regions, respectively. As ETFE copolymer is isomeric with poly(vinylidene fluoride) (18), the γ relaxations occur at about the same temperature. Activation energies are similar and are attributed to the motion of short amorphous segments. The β relaxation in PVF_2 is considered to be the main-chain amorphous relaxation and is analogous to the α relaxation in the ethylene–tetrafluoroethylene copolymer. However, the arrangement of dipoles in the all-trans conformation is more symmetrical.

Physical and Mechanical Properties. Modified ethylene–tetrafluoroethylene copolymer has a good combination of mechanical properties, including excellent cut-through and abrasion resistance, high flex life, and exceptional impact strength. As wire insulation, it withstands physical abuse during and after installation. Lightweight wire constructions are designed with a minimum diameter and are useful as single, general-purpose insulation and for multiple or composite constructions.

Modified ETFE is less dense, tougher, and stiffer and exhibits a higher tensile strength and creep resistance than PTFE, PFA, or FEP resins. It is ductile, and displays in various compositions the characteristic of a nonlinear stress–strain relationship. Typical physical properties of Tefzel products are shown in Table 1 (24,25). Properties such as elongation and flex life depend on crystallinity, which is affected by the rate of crystallization; values depend on fabrication conditions and melt cooling rates.

Light transmittance of 25-μm films in the visible-to-ir range varies from 91 to 95% for Tefzel 200 and from 89 to 93% for Tefzel 280. In the uv range transmittance increases from 50% at 200 nm to 90% at 400 nm.

Thermal Properties. Modified ETFE copolymer has a broad operating temperature range up to 150°C for continuous exposure (24). Cross-linking by radiation improves the high temperature capability further. However, prolonged exposure to higher temperatures gradually impairs the mechanical properties and results in discoloration.

The thermodynamic properties of Tefzel 200 and 280 are shown in Table 2; the annual rate of loss of weight with thermal aging for Tefzel 200 ranges from 0.0006 g/g at 135°C to 0.006 g/g at 180°C after an initial loss of absorbed gases of 0.0013 g/g at elevated temperature. The excellent thermal stability of ETFE is demonstrated by aging at 180°C; at this temperature, the annual weight loss of six parts per 1000, or a 1% weight loss, takes almost two years.

Friction and Bearing Wear of the Glass-Reinforced Copolymer. Glass reinforcement improves the frictional and wear properties of modified ETFE resins (HT-2004). For example, the dynamic coefficient of friction (689.5 kPa (100 psi) at > 3 m/min) for Tefzel 200 is 0.4, which drops to 0.3 for the 25% glass-reinforced product at these conditions (24). The wear factor also improves from 12×10^{-14} to 32×10^{-17} 1/Pa (6000×10^{-10} to 16×10^{-10} in.^{3} · min/ftlbf·h). These frictional and wear characteristics, combined with outstanding creep resistance, indicate suitability for bearing applications. Glass-reinforced ETFE is less abrasive on mating surfaces than most glass-reinforced polymers. Its static coefficient of friction depends on bearing pressure; for Tefzel HT-2004 the coefficient of friction changes from 0.51 at 68 Pa to 0.34 at 3.43 kPa (0.5 psi).

Dynamic friction depends on pressure and rubbing velocity (PV). The generation of frictional heat depends on the coefficient of friction and the PV factor. For the glass-reinforced product, temperature buildup begins at about PV 10,000 and thermal runaway occurs just below PV 20,000. High wear rates begin above

Table 1. Typical Properties of Tefzel[a]

Property	ASTM method	Tefzel 200, 280	Tefzel[b] HT-2004
ultimate tensile strength, MPa[c]	D638	44.8	82.7
ultimate elongation	D887-64T	200[d]	8
compressive strength, MPa[c]	D695	48.9	68.9
shear strength, MPa[c]		41.3	44.8
heat deflection temp, °C	D648		
at 0.45 MPa		104	265
at 1.8 MPa		74	210
max continuous use temp, no load, °C		150[e]	200
low temp embrittlement	D746	below −100°C	
tensile modulus, MPa[c]	D638	827	8270
flexural modulus, MPa[c]	D790	965	6550
impact strength notched Izod	D256		
at −54°C, J/m[f]		>1067	373
at 23°C		no break	485
deformation under load, 13.7 MPa[c] at 50°C, %	D621	4.11	0.68
coefficient of linear expansion per °C × 10^{-5}	D696-70		
20–30°C		9	3
50–90°C		9.3	1.7
104–180°C		14	3.2
specific gravity	D792	1.70	1.86
refractive index, n_D		1.4028	
flammability	UL 94	94 V-0	94 V-0
	D635	ATB[g] <5 s	
		ALB[g] 10 mm	
melting point, dta peak, °C		270	270
water absorption at saturation, %	D570	0.029	0.022
hardness			
Rockwell	D785	R50	R74
Durometer D		D75	
coefficient of friction[h]			
dynamic, 689 kPa (at >3 m/min)		0.4	0.3
static, 689 kPa			0.3

[a] At 23°C and 50% rh, unless otherwise specified.
[b] Reinforced with 25 wt % glass fiber.
[c] To convert MPa to psi, multiply by 145.
[d] Elongations between 100 and 300% are achieved with varying methods of sample fabrication.
[e] Long-term heat-aging tests on Tefzel 280 are in progress. It is expected that its continuous-use temperature will be above 150°C.
[f] To convert J/m to ftlbf/in., divide by 53.38.
[g] ATB, average time of burning to nearest 5 s; ALB: average length of burn to nearest 5 mm. Test bar thickness, 2.9 mm.
[h] Mating material AISI 1018 Steel, Rc20, 16AA; 689 kPa = 100 psi.

Table 2. Thermodynamic Properties of Modified ETFE

Property	Tefzel 200 and 280
melting point, °C	270
specific heat, J/(mol·K)[a]	0.46–0.47
heat of sublimation, kJ/mol[a]	50.2
heat of fusion,[b] J/g[a]	46.0
heat of combustion, kJ/g[a]	13.72
thermal conductivity, W/(m·K)	0.238
critical surface tension of molten resin, mN/m(=dyn/cm)	22

[a]To convert J to cal, divide by 4.184.
[b]Little dependence on temperature.

PV 15,000. The wear rate depends on the type of metal rubbing surface and finish, lubrication, and clearances. Lubrication, hard shaft surfaces, and high finishes improve wear rates. Table 3 gives wear factors for steel and aluminum. Because the wear rate of both ETFE and the metal is much higher for aluminum than for steel, an anodized surface is preferred with aluminum.

Electrical Properties. Modified ethylene–tetrafluoroethylene is an excellent dielectric (Table 4). Its low dielectric constant confers a high corona-ignition voltage. The dielectric constant does not vary with frequency or temperature. Both dielectric strength (ASTM D149) and resistivity are high. The loss characteristics

Table 3. Bearing Wear Rate[a] of Tefzel HT-2004

Pressure, kPa[b]	Velocity, cm/s	Wear factor, K × 10⁻¹⁷, 1/Pa[c]	
		Tefzel	Metal
On steel[d]			
6.8	2.5	32	8
6.8	5.1	28	12
6.8	7.6	38	26
6.8	8.9	60	32
6.8	10.2	fail	
On aluminum[e]			
2.0	5.1	2400	2400
0.68	25.4	960	780

[a]Thrust-bearing tester, no lubricants ambient air temperature, metal finish 406 nm.
[b]To convert kPa to psi, multiply by 0.145.
[c]To convert 1/Pa to (in.³·min)/(ftlbf·h), divide by 2×10^{-7}.
[d]AISI 1018.
[e]LM24M (English).

Table 4. Electrical Properties of ETFE Resins

Property	ASTM test	ETFE	Reinforced
dissipation factor, Hz	D150		
10^2		0.0006	0.004
10^3		0.0008	0.002
10^4			0.002
10^5			0.003
10^6		0.005	0.005
10^9		0.005	
10^{10}		0.010	0.012
volume resistivity, $\Omega \cdot cm$	D257	$>10^{16}$	10^{16}
surface resistivity, Ω/sq	D257	5×10^4	10^{15}
arc resistance, s		75	110

are minimum; the dissipation factor, although low, increases at higher frequencies. Glass reinforcement increases losses and the dielectric constant rises from 2.6 to 3.4 (from 10^2 to 10^{10} Hz); the dissipation factor is increased by tenfold. Exposure to radiation also increases losses. Dielectric strength is not reduced by thermal aging, unless a physical break occurs in the material. The short-time test of ASTM D149 gives values of 16–20 kV/mm with 3-mm thick specimens to 160–200 kV/mm with films 25–75 µm thick. Tracking resistance is about 70 s by ASTM D495. This is comparable to materials considered to be nontracking; under unusual conditions tracking occurs. When these resins are foamed they provide insulation with even lower dielectric constant (26).

Chemical Resistance and Hydrolytic Stability. Modified ethylene–tetrafluoroethylene copolymers are resistant to chemicals and solvents (Table 5) that often cause rapid degradation in other plastic materials. Performance is similar to that of perfluorinated polymers (27), which are not attacked by strong mineral acids, inorganic bases, halogens, and metal salt solutions. Organic compounds and solvents have little effect. Strong oxidizing acids, organic bases, and sulfonic acids at high concentrations and near their boiling points affect ETFE to varying degrees.

Physical properties remain stable after long exposure to boiling water. Tensile strength and elongation of Tefzel 200 are unaffected after 3000 h in boiling water. The higher molecular weight ETFE behaves similarly, whereas the glass-reinforced product shows a reduction of 25–35% in tensile strength with loss of reinforcement.

Water absorption of Tefzel is low (0.029% by weight), which contributes to its outstanding dimensional stability as well as to the stability of mechanical and electrical properties regardless of humidity.

High temperature resistance of ETFE and other fluoropolymers in automotive fuels and their permeation resistance have been discussed (28,29).

The ETFE copolymer can be cross-linked by radiation (30), despite the high content of tetrafluoroethylene units. Cross-linking reduces plasticity but en-

Table 5. Tefzel Resistance to Chemicals after Seven Days Exposure[a]

Chemical	Bp, °C	Test temperature, °C	Retained properties, %		
			Tensile strength	Elongation	Weight gain
organic acids and anhydrides					
acetic acid (glacial)	118	118	82	80	3.4
acetic anhydride	139	139	100	100	0
trichloroacetic acid	196	100	90	70	0
hydrocarbons					
mineral oil		180	90	60	0
naphtha		100	100	100	0.5
benzene	80	80	100	100	0
toluene	110	110			
amines					
aniline	185	120	81	99	2.7
aniline	185	180	95	90	
N-methylaniline	195	120	85	95	
N,N-dimethylaniline	190	120	82	97	
n-butylamine	78	78	71	73	4.4
di-n-butylamine	159	120	81	96	
di-n-butylamine	159	159	55	75	
tri-n-butylamine	216	120	81	80	
pyridine	116	116	100	100	1.5
solvents					
carbon tetrachloride	78	78	90	80	4.5
chloroform	62	61	85	100	4.0
dichloroethylene	77	32	95	100	2.8
methylene chloride	40	40	85	85	0
Freon 113	46	46	100	100	0.8
dimethylformamide	154	90	100	100	1.5
dimethyl sulfoxide	189	90	95	95	1.5
Skydrol		149	100	95	3.0
Aerosafe		149	92	93	3.9
A-20 stripper solution		140	90	90	
ethers, ketones, esters					
tetrahydrofuran	66	66	86	93	3.5
acetone	56	56	80	83	4.1
acetophenone	201	180	80	80	1.5
cyclohexanone	156	156	90	85	0
methyl ethyl ketone	80	80	100	100	0
n-butyl acetate	127	127	80	60	0
ethyl acetate	77	77	85	60	0
other organic compounds					
benzyl alcohol	205	120	97	90	
benzoyl chloride	197	120	94	95	
o-cresol	191	180	100	100	
decalin	190	120	89	95	
phthaloyl chloride	276	120	100	100	

Table 5. (*Continued*)

Chemical	Bp, °C	Test temperature, °C	Retained properties, %		
			Tensile strength	Elongation	Weight gain
inorganic acids					
hydrochloric (conc)	106	23	100	90	0
hydrobromic (conc)	125	125	100	100	
hydrofluoric (conc)		23	97	95	0.1
sulfuric (conc)		100	100	100	0
nitric, 70%	120	120	0	0	
chromic	125	125	66	25	
phosphoric (conc)		100			
halogens					
bromine (anhy)	59	23	90	90	1.2
chlorine (anhy)		120	85	84	7
bases, peroxides					
ammonium hydroxide		66	97	97	0
potassium hydroxide, 20%		100	100	100	0
sodium hydroxide, 50%		120	94	80	0.2
hydrogen peroxide, 30%		23	99	98	0
other inorganic compounds					
ferric chloride, 25%	104	100	95	95	0
zinc chloride, 25%	104	100	100	100	0
sulfuryl chloride	68	68	86	100	8
phosphoric trichloride	75	75	100	98	
phosphoric oxychloride	104	104	100	100	
silicon tetrachloride	60	60	100	100	

[a] Changes in properties <15% are considered insignificant; test performed on 250–1250-μm microtensile bars; tensile strength, elongation, and weight gain determined within 24 h after termination of exposure.

hances high temperature properties and nondrip performance. The irradiated resin withstands a 400°C solder iron for 10 min without noticeable effect.

Modified ETFE copolymer has excellent weather resistance; tensile strength and elongation are not affected. On the other hand, tensile and elongation properties of the glass-reinforced compound show a significant reduction.

Modified ETFE films are used as windows in greenhouses and conservatories due to their high transparency to both uv and visible light and excellent resistance to weathering (31).

Vacuum Outgassing and Permeability. Under vacuum, modified ethylene–tetrafluoroethylene copolymers give off little gas at elevated temperatures. The loss rate is about one-tenth of the acceptable maximum rates for spacecraft uses. Exposing 750-μm specimens for 24 h at 149°C to a high vacuum results in a maximum weight loss of 0.12%; volatile condensible material is less than 0.02%.

The following permeability values were determined on Tefzel film (100-μm, ASTM D1434) at 25°C (1 nmol/m·s·GPa = 0.5 cc·mil/100 in.^{2} d·atm):

Material	nmol/m·s·GPa
carbon dioxide	500
nitrogen	60
oxygen	200
helium	1800
water vapor (ASTM E96)	3.3

Fabrication

Modified ethylene–tetrafluoroethylene copolymers are commercially available in a variety of physical forms (Table 6) and can be fabricated by conventional thermoplastic techniques. Commercial ETFE resins are marketed in melt-extruded cubes, that are sold in 20-kg bags or 150-kg drums. In the United States, the 1992 price was $27.9–44.2/kg, depending on volume and grade; color concentrates are also available.

Like other thermoplastics, they exhibit melt fracture (32) above certain critical shear rates. In extrusion, many variables control product quality and performance (33).

Melt Processing. Articles are made by injection molding, compression molding, blow molding, transfer molding, rotational molding, extrusion, and coating. Films can be thermoformed and heat sealed (24). Because of high melt viscosity, ETFE resins are usually processed at high (300–340°C) temperatures.

Injection-molded articles shrink about 1.5–2.0% in the direction of resin flow and about 3.5–4.5% in the transverse direction under normal molding conditions. A 25% glass-reinforced composition shrinks only about 0.2–0.3% in the flow direction and about 3.0% in the transverse direction. Although shrinkage depends on shape and processing conditions, uniformity is excellent.

Molten ETFE polymers corrode most metals, and special corrosion-resistant alloys are recommended for long-term processing equipment; short-term prototype runs are possible in standard equipment.

Table 6. Forms of Modified ETFE Resins

Tefzel grade	Form	Melt flow,[a] g/10 min	Application
210	extruded cubes	45	injection molding, thin coating
200	extruded cubes	8	general-purpose, insulation, tubing, fasteners
280	extruded cubes	3	chemical resistance, jacketing, heavy-wall logging cables
HT-2000	compacted powder	8	compounded products
HT-2010	compacted powder	3	compounded products, coating lining
HT-2010	compacted powder	45	coating

[a] At 297°C and 45 N (5 kg) load.

Forming and Machining. Articles can be formed below the melting point with conventional metal-forming techniques. Tetrafluoroethylene–ethylene copolymers are readily machined with the same tools and feed rates as are used for nylon and acetal. For best dimensional stability, the article should be annealed at the expected use temperature before the final machine cut.

Coloring and Decorating. Commercial pigments that are thermally stable at the resin processing temperature may be used. Pigments may be dry-blended with the resin, or ETFE pellets may be blended with color concentrates, which are available in pellet form.

Nontreated surfaces can be hot-printed with special foils in a manner similar to a typewriter ribbon. The type is heated to about 321°C, and a printing pressure of 172–206 kPa (25–30 psi) is applied for about 0.25 s; no further treatment is required.

Stripes may be applied to wire coated with ETFE fluoropolymer over Du-Lite 817-5002 fluoropolymer clear enamel or other bases. Thermally stable pigments are required. Stripes may be applied by gravure-wheel-type applicators and oven-cured in-line.

Assembly. The success of many applications depends on the ability of ETFE fluoropolymer to be economically assembled.

Screw Assembly. Self-tapping screws are used for joining ETFE parts. For maximum holding power, the boss diameter should be about double the screw diameter, and the engagement length about 2.5 times the screw diameter; lubricants should be avoided. Threaded inserts can be molded in place, pressed in, or driven in ultrasonically.

Snap-Fit and Press-Fit Joints. Snap-fit joints offer the advantage that the strength of the joint does not diminish with time because of creep. Press-fit joints are simple and inexpensive, but lose holding power. Creep and stress relaxation reduce the effective interference, as do temperature variations, particularly with materials with different thermal expansions.

Cold or Hot Heading. Rivets or studs can be used in forming permanent mechanical joints. The heading is made with special tools and preferably with the rivet at elevated temperatures. Formed heads tend to recover part of their original shape if exposed to elevated temperatures, resulting in loose joints. Forming at elevated temperature reduces recovery.

Spin Welding. Spin welding is an efficient technique for joining circular surfaces of similar materials. The matching surfaces are rotated at high speed relative to each other and then brought into contact. Frictional heat melts the interface and, when motion is stopped, the weld is allowed to solidify under pressure.

Ultrasonic Welding. Ultrasonic welding has been applied to Tefzel with weld strength up to 80% of the strength of the base resin. Typical conditions include a contact pressure of 172 kPa (25 psi) and 1–2 s cycle time. The two basic designs, the shear and butt joints, employ a small initial contact area to concentrate and direct the high frequency vibrational energy.

Potting. Potting of wire insulated with Tefzel has been accomplished with the aid of a coating of a colloidal silica dispersion. The pots produced with a polysulfide potting compound meeting MIL-S-8516C Class 2 standards exhibit pullout strengths of 111–155 N (25–35 lbf).

Bonding. Surface treatment, such as chemical etch, corona, or flame treatments, is required for adhesive bonding of Tefzel. Polyester and epoxy compounds are suitable adhesives.

Ethylene–tetrafluoroethylene copolymers respond well to melt bonding to untreated aluminum, steel, and copper with peel strengths above 3.5 kN/m (20 lbf/in.). For melt bonding to itself, hot-plate welding is used. The material is heated to 271–276°C, and the parts are pressed together during cooling.

The plasma surface treatment of ETFE to improve adhesion has been studied (34).

Health and Safety

Large quantities of Tefzel have been processed and used in many demanding service applications. No cases of permanent injury have been attributed to these resins, and only limited instances of temporary irritation to the upper respiratory tract have been reported (35).

As with other melt-processable fluoropolymers, trace quantities of harmful gases, including hydrogen fluoride, diffuse from the resin even at room temperature. Therefore, the resins should be used in well-ventilated areas. Even though the resin is physiologically inert and nonirritating to the skin, it is recommended that spills on the skin be washed with soap and water. These resins are stable at 150°C and are recommended for continuous use at this temperature. Degradation, as measured by weight loss, is insignificant up to the melting point of 270°C. At processing temperatures sufficient quantities of irritating and toxic gases are generated to require removal of the gases by exhaust hoods over the die and at the hopper heater. For extrusion into water, a quench tank or partially filled container for purging is recommended. In extrusion operations proper procedures must be maintained to control temperature and pressure. The weight loss with increasing temperature is as follows:

Temperature, °C	Hourly weight loss, %
300	0.05
330	0.26
350	0.86
370	1.60

To remove all decomposition products, a "total-capture" exhaust hood is recommended.

Under normal processing conditions at 300–350°C, Tefzel resins are not subject to autocatalytic degradation. However, extended overheating can result in "blow-backs" through extruder feed hopper or barrel front.

Prolonged soldering in confined spaces with restricted air circulation requires ventilation. A small duct fan is recommended for hot-wire stripping. Tefzel articles should not be exposed to welding conditions.

The limiting oxygen index of Tefzel as measured by the candle test (ASTM D2863) is 30%. Tefzel is rated 94 V-0 by Underwriters' Laboratories, Inc., in their burning test classification for polymeric materials. As a fuel, it has a comparatively low rating. Its heat of combustion is 13.7 MJ/kg (32,500 kcal/kg) compared to 14.9 MJ/kg (35,000 kcal/kg) for poly(vinylidene fluoride) and 46.5 MJ/kg (110,000 kcal/kg) for polyethylene.

Bulk quantities of Tefzel fluoropolymer resins should be stored away from flammable materials. In the event of fire, personnel entering the area should have full protection, including acid-resistant clothing and self-contained breathing apparatus with a full facepiece operated in the pressure-demand or other positive-pressure mode. All types of chemical extinguishers may be used to fight fire involving Tefzel resins. Large quantities of water may be used to cool and extinguish the fire.

The Du Pont Haskell Laboratory for Toxicology and Industrial Medicine has conducted a study to determine the acute inhalation toxicity of fumes evolved from Tefzel fluoropolymers when heated at elevated temperatures. Rats were exposed to decomposition products of Tefzel for 4 h at various temperatures. The approximate lethal temperature (ALT) for Tefzel resins was determined to be 335–350°C. All rats survived exposure to pyrolysis products from Tefzel heated to 300°C for this time period. At the ALT level, death was from pulmonary edema; carbon monoxide poisoning was probably a contributing factor. Hydrolyzable fluoride was present in the pyrolysis products, with concentration dependent on temperature.

Testing and Standards

A description of modified ethylene–tetrafluoroethylene copolymers and their classification is given by the American Society for Testing and Materials under the designation D3159-83 (36). A comprehensive listing of industrial and military specifications is available (37).

Applications

Tefzel 200 is a general-purpose, high temperature resin for insulating and jacketing low voltage power wiring for mass transport systems, wiring for chemical plants, and control and instrumentation wiring for utilities. In injection-molded form, it is used for sockets, connectors, and switch components (38). Because of excellent mechanical properties it provides good service in seal glands, pipe plugs, corrugated tubing, fasteners, and pump vanes. In chemical service, it is used for valve components, laboratory ware, packing, pump impellers, and battery and instrument components.

Tefzel 210, the high melt-flow resin, provides a high speed processing product for use in coating of fine wire and injection molding of thin-walled or intricate shapes. It is also used for other fine-wire applications requiring high line speeds and mechanical strength, but where harsh environmental conditions are not anticipated.

For high temperature wiring with mechanical strength and stress-crack and chemical resistance, Tefzel 280 is preferred. Rated by UL at 150°C, it is widely used for insulating and jacketing heater cables and automotive wiring and for other heavy-wall application where temperatures up to 200°C are experienced for short periods of time or where repeated mechanical stress at 150°C is encountered. It is also suitable for oil-well logging cables and is used in transfer moldings and extrusions for lined chemical equipment. It is injection molded into articles with metal inserts, thick sections, and stock shapes.

BIBLIOGRAPHY

"Tetrafluoroethylene Copolymers With Ethylene" under "Fluorine Compounds, Organic" in *ECT* 3rd ed., Vol. 11, pp. 35–41, by R. L. Johnson, E. I. du Pont de Nemours & Co., Inc.

1. U.S. Pat. 2,468,664 (Apr. 26, 1949), W. E. Hanford and J. R. Roland (to E. I. du Pont de Nemours & Co., Inc.).
2. U.S. Pat. 2,479,367 (Aug. 16, 1949), R. M. Joyce, Jr. (to E. I. du Pont de Nemours & Co., Inc.).
3. Brit. Pat. 1,166,020 (Oct. 1, 1969), M. Modena and co-workers (to Montecatini Edison, SpA).
4. Jpn. Kokai 64 22,586 (Sept. 12, 1964), K. Hirose and co-workers (to Nitto Chemical Industry, Co., Ltd.).
5. Belg. Pat 725,356 (Feb. 14, 1969), Z. Kenkyusho (to Asahi Glass, Ltd.).
6. U.S. Pat. 2,946,763 (July 26, 1960), M. I. Bro and B. W. Sandt (to E. I. du Pont de Nemours & Co., Inc.).
7. D. P. Carlson, *Development of Tefzel Fluoropolymer Resins*, unpublished paper.
8. U.S. Pat. 3,624,250 (Nov. 30, 1971), D. P. Carlson (to E. I. du Pont de Nemours & Co., Inc.).
9. U.S. Pat. 4,123,602 (Oct. 31, 1978), H. Ukihashi and M. Yamake (to Asahi Glass Co.).
10. Brit. Pat. 1,353,535 (May 22, 1974), R. Hartwimmer (to Farbwerke Hoechst, AG).
11. U.S. Pat. 3,401,155 (Sept. 10, 1986), G. Borsini and co-workers (to Montecatini Edison SpA).
12. Y. Tabata, H. Shibano, and H. Sobue, *J. Polym. Sci. Part A* **2**(4), 1977 (1964).
13. U.S. Pat. 3,960,825 (June 1, 1976), D. N. Robinson and co-workers (to Pennwalt Corporation).
14. U.S. Pat. 4,338,237 (July 1982), R. A. Sulzbach and co-workers (to Hoechst Aktiengesellschaft).
15. M. Modena, C. Garbuglio, and M. Ragazzini, *Polym. Lett.* **10**, 153 (1972).
16. M. Modena, C. Garbuglio, and M. Ragazzini, *J. Polym. Sci. Part B* **10**, 153 (1972).
17. F. S. Ingraham and D. F. Wooley, Jr., *Ind. Eng. Chem.* **56**(9), 53 (1964).
18. S. Yano, *J. Polym. Sci. Part A-2* **8**, 1057 (1970).
19. H. W. Starkweather, *J. Polym. Sci. Part A-2* **11**, 587 (1973).
20. F. C. Wilson and H. W. Starkweather, *J. Polym. Sci. Part A-2* **11**, 919 (1973).
21. B. Chu and Chi Wu, *Macromolecules* **20**, 93–98 (1987).
22. B. Chu, C. Wu, and W. Buck, *Macromolecules* **22**, 371 (1989).
23. Z. Wang, A. Tontisakis, W. Tuminello, W. Buck, and B. Chu, *Macromolecules* **23**, 1444 (1990).
24. *Tefzel Fluoropolymer, Design Handbook*, E. I. du Pont de Nemours & Co., Inc., Wilmington, Del., 1973.
25. *Tefzel—Properties Handbook*, E-31301-3, E. I. du Pont de Nemours & Co., Inc., Wilmington, Del., Dec. 1991.
26. S. K. Randa, C. R. Frywald, and D. P. Reifschneider, *Proceedings of the 36th International Wire and Cable Symposium*, Arlington, Va., 1987, pp. 14–22.

27. Can. Pat. 900,075 (May 9, 1972), D. P. Carlson, J. A. Effenberger, and M. B. Polk (to E. I. du Pont de Nemours & Co., Inc.).
28. M. Carpenter, S. Chillons, and R. Will, Society of Automobile Engineers (SAE) International Technical Paper Series 910103, presented at the *International Congress and Exposition*, Detroit, Mich., Feb. 25–Mar. 1, 1991.
29. D. Goldsberry, S. Chillons and R. Will, SAE Technical Paper Series 910104, presented at the *International Congress and Exposition*, Detroit, Mich., Feb. 25–Mar. 1, 1991.
30. U.S. Pat. 3,738,923 (June 12, 1973), D. P. Carlson and N. E. West (E. I. du Pont de Nemours & Co., Inc.).
31. J. Emsley, *New Scientist*, 46 (April 22, 1989).
32. *Technical Information, Jacketing Rate Calculation*, Bulletin No. 10, Fluoropolymers Division Technical Service Laboratory, E. I. du Pont de Nemours & Co., Inc., Wilmington, Del., Aug. 1982.
33. *Extrusion Guide for Melt Processible Fluoropolymers*, Bulletin E-41337, E. I. du Pont de Nemours & Co., Inc., Wilmington, Del.
34. S. Kaplan, O. Kolluri, G. Hansen, R. Rushing, and R. Warren, technical paper presented at *Conference on Adhesives*, Society of Manufacturing Engineers, Atlanta, Ga., Sept. 12, 1989.
35. *Tefzel Fluoropolymers, Safe Handling Guide*, Bulletin E-85785, E. I. du Pont de Nemours & Co., Inc., Wilmington, Del., May 1986.
36. *Modified ETFE-Fluorocarbon Molding and Extrusion Materials*, ASTM D3159-83, American Society for Testing and Materials, Philadelphia, 1987.
37. *J. Teflon* **15**(1), 1974.
38. *Tefzel Fluoropolymers, Product Information* (Du Pont Materials for Wire and Cable), Bulletin E-81467, E. I. du Pont de Nemours & Co., Inc., Wilmington, Del., May 1986.

General Reference

S. V. Gangal, "Tetrafluoroethylene Polymers, Tetrafluoroethylene–Ethylene Copolymers," in J. I. Kroschwitz, ed., *Encyclopedia of Polymer Science and Engineering*, 2nd ed., Vol. 16, Wiley-Interscience, New York, 1989, pp. 626–642.

SUBHASH V. GANGAL
E. I. du Pont de Nemours & Co., Inc.

TETRAFLUOROETHYLENE–PERFLUOROVINYL ETHER COPOLYMERS

Perfluoroalkoxy (PFA) fluorocarbon resins are designed to meet industry's needs in chemical, electrical, and mechanical applications. These melt processable copolymers contain a fluorocarbon backbone in the main chain and randomly distributed perfluorinated ether side chains:

$$-CF_2-CF_2-CF-CF_2-$$
$$|$$
$$O$$
$$|$$
$$C_3F_7$$

A combination of excellent chemical and mechanical properties at elevated temperatures results in reliable, high performance service to the chemical processing and related industries. Chemical inertness, heat resistance, toughness and flexibility, stress-crack resistance, excellent flex life, antistick characteristics, little moisture absorption, nonflammability, and exceptional dielectric properties are among the characteristics of these resins.

The introduction of a perfluoromethyl side chain (Teflon FEP) greatly reduces the crystallinity of PTFE. Crystallinity is reduced even further by replacing the short side chain with a long side chain, such as perfluoropropyl ether. In contrast to Teflon FEP, only a small amount of vinyl ether is required to reduce crystallinity and develop adequate toughness.

Tetrafluoroethylene [116-14-3] and perfluorovinyl ether are copolymerized in aqueous (1,2) or nonaqueous (3) media. The polymer is separated and converted into various forms, such as extruded cubes, powders, beads, or dispersions. This family of products is manufactured by Du Pont, Daikin, and Hoechst and sold under the trade names of Teflon PFA, Neoflon AP, and Hostaflon TFA, respectively.

Monomers

Preparation. The preparation of tetrafluoroethylene has been described previously. Perfluorovinyl ethers (4–7) are prepared by the following steps. Hexafluoropropylene [116-15-4] (HFP) is oxidized to an epoxide HFPO [428-59-1] (5) which, on reaction with perfluorinated acyl fluorides, gives an alkoxyacyl fluoride.

$$CF_3-CF-CF_2 + R_F-C\overset{O}{\underset{F}{\diagup}} \longrightarrow R_F CF_2 OCF-C\overset{O}{\underset{F}{\diagup}}$$
$$\text{HFPO} \qquad\qquad\qquad\qquad\qquad\qquad\quad CF_3$$

The alkoxyacyl fluoride is converted to vinyl ethers by treatment with base at ca 300°C (8).

$$R_F CF_2 OCF-C\overset{O}{\underset{F}{\diagup}} + Na_2CO_3 \rightarrow R_F CF_2 OCF{=}CF_2 + 2\ CO_2 + 2\ NaF$$
$$\quad\ \ CF_3$$

where $R_F = F(CF_2)_n$

Alkoxyacyl fluorides are also produced by an electrochemical process (9).

Properties. Properties of perfluoropropyl vinyl ether [1623-05-8] (PPVE), a colorless, odorless liquid (mol wt 266) are shown in Table 1. Perfluoropropyl vinyl ether is an extremely flammable liquid and burns with a colorless flame. It is significantly less toxic than hexafluoropropylene; the average lethal concentration (ALC) is 50,000 ppm (10).

Table 1. Properties of Perfluoropropyl Vinyl Ether, $F_3C-CF_2-CF_2-O-CF=F_2$

Property	Value
critical temperature, K	423.58
critical pressure, MPa[a]	1.9
critical volume, cm^3/mol	435
surface tension, mN/m(=dyn/cm)	9.9
boiling point, °C	36
specific gravity at 23°C	1.53
vapor density at 75°C, g/cm^3	0.2
vapor pressure at 25°C, kPa[b]	70.3
solubility in water	0
odor	none
color	colorless
flash point, °C	−20
flammable limits in air,[c] % by vol	1

[a] To convert MPa to atm, divide by 0.1013.
[b] To convert kPa to psi (psia), multiply by 0.145.
[c] Extremely flammable.

Copolymerization

Tetrafluoroethylene–perfluoropropyl vinyl ether copolymers [26655-00-5] are made in aqueous (1,2) or nonaqueous media (3). In aqueous copolymerizations water-soluble initiators and a perfluorinated emulsifying agent are used. Molecular weight and molecular weight distribution are controlled by a chain-transfer agent. Sometimes a second phase is added to the reaction medium to improve the distribution of the vinyl ether in the polymer (11); a buffer is also added.

In nonaqueous copolymerization, fluorinated acyl peroxides are used as initiators that are soluble in the medium (12); a chain-transfer agent may be added for molecular weight control.

Temperatures range from 15 to 95°C, and the pressures from 0.45 to 3.55 MPa (65–515 psi). The temperatures used for the aqueous process are higher than those for the nonaqueous process.

Alkyl vinyl ethers tend to rearrange when exposed to free radicals (13). Temperatures must be kept low enough to prevent termination by free-radical coupling. In the aqueous process, temperatures below 80°C minimize the number of acid end groups derived from vinyl ether transfer. In the nonaqueous process, temperature must also be limited to avoid excessive vinyl ether transfer as well as reaction with the solvent. End groups are stabilized by treating the polymer with methanol, ammonia, or amines (14–16). Treatment of PFA with elemental fluorine generates CF_3 end groups and a very low level of contamination (17) which is important for the semiconductor industry (18).

The polymer is separated from the medium and converted to useful forms such as melt-extruded cubes for melt processible applications. Teflon PFA is also available as a dispersion, a fine powder, or in unmelted bead form.

Properties

The melting point of commercial Teflon PFA is 305°C, ie, between those of PTFE and FEP. Second-order transitions are at −100, −30, and 90°C, as determined by a torsion pendulum (21). The crystallinity of the virgin resin is 65–75%. Specific gravity and crystallinity increase as the cooling rate is reduced. An ice-quenched sample with 48% crystallinity has a specific gravity of 2.123, whereas the press-cooled sample has a crystallinity of 58% and a specific gravity of 2.157.

Mechanical Properties. Table 2 shows the physical properties of Teflon PFA (22,23). At 20–25°C the mechanical properties of PFA, FEP, and PTFE are

Table 2. Properties of Teflon PFA

Property	ASTM method	Teflon 340	Teflon 350
nominal melting point, °C		302–306	302–306
specific gravity	D3307	10.6	1.8
continuous use temp, °C		260	260
tensile strength, MPa[a]			
at 23°C	D1708	28	31
at 250°C		12	14
tensile yield, MPa[a]			
at 23°C	D1708	14	15
at 250°C		3.5	4.1
ultimate elongation, %			
at 23°C		300	300
at 250°C	D1708	480	500
flexural modulus, MPa[a]			
at 23°C	D790	655	690
at 250°C		55	69
creep resistance[b] tensile modulus, MPa[a]			
at 20°C	D695	270	270
at 250°C	D695	41	41
hardness Durometer	D2240	D60	D60
MIT folding endurance, 775–200 μm film thickness, cycles		50,000	500,000
water absorption, %	D570	0.03	0.03
coefficient of linear thermal expansion, per °C, $\times 10^{-5}$	D696		
20–100°C		12	12
100–150°C		17	17
150–210°C		20	20

[a]To convert MPa to psi, multiply by 145.
[b]Apparent modulus after 10 h: stress = 6.89 MPa at 20°C, 6.89 kPa at 250°C.

similar; differences between PFA and FEP become significant as the temperature is increased. The latter should not be used above 200°C, whereas PFA can be used up to 260°C. Tests at liquid nitrogen temperature indicate that PFA performs well in cryogenic applications (Table 3).

Unfilled Teflon PFA has been tested in mechanical applications using Teflon FEP-100 as a control (24). Tests were run on molded thrust bearings at 689.5 kPa (100 psi) against AISI 1080, Rc 20, 16AA steel, and at ambient conditions in air without lubrication. A limiting PV value of 5000 was found. Wear factors and dynamic coefficients of friction are shown in Table 4.

Table 3. Cryogenic Properties of Teflon PFA Resins

Property	ASTM method	At 23°C	At −196°C
yield strength, MPa[a]	D1708[b]	15	
ultimate tensile strength, MPa[a]	D1708[b]	18	129
elongation, %	D1708[b]	260	8
flexural modulus, MPa[a]	D790-71[c]	558	5790
impact strength, notched Izod, J/m[d]	D256-72a[e]	no break	64
compressive strength, MPa[a]	D695		414
compressive strain, %	D695		35
modulus of elasticity, MPa[a]	D695		4690

[a]To convert MPa to psi, multiply by 145.
[b]Crosshead speed B, 1.3 mm/min; used at both temperatures for more direct comparison.
[c]Method 1, procedure B.
[d]To convert J/m to ftlbf/in., divide by 53.38.
[e]Method A, head weight is 4.5 kg at 23°C and 0.9 kg at 160°C.

Table 4. Teflon PFA Fluorocarbon Resin Thrust-Bearing Wear-Test Results[a]

Velocity, m/min	Wear factor $K \times 10^{-17}$, 1/Pa[b]	Dynamic coefficient of friction	Test duration, h
Teflon PFA TE-9704			
0.91	3.12	0.210	103
3.05	3.67	0.214	103
9.1	1.96	0.229	103
15.24	1.38	0.289	103
Teflon FEP-100			
0.91	3.71	0.341	104
3.05	2.19	0.330	104
9.1	3.16	0.364	104
15.24	1.60	0.296	103

[a]Mating surface: AISA 1018 steel, Rc 20, 16AA; contact pressure; 689 kPa; at 20°C in air; no lubricant.
[b]To convert 1/Pa to (in.^{3}·min)/(ftlbf·h), divide by 2×10^{-7}.

Hardness (qv) is determined according to ASTM D2240 on $7.6 \times 12.7 \times 0.48$ cm injection-molded panels (25). Results on the D scale are 63–65 for Teflon PFA and 63–66 for Teflon FEP.

Chemical Properties. A combination of excellent chemical and mechanical properties at elevated temperatures result in high performance service in the chemical processing industry. Teflon PFA resins have been exposed to a variety of organic and inorganic compounds commonly encountered in chemical service (26). They are not attacked by inorganic acids, bases, halogens, metal salt solutions, organic acids, and anhydrides. Aromatic and aliphatic hydrocarbons, alcohols, aldehydes, ketones, ethers, amines, esters, chlorinated compounds, and other polymer solvents have little effect. However, like other perfluorinated polymers, they react with alkali metals and elemental fluorine.

Thermal Stability. Teflon PFA resins are very stable and can be processed up to 425°C. Thermal degradation is a function of temperature and time. A significant increase in melt flow rate indicates degradation after a short time above 425°C; at lower temperatures degradation takes longer. Degradation is not significant if the change in melt flow rate of the resin during molding is below 20%. Degradation is also indicated by the formation of small bubbles or discoloration; however, high stock temperatures may cause slight discoloration without adversely affecting properties.

Heat aging at 285°C, a temperature slightly below but near the melting point, increases the strength of Teflon PFA. Samples aged in a circulating air oven for 7500 h at 285°C show a decrease in melt flow number as defined by ASTM D2116. A decline in melt flow number indicates an increase in average molecular weight, which is also indicated by a 25% increase in tensile strength and enhanced ultimate elongation. Toughness is also measured by MIT flex life, which improves severalfold on heat aging at 285°C.

When exposed to fire, Teflon PFA contributes little in fuel value and is self-extinguishing when the flame is removed. The fuel value is approximately 5.4 MJ/kg (2324 Btu/lb). It passes the UL 83 vertical-flame test and is classified as 94VE-O according to UL 94. The limiting oxygen index (LOI) by ASTM D2863 is above 95%.

Electrical Properties. The electrical properties of Teflon PFA are given in Table 5. The dielectric constant of PFA resins is about 2.06 over a wide range of frequencies (10^2–2.4×10^{10} Hz), temperatures, and densities (ASTM D150). The values for PFA density vary only slightly, 2.13–2.17, and the dielectric constant varies only about 0.03 units over this range, among the lowest of all solid materials. Humidity has no measurable effect on the dielectric constant of PFA. The dielectric strength (short-term) of PFA resins is 80 kV/mm (0.25-mm films, ASTM D149); FEP films give similar results, whereas PTFE films are typically measured at 47 kV/mm. Like other fluoropolymer resins, PFA loses dielectric strength in the presence of corona discharge. The dissipation factor at low frequency (10^2–10^4 Hz) decreases with increasing frequency and decreasing temperature. Temperature and frequency have little influence on the dissipation factor over the frequency range 10^4–10^7 Hz. As frequencies increase to 10^{10} Hz, there is a steady increase in dissipation factor. Above 10^7 Hz, increases measured at room temperature are highest; a maximum at about 3×10^9 Hz is indicated. The higher dissipation factor with increasing frequency should be considered in electrical

Table 5. Electrical Properties of Teflon PFA

Property	Value
dielectric strength,[a] kV/m	79
volume resistivity,[b] ohm·cm	10^{18}
surface resistivity,[b] ohms/sq	10^{18}
dissipation factor[c]	
at 10^2 Hz	0.000027
at 10^6 Hz	0.000080
at 10^7 Hz	0.000145
at 10^9 Hz	0.00115
at 3×10^9 Hz	0.00144
at 1.4×10^{10} Hz	0.00131
at 2.4×10^{10} Hz	0.00124

[a] Short-term, 250-µm-thick sample.
[b] ASTM method D257.
[c] ASTM method D150.

insulation applications at high frequencies. The volume and surface resistivities of fluorocarbon resins are high and are not affected by time or temperature. When tested with stainless steel electrodes (ASTM D495), no tracking was observed for the duration of the test (180 s), indicating that PFA resin does not form a carbonized conducting path (27,28).

Optical Properties and Radiation Effects. Within the range of wavelengths measured (uv, visible, and near-ir radiation), Teflon PFA fluorocarbon film transmits slightly less energy than FEP film (29) (Table 6). In thin sections, the resin is colorless and transparent; in thicker sections, it becomes translucent. It is highly transparent to ir radiation; uv absorption is low in thin sections. Weather-O-Meter tests indicate unlimited outdoor life.

Like other perfluoropolymers, Teflon PFA is not highly resistant to radiation (30). Radiation resistance is improved in vacuum, and strength and elongation are increased more after low dosages (up to 30 kGy or 3 Mrad) than with FEP or PTFE. Teflon PFA approaches the performance of PTFE between 30 and 100 kGy

Table 6. Optical Properties of Teflon PFA Film

Property	ASTM method	Value
refractive index[a]	D542-50	1.350 ± 0.002
haze, %	D1003-52	4
light transmission, %		
uv,[b] 0.25–0.40 µm		55–80
visible, 0.40–0.70 µm		80–87
infrared, 0.70–2.1 µm		87–93

[a] Measured at 546 nm and 20°C.
[b] Cary Model Spectrophotometer.

Table 7. Effects of Radiation on Tensile Strength of PFA[a]

Exposure, kGy[b]	ASTM D1708 Tensile strength, MPa[c]	Elongation, %
0	30.27	358
5	28.20	366
10	24.96	333
20	21.24	302
50	14.55	35
200		<5
500		<5

[a]Sample: 250-μm compression-molded films of Teflon PFA 340 from G.E. resonance transformer 2 MeV capacity, at a current of 1 mA.
[b]To convert kGy to Mrad, multiply by 0.1.
[c]To convert MPa to psi, multiply by 145.

(3–10 Mrad) and embrittles above 100 kGy (10 Mrads). At 500 kGy (50 Mrad) PTFE, FEP, and PFA are degraded. The effect of radiation on tensile strength and elongation is shown in Table 7.

Fabrication

Teflon PFA resins are fabricated by the conventional melt-processing techniques used for thermoplastics. Processing equipment is constructed of corrosion-resistant materials and can be operated at 315–425°C. A general-purpose grade, PFA 340, is designed for a variety of molding and extrusion applications, including tubing, shapes, and molded components, in addition to insulation for electrical wire and cables. Because of the excellent thermal stability of PFA 350, a wide range of melt temperatures can be used for fabrication. Extrusion temperatures are 20–26°C above the melting point.

Teflon PFA 440 HP is a chemically modified form of PFA 340 that provides additional benefits such as enhanced purity and improved thermal stability. This product is suitable for producing tubing, pipe linings for production of ultrapure chemicals, semiconductor components, and fluid handling systems for high performance filters (31).

Extrusion. Like other thermoplastics, Teflon PFA resin exhibits melt fracture above certain critical shear rates. For example, samples at 372°C and 5-kg load show the following behavior:

Teflon PFA	Melt flow, g/10 min	Critical shear rate, s^{-1}
340	14	50
310	6	16
350	2	6

Because Teflon PFA melt is corrosive to most metals, special corrosion-resistant alloys must be used for the extrusion equipment, such as Hastelloy C, Monel 400, and Xaloy 306. Barrels, liners, screws, adapters, breaker plates, and dies are made of corrosion-resistant metals (32). Corrosion is promoted by resin degradation and high processing temperatures, long residence times, or dead spots. Extruders used with Teflon FEP are also suitable for PFA resins. Heaters and controllers capable of accurate operation in the range of 330–425°C are required. Extruder barrels should have three or four independently controlled heating zones, each equipped with its own thermocouple and temperature-indicating control.

The screw consists of a feed section, a rapid transition section, and a metering section; a rounded forward end prevents stagnation. The breaker plate that converts the rotary motion of the melt into smooth, straight flow should have as many holes as possible; both ends of each hole should be countersunk for streamlined flow.

The temperature of the melt downstream from the breaker plate may exceed the front barrel temperature, because of the mechanical work transmitted to the resin by the screw; it varies with screw speed and flow rate. The melt temperature is measured by a thermocouple inserted into the melt downstream from the breaker plate. A hooded exhaust placed over the extruder die and feed hopper removes decomposition products when the extrudate is heated.

High melt strength of Teflon PFA 350 permits large reductions in the cross section of the extrudate by drawing the melt in air after it leaves the die orifice (33). At a given temperature, the allowable flow rates are limited at the low end by resin degradation and at the high end by the onset of melt fracture. A broader range of specific gravities (2.13–2.17) may be obtained in articles fabricated from PFA 350 than with FEP 160. Unlike with polytetrafluoroethylene, higher crystallinity in PFA seems to have little effect on flex life.

Injection Molding. Any standard design plunger or reciprocating screw injection machine can be used for PFA 340, although a reciprocating screw machine is preferred (32). Slow injection into mold cavities avoids surface or internal melt fracture, and control of ram speed is important at low speed. Corrosion-resistant metals are used for parts in continuous contact with molten resin; Hastelloy C, and Xaloy 306 or 800 are recommended.

Because the mold is usually maintained at temperatures below the melting point of the resin, corrosion on the mold surface is less than in the molding machine. Nonreturn ball check valves and ring check valves are used; the latter is preferred for PFA. A streamlined flow must pass through the valve, preventing areas of stagnant flow or holdup and localized degradation.

A smear head causes less stagnation and overpacking than a nonreturn valve. A conventional-type reverse-tapered nozzle with the bore as large as possible without sudden changes in diameter is preferred. Independently controlled, zone-type heaters for heating the nozzle and at least two zones on the cylinder are used.

At a holdup time longer than 10–15 min at a high temperature, resin degradation is avoided by keeping the rear of the cylinder at a lower temperature than the front. At short holdup times (4–5 min), cylinder temperatures are the same in rear and front. If melt fracture occurs, the injection rate is reduced; pressures are in the range of 20.6–55.1 MPa (3000–8000 psi). Low back pressure and screw rotation rates should be used.

The cycle can usually be estimated on the basis of about 30 s/3 mm of thickness; most of it is devoted to ram-in-motion time (except for very thin sections). The mold temperature used with PFA 340 is often the highest temperature that allows the part to be ejected undamaged from the mold and retain its shape while cooling.

The resin must be of highest purity for optimum processing characteristics and properties. Degradation results in discoloration, bubbling, and change in melt flow rate.

Transfer Molding. Valve and fitting liners are made by a transfer-molding process (33), with the valve or fitting serving as the mold. Melted resin is forced into the fitting at a temperature above the melting point of the resin. The melt may be produced by an extruder or an injection molding machine or melted cubes contained in a melt pot and transferred by applying pressure to a piston in the pot. After the resin transfer is completed, the fitting is cooled under pressure. Stock temperatures of 350–380°C and fitting temperatures of 350–370°C are used to process PFA 350. A slight adjustment in the cooling cycles may be required for transfer molding PFA 350 because it has higher melting and freezing points than FEP.

Rotocasting Teflon PFA Beads. The resin has sufficient thermal stability for a commercial rotocasting operation; that is, TE-9738 has a melting point of about 303°C. In rotocasting trials, incoming flue gas temperatures of 355–365°C (34,35) and heat cycles of 90–180 min have been used. Conventional rotations for major and minor axes can be applied without modifications; Freecote 33 performs adequately as a mold release agent. Mold release instructions can be followed without modification. Heating cycles, including a preheat and a fusion stage, give consistent rotocasting. Preheating at 15–30°C below the fusion temperature takes 10–25 min. Heat-cycled Teflon PFA rotocastings are translucent white, often with bluish tinge. Rotocastings that have been heated too long may darken to a translucent brown. Uniform cooling is essential for undistorted, stress-free products; combinations of air and water are employed. The rotocasting is cooled below the resin melting temperature with air at ambient temperature and then with a water spray, and finally with a stream of air.

Dispersion Processing. A commercial aqueous dispersion of Teflon PFA 335 contains more than 50 wt % PFA particles, about 5 wt % surfactants and fillers. This dispersion is processed by the same technique as for PTFE dispersion. It is used for coating various surfaces, including metal, glass, and glass fabrics. A thin layer of Teflon PFA coating can also serve as an adhesive layer for PTFE topcoat.

Powder Coating. Teflon PFA is also available in a finely divided powder form. It can be used to produce thin layers on various surfaces by heating these surfaces above the melting point of PFA and then bringing the powder in contact with them. This allows a thin layer of the powder to melt on the surface of the substrate.

For some applications the powder is suspended in an aqueous medium or a solvent with the help of emulsifying agents and then sprayed onto the substrate.

The powder is also used as a filler to prepare sprayable compositions of PTFE dispersions, which then can be used to coat various substrates (36).

Pigmentation. Commercial color concentrates of Teflon PFA containing approximately 2% pigment can be easily dispersed in clear extruded cubes. The resin can also be dry-blended with stable inorganic pigments. At 0.1–1% concentration, the pigment has no appreciable effect on the dielectric strength and constant or mechanical properties. The dissipation factor of pigmented resin varies with the type and concentration of the pigment.

Pigment used for dry blending is dried overnight at 150°C in a vacuum oven to remove absorbed gases and moisture. It is screened through a 149-μm (100-mesh) screen directly onto the cubes, which are rolled or tumbled for at least 15 min. The pigmented resin is stored in an airtight container to prevent absorption of moisture.

Health and Safety

Safe practices employed for handling PTFE and FEP resins are adequate for Teflon PFA (37); adequate ventilation is required for processing above 330–355°C. In rotoprocessing, a vacuum (250–750 Pa or 1.8–5.6 mm Hg) in the oven ensures exhaust to the outside (36). Removal of end caps or opening of sealed parts in a well-ventilated area ensures ventilation of decomposition fumes. During rotoprocessing, molds should be vented.

Applications and Economic Aspects

The perfluorovinyl ether comonomer used for PFA is expensive, as is PFA. Most PFA grades are sold as extruded, translucent cubes in various colors at $47.9–60.3/kg. Some PFA types are also marketed in nonextruded forms.

Teflon PFA can be fabricated into high temperature electrical insulation and components and materials for mechanical parts requiring long flex life. Teflon PFA 350 is used as liner for chemical process equipment, specialty tubing, and molded articles for a variety of applications. Teflon PFA 340 is a general-purpose resin for tubing, shapes, primary insulation, wire and cable jacketing, injection- and blow-molded components, and compression-molded articles. Teflon PFA 440 HP is a chemically modified form of PFA-340 with enhanced purity and improved thermal stability while processing. This resin is suitable in semiconductor manufacturing, fluid handling systems for industry or life sciences, and instrumentation for precise measurements of fluid systems.

BIBLIOGRAPHY

"Tetrafluoroethylene Copolymers with Perfluorovinyl Ethers" under "Fluorine Compounds, Organic," in *ECT* 3rd ed., Vol. 11, pp. 42–49, by R. L. Johnson, E. I. du Pont de Nemours & Co., Inc.

1. U.S. Pat. 3,132,123 (May 5, 1964), J. F. Harris and D. I. McCane (to E. I. du Pont de Nemours & Co., Inc.).
2. U.S. Pat. 3,635,926 (Jan. 18, 1972) W. F. Fresham and A. F. Vogelpohl (to E. I. du Pont de Nemours & Co., Inc.).
3. U.S. Pat. 3,536,733 (Oct. 27, 1970), D. P. Carlson (to E. I. du Pont de Nemours & Co.).
4. U.S. Pat. 3,358,003 (Dec. 12, 1967), H. S. Eleuterio and R. W. Meschke (to E. I. du Pont de Nemours & Co., Inc.).
5. U.S. Pat. 3,180,895 (Apr. 27, 1965), J. F. Harris and D. I. McCane (to E. I. du Pont de Nemours & Co., Inc.).
6. U.S. Pat. 3,250,808 (Oct. 10, 1966), E. P. Moore, A. S. Milian, Jr., and H. S. Eleuterio (to E. I. du Pont de Nemours & Co., Inc.).
7. U.S. Pat. 4,118,421 (Oct. 3, 1978), T. Martini (to Hoechst Aktiengesellschaft).
8. U.S. Pat. 3,291,843 (Dec. 13, 1966), C. G. Fritz and S. Selman (to E. I. du Pont de Nemours & Co., Inc.).
9. U.S. Pat. 2,713,593 (July 1955), T. J. Brice and W. H. Pearlson (to Minnesota Mining and Manufacturing Co.).
10. A. H. Olson, E. I. du Pont de Nemours & Co., Inc., private communication, 1992.
11. U.S. Pat. 4,499,249 (Feb. 12, 1985), S. Nakagawa and co-workers (to Daikin Kogyo Co., Ltd.).
12. U.S. Pat. 2,792,423 (May 14, 1957), D. M. Young and W. N. Stoops (to Union Carbide and Carbon Corp.).
13. R. E. Putnam, in R. B. Seymour and G. S. Kirshenbaum, eds., *High Performance Polymers: Their Origin and Development*, Elsevier Scientific Publishing, Inc., New York, 1986, p. 279.
14. U.S. Pat. 3,674,758 (July 4, 1972), D. P. Carlson (to E. I. du Pont de Nemours & Co., Inc.).
15. U.S. Pat. 4,599,386 (July 8, 1986), D. P. Carlson and co-workers (to E. I. du Pont de Nemours & Co., Inc.).
16. PCT Int. Appl. WO 89,11,495 (1989), M. D. Buckmaster (to E. I. du Pont de Nemours & Co., Inc.).
17. U.S. Pat. 4,943,658 (1988), J. Imbalzano and D. Kerbow (to E. I. du Pont de Nemours & Co., Inc.).
18. C. J. Goodman and S. Andrews, *Solid State Technol.*, 65 (July 1990).
19. *PFA-Fluorocarbon Molding and Extrusion Materials*, ASTM 3307-86, American Society for Testing and Materials, Philadelphia, Pa., 1987.
20. *J. Teflon*, **15**(1) (1974).
21. R. A. Darby, E. I. du Pont de Nemours & Co., Inc., private communication, 1992.
22. *PFA Fluorocarbon Resins*, sales brochure, E08572, E. I. du Pont de Nemours & Co., Inc., Wilmington, Del.
23. M. I. Bro and co-workers, *29th International Wire and Cable Symposium*, Cherry Hill, N.J., Nov. 1980.
24. *Teflon PFA Fluorocarbon Resins: Wear and Frictional Data*, APD #2 bulletin, E. I. du Pont de Nemours & Co., Inc., Wilmington, Del., 1973.
25. *Teflon PFA Fluorocarbon Resins: Hardness*, APD #4 bulletin, E. I. du Pont de Nemours & Co., Inc., Wilmington, Del., 1973.
26. *Teflon PFA Fluorocarbon Resins: Chemical Resistance*, PIB #2 bulletin, E. I. du Pont de Nemours & Co., Inc., Wilmington, Del., 1972.
27. E. W. Fasig, D. I. McCane, and J. R. Perkins, paper presented at the *22nd International Wire and Cable Symposium*, Atlantic City, N.J., Dec. 1973.
28. *Handbook of Properties for Teflon PFA*, sales brochure, E46679, E. I. du Pont de Nemours & Co., Inc., Wilmington, Del., Oct. 1987.

29. *Teflon PFA Fluorocarbon Resins: Optical Properties*, APD #6 bulletin, E. I. du Pont de Nemours & Co., Inc., Wilmington, Del., 1973.
30. *Teflon PFA Fluorocarbon Resins: Response to Radiation*, APD #3 bulletin, E. I. du Pont de Nemours & Co., Inc., Wilmington, Del., 1973.
31. *Teflon PFA 440 HP*, product information, H-27760, E. I. du Pont de Nemours & Co., Inc., Wilmington, Del., 1990.
32. *Teflon PFA Fluorocarbon Resin: Injection Molding of Teflon PFA TE-9704*, PIB #4 bulletin, E. I. du Pont de Nemours & Co., Inc., Wilmington, Del., 1973.
33. *Teflon PFA Fluorocarbon Resins: Melt Processing of Teflon PFA TE-9705*, PIB #1 bulletin, E. I. du Pont de Nemours & Co., Inc., Wilmington, Del., 1973.
34. *Technical Information, No. 11, Processing Guidelines for Du Pont Fluoropolymer Rotocasting Powders of Tefzel and Teflon PFA*, E. I. du Pont de Nemours & Co., Inc., Wilmington, Del., 1982.
35. *Teflon PFA TE-9783 Rotation Molding Powder*, technical information, H-26600, E. I. du Pont de Nemours & Co., Inc., Wilmington, Del., June 1990.
36. Brit. Pat. 2,051,091B (Feb. 9, 1983), J. E. Bucino (to Fluorocoat Ltd.).
37. *Handling and Use of Teflon Fluorocarbon Resins at High Temperatures*, bulletin, E. I. du Pont de Nemours & Co., Inc., Wilmington, Del., 1961.

General Reference

S. V. Gangal, in J. I. Kroschwitz, ed., *Encyclopedia of Polymer Science and Engineering*, 2nd ed., Vol. 16, Wiley-Interscience, New York, 1989, pp. 614–626.

SUBHASH V. GANGAL
E. I. du Pont de Nemours & Co., Inc.

POLY(VINYL FLUORIDE)

Homopolymers and copolymers of vinyl fluoride are based on free-radical polymerization of vinyl fluoride and comonomers, usually under high pressure. Du Pont first commercialized a poly(vinyl fluoride)-based film in 1961 under the trade name Tedlar. Poly(vinyl fluoride) homopolymers and copolymers have excellent resistance to sunlight degradation, chemical attack, water absorption, and solvent, and have a high solar energy transmittance rate. These properties have resulted in the utilization of poly(vinyl fluoride) (PVF) film and coating in outdoor and indoor functional and decorative applications. These films are used where exceptional high temperature stability, outdoor longevity, stain resistance, adherence, and release properties are required.

Monomer

Vinyl fluoride [75-02-5] (VF) (fluoroethene) is a colorless gas at ambient conditions. It was first prepared by reaction of 1,1-difluoro-2-bromoethane [359-07-9] with zinc (1). Most approaches to vinyl fluoride synthesis have employed reactions of acetylene [74-86-2] with hydrogen fluoride (HF) either directly (2–5) or utilizing catalysts (3,6–10). Other routes have involved ethylene [74-85-1] and HF (11), pyrolysis of 1,1-difluoroethane [624-72-6] (12,13) and fluorochloroethanes (14–18), reaction of 1,1-difluoroethane with acetylene (19,20), and halogen exchange of vinyl chloride [75-01-4] with HF (21–23). Physical properties of vinyl fluoride are given in Table 1.

Table 1. Physical Properties of Vinyl Fluoride

Property	Value
molecular weight	46.04
boiling point, °C	−72.2
freezing point, °C	−160.5
critical temperature, °C	54.7
critical pressure, MPa^a	5.1
critical density, g/cm^3	0.320
liquid density at 21°C, g/cm^3	0.636
vapor pressure at 21°C, MPa^a	2.5
solubility in water at 80°C, g/100 g H$_2$O	
at 3.4 MPa^a	0.94
at 6.9 MPa^a	1.54

^{a}To convert MPa to atm, divide by 0.101.

Polymerization

Vinyl fluoride undergoes free-radical polymerization. The first polymerization involved heating a saturated solution of VF in toluene at 67°C under 600 MPa (87,000 psi) for 16 h (24). A wide variety of initiators and polymerization conditions have been explored (25–27). Examples of bulk (28,29) and solution (25,28,30,31) polymerizations exist; however, aqueous suspension or emulsion methods are generally preferred (26,32–40). VF volatility dictates that moderately high pressures be used. Photopolymerizations, usually incorporating free-radical initiators, are also known (26,28,29,35).

The course of VF polymerizations is dominated by the high energy and hence high reactivity of the propagating VF radical. The fluorine substituent provides little resonance stabilization, leading to a propagating intermediate which is indiscriminate in its reactions. Monomer reversals, branching, and chain-transfer reactions are common. The reactivity of the vinyl fluoride radical limits the choice of polymerization medium, surfactants, initiators, or other additives and makes impurity control important. Species which can participate in chain transfer or incorporate in the polymer can depress molecular weight or degrade the thermal stability characteristics of the final polymer.

The combination of triisobutylborane [1116-39-8] and oxygen has been used to polymerize VF at reduced temperature and pressure (41). Polymerization temperature was varied from 0 to 85°C with a corresponding drop in melting point from about 230°C (0°C polymerization) to about 200°C (85°C polymerization). This dependance of melting temperature, and degree of crystallinity, have been interpreted in terms of variations in the extent of monomer reversals during polymerization (42). Copolymers of VF with vinylidene fluoride [75-38-7] and tetrafluoroethylene [116-14-3] also have been prepared with this initiation system. VF tends toward alternation with tetrafluoroethylene and incorporates preferentially in copolymerization with vinylidene fluoride (see FLUORINE COMPOUNDS, ORGANIC–POLYTETRAFLUOROETHYLENE; POLY(VINYLIDENE FLUORIDE)).

Copolymers of VF and a wide variety of other monomers have been prepared (6,41–48). The high energy of the propagating vinyl fluoride radical strongly influences the course of these polymerizations. VF incorporates well with other monomers that do not produce stable free radicals, such as ethylene and vinyl acetate, but is sparingly incorporated with more stable radicals such as acrylonitrile [107-13-1] and vinyl chloride. An Alfrey-Price Q value of 0.010 ± 0.005 and an e value of 0.8 ± 0.2 have been determined (49). The low value of Q is consistent with little resonance stability and the e value is suggestive of an electron-rich monomer.

Polymer Properties

Poly(vinyl fluoride) [24981-14-4] (PVF) is a semicrystalline polymer with a planar, zig-zag configuration (50). The degree of crystallinity can vary significantly from 20–60% (51) and is thought to be primarily a function of defect structures. Wideline nmr and x-ray diffraction studies show the unit cell to contain two monomer units and have the dimensions of $a = 0.857$ nm, $b = 0.495$ nm, and $c = 0.252$ nm (52). Similarity to the phase I crystal form of poly(vinylidene fluoride) suggests an orthorhombic crystal (53).

The relationship of polymer structure to melting point and degree of crystallinity has been the subject of controversy. Head-to-head regio irregularities in PVF are known (51,54,55) and the concentration of such units has been suggested as the source of variations in melting point (42,47,56). Commercial PVF contains approximately 12% head-to-head linkages by ^{19}F-nmr and displays a peak melting point of about 190°C (47,48,57,58). Both nmr and ir studies have shown PVF to be atactic (47,51,54,55,59–62) and, as such, variations in stereoregularity are not thought to be a contributor to variations in melting point.

PVF with controlled amounts of head-to-head units varying from 0 to 30% have been prepared (47,48) by using a chlorine substituent to direct the course of polymerization of chlorofluoroethylenes and then reductively dechlorinating the products with tributyltin hydride. This series of polymers shows melting point distributions ranging from about 220°C for purely head-to-tail polymer down to about 160°C for polymer containing 30% head-to-head linkages. This study, however, does not report the extent of branching in these polymers. Further work has shown that the extent of branching has a pronounced effect upon melting temperature (57,58). Change of polymerization temperature from 90 to 40°C produces a change in branch frequency from 1.35 to 0.3%, while the frequency of monomer reversals is nearly constant ($12.5 \pm 1\%$). The peak melting point for this series varies from 186°C (90°C polymerization) to 206°C (40°C polymerization).

PVF displays several transitions below the melting temperature. The measured transition temperatures vary with the technique used for measurement. T_g (L) (lower) occurs at -15 to -20°C and is ascribed to relaxation free from restraint by crystallites. T_g (U) (upper) is in the 40 to 50°C range and is associated with amorphous regions under restraint by crystallites (63). Another transition at -80°C has been ascribed to short-chain amorphous relaxation and one at 150°C associated with premelting intracrystalline relaxation.

PVF has low solubility in all solvents below about 100°C (61). Polymers with greater solubility have been prepared using 0.1% 2-propanol polymerization modifier and were characterized in N,N-dimethylformamide solution containing 0.1 N LiBr. M_n ranged from 76,000 to 234,000 (osmometry), and M_s from 143,000 to 654,000 (sedimentation velocity). Sedimentation velocity molecular weights can be related to intrinsic viscosity using the Mark-Houwink equation:

$$\eta_{inh} = KM^a$$

Using an a value of 0.80, which is typical of an extended polar polymer in good solvent, K is determined to be 6.52×10^{-5} (64).

The conformational characteristics of PVF are the subject of several studies (53,65). The rotational isomeric state (RIS) model has been used to calculate mean square end-to-end distance, dipole moments, and conformational entropies. ^{13}C-nmr chemical shifts are in agreement with these predictions (66). The stiffness parameter (δ) has been calculated (67) using the relationship between chain stiffness and cross-sectional area (68). In comparison to polyethylene, PVF has greater chain stiffness which decreases melting entropy, ie, $(\Delta S)_m = 8.58$ J/(mol·K) [2.05 cal/(mol·K)] versus 10.0 J/(mol·K)[2.38 cal/(mol·K)].

A solubility parameter of 24.5–24.7 MPa$^{1/2}$ [12.0–12.1 (cal/cm^3)$^{1/2}$] has been calculated for PVF using room temperature swelling data (69). The polymer lost solvent to evaporation more rapidly than free solvent alone when exposed to air. This was ascribed to reestablishment of favorable dipole–dipole interactions within the polymer. Infrared spectral shifts for poly(methyl methacrylate) in PVF have been interpreted as evidence of favorable acid–base interactions involving the H from CHF units (70). This is consistent with the greater absorption of pyridine than methyl acetate despite a closer solubility parameter match with methyl acetate.

PVF is more thermally stable than other vinyl halide polymers. High molecular weight PVF is reported to degrade in an inert atmosphere, with concurrent HF loss and backbone cleavage occurring at about 450°C (71,72). In air, HF loss occurs at about 350°C, followed by backbone cleavage around 450°C.

More recent work reports the onset of thermal degradation at lower temperatures and provides a clearer picture of the role of oxygen (73–75). In the presence of oxygen, backbone oxidation and subsequent cleavage reactions initiate decomposition. In the absence of oxygen, dehydrofluorination eventually occurs, but at significantly higher temperatures.

PVF is transparent to radiation in the uv, visible, and near ir regions, transmitting 90% of the radiation from 350 to 2,500 nm. Radiation between 7,000 and 12,000 nm is absorbed (76). Exposure to low dose γ irradiation produces crosslinks in PVF and actually increases tensile strength and etching resistance, whereas the degree of crystallinity and melting point are reduced (77). PVF becomes embrittled upon exposure to electron-beam radiation of 10 MGy (10^9 rad), but resists breakdown at lower doses. It retains its strength at 0.32 MGy (32×10^6 rad) while polytetrafluoroethylene is degraded at 0.02 MGy (2×10^6 rad) (78).

Fabrication and Processing

Commercial PVF is insoluble at room temperature because of the large number of hydrogen bonds and high degree of crystallinity. Some latent solvents solvate PVF at temperatures above 100°C. PVF is converted to thin films and coatings. Processing of PVF, eg, by melt extrusion, depends on latent solvation of PVF in highly polar solvents and its subsequent coalescence. An example is plasticized melt extrusion of PVF into thin films (79). Pigments, stabilizers, plasticizers, and other additives can be incorporated in the film by dispersing them with the polymer in the latent solvent. The solvent is recovered by evaporation after extrusion. The extruded film can be biaxially oriented to varying degrees.

Poly(vinyl fluoride) can be applied to substrates with solvent-based or waterborne dispersions, or by powder-coating techniques. Viscosity modifiers are often needed to obtain a coatable dispersion. Dispersions can be applied by spraying, reverse roll coating, dip coating, and centrifugal casting. Other methods include casting on a continuous belt, extrusion into a hot liquid (80), and dipping a hot article into the dispersion (81).

Table 2 lists properties of PVF films. Various multilayer cast PVF films have been reported (82). Physical and tensile properties of the film depend on the extent of its orientation (83).

Adherability of the film may be enhanced by its treatment with flame, electric discharge, boron trifluoride gas, activated gas plasma, dichromate sulfuric acid, and a solution of alkali metal in liquid ammonia (84–87). A coating of polyurethane, an alkyl polymethacrylate, or a chlorinated adhesive can be applied to PVF surfaces to enhance adhesion (80,88,89).

Economic Aspects

Poly(vinyl fluoride) is available from Du Pont both as a resin and as transparent and pigmented films under the trademark Tedlar PVF film. Films are available in nonoriented and oriented grade in several tensile modifications and thicknesses, with either adherable or nonadherable release-grade surfaces. The 1992 prices ranged from $30 to $70/kg, vs $24 to $62/kg in 1988. Prices for specially tailored films were significantly higher.

Health and Environment

Acute inhalation exposure of rats to 200,000 ppm VF for 30 minutes or more produced weak anaesthesia and no deaths (90). In rats VF is only slightly metabolized at a rate of one-fifth that of vinyl chloride (91–95). An extensive program of toxicity testing of vinyl fluoride is in progress (96,97).

Vinyl fluoride is flammable in air between the limits of 2.6 and 22% by volume. Minimum ignition temperature for VF and air mixtures is 400°C. A small amount, < 0.2%, of terpenes is added to VF to prevent spontaneous polymeriza-

Table 2. Properties of Poly(vinyl fluoride) Film

Property	Value	ASTM[a] test method
Physical and thermal properties		
bursting strength, kPa[b]	200–450[c]	D774
coefficient of friction with metal	0.18–0.21	D1894-78
density, g/cm^3	1.38–1.72	weighed samples
impact strength, kJ/m[d]	43–90	D3420-80
refractive index, n_D	1.46	D542; Abbe refractometer, 30°C
tear strength, kJ/m[d]		
propagated	6–22	D1922-67
initial	129–196	D1004-66
tensile modulus, MPa[e]	44–110	D882
ultimate elongation, %	115–250	D882
ultimate yield strength, MPa[e]	33–41	D882
linear coefficient of expansion, cm/(cm·°C)	0.00005	air oven, 30 min
useful temperature range, °C		
continuous use	−70 to +107	
short cycle (1–2 h)	175	
zero strength, °C	260–300	hot bar
thermal conductivity (1°C/cm), W/(m·K)		
−30°C	0.14	
60°C	0.17	
self-ignition temperature, °C	390	D1929
solar energy transmittance, 359–2500 nm, %	90	E427-71
Permeability		
moisture absorption, %	0.5	D570-81
moisture vapor transmission,[f] nmol/(m^2·s)[g]		
at 7.0 kPa,[h] 39.5°C	4.65–29.4	E96-58T
gas permeability,[f] nmol/(m·s·GPa)[i] at		
98 kPa,[h] 23°C		D1434
carbon dioxide	22.4	
helium	302	
hydrogen	117	
nitrogen	0.5	
oxygen	6.6	
vapor transmission rate,[j] nmol/(m^2·s)[k] at 23.5°C		E96, modified
acetic acid	4.9	
acetone	1570	
benzene	13	
carbon tetrachloride	3.9	
ethyl acetate	13	
hexane	10	
water[l]	22	

Table 2. (*Continued*)

Property	Value	ASTM[a] test method
Electrical properties[m]		
corona endurance, h at 60 Hz, 40 V/μm	2.5–6.0	D2275
dielectric constant at 1 MHz, 23°C	6.2–7.7	D150-81
dielectric strength,		D150-81
short term ac, kV/μm	0.08–0.13	
short term dc, kV/μm	0.15–0.19	
dissipation factor, %		D150-81
1 MHz at 23°C	0.17–0.28	
1 MHz at 100°C	0.09–0.21	
10 kHz at 23°C	0.019–0.019	
10 kHz at 100°C	0.21–0.067	
volume resistivity for transparent film, GΩ·m		D257
23°C	2000–700	
100°C	0.7–2	
surface resistivity, GΩ·m		D257
23°C	60,000–20,000	
100°C	7–20	

[a]Unless otherwise noted.
[b]To convert kPa to psi, multiply by 0.145.
[c]Range dependent on composition and tensile modification.
[d]To convert kJ/m to ft·lbf/in., divide by 0.0534 (ASTM D256).
[e]To convert MPa to psi, multiply by 145.
[f]Measurements made on films of nominal 25 μm thickness.
[g]To convert nmol/(m^2·s) to g/(m^2·d), multiply by 1.94.
[h]To convert kPa to mm Hg, multiply by 7.5.
[i]To convert nmol/(m·s·GPa) to mL·mil/(m^2·d·atm), multiply by 7.725.
[j]At partial pressure of vapor at given temperature.
[k]To convert nmol/(m^2·s) to g/(m^2·d), multiply by 1.94 and by the density.
[l]At 39.5°C.
[m]Range of electrical properties is given; the first value refers to 54.8-μm transparent film, and the second value to 54.8-μm white pigmented film.

tion. The U.S. Department of Transportation has classified the inhibited VF as a flammable gas.

The self-ignition temperature of PVF film is 390°C. The limiting oxygen index (LOI) for PVF is 22.6% (98), which can be raised to 30% in antimony oxide-modified film (99). Hydrogen fluoride and a mixture of aromatic and aliphatic hydrocarbons (100) are generated from the thermal degradation of PVF. Toxicity studies, ie, survival and time to incapacitation, of polymers, cellulosics (101,102), and airplane interior materials (103) expose mice to pyrolysis products and show PVF thermal degradation products to have relatively low toxicity.

Uses

The uses of PVF depend on its weatherability, strength over a wide range of temperatures, and inertness toward a wide variety of chemicals, corrosives, and staining agents. It finds wide use as a protective or decorative coating. It can be applied as a preformed film in a laminating step or from a dispersion in a coating step. It may be transparent or pigmented in a variety of colors. Poly(vinyl fluoride) film is laminated to cellulosics, flexible vinyls, plastics, rubbers, and resin-impregnated felt. These laminated products are applied to exterior wall panels for buildings (104,105), highway sound barriers (106), automobile trim, truck and trailer siding (107), vinyl awnings, backlit signs (108), pipe covering (109), stain-resistant wall coverings, and aircraft cabin interiors (110,111).

On metal or plastic, PVF surfaces serve as a primer coat for painting, eg, automobile parts, or where improved adhesion is desired (112). Because of its moisture impermeability and wide operating temperature range, PVF film is used to fabricate bags to contain glass fiber mats for insulating exterior airplane walls and cargo space, and air conditioning ducts. PVF has long been used to construct bags for sampling gases (113).

Fiber-reinforced panels covered with PVF have been used for greenhouses. Transparent PVF film is used as the cover for flat-plate solar collectors (114) and photovoltaic cells (qv) (115). White PVF pigmented film is used as the bottom surface of photovoltaic cells. Nonadhering film is used as a release sheet in plastics processing, particularly in high temperature pressing of epoxy resins for circuit boards (116–118) and aerospace parts. Dispersions of PVF are coated on the exterior of steel hydraulic brake tubes and fuel lines for corrosion protection.

BIBLIOGRAPHY

"Poly(vinyl fluoride)" under "Fluorine Compounds, Organic," in *ECT* 2nd ed., Vol. 9, pp. 835–840, by L. E. Wolinski, E. I. du Pont de Nemours & Co., Inc.; in *ECT* 3rd ed., Vol. 11, pp. 57–64, by D. E. Brasure, E. I. du Pont de Nemours & Co., Inc.

1. F. Swarts, *Bull. Clin. Sci. Acad. Roy. Belg.* **7**, 383 (1901); F. Swarts, *J. Chem. Soc. Abstr.* **82**, 129 (1902).
2. U.S. Pat. 1,425,130 (Aug. 8, 1922), H. Plauson (to Plauson's Ltd.).
3. U.S. Pat. 2,118,901 (May 31, 1938), J. Soll (to I. G. Farbenindustrie AG).
4. A. V. Grossee and C. B. Linn, *J. Am. Chem. Soc.* **64**, 2289 (1942).
5. A. L. Henne, *Organic Reactions*, Vol. 2, John Wiley & Sons, Inc., New York, 1944.
6. U.S. Pat. 2,419,010 (Apr. 15, 1947), D. D. Coffman and T. A. Ford (to E. I. du Pont de Nemours & Co., Inc.).
7. U.S. Pat. 2,437,307 (Mar. 9, 1948), L. F. Salisburg (to E. I. du Pont de Nemours & Co., Inc.).
8. U.S. Pat. 2,674,632 (Apr. 1, 1954), B. F. Skiles (to E. I. du Pont de Nemours & Co., Inc.).
9. U.S. Pat. 3,178,483 (Apr. 13, 1965), C. M. Christy and G. Teufer (to E. I. du Pont de Nemours & Co., Inc.).
10. U.S. Pat. 3,607,955 (Sept. 21, 1971), L. E. Gardner (to Phillips Petroleum Co.).
11. Jpn. Kokai 77 122,310 (Oct. 14, 1977), T. Kuroda and T. Yamamoto (to Onoda Cement Co., Ltd.).
12. U.S. Pat. 2,442,993 (June 8, 1948), O. W. Cass (to E. I. du Pont de Nemours & Co., Inc.).
13. U.S. Pat. 2,461,523 (Feb. 15, 1949), D. D. Coffman and R. D. Cramer (to E. I. du Pont de Nemours & Co., Inc.).
14. U.S. Pat. 3,621,067 (Nov. 16, 1971), J. Hamersma (to Atlantic Richfield Co.).

15. Jpn. Pat. 46 21,607 (71 21,607) (June 18, 1971), B. Tatsutani and co-workers (to Electro Chemical Industrial Co., Ltd.).
16. U.S. Pat. 3,642,917 (Feb. 15, 1972), J. Hamersma (to Atlantic Richfield Co.).
17. Jpn. Pat. 47 11,728 (72 11,728) (Apr. 12, 1972), B. Tatsutani and co-workers (to Electro Chemical Industrial Co., Ltd.).
18. Jpn. Pat. 51 13,123 (76 13,123) (Apr. 26, 1976), B. Ryutani and co-workers (to Electro Chemical Industrial Co., Ltd.).
19. U.S. Pat. 3,317,619 (May 2, 1967), T. E. Hedge (to Diamond Shamrock Corp.).
20. T. S. Sirlibaev and co-workers, *Zh. Prikl. Khim.* **58** (7), 1666–1668 (1985).
21. Jpn. Pat. 47 11,726 (72 11,726), (Apr. 12, 1972), B. Tatsutani, I. Kobayashi, and K. Yamamoto (to Electro Chemical Industrial Co., Ltd.).
22. T. S. Sirlibaev, A. Akramkhodzhaev, A. A. Yul'chibaev, and K. U. Usmanov, *Synthesis of Vinyl Fluoride and 1,1-difluoroethane from Vinyl Chloride*, Deposited Document, VINITI 48575, USSR, 1975.
23. T. S. Sirlibaev and co-workers, *Uzb. Khim. Zh.* (1), 29–31 (1980).
24. H. W. Starkweather, *J. Am. Chem. Soc.* **56**, 1870 (1934).
25. G. H. Kalb and co-workers, *J. Appl. Polym. Sci.* **4**, 55 (1960).
26. A. E. Newkirk, *J. Am. Chem. Soc.* **68**, 2467 (1946).
27. D. Sianesi and G. Caporiccio, *J. Polym. Sci. Part A-1* **6**, 335 (1968).
28. D. Raucher and M. Levy, *J. Polym. Sci.* **13**(6) 1339–1346 (1975).
29. K. U. Usmanov and co-workers, *Russ. Chem. Rev.* **46** (5), 462–478 1977; trans. from Usp. Khim. **46,** 878–906 (1977).
30. T. S. Sirlibaev and I. Tirkashev, *Uzb. Khim. Zh.* **2**, 40–42 (1983).
31. D. Raucher and co-workers, *J. Polym. Sci.* **17**, 2825–2832 (1979).
32. Jpn. Pat. 74 027108 (July 15, 1974), H. Iwamichi and Y. Adachi (to Electro Chemical Industries Co.).
33. U.S. Pat. 3,627,744 (Dec. 14, 1971), R. A. Bonsall and B. Hopkins (to Monsanto Co.).
34. Jpn. Pat. 74 028,907 (July 30, 1974), S. Yashida, K. Tamaiima, and H. Kurovama (to Electro Chemical Industries Co., Ltd.).
35. V. G. Klaydin and co-workers, *Zh. Prikl. Khim.* **56**(2), 462–465 (1983).
36. A. K. Gafurov and co-workers, *Sb. Nauchn. Tr. Tashk. Gos. Univ. im. V. I. Lenina* **667**, 16–23 (1981).
37. U.S. Pat. 2,510,783 (June 6, 1950), F. L. Johnston (to E. I. du Pont de Nemours & Co., Inc.).
38. U.S. Pat. 3,129,207 (Apr. 14, 1964), V. E. James (to E. I. du Pont de Nemours & Co., Inc.).
39. Brit. Pat. 1,161,958 (Aug. 20, 1969), J. G. Frielink (to Deutsche Solvay-Werke Gesellschaft mit Beschrankter Haftung).
40. U.S. Pat. 3,265,678 (Aug. 9, 1966), J. L. Hecht (to E. I. du Pont de Nemours & Co., Inc.).
41. G. Natta and co-workers, *J. Polym. Sci. Part A-1* **3**, 4263 (1965).
42. D. Sianesi and G. Caporiccio, *J. Polym. Sci. Part A-1* **6**, 335 (1968).
43. U.S. Pat. 2,406,717 (Aug. 27, 1946), C. A. Thomas (to Monsanto Chemical Co.).
44. U.S. Pat. 2,847,401 (Aug. 12, 1958), E. W. Gluesenkamp and J. D. Calfee (to Monsanto Chemical Co.).
45. U.S. Pat. 3,057,812 (Oct. 9, 1962), J. R. Straughan, R. Stickl, Jr., and W. F. Hill, Jr. (to Union Carbide Corp.).
46. F. Z. Yusupbekova and co-workers, *Uzb. Khim. Zh.* (3), 47–51 (1987).
47. R. E. Cais and J. M. Kometani, *Polymer* **29**, 168–172 (1988).
48. R. E. Cais and J. M. Kometani, in J. C. Randall, ed., *NMR and Macromolecules*, ACS Symposium. Ser. No. 247, American Chemical Society, Washington, D.C., 1984, p. 153.

49. T. Alfrey and C. C. Price, *J. Polym. Sci.* **2**, 101 (1947).
50. G. Natta, *Makromol. Chem.* **35**, 94 (1960).
51. M. Goerlitz and co-workers, *Angew. Makromol. Chem.* **29/30**(371), 137 (1973).
52. G. Natta, I. W. Bassi, and G. Allegra, *Atti Accad. Naz. Lincei Cl. Sci. Fis. Mat. Natur. Rend.* **31**, 350–356 (1961).
53. J. B. Lando, H. G. Olf, and A. Peterlin, *J. Polym. Sci. Part A-1* **4**, 941–951 (1966).
54. G. Caporiccio, E. Strepparola, and D. Sianesi, *Chim. Ind. (Milan)* **52**, 28–36 (1970).
55. C. W. Wilson III and E. R. Santee, Jr., *J. Polym. Sci. Part C*, **8**, 97–112 (1965).
56. M. D. Hanes, PhD. dissertation, Case-Western Reserve University, Cleveland, Ohio, 1991.
57. D. W. Ovenall and R. E. Uschold, *Macromolecules* **24**, 3235 (1991).
58. L. L. Burger and M. T. Aronson, in *Polymer* **34**(12), pp. 25, 46 (1993).
59. M. D. Bruch, F. A. Bovey, and R. E. Cais, *Macromolecules* **17**, 2547–2551 (1984).
60. F. J. Weigert, *Org. Magn. Resonance* **3**, 373–377 (1977).
61. J. L. Koenig and J. J. Mannion, *J. Polym. Sci. Part A-2*, **4**, 401–414 (1966).
62. G. Zerbi and G. Cortili, *Spectrochim. Acta* **26**, 733–739 (1970).
63. R. F. Boyer, *J. Polym. Sci. Part C* **50**, 189–242 (1975).
64. M. L. Wallach and M. A. Kabayama, *J. Polym. Sci. Part A-1* **4**, 2667–2674 (1966).
65. A. E. Tonelli, *Macromelecules* **13**, 734–741 (1980).
66. A. E. Tonelli, F. C. Schiling, and R. E. Cais, *Macromolecules* **14**, 560 (1982).
67. H. Tianbai, *Yingyong Huaxue (Chinese Journal of Applied Chemistry)* **2**, 15–18 (1985).
68. R. F. Boyer and R. L. Miller, *Macromolecules* **10**, 1167–1169 (1977).
69. A. Chapira, Z. Mankowski, and N. Schmitt, *J. Polym. Sci. Part A-1* **20**, 1791–1796 (1982).
70. F. M. Fowkes, D. O. Tischler, J. A. Wolfe, L. A. Lannigan, C. M. Ademu-John, and M. J. Halliwell, *J. Polym. Sci. Part A-1* **22**, 547–566 (1984).
71. D. Raucher and M. Levy, *J. Polym. Sci. Part A-1* **17**, 2675–2680 (1979).
72. G. Montaudo, C. Puglisi, E. Scamporrino, and D. Vitalini, *J. Polym. Sci. Part A-1* **24**, 301–316 (1986).
73. M. L. O'Shea, C. Morterra, and M. J. D. Low, *Mater. Chem. Phys.* **25**, 501 (1990).
74. B. F. Mukhiddinov and co-workers, *Doki. Akad. Nauk SSSR*, **316**, 165–168 (1991).
75. W. E. Farneth, M. T. Aronson, and R. E. Uschold *Macromolecules* **26**(18), 4765 (Feb. 1993).
76. Technical Information Bulletin TD-31, E. I. du Pont de Nemours & Co., Inc., Wilmington, Del., 1979.
77. Y. Rosenberg, A. Siegmann, M. Narkis, and S. Shkolnik, *J. Appl. Polym. Sci.* **45**, 783 (1992).
78. R. Timmerman and W. Greyson, *J. Appl. Polym. Sci.* **6**, 456 (1962).
79. U.S. Pat. 2,953,818 (Sept. 27, 1960), L. R. Bartron (to E. I. du Pont de Nemours & Co., Inc.).
80. U.S. Pat. 3,723,171 (Mar. 27, 1973), O. Fuchs (to Dynamite Nobel AG).
81. U.S. Pat. 4,645,692 (Feb. 24, 1987), E. Vassiliou (to E. I. du Pont de Nemours & Co., Inc.).
82. U.S. Pat. 4,988,540 (Jan. 29, 1991), R. F. Davis, C. G. Bragaw, and T. P. Cancannon (to E. I. du Pont de Nemours & Co., Inc.).
83. U.S. Pat. 3,139,470 (June 30, 1964), R. S. Prengle and R. L. Richards, Jr. (to E. I. du Pont de Nemours & Co., Inc.).
84. U.S. Pat. 3,145,242 (Aug. 18, 1964), W. L. Bryan (to E. I. du Pont de Nemours & Co., Inc.).

85. U.S. Pat. 3,274,088 (Sept. 20, 1966), L. E. Wolinski (to E. I. du Pont de Nemours & Co., Inc.).
86. U.S. Pat. 3,122,445 (Feb. 25, 1964), R. O. Osborn (to E. I. du Pont de Nemours & Co., Inc.).
87. C. A. L. Westerdahl and co-workers, *Activated Gas Plasma Surface Treatment of Polymers for Adhesive Bonding, Part III*, Technical Report 4279, Picatinny Arsenal, Dover, N.J., 1972.
88. U.S. Pat. 4,215,177 (July 29, 1980), A. Strassel (to Produits).
89. U.S. Pat. 3,880,690 (Apr. 29, 1975), O. Fuchs (to Dynamite Nobel AG).
90. D. Lester and L. A. Greenberg, *Arch. Ind. Hyg. Occup. Med.* **2**, 335 (1950).
91. M. E. Andersen, *Neurobehav. Toxicol. Teratol.* **3**, 383 (1981).
92. M. E. Andersen, *Drug Metab. Rev.* **13**(5), 799 (1982).
93. J. G. Filser and H. M. Bolt, *Arch. Toxicol.* **42**, 123 (1979).
94. H. M. Bolt, *Arbeitsmed. Sozialmed. Praventivmed.* **15**, 49 (1980).
95. H. M. Bolt, R. J. Laib, and K. P. Klein, *Arch. Toxicol.* **47**, 71 (1981).
96. *Fed. Reg.* **52**(109), 21516 (June 8, 1987).
97. J. R. Fiddle, *Occupat. Health Safety News Dig.* **4**, 3 (1988).
98. C. P. Fennimore and F. J. Martin, *Combust. Flame* **10**, 135 (1966).
99. U.S. Pat. 3,963,672 (June 15, 1976), D. E. Brasure (to E. I. du Pont de Nemours & Co., Inc.).
100. I. N. Einhorn and co-workers, *Final Report, FRC/UU-41*, ETEC 75-022, NASA Contract No. NAS2-8244, National Technical Information System (NTIS), Springfield, Va., Dec. 14, 1974.
101. C. H. Hilado and co-workers, *J. Combust. Toxicol.* **3**, 157 (1976).
102. *Ibid.*, p. 270.
103. J. C. Spurgeon, *Report No. FAA-RD-78-131*, Federal Aviation Administration Contract No. 181-521-100, NTIS, Springfield, Va., Nov. 1978.
104. R. D. Leaversuch, *Mod. Plast.* **64**, 52 (July 1987).
105. V. M. Cassidy, *Mod. Met.* **41**, 36 (May 1985).
106. *Public Works* **III**, 78 (1980).
107. *Du Pont Mag.* **82**, 8–11 (Mar.–Apr. 1988).
108. *Du Pont Mag.* **82**, 13 (May–June 1988).
109. *Energy Managmt. Tech.* **8**, 22–23 (July–Aug. 1984).
110. R. A. Anderson and G. A. Johnson, *J. Fire Flammabil.* **8**, 364–381 (1977).
111. U.S. Pat. 5,137,775 (Aug. 11, 1992), S. Ebnesajjad and R. F. Davis (to E. I. du Pont de Nemours & Co., Inc.).
112. *Tedlar*, Du Pont Technical Bulletin TD-40, E. I. du Pont de Nemours & Co., Inc., Wilmington, Del., Feb. 1988.
113. J. C. Pau, J. E. Knoll, and M. R. Midgett, *J. Air Waste Managmt. Assoc.* **41**(8), 1095–1097 (1991).
114. B. Baum and M. Binette, *Report DOE/CS/35359-T1* (DE84011488), U.S. Dept. of Energy Contract No. AC04-78CS35359, NTIS, Springfield, Va., June 1983.
115. R. S. Sugimura, D. H. Otth, R. G. Ross, Jr., J. C. Arnett, and G. T. Samuelson, *IEEE Conference Report*, IEEE, New York, Oct. 21–25, 1985.
116. G. L. Schmutz, *Circuits Mfg.* **23**, 51 (Apr. 1983).
117. J. L. Wilson, C. L. Long, D. L. Mathews, and M. L. Wilson, *BDX-613-1657*, U.S. Department of Energy Contract No. EY-76-C-04-0613, NTIS, Springfield, Va., Jan. 1978.
118. Jpn. Pat. 62 214,939 (Jan. 21, 1987), S. Suzuki and S. Onari (to Matsushita Electric Works).

General References

D. E. Brasure and S. Ebnesajjad, in J. I. Kroschwitz, ed., *Concise Encyclopedia of Polymer Science and Engineering*, John Wiley & Sons, Inc., New York, 1990, pp. 1273–1275.

D. E. Brasure and S. Ebnesajjad, in J. I. Kroschwitz, ed., *Encyclopedia of Polymer Science and Engineering*, 2nd ed., Vol. 17, John Wiley & Sons, Inc., New York, 1989, pp. 468–491.

S. Ebnesajjad
L. G. Snow
Du Pont Company

POLY(VINYLIDENE FLUORIDE)

Poly(vinylidene fluoride) [24937-79-9] is the addition polymer of 1,1-difluoroethene [75-38-7], commonly known as vinylidene fluoride and abbreviated VDF or VF$_2$. The formula of the repeat unit in the polymer is —CH$_2$—CF$_2$—. The preferred acronym for the polymer is PVDF, but the abbreviation PVF$_2$ is also frequently used. The history and development of poly(vinylidene fluoride) technology has been reviewed (1–3).

PVDF is a semicrystalline polymer that contains 59.4 wt % fluorine and 3 wt % hydrogen and is commercially polymerized in emulsion or suspension using free-radical initiators. The spatial arrangement of the CH$_2$ and CF$_2$ groups along the polymer chain accounts for the unique polarity, unusually high dielectric constant, polymorphism, and high piezoelectric and pyroelectric activity of the polymer. It has the characteristic resistance of fluoropolymers to harsh chemical, thermal, ultraviolet, weathering, and oxidizing or high energy radiation environments. Because of these characteristics it has many applications in wire and cable products, electronic devices, chemical and related processing fields, as a weather-resistant binder for exterior architectural finishes, and in many specialized uses. The polymer is readily melt-processed using conventional molding or extrusion equipment; porous membranes are cast from solutions, and finishes are deposited from dispersions using specific solvents. PVDF contains an extremely low level of ionic contamination and does not require additives for stabilization during melt-processing, thereby qualifying it for applications such as ultrapure water systems where high purity is demanded from materials of construction.

There is growing commercial importance and escalating scientific interest in PVDF. The World Patent database, including the United States, lists 678 patents that cite the term poly(vinylidene fluoride) for the period 1963–1980 and 2052 patents for the period 1981–1992; *Chemical Abstracts* files covering the years 1967–1992 contain 5282 references for the same term. Thirty years ago there was only one commercial producer of PVDF in the world; now there are two in the United States, two in Japan, and three in Europe.

Monomer

Properties. Vinylidene fluoride is a colorless, flammable, and nearly odorless gas that boils at $-82°C$. Physical properties of VDF are shown in Table 1. It is usually polymerized above its critical temperature of 30.1°C and at pressures above 3 MPa (30 atm); the polymerization reaction is highly exothermic.

Preparation. Thermal elimination of HCl from 1-chloro-1,1-difluoroethane (HCFC-142b) [75-68-3] is the principal industrial route to VDF covered by numerous patents (8–19). Dehydrohalogenation of 1-bromo-1,1-difluoroethane (20), or 1,1,1-trifluoroethane (HFC-143a) (21–25), or dehalogenation of 1,2-dichloro-1,1-difluoroethane (26–28) are investigated alternative routes (see FLUORINE COMPOUNDS, ORGANIC–FLUORINATED ALIPHATIC COMPOUNDS).

The commercially preferred monomer precursor HCFC-142b has been prepared by hydrofluorination of acetylene (29), vinylidene chloride (30–32), or 1,1,1-trichloroethane (33–39).

$$CH\equiv CH \xrightarrow{+2\ HF} CH_3-CHF_2 \xrightarrow{+Cl_2} CH_3-CClF_2 + HCl$$

$$CH_2=CCl_2 \xrightarrow{+2\ HF} CH_3-CClF_2 + HCl$$

$$CH_3-CCl_3 \xrightarrow{+2\ HF} CH_3-CClF_2 + 2\ HCl$$

The monomer can also be continuously prepared by the pyrolysis of trifluoromethane (CHF_3) in the presence of a catalyst and either methane or ethylene (40–43). Passing 1,1-difluoroethane (CH_3-CHF_2), oxygen, and CO_2 over a catalyst gives a mixture of VDF and vinyl fluoride (44). Using either methanol or dichloromethane as a source of the carbene moiety, VDF can be continuously prepared from chlorodifluoromethane (HCFC-22) (CHF_2Cl) (45,46). Pyrolysis of dichlorodifluoromethane (CFC-12) (CCl_2F_2) with either methane (47) or methyl chloride

Table 1. Properties of Vinylidene Fluoride

Property	Value	Reference
molecular weight	64.038	
boiling point, °C	−84	
freezing point, °C	−144	
vapor pressure at 21°C, kPa[a]	3683	
critical pressure, kPa[a]	4434	4
critical temperature, °C	30.1	
critical density, kg/m^3	417	
explosive limits, vol % in air	5.8–20.3	5
heat of formation at 25°C, kJ/mol[b]	−345.2	6
heat of polymerization at 25°C, kJ/mol[b]	−474.21	7
solubility in water, cm^3/100 g at 25°C, 10 kPa[a]	6.3	

[a]To convert kPa to atm, divide by 101.3.
[b]To convert kJ to kcal, divide by 4.184.

yields the monomer (48,49). Copyrolysis of methane and either bromotrifluoro- (CBrF$_3$) or chlorotrifluoromethane (CClF$_3$) yields VDF (50). Deuterated VDF has also been prepared (51).

Storage and Shipment. VDF or HFC-1132a is stored and shipped in gas cylinders or high pressure tube trailers without polymerization inhibitor and is placarded as flammable compressed gas. Terpenes or quinones can be added to inhibit polymerization. Elf Atochem North America, Inc. and Ausimont USA, Inc. supply VDF in the United States; other producers are in Japan and Europe.

Health and Safety Factors. VDF is a flammable gas; its combustion products are toxic. Liquid VDF on contact with the skin can cause frostbite. Acute inhalation toxicity of VDF is low; median lethal concentrations (LC$_{50}$) for rats were 128,000 ppm after a single 4-h exposure (52) and 800,000 ppm after a 30-min exposure (53). Cumulative toxicity is low; exposure of rats and mice at levels of up to 50,000 ppm for 90 days did not cause any systemic toxicity (54,55). No teratogenic or reproductive effects were found in rats. VDF was positive in bacterial gene mutation assay but negative in mammalian gene mutation, chromosomal aberration, and cell transformation assays. In 1979, a paper reported that rats developed lipomas after being given over 52 weeks' oral doses of VDF dissolved in olive oil (56). More relevant, lifetime (18 months) inhalation studies on rats and mice have not detected chronic or carcinogenic effects up to 10,000 ppm VDF (57,58). Additional information is available (59,60). Toxicology test data on VDF were submitted to the EPA pursuant to a final test rule and consent order under the Toxic Substances Control Act (TSCA) (61).

Uses. Vinylidene fluoride is used for the manufacture of PVDF and for copolymerization with many fluorinated monomers. One commercially significant use is the manufacture of high performance fluoroelastomers that include copolymers of VDF with hexafluoropropylene (HFP) (62) or chlorotrifluoroethylene (CTFE) (63) and terpolymers with HFP and tetrafluoroethylene (TFE) (64) (see ELASTOMERS, SYNTHETIC–FLUOROCARBON ELASTOMERS). There is intense commercial interest in thermoplastic copolymers of VDF with HFP (65,66), CTFE (67), or TFE (68). Less common are copolymers with trifluoroethene (69), 3,3,3-trifluoro-2-trifluoromethylpropene (70), or hexafluoroacetone (71). Thermoplastic terpolymers of VDF, HFP, and TFE are also of interest as coatings and film. A thermoplastic elastomer that has an elastomeric VDF copolymer chain as backbone and a grafted PVDF side chain has been developed (72).

Polymer

Polymerization. The first successful polymerizations of VDF in aqueous medium using peroxide initiators at 20–150°C and pressures above 30 MPa were described in a patent issued in 1948 (73). About a year later, the first copolymerizations of VDF with ethylene and halogenated ethylenes were also patented (74). After a hiatus of over 12 years a commercially feasible process was developed and PVDF was ready for market introduction (2).

PVDF is manufactured using radical initiated batch polymerization processes in aqueous emulsion or suspension; operating pressures may range from 1 to 20 MPa (10–200 atm) and temperatures from 10 to 130°C. Polymerization method, temperature, pressure, recipe ingredients, the manner in which they are added to the reactor, the reactor design, and post-reactor processing are variables that influence product characteristics and quality.

Emulsion polymerization of VDF is a heterogeneous reaction that requires, as is typical with most fluorine-containing monomers, addition of a polyfluoroalkanoic acid salt as surfactant (75) to avoid radical scavenging reactions during polymerization. Sometimes chain-transfer agents or buffers, or both, are used in the emulsion process. Radical generators that initiate polymerization of VDF are either water-soluble, eg, persulfate salts (76–78), disuccinic acid peroxide (79), β-hydroxyalkyl peroxide (80,81), alkylperoxybutyric acid (82) or monomer soluble, eg, di-*tert*-butyl peroxide (83,84), dialkylperoxydicarbonate (85–88), or *tert*-butylperoxybutyrate (89). A radiotracer study found that the number of end groups formed in the polymer from primary radicals of the initiator decreased during emulsion polymerization of PVDF, whereas overall branching increased (90). Upon completion of the polymerization, the discharged reactor product is a milky white colloidal dispersion or latex that is subsequently filtered, coagulated, thoroughly washed and usually spray-dried to produce a very fine powder. It is typical of emulsion polymerization that the polymer solids in latex are spheres of about 250 nm in diameter and the dried powders contain agglomerates of about 2 to 5 μm in diameter. The powder is either packaged or processed as required for the intended use.

Suspension polymerization of VDF in water are batch processes in autoclaves designed to limit scale formation (91). Most systems operate from 30 to 100°C and are initiated with monomer-soluble organic free-radical initiators such as diisopropyl peroxydicarbonate (92–96), *tert*-butyl peroxypivalate (97), or *tert*-amyl peroxypivalate (98). Usually water-soluble polymers, eg, cellulose derivatives or poly(vinyl alcohol), are used as suspending agents to reduce coalescence of polymer particles. Organic solvents that may act as a reaction accelerator or chain-transfer agent are often employed. The reactor product is a slurry of suspended polymer particles, usually spheres of 30–100 μm in diameter; they are separated from the water phase thoroughly washed and dried. Size and internal structure of beads, ie, porosity, and dispersant residues affect how the resin performs in applications.

Solution polymerization of VDF in fluorinated and fluorochlorinated hydrocarbons such as CFC-113 and initiated with organic peroxides (99), especially bis(perfluoropropionyl) peroxide (100), has been claimed. Radiation-induced polymerization of VDF has also been investigated (101,102). Alkylboron compounds activated by oxygen initiate VDF polymerization in water or organic solvents (103,104). Microwave-stimulated, low pressure plasma polymerization of VDF gives polymer film that is <10 μm thick (105). Highly regular PVDF polymer with minimized defect structure was synthesized and claimed (106). Perdeuterated PVDF has also been prepared and described (107).

Polymer Properties. PVDF is a tough, semicrystalline engineering polymer. Compared to the softer and mechanically less robust perfluorocarbon polymers, PVDF has high mechanical and impact strength, and excellent resistance to both creep under long-term stress and fatigue upon cyclic loading (108,109). PVDF also has excellent abrasion resistance and thermal stability, and resists damage from most chemicals and solvents, as well as from ultraviolet and nuclear radiation. Typical PVDF design properties are shown in Table 2.

Properties of PVDF depend on molecular weight, molecular weight distribution, chain configuration, ie, the sequence in which the monomer units are linked together, including side groups or branching, and crystalline form. The morphology of PVDF reflects differences in both the utilized polymerization procedure and the thermomechanical treatment that followed polymerization. During radical-initiated polymerization, the head-to-tail addition of VDF molecules predominates, in which —CF_2— is denoted as "head" and —CH_2— as "tail," but reversed monomeric addition leading to head-to-head and tail-to-tail defects does occur; the extent of defects is influenced by polymerization process conditions, particularly temperature (110). The incidence of these defects is best determined by high resolution ^{19}F nmr (111,112); infrared (113) and laser mass spectrometry (114) are alternative methods. Typical commercial polymers show 3–6 mol % defect content. Polymerization methods have a particularly strong effect on the sequence of these defects. In contrast to suspension polymerized PVDF, emulsion polymerized PVDF forms a higher fraction of head-to-head defects that are not

Table 2. Properties of Poly(vinylidene fluoride)

Property	Method	Value
specific gravity	ASTM D792	1.75–1.80
water absorption, 24 h at 23°C, %	ASTM D570	0.04
refractive index, n_D	ASTM D542	1.42
melting peak, T_m, °C	ASTM D3418	156–180
crystallization peak, T_c, °C	ASTM D3418	127–146
glass transition, T_g, °C	ASTM D2236	−40
brittleness temperature, °C	ASTM D746	−62 to −64
deflection temperature at 1.82 MPa,[a] °C	ASTM D648	84–115
specific heat, kJ/kg·K[b]	DSC	1.26–1.42
thermal conductivity, W/K·m	ASTM D433	0.17–0.19
tensile stress at yield, MPa[a]	ASTM D638	28–57
tensile stress at break, MPa[a]	ASTM D638	31–52
elongation at break, %	ASTM D638	50–250
compressive strength, MPa[a]	ASTM D695	55–110
flexural strength, MPa[a]	ASTM D790	59–94
modulus of elasticity, MPa[a]		
in tension	ASTM D882	1040–2600
in flexure	ASTM D790	1140–2500
impact strength at 25°C, J/m[c]	ASTM D256	
unnotched		800–4270
notched		107–214
limiting oxygen index, %	ASTM D2863	43
vertical burn	UL 94	V-0
sand abrasion, m³/mm	ASTM D968	4.0

[a]To convert MPa to psi, multiply by 145.
[b]To convert kJ to kcal, divide by 4.184.
[c]To convert J/m to ft·lbf/in., divide by 53.38.

followed by tail-to-tail addition (115,116). Crystallinity and other properties of PVDF or copolymers of VDF are influenced by these defect structures (117).

Crystallinity affects toughness and mechanical strength as well as impact resistance. PVDF crystals are seen in the optical microscope as spherulites that are lamellae of polymer chain segments, which are packed crystallographically; the interposed amorphous regions consist of disordered chains. The crystallinity can range between 35 and 70%. Various parameters, including molecular weight, molecular weight distribution, polymerization method, thermal history, and cooling rates influence crystallization kinetics (118).

Unlike other synthetic polymers, PVDF has a wealth of polymorphs; at least four chain conformations are known and a fifth has been suggested (119). The four known distinct forms or phases are alpha (II), beta (I), gamma (III), and delta (IV). The most common α-phase is the trans-gauche (tgtg′) chain conformation placing hydrogen and fluorine atoms alternately on each side of the chain (120,121). It forms during polymerization and crystallizes from the melt at all temperatures (122,123). The other forms have also been well characterized (124–128). The density of the α polymorph crystals is 1.92 g/cm^3 and that of the β polymorph crystals 1.97 g/cm^3 (129); the density of amorphous PVDF is 1.68 g/cm^3 (130).

Relaxations of α-PVDF have been investigated by various methods including dielectric, dynamic mechanical, nmr, dilatometric, and piezoelectric and reviewed (3). Significant relaxation ranges are seen in the loss-modulus curve of the dynamic mechanical spectrum for α-PVDF at about 100°C (α′), 50°C (α″), −38°C (β), and −70°C (γ). PVDF relaxation temperatures are rather complex because the behavior of PVDF varies with thermal or mechanical history and with the testing methodology (131).

Suspension- and emulsion-polymerized PVDF exhibit dissimilar behavior in solutions. The suspension resin type is readily soluble in many solvents; even in good solvents, solutions of the emulsion resin type contain fractions of microgel, which contain more head-to-head chain defects than the soluble fraction of the resin (116). Concentrated solutions (15 wt %) and melt rheology of various PVDF types also display different behavior (132). The Mark-Houwink relation ($\eta = KM^a$) for PVDF in N-methylpyrrolidinone (NMP) containing 0.1 molar LiBr at 85°C, for the suspension (115) and emulsion (116) respectively is: $\eta = (4.5 \pm 0.3) \times 10^{-4} M^{0.70}$ and $\eta = 1.4 \times M^{0.96}$.

Unlike most crystalline polymers, PVDF exhibits thermodynamic compatibility with other polymers (133). Blends of PVDF and poly(methyl methacrylate) (PMMA) are compatible over a wide range of blend composition (134,135). Solid-state nmr studies showed that isotactic PMMA is more miscible with PVDF than atactic and syndiotactic PMMA (136). Miscibility of PVDF and poly(alkyl acrylates) depends on a specific interaction between PVDF and oxygen within the acrylate and the effect of this interaction is diminished as the hydrocarbon content of the ester is increased (137). Strong dipolar interactions are important to achieve miscibility with poly(vinylidene fluoride) (138). PVDF blends are the object of many papers and patents; specific blends of PVDF and acrylic copolymers have seen large commercial use.

PVDF cross-links readily when subjected to electron beam radiation (139) or gamma radiation (140). Cross-linking efficiency is proportional to molecular weight, molecular weight distribution, or extent of head-to-head chain defects (141). The cross-linked PVDF, when highly stressed or compressed above the melting point, exhibits thermodynamic and physical properties similar to polyethylene and polypropylene (142). Polyfunctional monomers having good solubility in PVDF increase the cross-linking rate (143,144). The effect of radiation on the structure and properties of PVDF has been reviewed (145).

Some electrical properties are shown in Table 3. Values of other parameters have been published (146). Polymorphism of the PVDF chains and the orientation of the two distinct dipole groups, —CF_2— and —CH_2—, rather than trapped space charges (147) contribute to the exceptional dielectric properties and the extraordinarily large piezoelectric and pyroelectric activity of the polymer (146,148,149).

Prolonged exposure of PVDF to processing temperatures exceeding 300°C could lead to discoloration and chemical reactions that present hazards. The primary reaction at high temperature is loss of hydrogen fluoride (HF) that results in conjugation, —CH=CF—CH=CF—, along the chains; this explains the observed discoloration. The extent and rate of discoloration is not homogeneous among PVDF resins and may be commensurate with chain perfection, ie, percentage of head-to-tail repeat units in the chain (150); reversed repeat units may interrupt dehydrofluorination (151). If the temperature exceeds 375°C in air, rapid thermal decomposition takes place and HF gas evolves. After 70 wt % loss, at about 480°C, the residue is char that eventually burns completely at higher temperature. The charring phenomenon is considered basic to the superior performance of PVDF in severe fire tests such as the Underwriters Laboratories UL 910 Modified Steiner Tunnel Test (152).

Fabrication and Processing. PVDF is available in a wide range of melt viscosities as powder or pellets to fulfill typical fabrication requirements; latices are also commercially available.

PVDF is readily molded in conventional compression, transfer, and injection-molding equipment (153–155); typical molding temperatures for the cylinder and nozzle are 180–240°C and molds are at 50–90°C. PVDF resins do not require drying because the resin does not absorb moisture. As a crystalline polymer, it shows a relatively high mold shrinkage of ca 3%. To obtain a high dimensional

Table 3. Electrical Properties of Poly(vinylidene fluoride)

Property	Method	Value
volume resistivity, $\Omega \cdot$cm	ASTM D257	$1.5–5 \times 10^{14}$
surface arc resistance, s	ASTM D495	50–60
dielectric strength, kV/mm	ASTM D149	63–67
dielectric constant at 25°C	ASTM D150	
1 kHz		8.15–10.46
10 kHz		8.05–9.90
100 kHz		7.85–9.61
dissipation factor		
1 kHz		0.005–0.026
10 kHz		0.015–0.021
100 kHz		0.039–0.058

stability, carbon-filled, mica-filled, or carbon–fiber-reinforced (156) grades are used. To achieve best results and avoid warping or voids, it is essential to coordinate the cooling rate with the crystallization of the resin or anneal the part at 140–150°C. For compression or transfer molding the PVDF pellets are preheated in an oven to 210–240°C and transferred to the mold that is heated to 190–200°C. The resin in the filled mold is placed under sufficient pressure to complete flow and fusion. Sufficient time must be allowed to cool the molded part under pressure to 90°C to prevent vacuum voids and distortion.

Smooth PVDF profiles of all types—film, sheet, rod, profile, pipe, tubing, fiber, monofilament, wire insulation, and cable-jackets—can be extruded; no heat stabilizers are needed. In both molding and extrusion operations, care must be exercised to eliminate hang-up zones in the equipment where molten resin (at 230–260°C) can stagnate and thermally decompose with time. Equipment built with material of construction used for processing polyolefins or PVC is adequate; for long-term or high shear processing, a highly wear-resistant alloy such as Xaloy 306 for barrel liner and SAE 4140 steel for the screw is suggested. Gradual transition-type screws having L/D ratios at least 20:1, ample metering sections, and compression ratio of about 3 are recommended. Temperature profiles vary from 190 to 290°C depending on resin grade and shape being extruded. Water quenching is practiced for wire insulation, tubing, and pipe, whereas sheet and flat film are melt-cast on polished steel rolls operating at 65–150°C (157).

PVDF sheets can be backed during extrusion-calendering using fabrics of glass (158), polyamide, or polyester fibers; they can also be press-laminated with the fabrics at 185 to 200°C (159). Nonvulcanized rubber can also be press-laminated with PVDF sheet at 150°C (160). Melt-cast PVDF sheets can be oriented uniaxially or biaxially to produce films with vastly increased mechanical strength, specular transmission, or ferroelectric activity (161). Blown-film equipment typically used for HDPE can also be used for extrusion of blown PVDF film. Monofilaments are usually extruded or spun at 240–260°C into a 30–50°C water bath and then reheated to 130–160°C, oriented using draw ratios of 3:1 to 5:1, and heat-set at elevated temperatures to produce high strength filament having tenacities of 350–440 mN/tex (4–5 gf/den) (162,163). Coextrusion of PVDF with other polymers is the subject of several patents. Interlayer adhesion is critical, although matching the coefficients of thermal expansion and melt viscosities are other important considerations (164). To promote interleaf bond, an adhesive "tie-layer" consisting of a polymer that is partially compatible both with PVDF and the incompatible polymer layer, such as ABS (165) and polyolefins (166), has been used in coextrusion.

Semifinished PVDF products can be machined and processed by methods used for other thermoplastics (155). PVDF parts can be joined by standard welding methods. Pipe, fittings, or sheets can be welded using a hot-air gun with a welding rod or a heated tool for butt or socket welding. Films can be bonded by heat sealing, high frequency welding, or ultrasonic welding.

Manufactured PVDF parts can be cross-linked using high energy radiation to produce high temperature wire insulation, and heat-shrinkable tubing or film.

Organosol dispersions of PVDF used extensively for exterior architectural finishes can be produced from the very fine powder obtained only by the emulsion polymerization method. These dispersions include the very fine PVDF powder, pigments, acrylate or methacrylate copolymer, and selected solvents (167–170); comparable water-based coating compositions can also be prepared (171–174). These dispersions are factory applied by spray or roller to primed steel or aluminum surfaces and oven-fused at 230–260°C to form continuous films that adhere firmly to the substrata. Other organic dispersions of PVDF are formulated for spray applications of relatively thick coatings to protect metals from corrosive environments. Powders for electrostatic spraying, fluidized-bed deposition, or rotomolding are obtained by melt compounding PVDF with appropriate ingredients, cryogenic grinding, and classification to desirable particle-size range for the application.

Microporous filtration membranes from VDF polymers are made by casting a polymer solution on a rigidly supported backing belt, then passing the belt through a bath to form the membrane, followed by extraction of any residual solvent from and drying of the membrane (175). Formation of microporous PVDF membranes has been reviewed (176). To improve performance, PVDF membranes are often chemically modified (177–181). Hollow fibers useful for microfiltration are produced by extruding a spinning solution of PVDF from an annular spinning orifice into coagulating liquids (182). Porous structures can also be made by sintering very fine granules under controlled conditions (183), from extruded compounds that contain leachable additives that upon extraction leave voids in the product, or by extrusion of compounds containing chemical blowing agents (184,185).

Economic Aspects

Because of its excellent combination of properties, processability, and relatively low price compared to other fluoropolymers, PVDF has become the largest volume fluoropolymer after PTFE; consumption in the United States has grown from zero in 1960 to about 6200 metric tons in 1991 (186). About 49% of the consumed volume is PVDF modified by copolymerization with 5–12-wt % HFP to enhance flexibility. In 1992, list price for homopolymer powders was $15.32/kg, and for pellets $15.42/kg; the reported market price was $14.09–14.22/kg (187). In the United States, almost all PVDF is supplied by Ausimont USA, Inc., Elf Atochem North America, Inc., and Solvay Polymers, Inc. Ausimont and Elf Atochem are producers; Solvay is an importer of the resin. Small amounts of resin are imported from Germany by Hüls America, Inc. and from Japan by Kureha Chemical Industry Co., Ltd. PVDF producers and their trademarks are listed in Table 4.

After 10 years of unabated rapid growth in the plenum wire and cable market, fluoropolymers including PVDF, primarily the flexible VDF/HFP copolymer, are beginning to lose market share to lower priced PVC-alloys. The loss of market share in the plenum market probably will be compensated by growth of PVDF in other fields; thus during the mid-1990s the total volume of PVDF may not grow (188).

Specifications and Standards. Commercial PVDF resin types and standards are defined in ASTM D3222. A list of military and industrial specifications covering applications, material suppliers, and PVDF resin grades can be found in Reference 189.

Table 4. Producers and Trademarks of Poly(vinylidene fluoride)

Producer	Country	Trademark
Ausimont USA, Inc.	United States	Hylar
Elf Atochem North America, Inc.	United States	Kynar
Elf Atochem, SA	France	Foraflon
Solvay & Cie, SA	Belgium	Solef/Vidar
Hüls, AG	Germany	Dyflor
Daikin Kogyo Co., Ltd.	Japan	Neoflon
Kureha Chemical Industry Co., Ltd.	Japan	KF Polymer

Health and Safety Factors

PVDF is a nontoxic resin and may be safely used in articles intended for repeated contact with food (190). Based on studies under controlled conditions, including acute oral, systemic, subchronic, and subacute contact; implantation; and tissue culture tests, no adverse toxicological or biological response has been found in test animals (191,192). PVDF is acceptable for use in processing and storage areas in contact with meat or poultry products prepared under federal inspection and it complies with the 3-A sanitary standards for dairy equipment.

PVDF is not hazardous under typical processing conditions. If the polymer is accidentally exposed to temperatures exceeding 350°C, thermal decomposition occurs with evolution of toxic hydrogen fluoride (HF).

Some silica-containing additives such as glass and titanium dioxide lower the thermal stability of PVDF and should be used with caution. Processors should consult the resin producer about safe processing practice.

Uses

PVDF is used in many diverse industrial applications for products that require high mechanical strength and resistance to severe environmental stresses. The most important fields of application for PVDF resins include electric and electronic industry products, architectural and specialty finishes, products for the chemical and related industries, and rapidly growing specialized uses.

In the electric and electronics field the largest usage of PVDF is for plenum wire and cables, plenum being the space between the suspended and structural ceiling in high rise buildings. PVDF-insulated wire and cables jacketed with the flexible VDF–HFP copolymer pass the UL 910 specification for low smoke generation and flame spread and are approved for remote-control, signaling and power-limited circuits, fire protective signaling systems, and communication systems. Other important wire constructions include cross-linked PVDF jackets (193) for commercial aircraft, industrial power control, and cathodic protection wires and cables. Self-limiting strip heaters consisting of a cross-linked conductive PVDF core, which separates two parallel conductors, and a fluoropolymer jacket are useful for heating pipes or other process fluid-handling equipment (194,195).

Cross-linked heat-shrinkable PVDF tubings (196) are used as connector sleeves for wires and cables, or to coat ordnance (197). Some sleeves incorporate a ring of solder, forming a so-called solder sleeve for power control, electronic, aircraft, and communication wiring.

Uniaxially or biaxially oriented PVDF film upon metallization and poling under a high dielectric field is a flexible, tough, light, and active transducer for many piezo- and pyroelectric applications (198,199). Current applications include infrared detectors (200); audio devices, eg, stereo speakers, microphones, headphones, phonograph cartridges, hydrophones for long-range tracking in ocean depth; pressure or stress sensors (201); contactless keyboards (202); motion detectors (203); and medical devices, eg, detectors for heartbeat and breathing rate, or sensors for ultrasonic imaging.

The largest commercial application for PVDF homopolymer powder is as a base for long-lasting decorative finishes on aluminum and galvanized steel siding, curtain-wall panels, roofing systems, aluminum extrusions and other building components (204) that are used on power plants, schools, airport buildings, department stores, high rise office and hotel buildings, sports stadiums, and, to a lesser extent, residential buildings. These organosol finishes, available in many colors from paint companies throughout the world, are factory-applied by conventional state-of-the-art coil or spray-coating procedure to the primed base metal (205). Usually, the coating consists of a suitable primer layer up to about 5 µm thick and a 20–30 µm finishing layer of a PVDF topcoat. Accelerated weathering tests along with the experience with buildings erected since the 1960s prove that these finishes are unique in durability in terms of film integrity, color retention, corrosion resistance, flexibility, sand-abrasion resistance, and chemical resistance (206). Similar PVDF organosol dispersions are also being used for corrosion-protection coating of automotive break-line tubings. Pigmented thin film that is continuously cast from PVDF solutions or dispersions is used for decorative laminates and has been specified for body trim by principal automobile manufacturers. PVDF-based powders analogous in composition to the liquid finishes have been proposed as decorative protective coatings for metallic substrata (207). PVDF-based powder for rotomolding, eg, for tanks, valves, or fittings, and for fluidized-bed deposition and electrostatic spraying are available (208).

Fluid-handling systems in the chemical processing and related fields are also large users of PVDF products such as solid or lined pipes, valves, pumps, tower packing, and tank and trailer linings (209,210). Because PVDF is manufactured by methods that assure extremely low ionic contamination, it has qualified for use in ultrapure water systems (211–213), including WFI (water for injection) and *U.S. Pharmacopoeia* (USP) standards (214). Blow-molded PVDF bottles are used for shipping or storing high purity chemicals in the semiconductor industry (215). Extruded monofilament woven into coarse fabric is used widely for drum filtration during bleaching of wood pulp with chemicals, eg, sodium hypochlorite or chlorine dioxide and caustic soda. Like other fluoropolymers, PVDF is used as a binder for asbestos-fiber-based diaphragms used in cells for the electrolysis of brine to produce chlorine and caustic soda (216,217).

PVDF-based microporous filters are in use at wineries, dairies, and electrocoating plants, as well as in water purification, biochemistry, and medical devices. Recently developed nanoselective filtration using PVDF membranes is 10 times more effective than conventional ultrafiltration (UF) for removing viruses from protein products of human or animal cell fermentations (218). PVDF protein-sequencing membranes are suitable for electroblotting procedures in protein research, or for analyzing the phosphoamino content in proteins under acidic and basic conditions or in solvents (219).

Pigmented PVDF and ABS laminates manufactured by coextrusion with a tie-coat exhibit excellent weather resistance resulting from the protective PVDF cap layer; they are used in Europe for thermoformed automotive dash panels, trailer and tractor roofs, motorcycle gas tank housings, and lawn-mower blade guards (220). A PVDF alloy which is a blend of PVDF and alkyl methacrylate homo- or copolymer is coextruded with acrylate or methacrylate resin blend to form a sheet for hydrosanitary components (221). Similar blends of PVDF and compatible resins can be coextruded both with PVC, to form home siding panels with outstanding resistance to weather (222), and with an engineering resin, ie, polycarbonate, polyurethane, polyamide, polyester, or ABS, or their compounds (223).

In Japan, PVDF monofilament for fishing lines for both commercial and sport fishing is a specialty in demand (224–226) because it displays no water absorption, is not visible in water, and has high knot strength and high specific gravity. PVDF as a processing aid eliminates melt fracture and other flow-induced imperfections in blown LLDPE and HDPE films (227). Optical disk memory devices utilize the decrease in transmittance on crystallization of PVDF and thus provide an overwritable memory (228). The exceptional dielectric properties of PVDF are utilized in electrophotographic carrier (toner) compositions (229).

BIBLIOGRAPHY

"Poly(vinylidene fluoride) under "Fluorine Compounds, Organic" in *ECT* 2nd ed., Vol. 9, pp. 840–847, by W. S. Barnhart and N. T. Hall, Pennsalt Chemicals Corp.; in *ECT* 3rd ed., Vol. 11, pp. 64–74, by J. E. Dohany and L. E. Robb, Pennwalt Corp.

1. J. E. Dohany and J. S. Humphrey, in J. I. Kroschwitz, ed., *Encyclopedia of Polymer Science and Engineering*, 2nd ed., Vol. 17, p. 532.
2. J. E. Dohany, in R. B. Seymour and G. S. Kirshenbaum, eds. *High Performance Polymers: Their Origin and Development*, Elsevier Science Publishing Co., New York, 1986, p. 287.
3. A. J. Lovinger in G. C. Bassett, ed., *Developments in Crystalline Polymers*, Vol. 1, Applied Science Publishers, Ltd., Barking, UK, 1982, pp. 195–273.
4. W. H. Mears and co-workers, *Ind. Eng. Chem.* **47**(7), 1449–1454 (1955).
5. A. N. Baratov and V. M. Kucher, *Zh. Prikl. Khim.* **38**(5), 1068–1072 (1965).
6. D. R. Stull, E. F. Westrum, and G. C. Sinke, *The Chemical Thermodynamics of Organic Compounds*, John Wiley & Sons, Inc., New York, 1969, p. 502.
7. W. D. Wood, J. L. Lacina, B. L. DePrater, and J. P. McCullough, *J. Phys. Chem.* **68**(3), 579 (1964).
8. U.S. Pat. 2,551,573 (Aug. 5, 1951), F. B. Downing, A. F. Benning, and R. C. McHarness (to E. I. du Pont de Nemours & Co., Inc.).
9. U.S. Pat. 2,774,799 (Dec. 18, 1956), R. Mantell and W. S. Barnhart (to M. W. Kellogg Co.).
10. USSR Pat. 216,699 (Apr. 26, 1968), B. P. Zverev, A. L. Goldinov, Yu. A. Panshin, L. M. Borovnev, and N. S. Shirokova.

11. U.S. Pat. 3,246,041 (Apr. 12, 1966), M. E. Miville and J. J. Earley (to Pennwalt Corp.).
12. Ger. Pat. 1,288,085 (Jan. 30, 1969), F. Kaess, K. Lienhard, and H. Michaud (to Sueddeutsche Kalkstickstoff-Werke A.G.).
13. Ger. Pat. 1,288,593 (Feb. 6, 1969), F. Kaess, K. Lienhard, and H. Michaud (to Sueddeutsche Kalkstickstoff-Werke A.G.).
14. Jpn. Pat. 58 217,403 (Dec. 17, 1983), (to Pennwalt Corp.).
15. H. Mueller, G. Emig, and H. Hofmann, *Chem. Ing. Tech.* **56**(8), 626–628 (1984); *Chem. Abstr.* **101**(19), 170373v (1984).
16. J. Wolfrum, M. Schneider, *Proc. SPIE Int. Soc. Opt. Eng.* **458**, 46–52 (1984); *Chem. Abstr.* **101**(13), 110151 (1984).
17. J. Wolfrum, *Laser Chem.* **6**(2), 125–147 (1986).
18. Z. F. Dong, M. Schneider, J. Wolfrum, *Int. J. Chem. Kinet.* **21**(6), 387–397 (1989); *Chem. Abstr.* **111**(23), 213982 (1989).
19. Can. Pat. 2,016,691 (Dec. 28, 1990), M. Y. Elsheikh (to Elf Atochem North America, Inc.).
20. Fr. Pat. 1,337,360 (Sept. 13, 1963), Produits Chimique Pechiney Saint-Gobain.
21. Jpn. Pat. 68 29,126 (Dec. 13, 1968), H. Ukihashi and M. Ichimura (to Asahi Glass Co., Ltd.).
22. Jpn. Pat. 62 169,737 (July 25, 1987), (to Pennwalt Corp.).
23. U.S. Pat. 4,818,513 (Apr. 4, 1989), F. C. Trager, J. D. Mansell, and W. E. Wimer (to PPG Industries, Inc.).
24. Eur. Pat. Appl. 402,652 (Dec. 19, 1990), M. Y. Elsheikh and M. S. Bolmer (to Elf Atochem North America, Inc.).
25. Eur. Pat. Appl. 407,711 (Jan. 16, 1991), M. Y. Elsheikh (to Elf Atochem North America, Inc.).
26. U.S. Pat. 2,401,897 (June 11, 1946), A. F. Benning, F. B. Downing, and R. J. Plunkett (to E. I. du Pont de Nemours & Co., Inc.).
27. Jpn. Pat. 68 11,202 (May 11, 1968), Kureha Chem. Ind. Co., Ltd.
28. U.S. Pat. 2,734,090 (Feb. 7, 1956), J. C. Calfee and C. B. Miller (to Allied Chemical Corp.).
29. Ger. Pat. 2,659,712 (July 6, 1976), N. Schultz, P. Martens and H-J. Vahlensieck (to Dynamit Nobel AG).
30. U.S. Pat. 3,600,450 (Aug. 17, 1971), F. Kaess and H. Michaud (to Sueddeutsche Kalkstickstoff-Werke AG).
31. Eur. Pat. 3,723,549 (Mar. 27, 1973), F. Kaess, K. Lienhard, and H. Michaud (to Sueddeutsche Kalkstickstoff-Werke AG).
32. Eur. Pat. Appl. 361,578 (Apr. 4, 1990), J. Franklin and F. Janssens (to Solvay & Cie.).
33. E. T. McBee and co-workers, *Ind. Eng. Chem.* **39**(3), 409–412 (1947).
34. U.S. Pat. 3,833,676 (Sept. 3, 1974), R. Ukaji and I. Morioka (to Daikin Industries, Ltd.).
35. Jpn. Pat. 58 217,403 (Dec. 17, 1983), (to Pennwalt Corp.).
36. Eur. Pat. Appl. 297,947 (Jan. 4, 1989), B. Cheminal and A. Lantz (to Elf Atochem SA).
37. Eur. Pat. Appl. 407,689 (Jan. 16, 1991), D. W. Wright and B. L. Wagner (to Elf Atochem North America, Inc.).
38. Eur. Pat. Appl. 421,830 (Apr. 10, 1991), M. Bergougnan, J. M. Galland, and S. Perdieux (to Elf Atochem SA).
39. Jpn. Pat. 03 151,335 (June 27, 1991), M. Iwasaki and T. Yoshida (to Toa Gosei Industry Co., Ltd.).
40. U.S. Pat. 3,047,637 (July 31, 1962), F. Olstowski (to The Dow Chemical Co.).
41. Fr. Pat. 1,330,146 (June 2, 1963), A. E. Pavlath and F. H. Walker (to Stauffer Chemical Co.).

42. U.S. Pat. 3,188,356 (June 8, 1965), M. Hauptschein and A. H. Feinberg (to Pennwalt Corp.).
43. Jpn. Pat. 65 22,453 (Oct. 5, 1965), S. Okazaki and N. Sakauchi (to Kureha Chemical Industry Co., Ltd.).
44. Eur. Pat. Appl. 461,297 (Dec. 18, 1991), M. S. Bolmer and M. Y. Elsheikh (to Elf Atochem North America, Inc.).
45. U.S. Pat. 3,073,870 (Jan. 15, 1963), D. M. Marquis (to E. I. du Pont de Nemours & Co., Inc.).
46. Jpn. Pat. 68 10,602 (May 4, 1968), Y. Kometani and M. Takemoto (to Daikin Kogyo Co., Ltd.).
47. Eur. Pat. Appl. 313,254 (Apr. 26, 1989), D. W. Edwards (to Imperial Chemical Industries, PLC).
48. Ger. Pat. 42,730 (Jan. 5, 1966), H. Madai.
49. U.S. Pat. 3,428,695 (Feb. 18, 1969), J. R. Soulen and W. F. Schwartz (to Pennwalt Corp.).
50. U.S. Pat. 3,089,910 (May 14, 1963), F. Olstowski and J. D. Watson (to The Dow Chemical Company).
51. R. E. Cais and J. M. Kometami, *Macromolecules* **17**, 1887–1889 (1984).
52. C. P. Carpenter, U. C. Pozzani, and H. F. Smith, *J. Ind. Hyg. Toxicol.* **31**, 343 (1949).
53. L. A. Greenberg and O. Lester, *Arch. Ind. Hyg. Occ. Med.* **2**, 335 (1950).
54. Litton Bionetics, Inc., LBI Project No. 12199-02, National Toxicology Program, Contract No. NO1-ES-28, 1984.
55. Litton Bionetics, Inc., LBI Project No. 12199-03, National Toxicology Program, Contract No. NO1-ES-2, 1984.
56. C. Maltoni and D. Tovoli, *Med. Lavoro*, **5**, 353 (1979).
57. TNO Nutrition and Food Research Project No. B 84-1408, Report No. 91.039, Netherlands Institute for Applied Scientific Research, Delft, the Netherlands, 1991.
58. Bio/Dynamics, Inc., Project 87-8022, CMA Reference NO. FIG-3.3-ONCO-BIO, 1991.
59. G. L. Kennedy, Jr., *Crit. Rev. Toxicol.* **21**, 149 (1990).
60. *Vinylidene Fluoride*, Toxicology Data, Elf Atochem North America, Inc., Philadelphia, Pa., July 1992.
61. U.S. *Fed. Regist.* **57**(3), 409 (1992).
62. U.S. Pat. 3,051,677 (Aug. 28, 1964), D. R. Rexford (to E. I. du Pont de Nemours & Co., Inc.).
63. U.S. Pat. 2,738,343 (Mar. 13, 1956); U.S. Pat. 2,752,331 (June 26, 1956), A. Dittman, H. J. Passino, and W. O. Teeters (to M. W. Kellogg Co.).
64. U.S. Pat. 2,968,649 (Jan. 17, 1961), J. R. Pailthorp and H. E. Schroeder (to E. I. du Pont de Nemours & Co., Inc.).
65. U.S. Pat. 3,178,399 (Apr. 13, 1965), E. S. Lo (to 3M Co.).
66. Eur. Pat. Appl. 456,019 (Nov. 13, 1991), L. A. Barber (to Elf Atochem North America, Inc.).
67. U.S. Pat. 4,851,479 (Jul. 25, 1989), J. Blaise and P. Kappler (to Elf Atochem SA).
68. Brit. Pat. 827,308 (Feb. 3, 1960), (to 3M Co.).
69. Eur. Pat. Appl. 320,344 (June 14, 1989), P. Kappler (to Elf Atochem SA).
70. U.S. Pat. 3,706,723 (Dec. 19, 1972), S. Chandrasekaran and M. B. Mueller (to Allied Signal Corp.).
71. U.S. Pat. 4,591,616 (May 27, 1986), S. Miyata and S. Kobayashi (to Central Glass Co., Ltd.).
72. U.S. Pat. 4,472,557 (Sept. 18, 1984), C. Kawashima and T. Yasumura (to Central Glass Co., Ltd.).
73. U.S. Pat. 2,435,537 (Feb. 3, 1948), T. A. Ford and W. E. Hanford (to E. I. du Pont de Nemours & Co., Inc.).

74. U.S. Pat. 2,468,054 (Apr. 26, 1949), T. A. Ford (to E. I. du Pont de Nemours & Co., Inc.).
75. U.S. Pat. 2,559,752 (July 10, 1951), K. L. Berry (to E. I. du Pont de Nemours & Co., Inc.).
76. U.S. Pat. 3,714,137 (Jan. 30, 1973), K. Lienhard and D. Ulmschneider (to Suddeutsche Kalkstickstoff-Werke AG).
77. U.S. Pat. 4,025,709 (May 24, 1977), J. Blaise and E. Grimaud (to Elf Atochem SA).
78. Eur. Pat. Appl. 387,938 (Sept. 19, 1990), X. Bacque and P. Lasson (to Solvay & Cie.).
79. U.S. Pat. 3,245,971 (Apr. 12, 1966), H. Iserson (to Pennwalt Corp.).
80. U.S. Pat. 3,640,985 (Feb. 8, 1972), H. C. Stevens (to PPG Industries, Inc.).
81. U.S. Pat. 3,708,463 (Jan. 2, 1973), J. P. Stallings (to Diamond Shamrock Corp.).
82. U.S. Pat. 3,642,755 (Feb. 15, 1972), J. A. Baxter, C. O. Eddy, and H. C. Stevens (to PPG Industries, Inc.).
83. U.S. Pat. 3,193,539 (July 6, 1965), M. Hauptschein (to Pennwalt Corp.).
84. U.S. Pat. 4,076,929 (Feb. 28, 1978), J. E. Dohany (to Pennwalt Corp.).
85. U.S. Pat. 3,475,396 (Oct. 28, 1969), G. H. McCain, J. R. Semancik, and J. J. Dietrich (to Diamond Shamrock Corp.).
86. U.S. Pat. 3,857,827 (Dec. 31, 1974), J. E. Dohany (to Pennwalt Corp.).
87. U.S. Pat. 4,360,652 (Nov. 23, 1982), J. E. Dohany (to Pennwalt Corp.).
88. U.S. Pat. 4,569,978 (Feb. 11, 1986), L. A. Barber (to Pennwalt Corp.).
89. U.S. Pat. 3,598,797 (Aug. 10, 1971), Y. Kometani, M. Okuda, and C. Okuno (to Daikin Industries, Ltd.).
90. L. Y. Madorskaya and co-workers, *Vysokomol. Soedin., Ser. B* **31**(10), 737–742 (1989).
91. Eur. Pat. Appl. 215,710 (Mar. 25, 1987), J. Blaise (to Elf Atochem SA).
92. U.S. Pat. 3,553,785 (Jan. 12, 1971), Y. Amagi and N. Bannai (to Kureha Chemical Co.).
93. U.S. Pat. 3,781,265 (Dec. 25, 1973), J. E. Dohany (to Pennwalt Corp.).
94. U. S. Pat. 4,542,194 (June 18, 1985), J. Dumoulin (to Solvay & Cie.).
95. Jpn. Pat. 01 129,005 (May 29, 1989), K. Ihara, Y. Noda, and T. Amano (to Daikin Industries, Ltd.).
96. Jpn. Pat. 02 029,402 (Jan. 31, 1990), J. Watanabe (to Shin-Etsu Chemical Industry Co., Ltd.).
97. U.S. Pat. 3,780,007 (Dec. 18, 1973), J. F. Stallings (to Diamond Shamrock Corp.).
98. Eur. Pat. Appl. 417,585 (Mar. 20, 1991) and 423,097 (Apr. 17, 1991), P. Lasson (to Solvay & Cie.).
99. Brit. Pat. 1,057,088 (Feb. 1, 1967), Kali-Chemie AG.
100. Ger. Pat. 1,806,426 (May 16, 1969); Fr. Pat. 1,590,301 (Apr. 14, 1970), D. P. Carlson (to E. I. du Pont de Nemours & Co., Inc.).
101. W. W. Doll and J. B. Lando, *J. Appl. Polym. Sci.* **14**, 1767 (1970).
102. U.S. Pat. 3,616,371 (Oct. 26, 1971), H. Ukihashi and M. Ichimura (to Asahi Glass Co., Ltd.).
103. Brit. Pat. 1,004,172 (Sept. 8, 1965), Deutsche Solvay-Werke GmbH.
104. R. Liepins, J. R. Surles, N. Morosoff, V. T. Stannett, M. L. Timmons, and J. J. Wortman, *J. Polym. Sci. Part A-1* **16**, 3039 (1978).
105. Eur. Pat. Appl. 403,915 (Dec. 27, 1990), J. Kammermaier and G. Rittmayer (to Siemens AG).
106. U.S. Pat. 4,438,247 (Mar. 20, 1984), R. E. Cais (to AT&T Technologies).
107. R. E. Cais and J. M. Kometani, *Macromolecules*, **17**, 1887 (1984).
108. P. E. Bretz, Ph.D. dissertation, Lehigh University, Bethlehem, Pa., 1980.
109. P. E. Bretz, R. W. Hertzberg, and J. A. Manson, *Polymer* **22**, 1272–1278 (1981).
110. M. Gorlitz, R. Minke, W. Trautvetter, and G. Weisgerber, *Angew. Makromol. Chem.* **29/30** 137 (1973).

111. R. C. Ferguson and E. G. Baume, Jr., *J. Phys. Chem.* **83**, 1379 (1979).
112. R. C. Ferguson and D. W. Ovenall, *Polymer Preprints, Div. Polym. Chem. Am. Chem. Soc.* **25**(1), 340 (1984).
113. M. A. Bachmann, W. Gordon, J. L. Koenig, and J. B. Lando, *J. Appl. Phys.* **50**, 6106 (1979).
114. D. E. Mattern, L. Fu-Tyan, and D. M. Hercules, *Anal. Chem.* **56**, 2762–2769 (1984).
115. G. Lutringer and G. Weill, *Polymer* **32**(5), 877 (1991).
116. G. Lutringer, B. Meurer, and G. Weill, *Polymer* **32**(5), 884 (1991).
117. A. J. Lovinger, D. D. Davis, R. E. Cais, and J. M. Kometani, *Polymer* **28**, 617–626 (1987).
118. S. Russel, K. L. McElroy, and L. H. Judovits, *Polym. Eng. Sci.* **32**(17), 1300 (1992).
119. A. J. Lovinger, *Macromolecules* **15**, 40 (1982).
120. J. Herschinger, D. Schaefer, H. W. Spiess, and A. J. Lovinger, *Macromolecules* **24**, 2428 (1991).
121. M. A. Bachmann and J. B. Lando, *Macromolecules* **14**, 40 (1981).
122. A. J. Lovinger, *J. Polym. Sci. Part A-2* **18**, 793–809 (1980).
123. Y. S. Yadav and P. C. Jain, *J. Macromol. Sci. Phys.* **B25**(3), 335 (1986).
124. T. Mizuno, K. Nakamura, N. Murayama, and K. Okuda, *Polymer* **26**(6), 853 (1985).
125. A. J. Lovinger, *Polymer* **21**(11), 1317 (1980).
126. C. C. Hsu and P. H. Geil, *Polymer Comm.* **27**, 105 (1986).
127. W. M. Prest and D. J. Luca, *J. Appl. Phys.* **49**(10), 5042 (1978).
128. G. T. Davis, J. E. McKinney, M. G. Broadhurst, and S. C. Roth, *J. Appl. Phys.* **49**, 4998 (1978).
129. R. Hasegawa, Y. Takahashi, Y. Chatani, and H. Tadokoro, *Polymer J.* **3**, 600 (1972).
130. K. Nakagawa and Y. Ishida, *Kolloid Z. Z. Polym.* **251**, 103 (1973).
131. A. J. Lovinger and T. T. Wang, *Polymer* **20**, 725 (1979).
132. K. F. Auyeung, *Polym. Eng. Sci.* **30**(7), 394 (1990).
133. D. R. Paul and J. W. Barlow, *J. Macromol. Sci. Rev. Macromol. Chem.* **C18**, 109 (1980).
134. J. S. Noland, N. N.-C. Hsu, R. Saxon, and J. M. Schmitt, *Advan. Chem. Ser.* **99**, 15 (1971).
135. J. Mijovic, H.-L. Luo, and C. D. Han, *Polym. Eng. Sci.* **22**(4), 234 (1982).
136. A. P. A. M. Eijkelenboom and co-workers, *Macromolecules*, **25**(18), 4511 (1992).
137. D. C. Wahrmund, R. E. Bernstein, J. W. Barlow, and D. R. Paul, *Polym. Eng. Sci.* **18**, 677 (1978).
138. G. Guerra, F. E. Karasz, and W. J. MacKnight, *Macromolecules* **19**, 1935 (1986).
139. R. Timmerman and W. Greyson, *J. Appl. Polym. Sci.* **6**(22), 456 (1962).
140. T. Yoshida, R. E. Florin, and L. A. Wall, *J. Polymer Sci.* **A3**, 1685 (1965).
141. K. Makuuchi, M. Asano, and T. Abe, *Nippon Nogei Kagaku Kaishi* (4), 686 (1976).
142. S.-H. Hyon and R. Kitamaru, *Bull. Inst. Chem. Res. Kyoto Univ.* **57**(2), 193 (1979).
143. K. Makuuchi, F. Yoshii, and T. Abe, *Nippon Nogei Kagaku Kaishi*, (10), 1828 (1975).
144. V. S. Ivanov, I. I. Migunova, and A. I. Mikhailov, *Radiat. Phys. Chem.* **37**(1), 119 (1991).
145. A. J. Lovinger in R. L. Clough and S. W. Shalaby, eds., *Radiation Effects on Polymers*, American Chemical Society Symposium Series 475, ACS, Washington, D.C., 1991, p. 84.
146. R. G. Kepler, *Ann. Rev. Phys. Chem.* **29**, 497 (1978).
147. D. K. Das-Gupta, *Ferroelectrics* **118**, 165 (1991).
148. A. J. Lovinger, *Science* **220**, 1115 (1983).
149. T. T. Wang, J. M. Herbert, and J. M. Glass, eds., *The Applications of Ferroelectric Polymers*, Chapman and Hall, New York, 1988.
150. A. J. Lovinger and D. J. Freed, *Macromolecules* **13**, 889 (1980).

151. H. Ishii, *Kobunshi Kagaku* **27**(307), 858 (1970).
152. U.S. Pat. 4,401,845 (Aug. 30, 1983), J. W. Michaud and O. R. Odhner (to Pennwalt Corp.).
153. *Kynar PVDF*, Technical Brochure, Elf Atochem North America, Inc., Philadelphia, Pa., 1990.
154. *Hylar PVDF*, Technical Brochure, Ausimont USA, Inc., Morristown, N.J., 1991.
155. *Solef PVDF*, Technical Brochure, Solvay & Cie. SA, Brussels, Belgium, 1987.
156. U.S. Pat. 4,328,151 (May 4, 1982), D. N. Robinson (to Pennwalt Corp.).
157. *Extrusion of Kynar and Kynar Flex Poly(vinylidene Fluoride) (PVDF)*, Technical Data, Elf Atochem North America Inc., Philadelphia, Pa., Apr. 1990.
158. U.S. Pat. 3,922,186 (Nov. 25, 1975), M. Segawa, Y. Kawakami, and I. Itoh (to Kureha Chemical Co., Ltd.).
159. U.S. Pat. 4,208,462 (June 17, 1980), R. Dauphin and N. Maquet (to Solvay & Cie.).
160. *New Mater. Jpn.,* 2 (Feb. 1990).
161. U.S. Pat. 4,481,158 (Nov. 6, 1984), P. Georlette and N. Maquet (to Solvay & Cie.).
162. U.S. Pat. 4,264,555 (Apr. 28, 1981), E. Lang, W. Nachtigall, and J. Stark (to Dynamit Nobel AG).
163. U.S. Pat. 4,302,556 (Nov. 24, 1981), H. Endo, H. Ohhira, and T. Sasaki (to Kureha Chemical Co., Ltd.).
164. U.S. Pat. 4,051,293 (Sept. 27, 1977), D. F. Wiley (to Cosden Oil & Chemical Co.).
165. U.S. Pat. 4,317,860 (Mar. 2, 1982), A. Strassel (to Atochem SA).
166. Eur. Pat. Appl. 484,053 (May 6, 1992), T. Ozu, K. Hayama, K. Abe, and K. Hata (to Mitsubishi Petrochemical Co., Ltd.).
167. U.S. Pat. 3,340,222 (Sept. 5, 1967), J. C. Fang (to E. I. du Pont de Nemours & Co., Inc.).
168. U.S. Pat. 4,314,004 (Feb. 2, 1982), to R. L. Stoneberg (to PPG Industries, Inc.).
169. U.S. Pat. 4,400,487 (Aug. 23, 1983), R. L. Stoneberg and R. R. Stec (to PPG Industries, Inc.).
170. U.S. Pat. 4,656,768 (Apr. 14, 1987), A. J. Tortorello and C. A. Higginbotham (to DeSoto, Inc.).
171. U.S. Pat. 4,022,737 (May 10, 1977), K. Sekmakas and R. O. Yates (to DeSoto, Inc.).
172. U.S. Pat. 4,141,873 (Feb. 27, 1979), J. E. Dohany (to Pennwalt Corp.).
173. U.S. Pat. 4,309,328 (Jan. 5, 1982), D. W. Carson, R. C. Gray, and G. W. Luckock (to PPG Industries, Inc.).
174. U.S. Pat. 4,383,075 (May 10, 1983), P. T. Abel (to SCM Corp.).
175. U.S. Pat. 4,203,847 and 4,203,848 (May 20, 1980), J. D. Grandine (to Millipore Corporation).
176. A. Bottino, G. Camera-Roda, G. Capannelli, and S. Munari, *J. Membr. Sci.* **57**, 1 (1991).
177. U.S. Pat. 4,340,482 (July 20, 1982), S. Sternberg (to Millipore Corp.).
178. F. F. Stengaard, *J. Membr. Sci.* **36**, 257 (1988); *Desalination* **70**, 207 (1988).
179. U.S. Pat. 4,849,106 (July 18, 1989), L. Mir (to Koch Membrane Systems, Inc.).
180. U.S. Pat. 4,954,256 (Sept. 4, 1990), P. J. Degen, I. Rothman, and T. C. Gsell (to Pall Corporation).
181. U.S. Pat. 5,137,633 (Aug. 11, 1992), D. Wang (to Millipore Corp.).
182. U.S. Pat. 4,399,035 (Aug. 16, 1983), T. Nohmi and T. Yamada (to Asahi Kasei Kogyo Kabushiki Kaisha).
183. U.S. Pat. 3,896,196 (July 22, 1975), C. A. Dickey and J. E. McDaniel (to Glasrock Products, Inc.).
184. U.S. Pat. 4,425,443 (Jan. 10, 1984), P. Georlette and J. Leva (to Solvay & Cie.).

185. U.S. Pat. 4,615,850 (Oct. 7, 1986) and 4,675,345 (June 23, 1987), R. L. Pecsok (to Pennwalt Corp.).
186. M. J. Haley with A. Leder and Y. Sakuma, *Chemical Economics Handbook*, SRI International, Menlo Park, Calif., 1992.
187. L. Manolis Sherman, *Plastics Technol.* **38**(13), 77 (1992).
188. *Chem. Mark. Rep.* **240**(23), 7,26 (1991); *ibid.* **237**(6), 7,21 (1990).
189. R. J. Martino, ed., *Modern Plastics Encyclopedia for '93*, McGraw-Hill Book Co., Inc., New York, 1992, pp. 211–212.
190. U.S. Federal Regulations, Title 21, Chapt. I, Part 177.2510.
191. W. L. Guess, and J. Autian, *J. Oral Therapeut. Pharm.* **3**(2), 116 (1966).
192. D. J. Yturraspe, W. V. Lumb, S. Young, and H. G. Gorman, *J. Neurosurg.* **42**(1), 47 (1975).
193. U.S. Pat. 3,269,862 (Aug. 30, 1966), V. L. Lanza and E. C. Stivers (to Raychem Corp.).
194. *Plastics Design Forum*, (Nov-Dec. 1976).
195. U.S. Pat. 4,318,881 (Mar. 9, 1982); U.S. Pat. 4,591,700 (May 27, 1986), U. K. Sopory (to Raychem Corp.).
196. U.S. Pat. 3,582,457 (June 1, 1971), F. E. Bartell (to Electronized Chemicals Corp.).
197. M. D. Heaven, *Prog. Rubber Plast. Tech.* **2**, 16 (1986).
198. N. Murayama, *J. Polym. Sci. Part A-2* **13**, 929 (1975).
199. M. G. Broadhurst, S. Edelman, and G. T. Davis, *Am. Chem. Soc. Org. Coat. Plast. Chem.* **42**, 241 (1980).
200. H. Meixner and G. Mader, *Phys. Unserer Zeit* **21**(5), 210 (1990).
201. M. U. Anderson and D. E. Wackerbarth, Sandia National Laboratories Report SAND-88-2327; Order No. DE89010529, Albuquerque, N.M., 1988.
202. G. T. Pearman, J. L. Hokanson, and T. R. Meeker, *Ferroelectrics* **28**, 311 (1980).
203. B. Andre, J. Clot, E. Partouche, J. J. Simonne, and F. Bauer, *Sens. Actuators* **A33**, 111 (1992).
204. *Buildings*, 78 (Oct. 1978); *Building Design & Construction*, 134 (May 1983).
205. J. E. Dohany and N. P. Murray, in J. A. Wilkes, ed, *Encyclopedia of Architecture, Design, Engineering and Construction*, John Wiley & Sons, Inc, New York, 1988, p. 478.
206. American Architectural Manufacturers Association, Palatine, Ill., Specification No. AAMA 605.
207. U.S. Pat. 4,770,939 (Sept. 13, 1988) and U.S. Pat. 5,030,394 (July 9, 1991), W. Sietses, T. M. Plantenga, and J.-P. Dekerk (to Labofina SA).
208. *Pulp Pap.* **63**(13), 167 (1989).
209. N. L. Maquet, *Proceedings of the AESF Annual Technical Conference*, Vol. 73, American Electroplater's and Surface Finishers Society, Orlando, Fla., 1986, pp. 1–3.
210. D. K. Heffner, *Mater. Perform.* **31**(7), 33–36 (1992).
211. S. P. Daly, J. E. Dohany, and J. S. Humphrey, *Proceedings, 32nd Annual Conference Institute of Environmental Sciences*, Mt. Prospect, Ill., 1986, p. 397.
212. J. S. Humphrey, J. E. Dohany, and C. Ziu, *1st Annual High Purity Water Conference Proceedings*, Philadelphia, Pa., 1987, p. 135
213. J. M. De Berraly, *Ultrapure Water J.* **4**(4), 36 (1987); *CPI Equip. Reporter*, (July-Aug. 1988).
214. D. Spann, C. Mitchell, and D. A. Toy, *Chem. Process.* 26 (Mar. 1988).
215. *Chem. Mark. Rep.* **239**(4), 49 (1991).
216. U.S. Pat. 4,093,533 (June 6, 1978), R. N. Beaver and C. W. Becker (to The Dow Chemical Company).
217. U.S. Pat. 4,341,596 (July 27, 1982), P. R. Mucenieks (to FMC Corporation).
218. *Chem. Eng.* **98**(7), 17,19 (1991).

219. *Biotech. News*, **10**(12), 6 (1991); *ibid.*, **9**(12), 7 (1990).
220. A. Strassel, *Kunststoffe*, **78**(9), 801 (1988).
221. Eur. Pat. Appl. 419,166 (Mar. 27, 1991), C. Sempio, A. Anghileri, M. Binaghi, T. Ronchetti, and I. Vailati (to Vedril S.p.A.).
222. U.S. Pat. 4,585,701 (Apr. 29, 1986), E. J. Bartoszek and S. F. Mones (to Pennwalt Corp.).
223. U.S. Pat. 4,563,393 (Jan. 7, 1986), Y. Kitagawa, A. Nishioka, Y. Higuchi, T. Tsutsumi, T. Yamaguchi, and T. Kato (to Japan Synthetic Rubber Co., Ltd.).
224. H. Endo and S. Ohira, *Sen-i Gakkaishi* **47**(6), 333 (1991).
225. Jpn. Pat. 92 91,215 (Mar. 24, 1990), Y. Nishikawa, H. Nakada, and T. Sato (to Toray K. K.).
226. Jpn. Pat. 87 25,0217 (Oct. 31, 1987), K. Nakagawa, K. Toma, S. Murakami, and T. Eguchi (to Unitika Ltd.).
227. *Modern Plast.* **69**(6), 133 (1992); *Plast. World*, **50**(7), (1992).
228. Jpn. Pat. 03 13,383 (Jan. 22, 1991), A. Tanaka, Y. Kojima (to Fujitsu Ltd.).
229. Jpn. Pat. 03 01,164 (Jan. 7, 1991), H. Okuno, E. Tominaga, R. Kimura, M. Takeda, and T. Aokit

JULIUS E. DOHANY
Consultant

POLYCHLOROTRIFLUOROETHYLENE

Many challenging industrial and military applications utilize polychlorotrifluoroethylene [9002-83-9] (PCTFE) where, in addition to thermal and chemical resistance, other unique properties are required in a thermoplastic polymer. Such has been the destiny of the polymer since PCTFE was initially synthesized and disclosed in 1937 (1). The synthesis and characterization of this high molecular weight thermoplastic were researched and utilized during the Manhattan Project (2). The unique combination of chemical inertness, radiation resistance, low vapor permeability, electrical insulation properties, and thermal stability of this polymer filled an urgent need for a thermoplastic material for use in the gaseous UF_6 diffusion process for the separation of uranium isotopes (see DIFFUSION SEPARATION METHODS).

Properties

The physical properties of PCTFE are primarily determined by a combination of molecular weight and percent crystallinity. Because of the lack of suitable solvents, a correlation between the number average molecular weight and zero-strength time (ZST: typical values of 200 to 400 s) has been developed (3,4). The high molecular weight thermoplastic has a melt temperature (T_m) of 211–216°C, a glass-transition temperature (T_g) of 71–99°C (5), and is thermally stable up to 250°C. The useful operational temperature range is considered to be from −240 to 200°C although an increase in service temperature can be achieved through selected fiber filling of the polymer (fiber glass, from 1 to 20% weight of the fiber).

The theoretical specific gravity of PCTFE for the amorphous and crystalline polymers has been calculated to range from 2.075 to 2.185, respectively (6–12). In reality, PCTFE molded parts have exhibited ranges of crystallinity from approximately 45% (specific gravity of 2.10) for quick-quenched parts to 65% (specific gravity of 2.13) for slow-cooled parts. The use of the terms amorphous and crystalline are relative but can be significant in the application. Basically, two types of crystallinity, micro and macro, exist in the polymer as a result of the synthesis and processing. The higher crystalline forms are less transparent, have higher tensile modulus, lower elongation, and have more resistance to liquids and vapors. The less crystalline form is optically clear, tough, and ductile, exhibiting higher elongation and lower modulus.

The typical mechanical properties that qualify PCTFE as a unique engineering thermoplastic are provided in Table 1; the cryogenic mechanical properties are recorded in Table 2. Other unique aspects of PCTFE are resistance to cold flow due to high compressive strength, and low coefficient of thermal expansion over a wide temperature range.

Table 1. Mechanical Properties of Polychlorotrifluoroethylene

Property	Value
tensile strength, MPa[a]	32–39
compressive strength, MPa[a]	38
modulus of elasticity, MPa[a]	1400
hardness, Shore D	76
deformation under load, at 25°C, 24 h, 7 MPa[a], %	0.3
heat deflection temperature, at 0.46 MPa[a], °C	126

[a]To convert MPa to psi, multiply by 145.

The high fluorine content contributes to resistance to attack by essentially all chemicals and oxidizing agents; however, PCTFE does swell slightly in halogenated compounds, ethers, esters, and selected aromatic solvents. Specific solvents should be tested. PCTFE has the lowest water-vapor transmission rate of any plastic (14,15), is impermeable to gases (see also BARRIER POLYMERS), and does not carbonize or support combustion.

PCTFE plastic is compatible with liquid oxygen, remains ductile at cryogenic temperatures (16–22), and retains its properties when exposed to either uv or gamma radiation. PCTFE exhibits a refractive index of 1.43 (ASTM D542) and an amorphous sheet can provide over 90% transmittance.

PCTFE exhibits very good electrical properties in terms of high insulation resistance, minimal tracking, corona formation, and surface flashover due to the polymer's nonwettable surface and ultralow moisture absorption (Table 3).

Table 2. Cryogenic Mechanical Properties of Polychlorotrifluoroethylene[a]

Property	PCTFE, % crystallinity	Temperature, °C	Value[a]
tensile: ultimate strength, MPa[b]	40	25	38.6
		−129	150
		−252	200
elongation, %	40	25	140
		−129	9
		−252	5
modulus of elasticity, MPa[b]	40	25	1520
		−129	5500
		−252	8700
impact strength notched Izod, J/m[c]	60	25	13.7
		−196	12.8
		−252	13.7

[a] ASTM D1430–89 Type 1, Grade 2 (13).
[b] To convert MPa to psi, multiply by 145.
[c] To convert J/m to ft·lbf/in., divide by 53.38 (see ASTM D256).

Table 3. Electrical Properties of PCTFE

Property	ASTM method	Value
dielectric strength,[a] V/μm	D149	20
arc resistance, s	D495	360
volume resistivity,[b] ohm cm^2/cm	D257	10^{18}
surface resistivity,[b] ohm	D257	10^{15}

[a] Short time. To convert to V/mil, multiply by 25.
[b] Fifty percent rh at 25°C.

Manufacture and Processing

The synthesis of the high molecular weight polymer from chlorotrifluoroethylene [79-38-9] has been carried out in bulk (23–27), solution (28–30), suspension (31–36), and emulsion (37–41) polymerization systems using free-radical initiators, uv, and gamma radiation. Emulsion and suspension polymers are more thermally stable than bulk-produced polymers. Polymerizations can be carried out in glass or stainless steel agitated reactors under conditions (pressure 0.34–1.03 MPa (50–150 psi) and temperature 21–53°C) that require no unique equipment.

After polymerization, the polymer is isolated from the latex or suspension. The suspension polymer, already in powder form, is washed to remove initiator residues and then dried. The emulsion polymer is coagulated from the latex by freezing or by the addition of salts, acids, and solvents (see LATEX TECHNOLOGY) and separated from the aqueous phase. The isolated powder is then washed and dried. The dried powder from either process additionally can be chemically treated to remove trace impurities that can result in chain degradation during further processing. Treatment with carboxylic acids (42), ozone in air (43), or chlorine (44) improves thermal stability, color, and light transmission of the final polymer. The polymer product can then be processed by plastic fabrication techniques in powder or melt-extruded pellet forms.

The lower molecular weight oils, waxes, and greases of PCTFE can be prepared directly by telomerization of the monomer or by pyrolysis of the higher molecular weight polymer (45–54).

PCTFE plastics can be processed by the standard thermoplastic fabrication techniques, eg, extrusion, injection, compression, and transfer molding. Specific corrosion-resistant alloys or chrome or nickel plating are recommended for equipment parts in contact with the polymer melt, such as molds, barrels, screws, etc (see PLASTICS TECHNOLOGY). The control of processing temperatures is paramount since prolonged overheating (above 260°C) can result in degradation of the polymer causing discoloration, voids, blisters, and loss of properties. The plastic can be easily machined from billets or rod stock on standard machining equipment to fabricate more precise part geometries, but sharp tools should be employed.

Economic Aspects

Several worldwide commercial manufacturers of PCTFE and vinylidene fluoride-modified copolymers [9010-75-7] offer a variety of products as shown in Table 4. PCTFE plastics have selling prices in the range of $40–100/kg, depending on the molecular weight, grade, product form, and supplier. As a result, PCTFE thermoplastics are used in high technology, specialty engineering areas where the unique combination of properties and part reliability demands a high performance thermoplastic polymer.

Table 4. PCTFE Manufacturers and Products

Trademark	Manufacturer	Product forms
homopolymers		
Daiflon	Daikin Koygo, Osaka, Japan	molding powder, pellets, dispersion oils, and greases
Kel-F 81	3M Co., St. Paul, Minn.	molding powders and pellets
Voltalef	Ugine Kuhlmann, Pierre-Benite, France	molding powders and pellets
Halocarbon oil	Halocarbon Products Corp., River Edge, N.J.	oils, waxes, and greases
copolymers		
Aclon, Aclar	Allied-Signal Chemical, Morristown, N.J.	molding powders, pellets, and film
Kel-F 800	3M Co., St. Paul, Minn.	molding powders

Specifications and Test Methods

PCTFE plastic is available in products that conform to ASTM 1430-89 Type I (Grades 1 and 2) and is suitable for processing into parts that meet MIL-P 46036

(Federal Specification LP-385C was canceled 1988). Standards for fabricated forms are available for compression molded heavy sections (AMS-3645 Class C), thin-walled tubing, rod, sheet, and molded shapes (AMS-3650). PCTFE plastics have been approved for use in contact with food by the FDA (55).

The test methods employed are the determination of molecular weight as measured by ZST (ASTM D1430); specific gravity (ASTM D792); tensile strength, elongation, and modulus (ASTM D638); compressive strength and modulus (ASTM D621); heat deflection (ASTM D648); impact strength (ASTM D256); flammability (ASTM D2863); hardness (ASTM D2240 and D785); and coefficient of linear expansion (ASTM D696).

Health and Safety Factors

In general, the PCTFE resins have been found to be low in toxicity and irritation potential under normal handling conditions. Specific toxicological information and safe handling procedures are provided by the manufacturer of specified PCTFE products upon request.

Uses

The principal uses of PCTFE plastics remain in the areas of aeronautical and space, electrical/electronics, cryogenic, chemical, and medical instrumentation industries. Applications include chemically resistant electrical insulation and components; cryogenic seals, gaskets, valve seats (56,57) and liners; instrument parts for medical and chemical equipment (58), and medical packaging; fiber optic applications (see FIBER OPTICS); seals for the petrochemical/oil industry; and electrodes, sample containers, and column packing in analytical chemistry and equipment (59).

The lower molecular weight PCTFE oils, waxes, and greases are used as inert sealants and lubricants for equipment handling oxygen and other oxidative or corrosive media. Other uses include gyroscope flotation fluids and plasticizers for thermoplastics.

BIBLIOGRAPHY

"Polychlorotrifluoroethylene" under "Fluorine Compounds, Organic" in *ECT* 3rd ed., Vol. 11, pp. 49–54, by A. C. West, 3M Co.

1. Brit. Pat. 465,520 (May 3, 1937), (to I. G. Farbenindustrie).
2. U.S. Pat. 2,564,024 (Aug. 14, 1951), W. T. Miller (to USAEC).
3. H. S. Kaufman, C. O. Kroncke, and C. K. Giannotta, *Mod. Plast.* **32**, 146 (1954).
4. E. K. Walsh and H. S. Kaufman, *J. Polym. Sci.* **26**, 1 (1957).
5. Y. P. Khanna and R. Kumar, *J. Polymer Sci.* **32**(11) 2010–2013 (1991).
6. T. Hashimoto, H. Kawasaki, and H. Kawai, *J. Polymer Sci.* **16**(2) 271–288 (1978).
7. E. Sacher, *J. Polymer Sci.* **18**(5), 333–337 (1980).
8. H. Matsuo, *J. Polymer Sci.* **21**, 331 (1956).

9. H. Matsuo, *J. Polymer Sci.* **25**, 234 (1957).
10. H. Matsuo, *Bull. Chem. Soc. Jpn.* **30**, 593 (1957).
11. J. D. Hoffman, *J. Am. Chem. Soc.* **74**, 1696 (1952).
12. J. D. Hoffman and J. J. Weeks, *J. Res. Nat. Bur. Stand.* **60**, 465 (1958).
13. R. E. Mowers *Cryogenic Properties of Poly(Chlorotrifluoroethylene), Technical Document Report No. RTD-TDR-63-11*, Air Force Contract No. AF04(611)-6354, 1962.
14. J. D. Hoffman and J. J. Weeks, *J. Chem. Phys.* **37**, 1723 (1962).
15. A. W. Myers and co-workers, *Mod. Plast.* **37**, 139 (1960).
16. N. Brown, B. D. Metzger, and Y. Imai, *J. Polym. Sci.* **16**, 1085 (1978).
17. Y. Imai and N. Brown, *Polymer* **18**, 298 (1977).
18. N. Brown and S. Fischer, *J. Poly. Sci. Polym. Phys.* **13**, 1315 (1975).
19. J. L. Currie, R. S. Irani, and J. Sanders, *Factors Affecting the Impact Sensitivity of Solid Polymer Materials in Contact with Liquid Oxygen*, ASTM Spec. Tech. Publ. 986, ASTM, Philadelphia, Pa., 1988, pp. 233–247; S. Chandrasekaran, in J. I. Kroschwitz, ed., *Encyclopedia of Polymer Science and Engineering*, 2nd ed., Vol. 3, John Wiley & Sons, Inc., 1985, pp. 463–480.
20. N. Schmidt and co-workers, *Ignition of Nonmetallic Materials by Impact of High-pressure Oxygen*, ASTM Spec. Tech. Publ. 1040, ASTM, Philadelphia, Pa., 1989, pp. 23–37.
21. B. J. Lockhart, M. D. Hampton, and C. J. Bryan, *The Oxygen Sensitivity/Compatability Ranking of Several Materials by Different Test Methods*, ASTM Spec. Tech. Publ. 1040, ASTM, Philadelphia, Pa., 1989, pp. 93–105.
22. M. L. Reath and R. S. Britton, *Plastics Manufacture and Processing*, 37-5 (1991).
23. U.S. Pat. 2,586,550 (Feb. 19, 1952), W. T. Miller, A. L. Dittman, and S. K. Reed (to USAEC).
24. U.S. Pat 2,792,377 (May 14, 1957), W. T. Miller (to 3M).
25. U.S. Pat. 2,636,908 (Apr. 28, 1953), A. L. Dittman and J. M. Wrightson (to M. W. Kellogg Co.).
26. Brit. Pat. 729,010 (Apr. 27, 1955), (to Farbenfabriken Bayer AG).
27. Fr. Pat. 1,419,741 (Dec. 3, 1965), (to Kureha Chemical Co.).
28. U.S. Pat. 2,700,662 (Jan. 25, 1955), D. M. Young and B. Thompson (to Union Carbide Co.).
29. U.S. Pat. 2,820,027 (Jan. 14, 1958), W. F. Hanford (to 3M).
30. M. Lazar, *J. Polym. Sci.* **29**, 573 (1958).
31. U.S. Pat. 2,613,202 (Oct. 7, 1952), G. F. Roedel (to General Electric Co.).
32. U.S. Pat. 2,600,202 (June 10, 1952), D. W. Caird (to General Electric Co.).
33. Fr. Pat. 1,155,143 (Apr. 23, 1958), (to Society d'Ugine).
34. U.S. Pat. 2,842,528 (July 8, 1958), R. L. Herbst and B. F. Landrum (to 3M).
35. U.S. Pat. 2,689,241 (Sept. 14, 1954) A. L. Dittman, H. J. Passino, and J. M. Wrightson, (to M. W. Kellogg Co.).
36. J. M. Hamilton, *Ind. Eng. Chem.* **45**, 1347 (1953).
37. U.S. Pat. 2,569,524 (Oct. 21, 1951), J. M. Hamilton (to E. I. du Pont de Nemours & Co., Inc.).
38. U.S. Pat. 2,744,751 (Dec. 19, 1956), H. J. Passino and co-workers (to M. W. Kellogg Co.).
39. Brit. Pat. 840,735 (July 6, 1960), F. Fahnoe and B. F. Landrum (to 3M).
40. U.S. Pat. 2,559,749 (July 10, 1951), A. F. Benning (to E. I. du Pont de Nemours & Co., Inc.).
41. U.S. Pat. 2,559,749 (July 10, 1951), K. L. Berry (to E. I. du Pont de Nemours & Co., Inc.).
42. U.S. Pat. 2,751,376 (July 19, 1956), R. M. Mantell and W. S. Barnhart (to 3M).
43. U.S. Pat. 2,902,477 (Sept. 1, 1959), E. Fischer, K. Weissermel, and G. Bier (to 3M).

44. U.S. Pat. 3,045,000 (July 17, 1962), R. R. Divis.
45. U.S. Pat. 2,770,659 (Nov. 13, 1956), W. S. Barnhart (to M. W. Kellogg Co.).
46. U.S. Pat. 2,786,827 (Mar. 26, 1957), W. S. Barnhart (to M. W. Kellogg Co.).
47. U.S. Pat. 2,664,449 (Dec. 29, 1953), W. T. Miller (to USAEC).
48. U.S. Pat. 2,902,477 (Apr. 28, 1953), W. T. Miller (to M. W. Kellogg Co.).
49. U.S. Pat. 2,636,908 (Apr. 28, 1953), A. L. Dittman and J. M. Wrighton (to M. W. Kellogg Co.).
50. U.S. Pat. 2,706,715 (Apr. 19, 1955), R. C. Conner (to M. W. Kellogg Co.).
51. U.S. Pat. 2,716,141 (Aug. 23, 1955), W. T. Miller (to 3M).
52. U.S. Pat. 2,854,490 (Sept. 30, 1958), E. Fischer and H. Frey (to Farbwerke Hoechst).
53. U.S. Pat. 2,992,988 (July 18, 1961), C. D. Dipner (to 3M).
54. U.S. Pat. 3,076,765 (Feb. 5, 1963), F. W. West, R. J. Seffl, and L. J. Reilly (to 3M).
55. *Code of Federal Regulations*, Title 21, Paragraph 177.1380, U. S. Government Printing Office, Washington, D.C., revised Apr. 1, 1979, p. 604.
56. W. Broadway, "A Pressure Sensitivity and Temperature Response Butterfly Valve for Cryogenic Service," paper presented at *Energy Technology Conference and Exhibition*, Houston, Nov. 5–9, 1978.
57. W. Broadway, "Development of Cryogenic Butterfly Valve Seat," paper presented at *AIChE 71st Annual Meeting, Cryogenic Equipment Session*, Miami, Nov. 12–16, 1978.
58. J. E. Harrar and R. J. Sherry, *Anal. Chem.* **47**, 601 (1975).
59. S. L. Petersen and D. E. Tallman, *Anal. Chem.* **62**(5) 459–465 (1990).

General References

R. P. Bringer, "Influence of Unusual Environmental Conditions on Fluorocarbon Plastics," paper presented at *SAMPE (Society of Aerospace Material and Process Engineers) Symposium*, St. Louis, Mo., May 7–9, 1962.

R. P. Bringer and C. C. Solvia, *Chem. Eng. Prog.* **56**(10), 37 (1960).

R. E. Schawmm, A. F. Clark, and R. P. Reed, *A Compilation and Evaluation of Mechanical, Thermal and Electrical Properties of Selected Polymers*, NBS Report, AEC SAN-70-113, SANL 807 Task 7, SANL Task 6, National Technical Information Service, U.S. Dept. of Commerce, Springfield, Va., Sept. 1973, pp. 335–443.

C. A. Harper, ed., *Handbook of Plastics Elastomers, and Composit*, 2nd ed. McGraw-Hill Book Co., New York, 1992.

W. T. Miller, "General Discussion of Chlorotrifluoroethylene Polymers," in C. Slesser and S. R. Schram, eds., *Preparation, Properties and Technology of Fluorine and Organic Fluoro Compounds*, McGraw-Hill Book Co., New York, 1951.

L. A. Wall, ed., *Fluoropolymers*, Vol. 25, Wiley-Interscience, New York, 1992, Chapts. 15 and 16.

S. Chandrasekaran, "Chlorotrifluoroethylene Homopolymer" under "Chlorotrifluoroethylene Polymers," in J. I. Kroschwitz, ed., *Encyclopedia of Polymer Science and Engineering*, 2nd ed., Vol. 3, John Wiley & Sons, Inc., New York, 1985, pp. 463–480.

G. H. MILLET
J. L. KOSMALA
3M Company

BROMOTRIFLUOROETHYLENE

Bromotrifluoroethylene is a valuable reagent for the synthesis of trifluorovinylic compounds by means of its intermediate organometallic compounds.

Physical Properties

The monomer, bromotrifluoroethylene [598-73-2], CF_2=CFBr, is a colorless gas; bp $-3.0°C$ at 101 kPa (754 mm Hg); 58°C at 790 kPa (100 psig); and d_4^{25} 1.86 g/cm^3. Since it is spontaneously flammable in air, its odor is that of its oxidation products, mixed carbonyl halides. The olefin can be distilled, but it polymerizes on standing at ambient temperature unless an inhibitor such as 0.1% tributylamine is added. If desired, the inhibitor can be readily removed by passing the gas through silica gel. Higher temperatures or uv light increase the polymerization rate. The nmr (1,2), ir (3), uv (4), and photoelectron spectra (5) of the monomer have been reported, and some thermochemical data have been calculated for it (3). Its dipole moment has been determined to be 2.54×10^{-30} C·m (0.76 D) (6).

Chemical Properties

Many reactions of bromotrifluoroethylene have been studied. Under basic conditions it adds alcohols such as methanol (7,8) or ethanol (8,9), forming ethers with the general formula $ROCF_2CFBrH$. Similarly, diethylamine adds to it giving $(C_2H_5)_2NCF_2CFBrH$ (10). This addition is faster than diethylamine additions to tetrafluoroethylene or chlorotrifluoroethylene. Vapor-phase photochemical bromination of bromotrifluoroethylene gives the expected adduct, $CF_2BrCFBr_2$ (11). On the other hand, photochemical chlorination results in only 60% of the expected adduct and 40% scrambled bromo and chloro products. Hydrogenation of the double bond can be accomplished either catalytically (12) or with sodium borohydride (13). The former method also gives some trifluoroethylene and the latter method gives only a 37% yield of the pure product, $CF_2HCFBrH$. Other reagents that add across the double bond include S_2Cl_2, to give mainly a disulfide (14), and aqueous sodium nitrite (15).

Another class of reactions that bromotrifluoroethylene undergoes is cycloaddition with acetylenes (16) or olefins (17). When heated, the pure monomer, in addition to polymerizing, dimerizes to cis- and trans-1,2,-dibromohexafluorocyclobutane, which can be debrominated with zinc to perfluorocyclobutene [697-11-0] (18). Bromotrifluoroethylene is a valuable reagent for the synthesis of trifluorovinylic compounds by means of its intermediate organometallics. Trifluorovinylzinc (19) and trifluoromagnesium bromides (20,21) may be prepared directly using the metals, whereas trifluorovinyllithium [683-78-3] results from the metathesis metal–bromine exchange (22). The lithium and magnesium derivatives are thermally unstable but readily convert aldehydes, ketones, and carbon dioxide to alcohols (22,23) and trifluoroacrylic acid [433-68-1] (20). The zinc derivative transforms iodo- or bromobenzenes to α,β,β-trifluorostyrenes in the presence of zero valent palladium complexes (24) and participates in a variety of alkylation, coupling, and acylation reactions upon the addition of cuprous salts (25).

Manufacture

Although bromotrifluoroethylene was first prepared (9) by the dehydrobromination of 1,2-dibromo-1,1,2-trifluoroethane [354-04-1], it is more conveniently prepared from chlorotrifluoroethylene [79-38-9] by the following high yield steps (26):

$$CF_2=CFCl + HBr \rightarrow CF_2BrCFClH \xrightarrow{Zn} CF_2=CHF + ZnBrCl$$

$$CF_2=HF + Br_2 \rightarrow CF_2BrCHFBr \xrightarrow{KOH} CF_2=CFBr + KBr + H_2O$$

Bromotrifluoroethylene is manufactured and sold in commercial quantities with a purity of 99.9% by the Halocarbon Products Corp. for about $100/kg as of 1992.

Polymers

The olefin can be polymerized in trichlorofluoromethane solution at $-5°C$ for 7 days with a halogenated acetyl peroxide such as 0.037% trichloroacetyl peroxide as the initiator (27). Alternatively, it may be polymerized in an aqueous suspension with 2 parts by weight distilled water, 0.01 part ammonium persulfate, 0.004 part sodium bisulfite, and 0.001 part hydrated ferrous sulfate present for each part of the monomer. Mixing for 24.5 h at 20°C gives a 52% conversion of the monomer to the homopolymer (27). Prepared either way, the homopolymer [55157-25-0] is a white powder soluble in acetone and useful as a hard, chemically resistant coating for metal or fabric surfaces. The addition of small amounts of chain-transfer agents such as chloroform, carbon tetrachloride, bromotrichloromethane, and in particular, 0.01 to 0.3% 1-dodecanethiol to the polymerization mixture gives a lower molecular weight homopolymer that is softer and more soluble. Copolymers of bromotrifluoroethylene with many other monomers such as chlorotrifluoroethylene (28), tetrafluoroethylene (29), or trifluoronitrosomethane (30) have been reported. Neither the homopolymer nor the copolymers have any commercial utility.

Telomers. Bromotrifluoroethylene telomers have been prepared using chain-transfer agents such as CF_3SSCF_3 (31), C_2F_5I (32), CBr_4 (32), or CBr_3F (33). For example, when the olefin is slowly added to tribromofluoromethane under light from sunlamps, a liquid is obtained which, after saturation and distillation, has a viscosity of 510 mm^2/s ($=cSt$) and a density of 2.65 g/cm^3 at 58.3°C. These and other bromotrifluoroethylene telomers are useful as flotation agents and damping fluids for gyroscopes and accelerometers in inertial guidance systems. These telomers are noncrystalline up to a higher degree of polymerization than those of chlorotrifluoroethylene, allowing the preparation of liquids of greater viscosity. The higher densities of the bromine-containing oils permit further miniaturization of the instruments floating in them. For these reasons, these oils complement their less expensive chloro analogues in this application. Commercially available bromotrifluoroethylene telomers have densities of 2.14–2.65 g/cm^3 and viscosities of 2–4000 mm^2/s(= cSt). These fluids are expensive but are made in small volume for the aerospace industry.

Toxicity

Rats exposed to 500 ppm of bromotrifluoroethylene died following a 4-h exposure. Since the monomer decomposes in air, the level of exposure to it was actually lower. The effects in rats of repeated exposure over a two-week period have been studied. At 50 ppm, the animals lost weight and renal damage was noted although the effect was reversible. Very mild testicular damage was seen at 50 but not 10 ppm. The amount of urinary fluoride excreted suggested that extensive metabolism was occurring (34).

BIBLIOGRAPHY

"Polybromotrifluoroethylene" under "Fluorine Compounds, Organic," in *ECT* 2nd ed., Vol. 9, pp. 833–835, by L. L. Ferstandig, Halocarbon Products Corp.; in *ECT* 3rd ed., Vol. 11, pp. 54–56, by G. Astrologes, Halocarbon Products Corp.

1. D. D. Elleman and S. L. Manatt, *J. Chem. Phys.* **36**, 1945 (1962).
2. J. Reuben, Y. Shvo, and A. Demiel, *J. Am. Chem. Soc.* **87**, 3995 (1965).
3. D. E. Mann, N. Acquista, and E. K. Plyler, *J. Chem. Phys.* **22**, 1199 (1954).
4. J. Schander and B. R. Russel, *J. Mol. Spectrosc.* **65**, 379 (1977).
5. K. Wittel and H. Bock, *Chem. Ber.* **107**, 317 (1974).
6. E. J. Gauss and T. S. Gilman, *J. Phys. Chem.* **73**, 3969 (1969).
7. U.S. Pat. 3,666,864 (May 30, 1972), R. C. Terrel (to Airco, Inc.).
8. A. Demiel, *J. Org. Chem.* **25**, 993 (1960).
9. F. Swarts, *Chem. Zentr.* **II**, 281 (1899).
10. R. N. Sterlin and co-workers, *Izv. Akad. Nauk SSSR Otd. Khim. Nauk.* **4**, 8,22 (1962).
11. J. R. Lacher, R. D. Burkhart, and J. D. Park, *Univ. Colo. Stud. Ser. Chem. Pharm.* **4**, 8,22 (1962).
12. J. R. Lacher and co-workers, *J. Phys. Chem.* **61**, 1125 (1957).
13. A. L. Anderson, R. T. Bogan, and D. J. Burton, *J. Fluorine Chem.* **1**, 121 (1971).
14. U.S. Pat. 2,451,411 (Oct. 12, 1948), M. S. Raasch (to E. I. du Pont de Nemours & Co., Inc.).
15. A. M. Krzhizhevskii, Y. A. Cheburkov, and I. L. Knunyants, *Izv. Akad. Nauk SSSR Ser. Khim.*, 2144 (1974).
16. J. C. Blazejewski, D. Cantacuzene, and C. Wakselman, *Tetrahedron Lett.*, 2055 (1974).
17. U.S. Pat. 3,954,893 (May 4, 1976), G. J. O'Neill, R. S. Holdsworth, and C. W. Simons (to W. R. Grace and Co.).
18. W. R. Cullen and P. Singh, *Can. J. Chem.* **41**, 2397 (1963).
19. S. W. Hansen, T. D. Spawn, and D. J. Burton, *J. Fluorine Chem.* **35**, 415 (1987).
20. J. L. Knunyants and co-workers, *Izv. Akad. Nauk SSSR Otdel Khim. Nauk.*, 1345 (1958).
21. H. D. Kaesz, S. L. Stafford, and F. G. A. Stone, *J. Am. Chem. Soc.* **81**, 6336 (1959).
22. P. Tarrant, P. Johncock, and J. Savory, *J. Org. Chem.* **28**, 839 (1963).
23. D. D. Denson, C. F. Smith, and C. Tamborski, *J. Fluorine Chem.* **3**, 247 (1974).

24. P. L. Heinze and D. J. Burton, *J. Org. Chem.* **53**, 2714 (1988).
25. D. J. Burton and S. W. Hansen, *J. Am. Chem. Soc.* **108**, 4229 (1986).
26. J. D. Park, W. R. Lycan, and J. R. Lacher, *J. Am. Chem. Soc.* **73**, 711 (1951).
27. U.S. Pat. 2,793,202 (May 21, 1957), J. M. Hoyt (to M. W. Kellogg Co.).
28. Brit. Pat. 593,605 (Oct. 21, 1947), (to E. I. du Pont de Nemours & Co., Inc.).
29. S. Afr. Pat. 69 05,518 (Mar. 3, 1970), J. Kuhls, H. Hahn, and A. Steininger (to Farbwerke Hoechst AG).
30. P. Tarrant, E. C. Stump, Jr., and C. D. Padgett, *Polym. Prepr. Am. Chem. Soc. Div. Polym. Chem.* **12**, 391 (1971).
31. R. E. A. Bear and E. E. Gilbert, *J. Fluorine Chem.* **4**, 107 (1974).
32. Ger. Pat. 2,235,885 (Feb. 7, 1974), J. Kuhls, H. Fitz, and P. Haasemann (to Farbwerke Hoechst AG).
33. U.S. Pat. 3,668,262 (June 6, 1972), A. L. Dittman (to Halocarbon Products Corp.).
34. G. L. Kennedy, Jr., *CRC Crit. Rev. Toxicol.* **21**, 149 (1990).

ARTHUR J. ELLIOTT
Halocarbon Products Corporation

POLY(FLUOROSILICONES)

The presence of carbon–fluorine bonds in organic polymers is known to characteristically impart polymer stability and solvent resistance. The poly(fluorosilicones) are siloxane polymers with fluorinated organic substituents bonded to silicon. Poly(fluorosilicones) have unique applications resulting from the combination provided by fluorine substitution into a siloxane polymer structure (see SILICON COMPOUNDS, SILICONES).

The incorporation of a single carbon–fluorine bond into a polymer cannot provide the stability and solvent resistance offered by multiple bonds or clusters of carbon–fluorine bonds available with substituents like the CF_3, C_2F_5, or C_3F_7 groups. Therefore, commercially interesting poly(fluorosilicones) have at least one CF_3 group per silicon in their structure. The proximity of silicon and fluorine in such compounds governs the stability of the structure. If fluorine is alpha to silicon the compounds are subject to thermal rearrangement (eq. 1). The thermodynamically more stable silicon–fluorine bond and difluoromethylene by-product form. The by-product undergoes further chemistry characteristic of divalent carbon compounds.

$$CF_3-\underset{\underset{R''}{|}}{\overset{\overset{R}{|}}{Si}}-R' \rightarrow F-\underset{\underset{R''}{|}}{\overset{\overset{R}{|}}{Si}}-R' + [:CF_2] \tag{1}$$

When fluorine is beta to silicon, compounds undergo a facile elimination of an ethylenic compound and again form the stable silicon–fluorine bond (eq. 2).

$$CF_3CH_2-\underset{\underset{R''}{|}}{\overset{\overset{R}{|}}{Si}}-R' \rightarrow F-\underset{\underset{R''}{|}}{\overset{\overset{R}{|}}{Si}}-R' + CF_2{=}CH_2 \tag{2}$$

Structures with the widest temperature range of demonstrated stability have fluorine in the gamma position relative to silicon (or further removed), as in $CF_3CH_2CH_2SiRR''R''''$. Longer hydrocarbon chains, with or without hetero atoms, are feasible, but oxidative stability is compromised and such materials are generally disfavored. Poly(3,3,3-trifluoropropyl)methylsiloxane [26702-40-9] demonstrates this structural principle. This polymer is one key member of the industrially important family of fluorosilicone materials.

Properties

Fluorosilicone elastomers can be formulated to provide specific durometer (hardness), tear strength, modulus, and solvent resistance properties (1). The specific gravity is 1.35 to 1.65. Durometer variation in the range of 20–80 Shore A-2 are attainable by formulation methods. Materials designed to resist tearing exhibit a tear strength of up to 52.5 kN/m (300 ppi). Modulus at 100% elongation can range from 0.4–6.2 MPa (60–900 psi). Elongation is 150–500%. Compression set as low as 6% (22 h/177°C) can be achieved, with the upper end being about 35%. Tensile strength in the range of 5.5–12.4 MPa (800–1800 psi) has been reported. Rubber resiliency (Bashore) is 14–30%.

Fluid and Chemical Resistance. Fluorosilicone elastomers and greases are especially suited for applications involving repeated exposure to fuels, oils, hydraulic fluids, and various chemicals (1). Fluid resistance is excellent to almost all solvents including alcohol–hydrocarbon mixtures currently being evaluated as alternative fuels. Even at elevated temperatures, prolonged immersion causes only slight elastomer swelling. Exceptions to this rule are highly polar solvents, such as esters and ketones.

Heat Resistance. Fluorosilicone elastomers have long-lasting dependability in static and dynamic applications over a wide range of temperature. Thermal cycling does not lead to embrittlement. Elastomer service temperatures range from −60 to 200°C. Some elastomers have service temperature excursions allowable (for hours) up to 250°C with little change in hardness; brief exposures up to 260°C result in retention of about 50% of the original tensile strength.

Low Temperature Properties. The property of solvent resistance makes fluorosilicone elastomers useful where alternative fluorocarbon elastomers cannot function. The ability to retract to 10% of their original extension after a 100% elongation at low temperature is an important test result. Fluorosilicones can typically pass this test down to −59°C. The brittle point is approximately −68°C.

Electrical Properties. Like unfluorinated silicone counterparts, fluorosilicone elastomers have inherently good electrical insulating properties. The dielectric properties remain relatively unchanged when the elastomer is exposed to severe environments.

Manufacture

Monomer Production. The key industrial monomer is 2,4,6-trimethyl-2,4,6-tris-(3,3,3-trifluoropropyl)cyclotrisiloxane [2374-14-3], which is produced by the hydrosilylation of 3,3,3-trifluoropropene [677-21-4] with methyldichlorosilane [75-54-7], catalyzed by various platinum and other noble metal compounds (eq. 3).

$$CF_3CH{=}CH_2 + CH_3SiHCl_2 \rightarrow CF_3CH_2CH_2Si(CH_3)Cl_2 \qquad (3)$$

The preparation of 3,3,3-trifluoropropene in high yield has been described (2). The hydrosilylation reaction can also be conducted using peroxides, radiation, or photochemical means. The hydrosilylation product, 3,3,3-trifluoropropylmethyldichlorosilane [675-62-7], is hydrolyzed with water to form a hydrolyzate siloxane mixture of cyclic siloxanes and linear hydroxyl end-blocked siloxanes (eq. 4). This hydrolyzate mixture is washed to remove residual acid and then made basic with sodium hydroxide or potassium hydroxide. Distillation of the resulting mixture under reduced pressure affords the cyclotrisiloxane monomer as the lowest boiling siloxane. Base catalysis continues to rearrange both linear and cyclic materials in the distillation pot to re-form additional cyclotrisiloxane, which can then be recovered until nearly all the siloxane material is converted to useable monomer.

$$CF_3CH_2CH_2Si(CH_3)Cl_2 \longrightarrow \text{hydrolyzate} \longrightarrow \underset{\substack{\text{cyclotrisiloxane}}}{\text{structure}} \qquad (4)$$

Preparation of other fluorosilicone monomers follows methods similar to that described above. For example, 2,4,6-trimethyl-2,4,6-tris(3,3,4,4,5,5,6,6,6-nonafluorohexyl)cyclotrisiloxane [38521-58-3] is produced from 3,3,4,4,5,5,6,6,6-nonafluorohexene [19430-93-4] and methyldichlorosilane in three steps (3).

Polymerization. Cyclotrisiloxanes are strained ring compounds. Polymerization is driven by relief of this ring strain. Acid- or base-catalyzed equilibration reactions of cyclotrisiloxane with a measured amount of end-blocking agent lead to fluid polymers with predictable molecular weight distributions. The amount of added end blocking agent controls polymer chain length. Various polar, aprotic solvents are known promoters, including tetrahydrofuran, acetonitrile, and dimethylformamide. The reactions with metallic base catalysts (MB) adhere to the following order of decreasing rate: K > Na > Li. After neutralization or removal of the catalyst, fluid polymers are used as is or to formulate grease compounds. If a reactive end blocking group was incorporated during polymerization, the fluid polymer is then useful for making a coating, a sealant, or a liquid rubber product.

High molecular weight polymers or gums are made from cyclotrisiloxane monomer and base catalyst. In order to achieve a good peroxide-curable gum, vinyl groups are added at 0.1 to 0.6% by copolymerization with methylvinylcyclosiloxanes. Gum polymers have a degree of polymerization (DP) of about 5000 and are useful for manufacture of fluorosilicone rubber. In order to achieve the gum state, the polymerization must be conducted in a kinetically controlled manner because of the rapid depolymerization rate of fluorosilicone. The expected thermodynamic end point of such a process is the conversion of cyclotrisiloxane to polymer and then rapid reversion of the polymer to cyclotetrasiloxane [429-67-4]. Careful control of the monomer purity, reaction time, reaction temperature, and method for quenching the base catalyst are essential for reliable gum production.

Compounding. Fluorosilicone gums are compounded generally with fumed or precipitated silica fillers, hydroxy-containing low viscosity silicone oils, and readily available peroxides to produce various rubber products.

These rubber products are based on high molecular weight ($M_w > 700,000$) polymers, but because of the freedom of rotation of the silicon–oxygen bond they are soft and easily processed in conventional mixers, water-cooled mills and calenders. Gum plasticities are generally 2.3–3.6 mm and fully compounded bases are in the 2.5–5.1 mm range. Processing characteristics can be modified with hydroxy-containing low viscosity silicone oils, silicone gums, 1–10 parts of polydimethylsilicone softeners, and 1–50 parts of fumed or precipitated silica fillers. Alternatively, fillers based on diatomaceous earth or crushed quartz can be utilized. Most fluorosilicone rubbers can also be colored by the addition of pigments.

Vulcanization. Fluorosilicone elastomers can be peroxide-vulcanized by a free-radical mechanism using vinyl side groups that have been incorporated into the basic polymer structure during the initial polymerization process. Peroxide initiated cross-linking results in a carbon–carbon bond formed by the reaction of free radicals. The free radicals are generated in the polymer via the peroxide radicals, either by abstracting hydrogen from a methyl group or by adding to a vinyl group.

During the vulcanization, the volatile species formed are by-products of the peroxide. Typical cure cycles are 3–8 min at 115–170°C, depending on the choice of peroxide. With most fluorosilicones (as well as other fluoroelastomers), a post-cure of 4–24 h at 150–200°C is recommended to maximize long-term aging properties. This post-cure completes reactions of the side groups and results in an increased tensile strength, a higher cross-link density, and much lower compression set.

Another form of vulcanization is the "addition reaction" of one polymer containing a SiH functionality with a second polymer containing a vinyl functionality. These two polymers are made individually by copolymerization methods and then formulated prior to vulcanizing into two-part or inhibited one-part systems. The vulcanization reaction is usually catalyzed by platinum compounds or other noble metal compounds. Heat may or may not be employed, depending on the catalyst or inhibitor selected. The resulting —CH_2CH_2— linkage between siloxane chains vulcanizes the polymers into the desired elastomeric form. Many liquid silicone rubber (LSR) systems use this type of vulcanization for the rapid-cycle injection molding of fluorosilicone parts.

Fabrication. Fluorosilicones can be molded, extruded, or calendered by any of the conventional methods employed in the industry. Compression molding is the most widely used method and is ideal for a great many fabrications at 115–170°C and 5.5–10.3 MPa (800–1500 psi). Injection molding becomes increasingly important for high production operations and generally requires higher temperatures and pressures than compression molding. Transfer-press molding is particularly useful for molding complex parts in a multicavity press. Where dimensional accuracy of molded parts is important, shrinkage of the parts must be considered in the design of the mold. Linear shrinkage of most fluorosilicones is 2.5–3.5%.

Extrusion techniques are used to make tubes, rods, gaskets, preforms, etc. Standard rubber equipment may be used to extrude fluorosilicone elastomers. The green strength of fluorosilicones is less than that of typical fluorocarbon elastomers, and this should be considered when designing the feed system.

Calendering is used to produce long, thin sheets of fluorosilicone elastomers and to coat fluorosilicones on reinforcing substrates, eg, certain polymers and metals, to provide the protection of fluorosilicones at a minimal cost. When very thin films are desired or when the vulcanizing temperature of typical fluorosilicones is too high for the supporting substrate, room temperature vulcanizing (RTV) dispersions are commonly available. Dispersions of fluorosilicone gums, bases, and fully compounded stocks, with or without catalysts, can be used to coat many fibrous metal or polymeric substrates. It is necessary to maintain dispersion neutrality to maximize the physical properties of the resulting coating. Usually Silastic A-4040 Primer is recommended to prepare a dense polymeric or metal substrate surface for bonding.

Economic Aspects

Globally, there is a small number of basic fluorosilicone producers: General Electric Co. and Dow Corning Corp. in the United States, ShinEtsu in Japan, and Wacker Chemie in Germany. Prices tend to be about $55–220/kg and higher depending on the physical form and the application.

Production capacity is not well understood because producers do not report their figures. Growth of fluorosilicones in automotive elastomer applications was estimated (3) at 15–20% per year through the 1980s. U.S. and western European fluorosilicone elastomer consumption (4) was estimated in 1982 to reach 1950 t by 1986, up from 1227 t in 1981. A 1989 market study (5) on elastomers reported its 1988 estimation of U.S. fluorosilicone production at 2770 t and U.S. capacity at 3270 t per year. Growth of U.S. fluorosilicone elastomers was estimated at an average of 7.5% per year for the period 1989 through 1993.

Health and Safety Factors

Information on fluorosilicone polymers is limited to safe handling information available in specific fluorosilicone product brochures. No known chronic health effects have been reported. Eye contact with fluorosilicone fluid materials may cause temporary eye discomfort with redness and dryness similar to wind burn. A single prolonged skin exposure (24–48 h) causes no known adverse effect. Small amounts transferred to the mouth by the fingers during incidental use should not cause injury. Swallowing large amounts of the fluid may cause digestive discomfort. Attempted inhalation of fluids showed no eye or respiratory passage irritation. Fluorosilicone sealant and rubber materials use a variety of curing agents. Curing to the final form usually releases small amounts of volatile by-products and unnecessary exposure during curing should be avoided. Ventilation to control vapor exposure is recommended. Some uses at elevated temperatures or in aerosol-spray applications may require added precautions.

Uses

Surface Protection. The surface properties of fluorosilicones have been studied over a number of years. The CF_3 group has the lowest known intermolecular force of polymer substituents. A study (6) of liquid and solid forms of fluorosilicones has included a comparison to fluorocarbon polymers. The low surface tensions for poly(3,3,3-trifluoropropyl)methylsiloxane and poly(3,3,4,4,5,5,6,6,6-nonafluorohexyl)methylsiloxane both resemble some of the lowest tensions for fluorocarbon polymers, eg, polytetrafluoroethylene.

Solutions of fluorosilicones impart oil and water repellent finishes to nylon–cotton fabrics. One series of C-1 through C-9 perfluoroalkyl substituents with varying structures were attached to silicon through amide or ether linkages. The fluorosilicones having perfluorinated straight-chain substituents with seven or more carbons gave the best repellencies (7) and exhibited durability toward repeated laundering, wear, and dry cleaning. The amide linkage to silicon was preferred over the ether linkage. Some nonfluorinated silicone can be tolerated in a fluorosilicone copolymer without affecting the repellency of the resulting treated fabric (see TEXTILES, FINISHING).

Foam Control. Whereas some silicones are known to be foam promoters, Dow Corning FS-1265 Fluid is a liquid fluorosilicone with effective antifoam properties. Petroleum industry application of fluids and dispersions in gas–oil separators on offshore drilling platforms has been successful. Their use peaked in the early 1980s, coinciding with constrained crude oil capacity and production. Diesel fuels are an excellent solvent for dimethylsilicones and render them ineffective as an antifoam. A new antifoam which does not require the use of added silica is formulated from a fluorosilicone copolymer. It has shown promise to antifoam (8) diesel fuel (see DEFOAMERS).

Fluids (Oils) and Greases. Fluorosilicone fluid polymers or oils are used as lubricants for pumps and compressors in harsh chemical service, such as those using acids, bases, and halogenated compounds and solvents. Unlike their hydrocarbon oil counterparts, fluorosilicones resist oxidation and degradation to form deposits, while providing a flatter viscosity–temperature profile than the hydrocarbon oils. Fluorosilicone fluids are chemically inert and do not corrode or react with most engineering materials of construction, including many metals, rubbers, and plastics. A serviceable temperature range is -40 to $204°C$ in open systems and up to $288°C$ in closed systems. The high autoignition temperature (above $480°C$) makes air service possible over a wider temperature range with fire-resistant properties.

Gels. Fluorosilicone fluids with vinyl functionality can be cured using the platinum catalyst addition reactions. The cure can be controlled such that a gel or a soft, clear, jelly-like form is achieved. Gels with low (12% after 7 d) swell in gasoline fuel are useful (9) to protect electronics or circuitry from dust, dirt, fuels, and solvents in both hot (up to $150°C$) and cold (down to $-65°C$) environments. Applications include automotive, aerospace, and electronic industries, where harsh fuel–solvent conditions exist while performance requirements remain high.

Sealants. Applications for sealants (qv) parallel the solvent-resistant applications cited below for rubbers. Sealants can be a one-part system utilizing a fluorosilicone with moisture-sensitive end groups like acetoxy groups. Hydrolysis of acetoxy end groups liberates acetic acid as a by-product. The compositions are called room temperature vulcanizing (RTV) sealants because they react with ambient moisture in air to hydrolyze the end group and cause cure by hydroxyl end group condensation of polymer ends. Cross-linking occurs through multifunctional moisture-sensitive cross-linker additives. Sealants can also be a two-part system with polymer cure depending on the reaction of a vinyl group with a silicon hydride. No volatile by-products are liberated. Application is primarily where thick section cure is desired.

Uses include both subsonic and supersonic aircraft fuel sealing applications, where temperatures of -57 to $232°C$ are experienced while in constant contact with aircraft fuel. Filleting sealants for integral fuel tanks built into the aircraft wings and noncurable channeling sealants comprise their primary uses. They are also used in bonding, sealing, caulking, encapsulating, and potting applications.

Rubber. Fluorosilicone rubber is used successfully as O-rings for fuel lines containing gasoline and aviation fuels. Its insulative properties allow its use as spark-plug boots and plug wire in transportation vehicles. The effect of alternative fuels on fluorosilicone rubbers has generated the largest interest in this class of materials. The use of alcohol–gasoline mixtures has been selected because it is a key transportation vehicle fuel of the future. Fluorosilicone rubbers exhibit stable physical properties (10) over the entire range of methanol–fuel blends for periods up to six months at 60°C. They also return to their original physical properties after dryout following immersion. Immersion tests have utilized methanol–ASTM Reference Fuel C (RFC) blends and blends of methyl *tert*-butyl ether (MTBE) with RFC. The greatest fuel swell (33%) is seen at a methanol–RFC ratio of 25:75. This behavior is identical to that shown by a fluorocarbon reference material. Similar resistant effects were seen using oxidized gasoline or "sour" gasoline as the immersion medium. Its low temperature properties are demonstrated as follows: fluorosilicone rubbers can be formulated to show the smallest value for compression set after immersion in 25:75 methanol–RFC for 22 hours at 177°C and $-30°C$ (38% vs fluorocarbon reference at 100%). MTBE is shown to have less effect on fluorosilicones than methanol.

Heat and oil resistance coupled with its low swell have led automotive applications into laminated tubing and hoses (11) with this material. This resistance to the effects of ASTM No. 3 oil at service temperatures of 200°C makes it competitive with fluorocarbons and with the tetrafluoroethylene–propylene copolymer. Fluorosilicones are used to make exhaust gas recirculation (EGR) diaphragms for some passenger cars.

Hydraulic fluid resistance makes fluorosilicones the preferred military aircraft choice for the manufacture of the flexible bellows (12) between the hydraulic fluid reservoir and the suction pump on Northrop Corp.'s T-38 trainers and T-5 fighters. Its use allows for fluid continuity during normal and inverted flight attitudes.

Resiliency provides another opportunity for the rubber functioning as a cushion between stainless steel loop clamps and fuel–hydraulic fluid lines in aircraft. Pratt and Whitney F-100 military jet engine use (12) provides vibration damping without the clamp abraiding the tube surfaces in normal service as well as at temperatures down to $-55°C$.

Electrically conductive rubber (13) can be achieved by incorporation of conductive fillers, eg, use of carbon or metal powders. These rubbers exhibit volume resistivities as low as 10^{-4} Ω·cm. Applications include use in dissipation of static charge and in conductive bridging between dissimilar electronic materials under harsh operating conditions.

BIBLIOGRAPHY

"Poly(fluorosilicones)" under "Fluorine Compounds, Organic" in *ECT* 3rd ed., Vol. 11, pp. 74–81 by Y. K. Kim, Dow Corning Corp.
1. D. J. Cornelius and C. M. Monroe, *Polym. Eng. Sci.* **25**(8), 467–473 (1985).
2. U.S. Pat. 4,465,786 (Aug. 14, 1984), M. F. Zimmer, W. E. Smith, and D. F. Malpass (to General Electric Co.); U.S. Pat. 4,798,818 (Jan. 17, 1989), W. X. Bajzer, R. L. Bixler, Jr., M. D. Meddaugh, and A. P. Wright (to Dow Corning Corp.).
3. T. Gabris, *Rubber World* **184**(1), 41, 59 (1981).
4. *Chem. Mark. Rep.*, 351 (Dec. 6, 1982).
5. J. L. Leone and J. P. Kelly, *Strategic Opportunities in Performance Elastomers U.S.A. 1989,* Kline & Company, Inc., Fairfield, N.J., 1989.
6. H. Kobayashi and M. J. Owen, *Macromolecules* **23**, 4929–4933 (1990).
7. J. W. Bovenkamp and B. V. LaCroix, *Ind. Eng. Chem. Prod. Res. Dev.* **20**, 130–133 (1981).
8. G. C. Sawicki and J. W. White, *Specialty Chemicals* **12**(2), 140–145 (1992).
9. M. T. Maxson and K. F. Benditt, SAE Technical Paper Series, Paper No. 88023, *SAEQ. Trans.* **97**(Part 2), 1–7 (1989).
10. M. S. Virant, L. D. Fiedler, T. L. Knapp, and A. W. Norris, SAE Technical Paper Series, Paper No. 910102, *SAEQ. Trans.* **100**(5), 37–48 (1991).
11. J. W. Horvath, *Rubber World* **197**(3), 21–29 (1987); R. E. Eggers, *Rubber World*, **204**(3), 24–35 (1991).
12. M. J. Dams, *Kautsch. Gummi Kunstst.* **38**(12), 1109–1113 (1985).
13. L. Kroupa, *Rubber World* **200**(3), 23–28 (1989).

WILLIAM X. BAJZER
YUNG K. KIM
Dow Corning Corporation

FLUORITE, FLUOROSPAR. See FLUORINE COMPOUNDS, INORGANIC–CALCIUM FLUORIDE.

FLUOROALUMINATES, FLUOROBERYLLATES, FLUOROPHOSPHATES, FLUOROSILICATES, AND SIMILAR ENTRIES. See FLUORINE COMPOUNDS, INORGANIC.

FLUOROCHEMICALS. See FLUORINE COMPOUNDS, ORGANIC.

FLUOROCARBON ELASTOMERS

Fluorocarbon elastomers are synthetic, noncrystalline polymers that exhibit elastomeric properties when cross-linked. They are designed for demanding service applications in hostile environments characterized by broad temperature ranges and/or contact with chemicals, oils, or fuels.

Military interest in the development of fuel and thermal resistant elastomers for low temperature service created a need for fluorinated elastomers. In the early 1950s, the M. W. Kellogg Co. in a joint project with the U.S. Army Quartermaster Corps, and 3M in a joint project with the U.S. Air Force, developed two commercial fluorocarbon elastomers. The copolymers of vinylidene fluoride, $CF_2=CH_2$, and chlorotrifluoroethylene, $CF_2=CFCl$, became available from Kellogg in 1955 under the trademark of Kel-F (1–3) (see FLUORINE COMPOUNDS, ORGANIC—POLYCHLOROTRIFLUOROETHYLENE; POLY(VINYLIDENE) FLUORIDE). In 1956, 3M introduced a polymer based on poly(1,1-dihydroperfluorobutyl acrylate) trademarked 3M Brand Fluororubber 1F4 (4). The poor balance of acid, steam, and heat resistance of the latter elastomer limited its commercial use.

In the late 1950s, the copolymers of vinylidene fluoride and hexafluoropropylene, $CF_2=CFCF_3$, were developed on a commercial scale by 3M (Fluorel) and by Du Pont (Viton) (5–8). In the 1960s, terpolymers of vinylidene fluoride, hexafluoropropylene, and tetrafluoroethylene, $CF_2=CF_2$, were developed (9) and were commercialized by Du Pont as Viton B. At about the same time, Montedison developed copolymers of vinylidene fluoride and 1-hydropentafluoropropylene as well as terpolymers of these monomers with tetrafluoroethylene, marketed as Tecnoflon polymers (10,11).

In the 1960s and 1970s, additional elastomers were developed by Du Pont under the Viton and Kalrez trademarks for improved low temperature and chemical resistance properties using perfluoro(methyl vinyl ether), $CF_2=CFOCF_3$, as a comonomer with vinylidene fluoride and/or tetrafluoroethylene (12,13) (see FLUORINE COMPOUNDS, ORGANIC—TETRAFLUOROETHYLENE POLYMERS AND COPOLYMERS).

Bromine- and iodine-containing fluoroolefins have been copolymerized with the above monomers in order to allow peroxide cure (14–21). The peroxide cure system does not require dehydrofluorination of the polymer backbone, resulting in an elastomer that shows improved properties after heat and fluid aging.

Copolymers of propylene and tetrafluoroethylene, which are sold under the Aflas trademark by 3M, have been added to the fluorocarbon elastomer family (21–26). Also 3M has introduced an incorporated cure copolymer of vinylidene fluoride, tetrafluoroethylene and propylene under the trademark Fluorel II (27). These two polymers (Aflas and Fluorel II) do not contain hexafluoropropylene. The substitution of hexafluoropropylene with propylene is the main reason why these polymers show excellent resistance toward high pH environments (28). Table 1 lists the principal commercial fluorocarbon elastomers in 1993.

Properties

Table 2 summarizes general characteristics of vulcanizates prepared from commercially available fluorocarbon elastomer gumstocks.

Table 1. Commercial Fluorocarbon Elastomers

Copolymer	CAS Registry Number	Trademark	Supplier
poly(vinylidene fluoride-co-hexafluoropropylene)	[9011-17-0]	Dai-el	Daikin
		Fluorel	3M
		Tecnoflon	Ausimont
		Viton	Du Pont
plus cure-site monomer[a]		Fluorel	3M
poly(vinylidene fluoride-co-hexafluoropropylene-co-tetrafluoroethylene) with and without cure-site monomer[a]	[25190-89-1]	Dai-el	Daikin
		Fluorel	3M
		Tecnoflon	Ausimont
		Viton	Du Pont
poly(vinylidene fluoride-co-tetrafluoroethylene-co-perfluorovinyl ether) plus cure-site monomer[a]		Viton	Du Pont
		Tecnoflon	Ausimont
poly(tetrafluoroethylene-co-perfluoro(methyl vinyl ether) plus cure-site monomer[b]		Kalrez	Du Pont
poly(tetrafluoroethylene-co-propylene)[a]	[27029-05-6]	Aflas	Asahi Glass
		Aflas	3M
poly(vinylidene fluoride-co-chlorotrifluoroethylene)[a]	[9010-75-7]	Kel-F	3M
poly(vinylidene fluoride-co-tetrafluoroethylene-co-propylene)[b]	[54675-89-7]	Fluorel II	3M
		Aflas	Asahi Glass
plus cure site		Aflas	3M

[a] Peroxide curable.
[b] Proprietary cure system.

Thermal Stability. The retention of elongation after thermal aging of fluorocarbon elastomers is an indication of their thermal stability. Figure 1 is a plot of percent retention of initial elongation vs days exposure to dry heat (150°C) for a number of oil-resistant elastomers (29), and shows that fluorocarbon elastomers are far superior to hydrocarbon elastomers. A more severe test at 205°C shows that a typical fluorocarbon molded goods compound retains 95% of initial elongation after one year. Retention of tensile strength is another important characteristic of fluorocarbon elastomers. Figure 2 shows the results of long-term heat aging on a typical O-ring compound made from vinylidene fluoride/hexafluoropropylene copolymer. Fifty percent of the initial tensile strength is retained after a period of one year at 205°C or after more than two months at 260°C.

Chemical Resistance. Fluorocarbon elastomer compounds show excellent resistance to automotive fuels and oils, hydrocarbon solvents, aircraft fuels and oils, hydraulic fluids, and certain chlorinated solvents, and may be used without reservation.

They show good to excellent resistance to highly aromatic solvents, polar solvents, water and salt solutions, aqueous acids, dilute alkaline solutions, oxidative environments, amines, and methyl alcohol. Care must be taken in choice of proper gum and compound. Hexafluoropropylene-containing polymers are not recommended for use in contact with: ammonia, strong caustic (50% sodium hydroxide above 70°C), and certain polar solvents such as methyl ethyl ketone and

Table 2. Fluorocarbon Elastomers Physical Property Ranges

Property	Value
Physical properties	
tensile strength, MPa[a]	7.00–20.00
100% modulus, MPa	2.00–16.00
elongation at break, %	100–500
hardness range, Shore A	50–95
compression-set[b]	
70 h at 25°C	9–16
70 h at 200°C	10–30
1000 h at 200°C	50–70
specific gravity (gumstock)	1.54–1.88
low temperature flexibility, °C[c]	0 to −30
brittle point (ASTM D746), °C	0 to −50
thermal degradation temperature, °C	400 to 550
General characteristics	
gas permeability	very low
flammability	self-extinguishing or nonburning (when properly formulated)
radiation resistance	good to fair
abrasion resistance	good and satisfactory for most uses
weatherability and ozone resistance	outstanding (unaffected after 200 h exposure to 150 ppm ozone)

[a]To convert MPa to psi, multiply by 145.
[b]ASTM method B, 3.5 mm O-ring.
[c]Highly dependent on grade of material used.

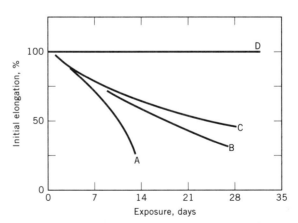

Fig. 1. Retention of elongation of vulcanized elastomers at 150°C. A, nitrile rubber, NBR; B, ethylene–propylene–diene rubber, EPDM; C, acrylic elastomer, AM; D, fluorocarbon elastomer.

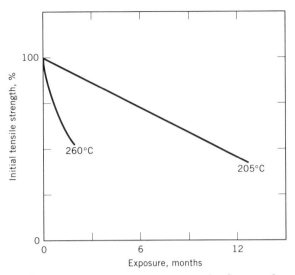

Fig. 2. Tensile strength retention, continuous service, for fluorocarbon elastomers, Compound I (see Table 4).

low molecular weight esters. However, perfluoroelastomers can withstand these fluids. Propylene-containing fluorocarbon polymers can tolerate strong caustic.

Recent innovations in fluorocarbon elastomer development have led to more highly fluorinated materials that possess greater solvent resistance. Included in this class of highly fluorinated materials are Fluorel FLS 2530 and 2650, Viton GF and GFLT, Dai-el G-912, and Kalrez. Figure 3 demonstrates the effect of high fluorine incorporation on volume swell resistance. These highly solvent resistant materials are expected to find wide applicability in the automotive, pollution control, and petrochemical markets.

Compression Set Resistant. One property of fluorocarbon elastomers that makes them uniquely valuable to the sealing industry is their extreme resistance to compression set. Figure 4 plots compression set vs time for compounds prepared especially for compression set resistance (O-ring grades).

Manufacture and Processing

Manufacture of Fluorocarbon Elastomers. Elastomers listed in Table 1 are typically prepared by high pressure, free-radical, aqueous emulsion polymerization techniques (30–33). The initiators (qv) can be organic or inorganic peroxy compounds such as ammonium persulfate [7727-54-0]. The emulsifying agent is usually a fluorinated acid soap, and the temperature and pressure of polymerization ranges from 30 to 125°C and 0.35 to 10.4 MPa (50–1500 psi). The molecular weight of the resultant polymers is controlled by the ratios of initiator to monomer, by the choice of chain-transfer reagents, or both. Typical chain-transfer

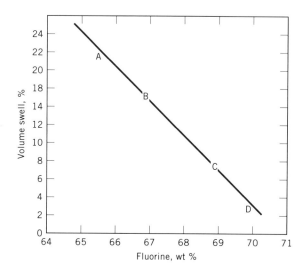

Fig. 3. The percent volume swell in benzene after seven days at 21°C compared with the wt % of fluorine on standard recommended compounds. A, copolymers of vinylidene fluoride–hexafluoropropylene; B, terpolymers of vinylidene fluoride–hexafluoropropylene–tetrafluoroethylene; C, terpolymers of vinylidene fluoride–hexafluoropropylene–tetrafluoroethylene-cure site monomer; D, copolymer of tetrafluoroethylene–perfluoro(methyl vinyl ether)-cure site monomer.

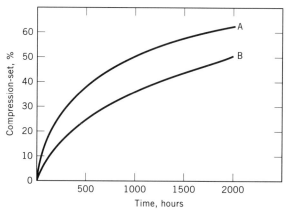

Fig. 4. Compression-sets of fluorocarbon elastomers at 200°C, 3.5 mm O-rings; A, Compound I (see Table 4); B, Compound II (see Table 4).

agents are 2-propanol, methanol, acetone, diethyl malonate [105-53-3], and dodecylmercaptan (34–36). A typical polymerization recipe is as follows.

Component	Amount, g
vinylidene fluoride	61
hexafluoropropylene	39
diethyl malonate	0.1
ammonium persulfate	0.5
ammonium perfluorooctanate	0.12
potassium phosphate, dibasic	1.0
water	304

The aqueous emulsion polymerization can be conducted by a batch, semibatch, or continuous process (Fig. 5). In a simple batch process, all the ingredients are charged to the reactor, the temperature is raised, and the polymerization is run to completion. In a semibatch process, all ingredients are charged except the monomers. The monomers are then added continuously to maintain a constant pressure. Once the desired solids level of the latex is reached (typically 20–40% solids) the monomer stream is halted, excess monomer is recovered and the latex is isolated. In a continuous process (37), feeding of the ingredients and removal of the polymer latex is continuous through a pressure control or relief valve.

The polymer latex is then coagulated by addition of salt or acid, a combination of both, or by a freeze–thaw process. The crumb is washed, dewatered, and dried. Since most fluorocarbon elastomer gums are sold with incorporated cure systems, the final step in the process involves incorporation of the curatives. This can be done on a two-roll mill, in an internal mixer, or in a mixing extruder.

Cross-Linking Chemistry. Like other thermosetting elastomers, fluorocarbon elastomers must be cured in order to get useful properties. Three distinct cross-linking systems have been developed to achieve this goal: diamine, bisphenol–onium, and peroxide curing agents (Table 3). Over the years, the bisphenol–onium cure system, which is the most practical in terms of processing latitude and cured properties, has become the most widely used.

These three cure systems have in common the need for a two-step cure cycle to generate the best cured properties. The first step is the application of heat and pressure in a mold to shape the article (press cure). The second step is a high temperature oven cycle at atmospheric pressure to obtain the final cured properties.

The manufacture of the majority of fluorocarbon elastomer gums includes the addition of an incorporated cure system comprising an organic onium cure accelerator, such as triphenylbenzylphosphonium chloride [1100-88-5], and a bisphenol cross-linking agent, such as hexafluoroisopropylidenediphenol [1478-61-1]. These incorporated cure systems offer improved compression set performance, processing safety, and fast cure cycles to fabricators, who need add only metal oxides as acid acceptors and reinforcing fillers for a complete formulation (38–43).

The chemistry of this cure system has been the subject of several studies (44–47). It is now generally accepted that the cure mechanism involves dehydro-

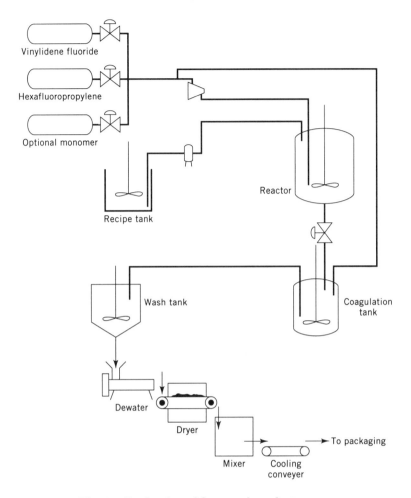

Fig. 5. Production of fluorocarbon elastomers.

fluorination adjacent to hexafluoropropylene monomer units. The subsequent fluoroolefin is highly reactive toward nucleophilic attack by a variety of curatives (eg, diamines, diphenols).

In addition to the incorporated cure gums, there are also raw gums that contain no curative, to which the fabricator adds cure ingredients, such as diamines, bisphenols, or peroxides (48), in addition to formulation (compound) ingredients. Although peroxide cure systems historically have suffered a poor reputation with respect to processibility (mold sticking and mold fouling), recent advancements in these areas have yielded greatly improved products. The use of iodine–bromine containing fluorinated chain-transfer agents has resulted in high fluorine grade polymers with iodine–bromine end groups. The rheology of these materials is characterized by very fast cure rates. In addition, these materials exhibit much better mold release properties than earlier grades (49–51). The advantages and disadvantages of the cure systems are given in Table 3.

Table 3. Cure Systems for Fluorocarbon Elastomers

Property	Diamine	Dihydroxy cross-linker/ accelerator	Peroxide/coagent
rheology			
cure	slow	excellent	good
scorch	poor	excellent	good
compression set	poor	excellent	intermediate
chemical resistance	typical	typical	typical[a]
thermal resistance	excellent	excellent	good
processing	poor	excellent	poor to good[b]
other	FDA approved good bondability	easy to modify	HAV[c] curable

[a] Improved steam and base resistance.
[b] Special grades required.
[c] HAV = hot air vulcanization.

Compounding. Owing to the number of ingredients required in a conventional rubber recipe, fluorocarbon elastomer compounding seems simple compared to typical hydrocarbon elastomer recipes. However, the apparent simplicity of such formulations makes a selection of appropriate ingredients especially important in order to obtain the excellent properties inherent in available gumstocks. A typical recipe in parts per hundred of rubber (phr) by weight is as follows:

Component	Amount, phr
rubber	100
inorganic base: magnesium oxide, calcium hydroxide	6–20
filler (reinforcing or nonreinforcing)	0–60
accelerators or curatives (if not included in base rubber)	0–6
process aids	0–2

With a clear idea of use requirements and rubber response to specific additives, a formulation may be selected. Uses generally fall into one of three classes: O-rings, molded goods, and extruded forms.

O-Rings. In O-ring applications, the primary consideration is resistance to compression set. A fluorocarbon elastomer gum is chosen for O-ring applications based on its gum viscosity, cross-link density, cure system, and chemical resistance so that the best combination of processibility and use performance is obtained. Sample formulations for such uses are given in Table 4.

Long-term compression set resistance is described in Figure 4. Lower set values are achievable by use of higher viscosity gumstock at comparable cross-link densities. Compression set resistance is also very dependent on the cure system chosen. The bisphenol cure system offers the best compression set resistance available today, as shown in Table 5.

Molded Goods. In molded goods compounding, the most important physical property in the final vulcanizate is usually elongation to break, with compression

Table 4. Fluorocarbon Elastomer O-Ring Compounds

Typical formulation	I[a]	II[b]	III[c]
Compound ingredients, phr			
MT black (N-990)	30	30	25
calcium hydroxide	6	6	
magnesium oxide	3	3	
hexafluoroisopropylidenediphenol	2.1	2.1	
triphenylbenzylphosphonium chloride	0.45	0.45	
triallyl isocyanurate			5
α,α-bis(*t*-butylperoxy)diisopropylbenzene			1
sodium stearate			1
Physical properties[d]			
tensile strength, MPa[e]	15.0	15.0	16.0
elongation at break, %	200	200	300
hardness, Shore A	75	75	71
compression-set (3.5 mm O-rings), %			
for 70 h at 200°C, (ASTM D395)	15	10	35
specific gravity	1.8	1.8	1.6

[a] 100 phr FKM2230 (ML 1 + 10 at 121°C = 40) where FKM2230 is poly(vinylidene fluoride-*co*-hexafluoropropylene); available from 3M.
[b] 100 phr FKM2178 (ML 1 + 10 at 121°C = 100) where FKM2178 is poly(vinylidene fluoride-*co*-hexafluoropropylene); available from 3M.
[c] 100 phr FKM100S (ML 1 + 10 at 121°C = 90) where FKM100S is poly(tetrafluoroethylene-*co*-propylene); available from 3M.
[d] Press cure: 5 min at 177°C; post-cure: 24 h at 230°C.
[e] To convert MPa to psi, multiply by 145.

set being a secondary consideration. Since complex shapes are often required, compound flow is also an important parameter. These objectives generally are best met by beginning with gum of the lowest initial viscosity that is consistent with good physical properties. This gum is then cured to lower cross-link density, compared to O-ring formulations, to permit high elongation.

Comparison of starting viscosities and properties of a molded goods compound (Table 6) with an O-ring compound (see Table 4) shows differences in elongation as a result of lower cross-link density. Even higher elongations are achievable with special formulations, or products designed for exceptionally high tear strengths in the press cured state.

The effects of specific fillers in molded goods applications are known (52), and are of special importance for water- and acid-resistant compounds. It is good policy to contact the suppliers for specific recommendations to meet the balance of properties required.

Extruded Articles. In extruded article compounding, the most important parameters are scorch safety and flow characteristics (53). The bisphenol cure system again offers the best scorch resistance of the available fluorocarbon elastomer cure systems. Good flow characteristics can be achieved through proper selection of gum viscosities. Also, the addition of process aids to the formulation can en-

Table 5. Effect of Cure System on Processing Safety and Compression Set Resistance for FKM2260[a]

Formulation	Amine	Bisphenol	Peroxide
Compound ingredients, phr			
MT black (N-990)	30	30	35
magnesium oxide	10	3	
calcium hydroxide		6	3
N,N'-dicinnamylidene-1,6-hexanediamine	2.5		
hexafluoroisopropylidenediphenol		2.1	
triphenylbenzylphosphonium chloride		0.45	
triallyl isocyanurate			2.5
2,5-dimethyl-2,5-di-*t*-butylperoxyhexane			2.5
Properties			
Mooney scorch[b] at 121°C			
minimum	68	66	48
point rise (25 min)	76	2	6
compression-set, %[c]	48	13	25

[a] 100 phr of this material, which has ML 1 + 10 at 121°C = 60 and is poly(vinylidene fluoride-*co*-hexafluoropropylene) plus cure site monomer; available from 3M.
[b] ASTM D1646.
[c] 3.5 mm O-rings for 70 h at 200°C (ASTM D395).

hance the flow characteristics. Typical formulations for extrusion grade fluorocarbon elastomers are given in Table 7.

Formulation Parameters. Gum viscosity is of primary importance to the determination of processability, as this factor affects vulcanizate properties, especially compression set. Gums are available with Mooney Viscosity (ML1 + 10 at 121°C) values of 5–160; a range of 20–60 is preferred for the optimum combination of flow and physical properties. Higher viscosities can cause excessive heat buildup during mixing without a compensatory gain in physical strength. Compound viscosity depends on gum viscosity and on filler selection (type and loading). A preferred range, as measured by Mooney scorch (MS at 121°C), is 25–60.

Compound stability and safety must also be considered when determining processibility, as they are strongly affected by compounding ingredients and cure systems. The data in Table 5 clearly show the effect that choice of cure system can have on scorch stability and processing minimums as determined by standard testing techniques (MS at 121°C). The most workable formulations are compounded with raw gums containing the bisphenol or incorporated cure systems. These elastomers offer the processor the best starting point for maximum processibility.

Mixing. Fluorocarbon elastomer formulations can be compounded by any standard rubber mixing technique. Open mill mixing can be used since most commercial gums mix well. Exceptions to this are very low viscosity gums that have a tendency to stick to the rolls, and very high viscosity gums that are excessively tough.

Table 6. Fluorocarbon Elastomer Molded Goods Compound

Typical formulation	I[a]	II[b]
Compound ingredients, phr		
MT black (N-990)	30	25
calcium hydroxide	6	
magnesium oxide	3	
triphenylbenzylphosphonium chloride	0.4	
hexafluoroisopropylidenediphenol	1.7	
triallyl isocyanurate		5
α,α-bis(*t*-butylperoxy)diisopropylbenzene		1
sodium stearate		1
Post-cured physical properties[c]		
tensile strength, MPa[d]	14.4	13.5
elongation at break, %	265	315
hardness, Shore A	74	71
compression set, (1.27 cm disk), %		
for 70h at 200°C (ASTM D395)	20	40
specific gravity	1.8	1.6

[a] 100 phr FKM2230 (ML 1 + 10 at 121°C = 40) where FKM2230 is poly(vinylidene fluoride-*co*-hexafluoropropylene); available from 3M.
[b] 100 phr FKM 150 P (ML 1 + 10 at 121°C = 75) where FKM 150 P is poly(tetrafluoroethylene-*co*-propylene); available from 3M.
[c] Press cure: 5 min at 177°C; post-cure: 24 h at 230°C.
[d] To convert MPa to psi, multiply by 145.

Internal mixing is widely used with fluorocarbon elastomers. Gumstocks and compounds that are particularly successful fall in the viscosity ranges discussed earlier, and use both incorporated bisphenol-type and peroxide cure systems. A typical internal mix cycle runs 6–8 min with a drop temperature of 90–120°C. The typical formulations in Tables 4 and 7 are readily mixed in an internal mixer.

Preforming. Extrusion preforming is easily accomplished if relatively cool barrel temperatures are used with either a screw or piston type extruder (Barwell). It is important that the gums be used in the appropriate viscosity ranges, and that scorching be avoided.

Calendering operations are done routinely, and warm rolls (40–90°C) are recommended for optimum sheet smoothness. A process aid, such as low molecular weight polyethylene wax, is often used. Sheet thicknesses of 0.5–1.3 mm (20–50 mils) can normally be produced.

Molding. Compression molding is generally used when it is desirable to conserve material, and when a molding operation is set up to allow preparation of large numbers of preforms with minimum labor costs. Flow requirements are minimal and high viscosity gums may be used.

Transfer molding minimizes preforming, and is usually used for the production of very small parts; however, this technique may generate excessive amounts

Table 7. Fluorocarbon Elastomer Extrusion Grade Compound

Formulation	I[a]	II[a]
Compound ingredients, phr		
hexafluoroisopropylidenediphenol	1.9	1.9
triphenylbenzylphosphonium chloride	0.45	0.45
MT black (N-990)	35	15
HAF black (N-326)		5
SRF black (N-762)		7
magnesium oxide	3	9
calcium hydroxide	6	
carnauba wax	1	1
Physical properties[b]		
tensile strength, MPa[c]	7.6	12.4
elongation at break, %	280	330
hardness, Shore A	75	75

[a] 100 phr FKM2145 (ML 1 + 10 at 121°C = 30) where FKM2145 is poly(vinylidene fluoride-co-hexafluoropropylene); available from 3M.
[b] Press cured 45 min at 160°C.
[c] To convert MPa to psi, multiply by 145.

of scrap material. Flow requirements can be quite high, but fluorocarbon elastomers are available that are effective in this application.

Injection molding is finding expanded usage in the rubber industry. Fluorocarbon elastomers can be successfully molded via this technique. Selection of the proper viscosity and cure rheology are very important due to the occurrence of high shear and fast cures.

All types of molding may be carried out at 150 to 200°C. This allows molding times of five minutes or less for most fluorocarbon elastomer parts, but this time is dependent on part size.

Extrusion. Extrusion techniques are used in the preparation of tubing, hose, O-ring cord, preforms and shaped gaskets. Typical extrusion conditions are 70 to 85°C for the barrel temperature and 95 to 110°C for the head temperature. The extruded forms are normally cured in a steam autoclave at 150 to 165°C. Some special grades of peroxide curable fluorocarbon elastomers can be hot air vulcanized.

Post-Curing. Post-curing at elevated temperatures develops maximum physical properties (tensile strength and compression-set resistance) in fluorocarbon elastomers. General post-cure conditions are 16 to 24 h at 200 to 260°C.

Economic Factors

Annual worldwide fluorocarbon elastomer usage totals about 7300 metric tons. Approximately 40% of this usage is in the United States, 30% in Europe, and 20% in Japan. Prices in 1991 were $30–110/kg.

Specifications

Commercially available fluorocarbon elastomers meet automotive specifications in the HK section of ASTM D2000 and SAE J-200. ASTM D1418 specifies designations of composition, eg, fluorocarbon elastomers are designated CFM, FKM, or FFKM. Commercially available fluorocarbon elastomers offer a balance of those properties needed to meet the major O-ring specifications, such as AMS 7276, AMS 7280A, AMS 7259, MIL 83248 Amendment 1 Type II, Class I and II.

Certain grades and formulations of the fluorocarbon elastomers are qualified under the code of Federal Regulations, 21, Food and Drugs, Part 177.2600 for use as rubber articles whose intended applications require repeated or continuous contact with food. Elastomer suppliers will provide assistance in formulating for specified uses.

Test Methods

The fluorocarbon elastomer raw gums provided for rubber molding are tested for Mooney viscosity (ASTM D1646) and for specific gravity (ASTM D297). When compounded as described above, the stocks are tested for Mooney cure (ASTM D1646), Mooney scorch (ASTM D1646), and oscillating-disk rheometer cure rate (ASTM D2084). The vulcanizates are evaluated regarding original physical properties (ASTM D412, D2240, and D1414), aged physical properties (ASTM D573), compression set (ASTM D395), and fluid aging (ASTM D471). Low temperature properties are measured by low temperature retraction TR10 (ASTM D1329) and brittle point (ASTM D2137) tests.

Health and Safety Factors

In general, under normal handling conditions, the fluorocarbon elastomers have been found to be low in toxicity and irritation potential. Specific toxicological, health, and safe handling procedures are provided by the manufacturer of each fluorocarbon elastomer product upon request.

Uses

About 60% of the United States usage is in ground transportation. Typical components include engine oil seals, fuel system components such as hoses and O-rings, and a variety of drive train seals. Growth in this area is expected to continue with the general strength of the U.S. automotive industry coupled with increased demands from higher underhood temperatures, alcohol containing fuels, and more aggressive lubricants. Other major U.S. segments include petroleum/petrochemical, industrial pollution control, and industrial hydraulic and pneumatic applications. These areas will be more dependent upon general industrial production and overall energy demands, and will show slower growth than the automotive segment.

The usage pattern in Europe and Japan is more dependent upon the automotive industry. However, with the recent concern about acid rain, the European and U.S. markets should show increased interest in fluorocarbon elastomers for pollution control applications. On the other hand, the Japanese market has a sizable outlet in electrical and general machinery manufacturing (eg, copiers). Petroleum applications are of little interest outside the United States.

The principal original use of fluorocarbon elastomers in the aircraft industry now accounts for less than 10% of the total fluorocarbon elastomer consumption.

BIBLIOGRAPHY

"Elastomers, Synthetic (Fluorinated)" in *ECT* 3rd ed., Vol. 8, pp. 500–515, by Arthur C. West and Allan G. Holcomb, 3M Co.

1. M. E. Conroy and co-workers, *Rubber Age* **76,** 543 (1955).
2. C. B. Griffis and J. C. Montermoso, *Rubber Age* **77,** 559 (1955).
3. W. W. Jackson and D. Hale, *Rubber Age* **77,** 865 (1955).
4. F. A. Bovey and co-workers, *J. Polym. Sci.* **15,** 520 (1955).
5. U.S. Pat. 3,051,677 (Aug. 28, 1962), D. R. Rexford (to E. I. du Pont de Nemours & Co., Inc.).
6. S. Dixon, D. R. Rexford, and J. S. Rugg, *Ind. Eng. Chem.* **49,** 1687 (1957).
7. U.S. Pat. 3,318,854 (May 9, 1967) F. J. Honn and W. M. Sims (to 3M Co.).
8. J. S. Rugg and A. C. Stevenson, *Rubber Age* **82,** 102 (1957).
9. U.S. Pat. 2,968,649 (Jan. 17, 1961), J. P. Pailthrop and H. E. Schroeder (to E. I. du Pont de Nemours & Co., Inc.).
10. U.S. Pat. 3,331,823 (July 18, 1967), D. Sianesi, G. Bernardi, and A. Regio (to Montedison).
11. U.S. Pat. 3,335,106 (Aug. 8, 1967), D. Sianesi, G. C. Bernardi, and G. Diotalleri (to Montedison).
12. U.S. Pat. 3,235,537 (Feb. 15, 1966), J. R. Albin and G. A. Gallagher (to E. I. du Pont de Nemours & Co., Inc.).
13. Ger. Offen. 2,457,102 (Aug. 7, 1975), R. Baird and J. D. MacLachlan (to E. I. du Pont de Nemours & Co., Inc.).
14. U.S. Pat. 4,035,565 (July 12, 1977), D. Apotheker and P. J. Krusic (to E. I. du Pont de Nemours & Co., Inc.).
15. U.S. Pat. 4,418,186 (Nov. 29, 1983), M. Yamabe and co-workers, (to Asahi Glass Co., Ltd.).
16. A. L. Barney, G. H. Kalb, and A. A. Kahn, *Rubber Chem. Technol.* **44,** 660 (1971).
17. A. L. Barney, W. J. Keller, and N. M. Van Gulick, *J. Polym. Sci. A-1* **8,** 1091 (1970).
18. G. H. Kalb, A. L. Barney, and A. A. Kahn, *Am. Chem. Soc. Dev. Polym. Chem.* **13,** 490 (1972).
19. S. M. Ogintz, *Lubric. Eng.* **34,** 327 (1978).
20. U.S. Pat. 4,251,399 (Feb. 17, 1981), M. Tomoda and Y. Ueta (to Daikin Kogyo Co., Ltd.).
21. U.S. Pat. 4,263,414 (Apr. 21, 1981), A. C. West (to 3M Co.).
22. Y. Tabata, K. Ishigure, and H. Sobue, *J. Polym. Sci.* **A2,** 2235 (1964).
23. G. Kojima and Y. Tabata, *J. Macromol. Sci. Chem.* **A5**(6), 1087 (1971).
24. K. Ishigure, Y. Tabata, and K. Oshima, *Macromolecules* **6,** 584 (1973).

25. G. Kojima and Y. Tabata, *J. Macromol. Sci. Chem.* **A6**(3), 417 (1972).
26. G. Kojima, H. Kojima, and Y. Tabata, *Rubber Chem. Technol.* **50,** 403 (1977).
27. U.S. Pat. 4,882,390 (Nov. 21, 1989), W. Grootaert and R. E. Kolb (to 3M Co.).
28. W. M. Grootaert, R. E. Kolb, and A. T. Worm, *Rubber Chem. and Technol.* **63**(4), 516–522 (1990).
29. J. R. Dunn and H. A. Pfisterer, *Rubber Chem. Technol.* **48,** 356 (1976).
30. U.S. Pat. 3,051,677 (Aug. 28, 1962), D. R. Rexford (to E. I. du Pont de Nemours & Co., Inc.).
31. U.S. Pat. 3,053,818 (Sept. 11, 1962), F. J. Honn and S. M. Hoyt (to 3M Co.).
32. U.S. Pat. 2,968,649 (Jan. 17, 1961), J. P. Pallthorp and H. E. Schroeder (to E. I. du Pont de Nemours & Co., Inc.).
33. A. L. Logothetis, *Prog. Polym. Sci.* **14,** 251–296 (1989).
34. U.S. Pat. 3,069,401 (Dec. 18, 1962), G. A. Gallagher (to E. I. du Pont de Nemours & Co., Inc.).
35. U.S. Pat. 3,080,347 (Mar. 5, 1963), C. L. Sandberg (to 3M Co.).
36. U.S. Pat. 3,707,529 (Dec. 26, 1972), E. K. Gladding and J. C. Wyce (to E. I. du Pont de Nemours & Co., Inc.).
37. U.S. Pat. 3,845,024 (Oct. 29, 1974), S. D. Weaver (to E. I. du Pont de Nemours & Co., Inc.).
38. U.S. Pat. 3,655,727 (Apr. 11, 1972), K. U. Patel and J. E. Maier (to 3M Co.).
39. U.S. Pat. 3,712,877 (Jan. 23, 1973), K. U. Patel and J. E. Maier (to 3M Co.).
40. U.S. Pat. 3,752,787 (Aug. 14, 1973), M. R. deBrunner (to E. I. du Pont de Nemours & Co., Inc.).
41. U.S. Pat. 3,857,807 (Dec. 31, 1974), Y. Kometani and co-workers (to Daikin Kogyo Co.).
42. U.S. Pat. 3,864,298 (Feb. 4, 1975), Y. Kometani and co-workers (to Daikin Kogyo Co.).
43. U.S. Pat. 3,920,620 (Nov. 18, 1975), C. Ceccato, S. Geri, and L. Calombo (to Montedison).
44. W. W. Schmiegel, *Kautsch. Gummikunst.* **31,** 137 (1971).
45. W. W. Schmiegel, *Angew. Makromolek. Chem.* **76/77,** 39 (1979).
46. P. Venkateswarlv and co-workers, *Paper 123 presented at the 136th ACS Rubber Division Meeting*, Detroit, Mich., Oct. 17–20, 1989.
47. V. Ardella and co-workers, *Paper 57 presented at the 140th ACS Rubber Division Meeting*, Detroit, Mich., Oct. 8–11, 1991.
48. J. E. Alexander and H. Omura, *Elastomerics* **2,** 19 (1978).
49. U.S. Pat. 4,243,770 (Jan. 6, 1981), M. Tatemoto and co-workers (to Daikin Kogyo Co.).
50. U.S. Pat. 4,501,869 (Feb. 26, 1985), M. Tatemoto and co-workers (to Daikin Kogyo Co.).
51. U.S. Pat. 4,745,165 (May 17, 1988), V. Arcella and co-workers (to Ausimont S.p.A.).
52. 3M Company product brochure, *FLUOREL 2170-Compounding with Various Fillers*, available from Commercial Chemicals Division, St. Paul, Minn., June 1984.
53. R. Christy, *Rubber World* **184**(6), 38 (1981).

General References

K. J. L. Paciorek and L. A. Wall, eds., *High Polymers*, Vol. 25, Wiley-Interscience, New York, 1972, pp. 291–313.

L. E. Creneshaw and D. L. Tabb, *The Vanderbilt Rubber Handbook*, 13th ed., R. T. Vanderbilt Comp., Inc., Norwalk, Conn., 1990, pp. 211–222.

M. Morton, ed., *Rubber Technology*, 3rd ed., Van Nostrand Reinhold Co., New York, 1987, pp. 410–437.

R. G. Arnold, A. L. Barney, and D. C. Thompson, *Rubber Chem. Technol.* **46,** 619 (1973).

J. C. Montermoso, *Rubber Chem. Technol.* **37,** 1521 (1961).

W. W. Schmiegel, *Makromekulare Chemie* **76/77,** 39 (1979).

D. Apotheker and co-workers, *Rubber Chemistry and Technology*, 1004 (1982).
J. C. Arthur, Jr., ed., *ACS Symposium Series 260, Polymers for Fibers and Elastomers*, American Chemical Society, Washington, D.C., 1984.
A. L. Logothetis, *Prog. Polym. Sci.* **14,** 251–296 (1989).

WERNER M. GROOTAERT
GEORGE H. MILLET
ALLAN T. WORM
3M Company

INDEX

A

Acetaldehyde trifluoroborane [306-73-0], 64
Acetic acid trifluoroborane [753-53-7], 64
Acetic anhydride trifluoroborane [591-00-4], 64
Acetone [67-64-1]
　perfluoro fluid solubility, 288
　reaction with fluorine, 6
Acetone trifluoroborane [661-27-8], 64
Acetophenone trifluoroborane [329-25-9], 64
Acetylene [74-86-2]
　vinyl fluoride from, 443
Acetylium tetrafluoroborate [2261-02-1], 64
Acifluorofen sodium [62476-59-9], 353
Aclar, 475
Aclon, 475
Acrylonitrile [107-13-1], 365
　in nitrile rubber, 505
　VF copolymers, 445
Adhesives
　fluorine polymer surfaces for, 21
Aerosol direct fluorination, 252
Aflas, 490
Aflon COP, 417
Alcoa Alzak process
　fluoroboric acid in, 71
Algoflon, 382
Algogrene, 273
Alkanes, fluorinated, 259
Alkylation
　use of fluorine cmpds, 30
　use of HF, 132
Allyl chlorodifluoroacetate [118337-48-7], 362
Alumina
　reaction with fluorine, 20

Alumina trihydrate
　reaction with HF, 133
Aluminum
　aluminum fluoride in mfg of, 35
　fluorine corrosion of, 16
　fluorspar in prdn of, 92
　production of, 71
　reaction with fluorine, 4
Aluminum difluoride [13569-23-8], 33
Aluminum fluoride [7784-18-1]
　HF in prdn of, 133
Aluminum fluoroborate [14403-54-4], 72
Aluminum industry
　use of HF in, 133
Aluminum metallurgy
　fluorozirconates for, 225
Aluminum monofluoride [13595-82-9], 33
Aluminum sulfate, 71
Aluminum trifluoride [7784-18-1], 33
Aluminum trifluoride monohydrate [12252-28-7],
　[15621-55-3], 35
Aluminum trifluoride trihydrate [15098-87-0], 34
2-Aminobenzotrifluoride [88-17-5], 352, 366
3-Aminobenzotrifluoride [98-16-3], 352
2-Amino-5-chlorobenzotrifluoride [445-03-4], 349, 355
3-Amino-4-chlorobenzotrifluoride [121-50-6], 355
3-Amino-4-ethylsulfonylbenzotrifluoride [382-85-4],
　355
Ammonia [7664-41-7]
　reaction with fluorine, 6
Ammonium acid difluoride [1341-49-7], 47
Ammonium bifluoride [1341-49-7], 47
Ammonium fluoride [12125-01-8], 47
Ammonium fluoroaluminate, 36
Ammonium fluorosilicate [16919-19-0], 48
Ammonium fluorosulfate, 207

Ammonium hexafluorantimonate
  antimony trifluoride from, 51
Ammonium hexafluorophosphate [16941-11-0], 171
Ammonium magnesium fluoride [35278-29-6], 144
Ammonium persulfate [7727-54-0]
  as initiator, 493
Ammonium salts
  or perfluorocarboxylic acids, 314
Analgesics
  fluorine compounds, 236
Anesthetics
  HF in prdn of, 131
Anilinium tetrafluoroborate [15603-97-1], 64
Anisole trifluoroborane [456-31-5], 64
Anodes
  for fluorine cells, 11
Antibacterial agents
  ring-fluorinated aromatics as, 335
Antifoam
  liquid fluorosilicone, 487
Antimicrobial agent
  use of silver fluoride, 183
Antimony
  antimony pentafluoride from, 52
Antimony(II) fluoroborate [14486-20-5], 78
Antimony pentafluoride [7783-70-2], 51
  as fluorinating agent, 230
  from fluorine, 21
  with fluorosulfuric acid, 203
Antimony pentafluoride dihydrate [65277-49-8], 52
Antimony pentoxide, 53
Antimony trifluoride [7783-56-4], 50
  fluorinating carbon compounds, 230
Antistick applications
  FEP in, 413
Antiviral agents
  fluorine compounds, 236
Appetite depressants
  fluorine compounds, 236
Aqueous dispersions
  of PTFE, 396
Argon, 2
Aromatic fluorine compounds, 325
Arsenic
  effect on hydrogen fluoride mfg, 125
Arsenic(III) fluoride [7784-35-2], 55
Arsenic pentafluoride [7784-36-3], 55
Arsenic trifluoride oxide [15120-14-6], 55
Arsenous fluoride [7784-35-2], 55
Asahiflon, 273
Automotive industry
  use of potassium tetrafluoroaluminate, 43

B

Balz-Schiemann reaction, 326
Barite
  from fluorspar ore, 89
Barium fluoride [7782-32-8], 58
Barium fluoroborate dihydrate [72259-09-7], 72
Barium fluoroborates, 72
Barium hexafluorogermanate [60897-63-4], 101
Barium tetrafluorobromate [35967-90-9], 108
Barrier properties
  of PTFE, 391
Basic zirconium carbonate, 224
Batteries
  copper(II) fluoride in, 98
  lead difluoride in, 139
  lithium fluorophosphate in, 172
  molten lithium fluoride in, 142
Bayer's alumina, 41
Bendroflumethiazide [73-48-3], 236
Benfluralin [1861-40-1], 236
Benzaldehyde trifluoroborane [456-30-4], 64
Benzal fluoride [455-31-2], 232
Benzalkonium chloride
  for HF burns, 134
Benzene [71-43-2]
  perfluoro fluid solubility, 288
  surface tensions of, 229
Benzenediazonium
  hexafluorophosphate, 171
Benzophenone trifluoroborane [322-21-4], 64
3,4-Benzopyrene
  boron trifluoride adduct, 63
Benzotrifluoride [98-08-8], 231
  enzymatic oxidation, 350
  perfluoro fluid solubility, 288
Benzotrifluorides
  by cyclization, 330
Benzyl alcohol [100-51-6]
  perfluoro fluid solubility, 288
Benzyl alcohol trifluoroborane [456-31-5], 64
Bergmann's reagent, 332
Beryllium fluoride [7787-49-7], 49
Biological oxidation
  of fluoroaromatics, 334
Bis(acetic acid) trifluoroborane [373-61-5], 64
Bis(2-chloroethanol) trifluoroborane [72985-81-0], 64
Bis(difluorobromine) hexafluorostannate [72229-86-8], 108
Bis(ethanol) trifluoroborane [373-59-1], 64
1,4-Bis(4-fluorobenzoyl)benzene [108464-88-6]
  by Friedel-Crafts synthesis, 332
1,1-Bis(4-fluorophenyl)-2,2,2-trichloroethane [475-26-3], 335
Bismuth pentafluoride [7787-62-4]
  as fluorinating agent, 244
Bis(pentafluorophenyl)phenylphosphine [5074-71-5], 344
Bisphenol A [80-05-7]
  trifluoromethylphenyl ethers from, 356
Bis(phenol) trifluoroborane [462-05-5], 64

INDEX

Bis(tetrafluoroammonium) hexafluoronickelate [63105-40-8], 151
Bis(trifluoroethyl) ether
   from 2,2,2-trifluoroethanol, 283
3,5-Bis(trifluoromethyl)aniline [328-74-5], 355
1,3-Bis(trifluoromethyl)benzene [402-31-3], 348
2,3-Bis(trifluoromethyl)perfluoro-2,3-epoxybutane [1708-78-7], 296
Bis(trifluorylmethyl)trioxide [1718-18-9], 159
Blowing agents
   fluorine compounds, 227
   HF in prdn of, 131
Blow molding
   use of fluorine, 7
Blow molding fluorination, 256
Boric acid, 60
   fluoroboric acid from, 70
Boron
   effect on hydrogen fluoride mfg, 125
   reaction with fluorine, 5
Boron nitride [10043-11-5]
   boron trifluoride in mfg of, 66
Boron trichloride [10294-34-5]
   boron trifluoride in mfg of, 66
Boron trifluoride [7637-07-2], 5, 60
   fluorosulfuric acid in prdn of, 206
Boron trifluoride dihydrate [13319-75-0], 63
Brass
   in presence of fluorine, 16
Brazing
   use of fluoroaluminates, 43
Breakdown voltage
   of sulfur hexafluoride, 191
Breeder reactors
   molten lithium fluoride in, 142
Brick
   fluorine cmpds in mfg of, 30
   fluorspar in prdn of, 92
Brominated fluoromethanes, 275
Bromine
   in bromine trifluoride, 103
Bromine monofluoride [13863-59-7], 102
Bromine pentafluoride [7789-30-2], 102
Bromine trifluoride [7787-71-5], 102
   from fluorine, 21
3-Bromobenzotrifluoride [401-78-5], 349
2-Bromo-1-chloro-1,1-difluoroethane [421-01-2], 276
Bromochlorodifluoromethane [353-59-3], 276
Bromochlorofluoromethane [593-98-6], 276
1-Bromo-2-chloro-1,1,2-trifluoroethane [354-06-3], 276
2-Bromo-2-chloro-1,1,1-trifluoroethane [151-67-7], 276
Bromodichlorofluoromethane [353-58-2], 276
2-Bromo-2,1-dichloro-1,1,1-trifluoroethane [354-50-7], 276
Bromodifluoroacetic acid [667-27-6], 306
2-Bromo-1,1-difluoroethane [359-07-9], 276

Bromodifluoromethane [1511-62-2], 276
4-Bromo-2-fluoroaniline [367-24-8], 346
p-Bromofluorobenzene [460-00-4], 333
4-Bromofluorobenzene [460-00-4], 333
4-Bromo-2-fluorobiphenyl [41604-19-7], 346
Bromofluorocarbons, 277
1-Bromo-2-fluoroethane [762-49-2]
   from 2-fluoroethanol, 281
Bromopentafluorobenzene [344-04-7], 344, 346
1-Bromo-1,1,2,2-tetrafluoroethane [354-07-4], 276
2-Bromo-1,1,1,2-tetrafluoroethane [124-72-1], 276
1-Bromo-2,4,5-trifluorobenzene [327-52-6], 342
2-Bromo-1,1,1-trifluoroethane [421-06-7], 276
Bromotrifluoroethylene [598-73-2], 479
Bromotrifluoromethane [75-63-8], 276
Burton's reagent, 330
1,3-Butadiene [106-99-0]
   in nitrile rubber, 505
2-Butanone [78-93-3]
   perfluoro fluid solubility, 288

C

Cadmium fluoroborate [14886-19-2], 78
Calcium carbonate
   effect on hydrogen fluoride mfg, 125
Calcium fluoride [7789-75-5], 1, 83
Calcium fluoroborate dihydrate [27860-81-7], 72
Calcium gluconate
   for HF burns, 134
   for hydrofluoric acid burns, 48
Calcium oxide phosphate [1306-01-0], 83
Calcium sulfate [7778-18-9]
   hydrogen fluoride by-product, 127
Caprolactam
   fluoroborates in mfg, 79
Carbate, 70
Carbon
   reaction with fluorine, 7
Carbon tetrachloride [56-23-5]
   perfluoro fluid solubility, 288
Carbon tetrafluoride [75-73-0], 7, 228, 261
Carboxylic acids
   boron trifluoride in mfg of, 66
Carboxylic acids, fluorinated, 311
β-Carotene
   boron trifluoride adduct, 63
Catalysts
   boron trifluoride as, 66
   fluoroborates as, 79
   fluorosulfuric acid as, 205
   hydrogen fluoride as, 120
   iron fluoride as, 137
   silver fluoroborate, 79
Cathode material
   use of silver fluoride, 183

Caustic potash
  reaction with fluorine, 20
Cement
  fluorine cmpds in mfg of, 30
  fluorspar in prdn of, 92
Ceramics
  lithium fluoride in mfg, 142
Cerium tetrafluoride [10060-10-3]
  as fluorinating agent, 244
Cesium trifluorogermanite [72121-41-6], 100
CFC alternatives, 271
CFCs. See *Chlorofluorocarbons.*
Chain conformations
  of PVDF, 459
Chain-transfer agents
  for fluorocarbon elastomers, 493
Chemically resistant electrical insulation
  PCTFE in, 476
Chemical resistance
  of Tefzel, 423
Chemical vapor deposition
  NF_3 cleaning of reactors, 156
  use of tungsten hexafluoride, 217
Chlorfluazuron [71422-67-8], 364
Chloride, vinyl [75-01-4]
  vinyl fluoride from, 443
Chlorine monofluoride [7790-91-2], 102
Chlorine pentafluoride [13637-63-3], 102
Chlorine trifluoride [7790-91-2], 2, 102
  from fluorine, 21
Chlorobenzene [108-90-7]
  perfluoro fluid solubility, 288
2-Chlorobenzotrifluoride [88-16-4], 352
4-Chlorobenzotrifluoride [98-15-7], 352
Chlorodifluoroacetic acid [76-04-0], 306
1-Chloro-1,1-difluoroethane [75-68-3], 266, 455
1-Chloro-1,2-difluoroethane [338-64-7], 266
2-Chloro-1,1-difluoroethane [338-65-8], 266
Chlorodifluoromethane [75-45-6], 266
  hexafluoropropylene from, 405
5-Chloro-2,3-difluoropyridine [89402-43-7], 362
4-Chloro-3,5-dinitrobenzotrifluoride [393-75-9], 351
2-Chloro-4-fluoroacetophenone [456-04-2]
  by Friedel-Crafts synthesis, 332
3-Chloro-4-fluoroaniline [367-21-5], 341
Chlorofluoroantimonates, 52
m-Chlorofluorobenzene [625-98-9], 333
4-Chloro-4-fluorobutyrophenone [3874-54-2]
  by Friedel-Crafts synthesis, 332
Chlorofluorocarbons (CFCs), 227, 259
  perfluoroether replacements, 292
1-Chloro-2-fluoroethane [762-50-5], 266
Chlorofluoromethane [593-70-4], 266
  hexafluorobenzene from, 346
2-Chloro-3-fluoro-5-methylpyridine [34552-15-3], 361
2-Chloro-3-fluoropyridine [17282-04-1], 361
2-Chloro-6-fluorotoluene [443-83-4], 341

Chloroform [67-66-3]
  perfluoro fluid solubility, 288
2-Chloro-5-nitrobenzotrifluoride [777-37-7], 350
4-Chloro-3-nitrobenzotrifluoride [121-17-5], 350
Chloropentafluorobenzene [344-07-0], 346
Chloropentafluoroethane [76-15-3], 265
3-Chloropentafluoro-1-propene [79-47-0], 405
1-Chloro-1,2,2,2-tetrafluoroethane [2837-89-0], 266
  hexafluoropropylene from, 405
2-Chloro-1,1,1-trifluoroethane [75-88-7]
  trifluoroethanol from, 283
1-Chloro-2,2,2-trifluoroethyl difluoromethyl ether [26675-46-7]
  from 2,2,2-trifluoroethanol, 283
Chlorotrifluoroethylene [79-38-9], 405
  bromotrifluoroethylene from, 480
  in fluorocarbon elastomers, 491
  polymerization of, 474
  from thermolysis, 235
2-Chloro-5-trifluoromethylpyridine [52334-81-3], 359
5-Chloro-2,4,6-trifluoropyrimidine [697-83-6], 359
  fiber-reactive dyes, 367
Cholinesterase inhibiting
  monofluorophosphate esters, 172
Christiansen colors, 39
CI Pigment Yellow 154 [88-17-5], 355
Ciprofloxacin [85721-33-1], 366
Circuit breakers
  sulfur hexafluoride in, 194
Cleaning
  use of CFCs, 268
Cleaning agents
  fluorine compounds, 227
Cloflucarban [369-77-7], 353
Coatings
  fluorine polymer surfaces for, 21
  PVDF in, 464
Cobalt, 96
Cobalt difluoride [10026-17-2], 96
Cobalt(II) fluoride [10026-17-2], 96
Cobalt(III) fluoride [10026-18-3], 97
Cobalt(II) fluoride dihydrate [13455-27-1], 96
Cobalt(II) fluoride tetrahydrate [13817-37-3], 96
Cobalt(II) fluoride trihydrate [13762-15-7], 96
Cobalt(II) fluoroborate [26490-63-1], 78
Cobalt trifluoride [10026-18-3], 6, 97
  as fluorinating agent, 244
Cocoa butter, 71
Coextrusion
  of PVDF, 461
Color centers
  in fluorite, 84
Compound 1080, 305
Conjunctivitis
  from HF, 134
Copper, 98
  fluorine corrosion of, 16

reaction with fluorine, 4
Copper cathode, 8
Copper(II) fluoride dihydrate [13454-88-1], 98
Copper(I) fluoroborate [14708-11-3], 79
Copper(II) fluoroborate [38465-60-0], 78
Copper hydroxyfluoride [13867-72-6], 98
Crop protection chemicals
  containing fluorine, 335
  fluoroaromatics in, 325
Cross-linking systems, 495
Crown ethers
  fluorination of, 253
Cryogenic seals
  PCTFE in, 476
Cryolite [15096-52-3], 29, 34
  HF in the prdn of, 133
Cupric fluoride [7789-19-7], 98
Curing agents
  boron trifluoride as, 66
  zinc fluoroborate, 78
Curing systems
  for fluorocarbon elastomers, 495
Cyanuric fluoride [675-14-9], 368
Cyclic perfluoroethers, 312
Cyclohexane [110-82-7]
  perfluoro fluid solubility, 288
Cyclohexanesulfonic acids, perfluorinated, 319
Cyclotetrasiloxane [429-67-4], 484
Cyclothiazide [2259-96-3], 333
Cyclotrisiloxanes polymerization, 484
Cyfluthrin [68359-37-5], 336

D

Daiflon, 273, 475
DDT fluoro analogue, 335
Decafluorobiphenyl [434-90-2], 346
Defoamers
  liquid fluorosilicone, 487
Dehydrating agent
  hydrogen fluoride as, 120
Demnum, 291, 301
Dental materials
  fluorozincates in, 222
Dentifrices
  sodium monofluorophosphate as, 168
  stannous flouride in, 211
2-Deoxy-5-fluorouridine [50-91-9], 367
Desflurane [57041-67-5], 236, 283
Detergent aklylation
  use of HF, 132
Dexamethasone [50-02-2], 236
3,5-Diaminobenzotrifluoride [368-53-6], 351, 356
Diazotization routes
  to fluoroaromatics, 326

Dibenzo crown ethers
  fluorination of, 253
Diborane, 61
Dibromochlorofluoromethane [353-55-9], 276
Dibromodifluoromethane [75-61-6], 276
Dibromofluoromethane [1868-53-7], 276
1,2-Dibromo-1,1,2,2-tetrafluoroethane [124-73-2], 276
1,2-Dibromo-1,1,2,-trifluoroethane [354-04-1]
  bromotrifluoroethylene from, 480
Dicarboxylic acids, perfluoro, 315
2,4-Dichlorobenzotrifluoride [320-60-5], 352
3,4-Dichlorobenzotrifluoride [328-84-7], 352
Dichlorobromofluoromethane [353-58-2], 276
1,1-Dichloro-2,2-difluoroethane [471-43-2], 266
1,2-Dichloro-1,1-difluoroethane [1649-08-7], 266
1,2-Dichloro-1,2-difluoroethane [431-06-1], 266
Dichlorodifluoromethane [75-71-8], 265
Dichlorofluoroacetonitrile [83620-05-7], 361
1,1-Dichloro-1-fluoroethane [1717-00-6], 266
1,2-Dichloro-1-fluoroethane [430-57-9], 266
Dichlorofluoroiodomethane [420-48-4], 276
Dichlorofluoromethane [75-43-4], 266
  hexafluorobenzene from, 346
2,4-Dichloro-5-fluorotoluene [86522-86-3]
  quinolone antibacterial from, 327
1,1-Dichloro-2,2,3,3,3-pentafluoropropane [422-56-0], 266
1,3-Dichloro-1,1,2,2,3-pentafluoropropane [507-55-1], 266
1,1-Dichloro-1,2,2,2-tetrafluoroethane [374-07-2], 265
1,2-Dichloro-1,1,2,2-tetrafluoroethane [76-14-2], 265
1,1-Dichloro-2,2,2-trifluoroethane [306-83-2], 266
1,2-Dichloro-1,1,2-trifluoroethane [354-23-4], 266
1,2-Dichloro-1,1,2-trifluoro-2-iodoethane [354-61-0], 276
2,3-Dichloro-5-trifluoromethylpyridine [69045-84-7], 365
2,2-Dichloro-3,3,3-trifluoropropionaldehyde [82107-24-2], 365
3,5-Dichloro-2,4,6-trifluoropyridine [1737-93-5], 362
Dielectric
  sulfur hexafluoride as, 190
N,N-Diethyldifluoroacetamide [56425-08-2], 306
Diethyl ether trifluoroborane [109-63-7], 64
Diethyl malonate [105-53-3]
  as chain-transfer agent, 495
Diflubenzuron [35367-38-5], 236, 339
Diflufenican [83164-33-4], 339
Diflunisal [22494-42-4], 236, 339, 346
Difluoroacetic acid [381-73-7], 306
  from 2,2-difluoroethanol, 282
2,4-Difluoroaniline [367-25-9], 341
  difluorobenzene from, 338
3,5-Difluoroaniline [372-39-4], 342
Difluoroaromatics, 338
1,2-Difluorobenzene [367-11-3], 340
1,3-Difluorobenzene [372-18-9], 340

1,4-Difluorobenzene [540-36-3], 340
2,6-Difluorobenzonitrile [1897-52-5], 341
4,4-Difluorobenzophenone [345-92-6]
 by Friedel-Crafts synthesis, 332
 PEEK from, 333
2,4-Difluorobiphenyl [2285-28-1], 346
Difluorobromine hexafluoroantimonate [19379-47-6], 108
Difluorobromine hexafluorobismuthate [36608-81-8], 108
Difluorobromine hexafluoroniobate [72229-87-9], 108
Difluorobromine hexafluorotantalate [35967-87-4], 108
Difluorobromine tetrafluoroborate [14282-83-8], 64
3,4-Difluorobromobenzene [348-61-8], 338
1,1-Difluoro-2-bromoethane [359-07-9]
 vinyl fluoride from, 443
Difluorochloromethane [75-45-6], 266
Difluorodiazine [10578-16-2], 152
1,5-Difluoro-2,4-dinitrobenzene [327-92-4], 332
sym-Difluorodiphosphoric acid [44801-72-1], 167
Difluorodisulfane [13709-35-8], 198
Difluoroethane
 vinyl fluoride from, 443
1,1-Difluoroethane [75-37-6], 232, 262
1,2-Difluoroethane [624-72-6], 262
2,2-Difluoroethanol [359-13-7], 230, 282
1,1-Difluoroethane [75-38-7], 454
2,2-Difluoroethyl difluoromethyl ether [32778-16-8]
 from 2,2-difluoroethanol, 282
2,2-Difluoroethyl ethyl ether [82907-09-3]
 from 2,2-difluoroethanol, 282
2,2-Difluoroethyl methyl ether [461-57-4]
 from 2,2-difluoroethanol, 282
Difluoroiodomethane [1493-03-4], 276
Difluoromethane [75-10-5], 228, 262
2-Difluoromethoxy-1,1,1,2-tetrafluoroethane [57041-67-5]
 from 2,2,2-trifluoroethanol, 283
Difluoromethyl fluoromethyl ether [461-63-2], 230
Difluoromonosulfane [13814-25-0], 198
1,4-Difluoronaphthalene [315-52-6], 347
2,4-Difluoronitrobenzene [446-35-5], 342
3,4-Difluoronitrobenzene[369-34-6], 338
Difluorophosphoric acid [13779-41-4], 167
1,3-Difluoropropane [462-39-5], 262
2,2-Difluoropropane [420-45-1], 262
2,4-Difluoropyridine [3491-90-7], 359
2,6-Difluoropyridine [1513-65-1], 359
Difluorotetrasulfane [31517-18-7], 198
Difluorotrisulfane [31517-17-6], 198
Dihydroxyfluoroboric acid [17068-89-2], 70
1,4-Diiodoperfluorobutane [375-50-8], 277
1,2-Diiodoperfluorethane [354-65-4], 277
Diisopropyl phosphorofluoridate [55-91-4], 172
Dinitramine [29091-05-2], 351
3,5-Dinitrobenzotrifluoride [401-99-0], 349
2,4-Dinitro-5-fluoroaniline [361-81-7], 332

3,5-Dinitro-2-fluoroaniline [18646-02-1], 332
2,4-Dinitrofluorobenzene [70-34-8], 332
Dioxygen difluoride [7783-44-0], 5, 160
Diphosphorus tetrafluoride [13537-32-1], 165
Direct fluorination, 243
 perfluorocarboxylic acids from, 311
Dispersion processing
 Teflon PFA, 440
Disulfane tetrafluoride [27245-05-2], 198
Disulfonyl fluorides, 322
Disulfur decafluoride [5714-22-7], 5, 197
Disulfur difluoride [13709-35-8], 5
Dithiopyr [97886-45-8], 364
Diuretics
 fluorine compounds, 236
Dodecafluoronorbornane [374-82-3], 249
Dosimeters
 lithium fluoride in, 142
Drimalene Dyestuffs, 368
Drimarene Dyestuffs, 368
Drugs
 fluoroaromatics in, 325
Dyes
 trifluoromethyl in, 353

E

ECF. See *Electrochemical fluorination*.
Electrical applications
 FEP in, 413
Electrically conductive rubber, 489
Electrochemical fluorination (ECF), 233, 244
 of alkanesulfonyl halides, 318
 perfluorocarboxylic acids from, 311
Electrochemical route
 to fluoroaromatics, 329
Electromagnetic shielding materials
 copper(II) fluoride in, 99
Electronics industry
 perfluorinated fluids in, 292
 use for aqueous HF, 133
 use of NF_3, 156
Electroplating
 stannous fluoroborate in, 212
 use of fluoroborates, 69, 79
 use of nickel tetrafluoroborate, 151
 use of potassium fluoride, 177
Emulsion polymerization
 of VDF, 457
Enamels
 aluminum fluoride in, 35
 fluorspar in prdn of, 92
Enflurane [13838-16-9], 236
Enoxacin [74011-58-8], 361, 366
Epichlorohydrin, 71

Etchant
  nitrogen trifluoride as, 156
Etching
  use of sulfur hexafluoride, 195
ETFE. See *Ethylenetetrafluoroethylene copolymers.*
Ethanol trifluoroborane [353-41-3], 64
Ethyl acetate [141-78-6]
  perfluoro fluid solubility, 288
Ethyl difluoroacetate [454-31-9], 306
Ethylene [74-85-1]
  TFE copolymers, 417
  vinyl fluoride from, 443
Ethylenepropylenediene monomer, 7
Ethylenetetrafluoroethylene copolymers (ETFE), 417
Ethyl ether [60-29-7]
  perfluoro fluid solubility, 288
Ethyl fluoride [353-36-6], 262
Ethyl fluoroacetate [459-72-3], 305, 361
Ethyl formate trifluoroborane [462-33-9], 64
Ethyl 4,4,4-trifluoroacetoacetate [372-31-6], 366
Exfluor process, 245
Extrusion
  of PFA resins, 438

F

Fenfluramine [458-24-2], 355
Fenfluramine hydrochloride [404-82-0], 236
Fentrifanil [62441-54-7], 354
FEP. See *Fluorinated ethylene–propylene copolymer.*
FEP fluorocarbon resin, 404
Fermentation
  aluminum fluoride in, 35
Fiber optic applications
  PCTFE in, 476
Fiber optics
  lead difluoride in, 139
  mercuric fluoride in, 147
  use of aluminum fluoride, 36
Fiber-reactive dyes
  fluorinated aromatics in, 368
Filled PTFE, 386
Fillers
  fluoroborates in, 75
Filters
  lead difluoride in, 139
Fine powder PTFE resins, 395
Fire extinguishants
  HF in prdn of, 131
Fire retardant
  potassium fluorozirconate as, 225
Fishing lines
  PVDF monofilament, 465
Flame retardants
  iron fluoride in mfg of, 137

Flamprop-isopropyl [52756-22-6], 336
Flazalone [21221-18-1], 337
Flazasulfuron [104040-78-0], 364
Flecainide acetate [54143-56-5], 358
Fleroxacin [79660-72-3], 366
Floctafenine [23779-99-9], 365
Flophemsyl reagents, 334
Florox reagent, 344
Floxacillin [5250-39-5], 337
Fluazilopbutyl [69806-50-4], 364
Fluazinam [79622-59-6], 364
Fluconazole [86386-73-4], 339
Flucythrinate [70124-77-5], 358
Fluellite, 35
Flufenamic acid [53-78-9], 355
Flufenoxuron [101463-69-8], 339
Flumethiazide [148-56-1], 355
Flumetramide [7125-73-7], 355
Fluocinonide [356-12-7], 236
Fluometuron [2164-17-2], 236, 353
Fluomine [6220-65-5], 334
Fluon, 382
Fluorapatite [1306-05-4], 28
Fluorel, 490
Fluorenert FC-75 [11072-16-5], 286
Fluoridation
  of potable water, 187
Fluoride glasses, 51, 224
  tantalum pentafluoride in, 210
Fluoride glass optical fiber, 51
Fluorides
  in drinking water, 28
Fluorinated aliphatic compounds, 259
Fluorinated epoxides, 294
Fluorinated ethylene–propylene (FEP) copolymer
  [25067-11-2], 404
Fluorinated O-rings, 255
Fluorinated plastics, 264
Fluorinated quinolone antibacterials, 366
Fluorinating agents, 7
  antimony pentafluoride as, 52
  antimony trifluoride as, 51
  cobalt trifluoride as, 97
  copper(II) fluoride, 98
  hydrogen fluoride as, 107, 120
  iron fluoride as, 137
  silver difluoride as, 183
  silver fluoride as, 182
  sulfur tetrafluoride as, 195
  tungsten hexafluoride as, 217
  zinc fluoride as, 221
Fluorination, 243
Fluorine [7782-42-4], **1**
  potassium bifluoride in mfg of, 178
Fluorine azide [14986-60-8], 152
Fluorine cell
  lithium fluoride in, 142

Fluorine compounds
  from HF, 133
Fluorine compounds, inorganic, **27**
  aluminum, 33
  ammonium, 47
  antimony, 50
  arsenic, 55
  barium, 58
  boron, 60
  calcium, 83
  cobalt, 96
  copper, 98
  germanium, 100
  halogens, 102
  hydrogen, 115
  iron, 136
  lead, 139
  lithium, 141
  magnesium, 143
  mercury, 146
  molybdenum, 148
  nickel, 150
  nitrogen, 152
  oxygen, 158
  phosphorus, 165
  potassium, 176
  rhenium, 180
  silver, 182
  sodium, 186
  sulfur, 188
  tantalum, 210
  tin, 211
  titanium, 214
  tungsten, 217
  zinc, 221
  zirconium, 223
Fluorine compounds, organic, **227**
  bromotrifluoroethylene, 479
  direct fluorination, 242
  fluorinated acetic acids, 304
  fluorinated aliphatic compounds, 259
  fluorinated aromatic compounds, 325
  fluorinated higher carboxylic acids, 311
  fluorocarbon elastomers, 490
  fluoroethanols, 281
  fluoroethers and fluoroamines, 285
  perfluorinated ethylene–propylene copolymers, 404
  perfluoroalkanesulfonic acids, 318
  perfluoroepoxides, 294
  polychlorotrifluoroethylene, 472
  poly(fluorosilicones), 482
  polytetrafluoroethylene, 381
  poly(vinyl fluoride), 443
  poly(vinylidene fluoride), 454
  tetrafluoroethylene–ethylene copolymers, 417
  tetrafluoroethylene–perfluorovinyl ether copolymers, 431

Fluorine-containing dyes, 353
Fluorine peroxide [7783-44-0], 28
Fluorinert FC-27, 6
Fluorinerts, 245
Fluorinols, 282
Fluorite
  crystal type, 84
Fluorite [14542-23-5], 28, 83
Fluoroacetaldehyde [1544-46-3]
  from 2-fluoroethanol, 281
Fluoroacetamide [640-19-7], 306
  potassium fluoride in mfg of, 177
Fluoroacetic acid [144-49-0], 304
  from 2-fluoroethanol, 281
Fluoroacetonitrile [503-20-8], 305
4-Fluoroacetophenone [403-42-9]
  by Friedel-Crafts synthesis, 332
Fluoroacetyl chloride [359-06-8], 305
Fluoroacrylate polymers, 279
m-Fluoroaniline [372-19-0]
  difluorobenzene from, 338
2-Fluoroaniline [348-54-9], 338, 341
3-Fluoroaniline [372-19-0], 341
4-Fluoroaniline [371-40-4], 341
  fluorodeoxygenation of nitrobenzene, 329
3-Fluoroanisole [456-49-5], 334
Fluoroapatite [1306-05-4], 165
Fluoroaromatics, 325
Fluorobenzene [462-06-6], 234
4-Fluorobenzeneboronic acid [1765-93-1], 334
2-Fluorobenzonitrile [394-47-8]
  by fluorodesulfonylation, 329
4-Fluorobenzoyl chloride [403-43-0]
  by Friedel-Crafts synthesis, 332
3-(4-Fluorobenzoyl)-2-naphthalenecarboxylic acid [91786-16-2]
  by Friedel-Crafts synthesis, 332
Fluorobiphenyls, 346
p-Fluoro-α, α-bis(trifluoromethyl)benzyl alcohol [2402-74-6], 330
Fluoroboric acid [16872-11-0], 69, 212
Fluoroboric acid and fluoroborates, 69
Fluorocarbon elastomers, 490
Fluorocarbon polymers
  from direct fluorination, 255
Fluorocarbons, 227
  HF in prdn of, 131
3-Fluorocatehol [363-52-0], 334
Fluorochemical Acid FC-24, 321
Fluorochemical surfactants, 322
Fluorochloroethanes
  vinyl fluoride from, 443
5-Fluorocytosine [2022-85-7], 367
Fluorodediazoniation, 326
Fluorodenitration
  of nitroaromatics, 328
Fluorodesulfonylation, 329

INDEX

Fluorodifen [15457-05-3], 236
5-Fluoro-2,6-dihydroxynicotinamide [655-13-0], 361
Fluorodiiodomethane [1493-01-2], 276
1-Fluoro-2,4-dinitrophenyl-5-L-alanine amide [95713-52-3], 332
Fluoroelastomers
   from HF, 133
2-Fluoroethanol [371-62-0], 281
Fluoroethene [75-02-5]
   polymerization of, 443
Fluoroform [75-46-7], 262
   hexafluoropropylene from, 405
Fluorogermanic acid [16950-43-9], 101
2-Fluoro-6-hydroxypridine [55758-32-2], 361
Fluoroimide [41205-21-4], 336
Fluoroiodomethane [373-53-5], 276
1-Fluoronaphthalene [321-38-0], 347
2-Fluoronaphthalene [323-09-1], 347
4-Fluoro-3-nitroaniline [364-76-1], 332
o-Fluoronitrobenzene [1493-27-2], 331
p-Fluoronitrobenzene [350-46-9], 331
4-Fluoro-3-nitrobenzenesulfonic acid [349-05-3], 332
4-Fluoro-3-nitrobenzoates, 332
4-Fluoro-7-nitrobenzofurazan [29270-56-2], 332
4-Fluoro-3-nitro-N,N-bis(hydroxyethyl) aniline [29705-38-2], 332
Fluoronitrofen [13738-63-1], 336
1-Fluoro-4-nitronaphthalene [341-92-4], 347
4-Fluoro-3-nitrophenylazide [28166-06-5], 332
4-Fluoro-3-nitrophenyl sulfone [51451-34-4], 332
4-Fluoro-3-nitrohphenyltrimethylammonium iodide [39508-27-5], 332
3-Fluoro-4-nitropyridine [13505-01-6], 361
3-Fluoro-4-nitropyridine oxide [769-54-0], 361
5-Fluoroorotic acid [703-95-7], 367
p-Fluorophenol [371-41-5], 333
2-Fluorophenol [367-12-4], 334
4-Fluorophenylacetonitrile [459-22-3], 347
2-Fluorophenyllithium [348-53-8], 334
4-Fluorophenyllithium [1493-23-8], 334
4-Fluorophenylmagnesium bromide [352-13-6], 333
Fluoroplast, 382
Fluoropolymer
   HF in prdn of, 131
1-Fluoropropane [460-13-9], 262
2-Fluoropropane [420-26-8], 262
2-Fluoropyridine [372-48-5], 359
3-Fluoropyridine [372-47-4], 359
4-Fluoropyridine [694-52-0], 359
5-Fluoro-3-pyridinemethanol [22620-32-2], 361
1-(3-Fluoro-2-pyridyl)piperazine hydrochloride [85386-84-1], 361
5-Fluoro-2, 4-pyrimidinedione, 7
Fluoropyrimidines, 366
2-Fluoroquinoline [580-21-2], 365
Fluororubber, 490
Fluorosalan [4776-06-1], 353

Fluorosilicic acid, 34
   fluoroboric acid from, 70
Fluorosilicone elastomers, 483
Fluorosilicone rubber, 484
Fluorosis, 20, 31, 92
Fluorosquaraine, 338
Fluorosulfonyldifluoroacetyl fluoride [677-67-8], 323
Fluorosulfuric acid [7789-21-1], 202
2-Fluorotoluene [95-52-3], 341
3-Fluorotoluene [352-70-5], 341
4-Fluorotoluene [352-32-9], 341
Fluorotriazines, 368
Fluorotrichloromethane [75-69-4], 265
5-Fluorouracil [51-21-8], 7, 235, 366
Fluorozirconic acid [12021-95-3], 223
Fluorspar [14542-23-5], 2, 28, 60, 83
   fluoroboric acid from, 70
   hydrogen fluoride from, 124
Fluosilicic acid [16961-83-4]
   hydrogen fluoride by-product, 126
   hydrogen fluoride from, 127
Fluosol, 264
Fluosol DA, 292
Fluoxetine [54910-89-3], 355
Fluphenazine hydrochloride [146-56-5], 236
Flurazepam hydrochloride [1172-18-5], 337
Flurbiprofen [5104-49-4], 346
Flurothyl, 283
Fluroxene [406-90-6], 236, 283
Fluroxypyr [81406-37-3], 236
Fluroxypyr-(1-methylheptyl) [81406-37-3], 362
Flurprimidol [56425-91-3], 358
Flurtamone [96525-23-4], 349
Flusilazole [85509-19-9], 236, 333
Fluspirilene [1841-19-6], 337
Flutec, 261
Flutriafol [76674-21-0], 236, 336
Fluvalinate [69409-94-5], 354
Fluxes
   fluoroborates in, 75
   magnesium fluoride in, 145
   use of aluminum fluoride, 35
   use of potassium bifluoride, 178
Fluxing agent
   fluorspar as, 92
FM, 275
Foam blowing agents, 268
Foam products
   fluorine compounds, 227
Foams
   fluorine treatment, 7
Fomblin, 291, 301
Forane, 273
Freon E, 301
Fridohna, 273
Friedel-Crafts reactions
   boron trifluoride in, 65

Frigen, 273
Ftoroplast, 110
Fungicides
　containing fluorine, 335
　copper(II) fluoride in, 99
　fluorine compounds, 236
Furosemide [54-31-9], 333

G

Gas chromatography
　in fluoride analysis, 31
Gas solubility, 288
Gauges
　for fluorine, 17
GB, 172. (See also *Sarin*.)
Gels
　from fluorosilicones, 487
Genetron, 273
Germanium difluoride [13940-63-1], 100
Germanium fluoride [39717-71-0], 100
Germanium tetrafluoride [7783-58-6], 101
Gifblaar, 305
Glass
　ammonium bifluoride frosting, 49
　fluorspar in prdn of, 92
　HF in etching of, 133
　lead difluoride in, 139
　mercuric fluoride in, 147
　stannous fluoride in mfg, 212
　zinc fluoride in, 221
Glass fibers
　fluorine cmpds in mfg of, 30
　fluorspar in prdn of, 92
Glaucoma
　treatment of, 172
Glazes
　aluminum fluoride in, 35
Global warming, 271
Gomberg-Bachmann arylation
　fluorinated biphenyls, 346
Granular PTFE, 385
Graphite [7782-42-5]
　antimony pentafluoride intercalation of, 52
　reaction with fluorine, 7
Gray iron
　copper(II) fluoride in casting of, 99
Greases
　from fluorosilicones, 487
Greenhouses
　PVF film for, 450
Greenhouse warming potential (GWP), 271
Grinding wheels
　fluorotitanates in, 216
　fluorozirconates for, 225

GWP. See *Greenhouse warming potential*.
Gyroscope flotation fluids
　PCTFE in, 476

H

Halazepam [23092-17-3]
　from 2,2,2-trifluoroethanol, 283
Halex reaction
　fluoroaromatics from, 327
Hall aluminum process
　fluorine cmpds in mfg of, 30
Hall-Haroult process
　aluminum fluoride in, 35
Halocarbon oil, 475
Halocarbons, 269
Halogen-exchange agents
　for fluorination, 232
Halogens, 102
Halon ET, 417
Halons, 270, 382
Haloperidol [52-86-8], 337
Halothane [151-67-7], 236, 276
Haloxydine [2693-61-0], 362
Haloxyfopmethyl [69806-40-2], 364
Hastelloy-C
　use with HF, 129
HCFC. See *Hydrochlorofluorocarbons*.
HCFC-142b
　vinylidene fluoride from, 455
Heat resistance
　of fluorosilicones, 488
Helium, 2
Heptadecafluorodecyl iodide [2043-53-0], 279
Heptadecafluorooctanesulfonic acid [1763-23-1], 320
1,1,1,2,3,3,3-Heptafluoropropane [431-89-0], 262
Heptafluoroquinoline [13180-38-6], 365
Heptane [142-82-5]
　perfluoro fluid solubility, 288
Herbicides
　containing fluorine, 335
　copper(II) fluoride in, 99
　fluorine compounds, 236
　hexafluoroarsenates as, 56
Hexamminenickel tetrafluoroborate [13877-20-8], 64
Hexaflumuron [86479-06-3], 358
Hexafluoroacetone [604-16-2]
　copolymers of, 254
Hexafluoroacetone copolymers
　fluorination of, 254
Hexafluoroantimonate ion [17111-95-4], 50
Hexafluoroantimonic acid [72121-43-8], 53
Hexafluoroarsenates, 56
Hexafluoroarsenic acid [17068-85-8], 56
Hexafluorobenzene [392-56-3], 229, 345

INDEX

Hexafluorohafnic acid [*12021-47-5*], 216
Hexafluoroisopropylidenediphenol [*1478-61-1*]
  in fluoroelastomer curing, 495
Hexafluorophosphoric acid [*16940-811*], 167
Hexafluorophosphoric acid hexahydrate [*40209-76-5*], 169
Hexafluoropropene [*116-15-4*]
  perfluoropolyethers from, 285
Hexafluoropropene oxide [*428-59-1*]
  perfluoropolyethers from, 285
Hexafluoropropylene [*116-15-4*], 432
  in FEP, 404
  in fluorocarbon elastomers, 491
  from TFE, 384
  from thermolysis, 235
Hexafluoropropylene oxide [*428-59-1*], 295
  perfluorovinyl ethers from, 432
Hexafluoropropylene–tetrafluoroethylene copolymers, 406
Hexafluorostannate anion [*21340-04-5*], 212
Hexafluorotitanic acid [*17439-11-1*], 215
Hexafluorozirconic acid [*12021-95-3*], 216, 225
HF. See *Hydrogen fluoride*.
HF-Bamberger reaction, 329
HFC. See *Hydrofluorocarbons*.
HFC-1132a, 456
High voltage coaxial lines
  sulfur hexafluoride in, 194
Hoopes cell
  aluminum fluoride in, 35
Hostaflon, 382
Hostaflon ET, 417
Hostaflon TFA, 432
Hostile environments
  elastomers for, 490
Hyamine 1622
  for hydrofluoric acid burns, 48
Hydramethylon [*67485-29-4*], 354
Hydraulic fluid resistance
  of fluorosilicones, 488
Hydraulic fluids
  perfluoropolyethers, 292
Hydroborofluoric acid, 69
Hydrocarbon polymers
  direct fluorination, 255
Hydrochlorofluorocarbons (HCFCs), 228, 259
Hydrofluoric acid [*7664-39-3*]
  fluorine from, 13
  reaction with fluorine, 5
Hydrofluoroboric acid, 69
Hydrofluorocarbons (HFCs), 259
Hydrogen [*1333-74-0*], 115
  from HF, 134
Hydrogenated nitrile rubber [*88254-10-8*], 505
Hydrogen blistering
  of steel, 129
Hydrogen fluoride (HF) [*7664-39-3*], 115
  antimony trifluoride from, 51
  from calcium fluoride, 84
  fluorinating carbon compounds, 229
  formation of cobalt difluoride, 96
Hydroxyapatite, 28
m-Hydroxybenzotrifluoride [*98-17-9*], 348
p-Hydroxybenzotrifluoride [*402-45-9*], 348
3-Hydroxybenzotrifluoride [*98-17-9*]
  biotransformation route to, 350
4-Hydroxytetrafluoropyridine [*2693-66-5*], 362
Hypnotics
  ring-fluorinated aromatics as, 335
Hypocalcemia
  from HF exposure, 134
Hypofluorites, 28

I

Imaging systems
  fluoroaromatics in, 325
Inconel
  fluorine corrosion of, 16
Indanthrene blue CLB [*6942-78-0*], 356
Indium fluoroborate [*27765-48-6*], 78
Inert fluids, perfluorinated, 287
Infrared spectroscopy
  in fluoride analysis, 31
Inhalation anesthetics
  fluorine compounds, 236
Initiators
  for fluorocarbon elastomers, 493
Injection molding
  of Teflon PFA, 439
Insecticides
  containing fluorine, 335
  fluorine compounds, 236
Intercalation
  fluorine into carbon anodes, 12
Iodine heptafluoride [*16921-96-3*], 102
Iodine pentafluoride [*7783-66-6*], 102
  from fluorine, 21
Iodofluorocarbons, 275
Iodopentafluorobenzene [*827-15-6*], 347
Ion implantation
  use of germanium tetrafluoride, 101
  use of phosphorus pentafluoride, 166
Ion-selective electrodes
  for fluoride, 91
  for fluoride analysis, 31
  for fluoroborates, 71
Iron [*7439-89-6*]
  fluorine corrosion of, 16
  reaction with fluorine, 4
Iron–boron–rare-earth alloy
  iron fluorides in, 136

Iron(II) fluoride [7789-28-8], 136
Iron(III) fluoride [7783-50-8], 137
Iron(II) fluoride tetrahydrate [13940-89-1], 137
Iron(III) fluoride trihydrate [15469-38-2], 137
Iron(II) fluoroborate [13877-16-2], 78
Iron oxide
    effect on hydrogen fluoride mfg, 124
Iron pentafluoride heptahydrate, 138
Isceon, 273
Isobutane [75-28-5], 132
    alkylation using HF, 132
Isobutene
    alkylation using HF, 133
Isobutylene
    dimerization of, 120
Isoflurane [26675-46-7], 236, 283
Isoflurophate USP, 172
Isopropyl alcohol [67-63-0]
    perfluoro fluid solubility, 288
Isopropyl fluoride [420-26-8], 262

J

Jet fluorination, 250

K

Kalrez, 490
Kel-F, 490
Kel-F 81, 475
Kel-F 800, 475
K7 fluid [59884-34-3], 286
Klea, 273
Knoop hardness
    of calcium fluoride, 86
Kobayashi's reagent, 356
Krypton, 2
Krypton difluoride [13773-81-4], 6
Krytox, 254, 291, 301
Kynar, 96

L

Laboratory ware
    FEP in, 413
Lactofen [77501-63-4], 354
LaMar process, 245
Laminated products
    PVF as, 450
Lanthanum fluoride
    in fluoride analysis, 91

Lanthanum nitrate [10099-59-9]
    in fluoride analysis, 31
Lasers
    nitrogen trifluoride in, 156
Laundries, 49
Laundry sour
    use of sodium fluoride, 187
Lead difluoride [7783-46-2], 139
Lead fluoride [7783-55-3], 166
Lead(II) fluoroborate [13814-96-5], 78
Lead sulfide
    fluorspar by-product, 89
Lead tetrafluoride [7783-59-7], 139
    as fluorinating agent, 244
Lenses
    of calcium fluorite, 92
Levafix, 368
Lewis acid, 60
Lidoflazine [3416-26-0], 337
Lime, 71, 89
    reaction with fluorine, 20
Limestone
    reaction with fluorine, 20
Liquid crystals
    fluoroaromatics in, 325, 336
Lithium batteries
    fluorinated carbon in, 7
Lithium borohydride, 61
Lithium fluoride [7789-24-4], 141
Lithium fluoroborate, 72
Lithium hexafluoroarsenate [29935-35-1], 56, 320
Lithium hexafluorophosphate [21324-40-3], 172
Lithium perchlorate [7791-03-9], 320
Lithium triflate [33454-82-9]
    electrical conductivity of, 320
Lomefloxacin [98079-51-7], 366
Loudspeakers
    sulfur hexafluoride in, 195
Lubricants
    boron trifluoride as, 66
    boron trifluoride on mfg of, 66
    fluorinated carbon in, 7
    fluorosilicone fluids, 487
    perfluoropolyethers, 292
Luminescene
    of fluorite, 84
Lycopene
    boron trifluoride adduct, 63

M

Magnesium
    fluorine corrosion of, 16
Magnesium alloy
    in fluorine mfg, 13

Magnesium fluoride [7783-40-6], 143
Magnesium fluoroborate hexahydrate [19585-07-0], 72
Magnetic resonance imaging
  role of fluorine, 336
Magnets
  iron fluorides in, 136
Manganese trifluoride [7783-53-1]
  as fluorinating agent, 244
Marfey's reagent, 333
Mass spectrometry
  in fluoride analysis, 31
Medical packaging
  PCTFE in, 476
Mefloquine [53230-10-7], 365
Melt fracture
  of PFA resins, 438
Membranes
  of PVDF, 464
Mercuric fluoride [7783-39-3], 147, 231
Mercurous fluoride [13967-25-4], 146, 231
Mercury
  with fluorosulfuric acid, 203
Mercury(I) fluoride [13967-25-4], 146
Mercury(II) fluoride [7783-39-3], 147
Metaboric acid
  fluoroboric acid from, 70
Metal finishing
  hexafluorozirconic acid, 225
  use of fluoroborates, 69
Metal fluorides
  as fluorinating agents, 244
Metallization
  of PVDF film, 464
Metallurgical flux
  zinc fluoride as, 221
Metal surface cleaners
  monofluorophosphates as, 170
Metal surface treatments
  use of fluorotitanic acid, 216
Methanol [67-56-1]
  perfluoro fluid solubility, 288
  reaction with fluorine, 6
Methoxyflurane [76-38-0], 236
Methyl bromide [74-83-9]
  surface tensions of, 229
Methyl tert-butyl ether
  fluorosulfuric acid in prpn of, 206
Methyldichlorosilane [75-54-7], 483
Methyl difluoroacetate [433-53-4], 230, 306
Methylene fluoride [75-10-5], 262
Methyl fluoride [593-53-3], 228, 262
2-Methyl-3-nitrobenzotrifluoride [6656-49-1], 356
Methyl nonafluorobutanesulfonate [6401-03-2]
  acetolysis of, 322
Methyl toluenesulfonate [80-48-8]
  acetolysis of, 322
Methyl trifluoromethanesulfonate [333-27-7]
  acetolysis of, 322
Methyl 3-(trifluoromethyl)benzoate [2557-13-3], 330
Metspar, 90
Microbial oxidation
  of fluorobenzene, 334
Microwave technology
  use of sulfur hexafluoride in, 192
Milk
  for HF poisoning, 134
Milk of magnesia
  for HF poisoning, 134
Mohs' hardness
  of calcium fluoride, 86
Molasses, 89
Molding PTFE, 394
Molecular sieves, 193
Molybdenum alloys
  with chlorine trifluoride, 105
Molybdenum dioxydifluoride [13824-57-2], 149
Molybdenum hexafluoride [7783-77-9], 4, 148
Molybdenum oxytetrafluoride [14459-59-7], 148
Molybdenum pentafluoride [13819-84-6], 149
Molybdenum–rhemium alloys, 149
Molybdenum tetrafluoride [23412-45-5], 149
Molybdenum trifluoride [20193-58-2], 149
Molybdenum–tungsten alloys, 149
Monel, 10
  fluorine corrosion of, 16
  use in fluorine cells, 12
  use with HF, 129
Monofluorinated aromatic drugs, 337
Monofluoroaromatic crop protection chemicals, 336
Monofluorophosphoric acid [13537-32-1], 167
Montreal Protocol, 228, 270
Mukaiyama's reagent, 360

N

Nafion, 323
2-Naphthol
  fluorosulfuric acid in prpn of, 206
Natural rubber
  fluorination of, 255
Neoflon, 404
Neoflon AP, 432
Neoflon EP, 417
Neon, 2
Neopentanal trifluoroborane [306-78-5], 64
Nerve gas
  fluorophosphate esters as, 172
Neutron counters
  boron trifluoride in, 66
Nickel
  fluorine corrosion of, 16
  reaction with fluorine, 4

Nickel anodes, 8
Nickel–copper alloy
  use with HF, 129
Nickel fluoride [10028-18-9], 150
Nickel fluoride tetrahydrate [13940-83-5], 150
Nickel(II) fluoroborate [14708-14-6], 78
Nickel–molybdenum–chromium alloy
  use with HF, 129
Nickel tetrafluoroborate [14708-14-6], 151
Niobium
  with chlorine trifluoride, 105
Nitrile rubber, 505
m-Nitrobenzotrifluoride [98-46-4], 349
o-Nitrobenzotrifluoride [384-22-5], 349
p-Nitrobenzotrifluoride [402-54-0], 349
3-Nitrobenzotrifluoride [98-46-4], 353
3-Nitro-2,6-difluoropyridine [5860-02-1], 362
Nitrogen, 152
  reaction with fluorine, 5
Nitrogen dioxide
  adduct with antimony pentafluoride, 52
Nitrogen fluoride
  from cobalt trifluoride, 97
Nitrogen oxides
  removal using copper(II) fluoride, 99
Nitrogen trifluoride [7783-54-2], 4, 152
Nitron
  for hexafluorophosphate analysis, 169
Nitrosonium hexafluororhenate [60447-76-9], 180
Nitrosyl tetrafluoroborate [14635-75-7], 64
4-Nitro-3-(trifluoromethyl)phenol [88-30-2], 353
Nocolok 100 Flux, 43
Nonafluorobutanesulfonic acid [59933-66-3], 320
3,3,4,4,5,5,6,6,6-Nonafluorohexene [19430-93-4], 484
Norfloxacin [70458-96-7], 366
Norflurazon [27314-13-2], 354
Nuclear magnetic resonance
  of boron trifluoride, 61
  in fluoride analysis, 31
Nuclear reactor fuel
  HF in mfg of, 132
Nuclear reactors
  boron trifluoride in, 66

O

Octafluoroisobutylene [382-21-8], 239
Octafluoronaphthalene [313-72-4], 347
Octafluorotoluene [434-64-0], 328
Octane [111-65-9]
  surface tensions of, 229
Ofloxacin [82419-36-1], 366
Oil resistance
  of fluorosilicones, 488
Oil wells
  use of halogen fluorides in, 111
Olah's reagent, 357
Olefins, light
  alkylation using HF, 132
Oleum
  hydrogen fluoride from, 125
Optical crystals, 145
Optical fiber preforms
  tantalum pentafluoride in, 210
Optical fibers
  zinc fluoride in, 221
  zirconium fluoride in, 224
Organic fluorine compounds, 227
Organometallics
  from bromotrifluoroethylene, 479
Organosol finishes
  PVDF in, 464
O-ring compounds, 498
O-rings
  surface fluorination, 255
Oxygen difluoride [7783-41-7], 158
  reaction with fluorine, 5

P

Parasympathomimetic agents
  monofluorophosphate esters, 172
Paschen's rule, 192
PCTFE. See *Poly(chlorotrifluoroethylene)*.
PEEK. See *Polyetheretherketones*.
Pentachlorofluoroethane [354-56-3], 265
Pentafluoroaniline [771-60-8], 343
Pentafluorobenzaldehyde [653-37-2], 344
Pentafluorobenzoic acid [602-94-8], 344
Pentafluorobenzoyl chloride [2251-50-8], 344
Pentafluorobenzyl alcohol [440-60-8], 344
2,3,4,5,6-Pentafluorobenzyl bromide [1765-40-8], 344
2,3,4,5,6-Pentafluorobenzyl chloroformate
  [53526-74-2], 344
Pentafluorochloroethane [76-15-3], 265
Pentafluoroethane [354-33-6], 262
Pentafluoroethanesulfonic acid [354-88-1], 320
Pentafluorophenol [771-61-9], 344
Pentafluorophenyllithium [1076-44-4], 344
Pentafluorophenylmagnesium bromide [879-05-0], 344
1,1,1,2,2-Pentafluoropropane [1814-88-6], 262
1,1,2,2,3-Pentafluoropropane [679-86-7], 262
Pentafluoropyridine [70-16-3], 359
Pentafluorotoluene [771-56-2], 235, 344
n-Pentane
  permeation rate, 7
Perfloxacin [70458-92-3], 366
Perfluorinated carboxylic acids, 311
Perfluorinated inert fluids, 287
Perfluoroadipic acid [336-08-3], 315

INDEX 521

Perfluoroaliphatic ethers, 285
Perfluoroalkanesulfonic acids, 318
Perfluoroalkoxy (PFA) fluorocarbon resins, 431
Perfluoroammonium salts, 152
Perfluorobenzene [392-56-3], 345
Perfluorobis(2-butoxyethoxy)methane [130085-23-3], 286
Perfluoro-1-bromooctane [423-55-2], 277
Perfluorobutane [355-25-9], 229, 261
Perfluorobutanesulfonic acid, 320
Perfluorobutanoic acid [375-22-4], 311
Perfluorobutyl ether [308-48-5], 286
Perfluorobutyl iodide [423-39-2], 277
Perfluoro-2-butyltetrahydrofuran [335-36-4], 286
Perfluorocarbons (PFCs), 4, 259
Perfluoro crown ethers, 253
Perfluorocryptand, 253
Perfluorocyclobutane [115-25-3], 261
Perfluorocyclobutene [697-11-0]
   from bromotrifluoroethylene, 479
Perfluorocyclohexane [355-68-0], 229
Perfluorocyclohexane carboxylic acid [374-88-9], 311
Perfluorodecahydronaphthalene [306-94-5], 347
Perfluorodecalin [306-94-5], 261
Perfluorodecanoic acid [335-76-2], 311
Perfluorodecyl iodine [423-62-1], 277
Perfluorodicarboxylic acids, 314
Perfluoro-1,2:9,10-diepoxydecane [13714-90-4], 297
Perfluoro-1,2:5,6-diepoxyhexane [140173-03-1], 296
Perfluoro-1,2:7,8-diepoxyoctane [13714-88-0], 297
Perfluoro(diethylamino)ethyl ether [108709-75-7], 286
Perfluoro-1,3-dimethylcyclohexane [335-27-3], 261
Perfluoro-1,4-dimethylcyclohexane [374-77-6]
   surface tensions of, 229
Perfluoroepoxides, 294
Perfluoro-1,2-epoxybutane [3709-90-8], 295
Perfluoro-2,3-epoxybutane [773-29-5], 295
Perfluoro-1,2-epoxycyclobutane [13324-28-2], 295
Perfluoro-1,2-epoxycyclohexane [5927-67-3], 297
Perfluoro-1,2-epoxycyclopentane [710-70-3], 296
Perfluoro-1,2-epoxyhexane [72804-48-9], 296
Perfluoro-5,6-epoxy-1-hexane [15453-10-8], 296
Perfluoro-1,2-epoxy-5-methyl-4-oxahexane [84424-45-3], 297
Perfluoro-2,3-epoxy-4-methylpentane [788-50-1], 296
Perfluoro-7,8-epoxy-1-octene [72264-78-9], 297
Perfluoro-1,2-epoxy-6-oxaheptane [71877-16-2], 296
Perfluoro-4,5-epoxy-1-pentene [15453-08-4], 295
Perfluoro-3,4-epoxy-2,3,5-trimethylhexane [2355-27-3], 297
Perfluoro-4,5-epoxyvaleryl fluoride [140173-04-2], 296
Perfluoroethane [76-16-4], 229, 261
Perfluoroethanesulfonic acid, 320
Perfluoroethyl bromide [354-55-2], 276
Perfluoroethyl chloride [76-15-3], 265
Perfluoroethyl iodide [354-64-3], 107, 234, 277

Perfluoro-4-ethylmorpholine [55716-11-5], 286
Perfluoro-2-ethyltetrahydrofuran [356-48-9], 286
Perfluoroglutaric acid [376-73-8], 315
Perfluoroheptane [335-57-9], 261
   surface tensions of, 229
Perfluoroheptanoic acid [375-58-85-9], 311
Perfluoroheptyl iodide [335-58-0], 277
Perfluorohexane [355-42-0], 261, 286
Perfluorohexanesulfonic acid, 320
Perfluorohexanoic acid [307-24-4], 311
Perfluorohexyl ether [424-20-4], 286
Perfluorohexyl iodide [355-43-1], 277
Perfluoro-1-iododecane [2043-53-0], 279
Perfluoroisobutylene oxide [707-13-1], 295
Perfluoro-4-isopropylmorpholine [1600-71-1], 286
Perfluoromalonic acid [1514-85-5], 315
Perfluoromethylcyclohexane [355-02-2]
   surface tensions of, 229
Perfluoro-2-methyldecalin [306-92-3], 261
Perfluoro-1-methyl-1,2:4,5-diepoxycyclohexane [130482-35-8], 297
Perfluoro-2-methyl-2,3-epoxypentane [788-67-0], 296
Perfluoro-4-methylmorpholine [382-28-5], 286
Perfluoro(methyl vinyl ether) [1187-93-5]
   in fluorocarbon elastomers, 490
Perfluoronaphthalene [313-72-4], 235
Perfluorononane [375-96-2], 286
Perfluorononyl iodide [558-97-4], 277
Perfluoronorbornane [374-82-3], 249
Perfluorooctadecanoic acid, 312
Perfluorooctane [307-34-6], 286
Perfluorooctanesulfonic acid [1763-23-1], 234, 320
Perfluorooctanesulfonyl fluoride [307-35-7], 234
Perfluorooctanoic acid [335-67-1], 234, 311
Perfluorooctanoyl fluoride [335-66-0], 234
Perfluorooctyl bromide [423-55-2], 277
Perfluorooctyl iodide [507-63-1], 277, 279
Perfluoro-1-oxaspiro[2,3]hexane [53389-66-5], 295
Perfluoropentane [678-26-2], 261
Perfluoropentanesulfonic acid, 320
Perfluoropentanoic acid [2706-90-3], 311
Perfluoropentyl iodide [638-79-9], 277
Perfluorophenylglycidyl ether [84329-68-0], 297
Perfluoropolyethers, 6, 254, 285
Perfluoropropane [76-19-7], 229, 261
Perfluoropropanoic acid [422-64-0], 311
Perfluoropropyl iodide [754-34-7], 277
Perfluoropropyl vinyl ether [1623-05-8]
   TFE copolymerization, 432
Perfluorosebacic acid [307-78-8], 315
Perfluorosuberic acid [678-45-5], 315
Perfluorosuccinic acid [377-35-8], 315
Perfluorotertiary amines, 285
Perfluorotetradecahydrophenanthrene [306-91-2], 261
Perfluorotetradecanedioic acid, 315
Perfluorotetradecanoic acid, 311
Perfluorotoluene [434-64-0], 348

Perfluorotributylamine [*311-89-7*], 286
Perfluorotriethylamine [*359-70-6*], 286
Perfluorotripropylamine [*338-83-0*], 286
Perfluorovinyl ether [*13269-86-8*]
   in fluorocarbon elastomers, 491
Peroxydisulfuryl difluoride [*13709-32-5*], 206
Peroxytrifluoroacetic acid [*359-48-8*], 308
Perrhenyl fluoride [*25813-73-4*], 181
Pesticides
   sodium fluoride in, 186
Petroleum alkylation
   use of HF, 132
Petroleum coke
   anodes from, 11
Petroleum ether [*8032-32-4*]
   perfluoro fluid solubility, 288
PFA. See *Perfluoroalkoxy fluorocarbon resins*.
PFA resins, 434
PFC. See *Perfluorocarbons*.
Pharmaceuticals
   fluorination of, 7
Phenyl acetate trifluoroborane [*30884-81-6*], 64
m-Phenylenediamine [*108-45-2*]
   difluorobenzene from, 338
o-Phenylenediamine [*95-54-5*]
   1,2-difluorobenzene from, 338
p-Phenylenediamine [*106-50-3*], 342
Phenyl trifluoromethyl ether [*456-55-3*], 356
Phosphate rock, 28, 87
Phosphoric acid [*7664-38-2*]
   hydrogen fluoride as by-product, 127
Phosphorus oxyfluoride [*13478-20-1*], 165
Phosphorus pentafluoride [*7647-19-0*], 165
Phosphorus pentoxide
   effect on hydrogen fluoride mfg, 125
Phosphorus thiofluoride [*2404-52-6*], 165
Phosphorus trifluoride [*7783-55-3*], 165
Photography
   lead difluoride in, 139
Photoinitiators
   hexafluoroantimonates as, 53
Photooxidation
   perfluorocarboxylic acids from, 311
Photopolymerization
   hexafluorophosphate catalysts for, 171
Photovoltaic cells
   PVF film for, 450
Pickling solutions
   use of ammonium bifluoride, 47
Poly(acrylonitrile-*co*-1,3-butadiene) [*9003-18-3*], 505
Polybromotrifluoroethylene [*55157-25-0*], 480
Poly(carbon monofluoride), 242
Poly(chlorotrifluoroethylene) (PCTFE) [*9002-83-9*], 472
Polyetheretherketones (PEEK)
   from 4,4-difluorobenzophenone, 333
Polyflon, 382

Poly(fluoromethylidyne), 242
Poly(fluorosilicones), 482
Polymerizing agent
   hydrogen fluoride as, 120
Polymorphs
   of PVDF, 459
Polyolefins
   fluorine treatment, 7
Polytetrafluoroethylene (PTFE) [*9002-84-0*], 264, 381
   in fluorine mfg, 13
Poly(tetrafluoroethylene-*co*-hexafluoropropylene) [*25067-11-2*], 264
Poly(tetrafluoroethylene-*co*-perfluoro(methyl vinyl ether)), 491
Poly(tetrafluoroethylene-*co*-propylene) [*27029-05-6*], 491
Poly(3,3,3-trifluoropropyl)methylsiloxane [*26702-40-9*], 483
Poly(vinyl fluoride) (PVF) [*24981-14-4*], 445
Poly(vinylidene fluoride) (PVDF) [*24937-79-9*], 454
   as ETFE isomer, 418
Poly(vinylidene fluoride-*co*-chlorotrifluoroethylene) [*9010-75-7*], 491
Poly(vinylidene fluoride-*co*-hexafluoropropylene) [*9011-17-0*], 491
Poly(vinylidene fluoride-*co*-hexafluoropropylene -*co*-tetrafluoroethylene) [*25190-89-1*], 491
Poly(vinylidene fluoride-*co*-tetrafluoroethylene-*co*-perfluorovinyl ether), 491
Poly(vinylidene fluoride-*co*-tetrafluoroethylene--*co*-propylene) [*54675-89-7*], 491
Porapak QS, 193
Positron emission tomography
   fluoroaromatics in, 325
   role of fluorine, 335
Potassium acid fluoride, 178
Potassium bifluoride [*7789-29-9*], 176
   fluorine from, 8
Potassium fluoride [*7789-23-3*], 176
   as fluorinating agent, 230
   replacement of mercury(II) fluoride, 147
Potassium fluorozincate [*13827-02-6*]
   dental material catalysts, 222
Potassium fluorozirconate [*16923-95-8*], 225
Potassium hexafluoroaluminate [*13775-52-5*], 43
Potassium hexafluoroarsenate [*17029-22-0*], 56
Potassium hexafluorobromate [*32312-22-4*], 108
Potassium hexafluorogermanate [*7783-73-5*], 101
Potassium hexafluoroiodate [*20916-97-6*], 108
Potassium hexafluorophosphate [*17084-13-8*], 171
Potassium hydrogen difluoride, 178
Potassium hydroxypentafluoroarsenate(V) [*17068-84-7*], 56
Potassium monofluorophosphate [*14104-28-0*], 170
Potassium octafluororhenate [*57300-90-0*], 180
Potassium pentafluoroaluminate monohydrate [*41627-26-3*], 43

Potassium perfluorooctanesulfonate [2795-39-3], 321
Potassium tetrafluoroaluminate [14484-69-6], 37
Potassium tetrafluoroborate [14075-53-7], 64
Potassium trifluorogermanite [72121-42-7], 100
Powder coating
  Teflon PFA, 440
Prickly pear, 56
Primsulfuronmethyl [86209-51-0], 358
Profluralin [26399-36-0], 236
Propellants, 268
Propenes
  alkylation using HF, 133
Propylene [115-07-1]
  in fluorocarbon elastomers, 491
n-Propyl fluoride [460-13-9], 262
Protective coatings
  PVF as, 450
PTFE. See *Polytetrafluoroethylene*.
PTFE fibers, 396
PVDF. See *Poly(vinylidene fluoride)*.
PVDF film
  metallization, 464
PVDF membranes, 464
PVDF monofilament, 465
PVF. See *Poly(vinyl fluoride)*.
PVF films, 447
Pyridines
  fluorinated, 357
Pyrimidines
  fluorinated, 357
Pyrosulfuryl fluoride [37240-33-8], 159
Pyroxofop [105512-06-9], 362

Q

Quinolines
  fluorinated, 357

R

Radiators
  of aluminum, 43
Radon, 2
  fluorination of, 106
Raman spectroscopy
  in fluoride analysis, 31
Reactive dyes, 368
Refrigerants, 268
  fluorine cmpds, 227
  HF in prdn of, 131
Repellant for termites
  copper(II) fluoride in, 99
Resins
  boron trifluoride in mfg of, 66

Rhenium
  reaction with fluorine, 21
Rhenium dioxytrifluoride [57246-89-6], 181
Rhenium heptafluoride [17029-21-9], 181
Rhenium hexafluoride [10049-17-9], 21, 180
Rhenium oxypentafluoride [23377-53-9], 181
Rhenium oxytetrafluoride [17026-29-8], 181
Rhenium pentafluoride [30937-52-1], 181
Rhenium tetrafluoride [15192-42-4], 181
Rhodium fluoroborate [30935-54-7], 79
Road aggregate, 127
Rodenticides, 305
Rotocasting
  of Teflon PFA, 440
Rubber
  fluorination of, 255
  fluorine treatment, 7
Rust
  removal using iron fluoride, 137

S

Sanger's reagent, 332
Sarin [107-44-8], 172. (See also *GB*.)
Scale
  removal using ammonium bifluoride, 48
Scintillation counters
  lead difluoride in, 139
Scotchban, 322
Scotchgard, 279, 322
Sealants
  from fluorosilicones, 487
Seals
  PCTFE in, 476
Sedatives
  ring-fluorinated aromatics as, 335
SELECTOFLUOR reagents, 326
Semiconductor materials
  boron trifluoride in, 66
Semiconductors
  arsenic pentafluoride in, 56
  arsenous fluoride in mfg, 55
  use of phosphorus pentafluoride, 166
Sevoflurane [28523-86-6], 236
Silanes
  copper(II) fluoride fluorination of, 98
Silica [7631-86-9]
  effect on hydrogen fluoride mfg, 124
Silicon
  chlorine trifluoride as etchant, 111
  reaction with fluorine, 5
Silicon chips
  tungsten hexafluoride, 219
Silicon tetrafluoride [7783-61-1], 5, 204
  fluorosulfuric acid in prdn of, 206

Silicon wafers
    ammonium bifluoride in mfg, 49
    HF as etchant, 133
Silver difluoride [7783-95-1], 183
    as fluorinating agent, 244
Silver fluoride [7775-41-9], 182
Silver fluoroborate [1404-20-2], 78
Silver hexafluoroantimonate [26042-64-8], 184
Silver hexafluorophosphate [26042-63-7], 172, 184
Silver monofluorophosphate [66904-72-1], 170
Silver solder fluxes
    use of potassium fluoride, 176
Silver subfluoride [1302-01-8], 182
Silver tetrafluoroborate [14104-20-2], 184
Silver tetrafluorobromate [35967-89-6], 108
Silver trifluoride [91899-63-7], 184
Silver trifluoroacetate [2966-50-9], 184
Silver trifluoromethane sulfonate [2923-28-6], 184
Simons' electrochemical cell, 244
Soda lime
    reaction with fluorine, 20
Sodium, 186
Sodium acid fluoride, 187
Sodium alizarine sulfonate
    in fluoride analysis, 91
Sodium aluminum fluoride
    HF in prdn of, 133
Sodium bifluoride [1333-83-1], 186
Soduim fluoride [7722-88-5], 133, 186
Sodium fluoroacetate [62-74-8], 305
    potassium fluoride in mfg of, 177
Sodium fluorozincate [18251-84-8]
    dental material catalysts, 222
Sodium heptafluorobutyrate [2218-84-4], 405
Sodium hexafluoroaluminate [15096-52-2], 37
Sodium hexafluoroammoniate [65176-04-7], 51
Sodium hydrogen fluoride, 187
Sodium hydroxide [1310-73-2]
    reaction with HF, 133
Sodium hydroxyfluoroborate [13876-97-6], 70
Sodium monofluorophosphate [7631-97-2], 168
Solar cells
    arsenous fluoride in mfg, 55
Solar collectors
    FEP film, 413
    PVF film for, 450
Solkane, 273
Solution polymerization
    of VDF, 457
Solvents
    HF in prdn of, 131
Sorbinil [68367-52-2], 337
Spiro compounds
    fluorination of, 254
Stainless steel
    HF in pickling of, 133
Stannic fluoride [7783-62-2], 212

Stannous fluoride [7783-47-3], 211
Stannous fluoroborate [13814-97-6], 212
Steam
    reaction with fluorine, 20
Stearic acid trifluoroborane [60274-92-2], 64
Steel
    boron trifluoride in prpn of, 66
    fluorine cmpds in mfg of, 30
    fluorine corrosion of, 16
    fluorspar flux for, 89
    use with HF, 129
Steel anode, 9
Steel cathodes
    use in fluorine cells, 12
Stockbarger process, 177
Strabismus
    treatment of, 172
Strontium fluoroborate tetrahydrate [27902-05-2], 72
Sulfamic acid
    fluorosulfuric acid in prpn of, 206
Sulfur, 203
    reaction with fluorine, 5
Sulfur difluoride, 198
Sulfur dioxide [7446-09-5]
    adduct with antimony pentafluoride, 52
Sulfur hexafluoride [2551-62-4], 4, 188
Sulfuric acid [7664-93-9]
    hydrogen fluoride from, 124
Sulfur pentafluoride [10546-01-7], 197
Sulfur tetrafluoride [7783-60-0], 5, 195, 232
Sulfuryl fluoride [2699-79-8], 159
Sulindac [38194-50-2], 236
Superacids
    antimony fluorides as, 53
    fluorosulfuric acid as, 203
    hydrogen fluoride as, 124
Superconductors
    copper(II) fluoride in, 98
Surfactants
    perfluoroalkanesulfonic acid salts, 321
Surgical gloves
    surface fluorination, 256
Suspension polymerizations
    of VDF, 457
Suva, 273
Swarts reactions
    use of antimony trifluoride, 51
Synthetic rubber
    fluorination of, 255

T

Takagi reagent, 356
Tamed fluorine
    aromatic fluorination, 326

INDEX 525

Tantalum pentafluoride [7783-71-3], 210
Tecnoflon polymers, 490
Tedlar, 447
Teeth
   sodium fluoride treatment, 187
Teflon, 70, 96, 264, 381
Teflon FEP, 264, 406
Teflon PFA, 432
Teflubenzuron [83121-18-0], 339
Tefluthrin [795-38-2], 343
Tefzel, 417
Tegafur [17902-23-7], 367
Telomerization
   fluorinated compounds, 234
Telomerization of tetrafluoroethylene
   perfluorocarboxylic acids from, 311
Telomers
   bromotrifluoroethylene, 480
Temafloxacin hydrochloride [105784-61-0], 366
Tennis balls
   sulfur hexafluoride in, 195
Tetracene
   boron trifluoride adducts, 63
1,1,1,2-Tetrachlorodifluoroethane [76-11-9], 265
1,1,2,2-Tetrachloro-1,2-difluoroethane [76-12-0], 265
1,1,1,2-Tetrachloro-2-fluoroethane [354-11-0], 266
1,1,2,2-Tetrachloro-1-fluoroethane [354-14-3], 266
Tetrachloroisophthalonitrile [1897-45-6], 342
2,3,5,6-Tetrachloropyrimidine [1780-40-1]
   fluorination of, 367
1,2,3,4-Tetrafluorobenzene [551-62-2], 340
1,2,3,5-Tetrafluorobenzene [2367-82-0], 340
1,2,4,5-Tetrafluorobenzene [327-54-8], 340
2,3,4,5-Tetrafluorobenzoic acid [1201-31-6], 343
2,3,5,6-Tetrafluorobenzotrifluoride [651-80-9], 356
Tetrafluoroboric acid [16872-11-0], 69, 151
1,1,1,2-Tetrafluoro-2-chloroethane [2837-89-0], 266
1,1,2,2-Tetrafluoro-1-chloroethane [354-25-6], 266
1,1,1,2-Tetrafluoroethane [811-97-2], 262
1,1,2,2-Tetrafluoroethane [359-35-3], 262
Tetrafluoroethylene (TFE) [116-14-3], 234
   cyclodimerization of, 263
   ethylene copolymers, 417
   in FEP, 404
   in fluorocarbon elastomers, 491
   manufacture of, 382
   perfluoropolyethers from, 285
   perfluorovinyl ether copolymers, 432
   polymerization of VF, 444
Tetrafluoroethylene–ethylene copolymers, 418
Tetrafluoroethylene oxide [694-117-7], 295
Tetrafluoroethylene–perfluoropropyl vinyl ether
   copolymers [26655-00-5], 433
1,1,2,2-Tetrafluoroethyl-2,2,2-trifluoroethyl ether
   [406-78-0]
   from 2,2,2-trifluoroethanol, 283
Tetrafluorohydrazine [10036-47-2], 152

Tetrafluoroiodine hexafluoroantimonate [41646-48-4],
   108
Tetrafluoromethane [75-73-0], 261
2,3,4,5-Tetrafluoronitrobenzene [5580-79-0], 343
Tetrafluorooxetane [765-63-9]
   perfluoropolyethers from, 285
1,1,1,3-Tetrafluoropropane [460-36-6], 262
2,3,5,6-Tetrafluoropyridine [2875-18-5], 359
2,4,5,6-Tetrafluoropyrimidine [767-79-3], 359
2,3,5,6-Tetrafluoro-4-trifluoromethylphenol
   [2787-79-3], 348
Tetrafluron [27954-37-6], 358
Tetrahydrofuran, 206
Tetrahydrofuran trifluoroborane [462-34-0], 64
Tetramethylammonium hexafluorophosphate
   [558-32-7], 172
Tetramethylammonium tetrafluoroborate [661-36-9],
   64
Tetraoxygen difluoride [12020-93-8], 5
Tetraphenylarsonium chloride
   for hexafluorophosphate analysis, 169
Textile finishes
   fluorosilicones in, 487
TFE. See Tetrafluoroethylene.
Thermocouples
   for fluorine, 17
Thin films
   rhenium hexafluoride in mfg of, 180
Thionyl fluoride [7783-42-8], 195
Thiothionyl fluoride [16860-99], 198
Thorium nitrate [13823-29-5]
   in fluoride analysis, 31, 91
3M Brand, 275
Tin
   with fluorosulfuric acid, 203
   reaction with fluorine, 5
Tin(II) fluoroborate [13814-96-5], 78
Tin tetrafluoride [7783-62-2], 5
Tire scrap
   fluorine treatment, 7
Titanium
   reaction with fluorine, 5
Titanium alloys
   with chlorine trifluoride, 105
Titanium(IV) fluoride dihydrate [60927-06-2], 215
Titanium(IV) oxyfluoride [13537-16-1], 215
Titanium tetrafluoride [7783-63-3], 5, 215
Titanium trifluoride [13470-08-1], 214
Toltrazuril [69004-03-1], 357, 358
Toluene [108-88-3]
   perfluoro fluid solubility, 288
Tosufloxacin [108138-46-1], 366
Tranquilizers
   fluorine cmpds, 236
   ring-fluorinated aromatics as, 335
Transfer molding
   of Teflon PFA, 440

Triallyl isocyanurate [1025-15-6]
  in fluoroelastomer curing, 498
Triazines
  fluorinated, 357
Tribromofluoromethane [353-54-8], 276
  hexafluorobenzene from, 346
2,4,6-Trichlorobenzotrifluoride [567-59-9], 348
1,1,1-Trichloro-2,2-difluoroethane [354-12-1], 266
1,1,2-Trichloro-1,2-difluoroethane [354-15-4], 266
1,2,2-Trichloro-1,1-difluoroethane [354-21-2], 266
1,1,2-Trichloro-1-fluoroethane [811-95-0], 266
1,1,2-Trichloro-2-fluoroethane [359-28-4], 266
Trichlorofluoromethane [75-69-4], 265
1,1,1-Trichloro-2,2,2-trifluoroethane [354-58-5], 265, 365
1,1,2-Trichloro-1,2,2-trifluoroethane [76-13-1], 6, 265
Tridecafluorohexanesulfonic acid [355-46-4], 320
Triflates, 320
Triflumizole [68694-11-1], 354
Triflumuron [64628-44-0], 358
Trifluoperazine hydrochloride [440-17-5], 236
Trifluoroacetaldehyde [75-90-1], 307
  from alcohol oxidation, 283
Trifluoroacetamide [354-38-1]
  trifluoroethanol from, 283
Trifluoroacetic acid [76-05-1], 235, 307
  from alcohol oxidation, 283
Trifluoroacetic anhydride [407-25-0], 235
  trifluoroethanol from, 283
4,4,4-Trifluoroacetoacetate [372-31-6], 365
2,4,5-Trifluoroacetophenone [129322-83-4], 342
Trifluoroacetyl chloride [354-32-5], 308
Trifluoroacetyl fluoride [354-34-7], 307
Trifluoroacrylic acid [433-68-1]
  from bromotrifluoroethylene, 479
2,3,5-Trifluoroaniline [363-80-4], 343
2,4,5-Trifluoroaniline [57491-45-9], 343
Trifluoroanisole [456-55-3], 356
1,2,3-Trifluorobenzene [1489-53-8], 340
1,2,4-Trifluorobenzene [367-23-7], 340
1,3,5-Trifluorobenzene [372-38-3], 340
2,4,5-Trifluorobenzoic acid [446-17-3], 342
Trifluoroborane [7637-07-2], 60
Trifluoroboroxine [13703-95-2], 61
1,1,1-Trifluoro-2-chloroethane [75-88-7], 266
1,1,2-Trifluoro-2-chloroethane [431-07-2], 266
Trifluorochloromethane [75-72-9], 265
1,1,1-Trifluoro-2,2-dichloroethane [306-83-2], 266
1,1,1-Trifluoroethane [420-46-2], 262
1,1,2-Trifluoroethane [430-66-0], 262
2,2,2-Trifluoroethanol [75-89-8], 282, 307
Trifluoroethyl ethyl ether [461-24-5]
  from 2,2,2-trifluoroethanol, 283
2,2,2-Trifluoroethyl trifluoroacetate [407-38-5]
  trifluoroethanol from, 283
2,2,2-Trifluoroethyl vinyl ether [406-90-6]
  from 2,2,2-trifluoroethanol, 283

Trifluoroglycidyl fluoride [24419-82-7], 295
1,1,1-Trifluoro-2-iodoethane [353-83-3], 276
Trifluoroiodomethane [2314-97-8], 276
Trifluoromethane [75-46-7], 228, 262, 307
Trifluoromethanesulfonamides, 321
Trifluoromethanesulfonic acid [1493-13-6], 320
Trifluoromethoxybenzene [456-55-3], 356
Trifluoromethyl aromatics, 347
m-Trifluoromethylbenzoic acid [454-92-2], 348
o-Trifluoromethylbenzoic acid [433-97-6], 348
3-(Trifluoromethyl)benzyl chloride [705-29-3], 349
Trifluoromethylcyclohexane [401-75-2], 348
Trifluoromethyl group
  in dyes, 353
Trifluoromethyl iodide [2314-97-8], 234
4-Trifluoromethylmercaptophenol [825-83-2], 357
3-(Trifluoromethylphenyl)acetonitrile [2338-76-3], 349
3-Trifluoromethylphenylisocyanate [329-01-1], 353
m-Trifluoromethylphenyllithium [368-49-0], 348
o-Trifluoromethylphenyllithium [49571-35-5], 348
p-Trifluoromethylphenyllithium [2786-01-8], 348
m-Trifluoromethylphenylmagnesium bromide [402-26-6], 348
o-Trifluoromethylphenylmagnesium bromide [395-47-1], 348
p-Trifluoromethylphenylmagnesium bromide [402-51-7], 348
p-Trifluoromethylphenylmagnesium chloride [2923-41-3], 348
N-(3-Trifluoromethylphenyl)-N,N-dimethylurea [2164-17-2], 353
2-Trifluoromethylpyridine [368-48-9], 359
3-Trifluoromethylpyridine [3796-23-4], 359
4-Trifluoromethylpyridine [3796-24-5], 359
2-Trifluoromethylpyridine-1-oxide [22253-71-0], 365
3,3,3-Trifluoropropene [677-21-4], 232
  hydrosilyation of, 483
3,3,3-Trifluoropropylmethyldichlorosilane [675-62-7], 484
2,4,6-Trifluoropyridine [3512-17-2], 359
2,4,6-Trifluoropyrimidine [696-82-2], 359
α,α,α-Trifluorotoluene [98-08-8], 350
2,4,6-Trifluoro-1,3,5-triazine [675-14-9], 359
Trifluorovinylic compounds
  from bromotrifluoroethylene, 479
Trifluorovinyllithium [683-78-3]
  from bromotrifluoroethylene, 479
Trifluperiodol [749-13-3], 355
Triflupromazine [146-54-3], 236, 355
Trifluralin [1582-09-8], 354
Trifluridine [70-00-8], 236
Triisobutylborane [1116-39-8]
  polymerization of VF, 444
2,4,6-Trimethyl-2,4,6-tris(3,3,4,4,5,5,6,6,6-nonafluorohexyl)cyclotrisiloxane [38521-58-3], 484

INDEX 527

2,4,6-Trimethyl-2,4,6-tris-(3,3,3-trifluoropropyl)
  cyclotrisiloxane [2374-14-3], 483
Trioxygen difluoride [16829-28-0], 5
Triphenylbenzylphosphonium chloride [1100-88-5]
  in fluoroelastomer curing, 495
Tris(perfluoro-n-butyl)amine [311-89-7], 234
Tris(perfluoroethyl)-s-triazine [858-46-8], 368
Tris(perfluoroheptyl)-s-triazine [21674-38-4], 368
Tris(perfluorononyl)-s-triazine [57104-59-4], 368
Tris(perfluoropropyl)-s-triazine [915-22-9], 368
Tris(trifluoromethyl)-s-triazine [368-66-1], 368
2,4,6-Tris(trifluoromethyl)-1,3,5-triazine [368-66-1], 359
Trouton constant
  for disulfur decafluoride, 197
Tubing cutters
  halogen fluorides as, 111
Tungsten carbide coatings
  from tungsten hexafluoride, 220
Tungsten(VI) fluoride [7783-82-6], 217
Tungsten hexafluoride [7783-82-6], 4, 217
Tungsten pentafluoride [19357-83-6], 217
Tungsten–rhenium alloys, 180
Tungsten tetrafluoride [13766-47-7], 217
Turpentine [8006-64-2]
  perfluoro fluid solubility, 288

U

Ullmann coupling reactions
  fluorinated biphenyls, 346
Ultramark, 344
Undecafluoropentanesulfonic acid [3872-25-1], 320
Uracil [66-22-8]
  fluorination of, 7, 366
Uranium
  with chlorine trifluoride, 105
Uranium hexafluoride [7783-81-5], 2
  from chlorine trifluoride, 111
  as fluorinating agent, 244
  HF in prdn of, 132
Uranium oxide
  conversion using HF, 132
Uranium reactor fuel
  HF in mfg of, 132
Uv windows
  use of magnesium fluoride, 143

V

Valves
  for fluoride, 17
VDF. See Vinylidene fluoride.
Verofix, 368

VF. See Vinyl fluoride.
Vinyl chloride [75-01-4]
  VF copolymers, 445
Vinyl fluoride (VF) [75-02-5], 232, 443
Vinylidene fluoride (VDF) [75-38-7], 454
  in fluorocarbon elastomers, 491
  polymerization of VF, 444
  from thermolysis, 235
Vinylidene fluoride-modified copolymers [9010-75-7], 475
Viton, 264, 490
Volhard method
  use in fluorine analysis, 18
Voltalef, 475

W

Wallach procedure, 327
Water [7732-18-5]
  perfluoro fluid solubility, 288
  reaction with fluorine, 5
  surface tensions of, 229
Water fluoridation
  use of sodium fluoride, 187
Water repellent finishes
  fluorosilicones in, 487
Willard-Winters method
  for fluoride, 91
Windows
  of calcium fluorite, 92
  of LiF for x-rays, 142
Wire covering
  FEP in, 413
Wood charcoal
  reaction with fluorine, 7

X

Xenon, 2, 183
  fluorination of, 106
Xenon difluoride [13709-36-9], 6
Xenon hexafluoride [13693-09-9], 6
Xenon tetrafluoride [13709-61-0], 6
X-ray fluorescence
  in fluoride analysis, 91
Xylene [1330-20-7]
  perfluoro fluid solubility, 288

Z

ZBLAN glasses, 36, 224
Zeolites

nitrogen fluoride purification, 154
Zepel, 279
Zephiran chloride
   for hydrofluoric acid burns, 48
Zinc fluoride [*7783-49-5*], 221
Zinc fluoride tetrahydrate [*13986-18-0*], 221
Zinc fluoroborate [*13826-88-5*], 78
Zinc sulfide
   fluorspar by-product, 89
Zirconium
   reaction with fluorine, 5
Zirconium difluoride [*7783-49-5*], 223
Zirconium hydroxy carbonate, 224
Zirconium tetrafluoride [*7783-64-4*], 5, 223
Zirconium tetrafluoride monohydrate [*14956-11-3*], 223
Zirconium trifluoride [*13814-22-7*], 223
Zonyl, 279